Lecture Notes in Computer S

Commenced Publication in 1973
Founding and Former Series Editors:
Gerhard Goos, Juris Hartmanis, and Jan van Leeuwen

Victor Shoup (Ed.)

Advances in Cryptology – CRYPTO 2005

25th Annual International Cryptology Conference
Santa Barbara, California, USA, August 14-18, 2005
Proceedings

 Springer

Volume Editor

Victor Shoup
New York University, Department of Computer Science
251 Mercer Street, New York, NY 10012, USA
E-mail: shoup@cs.nyu.edu

Library of Congress Control Number: Applied for

CR Subject Classification (1998): E.3, G.2.1, F.2.1-2, D.4.6, K.6.5, C.2, J.1

ISSN   0302-9743
ISBN-10   3-540-28114-2 Springer Berlin Heidelberg New York
ISBN-13   978-3-540-28114-6 Springer Berlin Heidelberg New York

Springer is a part of Springer Science+Business Media

springeronline.com

© International Association for Cryptologic Research 2005
Printed in Germany

Typesetting: Camera-ready by author, data conversion by Scientific Publishing Services, Chennai, India
Printed on acid-free paper   SPIN: 11535218   06/3142   5 4 3 2 1 0

Preface

These are the proceedings of Crypto 2005, the 25th Annual International Cryptology Conference. The conference was sponsored by the International Association for Cryptologic Research (IACR) in cooperation with the IEEE Computer Science Technical Committee on Security and Privacy and the Computer Science Department of the University of California at Santa Barbara. The conference was held in Santa Barbara, California, August 14–18, 2005.

The conference received 178 submissions, out of which the program committee selected 33 for presentation. The selection process was carried out by the program committee via an "online" meeting. The authors of selected papers had a few weeks to prepare final versions of their papers, aided by comments from the reviewers. However, most of these revisions were not subject to any editorial review.

This year, a "Best Paper Award" was given to Xiaoyun Wang, Yiqun Lisa Yin, and Hongbo Yu, for their paper "Finding Collisions in the Full SHA-1."

The conference program included two invited lectures. Ralph Merkle delivered an IACR Distinguished Lecture, entitled "The Development of Public Key Cryptography: a Personal View; and Thoughts on Nanotechnology." Dan Boneh gave an invited talk, entitled "Bilinear Maps in Cryptography."

We continued the tradition of a "rump session," featuring short, informal presentations (usually serious, sometimes entertaining, and occasionally both). The rump session was chaired this year by Phong Q. Nguyễn.

I would like to thank everyone who contributed to the success of this conference. First, thanks to all the authors who submitted papers: a conference program is no better than the quality of the submissions (and hopefully, no worse). Second, thanks to all the members of the program committee: it was truly an honor to work with a group of such talented and hard working individuals. Third, thanks to all the external reviewers (listed below) for assisting the program committee: their expertise was invaluable. Fourth, thanks to Matt Franklin, Dan Boneh, Jan Camenisch, and Christian Cachin for sharing with me their experiences as previous Crypto and Eurocrypt program chairs. Finally, thanks to my wife, Miriam, and my children, Alec and Nicol, for their love and support, and for putting up with all of this.

June 2005   Victor Shoup

CRYPTO 2005

August 14–18, 2005, Santa Barbara, California, USA

Sponsored by the
International Association for Cryptologic Research (IACR)

in cooperation with
IEEE Computer Society Technical Committee on Security and Privacy,
Computer Science Department, University of California, Santa Barbara

General Chair
Stuart Haber, HP Labs, USA

Program Chair
Victor Shoup, New York University, USA

Program Committee

Masayuki Abe .... NTT Information Sharing Platform Laboratories, Japan
Boaz Barak .... Institute for Advanced Study & Princeton University, USA
Amos Beimel .... Ben-Gurion University, Israel
Alex Biryukov .... Katholieke Universiteit Leuven, Belgium
John Black......................... University of Colorado at Boulder, USA
Alexandra Boldyreva .... Georgia Institute of Technology, USA
Jan Camenisch................. IBM Zurich Research Laboratory, Switzerland
Jean-Sébastien Coron .... University of Luxembourg, Luxembourg
Craig Gentry...................................... DoCoMo USA Labs, USA
Shai Halevi........................ IBM T. J. Watson Research Center, USA
Stanislaw Jarecki...................... University of California at Irvine, USA
Antoine Joux..................... DGA & Univ. Versailles St-Quentin, France
Jonathan Katz................................ University of Maryland, USA
Arjen Lenstra.. Lucent Technologies, USA & TU Eindhoven, The Netherlands
Yehuda Lindell...................................... Bar-Ilan University, Israel
Tal Malkin .... Columbia University, USA
Ilya Mironov....................................... Microsoft Research, USA
David Naccache..................... Gemplus, France & Royal Holloway, UK
Moni Naor............................ Weizmann Institute of Science, Israel
Leonid Reyzin....................................... Boston University, USA
Louis Salvail................................... Aarhus Universitet, Denmark
Alice Silverberg .... University of California at Irvine, USA
Adam Smith.......................... Weizmann Institute of Science, Israel
Rebecca Wright...................... Stevens Institute of Technology, USA

Advisory Members
Matt Franklin (Crypto 2004 Program Chair) .... UC Davis, USA
Cynthia Dwork (Crypto 2006 Program Chair)....... Microsoft Research, USA

External Reviewers

Luis von Ahn
Jesus F. Almansa
Michael Anshel
Frederik Armknecht
Michael Backes
Endre Bangerter
Paulo Barreto
Donald Beaver
Mihir Bellare
Daniel J. Bernstein
Bhargav Bhatt
Ian F. Blake
Daniel Bleichenbacher
Dan Boneh
Xavier Boyen
An Braeken
Eric Brier
Christian Cachin
Ran Canetti
Pascale Charpin
Melissa Chase
Benoit Chevallier-Mames
Martin Cochran
Nicolas Courtois
Ivan Damgård
Christophe De Cannière
Nenad Dedić
Michael de Mare
Claudia Diaz
Xuhua Ding
Hans Dobbertin
Yevgeniy Dodis
Iwan Duursma
Ariel Elbaz
Michael Engling
Marc Fischlin
Matthias Fitzi
Gerhard Frey
Eiichiro Fujisaki
Steven Galbraith
Juan Garay
Rosario Gennaro
Daniel Gottesman
Louis Goubin
Prateek Gupta

Helena Handschuh
Jonathan Herzog
Susan Hohenberger
Omer Horvitz
Nick Howgrave-Graham
Jim Hughes
Dae Hyun Yum
Yuval Ishai
Geetha Jagannathan
Marc Joye
Charanjit Jutla
Yael Tauman Kalai
Alexander Kholosha
Chiu-Yuen Koo
Hugo Krawczyk
Kaoru Kurosawa
Eyal Kushilevitz
Tanja Lange
Joseph Lano
Gregor Leander
Homin Lee
Wen-Ching Winnie Li
Anna Lysyanskaya
David M'Raihi
Phil Mackenzie
John Malone-Lee
Alexander May
Daniele Micciancio
Sara Miner More
Tal Moran
Shiho Moriai
Ryan Moriarty
Frédéric Muller
Kumar Murty
Steven Myers
Anderson Nascimento
Antonio Nicolosi
Jesper Buus Nielsen
Kobbi Nissim
Kazuo Ohta
Tatsuaki Okamoto
Siddika Berna Örs
Pascal Paillier
Matthew Parker
Rafael Pass

Thomas B. Pedersen
Krzysztof Pietrzak
Benny Pinkas
David Pointcheval
Joern-Mueller Quade
Tal Rabin
Zulfikar Ramzan
Omer Reingold
Pankaj Rohatgi
Guy Rothblum
Karl Rubin
Andreas Ruttor
Christian Schaffner
Berry Schoenmakers
Hovav Shacham
abhi shelat
Vitaly Shmatikov
Thomas Shrimpton
Hervé Sibert
Andrey Sidorenko
Nigel Smart
Dieter Sommer
Martijn Stam
Douglas R. Stinson
Koutarou Suzuki
Emmanuel Thomé
Eran Tromer
Fréderik Vercauteren
Eric Verheul
Emanuele Viola
Andrew Wan
Bogdan Warinschi
Hoeteck Wee
Benne de Weger
Enav Weinreb
Stephen Weis
Susanne Wetzel
Claire Whelan
Christopher Wolf
Nikolai Yakovenko
Shoko Yonezawa
Moti Yung
Sheng Zhong

Table of Contents

Efficient Collision Search Attacks on SHA-0

Xiaoyun Wang[1,*], Hongbo Yu[2], and Yiqun Lisa Yin[3]

[1] Shandong University, China
xywang@sdu.edu.cn
[2] Shandong University, China
yhb@mail.sdu.edu.cn
[3] Independent Security Consultant, Greenwich CT, US
yyin@princeton.edu

Abstract. In this paper, we present new techniques for collision search in the hash function SHA-0. Using the new techniques, we can find collisions of the full 80-step SHA-0 with complexity less than 2^{39} hash operations.

Keywords: Hash functions, Collision search attacks, SHA-0, SHA-1.

1 Introduction

The hash function SHA-0 was issued in 1993 as a federal standard by NIST. A revised version called SHA-1 was later issued in 1995 as a replacement for SHA-0. The only difference between the two hash functions is the additional rotation operation in the message expansion of SHA-1, which is supposed to provide more security. Both hash functions are based on the design principles of MD4.

In 1997, Wang found an attack on SHA-0 [14] which produces a collision with probability 2^{-58} by utilizing algebraic methods to derive a collision differential path. In 1998, Chabaud and Joux [6] independently found the same differential path through computer search. In August 2004, Joux [7] announced the first real collision of SHA-0, which consists of four message blocks (a pair of 2048-bit input messages). The collision search took about 80,000 hours of CPU time (three weeks of real time) and is estimated to have a complexity of about 2^{51} hash operations. To our knowledge, this is the best existing attack on the full 80-step SHA-0 prior to the work reported here.

The attacks in [14,6] found a differential path which is composed of certain 6-step local collisions. There is an obstacle to further improve these attacks, as finding a differential characteristic for two consecutive local collisions corresponding to two consecutive disturbances in the first round turns out to be impossible. This phenomenon makes it difficult to find a differential path which has a smaller number of local collisions in rounds 2-4 and no consecutive local collisions in the first round.

---

* Supported by the National Natural Science Foundation of China (NSFC Grant No.90304009) and Program for New Century Excellent Talents in University.

V. Shoup (Ed.): Crypto 2005, LNCS 3621, pp. 1–16, 2005.

In this paper, we introduce a new cryptanalytic method to cope with this difficulty. Our analysis includes the following techniques: Firstly, we identify an "impossible" differential path with few local collisions in rounds 2-4 and some consecutive local collisions in round 1. Secondly, we transform the impossible differential path into a possible one. Thirdly, we derive a set of conditions which guarantee that the modified differential path holds. Finally, we design message modifications to correct all the unfulfilled conditions in the first round as well as some such conditions in the second round. With these techniques, we can find collisions of the full SHA-0 with at most 2^{39} hash operations, which is a major improvement over existing attacks. The same techniques can be used to find near collisions of SHA-0 with complexity about 2^{33} hash operations.

We note that the new techniques have also been proven to be effective in the analysis of SHA-1[16].

The rest of the paper is organized as follows. In Section 2, we give a description of SHA-0. In Section 3, we provide an overview of the original attack on SHA-0 [14] and subsequent improvements [15,1,2,7,3]. In Section 4, we review the "message modification techniques" presented in [11,12,13] to break HAVA-128, MD5, MD4 and RIPEMD, and consider their effectiveness in improving existing attacks on SHA-0. In Section 5, we present our new collision search attacks on SHA-0. In Section 6, we give an example of real collision of SHA-0 found by computer search using the new techniques. We conclude the paper in Section 7.

2   Description of SHA-0

The hash function SHA-0 takes a message of length less than 2^{64} bits and produces a 160-bit hash value. The input message is padded and then processed in 512-bit blocks in the Damgård/Merkle iterative structure. Each iteration invokes a so-called compression function which takes a 160-bit chaining value and a 512-bit message block and outputs another 160-bit chaining value. The initial chaining value (called IV) is a set of fixed constants, and the final chaining value is the hash of the message.

In what follows, we describe the compression function of SHA-0. For each 512-bit block of the padded message, divide it into 16 32-bit words, $(m_0, m_1, ..., m_{15})$. The message words are first expanded as follows: for $i = 16, ..., 79$,

$$m_i = m_{i-3} \oplus m_{i-8} \oplus m_{i-14} \oplus m_{i-16}.$$

The expanded message words are then processed in four rounds, each consisting of 20 steps. The step function is defined as follows.

For $i = 1, 2, ..., 80$,

$$a_i = (a_{i-1} << 5) + f_i(b_{i-1}, c_{i-1}, d_{i-1}) + e_{i-1} + m_{i-1} + k_i$$
$$b_i = a_{i-1}$$
$$c_i = b_{i-1} << 30$$
$$d_i = c_{i-1}$$
$$e_i = d_{i-1}$$

The initial chaining value $IV = (a_0, b_0, c_0, d_0, e_0)$ is defined as:

$$(0x67452301, \ 0xefcdab89, \ 0x98badcfe, \ 0x10325476, \ 0xc3d2e1f0)$$

Each round employs a different Boolean function f_i and constant k_i, which is summarized in Table 1.

Table 1. Boolean functions and constants in SHA-0

rounds	steps		Boolean function f_i	constant k_i
1	$1 - 20$	IF:	$(x \wedge y) \vee (\neg x \wedge z)$	0x5a827999
2	$21 - 40$	XOR:	$x \oplus y \oplus z$	0x6ed6eba1
3	$41 - 60$	MAJ:	$(x \wedge y) \vee (x \wedge z) \vee (y \wedge z)$	0x8fabbcdc
4	$61 - 80$	XOR:	$x \oplus y \oplus z$	0xca62c1d6

3   Previous Attacks on SHA-0

In this section, we first describe the original collision attack on SHA-0 given by Wang in 1997 [14]. This sets up the basic framework for introducing our new techniques later on. For other independent attacks on SHA-0 the reader may wish to refer to [15,6,1,7,3].

3.1   Local Collisions of SHA-0

Informally, a local collision is a collision within a few steps of the hash function. A simple yet very important observation is that SHA-0 has a 6-step local collision that can start at any step i, and this type of local collision is the basic component in constructing full collisions.

Suppose a message difference in bit j first occurs in Step i (e.g., $\Delta m_{i-1,j} = 1$.) The difference will affect the chaining variables a, b, c, d, e consecutively in the next five steps. In order to offset these differences and reach a local collision, more message differences are introduced in subsequent message words. In Table 2, we illustrate the differential path of such a local collision. The chaining variable conditions under which the local collisions hold were given in [14,15].

The probability associated with the above local collision depends on the Boolean function, the bit position j, and some conditions on the message bits. The differential attack in [14] and [6] chooses $j = 2$ so that $j + 30$ becomes the MSB[1] to eliminate the carry effect in the last three steps. In addition, the following condition

$$m_{i,2} = \neg m_{i+1,7}$$

---

[1] Throughout this paper, we label the bit positions in a 32-bit word as $32, 31, 30, ..., 3, 2, 1$, where bit 32 is the most significant bit and bit 1 is the least significant bit. Please note that this is different from the convention of labelling bit positions from 31 to 0.

Table 2. A 6-step local collision of SHA-0 starting at step i. The measure of difference is $\oplus$. Addition in the exponents is modulo 32. "nc" stands for no carry. Δf is the output difference of the Boolean function

step	Δm	Δa	Δb	Δc	Δd	Δe	Conditions
i	2^j	2^j					nc
$i+1$	2^{j+5}		2^j				
$i+2$	2^j			2^{j+30}			nc, $\Delta f = 2^j$
$i+3$	2^{j+30}				2^{j+30}		nc, $\Delta f = 2^{j+30}$
$i+4$	2^{j+30}					2^{j+30}	nc, $\Delta f = 2^{j+30}$
$i+5$	2^{j+30}						nc

helps to offset completely the chaining variable difference in the second step of the local collision, where $x_{i,j}$ ($x = m$) denotes the j-th bit of message word x_i.

The message condition in round 3

$$m_{i,2} = \neg m_{i+2,2}$$

helps to offset the difference caused by the non-linear function in the third step of the local collision.

3.2   Differential Paths of SHA-0

At a high level, the differential path used in [14] is a sequence of local collisions joined together with possible overlaps. To construct such a path, we need to find a set of appropriate starting step for each local collision. We can use an 80-bit 0-1 vector $x = (x_0, ..., x_{79})$ to specify these starting steps, and the vector is called a *disturbance vector*. It is easy to show that the disturbance vector satisfies the same recursion defined by the message expansion. That is, for $i = 16, ...79$,

$$x_i = x_{i-3} \oplus x_{i-8} \oplus x_{i-14} \oplus x_{i-16}.$$

For the 80 variables x_i, any 16 consecutive ones determine the rest. So there are 16 free variables to be set for a total of 2^{16} possibilities.

In order for the disturbance vector to lead to a possible collision, several conditions on the disturbance vectors need to be imposed, and they are discussed in details in [14]. These conditions are summarized in Table 3.

From [15], we know condition 1 in Table 3 holds if and only if the following equations hold:

$$x_{11} = x_3 + x_8$$
$$x_{12} = x_4 + x_9$$
$$x_{13} = x_5 + x_{10}$$
$$x_{14} = x_0 + x_3 + x_6 + x_8$$
$$x_{15} = x_1 + x_4 + x_7 + x_9$$

Table 3. Conditions on disturbance vectors for SHA-0 with t steps

	Condition	Purpose
1	$x_i = 0$ for $i = 75, 76, 77, 78, 79$	to produce a collision in the last step 5
2	$x_i = 0$ for $i = -5, ..., -1$	to avoid truncated local collisions in first few steps
3	no consecutive ones in the first 17 variables	to avoid an impossible collision path due to a property of IF

Condition 2 in Table 3 holds if and only if

$$x_6 = x_0 + x_1 + x_2 + x_4$$
$$x_7 = x_0 + x_4$$
$$x_8 = x_0 + x_1 + x_5$$
$$x_9 = x_4$$
$$x_{10} = x_0 + x_5$$

We can also search for a disturbance vector using $(x_0, ..., x_{15})$ as the 16 variables. After imposing Conditions 1 and 2, there are 6 free variables remaining: $(x_0, ..., x_5)$. With Condition 3, only 3 choices are left for the 6 free variables, namely (001000) and (000100) and (000101), the first of which corresponds to the differential path given in [14].

We remark that the Hamming weight of the disturbance vector is closely related to the complexity of the attack. Given a disturbance vector x, we define $hw_{r+}(x)$ as the Hamming weight of x from step r to 80. To minimize the complexity, the Hamming weight $hw_{17+}(x)$ should as small as possible (although there are other more subtle conditions). The corresponding vector used in [14] have $hw_{17+} = 27$, and the complexity of collision search attack is about 2^{58}.

3.3   Existing Techniques for Improving the Attack

In the past year, there have been some major advances in the analysis of SHA-0. These latest attacks are built upon the differential attack by Chaband and Joux, while introducing new ideas for significant improvements. We summarize these techniques below.

- *Neutral bit techniques* [1]. This allows the collision search to start at a step $i > 17$. [2] Biham and Chen showed how to start the collision search of SHA-0 at step $i = 22$ [1] and reduce complexity of finding full collisions to 2^{56}.

  More interestingly, they were able to find near collisions of SHA-0 with complexity 2^{40}, and this provides a basis for finding multi-block collisions.

---

[2] Since the first 16 message words are independent, in general one can bypass the first 16 steps and start the search at $i = 17$.

Since a near collision does not require the first set of conditions on the disturbance vector, vectors with much lower Hamming weight can be found.
- Multi-block collision techniques [2,7,12]. The idea is to use near collisions in several message blocks to produce a collision. Using this technique together with the neutral bit technique, Joux reported the first real collision of the full 80-step SHA-0. The search complexity is estimated to be 2^{51} hash operations.

4   Message Modification Techniques and SHA-0

At the Rump Session of Crypto'04, Wang [10] announced collisions of several hash functions, including MD4, MD5, RIPEMD, and HAVEL-128. The collision search attacks on these hash functions [11,12,13] adopt a three-step approach: find a differential path leading to possible collisions, derive a set of sufficient conditions for the differential path to hold, and modify the message words to satisfy all conditions in the first round (as well as most conditions in the second round) so that the success probabilities can be greatly enhanced.

The "message modification" employed in the last step is a major innovation that makes these collision search attacks feasible. Message modification techniques have been introduced in attacking HAVAL-128, MD5, MD4, RIPEMD [11,12,13] and SHA-0 [15] (not gave the precise description in [15]). However, the more sophisticated hash functions such as SHA-0 and SHA-1 pose considerable new challenges, and require more powerful message modification techniques in their attack. We shall discuss various components of the message modification approach which, when suitably combined, can yield an effective attack. Full details will be omitted in this presentation.

4.1   Message Modification Techniques

In what follows, we provide a description of the more complicated message modification techniques for SHA-0. Following the terminology in [12], we also categorize the techniques into basic techniques and advanced techniques.

For the MD4-family of hash functions, including SHA, the step function F has the form of

$$a_i = F(\text{input chaining variables}, m_{i-1}),$$

where a_i is the output chaining variable and m_{i-1} is the message word applied in step i. Given a differential path that may lead to possible collisions, it is not hard to derive a set of sufficient conditions on a_i. The conditions are of the following forms:

- $a_{i,j} = 0$ or $a_{i,j} = 1$.
- $a_{i,j} = a_{i',j'}$ or $a_{i,j} = \neg a_{i',j'}$, for $i' < i$.

In fact, all these conditions can be combined into one general form:

$$a_{i,j} = v,$$

where v is a bit value that is fixed to be 0/1 or has been computed *before* step i (since $i' < i$). Therefore, we can treat them uniformly.

The main idea of the "'basic modification technique"is simply to set $a_{i,j}$ to the correct bit by modifying the corresponding bit of m_{i-1}. More specifically, the following operation is performed for each derived condition $a_{i,j} = v$.

- If $a_{i,j} \neq v$, then set $m_{i-1} = m_{i-1} \pm 2^{j-1}$ to correct the condition.

If there is a condition on $m_{i-1,j}$ in the collision differential path, the above modification isn't available. So, it needs to modify some message bits in the previous steps (maybe only one message bit of the previous step is modified) to correct the condition of $a_{i,j}$. These message modification techniques can be applied to a hash function up to the first 16 steps.

If the message word m_i is dependent on one or more of the earlier message words, then "advanced modification technique"are needed to deal with the complication. Roughly, a change in m_i will cause a change in m_t for some $t < 16$ and hence a change in a_{t+1}. The advanced technique can "correct" this change within the next few steps during which each of the chaining variables are updated once. Effectively, the correction process is the same as constructing a local collision. The process works if and only if modification of the message words in those steps does not affect any existing conditions on the chaining variables.

4.2   Application to SHA-0

Given a differential path of SHA-0 in any existing attacks, we can easily derive a set of sufficient conditions on the chaining variables by analyzing each local collision separately. Using the basic modification techniques, we can make all the conditions in the first 16 steps to hold in a systematic way.

The advanced modification techniques, however, do not seem to be directly applied to SHA-0 as in the cases of MD4, HAVAL-128 and MD5 etc. The effectiveness largely depends on how the conditions are "distributed" after step 16. For MD5, the conditions are very concentrated in steps 17 and 18, while for SHA-0 the conditions are spread out due to the local collisions. Another reason is the use of message expansion in SHA-0. As a result, the advanced modification techniques in [11,12,13] can only help to make a few conditions satisfied in steps $i > 16$. This would improve over the original attack on SHA-0 [14]. Given the neutral bit techniques [1] which already allow the bypass of the first 22 steps, the modification techniques, as they were used in MD5, are difficult to offer additional improvements over the best existing attacks on SHA-0. Therefore, new ideas are required in order to launch a more practical attack on SHA-0 and especially for extending the attack to SHA-1.

5   New Techniques for Searching Collisions in SHA-0

In this section, we present our new techniques for collision search in SHA-0. The techniques are quite effective for SHA-0 and can also be extended to attack SHA-1 [16].

5.1   Overview

The first key idea in our new techniques is to remove both Condition 2 and Condition 3 (see Table 3) on the disturbance vectors. Such relaxation provides a larger search space and allows us to find disturbance vectors whose hamming weights are much lower than those used in existing attacks, thereby greatly decreasing the complexity of the attack.

The cost, though, is much more complicated differential paths in the first round. In particular, the disturbance vector consists of consecutive ones in the first 16 steps as well as truncated local collision. We introduce several new techniques to construct a valid differential path given such a disturbance vector. This is the most difficult yet crucial part of the new analysis, without which it would be impossible to produce a *real* collision.

We also present a variation of the basic modification technique to deal with conditions in steps 17 through 20, effectively starting the collision search at step 21. Combining all these new techniques and some simple implementation tricks, we are able to reduce the collision search complexity of SHA-0 to below 2^{40}.

5.2   Finding Disturbance Vectors with Low Hamming Weight

In existing attacks, the difficulty of finding disturbance vectors of low Hamming weights is largely due to the following difference between the IF and XOR function: when c and d both change, the output of IF always changes, while the output of XOR never changes. For MAJ, the output changes with probability $1/2$. This motivates us to treat the first round differently so that Condition 3 can be relaxed.

We only impose Condition 1 in the search for good disturbance vectors. By doing so, we obtain many vectors with very small Hamming weights. Since we can use modification techniques to make all conditions in the first round to hold, we focus on vectors with small Hamming weight in rounds 2-4. Among the 2^{16} choices that satisfy Condition 1, about 30 of them have $17 \leq hw_{21+} \leq 19$, and four of which have Hamming weight 3 in the third round. We then picked the following one from the four candidates as the disturbance vector.

Table 4. A disturbance vector for producing a collision of SHA-0

step	vector
-5...-1	0 0 1 1 1
1...20	0 1 1 1 1 0 0 1 0 1 0 0 1 0 1 0 1 0 0 0
21...40	0 1 1 0 0 0 1 0 0 0 1 0 1 1 1 0 0 0 0 0
41...60	0 1 0 0 0 0 0 1 0 0 0 0 0 0 0 0 0 1 0 0
61...80	1 1 0 0 1 1 0 1 0 1 1 1 0 1 0 0 0 0 0 0

5.3   New Analysis Techniques

As we can see from the chosen vector, there are four consecutive 1s in the first 16 steps. In addition, there are three truncated local collisions since $x_{-3} = x_{-2} = x_{-1} = 1$. The corresponding message differences for the first 16 steps are given in Table 8 in the appendix. The most difficult part is to derive a differential path for the first 16 steps given the irregular message difference.

There are several techniques that we used to construct a valid differential path. Before diving into the details, we first present a few general ideas.

– Use "subtraction" instead of "exclusive-or" as the measure of difference to facilitate the precision of the analysis.
– Take advantage of special differential properties of IF. In particular, if there is a bit difference in one of the three inputs, the output will have a difference with probability 1/2. In addition, when the bit does flip, it can maintain or change the sign of the difference. Therefore, the function can either *preserve* or *absorb* an input difference, giving good flexibility for constructing differential paths.
– Take advantage of the carry effect. Since $2^j = -2^j - 2^{j+1} \ldots - 2^{j+k-1} + 2^{j+k}$ for any k, a single bit difference j can be expanded into several bits. This property makes it possible to introduce extra bit differences. To use the idea in a more sophisticated way, we can combine two sets of differences to produce one difference.
– Regroup the message differences. Some differences in local collisions shall remain unchanged to guarantee that the local collisions hold. Some other differences in a local collision will be reset to cancel out certain changed chaining variable bits – especially those bits produced by the message differences in the truncated collisions, and those arising from two consecutive local collisions.

5.4   Constructing the Specific Path

We first introduce some notation. Let $a_{i,j}$ denote the jth bit of variable a_i and $\Delta a_i = a_i' - a_i$ denote the difference. Note that we use *subtraction* difference rather than *exclusive-or* difference since keeping track of the signs is important in the analysis. Following the notation introduced in [11,12,13], we use $a_i[j]$ to denote $a_i[j] = a_i + 2^{j-1}$ with no bit carry, and $a_i[-j]$ to denote those $a_i[-j] = a_i - 2^{j-1}$ with no bit carry.

To construct a valid differential path, it is important to control the propagation of the differences in each chaining variables. At a high level, differences in b, c, d are mostly absorbed by the Boolean function IF. The differences in a and e need to be carefully controlled, and most of them are offset by using appropriate differences in b, c, d.

The complete differential path for the first 16 steps is given in Table 8 in the appendix. It may look quite complicated at a first glance, and so we provide a more concise description below which better illustrates the idea. Based

Table 5. Differences in a. The entries list the bit positions of the differences and their signs. For example, the difference 2^j is listed as $j+1$ and -2^j as $-(j+1)$. Bit positions in bold are expanded using the carry effect in the complete differential path given in Table 8

Δa	I	II	III	IV
a_1	$-2, \mathbf{7}, -32$			
a_2	$-7, \mathbf{12}, -5$			
a_3	$-12, \mathbf{17}, -10$	2		
a_4			**20**	9
a_5			25	**4**
a_6		2	$-\mathbf{10}, \mathbf{15}$	
a_7		2	-17	
a_8			$-\mathbf{12}$	
a_{10}		2		
a_{11}			10	
a_{13}		2		
a_{15}			-2	

on the step function of SHA-0, it is not hard to see that the differences in the chaining values are fully determined by the differences in a, which is given in Table 5 below. Bit differences that are expanded using the carry effect is shown in bold, and the expanded bits are not shown. The bit differences in a can be categorized into four groups as follows, and their rationale can then be better understood.

- Group I: differences due to Δm_0:
  These are message differences due to the "truncated" local collisions. Hence they are inherent from the chosen path and cannot be changed. They cause differences in e_5, e_6, e_7 that need to be cancelled. Most of them can be cancelled with existing differences in that step, except $e_5[-30]$, $e_6[-5]$, and $e_7[15]$.
- Group II: differences due to disturbance.
  These result in the usual 6-step local collisions.
- Group III: differences introduced to cancel $e_5[-30]$ and $e_7[15]$.
  Note that only $a_6[15]$ is for cancelling $e_7[15]$, and the rest are all for cancelling $e_5[30]$. This part is where the expansion using the carry effect is needed.
- Group IV: differences used for additional adjustments.
  These are $a_4[9]$ for producing $e_8[7]$ in order to cancel $m_9[-7]$ and $a_5[4]$ for both producing $e_9[2]$ to introduce the disturbance equivalent to $m_{10}[2]$ and producing $b_6[5]$ to cancel out $e_6[-5]$.

This is the most difficult yet crucial part of the new analysis, without which it would be impossible to produce a *real* collision. Furthermore, the analysis demonstrates some unexpected weaknesses in the design of the step update function. In particular, certain properties of Boolean function $(x \wedge y) \vee (\neg x \wedge z)$ and the carry effect actually facilitate, rather than prevent, differential attacks.

5.5 Deriving Conditions on m_i and a_i

As we discussed in Section 3, for each local collision starting at step i, the follow conditions on m should hold.

$$m_{i+1,7} = \neg m_{i,2} \tag{1}$$

$$m_{i+2,2} = \neg m_{i,2} \quad (For\ round\ 3) \tag{2}$$

The condition on the disturbance vector given in Table 4, there are a total of 19 disturbances in Rounds 2, 3 and 4. So equation (1) yields 19 conditions on message words. From 3 disturbances in round 3, there are another 3 conditions on message word corresponding to equation (2). From Table 8, there are another 9 necessary conditions on message word position bit 2 and 7. There are total 31 conditions on message positions 2 and 7, and by a straightforward search of the 2^{32} choices for two positions of $m_{0,2}, ..., m_{15,2}$ it turns out that several choices satisfy all the conditions.

After the conditions on the message words are determined, we can derive a set of sufficient conditions on a_i given the differential path. The derivation uses differential properties of the three Boolean functions as well as the carry propagation pattern of addition. The complete description of the conditions is given in Table 9 in the appendix.

5.6 A Variation of the Modification Techniques

There are a total of 45 conditions from step 17 to step 80. Here we introduce a variation of the message modification techniques to deal with the three conditions in step 17 through 20, and hence reducing the number of conditions to 42. The idea is better explained using an example, say the condition on $a_{17,32}$. Instead of modifying m_{16}, which is dependent on four earlier message words, we modify m_{15} in a way that will flip the bit $a_{16,27}$, which in turn flips the bit $a_{17,32}$ in step 17. The other two conditions are handled similarly.

5.7 Complexity Analysis

In this section, we analyze the complexity of our collision search attack. Since there are a total of 42 conditions after applying message modification, a straightforward implementation would yield a complexity of 2^{42} hash operations.

There are several simple techniques that we can use to further improve the efficiency of the attack. The idea is that we only need to compute a small number of steps of the 80-step hash operation. First, we can precompute a set of "good" message words that make all conditions satisfied in the first 14 steps and only leave the last two message words as free variables. Second, we can use an "early stopping technique". More specifically, we only need to carry out the computation until step 23 and then test whether the four conditions in steps 21 through 24 are satisfied. On average only a fraction of 2^{-4} of the messages will pass the test. Overall, we only need to compute from step 15 to 24, for a total of

10 steps. This immediately gives a factor of $80/10 = 8$ improvements in search complexity. Hence, the complexity of finding a full collision is at most 2^{39} hash operations.

Our analysis can also be used to find near collisions with much lower complexity. For near collisions, we have found quite a few disturbance vectors with $hw_{21+} = 14$, and an example is given in Table 6. For this vector, the total number of conditions in Rounds 2-4 is $(14 - 4) \times 2 + 4 \times 4 = 36$. Using early stopping techniques, we estimate that near collisions of SHA-0 can be found with complexity about 2^{33}.

Table 6. A disturbance vector for near collision of SHA-0

step	vector
1...20	1 0 0 1 1 0 1 1 1 1 1 0 1 0 1 1 0 1 1 1
21...40	0 0 0 1 0 0 0 1 0 1 0 1 0 0 0 0 0 0 0 0
41...60	0 1 0 0 1 0 0 0 0 1 0 0 0 0 0 1 0 0 0 0
61...80	1 0 0 1 0 1 1 0 0 0 0 0 0 1 0 0 0 1 0 0

Finally, we remark that using the multi-block technique for attacking MD5 [12], we can use near collisions to construct multi-block collisions with about the same complexity. Therefore, we expect multi-block collisions of SHA-0 can potentially be found with about 2^{33} hash operations.

6   A Collision Example of SHA-0

The two messages that collide are (M_0, M_1) and (M_0, M_1'), where

$$h_1 = compress(h_0, M_0)$$
$$h_2 = compress(h_1, M_1) = compress(h_1, M_1')$$

Note that the first message block M_0 is the same, and it is for producing an intermediate chaining value h_1 that satisfies the 14 conditions on a_0, b_0. (See Table 9). M_0 can be found with complexity 2^{14}. After that, the pair (M_1, M_1') can be found with complexity 2^{39}.

We remark that we can adjust the differential path under the conditions of the original initial value h_0 to find a one-block message collision differential path.

7   Conclusions

In this paper, we present a new collision search attack on SHA-0 with complexity 2^{39} hash operations. Compared with existing attacks on SHA-0, our method is much more efficient and real collisions can be found quickly on a typical PC.

The techniques developed in our analysis of SHA-0 are also applicable to SHA-1. As SHA-0 may be viewed as simpler variant of SHA-1, the analysis

Table 7. A collision of 80-step SHA-0. Padding rules are not applied to the input messages

h_0:	67452301 efcdab89 98badcfe 10325476 c3d2e1f0
M_0:	65c24f5c 0c0f89f6 d478de77 ef255245 83ae3a1f 2a96e508 2c52666a 0d6fad5a
	9d9f90d9 eb82281e 218239eb 34e1fbc7 5c84d024 f7ad1c2f d41d1a14 3b75dc18
h_1:	39f3bd80 c38bf492 fed57468 ed70c750 c521033b
M_1:	474204bb 3b30a3ff f17e9b08 3ffa0874 6b26377a 18abdc01 d320eb93 b341ebe9
	13480f5c ca5d3aa6 b9f3bd88 21921a2d 4085fca1 eb65e659 51ac570c 54e8aae5
M_1':	c74204f9 3b30a3ff 717e9b4a 3ffa0834 6b26373a 18abdc43 5320eb91 3341ebeb
	13480f1c 4a5d3aa6 39f3bdc8 a1921a2f 4085fca3 6b65e619 d1ac570c d4e8aaa5
h_2:	2af8aee6 ed1e8411 62c2f3f7 3761d197 0437669d

presented here serves to verify effectiveness of these new techniques for other SHA variants.

Our analysis demonstrates some weaknesses in the step updating function of SHA-0 and SHA-1. In particular, because of the simple step operation structure, certain properties of the Boolean function $(x \wedge y) \vee (\neg x \wedge z)$ combined with the carry effect actually facilitate, rather than inhibit, differential attacks. We hope that these insights can be useful in the design of more secure hash functions in the future.

Acknowledgements

It is a pleasure to acknowledge Arjen K. Lenstra for his important suggestions and corrections, and for spending his precious time on our research. We would like to thank Andrew C. Yao and Frances. Yao for their support and corrections on this paper. We also thank Ronald L. Rivest and many other anonymous reviewers for their important comments.

References

1. E. Biham and R. Chen. *Near Collisions of SHA-0*. Advances in Cryptology – Crypto'04, pp.290-305, Springer-Verlag, August 2004.
2. E. Biham and R. Chen. *New Results on SHA-0 and SHA-1*. Crypto'04 Rump Session, August 2004.
3. E. Biham, R. Chen, A. Joux, P. Carribault, W. Jalby and C. Lemuet. *Collisions in SHA-0 and Reduced SHA-1*. Advances in Cryptology–Eurocrypt'05, pp.36-57, May 2005.
4. NIST. *Secure hash standard*. Federal Information Processing Standard, FIPS-180, May 1993.
5. NIST. *Secure hash standard*. Federal Information Processing Standard, FIPS-180-1, April 1995.
6. F. Chabaud and A. Joux. *Differential Collisions in SHA-0*. Advances in Cryptology – Crypto'98, pp.56-71, Springer-Verlag, August 1998.

7. A. Joux. *Collisions for SHA-0*. Rump session of Crypto'04, August 2004.
8. K. Matusiewicz and J. Pieprzyk. *Finding Good Differential Patterns for Attacks on SHA-1*. IACR Eprint archive, December 2004.
9. V. Rijmen and E. Osward. *Update on SHA-1*. RSA Crypto Track 2005, February 2005.
10. X. Y. Wang, D. G. Feng, X. J. Lai, and H. B. Yu. *Collisions for Hash Functions MD4, MD5, HAVAL-128 and RIPEMD*. In Rump session of Crypto'04 and IACR Eprint archive, August 2004.
11. X. Y. Wang, D. G. Feng, X. Y. Yu. The Collision Attack on Hash Function HAVAL-128. In Chinese, Science in China, Series E, Vol. 35(4), pp.405-416, April, 2005.
12. X. Y. Wang and H. B. Yu. *How to Break MD5 and Other Hash Functions*. Advances in Cryptology–Eurocrypt'05, pp.19-35, Springer-Verlag, May 2005.
13. X. Y. Wang. X. J. Lai, D. G. Feng, H. Chen, X. Y. Yu. *Cryptanalysis for Hash Functions MD4 and RIPEMD*. Advances in Cryptology–Eurocrypt'05, pp.1-18, Springer-Verlag, May 2005.
14. X. Y. Wang. *The Collision attack on SHA-0*, In Chinese, to appear on www.infosec.edu.cn, 1997.
15. X. Y. Wang. *The Improved Collision attack on SHA-0*. In Chinese, to appear on www.infosec.edu.cn, 1998.
16. X. Y. Wang. Y. Lisa Yin, H. B. Yu. *Finding Collisions in the Full SHA-1*. These proceedings.

A   The Differential Path and Derived Conditions

In Table 8 we describe the details of the differential path that leads to a full collision of SHA-0. In Table 9, we list a set of sufficient conditions on the chaining variables a_i for the given differential path.

For the first 20 steps, since there are many conditions for each a_i, we use a compact representation for the conditions so that they can be easily visualized. More specifically, for the condition $a_{i,j} = v$ we put one symbol w in the row for a_i under bit position j, where w is defined as follows:

- If $v = 0$, then w = 0.
- If $v = 1$, then w = 1.
- If $v = a_{i-1,j}$, then w = a.
- If $v = \neg a_{i-1,j}$, then w = $\overline{a}$.
- If no condition on $a_{i,j}$, then w = -.

Table 8. A differential path for the first round of SHA-0. For ease of notation, the entries list the bit positions of the differences and their signs. For example, the difference 2^j is listed as $(j+1)$ and -2^j as $-(j+1)$

step i	x_{i-1}	Δm_{i-1}	Δa_i	Δb_i	Δc_i	Δd_i	Δe_i
1	0	$-2,7,32$	-2 $-7,-8,9$ -32				
2	1		$7,8,-9$ $-12,...,-21,22$ $5,-6$	Δa_1			
3	1	$2,7,32$	-12 $-17,-18,19$ -10 2	Δa_2	$\Delta a_1^{\lll 30}$		
4	1	-7	$-20,...,-24,25$ 9	$\ldots$	$\Delta a_2^{\lll 30}$	$\Delta a_1^{\lll 30}$	
5	1	-7	25 $-4,5$		$\ldots$	$\Delta a_2^{\lll 30}$	$\Delta a_1^{\lll 30}$
6	0	$2,7$	2 $10,11,-12$ $-15,16$			$\ldots$	$\Delta a_2^{\lll 30}$
7	0	$-2,32$	2 -17				$\ldots$
8	1	$2,32$	$12,...,22,-23$	$\ldots$			
9	0	-7			$\ldots$		
10	1	32	2			$\ldots$	
11	0	$7,32$	10	$\ldots$			$\ldots$
12	0	$2,32$		Δa_{11}	$\ldots$		
13	1	2	2		$\Delta a_{11}^{\lll 30}$	$\ldots$	
14	0	$-7,32$		Δa_{13}		$\Delta a_{11}^{\lll 30}$	$\ldots$
15	1	32	-2		$\Delta a_{13}^{\lll 30}$		$\Delta a_{11}^{\lll 30}$
16	0	$-7,32$		Δa_{15}		$\Delta a_{13}^{\lll 30}$	

The rest of the path consists of 19 6-step local collisions. The starting step of these collisions is specified by the disturbance vector given in Table 4.

Table 9. A set of sufficient conditions on a_i for the differential path given in Table 8

chaining variable	conditions on bits			
	$32 - 25$	$24 - 17$	$16 - 9$	$8 - 1$
a_0	--------	1-1100-1	--1--1-1	10------
b_0	--------	--------	------a-	----a-$\bar{a}$-
a_1	1-------	0a0011a0	aa1-10a0	11----1-
a_2	0-0-----	--011111	111111-1	0010a---
a_3	1-1--aaa	aa0-0011	0010101-	-010100-
a_4	1----a-0	11111000	--1111-0	110011--
a_5	0------0	-0001001	00100-0-	01-01---
a_6	-------0	-1011110	010-100-	-0--100-
a_7	0------1	a1011111	0100--00	0----10-
a_8	--------	-1000000	00000-11	1---1---
a_9	1-------	---00000	0011001-	----0---
a_{10}	--------	---11111	1111111-	------0-
a_{11}	0-------	--------	------0-	----1---
a_{12}	0-------	--------	--------	0---0---
a_{13}	--------	--------	--------	1---0-0-
a_{14}	1-------	--------	--------	----1---
a_{15}	0-------	--------	--------	----1-1-
a_{16}	1-------	--------	--------	----0---
a_{17}	0-------	--------	--------	------1-
a_{18}	1-------	--------	--------	--------
a_{19}	--------	--------	--------	--------
a_{20}	--------	--------	--------	--------

The conditions for the 19 local collisions in Rounds 2-4 are derived as follows. Note that the conditions depend on the bit $m_{i,2}$ which has been pre-determined.

- XOR rounds:

$$a_{i-1,4} = \neg a_{i-2,4} \; (or \; a_{i-1,4} = a_{i-2,4})$$
$$a_{i,2} = m_{i,2}.$$

- MAJ round:

$$a_{i-1,4} = \neg a_{i-2,4}$$
$$a_{i,2} = m_{i,2}$$
$$a_{i+1,32} = \neg a_{i-1,2}$$
$$a_{i+2,32} = \neg a_{i+1,2}$$

Finding Collisions in the Full SHA-1

Xiaoyun Wang[1,*], Yiqun Lisa Yin[2], and Hongbo Yu[3]

[1] Shandong University, Jinan 250100, China
xywang@sdu.edu.cn
[2] Independent Security Consultant, Greenwich CT, US
yyin@princeton.edu
[3] Shandong University, Jinan250100, China
yhb@mail.sdu.edu.cn

Abstract. In this paper, we present new collision search attacks on the hash function SHA-1. We show that collisions of SHA-1 can be found with complexity less than 2^{69} hash operations. This is the first attack on the full 80-step SHA-1 with complexity less than the 2^{80} theoretical bound.

Keywords: Hash functions, collision search attacks, SHA-1, SHA-0.

1 Introduction

The hash function SHA-1 was issued by NIST in 1995 as a Federal Information Processing Standard [5]. Since its publication, SHA-1 has been adopted by many government and industry security standards, in particular standards on digital signatures for which a collision-resistant hash function is required. In addition to its usage in digital signatures, SHA-1 has also been deployed as an important component in various cryptographic schemes and protocols, such as user authentication, key agreement, and pseudorandom number generation. Consequently, SHA-1 has been widely implemented in almost all commercial security systems and products.

In this paper, we present new collision search attacks on SHA-1. We introduce a set of strategies and corresponding techniques that can be used to remove some major obstacles in collision search for SHA-1. Firstly, we look for a near-collision differential path which has low Hamming weight in the "disturbance vector" where each 1-bit represents a 6-step local collision. Secondly, we suitably adjust the differential path in the first round to another possible differential path so as to avoid impossible consecutive local collisions and truncated local collisions. Thirdly, we transform two one-block near-collision differential paths into a two-block collision differential path with twice the search complexity. We show that, by combining these techniques, collisions of SHA-1 can be found with complexity less than 2^{69} hash operations. This is the first attack on the *full* 80-step SHA-1 with complexity less than the 2^{80} theoretical bound.

---

* Supported by the National Natural Science Foundation of China (NSFC Grant No.90304009) and Program for New Century Excellent Talents in University.

V. Shoup (Ed.): Crypto 2005, LNCS 3621, pp. 17–36, 2005.

In the past few years, there have been significant research advances in the analysis of hash functions. The techniques developed in these early works provide an important foundation for the attacks on SHA-1 presented in this paper. In particular, our analysis is built upon the original differential attack on SHA-0 [14], the near collision attack on SHA-0 [1], the multi-block collision techniques [12], as well as the message modification techniques used in the collision search attacks on HAVAL-128, MD4, RIPEMD and MD5 [11,13,12].

Our attack naturally is applied to SHA-0 and all reduced variants of SHA-1. For SHA-0, the attack is so effective that we are able to find real collisions of the full SHA-0 with less than 2^{39} hash operations [16]. We also implemented the attack on SHA-1 reduced to 58 steps and found real collisions with less than 2^{33} hash operations. In a way, the 58-step SHA-1 serve as a simpler variant of the full 80-step SHA-1 which help us to *verify* the effectiveness of our new techniques. Furthermore, our analysis shows that the collision complexity of SHA-1 reduced to 70 steps is less than 2^{50} hash operations.

The rest of the paper is organized as follows. In Section 2, we give a description of SHA-1. In Section 3, we provide an overview of previous work on SHA-0 and SHA-1. In Section 4, we present the techniques used in our new collision search attacks on SHA-1. In Section 5, we elaborate on the analysis details using the real collision of 58-step SHA-1 as a concrete example. We discuss the implication of the results in Section 6.

2   Description of SHA-1

The hash function SHA-1 takes a message of length less than 2^{64} bits and produces a 160-bit hash value. The input message is padded and then processed in 512-bit blocks in the Damgard/Merkle iterative structure. Each iteration invokes a so-called compression function which takes a 160-bit chaining value and a 512-bit message block and outputs another 160-bit chaining value. The initial chaining value (called IV) is a set of fixed constants, and the final chaining value is the hash of the message.

In what follows, we describe the compression function of SHA-1.

For each 512-bit block of the padded message, divide it into 16 32-bit words, $(m_0, m_1, ..., m_{15})$. The message words are first expanded as follows: for $i = 16, ..., 79$,

$$m_i = (m_{i-3} \oplus m_{i-8} \oplus m_{i-14} \oplus m_{i-16}) \lll 1.$$

The expanded message words are then processed in four rounds, each consisting of 20 steps. The step function is defined as follows.

For $i = 1, 2, ..., 80$,

$$a_i = (a_{i-1} \lll 5) + f_i(b_{i-1}, c_{i-1}, d_{i-1}) + e_{i-1} + m_{i-1} + k_i$$
$$b_i = a_{i-1}$$
$$c_i = b_{i-1} \lll 30$$

$$d_i = c_{i-1}$$
$$e_i = d_{i-1}$$

The initial chaining value $IV = (a_0, b_0, c_0, d_0, e_0)$ is defined as:

$$(0x67452301, \ 0xefcdab89, \ 0x98badcfe, \ 0x10325476, \ 0xc3d2e1f0)$$

Each round employs a different Boolean function f_i and constant k_i, which is summarized in Table 1.

Table 1. Boolean functions and constants in SHA-1

round	step		Boolean function f_i	constant k_i
1	$1-20$	IF:	$(x \wedge y) \vee (\neg x \wedge z)$	0x5a827999
2	$21-40$	XOR:	$x \oplus y \oplus z$	0x6ed6eba1
3	$41-60$	MAJ:	$(x \wedge y) \vee (x \wedge z) \vee (y \wedge z)$	0x8fabbcdc
4	$61-80$	XOR:	$x \oplus y \oplus z$	0xca62c1d6

3 Previous Work on SHA-0 and SHA-1

In 1997, Wang [14] presented the first attack on SHA-0 based on an algebraic method, and showed that collisions can be found with complexity 2^{58}. In 1998 Chabaud and Joux independently found the same collision differential path for SHA-0 by the differential attack. In the present work, as well as in the SHA-0 attack by [16], the algebraic method (see also Wang [15]) again plays an important role, as it is used to deduce message conditions both on SHA-0 and SHA-1 that should hold for a collision (or near-collision) differential path and be handled in advance.

3.1 Local Collisions of SHA-1

Informally, a local collision is a collision within a few steps of the hash function. A simple yet very important observation made in [14] is that SHA-0 has a 6-step local collision that can start at any step i. A kind of local collision can be referred to [16], and the chaining variable conditions for a local collision were taken from Wang [14].

The collision differential path on SHA-0 chooses $j = 2$ so that $j + 30 = 32$ becomes the MSB [1] to eliminate the carry effect in the last three steps. In addition, the following condition

$$m_{i,2} = \neg m_{i+1,7}$$

---

[1] Throughout this paper, we label the bit positions in a 32-bit word as $32, 31, 30, ..., 3, 2, 1$, where bit 32 is the most significant bit and bit 1 is the least significant bit. Please note that this is different from the convention of labelling bit positions from 31 to 0.

helps to offset completely the chaining variable difference in the second step of the local collision, where $m_{i,j}$ denotes the j-th bit of message word m_i.

The message condition in round 3

$$m_{i,2} = \neg m_{i+2,2}$$

helps to offset the difference caused by the non-linear function in the third step of the local collision.

Since the local collision of SHA-0 does not depend on the message expansion, it also applies to SHA-1. Hence, this type of local collision can be used as the basic component in constructing collisions and near collisions of the full 80-step SHA-0 and SHA-1.

3.2   Differential Paths of SHA-1

We start with the differential path for SHA-0 given in [14,15]. At a high level, the path is a sequence of local collisions joined together. To construct such a path, we need to find appropriate starting steps for the local collisions. They can be specified by an 80-bit 0-1 vector $x = (x_0, ..., x_{79})$ called a *disturbance vector*. It is easy to show that the disturbance vector satisfies the same recursion defined by the message expansion.

For the 80 variables x_i, any 16 consecutive ones determine the rest. So there are 16 free variables to be set for a total of 2^{16} possibilities. Then a "good" vector satisfying certain conditions can be easily searched with complexity 2^{16}.

In [2,9], the method for constructing differential paths of SHA-0 is naturally extended to SHA-1. In the case of SHA-1, each entry x_i in the disturbance vector is a 32-bit word, rather than a single bit. The vectors thus defined satisfy the SHA-1 message expansion.

That is, for $i = 16, ..., 79$,

$$x_i = (x_{i-3} \oplus x_{i-8} \oplus x_{i-14} \oplus x_{i-16}) \lll 1.$$

In order for the disturbance vector to lead to a possible collision, several conditions on the disturbance vectors need to be imposed, and they are discussed in details in [15] [6]. These conditions also extend to SHA-1 in a straightforward way, and we summarize them in Table 2.

In the case of SHA-0, 3 vectors are found among the 2^{16} choices, and two of them are valid when all three conditions are imposed.

In the case of SHA-1, it becomes more complicated to find a good disturbance vector with low Hamming weight due to large search space. Biham and Chen [2] used clever heuristics to search for such vectors for reduced step variants and they were able to find real collisions of SHA-1 up to 40 steps. They estimated that collisions of SHA-1 can be found up to 53-round reduced SHA-1 with about 2^{48} complexity, where the reduction is to the last 53 rounds of SHA-1. Rijmen and Osward [9] did a more comprehensive search using methods from coding theory, and their estimates on the complexity are similar.

Table 2. Conditions on disturbance vectors for SHA-1 with t steps

	Condition	Purpose
1	$x_i = 0$ for $i = t - 5, ..., t - 1$	to produce a collision in the last step t
2	$x_i = 0$ for $i = -5, ..., -1$	to avoid truncated local collisions in first few steps
3	no consecutive ones in same bit position in the first 16 variables	to avoid an impossible collision path due to a property of IF

Overall, since the Hamming weight of a valid disturbance vector grows quickly as the number of steps increases, it seems that finding a collision of the full 80-step SHA-1 is beyond the 2^{80} theoretical bound with existing techniques.

4   New Collision Search Attacks on SHA-1

In this section, we present our new techniques for search collisions in SHA-1. The techniques used in the attack on SHA-1 are largely built upon our new analysis of SHA-0 [16], in which we showed how to greatly reduces the search complexity to below the 2^{40} bound.

4.1   Overview

As we have seen in existing analysis of SHA-1, finding a disturbance vector with low Hamming weight is a necessary step in constructing valid differential paths that can lead to collision. On the other hand, the three conditions imposed on disturbance vectors seem to a major obstacle. There have been attempts to remove some of the conditions. For example, finding multi-block collisions using near collisions effectively relax the first condition, and finding collisions for SHA-1 without the first round effectively relax the second condition (although it is no longer SHA-1 itself). Even with both relaxation, the Hamming weight of the disturbance vectors is still too high to be useful for the full 80-step SHA-1.

A key idea of our new attack is to relax *all* the conditions on the disturbance vectors. In other words, we impose *no* condition on the vectors other than they satisfy the message expansion recursion. This allows us to find disturbance vectors whose Hamming weights are much lower than those used in existing attacks.

We then present several new techniques for constructing a valid differential path given such disturbance vectors. The resulting path is very complex in the first round due to consecutive disturbances as well as truncated local collisions that initiate from steps −5 through −1. This is the most difficult yet crucial part of new analysis, without which it would be impossible to produce a *real* collision.

Once a valid differential path is constructed, we apply the message modification techniques, first introduced by Wang et. al in breaking MD5 and other hash functions [15,11,12,13], to further reduce the search complexity. Such extension requires carefully deriving the exact conditions on the message words and chaining variables, which is much more involved in the case of SHA-1 compared with SHA-0 and other hash functions.

Besides the above techniques, we also introduce some new methods that are tailored to the SHA-1 message expansion. Combining all these techniques and a simple "early stopping" trick when implementing the search, we are able to present an attack on SHA-1 with complexity less than 2^{69}. These techniques are presented in more detail in Sections 4 and 5.

4.2   Finding Disturbance Vectors with Low Hamming Weight

Finding good disturbance vectors is the first important step in our analysis. Without imposing any conditions other than the message expansion recursion, the search becomes somewhat easier. However, since there are 16 32-bit free variables, the search space can be as large as 2^{512}. Instead of searching the entire space for a vector with minimum weight, we use heuristics to confine our search within a subspace that most likely contains good vectors.

We note that the 80 disturbance vectors $x_0, ..., x_{79}$ can be viewed as an 80-by-32 matrix where each entry is a single 0/1 bit. A simple observation is that for a matrix with low hamming weight, the non-zero entries are likely to concentrate in several consecutive columns of the matrix. Hence, we can first pick two entries $x_{i,j-1}$ and $x_{i,j}$ in the matrix and let two 16-bit columns starting at $x_{i,j-1}$ and $x_{i,j}$ to vary through all 2^{32} possibilities. There are 64 choices for i ($i = 0, 1, ..., 63$) and 32 choices for j ($j = 1, 2, ..., 32$). In fact, with the same i, different choices of j produce disturbance vectors that are rotations of each other, which would have the same Hamming weight. By setting $j = 2$, we can minimize the carry effect as discussed in Section 3.1. Overall, the size of the search space is at most $64 \times 2^{32} = 2^{38}$.

Using the above strategy, we first search for the best vectors predicting *one-block collisions*. For the full SHA-1, the best one is obtained by setting $x_{64,2} = 1$ and $x_{i,2} = 0$ for $i = 65, .., 79$. The resulting disturbance vector is given in Table 5. The best disturbance vectors for SHA-1 reduced to t-step is the same one with the first $80 - t$ vectors omitted. For SHA-1 variants up to 75 steps, the Hamming weight is still small enough up to allow an attack with complexity less than 2^{80}, and Table 7 summarizes the results for these variants.

In order to break the 2^{80} barrier for the full SHA-1, we continue to search for good disturbance vectors that predict *near collisions* and *two-block collisions*. To do so, we compute more vectors after step 80 using the same SHA-1 message expansion formula (also listed in Table 5).

Then we search all possible 80-vector intervals $[x_i, ..., x_{i+79}]$. *Any* set of 80 vectors with small enough Hamming weight can be used for constructing a near collision. In fact, we found a total of 12 good sets of vectors, and this gives us some freedom to pick the one that achieves the best complexity when taking into account other criteria and techniques (other than just the Hamming weight).

Table 3. Hamming weights (for Rounds 2-4) of best disturbance vectors for SHA-1 variants found by experiments. The comparison is made among different subsets of conditions listed in Table 2. The notation 1BC denotes one-block collision, 2BC is two-block collision, and NC implies near collision.

	Existing results				Our new results	
	SHA-1		SHA-1 w/o Round 1		SHA-1	
	conditions		conditions		conditions	
	1,2,3	2,3	1,2	2	1	-
step	1BC	NC,2BC	1BC	NC,2BC	1BC	NC,2BC
47	**26**	12	24	12	5	5
53	42	20	**16**	16	10	7
54	39	**24**	36	16	10	7
60					14	11
70					14	17
75					**26**	21
80					31	**25**

Finally, we compare the minimal Hamming weight of disturbance vectors found by experiments when different conditions are imposed. In Table 3, the last two columns are obtained from our new analysis and other data are from [2]. Provided that the average probability in 2-4 rounds is 2^{-3}, a valid disturbance vector should have a Hamming weight less than a threshold 27, because the corresponding collision (or near-collision) differential has the probability higher than 2^{-80} which can result in an attack faster than the 2^{80} theoretical bound. In the table, we mark the step in bold for which this threshold is reached. It is now easy to see that removing all the conditions has a significant effect in reducing the Hamming weight of the disturbance vectors.

4.3   Techniques for Constructing Differential Paths

In this section, we present our new techniques for constructing a differential path given a disturbance vector with low Hamming weight. Since the vector no longer satisfies the seemly required conditions listed in Table 2, constructing a valid differential path that leads to collisions becomes more difficult. Indeed, this is the most complicated part of our new attacks on SHA-1. It is also a crucial part of the analysis, since without a concrete differential path, we would not be able to search for *real* collisions.

Below, we describe the high-level ideas in these new analysis techniques.

– Use "subtraction" instead of "exclusive-or" as the measure of difference to facilitate the precision of the analysis.
– Take advantage of special differential properties of IF. In particular, when an input difference is 1, the output difference can be 1, −1 or 0. Hence, the function can *preserve*, *flip* or *absorb* an input difference, giving good flexibility for constructing differential paths.

- Take advantage of the carry effect. Since $2^j = -2^j - 2^{j+1}... - 2^{j+k-1} + 2^{j+k}$ for any k, a single bit difference j can be expanded into several bits. This property makes it possible to introduce extra bit differences.
- Use different message differences for the 6-step local collision. For example, $(2^j, 2^{j+5}, 0, 0, 0, 2^{j+30})$ is a valid message differences for a local collision in the first round.
- Introduce extra bit differences to produce the impossible bit-differences in the consecutive local collisions corresponding to the consecutive disturbances in the first 16 steps, or to offset the bit differences of chaining variables produced by truncated local collisions.

A near-collision differential path for the first message block is given in Table 11.

4.4   Deriving Conditions

Given a valid differential path for SHA-1 or its reduced variants, we are ready to derive conditions on messages and chaining variables. The derivation method was originally introduced in [14] for breaking SHA-0, and can be applied to SHA-1 since SHA-0 and SHA-1 have the same step update function. Most details can be found in our analysis of SHA-0 [16], and hence are omitted. Here we focus on the differences between SHA-0 and SHA-1 and discuss a new technique that is tailored to SHA-1.

Due to the extra shift operation in the message expansion of SHA-1, a disturbance can occur in bit positions other than bit 2 of the message words (as can be seen from Table 5), while for SHA-0, all disturbances initiate in bit 2. If this happens in the XOR rounds (round 2 and 4), the number of conditions will increase from 2 to 4 for each local collision. This can blow up the total number of conditions if not handled properly.

We describe a useful technique for utilizing two sets of message differences corresponding to two consecutive disturbances within the *same* step i to produce one 6-step local collision. For example, if there is a disturbance in both bit 1 and bit 2 of x_i, we can set the signs of the message differences Δm_i to be opposite in those two bits. This way, the actual message difference can be regarded as one difference bit in position 1, since $2^1 - 2^0 = 2^0$. Hence the number of conditions can be reduced from $4 + 2 = 6$ to 4.

The conditions for the near-collision path in Table 11 are given in Table 12.

4.5   Message Modification Techniques

Using the basic message modification techniques in [11,12,13], we can modify an input message so that all conditions on the chaining variables can hold in the first 16 steps. With some additional effort, we can modify the messages so that all conditions in step 17 to 22 also hold.

Note that message modification should keep all the message conditions to hold in order to satisfy the differential path. All the message conditions can be expressed as equations of bit variables in $m_0, m_1, ...m_{15}$ (message words before message expansion). Because of the 1-bit shift in message recursion, all the equations aren't contradictory. Suppose we would like to correct 10 conditions from step 17 to 22 by modifying the last 6 message words $m_{10}, m_{11}, ...m_{15}$. From Table 12, we know there are 32 chaining variable conditions, together with total 47 message equations from step 11 to step 16, the total number of conditions is 79 in step 11-16. Intuitively, this leaves a message space of size 2^{113}, which is large enough for modifying some message bits to correct 10 conditions.

4.6   Picking the Best Disturbance Vector

Once the conditions are derived and message modifications are applied, we can analyze the complexity in a very precise way, by counting the remaining number of conditions in Rounds 2 to 4. The counting rules depend on the Boolean function and locations of the disturbances occur in each round, and local collisions across boundaries of rounds need to be handled differently. The details are summarized in Table 8 in the appendix.

Given the disturbance vectors in Table 5, we find that for an 80-step near collision, the minimum Hamming weight is 25 using the 80 vectors with index [15,94]. However, the *minimum number of conditions* is 71 using the 80 vectors with index [17,96]. This is because the conditions in step 79 and 80 can be ignored for the purpose of near collisions, and the condition in step 21 can be made to hold (see Section 4.5). The step-by-step counting for the number of conditions for this vector is given in Table 9.

Using minimum number of conditions as the selection criteria, we pick the vectors with index [17,96] as the disturbance vectors for constructing an 80-step near collision.

4.7   Using Near Collisions to Find Collisions

Using the idea of multi-block collisions in [7,2,3,12], we can construct two-block collisions using near collisions. For MD5 [12], the complexity of finding the first block near-collision is higher than those of the second block near-collision because of the determination for the bit-difference positions and signs in the last several steps. Here we show that by keeping the bit-difference positions and the signs as free variables in the last two steps, we can maintain essentially twice the search complexity while moving from near collisions to two-block collisions. This idea is also applicable to MD5 to further improve its collision probability from 2^{-37} to 2^{-32}.

Let M_0 and M_0' be the two message blocks and $\Delta h_1 = h_1' - h_1$ be the output difference for the 80-step near collision. If we look closely at the disturbance vectors that we have chosen, there are 4 disturbances in the last 5 steps that will propagate to Δh_1, which become the input differences in the initial values for the second message block.

There are two techniques that we use to construct the differential path for the second message blocks M_1 and M_1'. First, we apply the techniques described in

Section 4.3 so that Δh_1 can be "absorbed" in the first 16 steps of the differential paths. Second, we set the conditions on M_1 so that the output difference Δh_2 will have opposite signs for each of the differences in Δh_1. In other words, we set the signs so that $\Delta h_2 + \Delta h_1 = 0$, meaning a collision after the second message block. We emphasize that setting these conditions on the message does *not* increase the number of conditions on the resulting differential path, and hence it does not affect the complexity.

To summarize, the near collision on the second message block can be found with the same complexity as the near collision for the first message block. Therefore, there is only a factor of two increase in the overall complexity for getting a two-block full collision.

4.8   Complexity Analysis and Additional Techniques

Using the modification techniques described in this section, we can correct the conditions of steps 17-22. Furthermore, message modification will not result in increased complexity if we use suitable implementation tricks such as "precomputation". First, we can precompute and fix a set of messages in the first 10 steps and leave the rest as free variables. By Table 9, we know that there are 70 conditions in steps 23-77. For three conditions in steps 23-24, we use the "early stopping technique". That is, we only need to carry out the computation up to step 24 and then test whether three conditions in steps 23-24 hold. This needs about 12 step operations including message modification for correcting conditions of steps 17-22. This is equivalent to about two SHA-1 operations. Hence, the total complexity of finding the near-collision for the full SHA-1 is about 2^{68} computations. Considering the complexity of finding the second near-collision differential path, the total complexity of finding a full SHA-1 collision is thus about 2^{69}.

The results for SHA-1 reduced variants are summarized in Table 6 and Table 7 in the appendix.

5   Detailed Analysis: a 58-Step Collision of SHA-1

When $t = 58$, our analysis suggests that collisions can be found with about 2^{33} hash operations, which is within the reach of computer search. In this section, we describe some details on how to find a real collision for this SHA-1 variant. The collision example is given in Table 4.

5.1   Constructing the Specific Differential Path

We first introduce some notation. Let $a_{i,j}$ denote the jth bit of variable a_i and $\Delta a_i = a_i' - a_i$ denote the difference. Note that we use *subtraction* difference rather than *exclusive-or* difference since keeping track of the signs is important in the analysis. Following the notation introduced in [12], we use $a_i[j]$ to denote $a_i[j] = a_i + 2^{j-1}$ with no bit carry, and $a_i[-j]$ to denote that $a_i[-j] = a_i - 2^{j-1}$ with no bit carry.

Table 4. A collision of SHA-1 reduced to 58 steps. Note that padding rules are not applied to the messages, and $compress(h_0, M_0) = compress(h_0, M'_0) = h_1$.

h_0:	67452301 efcdab89 98badcfe 10325476 c3d2e1f0
M_0:	132b5ab6 a115775f 5bfddd6b 4dc470eb 0637938a 6cceb733 0c86a386 68080139
	534047a4 a42fc29a 06085121 a3131f73 ad5da5cf 13375402 40bdc7c2 d5a839e2
M'_0:	332b5ab6 c115776d 3bfddd28 6dc470ab e63793c8 0cceb731 8c86a387 68080119
	534047a7 e42fc2c8 46085161 43131f21 0d5da5cf 93375442 60bdc7c3 f5a83982
h_1:	9768e739 b662af82 a0137d3e 918747cf c8ceb7d4

We use step 23 to step 80 of the disturbance vector in Table 5 to construct a 58-step differential path that leads to a collision. The specific path for the first 16 steps is given in Table 10, and the rest of the path consists of the usual local collisions.

As we discussed before, there are two major complications that we need to deal with in constructing a valid differential path in the first 16 steps. In what follows, we describe high-level ideas as how to deal with the above two problems, and some technical details are omitted.

1. Message differences from a disturbance initiated in steps -5 to -1. These differences are $m_0[30], m_1[-5, 6, -30, 31], m_2[-1, 30, -31]$.
2. Consecutive disturbances in the same bit position in the first 16 steps. There are two such sequences: (1) $x_{1,2}, x_{2,2}, x_{3,2}$ and (2) $x_{8,2}, x_{9,2}, x_{10,2}$.

It is more instructive to focus on the values of Δa_i without carry expansion, which is the left column for Δa_i in Table 10. We first consider the propagation of the difference $m_1[-5, 6]$. It produces the following differences:

$$a_2[5] \rightarrow a_3[10] \rightarrow a_4[15] \rightarrow a_5[20] \rightarrow a_6[25].$$

These differences in a propagate through b, c, d to the following differences in the chaining variable e:

$$e_6[3] \rightarrow e_7[8] \rightarrow e_9[13] \rightarrow e_9[18] \rightarrow e_{10}[23].$$

The differences in b, c, d are easy to deal with since they can be *absorbed* by the Boolean function. So we only need to pay attention to variables a and e. The difference $a_6[25]$ as well as the five differences in e_i are cancelled in the step *immediately after* the step in which they first occur. This way, they will not propagate further. The cancellation is done using either existing differences in other variables or extra differences from the carry effect. For example, we expand $a_8[-18]$ to $a_8[18, 19, ..., -26]$ so that $a_8[25, -26]$ can produce the bit difference $c_{10}[23, -24]$ to offset $e_{10}[23]$, and $a_8[-26]$ produce $b_9[-26]$ to cancel out $e_9[26]$.

The consecutive disturbances are handled in different ways. For the first sequence, the middle disturbance $m_2[2]$ is combined with $m_2[1]$ so that the disturbance is shifted from bit 2 to bit 1. For the second sequence, the middle disturbance $m_9[2]$ is offset by $c_9[2]$, which comes from the difference $a_7[4]$.

One might get too swamped with the technicality for deriving such a complicated differential path. It is helpful to summarize the flow in the main approach:
(1) analyze the propagation of differences, (2) identify wanted and un-wanted
differences, and (3) use the Boolean function and the carry effect to introduce
and absorb these differences.

5.2   Deriving Conditions on a_i and m_i

The method for deriving conditions on the chaining variables is essentially the
same as in our analysis of SHA-0 [16], and so the details are omitted here.

The method for deriving conditions on the messages is more complicated
since it involves more bit positions in the message words. To simplify the analysis, we first find a partial message (the first 12 words) that satisfies all the
conditions in the first 12 steps. This can be done using message modification
techniques in a systematic way. This leaves us with four free variables, namely
$m_{12}, m_{13}, m_{14}, m_{15}$. Next we can write each m_i ($i \geq 16$) as a function of the four
free variables using the message expansion recursion. Conditions on these m_i
then translate to conditions on $m_{12}, m_{13}, m_{14}, m_{15}$, and these bits will be fixed
during the collision search.

6   Conclusions

In this paper, we present the first attack on the full SHA-1 with complexity less
than 2^{69} hash operations. This attack is also available to find one-block collisions
for the SHA-1 reduced variants less than 76 rounds. For example, we can find a
collision of 75-round SHA-1 with complexity 2^{78}, and find a collision of 70-round
SHA-1 with complexity 2^{68}.

Some strategies of the attack can be utilized to further improve the attacks
on MD5 and SHA-0 etc. For example, applying the new technique of combining
near-collision paths into a collision path, we can improve the successful probability of the attack on MD5 from 2^{-37} to 2^{-32}.

At this point, it is worth comparing the security of the MD4 family of hash
functions against the best known attacks today. We can see that more complicated message preprocessing does provide more security. However, even for
SHA-1, the message expansion does not seem to offer enough avalanche effect
in terms of spreading the input differences. Furthermore, there seem to be some
unexpected weaknesses in the structure of all the step updating functions. In
particular, because of the simple step operation, the certain properties of some
Boolean functions combined with the carry effect actually facilitate, rather than
prevent, differential attacks.

We hope that the analysis on SHA-1 as well as other hash functions will
provide useful insight on design criteria for more security hash functions. We
anticipate that the design and analysis of new hash functions will be an important research topic in the coming years.

Acknowledgements

It is a pleasure to acknowledge Arjen K. Lenstra for his important suggestions, corrections, and for spending his precious time on our research. We would like to thank Andrew C. Yao and Frances. Yao for their support and corrections on this paper. We also thank Ronald L. Rivest and many other anonymous reviewers for their important comments.

References

1. E. Biham and R. Chen. *Near Collisions of SHA-0*. Advances in Cryptology – Crypto'04, pp.290-305, Springer-Verlag, August 2004.
2. E. Biham and R. Chen. *New Results on SHA-0 and SHA-1*. Crypto'04 Rump Session, August 2004.
3. E. Biham, R. Chen, A. Joux, P. Carribault, W. Jalby and C. Lemuet. *Collisions in SHA-0 and Reduced SHA-1*. Advances in Cryptology–Eurocrypt'05, pp.36-57, May 2005.
4. NIST. *Secure hash standard*. Federal Information Processing Standard, FIPS-180, May 1993.
5. NIST. *Secure hash standard*. Federal Information Processing Standard, FIPS-180-1, April 1995.
6. F. Chabaud and A. Joux. *Differential Collisions in SHA-0*. Advances in Cryptology – Crypto'98, pp.56-71, pringer-Verlag, August 1998.
7. A. Joux. *Collisions for SHA-0*. Rump session of Crypto'04, August 2004.
8. K. Matusiewicz and J. Pieprzyk. *Finding Good Differential Patterns for Attacks on SHA-1*. IACR Eprint archive, December 2004.
9. V. Rijmen and E. Osward. *Update on SHA-1*. RSA Crypto Track 2005, 2005.
10. X. Y. Wang, D. G. Feng, X. J. Lai, and H. B. Yu. *Collisions for Hash Functions MD4, MD5, HAVAL-128 and RIPEMD*. Rump session of Crypto'04 and IACR Eprint archive, August 2004.
11. X. Y. Wang, D. G. Feng, X. Y. Yu. *The Collision Attack on Hash Function HAVAL-128*. In Chinese, Science in China, Series E, Vol. 35(4), pp. 405-416, April, 2005.
12. X. Y. Wang and H. B. Yu. *How to Break MD5 and Other Hash Functions*. Advances in Cryptology–Eurocrypt'05, pp.19-35, Springer-Verlag, May 2005.
13. X. Y. Wang, X. J. Lai, D. G. Feng, H. Chen, X. Y. Yu. *Cryptanalysis for Hash Functions MD4 and RIPEMD*. Advances in Cryptology–Eurocrypt'05, pp.1-18, Springer-Verlag, May 2005.
14. X. Y. Wang. *The Collision attack on SHA-0*. In Chinese, to appear on www.infosec.edu.cn, 1997.
15. X. Y. Wang. *The Improved Collision attack on SHA-0*. In Chinese, to appear on www.infosec.edu.cn, 1998.
16. X. Y. Wang. H. B. Yu, Y. Lisa Yin, *Efficient Collision Search Attacks on SHA-0*. These proceedings. 2005.

A   Appendix: Tables

Table 5. Disturbance vectors of SHA-1. The 96 vectors x_i ($i = 0, ..., 95$) satisfy the SHA-1 message expansion recursion, but no other conditions. The second *italicized* index is only needed for numbering the 80 vectors that are chosen for constructing the best 80-step near collision.

index i	*index*	vector x_{i-1}	index i	*index*	vector x_{i-1}	index i	*index*	vector x_{i-1}
1		e0000000	33	*17*	80000002	65	*49*	2
2		2	34	*18*	0	66	*50*	0
3		2	35	*19*	2	67	*51*	0
4		80000000	36	*20*	0	68	*52*	0
5		1	37	*21*	3	69	*53*	0
6		0	38	*22*	0	70	*54*	0
7		80000001	39	*23*	2	71	*55*	0
8		2	40	*24*	2	72	*56*	0
9		40000002	41	*25*	1	73	*57*	0
10		2	42	*26*	0	74	*58*	0
11		2	43	*27*	2	75	*59*	0
12		80000000	44	*28*	2	76	*60*	0
13		2	45	*29*	1	77	*61*	0
14		0	46	*30*	0	78	*62*	0
15		80000001	47	*31*	0	79	*63*	0
16		0	48	*32*	2	80	*64*	0
17	*1*	40000001	49	*33*	3	81	*65*	4
18	*2*	2	50	*34*	0	82	*66*	0
19	*3*	2	51	*35*	2	83	*67*	0
20	*4*	80000002	52	*36*	2	84	*68*	8
21	*5*	1	53	*37*	0	85	*69*	0
22	*6*	0	54	*38*	0	86	*70*	0
23	*7*	80000001	55	*39*	2	87	*71*	10
24	*8*	2	56	*40*	0	88	*72*	0
25	*9*	2	57	*41*	0	89	*73*	8
26	*10*	2	58	*42*	0	90	*74*	20
27	*11*	0	59	*43*	2	91	*75*	0
28	*12*	0	60	*44*	0	92	*76*	0
29	*13*	1	61	*45*	2	93	*77*	40
30	*14*	0	62	*46*	0	94	*78*	0
31	*15*	80000002	63	*47*	2	95	*79*	28
32	*16*	2	64	*48*	0	96	*80*	80

Table 6. Search complexity for near collisions (NC) and two-block collisions (2BC) of SHA-1 reduced to t steps. "Start & end index" refers to the index for disturbance vectors in Table 5. The complexity estimation takes into account the speedup using early stopping techniques (see Section 4.8), and the estimation for 78-80 steps also takes into accounts the speedup by advanced modification techniques (see Section 4.5).

t-step SHA-1	start & end index of DV	HW in ro.2-4	# conditions in ro.2-4	complexity NC	2BC
80	17, 96	27	71	2^{68}	2^{69}
79	17, 95	26	71	2^{68}	2^{69}
78	17, 94	24	71	2^{68}	2^{69}
77	16, 92	23	71	2^{68}	2^{69}
76	19, 94	22	69	2^{66}	2^{67}
75	20, 94	21	65	2^{62}	2^{63}
74	21, 94	20	63	2^{60}	2^{61}
73	20, 92	20	61	2^{58}	2^{59}
72	23, 94	19	59	2^{56}	2^{57}
71	24, 94	18	55	2^{52}	2^{53}
70	25, 94	17	52	2^{49}	2^{50}
69	26, 94	16	50	2^{48}	2^{49}
68	27, 94	16	48	2^{46}	2^{47}
67	28, 94	16	45	2^{43}	2^{44}
66	29, 94	15	41	2^{39}	2^{40}
65	30, 94	13	40	2^{38}	2^{39}
64	29, 92	14	37	2^{35}	2^{36}
63	32, 94	12	35	2^{33}	2^{34}
62	33, 94	11	34	2^{32}	2^{33}
61	32, 92	11	31	2^{29}	2^{30}
60	29, 88	12	29	2^{27}	2^{28}
59	30, 88	10	28	2^{26}	2^{27}
58	29, 86	11	25	2^{23}	2^{24}
57	32, 88	9	23	2^{21}	2^{22}
56	33, 88	8	22	2^{20}	2^{21}
55	32, 86	8	19	2^{17}	2^{18}
54	33, 86	7	18	2^{16}	2^{17}
53	34, 86	7	18	2^{16}	2^{17}
52	32, 83	7	15	2^{13}	2^{14}
51	33, 83	6	14	2^{12}	2^{13}
50	34, 83	6	14	2^{12}	2^{13}

Table 7. Search complexity for one-block collisions of SHA-1 reduced to t steps. Explanation of the table is the same as that for 6.

SHA-1 reduced to t steps	start & end point of DV	HW in rounds 2-4	# conditions in rounds 2-4	search complexity
80	1, 80	31	96	2^{93}
79	2, 80	30	95	2^{92}
78	3, 80	30	90	2^{87}
77	4, 80	28	88	2^{85}
76	5, 80	27	83	2^{80}
75	6, 80	26	81	2^{78}
74	7, 80	25	79	2^{76}
73	8, 80	25	77	2^{74}
72	9, 80	25	77	2^{74}
71	10, 80	24	74	2^{71}
70	11, 80	24	71	2^{68}
69	12, 80	22	68	2^{66}
68	13, 80	21	62	2^{60}
67	14, 80	19	58	2^{56}
66	15, 80	19	55	2^{53}
65	16, 80	18	51	2^{49}
64	17, 80	18	48	2^{46}
63	18, 80	16	48	2^{46}
62	19, 80	16	45	2^{43}
61	20, 80	15	41	2^{39}
60	21, 80	14	39	2^{37}
59	22, 80	13	38	2^{36}
58	23, 80	13	35	2^{33}
57	24, 80	12	31	2^{29}
56	25, 80	11	28	2^{26}
55	26, 80	10	26	2^{24}
54	27, 80	10	24	2^{22}
53	28, 80	10	21	2^{19}
52	29, 80	9	17	2^{15}
51	30, 80	7	16	2^{14}
50	31, 80	7	14	2^{12}

Table 8. Rules for counting the number of conditions in rounds 2-4

step	disturb in bit 2	disturb in other bits	comments
19	0	1	For a_{21}
20	0	2	For a_{21}, a_{22}
21	1	3	Condition a_{20} is "truncated"
22-36	2	4	
37	3	4	
38-40	4	4	
41-60	4	4	
61-76	2	4	
77	2	3	Conditions are "truncated"
78	2	2	starting at step 77.
79	(1)	(1)	Conditions for step 79,80
80	(1)	(1)	can be ignored in analysis

Special counting rules:

1. If two disturbances start in both bit 2 and bit 1 in the same step, then they only result in 4 conditions (see Section 4.8).
2. For Round 3, two consecutive disturbances in the same bit position only account for 6 conditions (rather than 8). This is due to the property of the MAJ function.

Table 9. Example: Counting the number of conditions for the 80-step near collision. The "index" refers to the second italicized index in Table 5.

index	number of conditions	comments
21	$4 - 1 - 1 = 2$	4 cond's: $a_{20}, a_{21}, a_{22}, a_{23}$
		$-\ a_{20}$ due to truncation
		$-\ a_{21}$ using modification
23,24,27,28		
32,35,36	$2 \times 7 = 14$	
25,29,33,39	$4 \times 4 = 16$	
43,45,47,49	$4 \times 4 = 16$	
65,68,71,73,74	$4 \times 5 = 20$	
77	3	Truncation
79	0	2 conditions ignored
80	0	1 condition ignored
Total	71	

Table 10. The differential path for the 58-step SHA-1 collision. Note that x_i $(i = 0..15)$ are the disturbance vector for the first 16 steps, which correspond to the 16 vectors indexed by 23 through 38 in Table 5. The Δ entries list the positions of the differences and their signs. For example, the difference 2^j is listed as $(j + 1)$ and -2^j as $-(j + 1)$.

step i	x_{i-1}	Δm_{i-1}	Δa_i no carry	Δa_i with carry	Δb_i	Δc_i	Δd_i	Δe_i
1	80000001	30	30	$-30, 31$				
2	2	-2	2	$-2, 3$				
		$-5, 6$	5	5				
		-30	-30	-30				
		31	31	$-31, 32$	Δa_1			
3	2	$-1, -2$	1	1				
		-7	10	10				
		$30, -31$			Δa_2	$\Delta a_1^{\lll 30}$		
4	2	-7	-2	$2, -3$				
		30	15	$-15, 16$				
			-5	$5, -6$	...	$\Delta a_2^{\lll 30}$	$\Delta a_1^{\lll 30}$	
5	0	$-2, 7$	20	$-20, 21$				
		$30, 31$	28	$-28, 29$				
		32	-1	-1				
			-10	$10, 11, -12$	...		$\Delta a_2^{\lll 30}$	$\Delta a_1^{\lll 30}$
6	0	-2	25	25				
		$-30, -31$	15	$-15, 16$				
					...			$\Delta a_2^{\lll 30}$
7	1	$1, 32$	1	1				
			8	$-8, -9, 10$				
			$4, -21$	$4, -21$				...
8	0	-6	-18	$18, ..., -26$				
					...			
9	80000002	$1, 2$	$-2, 32$	$-2, 32$				
			-9	$9, ..., -19$		...		
10	2	-2						
		$-5, 7$						
		31					...	
11	80000002	$7, 31$	$2, -32$	$2, -32$				
			9	9	...			...
12	0	-2						
		$-5, -7$						
		-30						
		$31, -32$			Δa_{11}	...		
13	2	$-30, -32$	-2	-2				
						$\Delta a_{11}^{\lll 30}$	...	
14	0	$7, 32$						
					Δa_{13}		$\Delta a_{11}^{\lll 30}$	...
15	3	$1, 30$	1	1				
						$\Delta a_{13}^{\lll 30}$		$\Delta a_{11}^{\lll 30}$
16	0	$-6, -7$						
		30			Δa_{15}		$\Delta a_{13}^{\lll 30}$	

Table 11. The differential path for the 80-step SHA1 collision. Note that x_i $(i = 0..19)$ are the disturbance vector for the first 20 steps, which correspond to the 20 vectors indexed by 1 through 20 in Table 5. The Δ entries list the positions of the differences and their signs. For example, the difference 2^j is listed as $(j + 1)$ and -2^j as $-(j + 1)$.

step i	x_{i-1}	Δm_{i-1}	Δa_i no carry	Δa_i with carry	Δb_i	Δc_i	Δd_i	Δe_i
1	40000001	30	30, 31	30, 31				
2		$-2, -4$	2	$-2, 3$				
		6	6	$-6, -7, 8$				
		$-30, -31, 32$	30	$-30, -31, 32$	Δa_1			
3		2 $1, 2$	-1	-1				
		-7	4	4				
		30	11	$-11, -12, -13, 14$	Δa_2	$\Delta a_1^{\lll 30}$		
4	80000002	7	$-2, 9$	$-2, 9$				
		$29, -30$	16	$-16, -17, -18, 19$				
		-32	-32	-32	$...$	$\Delta a_2^{\lll 30}$	$\Delta a_1^{\lll 30}$	
5		1 $1, -2$	-5	$5, -6$				
		$-5, 7$	21	$-21, 22$				
		$29, 31, 32$	28	28		$...$	$\Delta a_2^{\lll 30}$	$\Delta a_1^{\lll 30}$
6		0 $-2, -6$	11	$-11, -12, 13$				
		$29, 31$	16	$-16, 17$				
		32	26	$-26, 27$			$...$	$\Delta a_2^{\lll 30}$
7	80000001	30	1	1				
			$-4, -6$	$-4, 6, -7$				
			32	32				$...$
8		2 $-2, -5, -6$	-19	$19, ..., -26$				
		$30, 31$			$...$			
9		2 $1, -2, -7$	-2	-2				
		$-30, -31$	-10	$10, ..., -20$		$...$		
10		2 7	2	2				
		-30					$...$	
11		0 $2, -7$	9	$-9, 10$				
		$-30, 31, -32$			$...$			$...$
12		0 2	-4	-4				
		$-30, -31$				$...$		
13		1 1	1	1				
		32					$...$	
14		0 -6						
								$...$
15	80000002	$-1, 2$	-32	-32				
16		2 $2, 5, -7$	2	2				
		-31			Δa_{15}			
17	80000002	-7	-2	-2				
		31	32	32	Δa_{16}	$\Delta a_{15}^{\lll 30}$		
18		0 $-2, -5, 7$						
		$30, 31, 32$				$\Delta a_{16}^{\lll 30}$	$\Delta a_{15}^{\lll 30}$	
19		2 30	2	2				
		32					$\Delta a_{16}^{\lll 30}$	$\Delta a_{15}^{\lll 30}$
20		0 -7						
		32			Δa_{19}			$\Delta a_{16}^{\lll 30}$

Table 12. A set of sufficient conditions on a_i for the differential path given in Table 11. The notation 'a' stands for the condition $a_{i,j} = a_{i-1,j}$ and 'b' denotes the condition $a_{19,30} = a_{18,32}$.

chaining variable	conditions on bits			
	$32-25$	$24-17$	$16-9$	$8-1$
a_1	a00-----	--------	1-----aa	1-0a11aa
a_2	01110---	------1-	0aaa-0--	011-001-
a_3	0-100---	-0-aaa0-	--0111--	01110-01
a_4	10010---	a1---011	10011010	10011-10
a_5	001a0---	--01-000	10001111	-010-11-
a_6	1-0-0011	1-1001-0	111011-1	a10-00a-
a_7	0---1011	1a0111--	101--010	-10-11-0
a_8	-01---10	000000aa	001aa111	---01-1-
a_9	-00-----	10001000	0000000-	---11-1-
a_{10}	0-------	1111111-	11100000	0-----0-
a_{11}	--------	------10	11111101	1-a--0--
a_{12}	0-------	--------	--------	10--11--
a_{13}	--------	--------	--------	11----10
a_{14}	-0------	--------	--------	----0-1-
a_{15}	10------	--------	--------	----1-0-
a_{16}	--1-----	--------	--------	----0-0-
a_{17}	0-0-----	--------	--------	------1-
a_{18}	--1-----	--------	--------	----a---
a_{19}	--b-----	--------	--------	------0-
a_{20}	--------	--------	--------	-----a--
a_{21}	--------	--------	--------	-------1

Pebbling and Proofs of Work

Cynthia Dwork[1], Moni Naor[2,*], and Hoeteck Wee[3,**]

[1] Microsoft Research, Silicon Valley Campus
dwork@microsoft.com
[2] Weizmann Institute of Science
moni.naor@weizmann.ac.il
[3] University of California, Berkeley
hoeteck@cs.berkeley.edu

Abstract. We investigate methods for providing easy-to-check proofs of computational effort. Originally intended for discouraging spam, the concept has wide applicability as a method for controlling denial of service attacks. Dwork, Goldberg, and Naor proposed a specific memory-bound function for this purpose and proved an asymptotically tight amortized lower bound on the number of memory accesses any polynomial time bounded adversary must make. Their function requires a large random table which, crucially, cannot be compressed.

We answer an open question of Dwork et al. by designing a compact representation for the table. The paradox, compressing an incompressible table, is resolved by embedding a time/space tradeoff into the process for constructing the table from its representation.

1 Introduction

In 1992 Dwork and Naor proposed that e-mail messages be accompanied by easy-to-check *proofs of computational effort* in order to discourage junk e-mail, now known as spam [12], and suggested specific CPU-bound functions for this purpose[1]. Noting that memory access speeds vary across machines much less than do CPU speeds, Abadi, Burrows, Manasse, and Wobber [1] initiated a fascinating new direction: replacing CPU-intensive functions with *memory-bound* functions, an approach that treats senders more equitably.

Memory-bound functions were further explored by by Dwork, Goldberg, and Naor [11], who designed a class of functions based on pointer chasing in a large shared random table, denoted T. We may think of T as part of the definition of their functions. Using hash functions modelled as truly random functions (i.e. 'random oracles'), they proved lower bounds on the amortized number of memory accesses that an adversary must expend per proof of effort, and gave a concrete implementation in which the size of the proposed table is $16MB$.

---

* Incumbent of the Judith Kleeman Professorial Chair. Research supported in part by a grant from the Israel Science Foundation.
** Work performed at Microsoft Research, Silicon Valley Campus.
[1] A similar proposal was later made by Back in 1997 [6].

There are two drawbacks to the use of a large random table in the definition of the function. If the proof-of-effort software is distributed bundled with other software, then the table occupies a large footprint. In addition, for users downloading the function (or a function update, possibly necessitated by a substantial growth in cache sizes), downloading such a large table might require considerable connect time, especially if the download is done on a common telephone line. Connection fees, i/o-boundedness, and the possibility of transmission errors suggest that five minutes of local computation, to be done once and for all (at least, until the next update), is preferable to five minutes of connect time. Thus a compact representation of T allows for easy distribution and frequent updates.

These considerations lead to the question of whether there might be a succinct representation of T. In other words, is it possible to distribute a *short program* for constructing T while still maintaining the lower bound on the amortized number of memory accesses? The danger is that the adversary (spammer) might be able to use the succinct description of T to generate elements in cache and on the fly, whenever they are needed, only rarely going to memory.

Roughly speaking, our approach is to generate T using a memory-bound *process*. Sources for such processes are time/space tradeoffs, such as those offered by *graph pebbling*, defined below, and *sorting*. We will use both of these: we exploit known dramatic time/space tradeoffs for pebbling in constructing a theoretical solution, with provable complexity bounds; the solution uses a hash function, modeled by a random oracle in the proof. We also describe a heuristic based on sorting. A very nice property of an algorithm whose most complex part is sorting is that it is easy to program, reducing a common source of implementation errors.

We will focus most of our discussion on the pebbling results. The heuristic based on sorting is described in Section 6. Although our work does not rely on computational assumptions, we nonetheless assume the adversary is restricted to polynomial time, or in any event that a spamming approach that requires superpolynomially many cpu cycles is not lucrative. This raises an interesting observation:

Remark 1. If the adversary were not restricted to polynomial time, then the proof of effort would have to be long. Otherwise, by Savitch's Theorem (relativized to a random oracle), the adversary could find the proof using space at most the square of the length of the proof (since guessing the proof has non-deterministic space complexity bounded by the proof length), which may be considerably smaller than the cache size.

Pebbling can be described as a game played on a directed acyclic graph $D(V, E)$ be with a set $S \subset V$ of inputs (nodes of indegree 0) and a set $T \subset V$ of outputs (nodes of outdegree 0). (Eventually we will identify the outputs of D with the elements of the table T.) The player has s pebbles. A pebble may be placed on an input at any time. A pebble may be placed on a node $v \in V \setminus S$ if and only if every vertex u such that $(u, v) \in E$ currently has a pebble. That is, a non-input may be pebbled if and only if all its immediate predecessors hold pebbles. Finally, a pebble may be removed from the graph at any time. Typically, the goal of the player is to pebble outputs using few moves and using few simultaneous pebbles; that is, efficiently and in such a way that at any time there are few pebbles on the graph.

Pebbling has been the subject of deep and extensive research, and it is in the context of proving lower bounds for computation on random access machines *superconcentrators* were invented (see [18]). These are graphs with large flow: for every set A of inputs and every set B of outputs of the same size, there are vertex-disjoint paths connecting A to B. Valiant [18] showed that every circuit for computing a certain type of transform contains a superconcentrator. Although these did not directly yield lower bounds, they eventually yielded time-space tradeoffs, via pebbling arguments.

Pebbling intends to capture time and space requirements for carrying out a particular computation, defined by the graph – we can think of a non-input as being associated with a function symbol, such as "+" or " ×". For example, a sum can be computed if its summands – as represented by the node's immediate predecessors – have been computed. We can think of placing a pebble on a node as tantamount to storing a (possibly newly computed) value in a register. Time/space tradeoffs for specific computation graphs are obtained by showing that no (time efficient) pebbling strategy exists that uses few simultaneous pebbles. Time/space tradeoffs for problems are obtained by showing that every computation graph for the problem yields a tradeoff.

As noted, in our case, the outputs of the graph will correspond to elements of T. If an output cannot be pebbled (in reasonable time) using few pebbles, i.e., little space, we would like to conclude that considerable time or memory accesses were devoted to finding the value associated with the corresponding output of the graph computation. But perhaps there is a *different* computation that yields the same outputs – a computation unrelated to the computation determined by the dag. In this case obtaining a function output does not imply that significant resources have been expended.

We force the adversary to adhere to the computation schema described by the graph by associating a random oracle with each node. That is, we *label* each non-input with the value obtained by applying the hash function to the labels of the predecessor vertices. (The inputs are numbered 1 through N, and the label of input i is the hash of i.) Using this we show, rigorously, how to convert the adversary's behavior to a pebbling.

Although this intuition is sound, our situation is more challenging: while we associate placing a pebble with storing a value in cache, our adversary is not limited to cache memory, but may use main memory as well; moreover, main memory is very large. How can we adapt the classical time/space tradeoffs to this setting?

We address this by constructing a dag D that, in addition to being hard to pebble, remains hard to pebble even when many nodes and their incident edges are removed. Roughly speaking, given the cache contents, we "knock out" those nodes whose labels can be (mostly) determined by the cache contents, together with their incident edges. We need that the "surviving" graph is still hard to pebble. We will also introduce the concept of *spontaneously generated* pebbles to capture memory reads. We show that pebbling remains hard unless the number of spontaneously generated pebbles is large.

We must ensure that no cut in the graph is knocked out. Intuitively, knocking out a cut in the graph corresponds to reducing the depth of the graph, and therefore reducing the difficulty of pebbling the graph. This leads us to a graph with slightly special structure: it is the concatenation of two pieces. The first is a stack of wide (no small cut) superconcentrators (by stack we mean a sequence of DAGs where the outputs of one are the inputs of the next DAG in the sequence). The description of the second is quite technical, and we defer it until the requirements have been better motivated (Section 4).

2   Complete Description of Our Abstract Algorithm

We now describe our algorithm, *Compact_MBound*, postponing the construction of the graph D to Section 4. The algorithm adds a table generation phase to Algorithm MBound of [11]. Thus the two algorithms are identical except that in the new algorithm the table T is generated by the procedure outlined below, while in the original algorithm the table T is completely random. For the concrete implementation, we use a heuristic for generating the table T, described in Section 6. We then combine this with the concrete implementation of Algorithm MBound proposed in [11].

Algorithm Compact_Mbound uses a collection of hash functions $\mathcal{H}' = \{H_0, H_1, H_2, H_3, H_4\}$. The function H_4 has been described in the Introduction; its role is to force the adversary to adhere to a computation defined by the graph D, and its output is w bits long, where w is the word size. The remaining hash functions are those used in the original Algorithm MBound. In our analysis, we will treat each hash function as a random oracle.

2.1   Building Table T

Both the (legitimate) sender and the receiver must build the table T, but this is done only once and the table T is then stored in main memory. After the table has been built, proofs of effort are constructed and checked using the algorithms of Dwork, Goldberg, and Naor [11], reproduced in Section 2.2 for completeness. We will provide an explicit construction of the graph D used in building T. This means that we may incorporate the algorithm for computing D (and thus T) into the proof-of-effort software.

The algorithm for constructing T first computes the graph D which has N inputs, N outputs and constant indegree d, and then numbers the input vertices $1, 2, \ldots, N$ and the output vertices $N+1, N+2, \ldots, 2N$. Next, each vertex of V is labeled with a w-bit string in an inductive fashion, beginning with the input vertices:

1. Input i is labeled $H_4(i)$, for $1 \leq i \leq N$.
2. For vertex $j \notin [N]$, let vertices $i_1 < \cdots < i_d$ be the predecessors of vertex j. Then vertex j is labeled $H_4(label(i_1), \ldots, label(i_d), j)$.

The entries of T correspond to the labels of the output vertices, namely:

$$T[i] = label(N + i), \; i = 1, 2, \ldots, N.$$

Once the table T has been computed and stored, the graph D and the node labels may be discarded.

2.2 Computing and Checking Proofs of Effort

The algorithm described here is due to [11]. It uses a modifiable array A, initialized for each trial. The adversary's model, described in Section 2.3, restricts the size of A: if w is the size of a memory word and b is the number of bits in a memory block (or cache line), then the algorithm requires that $|A|w > b$ bits. A word on notation: For arrays A and T, we denote by $|A|$ (respectively, $|T|$) the number of elements in the array. Since each element is a word of w bits, the numbers of *bits* in these arrays are $|A|w$ and $|T|w$, respectively.

At a high level, the algorithm is designed to force the sender of a message to take a random walk "through T," that is, to make a series of random-looking accesses to T, each subsequent location determined, in part, by the contents of the current location. Such a walk is called a *path*. Typically, the sender will have to explore many different paths until a path with certain desired characteristics is found. Such a path is called *successful*, and each path exploration is called a *trial*. Once a successful path has been identified, information enabling the receiver to check that a successful path has been found is sent along with the message.

The algorithm for computing a path in a generic trial is specified by two parameters ℓ (path length) and e (effort), and takes as input a message m, sender's name (or address) S, receiver's name (or address) R, and date d, together with a trial number k:

Initialization:
$\quad A = H_0(m, R, S, d, k)$
Main Loop: Walk for ℓ steps (ℓ is the path length):
$\quad c \leftarrow H_1(A)$
$\quad A \leftarrow H_2(A, T[c])$
Success occurs if after ℓ steps the last e bits of $H_3(A)$ are all zero.

A legitimate proof of effort is a 5-tuple (m, R, S, d, k) along with the value $H_3(A)$ for which success occurs. This may be verified with $O(\ell)$ work by just exploring the path specified by k and checking that the reported hash value $H_3(A)$ is correct and ends with e zeroes. The value of $H_3(A)$ is added to prevent the adversary from simply guessing k, which has probability $1/2^e$ of success.

An honest sender computes a proof of effort by repeating path exploration for $k = 1, 2, \ldots$ until success occurs. The probability of success for each trial is $1/2^e$, so the expected amount of work for the honest sender is $O(2^e \ell)$. The main technical component in [11] is showing that $\Omega(2^e \ell)$ work is also *necessary* (for a random T).

2.3   The Adversary's Model

We assume an adversary's computational model as in [11]. The adversary is assumed to be limited to a "standard architecture" as specified below:

1. There is a large memory, partitioned into m blocks (also called cache lines) of b bits each;
2. The adversary's cache is small compared to the memory. The cache contains at most s (for "space") *words*; a cache line typically contains a small number (for example, 16) of words;
3. Although the memory size is large compared to the cache, we assume that m is still only polynomial in the largest feasible cache size s;
4. Each word contains w bits (commonly, $w = 32$);
5. To access a location in the memory, if a copy is not already in the cache (a *cache miss* occurs), the contents of the block containing that location must be brought into the cache – a *fetch*; every cache miss results in a fetch;
6. We charge one unit for each fetch of a memory block. Thus, if two adjacent blocks are brought into cache, we charge two units (there is no discount for proximity at the block level);
7. Computation on data *in the cache* is essentially *free*. By not (significantly) charging the adversary for this computation, we are increasing the power of the adversary; this strengthens the lower bound.

3   Pebbling

The goal in pebbling is to find a strategy to pebble all the outputs while using only a few pebbles simultaneously and not too many steps (pebble placements). Pebbling has received much attention, in particular in the late seventies and early eighties, as a model for space bounded computation (as well as other applications, such as the relative power of programming languages) [9,13,14,16].

A directed acyclic graph with bounded indegree, N inputs, and N outputs is an N-*superconcentrator* if for any $1 \le k \le N$ and any sets $\mathcal{S}'$ of inputs and $\mathcal{T}'$ of outputs, both of size k, there are k vertex-disjoint paths connecting $\mathcal{S}'$ to $\mathcal{T}'$. Thus, superconcentrators are graphs with excellent flow. (Note that we do not assume the need to specify which input is connected to which output.)

The following classical results are relevant to our work. Here m denotes the number of vertices in the graph.

– Stacks of superconcentrators yield graphs with a very sharp tradeoffs: To pebble all the outputs of these graphs with fewer than N pebbles requires time exponential in the depth, *independent of the initial configuration of the pebbles* [13].

- Constant-degree constructions of linear-sized superconcentrators of small depth were given in [18,15,10]. The construction of minimum known density is by Alon and Capalbo [4].

The Basic Lower Bound Argument. Many proofs of pebbling results rely on the following so-called *Basic Lower Bound Argument* [16,13]. The claim of the Basic Lower Bound Argument is that to pebble $s + 1$ outputs of a superconcentrator with *any* initial placement of at most s pebbles requires the pebbling of $N - s$ inputs, independent of the initial configuration of the s pebbles.

To see this, suppose that fewer than $N - s$ inputs are pebbled. Then there exists a set S'' of $s + 1$ inputs that do not receive pebbles. By the superconcentrator property these $s + 1$ inputs are connected via vertex-disjoint paths to the target set of $s+1$ outputs that should be pebbled. Every one of these paths must at some point receive a pebble, else not all the target outputs can be pebbled. Since a node cannot be pebbled without pebbling all its ancestors, it follows that every input in our set of size $s + 1$ must receive a pebble at some point, contradicting the assumption.

3.1   Converting the Adversary's Moves to a Pebbling

The adversary does not define its operation in terms of pebbling but instead we assume that we (the provers of the lower bound) can follow its memory accesses and the applications of the functions of $\mathcal{H}'$ and in particular H_4. We now describe how the adversary's actions yield a pebbling of the graph. The pebbling is determined by an *off-line* inspection of the adversary's moves, i.e., following an execution of the adversary it is possible to describe the pebbling that occurred. Hence we call it *ex post facto* pebbling:

Placing Initial Pebbles. If H_4 is applied with $label(j')$ as an argument, and $label(j')$ was not computed via H_4, then we consider j' to have a pebble in an initial configuration. We sometimes refer to these as *spontaneously generated* pebbles.

Placing a Pebble. If H_4 is applied to i for some $1 \leq i \leq N$, then place a pebble on node i (recall the inputs are vertices $1, \ldots, N$, so node i in this case is an input vertex). Let j be a non-input vertex (so $j > N$), and let $i_1 < \cdots < i_d$ be the predecessors of vertex j. If H_4 is applied to $(label(i_1), \ldots, label(i_d), j)$, where $label(i_b)$ is the correct label of vertex i_b, $1 \leq b \leq d$, then place a pebble on vertex j.

Removing a Pebble. A pebble is removed as soon as it is not needed anymore. Here we use our clairvoyant capabilities and remove the pebble on node j' right after a call for H_4 with the correct value of $label(j')$ as one of the arguments if (i) $label(j')$ is not used anymore or (ii) $label(j')$ is computed again before it is used as an argument to H_4. That is, before the next time $label(j')$

appears as an argument to H_4 it also appears as the result of computing H_4 (the output of H_4).

We may relate the ex post facto pebbling strategy to the adversary's strategy as follows:

- Placing a pebble corresponds to making an oracle call to H_4. Hence, a lower bound on the number of (placement) moves in the pebbling game yields a lower bound on the number of oracle calls and thus the amount of work done by the adversary.
- The initial (spontaneously generated) pebbles correspond to the values of H_4 that the adversary "learns" without invoking H_4. Intuitively, this information must come from the cache contents and memory fetches. Therefore, we would expect that if the adversary has a cache of s words and fetches z bits from memory, then the adversary is limited to at most $s + z/w$ initial pebbles, since each pebble corresponds to a w-bit string.

The following lemma formalizes our intuition relating the number of pebbles used in the ex post facto pebbling to the cache size of the adversary and the number of bits the adversary fetches from memory. It says that with very high probability the ex post fact pebbling uses only $s + z/w$ simultaneous pebbles. The intuition is that if more pebbles are used, then somehow the sw bits in the cache and the additional z bits obtained from memory are being used to reconstruct $s + z/w + 1$ labels, or $sw + z + w$ bits.

Lemma 1. *Consider an adversary that operates for a certain number of steps and where:*

- *the adversary is using a standard architecture as specified in Section 2.3 with a cache of s words of size w; and*
- *the adversary brings from memory at most z bits.*

Then with probability at least $1 - 2^{-w}$ the maximum number of pebbles at any given point in the ex post facto pebbling is bounded by $s + z/w$. The probability is over $\mathcal{H}'$.

Proof. We need the following simple observation:

Claim 1. *Let $b_1 \ldots b_u$ be independent unbiased random bits and let $k \leq u$. Suppose we have a (randomized) reconstruction procedure that, given a hint of length $B < k$ (which may be based on the value of $b_1 \ldots b_u$), produces a subset $S \subset \{1, \ldots u\}$ of k indices and a guess of the values of $\{b_i \mid i \in S\}$. Then the probability that all k guesses are correct is at most $2^B/2^k$, where the probability is over the random variables and the coin flips of the hint generation and the reconstruction procedures.*

Proof (of claim). Fix an arbitrary sequence of random choices for the reconstruction procedure. Each fixed hint yields a choice of S and a guess of the bits of S. For any fixed hint, the probability, over choice of $b_1, \ldots, b_u$, that all the guessed values are consistent with the values of the elements of S is 2^{-k}. Therefore, by summing over all hints, the probability that there exists a hint yielding a guess consistent with the actual value of $b_1, \ldots, b_u$ is 2^{B-k}. □

Suppose there are $s + z/w + 1$ pebbles at some point in the ex post facto pebbling. We apply the claim to bound the probability that this occurs, as follows: the independent unbiased random bits $b_1, \ldots, b_u$ are from the truth table of $\mathcal{H}'$; the hint consists of three parts:

1. The cache contents $C \in \{0,1\}^{sw}$
2. The bits brought from main memory (at most z)
3. The values of the functions in $\mathcal{H}'$ needed to simulate the adversary *except those values of H_4 corresponding to the initial pebbles.*

The output of the reconstruction procedure is all the values of the functions in $\mathcal{H}'$ given in the hint, plus the labels of the $s + z/w + 1$ pebbled nodes. The reconstruction procedure works by simulating the adversary (up to the point where there are $s + z/w + 1$ pebbles) and outputting the labels of the pebbled nodes as the values are generated. The difference between the length of the output (in bits) and the length of the hint (in bits) is w, so the probability is bounded above by 2^{-w}. □

4   Description of Our Graphs

Call our graph D and let it be composed from two pieces D_1 and D_2. We are interested in a graph with a small number of nodes and edges, since each node corresponds to an invocation of H_4 and the number of edges corresponds to the total size of inputs in the H_4 calls. We are less concerned with the depth of the graph.

The dag D has $N = |T|$ inputs and outputs. It is constructed from two dags D_1 and D_2 via concatenation; that is:

- Inputs of D_1 are the inputs of D.
- The outputs of D_1 are the inputs of D_2.
- Outputs of D_2 are the outputs of D.

The properties we require from the two dags are different. In particular we allow the spammer more pebbles for the D_1 part. For each of D_1 and D_2 we first describe the properties needed (in terms of pebbling) and then mention constructions of graphs that satisfy these requirements.

The Dag $D_1 = (V_1, E_1)$: We require an almost standard pebbling lower bound property: for a certain β to be determined later, pebbling any $m > s + 2\beta b/w$ outputs of D_1 is either impossible or at the very least requires exponential (in the depth of D_1) time (making it infeasible) provided we are constrained to:

1. start from any initial setting of at most $s' = s + 2\beta b/w$ pebbles. The number of pebbles in the initial configuration comes from two sources: the cache size in words (s) and the bits brought from memory during a given interval, divided by the word size w. In Section 5 the number of blocks brought from memory will be denoted 2β, containing a total of $2\beta b$ bits. (In the language of Lemma 1, $z = 2\beta b$.) Hence $s' = s + 2\beta b/w$.

2. use at most s pebbles during the pebbling itself; that is, of the initially placed pebbles, some are designated as *permanent*, and do not move. Only s pebbles may be moved, and these s may be moved repeatedly.

Constructing D_1: One way to obtain such a graph is to consider a stack of $\ell_1 \in \omega(\log |T|)$ N-superconcentrators (that is, N inputs and N outputs). Then following the work of [13] (Section 4) we know that D_1 has the desired property: independent of the initial configuration, the time to pebble $m > s'$ outputs requires time at least exponential in the depth, hence, superpolynomial in $|T|$. This means that when we consider in Section 5 an interval of computation by the spammer, then by appealing to Lemma 1 we can argue that either (i) fewer than m outputs of D_1 were pebbled or (ii) at least $mw - 2\beta b$ bits were brought from main memory during this interval (which suffices to show high amortized memory accesses). This follows from the randomness of the labels of V_1 (that is, the randomness of H_4), as discussed above.

Since there are linear-sized superconcentrators [18,15,4], this means the size of V_1 can be some function in $\omega(|T| \log |T|)$. Also these constructions are explicit, so we have an *explicit* construction of D_1.

Remark 2. An alternative graph to the stack of superconcentrators is that of Paul, Tarjan and Celoni [14], where the number of nodes is $O(N \log N)$.

The Dag $D_2 = (V_2, E_2)$: The property needed for D_2 is that even if a significant fraction of the vertices fail it should be very hard to disconnect small sets of surviving outputs from the surviving inputs. (For now, think of a failed node as one whose label is largely determined by the cache contents.)

The vertices V_2 are partitioned into layers $L_1, L_2, \ldots L_{\ell_2}$ of size N. Suppose that nodes in V_2 fail but we are guaranteed that from each layer L_i at least a δ fraction of nodes survives. The condition on the surviving graph can be expressed as follows: There exists a set S' of the inputs and a set T' of the outputs both of size $\Omega(N)$, such that, for any set $U \subset T'$ of x outputs, where the bound x is derived from the proof of Algorithm Mbound, to completely disconnect U from all of S' by removing nodes requires either cutting $\Omega(x)$ vertices in *some* level not including the input, or cutting $\Omega(N)$ inputs.

Constructing D_2: One way to construct D_2 is using a stack of bipartite expanders on N nodes, where we identify the left set of one expander with the right set of the other, except for the inputs and outputs of D_2, which are identified with the leftmost and rightmost sets respectively; the orientation of the edges is from the inputs to the outputs. Expanders are useful for us for two reasons: (i) they do not have small cuts and (ii) they have natural fault tolerance properties. In particular, Alon and Chung [5] have shown that in any good enough expander if up to some constant (related to the expansion) fraction of nodes are deleted, then one can still find a smaller expander (of linear size) in the surviving graph (see also Upfal [17]).

Consider now the dag obtaining by stacking $\ell_2 = O(\log |T|)$ bipartite expanders where each side has $N = |T|$ nodes. We give here an intuitive explanation of why disconnecting a relatively small set U of surviving outputs from the surviving inputs requires deleting $|U|/2$ vertices at some level. Following the argument in [5], there exists a subgraph F of the surviving graph with the following property: every layer of F contains $\delta N/2$ vertices, and the bipartite graph induced by any two consecutive layers in F satisfies a vertex expansion property (in the direction from the output nodes towards the input nodes) with expansion factor 2 for subsets of size at most $\delta N/16$, say. Consider any set U of size $o(N)$ outputs in F. Clearly, by deleting U we can disconnect it from the inputs. Suppose we delete at most $|U|/2$ output vertices in F. Then, there are at least $|U|/2$ vertices left amongst the output nodes, which are connected to at least $|U|$ vertices in level $\ell_2 - 1$ in F. Again, if we delete at most $|U|/2$ output vertices in level $\ell_2 - 1$, then U must be connected to at least $|U|$ vertices in level $\ell_2 - 2$. Continuing this argument, we may deduce that U must be connected to at least $|U|/2$ input nodes in F, unless we delete at least $|U|/2$ nodes at some level. To ensure that disconnecting $|U|$ inputs is insufficient we use an additional property of D_2: the surviving subgraph D_2' contains a substantial superconcentrator. Restricting out attention to output sets U of the surviving superconcentrator suffices for our lower bound proof.

We conclude that the total number of nodes in V_2 can be $O(|T| \log |T|)$ and thus the dominating part is D_1. Also we have explicit construction of expanders and hence of D_2.

5   An Amortized Lower Bound on Cache Misses

In this section we prove that any spammer limited to a standard architecture (as specified in Section 2.3) and trying to generate many different proofs of computational efforts according to Algorithm Compact_Mbound presented in Section 2 (i.e. the verifier follows the algorithm there, while the spammer is free to apply any algorithm), will, with high probability over choices of the random oracles and the choices made by the adversary, have a large amortized number of cache misses.

5.1   The Lower Bound

We are now ready to state the main theorem:

Theorem 1. *Fix an adversary spammer $\mathcal{A}$. Consider an arbitrarily long but finite execution of $\mathcal{A}$'s program – we don't know what the program is, only that*

$\mathcal{A}$ is constrained to use an architecture as described in Section 2 and that its computation of H_1 and H_2 has to be via oracle calls.

Under the following additional conditions, the expectation, over choice of the hash functions $\mathcal{H}'$, and the coin flips of $\mathcal{A}$, of the amortized complexity of generating a proof of effort that will be accepted by a verifier is $\Omega(2^e \ell)$. Note that in [11], "choice of T" meant choice of the random table T. In our case, "choice of T" means choice of the hash function H_4.

- $|T| \geq 2s$ (recall that the cache contains s words of w bits each)
- $|A|w \geq bs^{1/5}$ (recall that b is the block size, in bits).
- $\ell > 8|A|$
- The total amount of work by the spammer (measured in oracle calls) per successful path is no more than $2^{o(w)}2^e \ell$ and no more than 2^{ℓ_1}, where ℓ_1 is the depth of the dag D_1.
- ℓ is large enough so that the spammer cannot call the oracle 2^ℓ times.

The amortized cost of a proof of effort is the sum of the costs of the individual proofs divided by the number of proofs. The theorem says that

$$E_{\mathcal{H}',\mathcal{A}}[\text{amortized cost of proof of effort}] = \Omega(2^e \ell) \qquad (1)$$

Remark 3. As noted in [11], if (for some reason) it must be the case that $|A| \leq O(b/w)$, then the lower bound obtained is $\Omega(2^e \ell / \log s)$. Also, as noted in [11], the theorem holds if expected amortized cost (over $\mathcal{H}'$ and flips) is replaced with "with high probability."

Our proof follows the structure of the proof in [11]; naturally, however, we must make several modifications since T is no longer random. We will describe the key lemma in the original proof, and sketch the proof given in [11]. We will then state and sketch the proof of the new version of the key lemma, yielding a proof of Theorem 1. We start with a simple lemma from [11]:

Lemma 2 ([11]). The expected amortized number of calls to H_1 and H_2 per proof of effort that will be accepted by a verifier is $\Omega(2^e \ell)$. The expectation is taken over T, $\mathcal{A}$, and $\mathcal{H} = \mathcal{H}' \setminus \{H_4\}$.

In our case an analogous lemma holds (with exactly the same proof). This time, the expectation is taken over $\mathcal{A}$ and $\mathcal{H}'$ (recall that in the current work T is defined by H_4 and the dag D).

The execution is broken into intervals in which, it will be shown, the adversary is forced to learn a large number of elements of T. That is, there will be a large number of scattered elements of T which the adversary will need in order to make progress during the interval, and very little information about these elements is in the cache at the start of the interval. The proof holds even if the adversary is allowed during each interval, to remember "for free" the contents of all memory locations fetched during the interval, provided that at the start of the subsequent interval the cache contents are reduced to sw bits once again.

For technical reasons the proof focuses on the values of A only in the second half of a path. Recall that A is modified at each step of the Main Loop; intuitively, since these modifications require many elements of T, these "mature" A's cannot be compressed. The definition of an interval allows focusing on progress on paths with "mature" A's.

A *progress call* to H_1 is a call in which the arguments have not previously been given to H_1 in the current execution. Let $n = s/|A|$. A progress call is *mature* if it is the jth progress call of the path, for $j > \ell/2$ (recall that ℓ is the length of a path).

Let k be a constant determined in [11]. An *interval* is defined by fixing an arbitrary starting point in an execution of the adversary's algorithm (which may involve the simultaneous exploration of many paths), and running the execution until kn mature progress calls (spread over any number of paths) have been made to oracle H_1.

At any point in the computation, the *view* of the spammer is T together with the parts of the oracles $\mathcal{H}$ that the spammer has explicitly invoked. Intuitively, the view contains precisely that information which can have affected the memory of the spammer. Since T is (when T is random, and in any case, could be) stored in memory, we consider it part of the view.

We now state the key lemma from [11]; recall that in that setting T is random.

Lemma 3. *There is a constant $k \geq 1$ where the following is true. Fix any integer i, the "interval number". Choose T and $\mathcal{H}$, and coin flips for the spammer. Run the spammer's algorithm, and consider the ith interval. The expected number of memory accesses made during this interval is $\Omega(n)$, where the expectation is taken over the choice of T, the functions $\mathcal{H}$, and the coin flips of the spammer. That is,*

$$E_{T,\mathcal{H},\mathcal{A}}[\text{number of memory accesses}] = \Omega(n) \qquad (2)$$

Note that between intervals the adversary is allowed to store whatever it wishes into the cache, taking into account all information it has seen so far, in particular, the table T and the calls it has made to the hash functions.

It is an easy consequence of this lemma that the amortized number of memory accesses to find a successful path is $\Omega(2^e \ell)$. This is true since by Lemma 2, success requires an expected $\Omega(2^e \ell)$ mature progress calls to H_1, and the number of intervals is the total number of mature progress calls to H_1 during the execution, divided by kn, which is $\Omega(2^e \ell/n)$. (Note that we have made no attempt to optimize the constants involved.)

5.2   Sketch of Proof of Key Lemma for Random T

We give here a slightly inaccurate but intuitive sketch of the key steps of the proof in [11] of Lemma 3.

The spammer's problem is that of *asymmetric communication complexity* between memory and the cache. Only the cache has access to the functions H_1 and H_2 (the arguments must be brought into cache in order to carry out the function calls). The goal of the (spammer's) cache is to perform *any kn* mature progress calls. Since by definition the progress calls to H_1 are calls in which the arguments *have not previously been given to H_1 in the current execution*, we can assume the values of H_1's responses on these calls are uniform over $1, \ldots, |T|$ given all the information currently in the system (memory and cache contents and queries made so far). The cache must tell the memory which blocks are needed for the subsequent call to H_2. The cache sends the memory $\beta \log m$ bits to specify the block numbers (which is by assumption $o(n \log m)$ bits), and gets in return βb bits altogether from the memory. The key to the proof is, intuitively, that the relatively few possibilities in requesting blocks by the cache imply that many different elements of T specified by the indices returned by the kn mature calls to H_1 must be derived using the same set of blocks. This is shown to imply that more than s elements of T can be reconstructed from the cache contents alone, which is a contradiction given the randomness of T.

It is first argued that a constant δ fraction of elements of T are largely undetermined by the contents of the cache. This is natural, since T is random and the cache can hold at most half the bits needed to represent T. For simplicity, assume that elements that are largely undetermined are in fact completely undetermined, that is, there is simply no information about these in the cache. Call these completely undetermined elements T'.

Simplifying slightly, it is next argued that, for a constant k to be determined later, if one fixes any starting point in the execution of the spammer's algorithm, and considers all oracle calls from the starting point until the knth call to H_1, there will be at least $5n$ pairs of calls to H_1 and H_2 on the same path; that is, H_2 is called on the index determined by the call to H_1.

Intuitively, this observation implies that, since the calls to H_1 return random indices into T, many of the elements of T selected by these invocations will be in T'. That is, there will be no information about them in the cache, and the spammer will have to go to memory to resolve them.

Let β be the average number of blocks sent by the main memory to the cache during an interval. Assume for the sake of contradiction that for at least half the kn-*tuples* of elements selected by H_1, the spammer makes only $2\beta = o(n)$ memory accesses, even though it needs $\Omega(n)$ elements in T' and about which it has no knowledge.

Unfortunately, this assumption of $o(n)$ memory accesses does not yield a contradiction: as a memory access fetches an entire block, which contains multiple words, the total number of bits retrieved (βb) is not necessarily less than the total number of bits needed (at least nw). Indeed, if $\beta \geq nw/b$ there is no contradiction, and it is *not* assumed that w/b is a fixed constant.

The contradiction is derived by using the fact that some set of 2β blocks suffices to reconstruct *many different* possible kn-tuples. This is immediate from a pigeonhole argument (since there are roughly $|T|^{kn}$ kn-tuples and $m^\beta = |T|^{O(\beta)}$ choices of β blocks, since we assume that the memory contains poly($|T|$) words). Let $\mathcal{G}'$ denote the largest such set of kn-tuples.

Let Σ denote the union over tuples in $\mathcal{G}'$, of the set of elements in the tuple. That is, Σ contains every element that appears in $\mathcal{G}'$. It is possible to show that

$|\Sigma|$ is large, i.e. when the entries are measured in terms of the missing bits, then the total number is $\omega(\beta b)$.

Note that if $\mathcal{G}'$ is known to the cache party, then intuitively, by sending the 2β blocks the memory transmits all $p \in \Sigma$ to the cache: to reconstruct any given p, the cache chooses any kn-tuple in $\mathcal{G}'$ containing p, acts as if the calls to H_1 returned the indices (in T) of the elements in this tuple, and runs the spamming program, extracting p in the process. However, $\mathcal{G}'$ is not known to the cache, since it may depend on the full memory content.

At this point, [11] argues that there is a small collection of "*mighty*" tuples, with the property that each tuple in the collection enables the transmission of "too many" elements in Σ. That is, there exists a too large set U of elements reconstructible from too few mighty tuples. This yields an information-theoretic argument that too many bits are obtained from too few. In the sequel, we let $x = |U|$. In [11] it is shown that setting $x = 4\beta b/w$ yields the desired contradiction. We do not repeat that proof here, but we use the same proof, and so, the same value of x.

This concludes the high-level sketch of the proof in [11] for the key lemma in the case that T is random.

5.3   Lower Bound When T Has a Compact Representation

The new key lemma for the lower bound proof is given below. We then sketch those aspects of the proof germane to the case of the compact representation of T. We let $s = |S|$. This is the size of the cache, in words.

Lemma 4. *There is a constant $k \geq 1$ where the following is true. Fix any integer i, the "interval number". Choose $\mathcal{H}'$ and coin flips for the spammer. Run the spammer's algorithm, and consider the ith interval. The expected number of memory accesses made during this interval is $\Omega(n)$, where the expectation is taken over the choice of $\mathcal{H}'$ and the coin flips of the spammer. That is,*

$$E_{\mathcal{H}',\mathcal{A}}[\text{number of memory accesses}] = \Omega(n) \qquad (3)$$

Consider an ex post facto pebbling on D induced by the adversary's execution during an interval, obtained as described in Section 3.1.

Either $\Omega(N)$ of the nodes at the level common to D_1 and D_2 been pebbled during the interval or not. In the first case, i.e., the case in which a constant fraction of the nodes at this level have been pebbled, by the discussion in Section 4, many blocks must be brought from memory. So if we are at an execution of a *good* tuple (one for which the adversary goes to memory at most $2\beta b/w$ times) we can conclude that this did not happen.

In the second case, i.e., the case in which $o(N)$ inputs of D_2 are pebbled, we will use the fault-tolerant flow property of D_2 to show that much information will have to be brought from memory *for some layer of D_2*.

As in the previous section, fix the cache contents C and consider the large set $\mathcal{G}'$ of tuples all utilizing the same set of of blocks from memory. The next claim (proved using Sauer's Lemma) guarantees that in each layer of D_2, there is a constant fraction δ of nodes for each which the label is mostly unknown:

Claim 2. *Fix cache contents* $C \in \{0,1\}^{sw}$. *For any* $1 \leq j \leq \ell_2$, *for* $\gamma, \delta \geq 0$, *consider the event that there exists a subset of the entries of* L_j, *called* L'_j, *of size at least* $\delta|T|$, *such that for each node* i *in* L'_j *there is a set* S_i *of* $2^{\gamma w}$ *possible values for label*(i) *and all the* S_i's *are mutually consistent with the cache contents* C. *Then there exist constants* $\gamma, \delta > 0$ *such that the probability of this event, over choice of the hash functions* $\mathcal{H}'$, *is high.*

Note that we applied the Lemma for each level individually and we cannot assume that the missing labels of different layers are necessarily independent of each other. To derive a contradiction we need to argue that there is some level where it is possible learn too many undetermined values from too few hints.

The fault tolerance property of D_2 tells us that even if we delete the remaining $(1-\delta)$ fraction of nodes (whose values may be determined from the cache contents and without access to H_4) from each layer, the surviving graph contains a large surviving superconcentrator, call it F, with excellent flow: to disconnect any set X of x outputs from all the inputs requires removing $\Omega(x)$ vertices at *some* level other than the input level to F, or disconnecting $\Omega(N)$ nodes from the input level (in the latter case we appeal to the properties of D_1).

The set X of outputs is obtained as in Section 5.2, from the union of the collection of *mighty* tuples in $\mathcal{G}'$. The collection covers a set X of unknown entries in T, and that X contains more bits than the $2\beta b$ bits of information brought from memory, ($|X| = x$ is roughly the size of $|\Sigma|$). Since each element in X is covered by some tuple in the collection, when the spammer $\mathcal{A}$ is initiated with that tuple, the ex post facto pebbling process must place a pebble on all paths from the inputs of F to that node. Therefore the union of the pebbles placed by $\mathcal{A}$ on all tuples in the collection disconnects X from the inputs F. By the properties of D_2 this means that $\Omega(x)$ spontaneously generated pebbles were placed at some level. A careful choice of x, following the argument in [11], yields a contradiction: too many bits from too few.

6   A Heuristic Based on Sorting

We now present an alternative construction of the table T, designed with an eye toward simplicity of definition. Our concrete heuristic is based, intuitively, on the known time/space lower bound tradeoff for sorting of Borodin and Cook [8]. However, as opposed to the pebbling results which we were able to convert into lower bound proofs, here we are left with a scheme with a conjectured lower bound only.

1. T is initialized to $T[i] = H_4(i)$, $1 \leq i \leq N$.
2. Repeat m times:
   Sort T.
   $T[i] := H_4(i, T[i])$.

Inserting i guarantees that collisions do not continue to be mapped to the same value (otherwise the number of distinct values in T could dwindle in successive applications).

The number of iterations m of the loop should be as large as possible while the loop can still be considered to take a "reasonable" amount of time on a relatively slow machine. Given current technology, to defeat a 16MB cache, we take the number of elements in T to be $n = 2^{23}$, where each element is a single 32-bit word. Thus T requires 2^{25} bytes $= 32$MB.

Note that after applying $H_4(i, T[i])$ we have a pretty good idea where this this value will end up after the next sorting phase, up to $\sqrt{|T|}$ roughly. However without actually sorting there does not seem to be a way to find the exact location, and we conjecture that the uncertainty increases with the number of iterations. We would therefore like the number of iterations of the loop to well exceed $\log_2 |T|$, say 40 for a 32MB table. The hash function H_4 may be instantiated with whichever function is considered 'secure' at the time of deployment. If this is considered too costly, then the "best" function that will allow 40 iterations in the desired running time should be used (in general we prefer more iterations than a more secure function).

It would be very interesting to see whether the time/space lower bounds known for sorting [8] and recent advances in space lower bounds [2,3,7] can be applied for the sorting heuristic in order to obtain lower bounds on the spammer's work.

References

1. M. Abadi, M. Burrows, M. Manasse, and T. Wobber. Moderately hard, memory-bound functions. In *Proc. 10th NDSS*, 2003.
2. M. Ajtai. Determinism versus non-determinism for linear time RAMs. In *Proc. 31st STOC*, pages 632–641, 1999.
3. M. Ajtai. A non-linear time lower bound for boolean branching programs. In *Proc. 40th FOCS*, pages 60–70, 1999.
4. N. Alon and M. Capalbo. Smaller explicit superconcentrators. *Internet Mathematics*, 1(2):151–163, 2003.
5. N. Alon and F. Chung. Explicit constructions of linear-sized tolerant networks. *Discrete Math*, 72:15–20, 1988.
6. A. Back. Hashcash – a denial of servic counter-measure. Available at http://www.cypherspace.org/hashcash/hashcash.pdf.
7. P. Beame, M. E. Saks, X. Sun, and E. Vee. Super-linear time-space tradeoff lower bounds for randomized computation. In *Proc. 41st FOCS*, pages 169–179, 2000.
8. A. Borodin and S. Cook. A time-space tradeoff for sorting on a general sequential model of computation. *SIAM J. Comput.*, 11(2):287–297, 1982.
9. S. Cook. An observation on time-storage trade off. *JCSS*, 9(3):308–316, 1974.
10. D. Dolev, C. Dwork, N. Pippenger, and A. Wigderson. Superconcentrators, generalizers, and generalized connectors with limited depth. In *Proc. 15th STOC*, pages 42–51, 1983.
11. C. Dwork, A. V. Goldberg, and M. Naor. On memory-bound functions for fighting spam. In *Advances in Cryptology – CRYPTO'03*, pages 426–444, 2003.
12. C. Dwork and M. Naor. Pricing via processing, or, combatting junk mail. In *Advances in Cryptology – CRYPTO'92*, pages 139–147, 1993.
13. T. Lengauer and R.E. Tarjan. Asymptotically tight bounds on time-space trade-offs in a pebble game. *JACM*, 29(4):1087–1130, 1982.

14. W. Paul, R. E. Tarjan, , and J. R. Celoni. Space bounds for a game on graphs. In *Proc. 8th STOC*, pages 149–160, 1976.
15. N. Pippenger. Superconcentrators. *SIAM J. Comput.*, 6(2):298–304, 1977.
16. M. Tompa. Time-space tradeoffs for computing functions, using connectivity properties of their circuits. *JCSS*, 20(2):118–132, 1980.
17. E. Upfal. Tolerating linear number of faults in networks of bounded degree. In *Proc. 11th PODC*, pages 83–89, 1992.
18. L. G. Valiant. On non-linear lower bounds in computational complexity. In *Proc. 7th STOC*, pages 45–53, 1975.

Composition Does Not Imply Adaptive Security

Krzysztof Pietrzak[*]

ETH Zürich, Department of Computer Science, CH-8092 Zürich Switzerland
pietrzak@inf.ethz.ch

Abstract. We study the question whether the sequential or parallel composition of two functions, each indistinguishable from a random function by non-adaptive distinguishers is secure against adaptive distinguishers. The sequential composition of $F(.)$ and $G(.)$ is the function $G(F(.))$, the parallel composition is $F(.) \star G(.)$ where $\star$ is some group operation. It has been shown that composition indeed gives adaptive security in the information theoretic setting, but unfortunately the proof does not translate into the more interesting computational case.

In this work we show that in the computational setting composition does not imply adaptive security: If there is a prime order cyclic group where the decisional Diffie-Hellman assumption holds, then there are functions F and G which are indistinguishable by non-adaptive polynomially time-bounded adversaries, but whose parallel composition can be completely broken (i.e. we recover the key) with only three adaptive queries. We give a similar result for sequential composition. Interestingly, we need a standard assumption from the asymmetric (aka. public-key) world to prove a negative result for symmetric (aka. private-key) systems.

1 Sequential and Parallel Composition

We continue to investigate the question whether composition of (pseudo) random functions yields a function whose security is in some sense superior to any of it's components. The two most natural ways to compose functions is to either apply them sequentially or in parallel. For two function F and G we denote by $G \circ F$ the *sequential composition*: $G \circ F(x) \stackrel{\text{def}}{=} G(F(x))$. And by $F \star G$ the *parallel composition*: $F \star G(x) \stackrel{\text{def}}{=} F(x) \star G(x)$ where $\star$ is some group operation defined on the range of F and G.

In the information theoretic model one considers computationally unbounded adversaries and only bounds the number of queries they are allowed to make. In this model Vaudenay [9] shows that if a permutation F cannot be distinguished from random with advantage more than ϵ by any adaptive (resp. non-adaptive)[1] distinguisher making q queries, then the sequential composition of k such permutations has security $2^{k-1}\epsilon^k$ against adaptive (resp. non-adaptive) distinguishers.

---

[*] Supported by the Swiss National Science Foundation, project No. 200020-103847/1.

[1] Adaptive means that the distinguisher can choose the $(i+1)$'th query after seeing the output to the i'th query. A non-adaptive distinguisher must decide which q queries he wants to make beforehand.

The same holds for parallel composition where F can be a function and doesn't have to be a permutation. For the computational case, where one considers polynomially time-bounded adversaries a similar amplification result was proven by Luby and Rackoff [3].[2] So if we have a function with some security against adaptive (resp. non-adaptive) distinguishers we can amplify this security for *the same* class of distinguishers in both models.

Another question is whether we always get adaptive security by the composition of non-adaptively secure functions. This is in fact true in the information theoretic model: Maurer and Pietrzak [4] show that if F and G both have security ϵ against *non-adaptive* distinguishers, then $F \star G$ has security $2\epsilon(1 + \ln \epsilon^{-1})$ against *adaptive* distinguishers (the same holds for $G \circ F$ if F and G are permutations). But no such result is known for the computational case. In fact, Myers [6] showed that there is an oracle relative to which non-adaptively secure permutations exist, but their sequential composition is not adaptively secure. This means that if it was indeed true that composition would always imply adaptive security, no relativizing proof for that does exist. As only very few non-relativizing proofs are known (not only in cryptography, but in complexity theory in general), Myers argues that this might be the reason for the lack of formal evidence that composition increases security even though this belief is shared by many cryptographers (including myself until recently).

Here we show that composition does not imply adaptive security in general if there is a group where the decisional Diffie-Hellman assumption holds. We will construct functions F and G which are indistinguishable by non-adaptive (polynomial time) distinguishers if the DDH assumption holds. But where a simple adaptive strategy exists to get the whole key out of $F \star G$ with only three adaptive queries. We then construct F and G such that the same holds for $G \circ F$.

1.1   Notation and Definitions

EFFICIENT/NEGLIGIBLE/INDISTINGUISHABLE. We denote by $\kappa \in \mathbb{N}$ our security parameter. An efficient algorithm is an algorithm whose running time is polynomial in κ. A function $\mu : \mathbb{N} \to [0,1]$ is negligible if for any $c > 0$ there is an n_0 such that $\mu(n) \leq 1/n^c$ for all $n \geq n_0$. Two families of distributions (indexed with κ) are indistinguishable if any efficient algorithm has negligible advantage (over a random guess) in distinguishing those distributions.

THE DDH ASSUMPTION. The DDH assumption for a prime order cyclic group $\mathcal{G} = \mathcal{G}(\kappa)$ states that for a generator g of $\mathcal{G}$ and random x, y the triplet g^x, g^y, g^{xy}

---

[2] Unlike in Vaudenay's information theoretic result, where k, the number of components in the cascade, can be arbitrary (in particular any function of n), the computational amplification proven in [3] requires k to be a constant and independent of the security parameter. Myers [5] proves a stronger amplification for PRFs (which unlike [3] allows to turn a weak PRF into a strong one) for a construction which is basically parallel composition with some extra random values XOR-ed to the inputs.

is indistinguishable from random. We denote the maximal advantage of any algorithm A running in time t for the DDH problem as

$$\mathbf{Adv}_{DDH}(t) \overset{\text{def}}{=} \max_A |\Pr_{x,y}[A(g^x, g^y, g^{xy}) \to 1] - \Pr_{a,b,c}[A(g^a, g^b, g^c) \to 1]|$$

For example the DDH assumption is believed to be true for the following groups: Let Q be a prime such that $Q - 1 = rP$ for some large prime P (say $\log(P) \geq \kappa$). Let h be a generator of $\mathbb{Z}_Q^*$, then $g = h^r$ is a generator of the subgroup $\mathcal{G} \overset{\text{def}}{=} \langle g \rangle$ of order P. In $\mathcal{G}$ any $a \neq 1$ is a generator, here 1 denotes the identity element.

THE EL-GAMAL CRYPTOSYSTEM. Let $\mathcal{G}, g, P$ be like above. The El-Gamal public-key cryptosystem [2] over $\mathcal{G}$ with generator g is defined as follows: The private-key is a random $x \in \mathbb{Z}_P$, and the public key is g^x. To encrypt $m \in \mathcal{G}$ with the public key g^x we choose $r \in \mathbb{Z}_P$ at random and compute the ciphertext in $\mathcal{G}^2$ as

$$\mathsf{Enc}_{g^x}(m, r) = (mg^{xr}, g^r)$$

The decryption of a ciphertext (a, b) with secret key x goes as

$$\mathsf{Dec}_x(a, b) = a/b^x$$

This scheme has some nice properties we will use. It is multiplicative homomorphic: Given an encryption $(mg^{xr}, g^r) = \mathsf{Enc}_{g^x}(m, r)$ of m we can compute an encryption of ℓm as $(\ell m g^{xr}, g^r) = \mathsf{Enc}_{g^x}(\ell m, r)$ even without knowing m or even the public key g^x. In particular given an encryption $(mg^{xr}, g^r) = \mathsf{Enc}_{g^x}(m, r)$ of a known message m we can compute $(g^{xr}, g^r) = \mathsf{Enc}_{g^x}(1, r) = \mathsf{Enc}_{g^x}(1, r)$, an encryption of 1, without even knowing the public key g^x. And further we can rerandomise this encryption by exponentiating with some random r' as $(g^{xrr'}, g^{rr'}) = \mathsf{Enc}_{g^x}(1, rr')$.[3]

DISTINGUISHER. By *distinguisher* we denote an efficient oracle algorithm which at the end of the computation outputs a decision bit. A distinguisher is *non-adaptive* if he generates all his queries before reading any inputs.

A function $\mathsf{R} : \mathcal{K} \times \mathcal{X} \to \mathcal{Y}$ is pseudorandom if for a random key $k \in \mathcal{K}$ the function $\mathsf{R}_k(.) \overset{\text{def}}{=} \mathsf{R}(k, .)$ is indistinguishable from a random function $\mathcal{R} : \mathcal{X} \to \mathcal{Y}$. We denote the distinguishing advantage for R form $\mathcal{R}$ of any distinguisher which runs in time t and makes at most q queries by

$$\mathbf{Adv}_\mathsf{R}(q, t) \overset{\text{def}}{=} \max_A |\Pr_k[A^{\mathsf{R}_k(.)} \to 1] - \Pr[A^{\mathcal{R}(.)} \to 1]|$$

We write $\mathbf{Adv}_\mathsf{R}^{non-adaptive}(q, t)$ if the maximum is only taken over all non-adaptive distinguishers.

---

[3] This is not the standard way of randomising El-Gamal encryptions, where one multiplies (mg^{xr}, g^r) with $(g^{xr'}, g^{r'})$ for a random r' to get $(mg^{x(r+r')}, g^{r+r'}) = \mathsf{Enc}_{g^x}(m, r + r')$. This randomisation works for any encrypted message (not just for $m = 1$), but it requires knowledge of the public key and because of that is not useful for our purpose.

1.2   Some Intuition

For the counterexamples to the conjecture that parallel (resp. sequential) composition does imply adaptive security we will define functions F and G whose output looks random with high probability for any fixed sequence of queries. But if we can query $F \star G$ (reps. $G \circ F$) adaptively we can somehow "convince" F and G of the fact that they are queried adaptively. We then define F, G such that they output their key when they are convinced. We achieve this by letting F and G "communicate" using a semantically secure public-key cryptosystem. The El-Gamal cryptosystem has all the additional features we will need.

THE PARALLEL COMPOSITION COUNTEREXAMPLE. We will now sketch our counterexample of two non-adaptively secure functions F, G where $F \star G$ can be broken with three adaptive queries. The full proof is given in Section 2. Let R be any adaptively secure pseudorandom function. The keyspace of F is a (El-Gamal) public/secret-key pair (pk_F, sk_F) and a key k_F for R (G's key is $(pk_G, sk_G), k_G$). The first thing F/G do on any input is to run it through R_{k_F}/R_{k_G} to produce some pseudorandomness.

We define F and G such that on one particular input α they output their public keys. So if we query $F \star G$ with α we get $pk_F pk_G$.

$$\alpha \to \left\{ \begin{array}{c} \xrightarrow{F} pk_F \\ \xrightarrow{G} pk_G \end{array} \right\} \to pk_F pk_G$$

We further define F and G such that for some fixed β on all inputs of the form (u, β) F computes $pk = u/pk_F$ and then outputs the encryption (using the randomness generated by R) of some fixed value γ under pk. G does the same thing. So if we now feed the output from the first query back into $F \star G$ we get

$$(pk_F pk_G, \beta) \to \left\{ \begin{array}{c} \xrightarrow{F} \mathsf{Enc}_{pk_G}(\gamma, r) \\ \xrightarrow{G} \mathsf{Enc}_{pk_F}(\gamma, r') \end{array} \right\} \to \mathsf{Enc}_{pk_G}(\gamma, r)\mathsf{Enc}_{pk_F}(\gamma, r')$$

And finally on general input (u, v) we define F as follows: First F divides v by the output it would have produced on input (u, β). If this value is an encryption of γ under pk_F, F is "convinced" that it is in an adaptive setting and outputs his key, otherwise F just outputs some pseudorandom stuff. G does the same thing. Let's see what happens if our third query consists of the outputs from the two first queries we made, i.e. $(pk_F pk_G, \mathsf{Enc}_{pk_G}(\gamma, r)\mathsf{Enc}_{pk_F}(\gamma, r'))$. Here F checks if the value computed as

$$\frac{\mathsf{Enc}_{pk_G}(\gamma, r)\mathsf{Enc}_{pk_F}(\gamma, r')}{\mathsf{Enc}_{pk_G}(\gamma, r) \leftarrow F(pk_F pk_G, \beta)} = \mathsf{Enc}_{pk_F}(\gamma, r')$$

is an encryption of γ, as is the case here F outputs its key, and so will G. To prove the non-adaptive security of F and G, we first observe that for a fixed input the above check will fail almost certainly. So we must only care about the α query, which gives the random pk_F and queries of the form (u, β) where we get $\mathsf{Enc}_{u/pk_\mathsf{F}}(\gamma, r)$. We show that if (given pk_F) we could distinguish such $\mathsf{Enc}_{u/pk_\mathsf{F}}(\gamma, r)$ from random, then DDH cannot be hard in $\mathcal{G}$.

THE SEQUENTIAL COMPOSITION COUNTEREXAMPLE. We will now sketch our counterexample of two non-adaptively secure functions F, G where $\mathsf{G} \circ \mathsf{F}$ can be broken with three adaptive queries. The full proof is given in Section 3. As in the previous section, let R be an adaptively secure pseudorandom function. Again, F's key is $(pk_\mathsf{F}, sk_\mathsf{F}), k_\mathsf{F}$ and G's key is $(pk_\mathsf{G}, sk_\mathsf{G}), k_\mathsf{G}$.

We define F such that it outputs his public key on some special input α. G first checks if the input is an encryption of 1 (using sk_G): if this is the case G is "convinced" and outputs his key. Otherwise the output is simply an encryption of the input. If the first query we make to $\mathsf{G} \circ \mathsf{F}$ is α then

$$\alpha \xrightarrow{\mathsf{F}} pk_\mathsf{F} \xrightarrow{\mathsf{G}} \mathsf{Enc}_{pk_\mathsf{G}}(pk_\mathsf{F}, r)$$

except in the unlikely case where by chance pk_F happens to be an encryption of 1 under sk_G.

F on inputs $u \neq \alpha$ "treats" u as if it was $\mathsf{Enc}_{pk}(pk_\mathsf{F}, r)$, i.e. an encryption of his public key pk_F under some key pk. Now (as described earlier in this section) F computes $\mathsf{Enc}_{pk}(1, rr')$, an encryption of 1 with some fresh randomness rr'. If we now feed back the output of the first query into $\mathsf{G} \circ \mathsf{F}$

$$\mathsf{Enc}_{pk_\mathsf{G}}(pk_\mathsf{F}, r) \xrightarrow{\mathsf{F}} \mathsf{Enc}_{pk_\mathsf{G}}(1, rr') \xrightarrow{\mathsf{G}} \mathsf{G}\text{'s key } (sk_\mathsf{G}, k_\mathsf{G})$$

and we get G's key. With the third query, which we will not sketch here we then can get F's key as well. Again the non-adaptive indistinguishability of F and G can be shown under the DDH assumption.

2   Parallel Composition Does Not Imply Adaptive Security

In this section we prove that there are two functions F and G, both $\mathcal{K} \times \mathcal{G}^3 \rightarrow \mathcal{G}^3$ ($\mathcal{K}$ denotes the keyspace) which are indistinguishable from a random function $\mathcal{G}^3 \rightarrow \mathcal{G}^3$ by any *non-adaptive* distinguisher if the DDH-assumption is true in $\mathcal{G}$. But the parallel composition $\mathsf{F} \star \mathsf{G}$ can be completely broken (i.e. we recover the keys of F and G) with only 3 *adaptive* queries.

The systems F and G are almost identically defined, we first define F and then make a small change to get G. Let $\mathsf{R} : \mathcal{K}_\mathsf{R} \times \mathcal{G}^3 \rightarrow \mathbb{Z}_P^3$ be any pseudorandom function with keyspace $\mathcal{K}_\mathsf{R}$. The keyspace of F and G is $\mathcal{K}_\mathsf{R} \times \mathbb{Z}_P$.

There is one annoying technicality we must consider; Because we do not only want to distinguish $\mathsf{F}\star\mathsf{G}$ from a random function in the adaptive case, but recover the keys of F and G, we must somehow encode the keys $(\mathcal{K}_\mathsf{R} \times \mathbb{Z}_P)^2$ into the range $\mathcal{G}^3$ of $\mathsf{F} \star \mathsf{G}$. For simplicity we will simply assume that this is possible, i.e. there are two mappings $\phi_1, \phi_2 : \mathcal{K}_R \times \mathbb{Z}_P \to \mathcal{G}^3$ such that from $\phi_1(k_1, x_1)\phi_2(k_2, x_2)$ we can recover k_1, k_2, x_1, x_2.[4]

F with key $(x \in \mathbb{Z}_P, k_F \in \mathcal{K}_R)$ on input (u, v, w) first computes some pseudorandom values.

$$(r_1, r_2, r_3) \leftarrow \mathsf{R}_{k_F}(u, v, w) \tag{1}$$

Now the output is computed as (we set the values α, β and γ as described in Section 1.2 to $\alpha = (1, 1, 1), \beta = (1, 1)$ and $\gamma = 1$)

$$\mathsf{F}(1, 1, 1) \to (g^x, g^{r_2}, g^{r_3})$$
$$\mathsf{F}(u \neq 1, 1, 1) \to ((u/g^x)^{r_1}, g^{r_1}, g^{r_3})$$
$$\mathsf{F}(u \neq 1, v \neq 1, w \neq 1) \to (a, b, c) \quad \text{where}$$
$$(d, e, f) \leftarrow \mathsf{F}(u, 1, 1) \tag{2}$$
$$\text{if } (v/d) = (w/e)^x \text{ then} \tag{3}$$
$$(a, b, c) = \phi_1(k_F, x) \tag{4}$$
$$\text{otherwise } (a, b, c) = (g^{r_1}, g^{r_2}, g^{r_3})$$
$$\mathsf{F}(\text{ all other cases }) \to (g^{r_1}, g^{r_2}, g^{r_3})$$

G with key (y, k_G) is defined similarly, but with (x, k_F) replaced with (y, k_G) and (4) replaced with

$$(a, b, c) = \phi_2(k_G, y)$$

2.1   Breaking F ⋆ G With 3 Adaptive Queries

We will now describe how to get the key out of F⋆G with 3 adaptive queries. The attack below is successful with probability almost 1. It only fails if by chance P divides one of the random values which appear in the exponent of g below. Below we denote with $r_{(i,j)}$ the pseudorandom value r_i computed by F in step (1) on the j'th input. We define $s_{(i,j)} \stackrel{\text{def}}{=} g^{r_{(i,j)}}$. Similarly the r', s' are defined for G. We will use s and s' for uninteresting terms whose only raison d'être is to pad the output to the right length. The first query we make is $(1, 1, 1)$

$$(1, 1, 1) \to \left\{ \begin{array}{l} \xrightarrow{\mathsf{F}} (g^x, s_{(2,1)}, s_{(3,1)}) \\ \xrightarrow{\mathsf{G}} (g^y, s'_{(2,1)}, s'_{(3,1)}) \end{array} \right\} \to \{g^{x+y}, s_{(2,1)}s'_{(2,1)}, s_{(3,1)}s'_{(3,1)})$$

Four our second query we use the first value from the above output.

$$(g^{x+y}, 1, 1) \to$$

$$\left\{ \begin{array}{l} \xrightarrow{\mathsf{F}} (g^{yr_{(1,2)}}, g^{r_{(1,2)}}, s_{(3,2)}) \\ \xrightarrow{\mathsf{G}} (g^{xr'_{(1,2)}}, g^{r'_{(1,2)}}, s'_{(3,2)}) \end{array} \right\} \to (g^{yr_{(1,2)}+xr'_{(1,2)}}, g^{r_{(1,2)}+r'_{(1,2)}}, s_{(3,2)}s'_{(3,2)})$$

---

[4] One could also easily solve this problem without this assumption by simply extending the range of F, G and R with a term $\{0, 1\}^{2\ell}$ for an ℓ such that $\mathbb{Z}_P \times \mathcal{K}_R$ can be encoded with ℓ bits (the group operation on this term is bitwise XOR). If F or G must output their key, they encode it into this term (F into the first, and G into the second half). In all other cases this term is simply filled with a pseudorandom value generated by R.

Our last query is a combination of the two outputs we have seen.

$$(g^{x+y}, g^{yr_{(1,2)}+xr'_{(1,2)}}, g^{r_{(1,2)}+r'_{(1,2)}})) \rightarrow \left\{ \begin{matrix} \overset{F}{\rightarrow} \phi_1(k_F, x) \\ \overset{G}{\rightarrow} \phi_2(k_G, y) \end{matrix} \right\} \rightarrow \phi_1(k_F, x)\phi_2(k_G, y)$$

Thus we learn the whole key! Let's see what happened in the last query. F on this input first by (2) simulated itself on the input $(g^{x+y}, 1, 1)$, which was exactly the input in the second query.

$$(g^{yr_{(1,2)}}, g^{r_{(1,2)}}, s_{(3,2)}) \leftarrow F(g^{x+y}, 1, 1)$$

Next by (3) F checked whether

$$g^{yr_{(1,2)}+xr'_{(1,2)}}/g^{yr_{(1,2)}} = (g^{r_{(1,2)}+r'_{(1,2)}}/g^{r_{(1,2)}})^x$$

and as this is true, F proceeds with (4) and outputs its key $\phi_1(k_F, x)$. Similarly G outputs its key $\phi_2(k_G, y)$.

2.2 Non-adaptive Indistinguishability of F and G

We will prove that

$$\mathbf{Adv}_F^{non-adaptive}(q, t) \leq \mathbf{Adv}_R(q, t') + \frac{2q}{P} + q\mathbf{Adv}_{DDH}(t') \qquad (5)$$

Where $t' = t + poly(\log P, q)$ for some polynomial *poly* which accounts for the overhead implied by the reduction we make. The same bound holds for G. Below we will treat R_{k_F} as if it was a truly random function, the $\mathbf{Adv}_R(q, t')$ term in (5) does account for this inaccuracy.

Assume that the non-adaptive distinguisher A chooses to make q queries (u_i, v_i, w_i) for $i = 1, \ldots, q$. We must only consider inputs of the form $(u, 1, 1)$ and $(u \neq 1, v \neq 1, w \neq 1)$ as in all other cases the output is simply computed by R_{k_F} and thus is random.

If we make a $(u \neq 1, v \neq 1, w \neq 1)$ query the output is also computed by R_{k_F}, except when $(v/d) = (w/e)^x$ for random d, e, x (here and below we say an element is random if its distribution is uniform over his domain. So here e, f are uniform over $\mathcal{G}$ and x over $\mathbb{Z}_P$). Now

$$\Pr_{d,e,x}[(v/d) = (w/e)^x] \leq 2P^{-1}$$

holds. To see this first note that we have $\Pr[w/e = 1] = \Pr[e = w] = P^{-1}$. Now as in $\mathcal{G}$ any element except 1 is a generator, conditioned on $w \neq e$ the $(w/e)^x$ is random and thus equal to v/d with probability P^{-1}.

So probability that for any of the $t \leq q$ queries of the form $(u \neq 1, v \neq 1, w \neq 1)$ will satisfy $(v/e) = (w/f)^x$ is at most $2tP^{-1}$, the $2qP^{-1}$ term in (5) does account for this probability.

Now we must only consider the case where all q queries of the from $(u_i, 1, 1)$. We make a deal with A: he will only make queries where $u_i \neq 1$ for all $1 \leq i \leq q$

but for this we allow A a $(q + 1)$'th query which must be $(1, 1, 1)$, clearly this can only help A.

Moreover we assume that A knows the discrete logarithm z_i of all his u_i's (i.e. $g^{z_i} = u_i$). Of course this can not be guaranteed, but not knowing them can only decrease A's advantage in the analysis below. So the output A gets on his query $(g^{z_1}, 1, 1), \ldots, (g^{z_q}, 1, 1), (1, 1, 1)$ is

$$(g^{(z_1-x)r_1}, g^{r_1}, *), \ldots, (g^{(z_q-x)r_q}, g^{r_q}, *), (g^x, *, *) \qquad (6)$$

where the $*$'s denote random values which are independent of all the other terms. Now distinguishing (6) from random is equivalent to distinguishing

$$(g^x, g^{r_1 x}, g^{r_1}), \ldots, (g^x, g^{r_q x}, g^{r_q}) \qquad (7)$$

for random $x, r_1, \ldots, r_q$ from

$$(a, b_1, c_1), \ldots, (a, b_q, c_q) \qquad (8)$$

where $a, b_1, \ldots, b_q, c_1, \ldots, c_q$ are random. To see this consider the (randomised) mapping τ (here $*$ are random values)

$$\tau[(\alpha, \beta_1, \gamma_1), \ldots, (\alpha, \beta_q, \gamma_q)] \to [(\gamma_1^{z_1} \beta_1^{-1}, \gamma_1, *), \ldots, (\gamma_q^{z_q} \beta_q^{-1}, \gamma_q, *), (\alpha, *, *)]$$

We get the distribution (6) if we apply τ to (7) and the uniform distribution over $(\mathcal{G}^3)^q$ if we apply τ to (8).

So assume A could distinguish (7) from (8) with probability ϵ, then we can construct an algorithm A' which can distinguish (g^x, g^{xr}, g^r) from a random (a, b, c) with advantage ϵ/q using a hybrid argument. The distribution of the i'th hybrid is

$$(g^x, g^{r_1 x}, g^{r_q}), \ldots, (g^x, g^{r_i x}, g^{r_i}), (g^x, b_{i+1}, c_{i+1}), \ldots, (g^x, b_q, c_q)$$

for random $x, r_1, \ldots, r_q, b_{i+1}, \ldots, b_q, c_{i+1}, \ldots, c_q$. Our A' on input (α, β, γ) chooses a random $i, 1 \leq i \leq q$ and generates the distribution

$$(\alpha, \alpha^{r_1}, g^{r_1}), \ldots, (\alpha, \alpha^{r_{i-1}}, g^{r_{i-1}}), (\alpha, \beta, \gamma), (\alpha, b_{i+1}, c_{i+1}), \ldots, (\alpha, b_q, c_q)$$

whose distribution is equal to the i'th hybrid if (α, β, γ) was generated as (g^x, g^{xr}, g^r) for random x, r, and equal to the $i - 1$'th hybrid if (α, β, γ) are three random values.

3   Sequential Composition Does Not Imply Adaptive Security

We will define two functions F and G, both $\mathcal{K} \times \mathcal{G}^3 \to \mathcal{G}^3$ which are indistinguishable from a random function $\mathcal{G}^3 \to \mathcal{G}^3$ by any *non-adaptive* distinguisher if the DDH-assumption is true in $\mathcal{G}$. But the sequential composition G ∘ F can be completely broken (i.e. we recover the keys of F and G) with only 3 *adaptive* queries. Unlike in the previous section, here F and G are defined somewhat differently.

Let $R : \mathcal{K}_R \times \mathcal{G}^3 \to \mathbb{Z}_P^3$ be any adaptively secure pseudorandom function. The keyspace of F and G is $\mathcal{K}_R \times \mathbb{Z}_P$. Let $\phi : \mathcal{K}_R \times \mathbb{Z}_P \to \mathcal{G}^2$ be some encoding of the keyspace of G into a subset of the range of $G \circ F$.

F with key $(x \in \mathbb{Z}_P, k_F \in \mathcal{K}_R)$ on input (u, v, w) first computes the pseudorandom values

$$(r_1, r_2, r_3) \leftarrow R_{k_F}(u, v, w)$$

Then the output is computed as (we set the value α as described in Section 1.2 to $(1, 1, 1)$)

$$F(1, 1, 1) \to (g^x, g^{r_2}, g^{r_3})$$
$$F(u \neq 1, v \neq 1, w) \to \text{ if } u = g^x \text{ then } (v, \phi(k_F, x))$$
$$\text{else } ((u/g^x)^{r_1}, v^{r_1}, g^{r_3})$$
$$F(\text{ all other cases }) \to (g^{r_1}, g^{r_2}, g^{r_3})$$

G with key $(y \in \mathbb{Z}_P, k_G \in \mathcal{K}_R)$ on input (u, v, w) first computes the pseudorandom values

$$(r_1, r_2, r_3) \leftarrow R_{k_G}(u, v, w)$$

Then the output is computed as

$$G(u \neq 1, v \neq 1, w) \to \text{ if } u = g^y \text{ then } (u, v, w)$$
$$\text{elseif } u = v^y \text{ then } (\phi(k_G, y), 1)$$
$$\text{else } (ug^{yr_1}, g^{r_1}, g^{r_3})$$
$$G(\text{ all other cases }) \to (g^{r_1}, g^{r_2}, g^{r_3})$$

3.1   Breaking $G \circ F$ With 3 Adaptive Queries

We will now describe how to get the key out of $G \circ F$ with three adaptive queries. The attack below is successful with probability almost 1. It only fails if by chance P divides one of the random values which appears in the exponent of g below. Let r, r', s, s' be like in the previous section. The first query we make is $(1, 1, 1)$

$$(1, 1, 1) \xrightarrow{F} (g^x, s_{(2,1)}, s_{(3,1)}) \xrightarrow{G} (g^x g^{yr'_{(1,1)}}, g^{r'_{(1,1)}}, s'_{(3,1)})$$

For the next query we use the first two terms of this output

$$(g^x g^{yr'_{(1,1)}}, g^{r'_{(1,1)}}, 1) \xrightarrow{F} (g^{yr'_{(1,1)}r_{(1,2)}}, g^{r'_{(1,1)}r_{(1,2)}}) \xrightarrow{G} (\phi(k_G, y), 1)$$

And we get G's key. Now with the y we just learned and the first output we can compute $g^x = g^x g^{yr'_{(1,1)}}/(g^{r'_{(1,1)}})^y$ and get F's key with the query

$$(g^x, g^y, 1) \xrightarrow{F} (g^y, \phi(k_F, x)) \xrightarrow{G} (g^y, \phi(k_F, x))$$

3.2   Non-adaptive Indistinguishability of F and G

The security of F and G can be reduced to the indistinguishability of R and the hardness of the DDH problem in $\mathcal{G}$ as in section 2.2, i.e.

$$\mathbf{Adv}_{\mathsf{F}}^{non-adaptive}(q,t) \leq \mathbf{Adv}_{\mathsf{R}}(q,t') + \frac{2q}{P} + q\mathbf{Adv}_{DDH}(t') \qquad (9)$$

And the same holds for G. Again we will treat $\mathsf{R}_{k_\mathsf{F}}$ as if it was a truly random function, the $\mathbf{Adv}_{\mathsf{R}(q,t')}$ term in (9) does account for that.

A query to G of the form (u,v,w) where $u=1$ or $v=1$ will just produce a random output. If the i'th query is of the form $(u_i \neq 1, v_i \neq 1, w_i)$ we get as output $(u_i g^{y r_i}, g^{r_i}, *)$ (again $*$ stands for a random value which is independent of all other terms) unless $u_i = v_i^y$ or $u_i = g^y$ for some i, the probability of each such event is exactly P^{-1} as y is random. With the union bound over all $i, 1 \leq i \leq q$ we get an upper bound $2qP^{-1}$ for the probability that any such event happens.

Thus we can assume that the queries are all of the form $(u_i \neq 1, v_i \neq 1, w_i)$ for $i = 1,\ldots,q$ and the output on the i'th query is $(u_i g^{y r_i}, g^{r_i}, *)$ for some random r_i. The distinguisher must now distinguish those $(u_i g^{y r_i}, g^{r_i})$ from sequence of random pairs (b_i, c_i) for $i = 1,\ldots,q$. Or equivalently (as the u_i's are known values) he must distinguish the sequence $(g^{y r_i}, g^{r_i})$ from random. We are generous and give g^y to the distinguisher. Now we can state that problem as distinguishing a sequence $(g^y, g^{y r_i}, g^{r_i})$ from (g^y, b_i, c_i) for $i = 1,\ldots,q$, those are exactly the sequences (7) and (8) for which we already proved that they cannot be distinguished with advantage more than $q\mathbf{Adv}_{DDH}(t')$.

Similarly the non-adaptive security of F can be reduced to the task of distinguishing $(u_i^{r_i} g^{-x r_i}, v_i^{r_i})$ for $i = 1,\ldots,q$ from random given g^x. We can assume that the adversary knows $s_i = \log_{v_i}(g), t_i = \log_{v_i}(u_i)$. Then he can map those tuples to $(g^x, g^{x r_i}, g^{r_i}) = (g^x, (u_i^{r_i} g^{-x r_i}/(v_i^{r_i})^{t_i})^{-1}, (v_i^{r_i})^{s_i})$. So again this is equivalent to distinguish the distribution (7) from (8).

4   Conclusions and Further Work

We showed that the sequential or parallel composition of pseudorandom functions with non-adaptive security is not adaptively secure in general if the DDH assumption is true in any group. Some interesting remaining question we're currently looking at are the following:

- Can we prove the same thing with pseudorandom *permutations*, ideally for efficiently invertible ones. This would show that cascading non-adaptively secure block-ciphers will not give adaptive security in general.
- We only gave counterexamples for the composition of two functions. How does this scale to the composition of $n > 2$ functions? This question has been partially answered in [8], where a non-adaptively secure PRF is constructed (under an assumption which is implied by the DDH-assumption) such that the *sequential* composition of *any* number of this PRFs can be distinguished with 2 adaptive queries with high probability. Unfortunately the approach used there seems not to generalise to parallel composition.

- Can we give this result under weaker assumptions or even unconditionally? Of course we may always assume that one-way functions exist (they imply pseudorandomness and vice-versa) as otherwise there's nothing to prove. We give a negative result in this direction in [7]. There we show that if a non-adaptively secure PRF exists where the sequential composition can be distinguished with two queries (as constructed in this paper[5]), then a secure key-agreement protocol exists. Thus any construction of such PRFs must either assume or unconditionally prove the existence of key-agreement (the DDH-assumption we use is know to imply key agreement [1]).
- The domain and the range for our counterexamples is a product of subgroups of $\mathbb{Z}_Q^*$. This is not what one usually does, can we adapt this such that the range and domain are $\{0,1\}^\ell$, ideally with standard bitwise XOR as group operation for parallel composition.

References

1. Whitfield Diffie and Martin E. Hellman. New directions in cryptography. *IEEE Transactions on Information Theory*, IT-22(6):644–654, 1976.
2. Taher El-Gamal. A public key cryptosystem and a signature scheme based on discrete logarithms. *IEEE Transactions on Information Theory*, 31(4):469–472, 1985.
3. Michael Luby and Charles Rackoff. Pseudo-random permutation generators and cryptographic composition. In *Proc, 18th ACM Symposium on the Theory of Computing (STOC)*, pages 356–363, 1986.
4. Ueli Maurer and Krzysztof Pietrzak. Composition of random systems: When two weak make one strong. In *Theory of Cryptograpy — TCC '04*, volume 2951 of *Lecture Notes in Computer Science*, pages 410–427, 2004.
5. Steven Myers. Efficient amplification of the security of weak pseudo-random function generators. *Journal of Cryptology*, 16(1):1–24, 2003.
6. Steven Myers. Black-box composition does not imply adaptive security. In *Advances in Cryptology — EUROCRYPT 04*, volume 3027 of *Lecture Notes in Computer Science*, pages 189–206, 2004.
7. Krzysztof Pietrzak. Exploring minicrypt, 2005. Manuscript.
8. Patrick Pletscher. Adaptive security of composition, 2005. Semester Thesis. Advisors K. Pietrzak and U. Maurer.
9. Serge Vaudenay. Decorrelation: A theory for block cipher security. *Journal of Cryptology*, 16(4):249–286, 2003.

---

[5] Actually we make three queries, but the third query is only needed to get the key of the second component, we can distinguish already after the second.

On the Discrete Logarithm Problem on Algebraic Tori*

R. Granger[1] and F. Vercauteren[2]

[1] University of Bristol, Department of Computer Science,
Merchant Venturers Building, Woodland Road,
Bristol, BS8 1UB, United Kingdom
granger@cs.bris.ac.uk
[2] Department of Electrical Engineering,
University of Leuven,
Kasteelpark Arenberg 10, B-3001 Leuven-Heverlee, Belgium
frederik.vercauteren@esat.kuleuven.ac.be

Abstract. Using a recent idea of Gaudry and exploiting rational representations of algebraic tori, we present an index calculus type algorithm for solving the discrete logarithm problem that works directly in these groups. Using a prototype implementation, we obtain practical upper bounds for the difficulty of solving the DLP in the tori $T_2(\mathbb{F}_{p^m})$ and $T_6(\mathbb{F}_{p^m})$ for various p and m. Our results do not affect the security of the cryptosystems LUC, XTR, or CEILIDH over prime fields. However, the practical efficiency of our method against other methods needs further examining, for certain choices of p and m in regions of cryptographic interest.

1 Introduction

The first instantiation of public key cryptography, the Diffie-Hellman key agreement protocol [5], was based on the assumption that discrete logarithms in finite fields are hard to compute. Since then, the discrete logarithm problem (DLP) has been used in a variety of cryptographic protocols, such as the signature and encryption schemes due to ElGamal [6] and its variants. During the 1980's, these schemes were formulated in the full multiplicative group of a finite field $\mathbb{F}_p$. To speed-up exponentiation and obtain shorter signatures, Schnorr [24] proposed to work in a small prime order subgroup of the multiplicative group $\mathbb{F}_p^\times$ of a prime finite field. Most modern DLP-based cryptosystems, such as the Digital Signature Algorithm (DSA) [9], follow Schnorr's idea.

Lenstra [15] showed that by working in a prime order subgroup G of $\mathbb{F}_{p^m}^\times$, for extensions that admit an optimal normal basis, one can obtain a further

---

* The work described in this paper has been supported in part by the European Commission through the IST Programme under Contract IST-2002-507932 ECRYPT. The information in this document reflects only the authors' views, is provided as is and no guarantee or warranty is given that the information is fit for any particular purpose. The user thereof uses the information at its sole risk and liability.

speed-up. Furthermore, Lenstra proved that when $|G| \mid \Phi_m(p)$ with $\Phi_m(x)$ the m-th cyclotomic polynomial and $|G| > m$, the minimal surrounding field of G truly is $\mathbb{F}_{p^m}$ and not a proper subfield. Lacking any knowledge to the contrary, the security of this cryptosystem has been based on two assumptions: firstly, the group G should be large enough such that square root algorithms [18] are infeasible and secondly, the minimal finite field in which G embeds should be large enough to thwart index calculus type attacks [18]. In these attacks one does not make any use of the particular form of the minimal surrounding finite field, i.e., $\mathbb{F}_{p^n}$, but only its size and the size of the subgroup of cryptographic interest.

More recent proposals, such as LUC [25], XTR [16] and CEILIDH [22], improve upon Schnorr's and Lenstra's idea, the latter two working in a subgroup $G \subset \mathbb{F}_{q^6}^\times$ with $|G| \mid \Phi_6(q) = q^2 - q + 1$, where q is a prime power. Brouwer, Pellikaan and Verheul [2] were the first to give a cryptographic application of effectively representing elements in G using only two $\mathbb{F}_q$-elements, instead of six, effectively reducing the communication cost by a factor of three.

Rubin and Silverberg [22] showed how to interpret and generalise the above cryptosystems using the algebraic torus $T_n(\mathbb{F}_q)$ which is isomorphic to the subgroup $G_{q,n} \subset \mathbb{F}_{q^n}^\times$ of order $\Phi_n(q)$. For "rational" tori, elements of $T_n(\mathbb{F}_q)$ can be compactly represented by $\varphi(n)$ elements of $\mathbb{F}_q$, obtaining a compression factor of $n/\varphi(n)$ over the field representation.

In this paper we develop an index calculus algorithm that works directly on rational tori $T_n(\mathbb{F}_q)$ and consequently show that the hardness of the DLP can depend on the form of the minimal surrounding finite field. The algorithm is based on the purely algebraic index calculus approach by Gaudry [10] and exploits the compact representation of elements of rational tori. The very existence of such an algorithm shows that the lower communication cost offered by these tori, may also be exploited by the cryptanalyst.

In practice, the DLP in T_2 and T_6 are most important, since they determine the security of the cryptosystems LUC [25], XTR [16], CEILIDH [22], and MNT curves [19]. We stress that when defined over prime fields $\mathbb{F}_p$, the security of these cryptosystems is not affected by our algorithm. Over extension fields however, this is not always the case. In this paper, we provide a detailed description of our algorithm for $T_2(\mathbb{F}_{q^m})$ and $T_6(\mathbb{F}_{q^m})$. Note that this includes precisely the systems presented in [17], and also those described in [28,27] via the inclusion of $T_n(\mathbb{F}_p)$ in $T_2(\mathbb{F}_{p^{n/2}})$ and $T_6(F_{p^{n/6}})$ when n is divisible by two or six, respectively, which for efficiency reasons is always the case. Our method is fully exponential for fixed m and increasing q. From a complexity theoretic point of view, it is noteworthy that for certain very specific combinations of q and m, for example when $m! \approx q$, the algorithms run in expected time $L_{q^m}(1/2, c)$, which is comparable to the index calculus algorithm by Adleman and DeMarrais [1]. However, our focus will be on parameter ranges of practical cryptographic interest rather than asymptotic results.

A complexity analysis and prototype implementation of these algorithms, show that they are faster than Pollard-Rho in the *full* torus $T_2(\mathbb{F}_{q^m})$ for $m \geq 5$

and in the *full* torus $T_6(\mathbb{F}_{q^m})$ for $m \geq 3$. However, in cryptographic applications one would work in a prime order subgroup of $T_n(\mathbb{F}_{q^m})$ of order around 2^{160}; in this case, our algorithm is only faster than Pollard-Rho for larger m.

From a practical perspective, our experiments show that in the cryptographic range, the algorithm for $T_6(\mathbb{F}_{q^m})$ outperforms the corresponding algorithm for $T_2(\mathbb{F}_{q^{3m}})$ and that it is most efficient when $m = 4$ or $m = 5$. Furthermore, for $m = 5$, both algorithms in practice outperform Pollard-Rho in a subgroup of $T_6(\mathbb{F}_{q^5})$ of order 2^{160}, for q^{30} up to and including the 960-bit scheme based in $T_{30}(\mathbb{F}_p)$ proposed in [27]. Compared to Pollard ρ our method seems to achieve in practice a 1000 fold speedup; its practical comparison with Adleman-DeMarrais is yet to be explored. Our experiments show that it is currently feasible to solve the DLP in $T_{30}(\mathbb{F}_p)$ with $\lceil \log_2 p \rceil = 20$, where we assume that a computation of around 2^{45} seconds is feasible.

The remainder of this paper is organised as follows. In Section 2 we briefly review algebraic tori and the notion of rationality. In Section 3 we present the philosophy of our algorithm and explain how it is related to classical index calculus algorithms. In Sections 4 and 5 we give a detailed description of the algorithm for $T_2(\mathbb{F}_{q^m})$ and $T_6(\mathbb{F}_{q^m})$ respectively. Finally, we conclude and give pointers for further research in Section 6.

2 Discrete Logs in Extension Fields and Algebraic Tori

Extension fields possess a richer algebraic structure than prime fields, in particular those with highly composite extension degrees. This has led some researchers to suspect that such fields may be cryptographically weak. For instance, in 1984 Odlyzko stated that fields with a composite extension degree 'may be very weak' [21]. The main result of this paper shows that these concerns may indeed be valid. A naive attempt to exploit the available subfield structure of extension fields in solving discrete logarithms, naturally leads one to consider the DLP on algebraic tori, as we show below.

2.1 A Simple Reduction of the DLP

Let $k = \mathbb{F}_q$ and let $K = \mathbb{F}_{q^n}$ be an extension of k of degree $n > 1$. Assume that $g \in K$ is a generator of $K^\times$ and let $h = g^s$ with $0 \leq s < q^n - 1$ be an element we wish to find the discrete logarithm of with respect to g.

Then by applying to g and h the norm maps N_{K/k_d} with respect to each intermediate subfield k_d of K, and solving the resulting discrete logarithms in these subfields, a simple argument shows that one can determine $s \bmod \mathrm{lcm}\{\Phi_d(q)\}_{d|n,d\neq n}$, where $\Phi_d(q)$ is the d-th cyclotomic polynomial evaluated at q. Modulo a cryptographically negligible factor, the remaining modular information required to determine the full discrete logarithm comes from the order $\Phi_n(q)$ subgroup of $K^\times$. As observed by Rubin and Silverberg [22], this subgroup is precisely the algebraic torus $T_n(\mathbb{F}_q)$.

2.2   The Algebraic Torus

In their CRYPTO 2003 paper [22], Rubin and Silverberg introduced the notion of torus-based cryptography. Their central idea was to interpret the subgroups of $K^\times$ as algebraic tori, and by exploiting birational maps from these groups to affine space, they obtained an efficient compression mechanism for elements of extension fields. Along with the existing public key cryptosystems XTR [16] and LUC [25], their method provides a reduction in bandwidth requirements for finite field discrete logarithm based protocols, which is becoming increasingly relevant as key-size recommendations become larger in order to maintain security levels.

Definition 1. *Let $k = \mathbb{F}_q$ and let $K = \mathbb{F}_{q^n}$ be an extension of k of degree $n > 1$. We define the algebraic torus $T_n(\mathbb{F}_q)$ as*

$$T_n(\mathbb{F}_q) = \{\alpha \in K \mid N_{K/k_d}(\alpha) = 1 \text{ for all subfields } k \subseteq k_d \subsetneq K\}.$$

Strictly speaking, $T_n(\mathbb{F}_q)$ refers only to the $\mathbb{F}_q$-rational points on the affine algebraic variety T_n, rather than the torus itself (see [22] for the exact construction).

Note that since $T_n(\mathbb{F}_q)$ is simply a subgroup of $\mathbb{F}_{q^n}^\times$, the group operation can be realised as ordinary multiplication in the field $\mathbb{F}_{q^n}$. The dimension of the variety T_n is $\phi(n) = \deg(\Phi_n(x))$, with $\phi(\cdot)$ the Euler totient function.

Let $G_{q,n}$ denote the subgroup of $\mathbb{F}_{q^n}^\times$ of order $\Phi_n(q)$. The following lemma from [22] provides some useful properties of T_n.

Lemma 1.

1. $T_n(\mathbb{F}_q) \cong G_{q,n}$ *and hence* $\#T_n(\mathbb{F}_q) = \Phi_n(q)$.
2. *If $h \in T_n(\mathbb{F}_q)$ is an element of prime order not dividing n, then h does not lie in a proper subfield of $\mathbb{F}_{q^n}/\mathbb{F}_q$.*

It follows that $T_n(\mathbb{F}_q)$ may be regarded as the 'primitive' subgroup of $\mathbb{F}_{q^n}^\times$, since by Lemma 1 it does not embed into a proper subfield. Hence in practice, one always uses a subgroup of $T_n(\mathbb{F}_q)$ in cryptographic applications, since otherwise a given DLP embeds into a proper subfield of $\mathbb{F}_{q^n}$ (see also [15]). In fact, using the decomposition

$$x^n - 1 = \prod_{d \mid n} \Phi_d(x)$$

in $\mathbb{Z}[x]$, the group $\mathbb{F}_{q^n}^\times$ can be seen to be almost the same as the direct product $\prod_{d \mid n} T_n(\mathbb{F}_q)$. Hence finding an efficient algorithm to solve the DLP on algebraic tori enables one to solve DLPs in extension fields, as well as vice versa.

2.3   Rationality of Tori over $\mathbb{F}_q$

In order to compress elements of the variety T_n, we make use of rationality, for particular values of n. The rationality of T_n means there exists a birational map from T_n to $\phi(n)$-dimensional affine space $\mathbb{A}^{\phi(n)}$. This allows one to represent nearly all elements of $T_n(\mathbb{F}_q)$ with just $\phi(n)$ elements of $\mathbb{F}_q$, providing an effective compression factor of $n/\phi(n)$ over the embedding of $T_n(\mathbb{F}_q)$ into $\mathbb{F}_{q^n}$. Since T_n has dimension $\phi(n)$, this compression factor is optimal. T_n is known to be rational when n is either a prime power, or is a product of two prime powers, and is conjectured to be rational for all n [22].

Formally, rationality can be defined as follows.

Definition 2. *Let T_n be an algebraic torus over $\mathbb{F}_q$ of dimension $d = \phi(n)$, then T_n is said to be rational if there is a birational map $\rho : T_n \to \mathbb{A}^{\phi(n)}$ defined over $\mathbb{F}_q$.*

This means that there are subsets $W \subset T_n$ and $U \subset \mathbb{A}^{\phi(n)}$, and rational functions $\rho_1, \ldots, \rho_{\phi(n)} \in \mathbb{F}_q(x_1, \ldots, x_n)$ and $\psi_1, \ldots, \psi_n \in \mathbb{F}_q(y_1, \ldots, y_{\phi(n)})$ such that $\rho = (\rho_1, \ldots, \rho_{\phi(n)}) : W \to U$ and $\psi = (\psi_1, \ldots, \psi_n) : U \to W$ are inverse isomorphisms. Furthermore, the differences $T \setminus W$ and $\mathbb{A}^{\phi(n)} \setminus U$ should be algebraic varieties of dimension $\leq (d-1)$, which implies that W (resp. U) is 'almost the whole' of T (resp. $\mathbb{A}^{\phi(n)}$).

The public key cryptosystem CEILIDH [22] is based on the algebraic torus T_6, which achieves a compression factor of three over the extension field representation. Rationality whilst useful, is not essential, since Van Dijk and Woodruff [28] showed that one can obtain key-agreement, signature and encryption schemes with bandwidth compressed by this factor asymptotically with the number of keys/signatures/messages, without relying on the conjecture stated above. Indeed, their result applies to any torus T_n, which helps explain the recent and increasing interest in torus-based cryptography.

3   Algorithm Philosophy

The algorithm as presented in Sections 4 and 5 is based on an idea first proposed by Gaudry [10], in reference to the DLP on general abelian varieties. While Gaudry's method is in principle an index calculus algorithm, the ingredients are very algebraic: for instance one need not rely on unique factorisation to obtain a notion of 'smoothness', as in finite field discrete logarithm algorithms.

As an introduction, in this section we consider Gaudry's idea in the context of computing discrete logarithms in $\mathbb{F}_{q^m}^{\times}$, and show how it is related to classical index calculus.

3.1   Classical Method

Let $\mathbb{F}_{q^m} = \mathbb{F}_q[t]/(f(t))$ for some monic irreducible degree m polynomial and let the basis be $\{1, t, \ldots, t^{m-1}\}$. Let g be a generator of $\mathbb{F}_{q^m}^{\times}$ and let $h \in \langle g \rangle$ be an element we are to compute the logarithm of w.r.t. g. Suppose also, for this example, that we are able to deal with a factor base of size q.

Classically, one would first reduce the problem to considering only monic polynomials, i.e., one considers the quotient $\mathbb{F}_{q^m}^{\times}/\mathbb{F}_q^{\times}$, and defines a factor base

$$\mathcal{F} = \{t + a : a \in \mathbb{F}_q\}.$$

Then for random $j, k \in \mathbb{Z}/((q^m - 1)/(q - 1))\mathbb{Z}$ one computes $r = g^j h^k$ and tests whether $r/\mathrm{lc}(r)$ decomposes over $\mathcal{F}$, with $\mathrm{lc}(r)$ the leading coefficient of r. This occurs with probability approximately $1/(m - 1)!$ for large q since the set of all products of $m - 1$ elements of $\mathcal{F}$ generates roughly $q^{m-1}/(m - 1)!$ elements of $\mathbb{F}_{q^m}^{\times}/\mathbb{F}_q^{\times}$.

Computing more than q such relations allows one to compute $\log_g h \bmod (q^m - 1)/(q - 1)$ as usual with a linear algebra elimination (and one applies the norm $N_{\mathbb{F}_{q^m}/\mathbb{F}_q}$ to g and h and solves the corresponding DLP in $\mathbb{F}_q^{\times}$ to recover the remaining modular information).

3.2   Gaudry's Method

Two essential points taken for granted in the above description are that there exist efficient procedures to compute:

- whether a given r decomposes over $\mathcal{F}$; this happens precisely when $r \in \mathbb{F}_q[t]$ splits over $\mathbb{F}_q$ or equivalently when $\gcd(t^q - t, r/\mathrm{lc}(r)) = r/\mathrm{lc}(r)$,
- the actual decomposition of r, i.e., to compute the roots of $r \in \mathbb{F}_q[t]$ in $\mathbb{F}_q$.

One may equivalently consider the following problem: determine whether the system of equations obtained by equating powers of t in the equality

$$\prod_{i=1}^{m-1} (t + a_i) = r/\mathrm{lc}(r) = r_0 + r_1 t + \cdots + r_{m-2} t^{m-2} + t^{m-1}, \qquad (1)$$

has a solution $(a_1, \ldots, a_{m-1}) \in \mathbb{F}_q^{m-1}$ and if so, to compute one such solution. Of course, in this trivial example the roots a_i can be read off from the factorisation of $r/\mathrm{lc}(r)$. However, one obtains a non-trivial example if the group operation on the left is more sophisticated than polynomial multiplication, such as elliptic curve point addition, which was Gaudry's original motivation for developing the algorithm. In this case the decomposition of a group element over the factor base can become more sophisticated, but the principle remains the same.

The central benefit of this perspective is that it can be applied in the absence of unique factorisation, since with a suitable choice of factor base, or more accurately a decomposition base, one can simply induce relations algebraically. For example, approaching the above problem from this slightly different perspective gives an algorithm for working directly in $\mathbb{F}_{q^m}^{\times}$, which is perhaps more natural than the stated quotient, $\mathbb{F}_{q^m}^{\times}/\mathbb{F}_q^{\times}$. Define a decomposition base

$$\mathcal{F} = \{1 + at : a \in \mathbb{F}_q\},$$

and again associate to the equality

$$\prod_{i=1}^{m} (1 + a_i t) \equiv r \equiv r_0 + r_1 t + \cdots + r_{m-1} t^{m-1} \pmod{f(t)}, \qquad (2)$$

the algebraic system obtained by equating powers of t.

Note that in (2) one must multiply m elements of $\mathcal{F}$ in order to obtain a probability of $1/m!$ for obtaining a relation, rather than the $m-1$ elements (and probability $1/(m-1)!$) of (1). The reason these probabilities differ is simply that the algebraic groups $\mathbb{F}_{q^m}^{\times}/\mathbb{F}_q^{\times}$ and $\mathbb{F}_{q^m}^{\times}$ over $\mathbb{F}_q$ are $m-1$ and m-dimensional respectively.

Ignoring for the moment that $\mathcal{F}$ essentially consists of degree one polynomials, and assuming that we want to solve this system without factoring $r/\mathrm{lc}(r)$, we are faced with finding a solution to a non-linear system, which would ordinarily require a Gröbner basis computation to solve. However writing out the left hand side in the polynomial basis $\{1, \ldots, t^{m-1}\}$ gives

$$\prod_{i=1}^{m}(1 + a_i t) = 1 + \overline{\sigma}_1 t + \cdots + \overline{\sigma}_m t^m$$

$$\equiv 1 + \overline{\sigma}_1 t + \cdots + \overline{\sigma}_{m-1} t^{m-1} + \overline{\sigma}_m(t^m - f(t)) \pmod{f(t)},$$

with $\overline{\sigma}_i$ the i-th elementary symmetric polynomial in the a_i. Equating powers of t then gives a linear system of equations in the $\overline{\sigma}_i$ for $i = 1, \ldots, m$. Given a solution $(\sigma_1, \ldots, \sigma_m)$ to this system of equations, r will decompose over $\mathcal{F}$ precisely when the polynomial

$$p(x) := x^m - \sigma_1 x^{m-1} + \sigma_2 x^{m-2} - \cdots + (-1)^m \sigma_m$$

splits over $\mathbb{F}_q$. Thus exploiting the symmetry in the construction of the algebraic system makes solving it much simpler. Although in this contrived example, solving the system directly and solving it using its symmetry are essentially the same, in general the latter makes infeasible computations feasible.

Following from this example, a simple observation is that for an algebraic group over $\mathbb{F}_q$ whose representation is m-dimensional, then using a decomposition base $\mathcal{F}$ of q elements, one must multiply m elements of $\mathcal{F}$ to obtain a constant probability of decomposition $1/m!$. Therefore, we conclude that the more efficient the representation of the group is, the higher the probability of obtaining a relation, and thus the corresponding index calculus algorithm will be more efficient.

In the following two sections, we apply this idea to rational representations of algebraic tori, and show that the above probability of $1/m!$ can be reduced significantly to $1/(m/2)!$ when m is divisible by 2 and to $1/(m/3)!$ when m is divisible by 6.

4   An Index Calculus Algorithm for $T_2(\mathbb{F}_{q^m}) \subset \mathbb{F}_{q^{2m}}^{\times}$

For q any odd prime power, we describe an algorithm to compute discrete logarithms in $T_2(\mathbb{F}_{q^m})$.

4.1   Setup

With regard to the extension $\mathbb{F}_{q^{2m}}/\mathbb{F}_{q^m}$, by Lemma 1 we know that

$$\#T_2(\mathbb{F}_{q^m}) = \Phi_2(q^m) = q^m + 1,$$

and hence we presume the DLP we consider is in the subgroup of this order. By applying the reduction of the DLP via norms as in Section 2, it is clear that the hard part actually is $T_{2m}(\mathbb{F}_q) \subsetneq T_2(\mathbb{F}_{q^m})$. Since in this section we use the properties of T_2 rather than T_{2m}, we only consider $T_2(\mathbb{F}_{q^m})$, or more accurately $(\mathrm{Res}_{\mathbb{F}_{q^m}/\mathbb{F}_q} T_2)(\mathbb{F}_q)$, where here Res denotes the Weil restriction of scalars (see also [22]).

Let $\mathbb{F}_{q^m} \cong \mathbb{F}_q[t]/(f(t))$ with $f(t) \in \mathbb{F}_q[t]$ an irreducible monic polynonmial of degree m and take the polynomial basis $\{1, t, \ldots, t^{m-1}\}$. Assuming that q is an odd prime power, we let $\mathbb{F}_{q^{2m}} = \mathbb{F}_{q^m}[\gamma]/(\gamma^2 - \delta)$ with basis $\{1, \gamma\}$, for some non-square $\delta \in \mathbb{F}_{q^m} \setminus \mathbb{F}_q$. Then using Definition 1, we see that

$$T_2(\mathbb{F}_{q^m}) = \{(x, y) \in \mathbb{F}_{q^m} \times \mathbb{F}_{q^m} : x^2 - \delta y^2 = 1\}.$$

This representation uses two elements of $\mathbb{F}_{q^m}$ to represent each point. The torus T_2 is one-dimensional, rational, and has the following equivalent affine representation:

$$T_2(\mathbb{F}_{q^m}) = \left\{ \frac{z - \gamma}{z + \gamma} : z \in \mathbb{F}_{q^m} \right\} \cup \{O\}, \tag{3}$$

where O is the point at infinity.

Here a point $g = g_0 + g_1\gamma \in T_2(\mathbb{F}_{q^m})$ in the $\mathbb{F}_{q^{2m}}$ representation has a corresponding representation as given above by the rational function $z = -(1 + g_0)/g_1$ if $g_1 \neq 0$, whilst the elements -1 and 1 map to $z = 0$ and $z = O$ respectively. The representation (3) thus gives a compression factor of two for the elements of $\mathbb{F}_{q^{2m}}$ that lie in $T_2(\mathbb{F}_{q^m})$. Furthermore since $T_2(\mathbb{F}_{q^m})$ has $q^m + 1$ elements, this compression is optimal (since for this example, including the point at infinity, we really have a map from $T_2(\mathbb{F}_{q^m}) \to \mathbb{P}^1(\mathbb{F}_{q^m})$).

4.2   Decomposition Base

As with any index calculus algorithm, we need to define a factor base, or in the case of Gaudry's algorithm, a decomposition base. Let

$$\mathcal{F} = \left\{ \frac{a - \gamma}{a + \gamma} : a \in \mathbb{F}_q \right\} \subset T_2(\mathbb{F}_{q^m}),$$

which contains q elements, since the map, given above, is a birational isomorphism from T_2 to $\mathbb{A}^1$. Note that if $\delta \in \mathbb{F}_q$, then $\mathcal{F}$ would lie in the subvariety $T_2(\mathbb{F}_q)$ and would not aid in our attack, which is why we ensured that $\delta \in \mathbb{F}_{q^m} \setminus \mathbb{F}_q$ during the setup.

4.3   Relation Finding

Writing the group operation additively, let P be a generator, and let $Q \in \langle P \rangle$ be a point we wish to find the discrete logarithm of with respect to P. For a given $R = [j]P + [k]Q$, we test whether it decomposes as a sum of m points in the decomposition base:

$$P_1 + \cdots + P_m = R, \tag{4}$$

with $P_1, \ldots, P_m \in \mathcal{F}$. From the representation we have chosen for T_2 we may equivalently write this as

$$\prod_{i=1}^{m} \left(\frac{a_i - \gamma}{a_i + \gamma} \right) = \frac{r - \gamma}{r + \gamma},$$

where the a_i are unknown elements in $\mathbb{F}_q$, and $r \in \mathbb{F}_{q^m}$ is the affine representation of R. Note that the left hand side is symmetric in the a_i. Upon expanding the product for both the numerator and denominator, we obtain two polynomials of degree m in γ whose coefficients are just plus or minus the elementary symmetric polynomials $\sigma_i(a_1, \ldots, a_m)$ of the a_i:

$$\frac{\sigma_m - \sigma_{m-1}\gamma + \cdots + (-1)^m \gamma^m}{\sigma_m + \sigma_{m-1}\gamma + \cdots + \gamma^m} = \frac{r - \gamma}{r + \gamma}.$$

Therefore, when we reduce modulo the defining polynomial of γ, we obtain an equation of the form

$$\frac{b_0(\sigma_1, \ldots, \sigma_m) - b_1(\sigma_1, \ldots, \sigma_m)\gamma}{b_0(\sigma_1, \ldots, \sigma_m) + b_1(\sigma_1, \ldots, \sigma_m)\gamma} = \frac{r - \gamma}{r + \gamma},$$

where b_0, b_1 are linear in the σ_i and have coefficients in $\mathbb{F}_{q^m}$. More explicitly, since $\gamma^2 = \delta \in \mathbb{F}_{q^m}$, these polynomials are given by

$$b_0 = \sum_{k=0}^{\lfloor m/2 \rfloor} \sigma_{m-2k} \delta^k \quad \text{and} \quad b_1 = \sum_{k=0}^{\lfloor (m-1)/2 \rfloor} \sigma_{m-2k-1} \delta^k,$$

where we define $\sigma_0 = 1$.

In order to obtain a simple set of algebraic equations amongst the σ_i, we first reduce the left hand side to the affine representation (3) and obtain the equation

$$b_0(\sigma_1, \ldots, \sigma_m) - b_1(\sigma_1, \ldots, \sigma_m)r = 0.$$

Since the unknowns σ_i are elements of $\mathbb{F}_q$, we express the above equation on the polynomial basis of $\mathbb{F}_{q^m}$ to obtain m linear equations over $\mathbb{F}_q$ in the m unknowns $\sigma_i \in \mathbb{F}_q$. This gives an $m \times m$ matrix M over $\mathbb{F}_q$ such that

- the $(m - 2k)$-th column contains the coefficients of δ^k,
- the $(m - 2k - 1)$-th column contains the coefficients of $-r\delta^k$.

Furthermore, let V be the $m \times 1$ vector containing the coefficients of $r\delta^{(m-1)/2}$ when m is odd or $-\delta^{m/2}$ when m is even, then $\Sigma = (\sigma_1, \ldots, \sigma_m)^T$ is a solution of the linear system of equations

$$M\Sigma = V.$$

If there is a solution Σ, to see whether this corresponds to a solution of (4) we test whether the polynomial

$$p(x) := x^m - \sigma_1 x^{m-1} + \sigma_2 x^{m-2} - \cdots + (-1)^m \sigma_m$$

splits over $\mathbb{F}_q$ by computing $g(x) := \gcd(x^q - x, p(x))$. If $g(x) = p(x)$, then the roots $a_1, \ldots, a_m$ will be the affine representation of the elements of the factor base which sum to R and we have found a relation.

4.4   Complexity Analysis and Experiments

The number of elements of $T_2(\mathbb{F}_{q^m})$ generated by all sums of m points in $\mathcal{F}$ is roughly $q^m/m!$, assuming no repeated summands and that most points admit a unique factorisation over the factor base. Hence the probability of obtaining a relation is approximately $1/m!$. Therefore in order to obtain q relations we must perform roughly $m!q$ such decompositions. Each decomposition consists of the following steps:

- computing the matrix M and vector V takes $O(m^3)$ operations in $\mathbb{F}_q$, using a naive multiplication routine,
- solving for Σ also requires $O(m^3)$ operations in $\mathbb{F}_q$,
- computing the polynomial $g(x)$ requires $O(m^2 \log q)$ operations in $\mathbb{F}_q$,
- if the polynomial $p(x)$ splits over $\mathbb{F}_q$, then we have to find the roots $a_1, \ldots, a_m$ which requires $O(m^2 \log m(\log q + \log m))$ operations in $\mathbb{F}_q$.

Note that the last step only has to be executed $O(q)$ times. The overall complexity to find $O(q)$ relations is therefore

$$O(m! \cdot q \cdot (m^3 + m^2 \log q)).$$

operations in $\mathbb{F}_q$.

Since in each row of the final relations matrix there will be $O(m)$ non-zero elements, we conclude that finding a kernel vector using sparse matrix techniques [13] requires $O(mq^2)$ operations in $\mathbb{Z}/(q^m + 1)\mathbb{Z}$ or about $O(m^3q^2)$ operations in $\mathbb{F}_q$. This proves the following theorem.

Theorem 1. *The expected running time of the T_2-algorithm to compute DLOGs in $T_2(\mathbb{F}_{q^m})$ is*

$$O(m! \cdot q \cdot (m^3 + m^2 \log q) + m^3q^2)$$

operations in $\mathbb{F}_q$.

Note that when $m > 1$ and the q^2 term dominates, by reducing the size of the decomposition base, the complexity may be reduced to $O(q^{2-2/m})$ for $q \to \infty$ using the results of Thériault [26], and a refinement reported independently by Gaudry and Thomé [11] and Nagao [20].

The expected running time of the T_2-algorithm is minimal when the relation stage and the linear algebra stage take comparable time, i.e. when $m! \cdot q \cdot (m^3 + m^2 \log q) \simeq m^3q^2$ or $m! \simeq q$. The complexity of the algorithm then becomes $O(m^3q^2)$, which can be rewritten as

$$
\begin{aligned}
O(m^3q^2) &= O\big(\exp(3\log m + 2\log q)\big) \\
&= O\big(\exp(2(\log q)^{1/2}(\log q)^{1/2})\big) \\
&= O\big(\exp(2(m\log m)^{1/2}(\log q)^{1/2})\big) \\
&= O\big(L_{q^m}(1/2, c)\big)
\end{aligned}
$$

with $c \in \mathbb{R}_{>0}$. Note that for the second and third equality we have used that $m! \simeq q$, and thus by taking logarithms $\log q \simeq m \log m$.

To assess the practicality of the T_2 algorithm, we ran several experiments using a simple Magma implementation, the results of which are given in Table 1. This table should be read as follows: the size of the torus cardinality, i.e., $\log_2(q^m)$, is constant across each row; for a given q^m, the table contains for $m = 1, \ldots, 15$, the $\log_2$ of the expected running times in seconds for the entire algorithm, i.e. both relation collection stage and linear algebra. For instance, for $q^m \cong 2^{300}$ and $m = 15$, the total time would be approximately 2^{51} seconds on one AMD 1700+ using our Magma implementation. For the fields where the torus is less than 160 bits in size, we use the full torus otherwise we use a subgroup of 160 bits to estimate the Pollard ρ costs.

Note that Table 1 does not take into account *memory* constraints imposed by the linear algebra step; since the number of relations is approximately q, we conclude that the algorithm is currently only practical for $q \leq 2^{23}$. Assuming that 2^{45} seconds, which is about 1.1×10^6 years, is feasible and assuming it is possible to find a kernel vector of a sparse matrix of dimension 2^{23}, Table 1 contains, in bold, the combinations of q and m which can be handled using our Magma implementation.

Table 1. $\log_2$ of expected running times (s) of the T_2-algorithm and Pollard-Rho in a subgroup of size 2^{160}

									m								
$\log_2 |\mathbb{F}_{q^{2m}}|$	$\log_2 |T_2(\mathbb{F}_{q^m})|$	ρ	1	2	3	4	5	6	7	8	9	10	11	12	13	14	15
200	100	34	88	40	52	36	**26**	**20**	**16**	**17**	**18**	**21**	**23**	**26**	**31**	**33**	**37**
300	150	59	138	66	87	62	48	38	**31**	**26**	**25**	**26**	**28**	**31**	**34**	**37**	**40**
400	200	65	188	92	121	88	68	55	46	39	**34**	**32**	**33**	**35**	**38**	**41**	**44**
500	250	66	238	117	155	114	89	73	61	52	45	40	**38**	**40**	**42**	**44**	**47**
600	300	66	289	142	189	139	110	90	76	65	57	51	45	44	46	48	51
700	350	66	339	168	223	165	130	107	91	78	69	61	55	50	50	52	54
800	400	66	389	193	256	190	150	124	105	91	80	71	64	58	56	55	58
900	450	68	439	219	290	215	171	141	120	104	92	82	74	67	62	61	62
1000	500	69	489	244	324	241	191	158	134	117	103	92	83	76	69	66	67

4.5 Comparison with Other Methods

In this section we compare the T_2-algorithm with the Pollard-Rho and index calculus algorithms.

Pollard-Rho in the Full Torus. Using the Pohlig-Hellman reduction, the overall running time is determined by executing the Pollard-Rho algorithm in the subgroup of $T_2(q^m)$ of largest prime order l. Since $\#T_2(q^m) = q^m + 1$, we have to analyse the size of the largest prime factor l. Note that the factorisation of $x^m + 1$ over $\mathbb{Z}[x]$ is given by

$$x^m + 1 = \frac{x^{2m} - 1}{x^m - 1} = \frac{\prod_{d|2m} \Phi_d(x)}{\prod_{d|m} \Phi_d(x)} = \prod_{d|2m, d \nmid m} \Phi_d(x),$$

which implies that the maximum size of the prime l is $O(q^{\phi(2m)})$, since the degree of $\Phi_{2m}(x)$ is $\phi(2m)$. The overall worst case complexity of this method is therefore $O(q^{\phi(2m)/2})$ operations in $\mathbb{F}_{q^{2m}}$ or $O(m^2 \cdot q^{\phi(2m)/2})$ operations in $\mathbb{F}_q$.

From a complexity theoretic point of view, we therefore conclude that for $m! \leq q$, our algorithm is as fast as Pollard-Rho whenever $m \geq 5$, since then $\phi(2m)/2 > 2$. As a consequence, we note that the T_2 algorithm does not lead to an improvement over existing attacks on LUC [25], XTR [16] or CEILIDH [22] over $\mathbb{F}_p$. Furthermore, also the security of MNT curves [19] defined over $\mathbb{F}_p$, where p is a large prime remains unaffected.

Pollard-Rho in a Subgroup of Prime Order $\simeq 2^{160}$. In cryptographic applications however, one would work in a subgroup of $T_2(\mathbb{F}_{q^m})$ of prime order l with $l \simeq 2^{160}$. To this end, we measured the average time taken for one multiplication for the various fields in Magma, and multiplied this time by the expected 2^{80} operations required by the Pollard-Rho algorithm. The results can be found in the third column of Table 1. The column for $m = 15$ is especially interesting since this determines the security of the T_{30} cryptosystem introduced in [27]. In this case, the T_2 is always faster than Pollard-Rho, and the matrices occurring in the linear algebra step would be feasible up to 700-bit fields.

Adleman/Demarrais in $\mathbb{F}_{q^{2m}}^{\times}$. The alternative approach would be to embed $T_2(\mathbb{F}_{q^m})$ into $\mathbb{F}_{q^{2m}}^{\times}$ and to apply a subexponential algorithm, which for all m and q can attain a complexity of $L_{q^{2m}}(1/2, c)$ as shown by Adleman and Demarrais [1]. Clearly, using the T_2 algorithm this is only possible for certain combinations of m and q, e.g. for $q \simeq m!$, which is also indicated by Table 1. Of course, when $q = p^n$ for p a prime, then we can choose a different $\bar{m}$ with $\bar{m}|n \cdot m$ such that $\bar{m}! \simeq p^{nm/\bar{m}}$. We do not know how the Adleman-DeMarrais algorithms performs.

Remark 1. The linearity of the decomposition method in fact holds for any torus T_{p^r}. However the savings are optimal for T_{2^r}, since $p^r/\phi(p^r)$ is maximal in this case. When one considers T_n for which n is divisible by more than one distinct prime factor, the rational parametrisation becomes non-linear, and hence so does the corresponding decomposition, as we see in the following section.

5   An Index Calculus Algorithm for $T_6(\mathbb{F}_{q^m}) \subset \mathbb{F}_{q^{6m}}^{\times}$

In this section we detail our algorithm to compute discrete logarithms in $T_6(\mathbb{F}_{q^m})$. The main difference with the T_2-algorithm is the non-linearity of the equations involved in the decomposition step.

5.1   Setup

Again, let $\mathbb{F}_{q^m} \cong \mathbb{F}_q[t]/(f(t))$, with $f(t)$ an irreducible polynomial of degree m and where we use the polynomial basis $\{1, t, t^2, \ldots, t^{m-1}\}$. Since T_6 is two-dimensional and rational, it is an easy exercise to construct a birational map from T_6 to $\mathbb{A}^2$ for a given representation of $\mathbb{F}_{q^{6m}}$. For the following exposition we make use of the the CEILIDH field representation and maps, as described in [22].

Let $q^m \equiv 2$ or $5 \bmod 9$, and for $(r, q) = 1$ let ζ_r denote a primitive r-th root of unity in $\overline{\mathbb{F}}_{q^m}$. Define $x = \zeta_3$ and let $y = \zeta_9 + \zeta_9^{-1}$, then clearly $x^2 + x + 1 = 0$ and $y^3 - 3y + 1 = 0$. Let $\mathbb{F}_{q^{3m}} = \mathbb{F}_{q^m}(y)$ and $\mathbb{F}_{q^{6m}} = \mathbb{F}_{q^{3m}}(x)$, then the bases we use are $\{1, y, y^2 - 2\}$ for the degree three extension and $\{1, x\}$ for the degree two extension.

Let $V(f)$ be the zero set of $f(\alpha_1, \alpha_2) = 1 - \alpha_1^2 - \alpha_2^2 + \alpha_1\alpha_2$ in $\mathbb{A}^2(\mathbb{F}_{q^m})$, then we have the following inverse birational maps:

- $\psi : \mathbb{A}^2(\mathbb{F}_{q^m}) \setminus V(f) \overset{\sim}{\longrightarrow} T_6(\mathbb{F}_{q^m}) \setminus \{1, x^2\}$, defined by

$$\psi(\alpha_1, \alpha_2) = \frac{1 + \alpha_1 y + \alpha_2(y^2 - 2) + (1 - \alpha_1^2 - \alpha_2^2 + \alpha_1\alpha_2)x}{1 + \alpha_1 y + \alpha_2(y^2 - 2) + (1 - \alpha_1^2 - \alpha_2^2 + \alpha_1\alpha_2)x^2}, \tag{5}$$

- $\rho : T_6(\mathbb{F}_{q^m}) \setminus \{1, x^2\} \overset{\sim}{\longrightarrow} \mathbb{A}^2(\mathbb{F}_{q^m}) \setminus V(f)$, which is defined as follows: for $\beta = \beta_1 + \beta_2 x$, with $\beta_1, \beta_2 \in \mathbb{F}_{q^{3m}}$, let $(1 + \beta_1)/\beta_2 = u_1 + u_2 y + u_3(y^2 - 2)$, then $\rho(\beta) = (u_2/u_1, u_3/u_1)$.

5.2   Decomposition Base

In this case the decomposition base consists of $\psi(at, 0)$, where a runs through all elements of $\mathbb{F}_q$ and t generates the polynomial basis, i.e.

$$\mathcal{F} = \left\{ \frac{1 + (at)y + (1 - (at)^2)x}{1 + (at)y + (1 - (at)^2)x^2} : a \in \mathbb{F}_p \right\}$$

which clearly contains q elements, for much the same reason as given in Section 4. The reason for considering $\psi(at, 0)$ instead of $\psi(a, 0)$ is that the minimal polynomials of x and y are defined over $\mathbb{F}_q$. Note that this implies that $\psi(a, 0) \in T_6(\mathbb{F}_q)$ for $a \in \mathbb{F}_q$ and so does not generate a fixed proportion of $T_6(\mathbb{F}_{q^m})$, as is needed.

5.3   Relation Finding

Since $(\text{Res}_{\mathbb{F}_{q^m}/\mathbb{F}_q} T_6)(\mathbb{F}_q)$ is $2m$-dimensional, we need to solve

$$P_1 + \cdots + P_{2m} = R, \tag{6}$$

with $P_1, \ldots, P_{2m} \in \mathcal{F}$. Assuming that R is expressed in its canonical form, i.e. $R = \psi(r_1, r_2)$, we get

$$\prod_{i=1}^{2m} \left(\frac{1 + (a_i t)y + (1 - (a_i t)^2)x}{1 + (a_i t)y + (1 - (a_i t)^2)x^2} \right)$$
$$= \frac{1 + r_1 y + r_2(y^2 - 2) + (1 - r_1^2 - r_2^2 + r_1 r_2)x}{1 + r_1 y + r_2(y^2 - 2) + (1 - r_1^2 - r_2^2 + r_1 r_2)x^2}.$$

After expanding the product of the numerators and denominators, the left hand side becomes the fairly general expression

$$\frac{b_0 + b_1y + b_2(y^2 - 2) + \left(c_0 + c_1y + c_2(y^2 - 2)\right)x}{b_0 + b_1y + b_2(y^2 - 2) + \left(c_0 + c_1y + c_2(y^2 - 2)\right)x^2} \tag{7}$$

with b_i, c_i polynomials over $\mathbb{F}_{q^m}$ of degree $4m$ in $a_1, \ldots, a_{2m}$. In general, these polynomials are rather huge and thus difficult to work with.

Example 1. For $m = 5$, the number of terms in the b_i (resp. c_i) is given by $B = [35956, 30988, 25073]$ (resp. $C = [35946, 31034, 24944]$) for finite fields of large characteristic.

However, note that these polynomials are by construction symmetric in the $a_1, \ldots, a_{2m}$ so we can rewrite the b_i and c_i in terms of the $2m$ elementary symmetric polynomials $\sigma_j(a_1, \ldots, a_{2m})$ for $j = 1, \ldots, 2m$. This has quite a dramatic effect on the complexity of these polynomials, i.e., the degree is now only quadratic and the number of terms is much lower, since the maximum number of terms in a quadratic polynomial in $2m$ variables is $4m + \binom{2m}{2} + 1$.

Example 2. For $m = 5$, when we rewrite the equations using the symmetric functions σ_i, the number of terms of the polynomials b_i and c_i reduces to $B = [16, 19, 18]$ and $C = [20, 16, 16]$.

Note that the polynomials b_i and c_i only have to be computed once and can be reused for each random point R.

To generate the system of non-linear equations, we use the embedding of $T_6(\mathbb{F}_{q^m})$ into $T_2(\mathbb{F}_{q^{3m}})$ and consider the Weil restriction of the following equality:

$$\frac{b_0 + b_1y + b_2(y^2 - 2)}{c_0 + c_1y + c_2(y^2 - 2)} = \frac{1 + r_1y + r_2(y^2 - 2)}{1 - r_1^2 - r_2^2 + r_1r_2}.$$

The above equation leads to 3 non-linear equations over $\mathbb{F}_{q^m}$ or equivalently, to $3m$ non-linear equations over $\mathbb{F}_q$ in the $2m$ unknowns $\sigma_1, \ldots, \sigma_{2m}$. Note that amongst the $3m$ equations, there will be at least m dependent equations, caused by the fact that we only considered the embedding in T_2 and not strictly in T_6.

The efficiency with which one can find the solutions of this system of non-linear equations depends on many factors such as the multiplicities of the zeros or the number of solutions at infinity. For each random R, the resulting system of equations has the same structure, since only the value of some coefficients changes, but for finite fields of large enough characteristic, not the degrees nor the numbers of terms. To determine the properties of these systems of equations we computed the Gröbner basis w.r.t. the lexicographic ordering using the Magma implementation of the F4-algorithm [7] and concluded the following:

- The ideal generated by the system non-linear equations is zero-dimensional, which implies that there is only a finite number of candidates for the σ_i.
- After homogenizing the system of equations, we concluded that there is only a finite number of solutions at infinity. This property is quite important, since we can then use an algorithm by Lazard [14] with proven complexity.

- The Gröbner basis w.r.t. the lexicographic ordering satisfies the so called Shape Lemma, i.e. the basis has the following structure:

$$\sigma_1 - g_1(\sigma_{2m}),\ \sigma_2 - g_2(\sigma_{2m}),\ \ldots,\ \sigma_{2m-1} - g_{2m-1}(\sigma_{2m}),\ g_{2m}(\sigma_{2m})\,,$$

  where $g_i(\sigma_{2m})$ is a univariate polynomial in σ_{2m} for each i. By reducing modulo g_{2m} we can assume that $\deg(g_i) < \deg(g_{2m})$ and by Bezout's theorem we have $\deg(g_{2m}) \le 2^{2m}$, since the non-linear equations are quadratic. However, our experiments show that in all cases we have $\deg(g_{2m}) = 3^m$.
- The polynomial $g_{2m}(\sigma_{2m})$ is squarefree, which implies that the ideal is in fact a radical ideal.

To test if a random point decomposes over the factor base, we first find the roots of $g_{2m}(\sigma_{2m})$ in $\mathbb{F}_q$, and then substitute these in the g_i to find the values of the σ_i for $i = 1, \ldots, 2m - 1$. For each such $2m$-tuple, we then test if the polynomial

$$p(x) := x^{2m} - \sigma_1 x^{2m-1} + \sigma_2 x^{2m-2} - \cdots + (-1)^{2m}\sigma_{2m}$$

splits completely over $\mathbb{F}_q$. If it does, then the roots a_i for $i = 1, \ldots, 2m$ lead to a possible relation of the form (6).

5.4   Complexity Analysis and Experiments

The probability of obtaining a relation is now $1/(2m)!$ and since the factor base again consists of q elements, we need to perform $(2m)!q$ decompositions. Each decomposition consists of the following steps:

- Since the polynomials b_i and c_i only need to be computed once, generating the system of non-linear equations requires $O(1)$ multiplications of multivariate polynomials with $O(m^2)$ terms with an $\mathbb{F}_{q^m}$-element. Using a naive multiplication routine, the overall time to generate one such system is therefore $O(m^4)$ operations in $\mathbb{F}_q$.
- Computing the Gröbner basis using the F5-algorithm algorithm [8] requires $O\left(\binom{4m}{2m}^{\omega}\right)$ operations in $\mathbb{F}_q$, with ω the complexity of matrix multiplication, i.e. $\omega = 3$ using a naive algorithm. Using the fact that

$$\binom{2n}{n} \cong \sqrt{\frac{\pi}{2}}(2n)^{-1/2}2^{2n} \in O(2^{2n})$$

  we obtain a complexity of $O(2^{12m})$ operations in $\mathbb{F}_q$.
- Since $\deg(g_{2m}) = 3^m$, computing $\gcd(g_{2m}(z), z^q - z)$ requires $O(3^{2m}\log q)$ operations in $\mathbb{F}_q$. On average, the polynomial will have one root in $\mathbb{F}_q$, so finding the actual roots takes negligible time.
- Testing if the polynomial $p(x)$ has roots in $\mathbb{F}_q$ requires $O(m^2 \log q)$ operations in $\mathbb{F}_q$. Since this only happens with probability $1/(2m)!$, when it does split, finding the actual roots is negligible.

The overall time complexity to generate sufficient relations therefore amounts to

$$O\left((2m)! \cdot q \cdot (2^{12m} + 3^{2m} \log q)\right)$$

operations in $\mathbb{F}_q$.

Finding an element in the kernel of a matrix of dimension q with $2m$ non-zero elements per row requires $O(mq^2)$ operations in $\mathbb{Z}/(\Phi_6(q^m)\mathbb{Z})$, which finally justifies the following complexity estimate:

Run Time Heuristic 1. *The expected running time of the T_6-algorithm to compute DLOGs in $T_6(\mathbb{F}_{q^m})$ is*

$$O((2m)! \cdot q \cdot (2^{12m} + 3^{2m} \log q) + m^3 q^2)$$

operations in $\mathbb{F}_q$.

Again, the results of [26,11,20] imply that the complexity can be reduced to $O(q^{2-1/m})$ as $q \to \infty$, since in this case the dimension is $2m$.

The expected running time of the T_6-algorithm is minimal precisely when the relation collection stage takes about the same time as the linear algebra stage, i.e. when $(2m)! \cdot 2^{12m} \simeq q$. Note that for such q and m, the term $3^{2m} \log q$ is negligible compared to 2^{12m}. The overall running time then again becomes

$$\begin{aligned}
O(m^3 q^2) &= O\left(\exp(3 \log m + 2 \log q)\right) \\
&= O\left(\exp(2(\log q)^{1/2}(\log q)^{1/2})\right) \\
&= O\left(\exp(2(2m \log 2m + 12m)^{1/2}(\log q)^{1/2})\right) \\
&= O\left(L_{q^m}(1/2, c)\right)
\end{aligned}$$

with $c \in \mathbb{R}_{>0}$. Note that for the second and third equality we have used $\log q \simeq 2m \log m + 12m \log 2$.

The practicality of the T_6-algorithm clearly depends on the efficiency of the Gröbner basis computation. Note that for small m, the complexity of the Gröbner basis computation is greatly overestimated by the $O(2^{12m})$ operations in $\mathbb{F}_q$.

Due to the use of the symmetric polynomials, the input polynomials are only quadratic instead of degree $4m$. As one can see from Table 2, this makes the algorithm quite practical. The table should be interpreted as for Table 1, i.e., the torus size is constant across each row and for a given size q^m, the table contains for $m = 1, \ldots, 5$, the $\log_2$ of the expected running times in seconds for the entire algorithm. Taking into account the memory restrictions on the matrix, i.e., the dimension should be limited by 2^{23}, the timings given in bold are feasible with the current Magma implementation.

Remark 2. Note that the column for $m = 5$ provides an upper bound for the hardness of the DLP in $T_{30}(\mathbb{F}_q)$, since this can be embedded in $T_6(\mathbb{F}_{q^5})$. This group was recently proposed [27] and also in [15] for cryptographic use where keys of length 960 bits were recommended, i.e., with q of length 32 bits. The above table shows that even with a Magma implementation it would be feasible

Table 2. $\log_2$ of expected running times (s) of the T_6-algorithm and Pollard-Rho in a subgroup of size 2^{160}

			m				
$\log_2 |\mathbb{F}_{p^{6m}}|$	$\log_2 |T_6(\mathbb{F}_{p^m})|$	ρ	1	2	3	4	5
200	67	18	25	18	14	20	29
300	100	34	42	36	21	24	32
400	134	52	59	54	32	29	36
500	167	66	75	71	44	33	39
600	200	66	93	88	55	40	42
700	234	66	109	105	67	48	46
800	267	66	127	122	78	57	51
900	300	68	144	139	90	65	56
1000	334	69	161	156	101	74	60

to compute discrete logarithms in $T_{30}(\mathbb{F}_p)$ with p a prime of around 20 bits. The embedding in $T_2(\mathbb{F}_{p^{15}})$ is about 2^{10} times less efficient as can be seen from the column for $m = 15$ in Table 1. In light of this attack, the security offered by the DLP in finite fields of the form $\mathbb{F}_{q^{30}}$ should be completely reassessed. Note that by simply comparing the complexities given in Theorem 1 and the above run time heuristic, it is a priori not clear that the T_6-algorithm is in fact faster than the corresponding T_2-algorithm. This phenomenon is caused by the overestimating the complexity of the Gröbner basis computation.

5.5   Comparison with Other Methods

In this section we compare the T_6-algorithm with the Pollard-Rho and index calculus algorithms.

Pollard-Rho in the Full Torus. Since the size of $T_6(\mathbb{F}_{q^m})$ is given by $\Phi_6(q^m) \simeq q^{2m}$, we conclude that the Pollard-Rho algorithm takes, in the worst case, $O(q^m)$ operations in $T_6(\mathbb{F}_{q^m})$ or $O(m^2 q^m)$ operations in $\mathbb{F}_q$. If we assume that q is large enough such that the term q^2 determines the overall running time, i.e., $(2m)!2^{12m} \leq q$, then the T_6-algorithm will be at least as fast as Pollard-Rho whenever $m \geq 3$. Again we note that the T_6 algorithm does not lead to an improvement over the existing attacks on LUC [25], XTR [16], CEILIDH [22] or MNT curves [19] as long as these systems are defined over $\mathbb{F}_p$. However, the security of XTR over extension fields, as proposed in [17] or of the recent proposal that works in $T_{30}(\mathbb{F}_p)$ [27], needs to be reassessed as shown below.

Pollard-Rho in a Subgroup of Prime Order $\simeq 2^{160}$. As for the T_2-algorithm, the third column of Table 2 contains the expected running time of the Pollard-Rho algorithm in a subgroup of $T_6(\mathbb{F}_{q^m})$ of prime order l with $l \simeq 2^{160}$. In this case, the column for $m = 5$ gives an upper bound of the security of the T_{30} cryptosystem introduced in [27]. As is clear from Table 2, for $m = 5$, our algorithm is always faster than Pollard-Rho, and the matrices occurring in the linear algebra step would be feasible up to 700-bit fields.

Adleman/Demarrais in $\mathbb{F}_{q^{6m}}^{\times}$. Using the embedding of $T_6(\mathbb{F}_{q^m})$ into $\mathbb{F}_{q^{6m}}^{\times}$ one can apply the subexponential algorithm of Adleman-Demarrais [1] which runs, for all m and q, in time $L_{q^{6m}}(1/2, c)$. Using the T_6 algorithm, it is possible to obtain a complexity of $L_{q^m}(1/2, c')$, but only when m and q grow according to a specific relation such as $(2m)!2^{12m} \simeq q$. Again, when $q = p^n$ with p a prime, we could choose a different $\bar{m}$ with $\bar{m}|n \cdot m$ such that $(2\bar{m})!2^{12\bar{m}} \simeq p^{mn/\bar{m}}$.

However, as was the case for the T_2-algorithm, the importance of Table 2 is that it contains the first practical upper bounds for the hardness of the DLP in extension fields $\mathbb{F}_{q^{6m}}^{\times}$, since there are no numerical experiments available based on the existing subexponential algorithms.

6   Conclusion and Future Work

In this paper we have presented an index calculus algorithm, following ideas of Gaudry, to compute discrete logarithms on rational algebraic tori. Our algorithm works directly in the torus and depends fundamentally on the compression mechanisms previously used in a constructive context for systems such as LUC, XTR and CEILIDH.

We have also provided upper bounds for the difficulty of solving discrete logarithms on the tori $T_2(\mathbb{F}_{q^m})$ and $T_6(\mathbb{F}_{q^m})$ for various q and m in the cryptographic range. These upper bounds indicate that if the techniques in this paper can be made fully practical and optimized, then they may weaken the security of practical systems based on T_{30}.

In the near future we wish to investigate the approach by Diem [4], who allows a larger decomposition base when necessary. The disadvantage of this approach is that it destroys the symmetric nature of the polynomials defining the decomposition of a random element over the factor base, which makes Gröbner basis techniques virtually impossible.

It is clear that the Magma implementations described in this paper are not optimised and many possible improvements exist. Two factors mainly determine the running time of the algorithm: first of all, the probability that a random element decomposes over the factor base and secondly, the time it takes to solve a system of non-linear equations over a finite field. The first factor could be influenced by designing some form of sieving, if at all possible, whereas the second factor could be improved by exploiting the fact that many very similar Gröbner bases have to be computed.

In addition the method needs to be compared in practice to the method of Adleman and DeMarrais.

Acknowledgements

The authors would like to thank Daniel Lazard for his invaluable comments regarding the details of the complexity of the Gröbner basis computation in the T_6-algorithm, and anonymous referees for constructive comments on earlier versions of this paper.

References

1. L. M. Adleman and J. DeMarrais. *A subexponential algorithm for discrete logarithms over all finite fields.* Math. Comp., **61 (203)**, 1–15, 1993.
2. A. E. Brouwer, R. Pellikaan and E. R. Verheul. *Doing more with fewer bits.* In Advances in Cryptology (ASIACRYPT 1999), Springer LNCS 1716, 321–332, 1999.
3. B. Buchberger. *A theoretical basis for the reduction of polynomials to canonical forms.* ACM SIGSAM Bull., **10 (3)**, 19–29, 1976.
4. C. Diem. *On the discrete logarithm problem in elliptic curves over non-prime fields.* Preprint 2004. Available from the author.
5. W. Diffie and M. E. Hellman. *New directions in cryptography.* IEEE Trans. Inform. Theory **22 (6)**, 644–654, 1976.
6. T. ElGamal. *A public key cryptosystem and a signature scheme based on discrete logarithms.* In Advances in Cryptology (CRYPTO 1984), Springer LNCS 196, 10–18, 1985.
7. J.-C. Faugère. *A new efficient algorithm for computing Gröbner bases (F_4),* J. Pure Appl. Algebra **139 (1-3)**, 61-88, 1999.
8. J.-C. Faugère. *A new efficient algorithm for computing Gröbner bases without reduction to zero (F_5),* In Proceedings of the 2002 International Symposium on Symbolic and Algebraic Computation, 75–83, 2002.
9. FIPS 186-2, *Digital signature standard.* Federal Information Processing Standards Publication 186-2, February 2000.
10. P. Gaudry. *Index calculus for abelian varieties and the elliptic curve discrete logarithm problem.* Cryptology ePrint Archive, Report 2004/073. Available from http://eprint.iacr.org/2004/073.
11. P. Gaudry and E. Thomé. *A double large prime variation for small genus hyperelliptic index calculus.* Cryptology ePrint Archive, Report 2004/153. Available from http://eprint.iacr.org/2004/153.
12. R. Granger, D. Page and M. Stam. *A comparison of CEILIDH and XTR.* In Algorithmic Number Theory Symposium (ANTS-VI), Springer LNCS 3076, 235–249, 2004.
13. B. A. LaMacchia and A. M. Odlyzko. *Solving large sparse linear systems over finite fields.* In Advances in Cryptology (CRYPTO 1990), Springer LNCS 537, 109–133, 1991.
14. D. Lazard. *Résolution des systèmes d'équations algébriques,* Theoret. Comput. Sci., **15 (1)**, 77–110, 1981.
15. A. K. Lenstra. *Using cyclotomic polynomials to construct efficient discrete logarithm cryptosystems over finite fields.* In Proceedings of ACISP97, Springer LNCS 1270, 127–138, 1997.
16. A. K. Lenstra and E. Verheul. *The XTR public key system.* In Advances in Cryptology (CRYPTO 2000), Springer LNCS 1880, 1–19, 2000.
17. S. Lim, S. Kim, I. Yie, J. Kim and H. Lee. *XTR extended to $GF(p^{6m})$.* In Selected Areas in Cryptography (SAC 2001), Springer LNCS 2259, 301–312, 2001.
18. A. J. Menezes, P. van Oorschot and S. A. Vanstone. *The Handbook of Applied Cryptography,* CRC press, 1996.

19. A. Miyaji, M. Nakabayashi and S. Takano. *New explicit conditions of elliptic curve traces for FR-reduction*. IEICE Trans. Fundamentals **E84-A (5)**, 1234–1243, 2001.

20. K. Nagao. *Improvement of Thériault algorithm of index calculus for Jacobian of hyperelliptic curves of small genus*. Cryptology ePrint Archive, Report 2004/161. Available from `http://eprint.iacr.org/2004/161`.

21. A. M. Odlyzko. *Discrete logarithms in finite fields and their cryptographic significance*. In Advances in Cryptology (EUROCRYPT 1984), Springer LNCS 209, 224–314, 1985.

22. K. Rubin and A. Silverberg. *Torus-based cryptography*. In Advances in Cryptology (CRYPTO 2003), Springer LNCS 2729, 349–365, 2003.

23. K. Rubin and A. Silverberg. *Using primitive subgroups to do more with fewer bits*. In Algebraic Number Theory Symposium (ANTS-VI), Springer LNCS 3076, 18–41, 2004.

24. C. P. Schnorr. *Efficient signature generation by smart cards*. J. Cryptology, **4**, 161–174, 1991.

25. P. Smith and C. Skinner. *A public-key cryptosystem and a digital signature system based on the Lucas function analogue to discrete logarithms*. In Advances in Cryptology (ASIACRYPT 1995), Springer LNCS 917, 357–364, 1995.

26. N. Thériault. *Index calculus attack for hyperelliptic curves of small genus*. In Advances in Cryptology (ASIACRYPT 2003), Springer LNCS 2894, 75–92, 2003.

27. M. van Dijk, R. Granger, D. Page, K. Rubin, A. Silverberg, M. Stam and D. Woodruff. *Practical cryptography in high dimensional tori*. In Advances in Cryptology (EUROCRYPT 2005), Springer LNCS 3494, 234–250, 2005.

28. M. van Dijk and D. P. Woodruff. *Asymptotically optimal communication for torus-based cryptography*. In Advances in Cryptology (CRYPTO 2004), Springer LNCS 3152, 157–178, 2004.

A Practical Attack on a Braid Group Based Cryptographic Protocol

Alexei Myasnikov[1], Vladimir Shpilrain[2], and Alexander Ushakov[3]

[1] Department of Mathematics, McGill University, Montreal, Quebec H3A 2T5
alexeim@math.mcgill.ca*
[2] Department of Mathematics, The City College of New York, New York, NY 10031
shpilrain@yahoo.com**
[3] Department of Mathematics, CUNY Graduate Center, New York, NY 10016
aushakov@mail.ru***

Abstract. In this paper we present a practical heuristic attack on the Ko, Lee et al. key exchange protocol introduced at Crypto 2000 [11]. Using this attack, we were able to break the protocol in about 150 minutes with over 95% success rate for typical parameters. One of the ideas behind our attack is using Dehornoy's handle reduction method as a counter measure to diffusion provided by the Garside normal form, and as a tool for simplifying braid words. Another idea employed in our attack is solving the *decomposition problem* in a braid group rather than the *conjugacy search problem*.

1 Introduction

Braid group cryptography has attracted a lot of attention recently due to several suggested key exchange protocols (see [1], [11]) using braid groups as a platform. We refer to [2], [6] for more information on braid groups.

Here we start out by giving a brief description of the Ko, Lee et al. key exchange protocol (subsequently called just the Ko-Lee protocol).

Let B_{2n} be the group of braids on $2n$ strands and $x_1, \ldots, x_{2n-1}$ its standard generators. Define two subgroups L_n and R_n of B_{2n} as follows:

$$L_n = \langle x_1, \ldots, x_{n-1} \rangle$$

and

$$R_n = \langle x_{n+1}, \ldots, x_{2n-1} \rangle.$$

Clearly, L_n and R_n commute elementwise. The Ko-Lee protocol [11] is the following sequence of operations:

---

* Partially supported by the NSF grant DMS-0405105.
** Partially supported by the NSF grant DMS-0405105.
*** Partially supported by Umbanet Inc. through an award from the U.S. Department of Commerce NIST, Advanced Technology Program, Cooperative Agreement No. 70NANB2H3012.

V. Shoup (Ed.): Crypto 2005, LNCS 3621, pp. 86–96, 2005.

(0) One of the parties (say, Alice) publishes a random element $w \in B_{2n}$ (the "base" word).

(1) Alice chooses a word a as a product of generators of L_n and their inverses. The word a is Alice's private key.

(2) Bob chooses a word b as a product of generators of R_n and their inverses. The word b is Bob's private key.

(3) Alice sends a normal form of the element $a^{-1}wa$ to Bob and Bob sends a normal form of the element $b^{-1}wb$ to Alice.

(4) Alice computes a normal form of

$$K_a = a^{-1}b^{-1}wba$$

and Bob computes a normal form of

$$K_b = b^{-1}a^{-1}wab.$$

Since $ab = ba$ in B_{2n}, the normal forms of K_a and K_b coincide. Thus Alice and Bob have the same normal form called their *shared secret key*.

We note that a particular normal form used in [11] is called the *Garside normal form* (see our Section 2).

Initially, the security of this problem was claimed to depend on the complexity of the *conjugacy search problem* in B_{2n} which is the following: for a given pair of words w_1, w_2 such that w_1 is conjugate to w_2 in B_{2n}, find a particular conjugator, i.e. a word x such that $w_1 = x^{-1}w_2x$. However, it was shown in [13] that solving the conjugacy search problem is not necessary to break the Ko-Lee protocol. More precisely, it was shown that for an adversary to get the shared secret key, it is sufficient to find a pair of words $a_1, a_2 \in L_n$ such that $w_1 = a_1wa_2$. Then $K_a = K_b = a_1b^{-1}wba_2$, where the element $b^{-1}wb$ is public because it was transmitted at step 3. The latter problem is usually called the *decomposition problem*. The fact that it is sufficient for the adversary to solve the decomposition problem to get the shared secret key was also mentioned, in passing, in the paper [11], but the significance of this observation was downplayed there by claiming that solving the decomposition problem does not really give a computational advantage over solving the conjugacy search problem.

In this paper, we show (experimentally) that a particular heuristic attack on the Ko-Lee protocol based on solving the decomposition problem is, in fact, by far more efficient than all known attacks based on solving the conjugacy search problem. With the running time of 150 minutes (on a cluster of 8 PCs with 2GHZ processor and 1GB memory each), the success rate of our attack program was over 96%; see Section 5 for more details.

We note that there is a polynomial-time *deterministic* attack on the Ko-Lee protocol based on solving a variant of the conjugacy search problem [3], but the authors of [3] acknowledge themselves that their attack is not practical and, in fact, has not been implemented.

Another idea employed in our attack is using Dehornoy's forms [4] for recovering words from Garside normal forms and for solving the decomposition problem. In the Ko-Lee protocol, Garside's algorithm for converting braid words into normal forms plays the role of a diffusion algorithm. We show (experimentally) that Dehornoy's algorithm can be used to weaken the diffusion and make the protocol vulnerable to a special kind of length based attack (see [7], [8], [9] for different versions of length based attacks).

To conclude the introduction, we note that several other, less efficient, attacks on the Ko-Lee protocol were suggested before; we refer to [5] for a comprehensive survey of these attacks as well as for suggestions on countermeasures.

Acknowledgments. We are grateful to R. Haralick for making a computer cluster in his lab available for our computer experiments.

2   Converting Garside Normal Forms to Words

The Garside normal form of an element $a \in B_n$ is the pair $(k, (\xi_1, \ldots, \xi_m))$, where $k \in \mathbb{Z}$ and $(\xi_1, \ldots, \xi_m)$ is a sequence of permutations (permutation braids) satisfying certain conditions (see [6] for more information). The braid a can be recovered from its normal form $(k, (\xi_1, \ldots, \xi_m))$ as a product of the kth power of the half twist permutation braid Δ and permutation braids $\xi_1, \ldots, \xi_m$:

$$a = \Delta^k \xi_1 \ldots \xi_m.$$

In this section we describe an algorithm which, given a Garside normal form of an element a, tries to find a geodesic braid word representing a. (A geodesic braid word of a given braid is a braid word of minimum length representing this braid.) Since all information transmitted by Alice and Bob is in Garside normal forms, we need this algorithm for our attack.

Note that for permutation braids it is easy to find geodesic braid words. Therefore, to convert a given Garside normal form $(k, (\xi_1, \ldots, \xi_m))$ to a word, one can find geodesic braid words $w_\Delta, w_{\xi_1}, \ldots, w_{\xi_m}$ for Δ and $\xi_1, \ldots, \xi_m$, respectively, and compose a word

$$w = w_\Delta^k w_{\xi_1} \ldots w_{\xi_m}$$

which represents the same word as the given normal form. The length of the obtained word w is

$$|k||w_\Delta| + |w_{\xi_1}| + \ldots + |w_{\xi_m}| = |k|\frac{n(n-1)}{2} + |w_{\xi_1}| + \ldots + |w_{\xi_m}|.$$

If $k \geq 0$ then the given braid a is positive and the word w is geodesic in the Cayley graph of B_n.

Before we proceed in the case $k < 0$, recall one property of the elemet Δ (see [2]). For any braid word $w = x_{i_1}^{\varepsilon_1} \ldots x_{i_k}^{\varepsilon_k}$, one has

$$\Delta^{-1} w \Delta = x_{n-i_1}^{\varepsilon_1} \ldots x_{n-i_k}^{\varepsilon_k}.$$

The result of conjugation of w by Δ will be denoted by w^Δ.

Consider now the case $k < 0$. Denote $-k$ by p. One can rewrite the normal form $\Delta^{-p}\xi_1\xi_2\ldots\xi_m$ in the following way:

$$\Delta^{1-p}(\Delta^{-1}\xi_1)\Delta^{p-1} \cdot \Delta^{2-p}(\Delta^{-1}\xi_2)\Delta^{p-2} \cdot \Delta^{3-p}(\Delta^{-1}\xi_3)\Delta^{p-3} \cdot \ldots \qquad (1)$$

Depending on the values of k and m the obtained decomposition (1) will end up either with $\Delta^{m-p}(\Delta^{-1}\xi_m)\Delta^{p-m}$ when $p > m$ or with ξ_m when $p \leq m$.

Note that the expressions $\Delta^{-1}\xi_i$ in brackets are inverted permutation braids and the length of a geodesic for $\Delta^{-1}\xi_i$ is $|\Delta| - |\xi_i|$. Compute a geodesic braid word w_i for each $\Delta^{-1}\xi_i$ in (1). Since Δ^2 generates the center of B_n, the conjugation by Δ^{i-p} either does not change the word (when $i - p$ is even) or acts the same way as the conjugation by Δ does. We have mentioned above that the conjugation by Δ does not increase the length of the word. Finally, conjugate the obtained words $w_1, \ldots, w_k$ by powers of Δ and denote the results by $w_1', \ldots, w_k'$. Clearly, the product

$$w' = w_1' \ldots w_k'$$

defines the same element of B_n as the given normal form does, but the word w' is shorter than w:

$$|w'| = \begin{cases} |k|\frac{n(n-1)}{2} - \sum_{i=1}^{m}|w_{\xi_i}|, & \text{if } -k > m \\ |k|\frac{n(n-1)}{2} - \sum_{i=1}^{|k|}|w_{\xi_i}| + \sum_{i=|k|+1}^{m}|w_{\xi_i}| & \text{if } -k \leq m \end{cases}$$

We performed a series of experiments in which we generated words of length l in generators of B_n and computed their Garside normal forms $(k, (\xi_1, \ldots, \xi_m))$. In the experiments, l was chosen to be sufficiently greater than n, e.g. $l > n^2$. In all cases k was approximately $-\frac{3l}{4n}$ while m was approximately $\frac{3l}{2n}$. Thus, almost in all cases the word w is longer than w'.

3   Minimization of Braids

Let B_n be the group of braids on n strands and let

$$\langle x_1, \ldots, x_{n-1} \; ; \; [x_i, x_j] = 1 \text{ (where } |i - j| > 1), \; x_i x_{i+1} x_i = x_{i+1} x_i x_{i+1} \rangle$$

be its standard presentation. Let w be a word in generators of B_n and their inverses. The problem of computing a geodesic word for w in B_∞ was shown to be NP-complete in [12]. It is known however (see e.g. [10], [15]) that many NP-complete problems have polynomial time generic- or average-case solutions, or have good approximate solutions. In this section we present heuristic algorithms for approximating geodesics of braids and cyclic braids.

By Dehornoy's form of a braid we mean a braid word without any "handles", i.e. a completely reduced braid word in the sense of [4]. The procedure that computes Dehornoy's form for a given word chooses a specific ("permitted") handle inside of the word and removes it. This can introduce new handles but the main result about Dehornoy's forms states that any sequence of handle reductions eventually terminates. Of course, the result depends on how one chooses the handles at every step. Let us fix any particular strategy for selecting handles. For a word $w = w(x_1, \ldots, x_{n-1})$ we denote by $D(w)$ the corresponding Dehornoy's form (i.e., the result of handle reductions where handles are chosen by the fixed strategy).

The following algorithm tries to minimize the given braid word. It exploits the property of Dehornoy's form that for a "generic" braid word one has $|D(w)| < |w|$.

Algorithm 1 *(Minimization of braids)*
SIGNATURE. $w' = Shorten(w)$.
INPUT. *A word $w = w(x_1, \ldots, x_{n-1})$ in generators of the braid group B_n.*
OUTPUT. *A word w' such that $|w'| \leq |w|$ and $w' = w$ in B_n.*
INITIALIZATION. *Put $w_0 = w$ and $i = 0$.*
COMPUTATIONS.

A. *Increment i.*
B. *Put $w_i = D(w_{i-1})$.*
C. *If $|w_i| < |w_{i-1}|$ then*
    1) Put $w_i = w_i^\Delta$.
    2) Goto A.
D. *If i is even then output w_{i+1}^Δ.*
E. *If i is odd then output w_{i+1}.*

The following simple example illustrates why the idea with conjugation by Δ works. Consider the braid word $w = x_2^{-1}x_1x_2x_1$. This braid is in Dehornoy's form, but the geodesic for the corresponding braid is x_1x_2, hence w is not geodesic. Now, the word $w^\Delta = x_1^{-1}x_2x_1x_2$ is not in Dehornoy's form. It contains one handle, removing of which results in the word x_2x_1 which is shorter than the initial word. If we call handles introduced by Dehornoy *left handles* and define *right handles* as subwords symmetric to left handles with respect to the direction of a braid, then the computation of Dehornoy's form of a word conjugated by Δ and conjugating the obtained result by Δ is essentially a process of removing right handles. We note that removing left handles might introduce right handles and vice versa, and the existence of forms without both left and right handles is questionable.

We would like to emphasize practical efficiency of Algorithm 1. We performed a series of experiments to test it; one of the experiments was the following sequence of steps:

1) generate a random freely reduced braid word $w \in B_{100}$ of length 4000;
2) compute its Garside normal form ξ;
3) transform ξ back into a word w' as described in Section 2;
4) apply Algorithm 1 to w'. Denote the obtained word by w''.

In all experiments the length of the obtained words w'' varied in the interval $[2500, 3100]$. Thus, the result was shorter than the input. It is possible that for a longer initial word w we would not get the same results, but the length 4000 is more than is used in the Ko-Lee protocol anyway.

The next algorithm is a variation of Algorithm 1 for cyclic braid words.

Algorithm 2 *(Minimization of cyclic braids)*
SIGNATURE. $w' = CycShorten(w)$.
INPUT. *A word* $w = w(x_1, \ldots, x_{n-1})$ *in generators of the braid group* B_n.
OUTPUT. *A word* w' *such that* $|w'| \leq |w|$ *and* $w' = w$ *in* B_n.
INITIALIZATION. *Put* $w_0 = w$ *and* $i = 0$.
COMPUTATIONS.

A. *Increment* i.
B. *Put* $w_i = w_{i-1}$.
C. *If* $|D(w_i)| < |w_i|$ *then put* $w_i = D(w_i)$.
D. *If* $w_i = w_i' \circ w_i''$ *(where* $|w_i'| - |w_i''| \leq 1$*) and* $|D(w_i''w_i')| < |w_i|$ *then put* $w_i = D(w_i''w_i')$.
E. *If* $|w_i| < |w_{i-1}|$ *then Goto A.*
F. *Output* w_i.

4   The Attack

In this section we describe a heuristic algorithm for solving the decomposition problem for a pair of words w_1 and w_2 as in the Ko-Lee protocol.

First we describe two auxiliary algorithms. The first algorithm decomposes a given word w into a product usv, where $u, v \in L_n$, trying to to make s as short as possible.

Algorithm 3 *(Decomposition 1)*
INPUT. *A braid word* $w = w(x_1, \ldots, x_{n-1})$.
OUTPUT. *A triple of words* (u, s, v) *such that* $u, v \in L_n$, $|s| \leq |w|$, *and* $usv = w$ *in* B_n.
INITIALIZATION. *Put* $u_0 = v_0 = \varepsilon$ *and* $s_0 = w$ *and* $i = 0$.
COMPUTATIONS.

A. *Increment* i.
B. *Put* $u_i = u_{i-1}$, $s_i = s_{i-1}$, *and* $v_i = v_{i-1}$.
C. *For each* $j = 1, \ldots, n - 1$ *check:*
    1) If $|D(x_j s_i)| < |s_i|$ *then*
        − *put* $u_i = u_i x_j^{-1}$;
        − *put* $s_i = D(x_j s_i)$;
        − *goto A.*
    2) If $|D(x_j^{-1} s_i)| < |s_i|$ *then*
        − *put* $u_i = u_i x_j$;
        − *put* $s_i = D(x_j^{-1} s_i)$;
        − *goto A.*

3) If $|D(s_i x_j)| < |s_i|$ then
  - put $v_i = x_j^{-1} v_i$;
  - put $s_i = D(s_i x_j)$;
  - goto A.
4) If $|D(s_i x_j^{-1})| < |s_i|$ then
  - put $v_i = x_j v_i$;
  - put $s_i = D(s_i x_j^{-1})$;
  - goto A.

D. Output the triple (u_i, s_i, v_i).

The next algorithm decomposes two given braid words w_1 and w_2 into products $u s_1 v$ and $u s_2 v$, respectively, where $u, v \in R_n$, trying to make s_1 and s_2 as short as possible.

Algorithm 4 *(Decomposition 2)*
INPUT. *Braid words w_1 and w_2.*
OUTPUT. *A quadruple of words (u, s, t, v) such that $u, v \in R_n$, $|s| \leq |w_1|$, $|t| \leq |w_2|$, $utv = w_2$ in B_n, and $usv = w_1$ in B_n.*
INITIALIZATION. *Put $u_0 = v_0 = \varepsilon$, $s_0 = w_1$, $t_0 = w_2$, and $i = 0$.*
COMPUTATIONS.

A. *Increment i.*
B. *Put $u_i = u_{i-1}$, $s_i = s_{i-1}$, $t_i = t_{i-1}$, and $v_i = v_{i-1}$.*
C. *For each $j = n+1, \ldots, 2n-1$ check:*
  1) If $|D(x_j s_i)| < |s_i|$ and $|D(x_j t_i)| < |t_i|$ then
    - put $u_i = u_i x_j^{-1}$;
    - put $s_i = D(x_j s_i)$;
    - put $t_i = D(x_j t_i)$;
    - goto A.
  2) If $|D(x_j^{-1} s_i)| < |s_i|$ and $|D(x_j^{-1} t_i)| < |t_i|$ then
    - put $u_i = u_i x_j$;
    - put $s_i = D(x_j^{-1} s_i)$;
    - put $t_i = D(x_j^{-1} t_i)$;
    - goto A.
  3) If $|D(s_i x_j)| < |s_i|$ and $|D(t_i x_j)| < |t_i|$ then
    - put $v_i = x_j^{-1} v_i$;
    - put $s_i = D(s_i x_j)$;
    - put $t_i = D(t_i x_j)$;
    - goto A.
  4) If $|D(s_i x_j^{-1})| < |s_i|$ and $|D(t_i x_j^{-1})| < |t_i|$ then
    - put $v_i = x_j v_i$;
    - put $s_i = D(s_i x_j^{-1})$;
    - put $t_i = D(t_i x_j^{-1})$;
    - goto A.

D. *Output (u_i, s_i, t_i, v_i).*

Now let w_1, w_2 be braid words in B_{2n} for which there exist words a_1, a_2 in L_n such that $w_1 = a_1 w_2 a_2$ in B_{2n}. Denote by $\overline{S}_{(w_1,w_2)}$ the solution set for the decomposition problem for the pair (w_1, w_2), i.e.,

$$\overline{S}_{(w_1,w_2)} = \{(q_1, q_2) \in L_n \times L_n \mid q_1 w_1 q_2 = w_2\} \text{ in } B_{2n}.$$

Let the triple (u_i, s_i, v_i) be the result of applying Algorithm 3 to the word w_i (where $i = 1, 2$) and (u, s, t, v) the result of applying Algorithm 4 to the pair (s_1, s_2). We will say that the pair (s, t) is a *simplified pair* of (w_1, w_2).

Lemma 1. *For a simplified pair (s, t) of (w_1, w_2) the following holds:*

$$\overline{S}_{(w_1,w_2)} = \{(u_2 q_1 u_1^{-1}, v_1^{-1} q_2 v_2) \mid (q_1, q_2) \in \overline{S}_{(s,t)}\}.$$

Proof. We have $s = u^{-1} u_1^{-1} w_1 v_1^{-1} v^{-1}$ and $t = u^{-1} u_2^{-1} w_2 v_2^{-1} v^{-1}$, where $u_1, u_2, v_1, v_2 \in L_n$ and $u, v \in R_n$. By the definition of $\overline{S}_{(s,t)}$, one has $(q_1, q_2) \in \overline{S}_{(s,t)}$ if and only if $q_1 s q_2 =_{B_{2n}} t$ in B_{2n}, or if and only if

$$q_1 u^{-1} u_1^{-1} w_1 v_1^{-1} v^{-1} q_2 = u^{-1} u_2^{-1} w_2 v_2^{-1} v^{-1}.$$

Since $q_1, q_2 \in L_n$ and $u, v \in R_n$, the last equality holds if and only if

$$q_1 u_1^{-1} w_1 v_1^{-1} q_2 = u_2^{-1} w_2 v_2^{-1},$$

or if and only if

$$u_2^{-1} q_1 u_1^{-1} w_1 v_1^{-1} q_2 v_2^{-1} = w_2,$$

or if and only if $(u_2^{-1} q_1 u_1^{-1}, v_1^{-1} q_2 v_2^{-1}) \in \overline{S}_{(w_1,w_2)}$.

Now represent the set of possible solutions $S = S_{(w_1,w_2)}$ of the decomposition problem for (w_1, w_2) as a directed graph with the vertex set

$$V = L_n \times L_n$$

and the edge set E containing edges of the following two types:

- $(q_1, q_2) \rightarrow (q_3, q_4)$ if $q_1 = q_3$ and $q_4 = \overline{q_2' \circ x_j^{\varepsilon} \circ q_2''}$ (where $q_2 = q_2' \circ q_2''$, $j \in \{1, \ldots, n-1\}$, and $\varepsilon = \pm 1$);
- $(q_1, q_2) \rightarrow (q_3, q_4)$ if $q_2 = q_4$ and $q_3 = \overline{s_1' \circ x_j^{\varepsilon} \circ q_1''}$ (where $q_1 = q_1' \circ q_1''$, $j \in \{1, \ldots, n-1\}$, and $\varepsilon = \pm 1$).

Define a function $\omega : S \rightarrow \mathbb{N}$ as follows:

$$(q_1, q_2) \overset{\omega}{\mapsto} |CycShorten(q_1 w_1 q_2 w_2^{-1})|.$$

(cf. our Algorithm 2).

Let w_1 be the base word in the Ko-Lee protocol and w_2 a word representing the normal form $\xi = a^{-1} w_1 a$ transmitted by Alice. In this notation we can formulate the problem of finding Alice's keys as a search problem in $S_{(w_1,w_2)}$. Clearly

$$\overline{S}_{(w_1,w_2)} = \{(q_1, q_2) \in S_{(w_1,w_2)} \mid \omega(q_1, q_2) = 0\}$$

and, therefore, the problem is to find a pair of braid words (q_1, q_2) such that $\omega(q_1, q_2) = 0$.

We want to stress that in some cases the set $S_{(w_1,w_2)}$ can be reduced. For example, let m be the smallest index of a generator in both words w_1 and w_2. Then we can impose a restriction $j \in \{m, \ldots, n-1\}$ and solve the search problem in a smaller space. This situation where $m > 1$ was very often the case in our computations.

The next algorithm is an attack on Alice's private key. The input of the algorithm is the base word w and the Garside normal form ξ of the braid word $a^{-1}wa$ transmitted by Alice. The algorithm finds a pair of words (α, β) in generators of L_n such that $\alpha w \beta = u$ in B_{2n}, where u is a braid word with the Garside normal form ξ. At step A, the algorithm transforms ξ into a word $\overline{w}$. At steps B and C, it computes a simplified pair (s, t) for $(w, \overline{w})$. At steps D-F, the algorithm performs a heuristic search in the key space $S_{(s,t)}$. The search starts at the point $(\varepsilon, \varepsilon)$, where ε is the empty word. In each iteration we choose an unchecked vertex with the minimum ω value and construct its neighborhood. The search stops when the point with zero ω value is found.

Algorithm 5 *(Attack on Alice's key)*
INPUT. *A braid word w and a Garside normal form ξ corresponding to a braid u for which there exists $a \in L_n$ such that $a^{-1}wa = u$ in B_{2n}.*
OUTPUT. *A pair of words $\alpha, \beta \in L_n$ such that $\alpha w \beta = u$ in B_{2n}.*
INITIALIZATION. *Put $u_0 = v_0 = \varepsilon$, $s_0 = w_1$, and $i = 0$.*
COMPUTATIONS.

A. *Convert a normal form ξ to a word $\overline{w}$.*
B. *Apply Algorithm 1 to words w and $\overline{w}$.*
B. *Let (u_1, s_1, v_1) be the result of applying Algorithm 3 to the word w and (u_2, s_2, v_2) the result of applying Algorithm 3 to the word $\overline{w}$.*
C. *Let (u, s, t, v) be the result of applying Algorithm 4 to the pair of words (s_1, s_2).*
D. *Let $Q = \{(\varepsilon, \varepsilon)\} \subset S_{(s,t)}$.*
E. *Choose an unchecked pair (q_1, q_2) from the set Q with the minimum ω value.*
F. *For each edge $(q_1, q_2) \to (q_1', q_2') \in S_{(s,st)}$ add a pair (q_1', q_2') to Q. If ω-value of some new pair (q_1', q_2') is 0, then output $(u_2 q_1' u_1^{-1}, v_1^{-1} q_2' v_2)$. Otherwise goto E.*

5   Experiments and Conclusions

We have performed numerous experiments of two types. Experiments of the first type tested security of the original Ko-Lee protocol, whereas experiments of the second type tested security of a protocol similar to that of Ko-Lee, but based on the decomposition problem.

An experiment of the first type is the following sequence of steps:

1) Fix the braid group B_{100}.
2) Randomly generate a base word w as a freely reduced word of length 2000 in the generators of B_{100}.

3) Randomly generate a word $a = a(x_1, \ldots, x_{49})$ as a freely reduced word of length 1000 in the generators of B_{100}.
4) Compute Garside normal forms ρ_1 and ρ_2 of w and $a^{-1}wa$, respectively.
5) Transform normal forms back into words w_1 and w_2 (see Section 2).
6) Apply Algorithm 1 to words w_1 and w_2.
7) Finally, apply Algorithm 5 to the pair (w_1, w_2).

We say that an experiment is successful if all of the above steps were performed in a reasonable amount of time (we allowed 150 minutes); otherwise we stopped the program. We performed 2466 such experiments and had success in 2378 of them, which means the success rate was 96.43%.

Experiments of the second type have different steps 3) and 4). They are as follows:

3') Randomly generate two words $a_1 = a_1(x_1, \ldots, x_{49})$ and $a_2 = a_2(x_1, \ldots, x_{49})$ as freely reduced words of length 1000.
4') Compute Garside normal forms ρ_1 and ρ_2 of w and a_1wa_2, respectively.

We performed 827 experiments of the second type and had success in 794 of them. This gives the success rate of 96.00%, so that the difference in the success rates of two types of experiments is statistically insignificant.

The conclusion therefore is that we were able to break the Ko-Lee protocol in about 150 minutes with over 95.00% success rate for typical parameters.

Finally, we note that there are several ways to improve the success rate. The easiest way is simply to increase the time allocated to experiments. Also, one can improve the algorithms themselves, in particular, Algorithm 1. With a better minimization algorithm the attack is likely to be more efficient. One can also somewhat narrow down the search space, etc.

6   Suggestions on Improving the Key Exchange Protocol

In this section, we briefly sketch a couple of ideas that may help to enhance security of the Ko-Lee protocol and, in particular, make it less vulnerable to the attack described in the previous sections.

1) Either increase the length of the private keys and the base or decrease the rank of the group. With the parameters suggested in [11], transmitted braids are sort of "sparse" which allows the adversary to simplify the initial braids substantially. The lengths of the transmitted braids should be at least on the order of n^2 (where n is the rank of the braid group) to prevent fast reconstruction of a short braid word from its normal form.

We note however that increasing the key length is a trade-off between security and efficiency. By comparison, the current key size used in the RSA cryptosystem is 512 bits, whereas to store a braid word of length l from the group B_n, $l\lfloor \log_2(2n) \rfloor$ bits are required. This number is approximately 8000 for $l = 1000$ and $n = 100$.

2) Choosing a "base" word w requires special attention. It might be a good idea to generate w as a geodesic in the Cayley graph of B_{2n} starting and terminating with the generator x_n or its inverse (the one which does not belong to L_n or R_n) such that any other geodesic representing w starts and terminates with $x_n^{\pm 1}$. Observe that for such w Algorithm 3 stops with the result $(\varepsilon, w, \varepsilon)$. Also, for such w and an arbitrary braid word w', Algorithm 4 applied to (w, w') stops with the result $(\varepsilon, w, w', \varepsilon)$.

3) Choose different commuting subgroups instead of L_n and R_n. This looks like the most promising suggestion at the moment; we refer to [14] for more details.

References

1. I. Anshel, M. Anshel, D. Goldfeld, *An algebraic method for public-key cryptography*, Math. Res. Lett. **6** (1999), 287–291.
2. J. S. Birman, *Braids, links and mapping class groups*, Ann. Math. Studies **82**, Princeton Univ. Press, 1974.
3. J. H. Cheon and B. Jun, *A polynomial time algorithm for the braid Diffie-Hellman conjugacy problem*, Crypto 2003, Lecture Notes in Comput. Sci. **2729** (2003), 212–225.
4. P. Dehornoy, *A fast method for comparing braids*, Adv. Math. **125** (1997), 200–235.
5. P. Dehornoy, *Braid-based cryptography*, Contemp. Math., Amer. Math. Soc. **360** (2004), 5–33.
6. D. B. A. Epstein, J. W. Cannon, D. F. Holt, S. V. F. Levy, M. S. Paterson, W. P. Thurston, *Word processing in groups*. Jones and Bartlett Publishers, Boston, MA, 1992.
7. D. Garber, S. Kaplan, M. Teicher, B. Tsaban, U. Vishne, *Probabilistic solutions of equations in the braid group*, preprint. http://arxiv.org/abs/math.GR/0404076
8. D. Hofheinz and R. Steinwandt, *A practical attack on some braid group based cryptographic primitives*, in Public Key Cryptography, 6th International Workshop on Practice and Theory in Public Key Cryptography, PKC 2003 Proceedings, Y.G. Desmedt, ed., Lecture Notes in Computer Science **2567**, pp. 187–198, Springer, 2002.
9. J. Hughes and A. Tannenbaum, *Length-based attacks for certain group based encryption rewriting systems*, Workshop SECI02 Securitè de la Communication sur Intenet, September 2002, Tunis, Tunisia. http://www.storagetek.com/hughes/
10. I. Kapovich, A. Myasnikov, P. Schupp and V. Shpilrain, *Average-case complexity for the word and membership problems in group theory*, Advances in Math. **190** (2005), 343–359.
11. K. H. Ko, S. J. Lee, J. H. Cheon, J. W. Han, J. Kang, C. Park, *New public-key cryptosystem using braid groups*, Advances in cryptology—CRYPTO 2000 (Santa Barbara, CA), 166–183, Lecture Notes in Comput. Sci. **1880**, Springer, Berlin, 2000.
12. M. S. Paterson, A. A. Razborov, *The set of minimal braids is co-NP-complete*, J. Algorithms **12** (1991), 393–408.
13. V. Shpilrain and A. Ushakov, *The conjugacy search problem in public key cryptography: unnecessary and insufficient*, Applicable Algebra in Engineering, Communication and Computing, to appear. http://eprint.iacr.org/2004/321/
14. V. Shpilrain and G. Zapata, *Combinatorial group theory and public key cryptography*, Applicable Algebra in Engineering, Communication and Computing, to appear. http://eprint.iacr.org/2004/242
15. J. Wang, *Average-case computational complexity theory*, Complexity Theory Retrospective, II. Springer-Verlag, New York, 1997, 295–334.

The Conditional Correlation Attack:
A Practical Attack on Bluetooth Encryption

Yi Lu[1], Willi Meier[2], and Serge Vaudenay[1]

[1] EPFL, CH-1015 Lausanne, Switzerland
http://lasecwww.epfl.ch
[2] FH Aargau, CH-5210 Windisch, Switzerland
meierw@fh-aargau.ch

Abstract. Motivated by the security of the nonlinear filter generator, the concept of correlation was previously extended to the conditional correlation, that studied the linear correlation of the inputs conditioned on a given (short) output pattern of some specific nonlinear function. Based on the conditional correlations, conditional correlation attacks were shown to be successful and efficient against the nonlinear filter generator. In this paper, we further generalize the concept of conditional correlations by assigning it with a different meaning, i.e. the correlation of the output of an arbitrary function conditioned on the unknown (partial) input which is uniformly distributed. Based on this generalized conditional correlation, a general statistical model is studied for dedicated key-recovery distinguishers. It is shown that the generalized conditional correlation is no smaller than the unconditional correlation. Consequently, our distinguisher improves on the traditional one (in the worst case it degrades into the traditional one). In particular, the distinguisher may be successful even if no ordinary correlation exists. As an application, a conditional correlation attack is developed and optimized against Bluetooth two-level E0. The attack is based on a recently detected flaw in the resynchronization of E0, as well as the investigation of conditional correlations in the Finite State Machine (FSM) governing the keystream output of E0. Our best attack finds the original encryption key for two-level E0 using the first 24 bits of $2^{23.8}$ frames and with 2^{38} computations. This is clearly the fastest and only practical known-plaintext attack on Bluetooth encryption compared with all existing attacks. Current experiments confirm our analysis.

Keywords: Stream Ciphers, Correlation, Bluetooth, E0.

1 Introduction

In stream ciphers, correlation properties play a vital role in correlation attacks (to name a few, see [7,8,9,15,18,19,26,27,30]). For LFSR-based[1] keystream generators, such as the nonlinear filter generator or the combiner, *correlation* commonly means a statistically biased relation between the produced keystream and

---

[1] LFSR refers to Linear Feedback Shift Register, see [28] for more.

V. Shoup (Ed.): Crypto 2005, LNCS 3621, pp. 97–117, 2005.

the output of certain LFSR sequences. In [1,21,22], the concept of (ordinary) correlations was further extended to the *conditional correlation* to describe the *linear* correlation of the inputs conditioned on a *given* (short) output pattern of a nonlinear function (with small input size). Based on conditional correlations, the conditional correlation attack received successful studies towards the nonlinear filter generator in [1,21,22]. In this paper, we assign a different meaning to conditional correlations, i.e. the correlation of the output of an arbitrary function (with favorable small input size) conditioned on the *unknown* (partial) input which is uniformly distributed. This might be viewed as the generalized opposite of [1,21,22]. As a useful application of our conditional correlations, imagine the attacker not only observes the keystream, but also has access to an intermediate computation process controlled partly by the key, which outputs a hopefully biased sequence for the right key and (presumably) unbiased sequences for wrong keys. If such side information is available, the *conditional correlation attack* may become feasible, which exploits correlations of the intermediate computation output conditioned on (part of) the inputs. In general, as informally conjectured in [22], conditional correlations are different and often larger than ordinary (unconditional) correlations, which effects reduced data complexity of conditional correlation attacks over ordinary correlation attacks.

Our first contribution consists of extracting a precise and general statistical model for dedicated key-recovery distinguishers based on the generalized conditional correlations. This framework deals with a specific kind of smart distinguishers that exploit correlations conditioned on the (partial) input, which is not restricted to keystream generators and is also applicable to other scenarios (e.g. side channel attacks like fault attacks in [4]). As the ordinary correlation serves as the criterion for the data complexity of the traditional distinguisher (that only exploits ordinary correlations), our result based on the sound theory of traditional distinguisher [5] tells that the conditional correlation serves similarly as the criterion for the data complexity of the smart distinguisher. The construction of the smart distinguisher also solves the unaddressed problem in [1,21,22] on how to make the best use of all the collected data, which can be transformed in the context of [1,21,22]. We prove that the smart distinguisher improves on the traditional one (in the worst case the smart distinguisher degrades into the traditional one), because our generalized conditional correlation is no smaller than the unconditional correlation. In particular, the smart distinguisher can still work efficiently even though the traditional one fails thoroughly. Meanwhile, we also study the computational complexity of the deterministic smart distinguisher for a special case, in which the essence of the major operation done by the distinguisher is identified to be *nothing but the regular convolution*. Thanks to Fast Walsh Transform[2] (FWT), when the key size is not too large, the smart distinguisher is able to achieve the optimal complete information set decoding and becomes a very powerful computing machine. Nonetheless, in general, with a very large key size, it is unrealistic to use the deterministic distinguisher as

---

[2] Note that most recently FWT was successfully applied in [9,23] to optimize different problems in correlation attacks.

complete information set decoding is impractical; many other efficient decoding techniques (e.g. the probabilistic iterative decoding) such as introduced in the previous conditional correlation attacks [22] or the correlation attacks will also apply to our smart distinguisher.

As a second contribution, we apply our smart distinguisher to a conditional correlation attack[3] on two-level E0, the keystream generator that is used in the short-range wireless technology Bluetooth [6]. The attack exploits the resynchronization flaw recently detected in [24]. Whereas in [24], this flaw is used for a traditional distinguisher based on results [12,16,17,23] of ordinary correlations, our conditional correlation attack relies on the systematic investigation of correlations conditioned on the inputs to the FSM in E0. These correlations extend a specific conditional correlation found in [23], which relates to one of the largest known biases in E0 as proved in [23]. The time complexity of our attack is optimized as the smart distinguisher works particularly well in this favorable case. Our best attack recovers the original encryption key for two-level E0 using the first 24 bits of $2^{23.8}$ frames after 2^{38} computations. Note that the number of necessary frames is below the maximum number 2^{26} of resynchronizations with the same user key as specified by Bluetooth [6]. Compared with all existing attacks [13,14,16,20,24,29] on two-level E0, our attack is clearly the fastest and only practical resynchronization attack[4] so far. Note that the resynchronization attacks on one-level E0 were well studied in [3,14,24] to be much more efficient.

The rest of the paper is structured as follows. In Section 2 we introduce some notations and give preliminaries. In Section 3, based on the generalized conditional correlation, the practical statistical model on smart distinguishers with side information is formalized and analyzed. In Section 4 we review the description of Bluetooth two-level E0 as well as the resynchronization flaw. In Section 5, correlations conditioned on input weights of E0 FSM are investigated. In Section 6, a key-recovery attack on two-level E0 is developed and optimized together with experimental results. Finally, we conclude in Section 7.

2 Notations and Preliminaries

Given the function $f : \mathcal{E} \to GF(2)^{\ell}$, define the distribution D_f of $f(X)$ with X uniformly distributed, i.e. $D_f(a) = \frac{1}{|\mathcal{E}|} \sum_{X \in \mathcal{E}} \mathbf{1}_{f(X)=a}$ for all $a \in GF(2)^{\ell}$. Following [5], recall that the Squared Euclidean Imbalance (SEI) of the distribution D_f is defined by

$$\Delta(D_f) = 2^{\ell} \sum_{a \in GF(2)^{\ell}} \left(D_f(a) - \frac{1}{2^{\ell}} \right)^2. \tag{1}$$

---

[3] For the conditional correlation attack related to the previous work [1,21,22] on Bluetooth E0, see [16].

[4] A resynchronization attack on stream cipher (a.k.a. the related-key attack) refers to the one that needs many frames of keystreams produced by different IVs (i.e. the public frame counter) and the same key in order to recover the key given the IVs.

For $\ell = 1$, it's easy to see that $\Delta(D_f)$ is closely related to the well known term *correlation*[5] $\epsilon(D_f)$ by $\Delta(D_f) = \epsilon^2(D_f)$. For brevity, we adopt the simplified notations $\epsilon(f), \Delta(f)$ to denote $\epsilon(D_f), \Delta(D_f)$ respectively hereafter. From the theory of hypothesis testing and Neyman-Pearson likelihood ratio (see [5]), $\Delta(f)$ tells us that the minimum number n of samples for an optimal distinguisher to effectively distinguish a sequence of n output samples of f from $(2^L - 1)$ truly random sequences of equal length is

$$n = \frac{4L \log 2}{\Delta(f)}. \tag{2}$$

Note that the result in Eq.(2) with $\ell = 1$ has long been known up to a constant factor $\frac{1}{2}$ in the theory of channel coding. In fact, correlation attacks has been very successful for almost two decades to apply the distinguisher that analyzes the biased sample of a single bit (i.e. the case $\ell = 1$) in order to reconstruct the L-bit key (or subkey), where only the right key can produce a biased sequence while all the wrong keys produce unbiased sequences. More recently, on the sound theoretical basis [5] of the generalized distinguisher, it was shown that this generalized distinguisher helps to improve the correlation attack when considering multi-biases simultaneously (for details see the key-recovery attack [23] on one-level E0 which halves the time and data complexities).

3   A Smart Distinguisher with Side Information

Given a function $f : GF(2)^u \times GF(2)^v \to GF(2)^r$, let $f_\mathcal{B}(X) = f(\mathcal{B}, X)$ for $\mathcal{B} \in GF(2)^u$ and $X \in GF(2)^v$, where the notation $f_\mathcal{B}(\cdot)$ is used to replace $f(\cdot)$ whenever $\mathcal{B}$ is given. Consider such a game between a player and an oracle. Each time the oracle secretly generates $\mathcal{B}, X$ independently and uniformly to compute $f_\mathcal{B}(X)$; the player, in turn, sends a guess on the current value of the partial input $\mathcal{B}$. Only when he guesses correctly, the oracle would output the value of $f_\mathcal{B}(X)$, otherwise, it would output a random and uniformly distributed $Z \in GF(2)^r$. Suppose the player somehow manages to collect 2^L sequences of n interaction samples with the following characteristics: one sequence has n samples $(f_{\mathcal{B}_i^\mathcal{K}}(X_i), \mathcal{B}_i^\mathcal{K})$ $(i = 1, \ldots, n)$ where $\mathcal{B}_i^\mathcal{K}$'s and X_i's are independently and uniformly distributed; the remaining $(2^L - 1)$ sequences all consist of n independently and uniformly distributed random variables $(Z_i^K, \mathcal{B}_i^K)$ $(i = 1, \ldots, n)$ for $K \neq \mathcal{K}$. One interesting question to the player is how to distinguish the biased sequence from the other sequences using the minimum number n of samples.

Note that the above problem is of special interest in key-recovery attacks, including the related-key attacks, where $\mathcal{B}_i^\mathcal{K}$'s are the key-related material (i.e. computable with the key and other random public parameters) and the oracle can be viewed as an intermediate computation process accessible to the attacker with only a limited number of queries. Thus, when the attacker knows the right

---

[5] Correlation is commonly defined by $D_f(1) = \frac{1}{2} + \frac{\epsilon(D_f)}{2}$; and $|\epsilon(D_f)| \leq 1$ by this definition.

key $\mathcal{K}$ he can collect n (hopefully biased) samples of f; on the other hand, if he uses the wrong key, he will only collect an unbiased sequence.

From Section 2, we know that the minimum number n of samples for the basic distinguisher which doesn't use the partial input $\mathcal{B}_i$'s is $n = 4L\log 2/\Delta(f)$. When the samples are incorporated with the $\mathcal{B}_i$'s, we can prove the following stronger result.

Theorem 1. *The smart distinguisher (in Algorithm 1) solves our above problem with*

$$n = \frac{4L\log 2}{E[\Delta(f_{\mathcal{B}})]} \tag{3}$$

and the time complexity $O(n \cdot 2^L)$, where the expectation is taken over all the uniformly distributed $\mathcal{B}$. Moreover, the distinguisher can achieve the optimal time complexity $O(n + L \cdot 2^{L+1})$ with precomputation $O(L \cdot 2^L)$ when $\mathcal{B}_i^K$'s and Z_i^K's can be expressed by:

$$\mathcal{B}_i^K = \mathcal{L}(K) \oplus c_i, \tag{4}$$

$$Z_i^K = \mathcal{L}'(K) \oplus c_i' \oplus g(\mathcal{B}_i^K), \tag{5}$$

for all L-bit K and $i = 1, 2, \ldots, n$, where g is an arbitrary function, $\mathcal{L}, \mathcal{L}'$ are $GF(2)$-linear functions, and c_i's, c_i''s are independently and uniformly distributed which are known to the distinguisher.

Algorithm 1. The smart distinguisher with side information

Parameters:
  1: n set by Eq.(3)
  2: $D_{f_{\mathcal{B}}}$ for all $\mathcal{B} \in GF(2)^u$

Inputs:
  3: uniformly and independently distributed u-bit $\mathcal{B}_1^K, \ldots, \mathcal{B}_n^K$ for all L-bit K
  4: $Z_1^{\mathcal{K}}, \ldots, Z_n^{\mathcal{K}} = f_{\mathcal{B}_1^{\mathcal{K}}}(X_1), \ldots, f_{\mathcal{B}_n^{\mathcal{K}}}(X_n)$ for one fixed L-bit $\mathcal{K}$ with uniformly and independently distributed v-bit vectors $X_1, \ldots, X_n$
  5: uniformly and independently distributed sequences $Z_1^K, Z_2^K, \ldots, Z_n^K$ for all L-bit K such that $K \neq \mathcal{K}$

Goal: find $\mathcal{K}$

Processing:
  6: **for all** L-bit K **do**
  7:     $G(K) \leftarrow 0$
  8:     **for** $i = 1, \ldots, n$ **do**
  9:         $G(K) \leftarrow G(K) + \log_2\left(2^r \cdot D_{f_{\mathcal{B}_i^K}}(Z_i^K)\right)$
 10:     **end for**
 11: **end for**
 12: output $\mathcal{K}$ that maximizes $G(\mathcal{K})$

Remark 2. Our smart distinguisher (Algorithm 1) turns out to be a derivative of the basic distinguisher in [5] and the result Eq.(3) for the simple case $r = 1$ was already pointed out (without proof) in [16] with a mere difference of a negligible constant term $2 \log 2 \approx 2^{0.47}$. Also note that the quantity $E[\Delta(f_\mathcal{B})]$ in Eq.(3) measures the correlation of the output of an arbitrary function conditioned on the (partial) input which is uniformly distributed and *unknown*[6]. In contrast, prior to our work, the conditional correlation, that refers to the linear correlation of the inputs conditioned on a given (short) output pattern of a nonlinear function, was well studied in [1,21,22] based on a different statistical distance other than SEI. Highly motivated by the security of the nonlinear filter generator, their research focused on the case where the nonlinear function is the augmented nonlinear filter function (with small input size) and the inputs are the involved LFSR taps. Obviously, the notion of our conditional correlation can be seen as the generalized opposite of [1,21,22], that addresses the issue of how to make the most use of all the data for the success. In Section 6, Theorem 1 is directly applied to Bluetooth two-level E0 for a truly practical attack.

Proof (sketch). Let us introduce a new distribution D over $GF(2)^{r+u}$ from $D_{f_\mathcal{B}}$ defined by

$$D(\mathcal{B}, Z) = \frac{1}{2^u} D_{f_\mathcal{B}}(Z), \tag{6}$$

for all $\mathcal{B} \in GF(2)^u, Z \in GF(2)^r$. We can see that our original problem is transformed into that of the basic distinguisher to distinguish D from uniform distribution. According to Section 2, we need minimum $n = 4L \log 2 / \Delta(D)$. So we compute $\Delta(D)$ by Eq.(1,6):

$$
\begin{aligned}
\Delta(D) &= 2^{r+u} \sum_{\mathcal{B} \in GF(2)^u} \sum_{Z \in GF(2)^r} \left(D(\mathcal{B}, Z) - \frac{1}{2^{r+u}} \right)^2 \\
&= 2^{r+u} \sum_{\mathcal{B} \in GF(2)^u} \sum_{Z \in GF(2)^r} \left(\frac{1}{2^u} D_{f_\mathcal{B}}(Z) - \frac{1}{2^{r+u}} \right)^2 \\
&= 2^{-u} \sum_{\mathcal{B} \in GF(2)^u} 2^r \sum_{Z \in GF(2)^r} \left(D_{f_\mathcal{B}}(Z) - \frac{1}{2^r} \right)^2 \\
&= E[\Delta(f_\mathcal{B})]. \tag{7}
\end{aligned}
$$

Meanwhile, the best distinguisher tries to maximize the probability $\prod_{i=1}^n D(\mathcal{B}_i, Z_i)$, i.e. the conditioned probability $\prod_{i=1}^n D_{f_{\mathcal{B}_i}}(Z_i)$. As the conventional approach, we know that this is equivalent to maximize $G = \sum_{i=1}^n \log_2(2^r \cdot D_{f_{\mathcal{B}_i}}(Z_i))$ as shown in Algorithm 1. The time complexity of the distinguisher[7] is obviously $O(n \cdot 2^L)$.

---

[6] According to the rule of our game, it's unknown to the distinguisher whether the sample $\mathcal{B}$ is the correct value used for the oracle to compute $f_\mathcal{B}(X)$ or not.

[7] In this paper, we only discuss the deterministic distinguisher. For the probabilistic distinguisher, many efficient and general decoding techniques (e.g. the probabilistic iterative decoding), which are successful in correlation attacks, were carefully presented in the related work [22] and such techniques also apply to our distinguisher.

Now, to show how to optimize the time complexity of the smart distinguisher when $\mathcal{B}_i^K$'s and Z_i^K's exhibit the special structure of Eq.(4, 5) for the second part of the theorem, let us first introduce two functions $\mathcal{H}, \mathcal{H}'$:

$$\mathcal{H}(K) = \sum_{i=1}^{n} \mathbf{1}_{\mathcal{L}(K)=c_i \text{ and } \mathcal{L}'(K)=c_i'} \tag{8}$$

$$\mathcal{H}'(K) = \log_2 \left(2^r \cdot D_{f_{\mathcal{L}(K)}} \left(\mathcal{L}'(K) \oplus g\left(\mathcal{L}(K)\right)\right)\right) \tag{9}$$

for $K \in GF(2)^L$. We can see that $G(K)$ computed in Line 7 to 10, Algorithm 1 is nothing but a simple convolution (denoted by $\otimes$) between $\mathcal{H}$ and $\mathcal{H}'$:

$$G(K) = (\mathcal{H} \otimes \mathcal{H}')(K) \stackrel{\text{def}}{=} \sum_{K' \in GF(2)^L} \mathcal{H}(K')\mathcal{H}'(K \oplus K'), \tag{10}$$

for all $K \in GF(2)^L$. It's known that convolution and Walsh transform (denoted by the hat symbol) are transformable, so we have

$$G(K) = \frac{1}{2^L} \widehat{\mathcal{H} \otimes \mathcal{H}'}(K) = \frac{1}{2^L} \widehat{\mathcal{H}''}(K), \tag{11}$$

where $\mathcal{H}''(K) = \widehat{\mathcal{H}}(K) \cdot \widehat{\mathcal{H}'}(K)$. This means that after computing $\mathcal{H}$ and $\mathcal{H}'$, the time complexity of our smart distinguisher would be dominated by three times of FWT, i.e. $\widehat{\mathcal{H}}, \widehat{\mathcal{H}'}, \widehat{\mathcal{H}''}$ in $O(3L \cdot 2^L)$. Moreover, since only c_i's, c_i''s may vary from one run of the attack to another, which are independent of $\mathcal{H}'$, we can also precompute $\widehat{\mathcal{H}'}$ and store it in the table; finally, the real-time processing only takes time $O(n + L \cdot 2^{L+1})$. □

Property 3. We have
$$\mathrm{E}[\Delta(f_\mathcal{B})] \geq \Delta(f),$$
where equality holds if and only if (iff) $D_{f_\mathcal{B}}$ is independent of $\mathcal{B}$.

For $r = 1$, this can be easily shown as follows. From Section 2, we have $\mathrm{E}[\Delta(f_\mathcal{B})] = \mathrm{E}[\epsilon^2(f_\mathcal{B})] \geq E^2[\epsilon(f_\mathcal{B})] = \epsilon^2(f) = \Delta(f)$ where equality holds iff $\epsilon(f_\mathcal{B})$ is independent of $\mathcal{B}$. In Appendix, we give the complete proof for the general case $\mathrm{E}[\Delta(f_\mathcal{B})] \geq \Delta(f)$.

Remark 4. As $\mathrm{E}[\Delta(f_\mathcal{B})], \Delta(f)$ measures the conditional correlation and the unconditional correlation respectively, this property convinces us that the former is no smaller than the latter. This relationship between the conditional correlation and the unconditional correlation was informally conjectured in [22]. We conclude from Eq.(3) that the smart distinguisher having partial (or side) information (i.e. $\mathcal{B}$ herein) about the biased source generator (i.e. $f_\mathcal{B}$ herein) always works better than the basic distinguisher governing no knowledge of that side information, as long as the generator is statistically dependent on the side information. Our result verifies the intuition that the more the distinguisher knows about the generation of the biased source, the better it works. In particular,

Property 3 implies that even if the fact that $\Delta(f) = 0$ causes the basic distinguisher to be completely useless as it needs infinite data complexity, in contrast, the smart distinguisher would still work as long as D_{f_B} is dependent on $\mathcal{B}$, i.e. $E[\Delta(f_B)] > 0$. In Section 5, we give two illustrative examples $E[\Delta(f_B)]$ on the core of Bluetooth E0 to be compared with their counterparts $\Delta(f)$.

4   Review on Bluetooth Two-Level E0

The core (Fig. 1) of Bluetooth keystream generator E0 (also called one-level E0) consists of four regularly-clocked LFSRs of a total 128 bits and a Finite State Machine (FSM) of 4 bits. Denote $B_t \in GF(2)^4$ the four output bits of LFSRs at time instance t, and $X_t \in GF(2)^4$ the FSM state at time instance t. Note that X_t contains the bit c_t^0 as well as the bit c_{t-1}^0 (due to the effect of a delay cell inside the FSM). Also note that the computation of the FSM next state X_{t+1} only depends on its current state X_t together with the Hamming weight $w(B_t)$ of B_t. At each time instance t, the core produces one bit $s_t = (w(B_t) \mod 2) \oplus c_t^0$, and then updates the states of LFSRs and FSM.

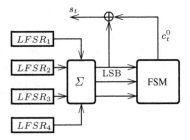

Fig. 1. The core of Bluetooth keystream generator E0

According to the Bluetooth standard [6], this core is used with a two-level initialization scheme to produce the keystream for encryption. That is, after a first initialization of LFSRs by an affine transformation of the effective encryption key $\mathcal{K}$ and the public nonce[8] $\mathcal{P}^i$ for the i-th frame, E0 runs at level one, whose last 128 output bits are permuted into LFSRs at level two for reinitialization; then E0 runs at level two to produce the final keystream $z_{t'}^i$ for $t' = 1, 2, \ldots, 2745$ (for clarity, we refer the time instance t and t' to the context of E0 level one and E0 level two respectively).

In order to review the reinitialization flaw discovered in [24], we first introduce some notations. Define the binary vector $\gamma = (\gamma_0, \gamma_1, \ldots, \gamma_{\ell-1})$ of length $|\gamma| = \ell \geq 3$ with $\gamma_0 = \gamma_{\ell-1} = 1$ and let $\bar{\gamma} = (\gamma_{\ell-1}, \gamma_{\ell-2}, \ldots, \gamma_0)$ represent the vector in reverse order of γ. Given ℓ and t, for the one-level E0, we define $\mathcal{B}_{t+1} = B_{t+1}B_{t+2} \ldots B_{t+\ell-2}$ and $C_t = (c_t^0, \ldots, c_{t+\ell-1}^0)$. Then, the function

---

[8] $\mathcal{P}^i$ includes a 26-bit counter and some user-dependent constant.

$h^\gamma_{\mathcal{B}_{t+1}} : X_{t+1} \mapsto \gamma \cdot C_t$ is well defined[9] for all t, where the dot operator between two vectors represents the inner $GF(2)$-product. Now, we let $(\mathcal{B}^i_{t+1}, X^i_{t+1})$ (resp. $(\mathcal{B}^i_{t'+1}, X^i_{t'+1})$) control the FSM to compute C^i_t (resp. $C^i_{t'}$) at E0 first (resp. second) level for the i-th frame. Note that initialization of LFSRs at E0 level one by an affine transformation of $\mathcal{K}, \mathcal{P}^i$ can be expressed by

$$\mathcal{B}^i_t = \mathcal{G}_t(\mathcal{K}) \oplus \mathcal{G}'_t(\mathcal{P}^i), \tag{12}$$

where $\mathcal{G}_t, \mathcal{G}'_t$ are public linear functions (which are dependent on ℓ but omitted from notations for simplicity). Moreover, we let $Z^i_{t'} = (z^i_{t'}, \dots, z^i_{t'+\ell-1})$. Then, as pointed out and detailed in [24], the critical reinitialization flaw of Bluetooth two-level E0 can be expressed as

$$\bar{\gamma} \cdot (Z^i_{t'} \oplus \mathcal{L}_{t'}(\mathcal{K}) \oplus \mathcal{L}'_{t'}(\mathcal{P}^i)) = \bigoplus_{j=1}^{4} (\gamma \cdot C^i_{t_j}) \oplus (\bar{\gamma} \cdot C^i_{t'}), \tag{13}$$

for any i and γ of length ℓ such that $3 \le \ell \le 8$, and $t' \in \bigcup_{k=0}^{2}\{8k+1, \dots, 8k + 9 - \ell\}$, where $t_1, \dots, t_4$ are functions[10] in terms of t' only, and $C^i_{t_1}, \dots, C^i_{t_4}$ share no common coordinate, and $\mathcal{L}_{t'}, \mathcal{L}'_{t'}$ are fixed linear functions which can be expressed by t', ℓ from the standard. By definition of h, Eq.(13) can be put equivalently as:

$$\bar{\gamma} \cdot (Z^i_{t'} \oplus \mathcal{L}_{t'}(\mathcal{K}) \oplus \mathcal{L}'_{t'}(\mathcal{P}^i)) = \bigoplus_{j=1}^{4} h^\gamma_{\mathcal{B}^i_{t_j+1}} (X^i_{t_j+1}) \oplus h^{\bar\gamma}_{\mathcal{B}^i_{t'+1}} (X^i_{t'+1}), \tag{14}$$

for any i, any γ with $3 \le \ell \le 8$ and $t' \in \bigcup_{k=0}^{2}\{8k+1, \dots, 8k + 9 - \ell\}$. Note that the usage of the bar operator actually reflects the fact that the loading of LFSRs at E0 level two for reinitialization is in reverse order of the keystream output at E0 level one.

5   Correlations Conditioned on Input Weights of FSM

Recall it has been observed in [23] that if $w(B_t)w(B_{t+1})w(B_{t+2})w(B_{t+3}) = 2222$ is satisfied, then, we always have

$$c^0_t \oplus c^0_{t+1} \oplus c^0_{t+2} \oplus c^0_{t+3} \oplus c^0_{t+4} = 1. \tag{15}$$

Let $\alpha_t = \gamma \cdot C_t$ with $\gamma = (1,1,1,1,1)$ and $\ell = 5$. Thus α_t is the sum on the left-hand side of Eq.(15). From Section 4 we know that given $\mathcal{B}_{t+1} = B_{t+1}B_{t+2}B_{t+3} \in GF(2)^{12}$, the function $h^\gamma_{\mathcal{B}_{t+1}} : X_{t+1} \mapsto \alpha_t$ is well defined for all

---

[9] Because c^0_t, c^0_{t+1} are contained in X_{t+1} already and we can compute $c^0_{t+2}, \dots, c^0_{t+\ell-1}$ by $\mathcal{B}_{t+1}, X_{t+1}$. Actually, the prerequisite $\gamma_0 = \gamma_{\ell-1} = 1$ on γ is to guarantee that knowledge of $\mathcal{B}_{t+1}, X_{t+1}$ is necessary and sufficient to compute $\gamma \cdot C_t$.

[10] Additionally, given t', the relation $t_1 < t_2 < t_3 < t_4$ always holds that satisfies $t_2 - t_1 = t_4 - t_3 = 8$ and $t_3 - t_2 \ge 32$.

t. Let $W(\mathcal{B}_{t+1}) \overset{\text{def}}{=} w(\mathcal{B}_{t+1})w(\mathcal{B}_{t+2}) \cdots w(\mathcal{B}_{t+\ell-2})$. Thereby, we deduct from [23] that $\alpha_t = 1$ whenever $W(\mathcal{B}_{t+1}) = 222$. In contrast to the (unconditional) correlation as mentioned in Section 2, we call it a conditional correlation[11], i.e. the correlation $\epsilon(h^\gamma_{\mathcal{B}_{t+1}}) = 1$ conditioned on $W(\mathcal{B}_{t+1}) = 222$.

This motivates us to study the general correlation $\epsilon(h^\gamma_{\mathcal{B}_{t+1}})$ conditioned on $\mathcal{B}_{t+1}$, or more precisely $W(\mathcal{B}_{t+1})$, when X_{t+1} is uniformly distributed. All the non-zero conditional correlations $\epsilon(h^\gamma_{\mathcal{B}_{t+1}})$ are shown in Table 1 in descending order of the absolute value, where $|\mathcal{B}_{t+1}|$ denotes the cardinality of $\mathcal{B}_{t+1}$ admitting any weight triplet in the group. As the unconditioned correlation $\epsilon(h^\gamma)$ of the bit α_t always equals the mean value[12] $\mathrm{E}[\epsilon(h^\gamma_{\mathcal{B}_{t+1}})]$ over the uniformly distributed $\mathcal{B}_{t+1}$, we can use Table 1 to verify $\epsilon(h^\gamma) = \frac{25}{256}$ (denote this value[13] by λ). Let $f_\mathcal{B} = h^\gamma_{\mathcal{B}_{t+1}}$ with $\gamma = (1,1,1,1,1)$. Now, to verify Property 3 in Section 3 we compute $\mathrm{E}[\Delta(f_\mathcal{B})] = \frac{544}{2^{12}} \approx 2^{-2.9}$, which is significantly larger than $\Delta(f) = \lambda^2 \approx 2^{-6.67}$. As another example, consider now $f_\mathcal{B} = h^\gamma_{\mathcal{B}_{t+1}}$ with $\gamma = (1,1,0,1)$ and $u = 8, v = 4, r = 1$. Similarly, the conditioned correlation of the corresponding sum $c_t^0 \oplus c_{t+1}^0 \oplus c_{t+3}^0$ (denoted by α'_t) is shown in Table 2. From Table 2, we get a quite large $\mathrm{E}[\Delta(f_\mathcal{B})] = 2^{-3}$ as well; in contrast, we can check that as already pointed out in Section 3, the unconditional correlation[14] $\Delta(f) = 0$ from Table 2.

6   Key-Recovery Attack on Bluetooth Two-Level E0

6.1   Basic Idea

Given the binary vector γ (to be determined later) with $3 \leq \ell \leq 8$, for all $\mathcal{B} \in GF(2)^{4(\ell-2)}$ such that $\epsilon(h^\gamma_\mathcal{B}) \neq 0$, define the function

$$g^\gamma(\mathcal{B}) = \begin{cases} 1, & \text{if } \epsilon(h^\gamma_\mathcal{B}) > 0 \\ 0, & \text{if } \epsilon(h^\gamma_\mathcal{B}) < 0 \end{cases}$$

to estimate the effective value of $h^\gamma_\mathcal{B}(X)$ (defined in Section 4) for some unknown $X \in GF(2)^4$. For a fixed $t' \in \bigcup_{k=0}^{2}\{8k+1, \ldots, 8k+9-|\gamma|\}$, let us guess the

---

[11] Note that earlier in [16], correlations conditioned on keystream bits (both with and without one LFSR outputs) were well studied for one-level E0, which differ from our conditional correlations and do not fit in the context of two-level E0 if the initial state of E0 is not recovered level by level.

[12] Note that $\mathrm{E}[\epsilon(h^\gamma_{\mathcal{B}_{t+1}})]$ is computed by an exhaustive search over all possible $X_{t+1} \in GF(2)^4$, $\mathcal{B}_{t+1} \in GF(2)^{12}$ and thus does not depend on t.

[13] This unconditional correlation was discovered by [12,16] and proved later on by [23] to be one of the two largest unconditioned correlations up to 26-bit output sequence of the FSM.

[14] Note that on the other hand the unconditional correlation $\epsilon(h^\gamma) = 2^{-4}$ with $\gamma = (1,0,1,1)$ (denote this value by λ'), shown first in [17], was proved by [23] to be the only second largest unconditioned correlations up to 26-bit output sequence of the FSM.

Table 1. Weight triplets to generate the biased bit α_t with $\gamma = (1, 1, 1, 1, 1)$ and $\ell = 5$

bias of α_t $\epsilon(h^\gamma_{\mathcal{B}_{t+1}})$	weight triplet(s) $W(\mathcal{B}_{t+1})$	cardinality $|\mathcal{B}_{t+1}|$
-1	220, 224	72
1	222	216
-0.5	120, 124, 210, 214 230, 234, 320, 324	192
0.5	122, 212, 322, 232	576
-0.25	110, 111, 114, 130 131, 134, 310, 311 314, 330, 331, 334	384
0.25	112, 113, 132, 133 312, 313, 332, 333	640

Table 2. Weight pairs to generate the biased bit α'_t with $\gamma = (1, 1, 0, 1)$ and $\ell = 4$

bias of α'_t $\epsilon(h^\gamma_{\mathcal{B}_{t+1}})$	weight pairs $W(\mathcal{B}_{t+1})$	cardinality $|\mathcal{B}_{t+1}|$
-1	01, 43	8
1	03, 41	8
-0.5	11, 33	32
0.5	13, 31	32

subkey $K_1 \overset{\text{def}}{=} (\mathcal{G}_{t_1}(\mathcal{K}), \ldots, \mathcal{G}_{t_4}(\mathcal{K}))$ of $16(\ell - 2)$ bits by $\widehat{K_1}$ and the one-bit subkey $K_2 \overset{\text{def}}{=} \bar{\gamma} \cdot \mathcal{L}_{t'}(\mathcal{K})$ by $\widehat{K_2}$. We set $K = (K_1, K_2)$, $\widehat{K} = (\widehat{K_1}, \widehat{K_2})$. As $\mathcal{P}^i$'s are public, for every frame i, we can use Eq.(12) to compute the estimate $\widehat{\mathcal{B}^i_{t_j+1}}$ for $\mathcal{B}^i_{t_j+1}$ for $j = 1, \ldots, 4$ with $\widehat{K_1}$. Denote

$$\mathcal{B}^i = (\mathcal{B}^i_{t_1+1}, \mathcal{B}^i_{t_2+1}, \mathcal{B}^i_{t_3+1}, \mathcal{B}^i_{t_4+1}),$$
$$\mathcal{X}^i = (X^i_{t_1+1}, X^i_{t_2+1}, X^i_{t_3+1}, X^i_{t_4+1}, X^i_{t'+1}, \mathcal{B}^i_{t'+1}, \widehat{K}).$$

Define the probabilistic mapping $\mathcal{F}^\gamma_{\mathcal{B}^i}(\mathcal{X}^i)$ to be a truly random bit with uniform distribution for all i such that $\prod_{j=1}^4 \epsilon(h^\gamma_{\mathcal{B}^i_{t_j+1}}) = 0$; otherwise, we let

$$\mathcal{F}^\gamma_{\mathcal{B}^i}(\mathcal{X}^i) = \bigoplus_{j=1}^4 \left(h^\gamma_{\mathcal{B}^i_{t_j+1}}(X^i_{t_j+1}) \oplus g^\gamma(\widehat{\mathcal{B}^i_{t_j+1}}) \right) \oplus h^{\bar{\gamma}}(\mathcal{B}^i_{t'+1}, X^i_{t'+1}). \tag{16}$$

Note that given $\widehat{K_2}$, $\mathcal{F}^\gamma_{\mathcal{B}^i}(\mathcal{X}^i)$ is accessible in the latter case as we have

$$\mathcal{F}^\gamma_{\mathcal{B}^i}(\mathcal{X}^i) = \bar{\gamma} \cdot (Z^i_{t'} \oplus \mathcal{L}'_{t'}(\mathcal{P}^i)) \oplus \widehat{K_2} \oplus \bigoplus_{j=1}^4 g^\gamma(\widehat{\mathcal{B}^i_{t_j+1}}),$$

for all i such that $\prod_{j=1}^{4} \epsilon(h^\gamma_{\overbrace{\mathcal{B}^i_{t_j+1}}}) \neq 0$ according to Eq.(14). With the correct guess $\widehat{K} = K$, Eq.(16) reduces to

$$\mathcal{F}^\gamma_{\mathcal{B}^i}(\mathcal{X}^i) = \bigoplus_{j=1}^{4} \left(h^\gamma_{\mathcal{B}^i_{t_j+1}}(X^i_{t_j+1}) \oplus g^\gamma(\mathcal{B}^i_{t_j+1}) \right) \oplus h^{\bar\gamma}(\mathcal{B}^i_{t'+1}, X^i_{t'+1}), \qquad (17)$$

for all i such that $\prod_{j=1}^{4} \epsilon(h^\gamma_{\mathcal{B}^i_{t_j+1}}) \neq 0$. As the right-hand side of Eq.(17) only involves the unknown $X^i = (X^i_{t_1+1}, X^i_{t_2+1}, X^i_{t_3+1}, X^i_{t_4+1}, X^i_{t'+1}, \mathcal{B}^i_{t'+1})$, we denote the mapping in this case by $f^\gamma_{\mathcal{B}^i}(X^i)$. With appropriate choice of γ as discussed in the next subsection, we can have $\mathrm{E}[\Delta(f^\gamma_{\mathcal{B}^i})] > 0$. With each wrong guess $\widehat{K} \neq K$, however, as shown in Appendix, we estimate $\mathcal{F}^\gamma_{\mathcal{B}^i}(\mathcal{X}^i)$ to be uniformly and independently distributed for all i (i.e. $\mathrm{E}[\Delta(\mathcal{F}^\gamma_{\mathcal{B}^i})] = 0$).

As we are interested in small ℓ for low time complexity, e.g. $|\ell| < 6$ as explained immediately next, we can assume from this constraint[15] that X^i's are uniformly distributed and that all X^i's, $\mathcal{B}^i$'s are independent. Submitting 2^L sequences of n pairs $(\mathcal{F}^\gamma_{\mathcal{B}^i}(\mathcal{X}^i), \widehat{\mathcal{B}^i})$ (for $i = 1, 2, \ldots, n$) to the distinguisher, we can fit in the smart distinguisher of Section 3 with $L = 16(\ell - 2) + 1, u = 16(\ell - 2), v = 20 + 4(\ell - 2), r = 1$ and expect it to successfully recover L-bit K with data complexity n sufficiently large as analyzed later. Note that the favourable $L < 64$ necessitates that $\ell < 6$.

6.2   Complexity Analysis and Optimization

From Eq.(3) in Section 3, the smart distinguisher needs data complexity

$$n = \frac{4L \log 2}{\mathrm{E}\left[\Delta\left(f^\gamma_{\mathcal{B}^i}\right)\right]}. \qquad (18)$$

To compute n, we introduce another probabilistic mapping $f'^\gamma_{\mathcal{B}^i}$ similar to $f^\gamma_{\mathcal{B}^i}$:

$$f'^\gamma_{\mathcal{B}^i}(\mathcal{X}^i) \overset{\text{def}}{=} \bigoplus_{j=1}^{4} h^\gamma_{\mathcal{B}^i_{t_j+1}}(X^i_{t_j+1}) \oplus h^{\bar\gamma}(\mathcal{B}^i_{t'+1}, X^i_{t'+1}). \qquad (19)$$

Theorem 5. *For all $\mathcal{B}^i = (\mathcal{B}^i_{t_1+1}, \mathcal{B}^i_{t_2+1}, \mathcal{B}^i_{t_3+1}, \mathcal{B}^i_{t_4+1}) \in GF(2)^{16(\ell-2)}$, we always have*

$$\Delta(f^\gamma_{\mathcal{B}^i}) = \Delta(f'^\gamma_{\mathcal{B}^i}).$$

---

[15] However, the assumption does not hold for $\ell = 7, 8$: with $\ell = 8$, we know that $X^i_{t_2+1}$ is fixed given $X^i_{t_1+1}$ and $\mathcal{B}^i_{t_1+1}$ as we have $t_2 = t_1 + 8$ from Section 4; with $\ell = 7$, two bits of $X^i_{t_2+1}$ are fixed given $X^i_{t_1+1}$ and $\mathcal{B}^i_{t_1+1}$. Similar statements hold concerning $X^i_{t_3+1}, \mathcal{B}^i_{t_3+1}$ and $X^i_{t_4+1}$.

Proof. This is trivial for the case where $\prod_{j=1}^{4} \epsilon(h^{\gamma}_{\mathcal{B}^i_{t_j+1}}) = 0$, because by definition $D_{f^{\gamma}_{\mathcal{B}^i}}$ is a uniform distribution and so is $D_{f'^{\gamma}_{\mathcal{B}^i}}$ by the famous Piling-up lemma (see [25]). Let us discuss the case where $\prod_{j=1}^{4} \epsilon(h^{\gamma}_{\mathcal{B}^i_{t_j+1}}) \neq 0$. In this case we know that given $\mathcal{B}^i$, $\bigoplus_{j=1}^{4} g^{\gamma}(\mathcal{B}^i_{t_j+1})$ is well-defined and it is a fixed value that doesn't depend on the unknown X^i. Consequently, we have $\Delta(f^{\gamma}_{\mathcal{B}^i}) = \Delta(f'^{\gamma}_{\mathcal{B}^i} \oplus \text{const.}) = \Delta(f'^{\gamma}_{\mathcal{B}^i})$.   □

We can use Theorem 5 to compute $\frac{4L\log 2}{n}$ from Eq.(18) as $\frac{4L\log 2}{n} = \mathrm{E}[\Delta(f^{\gamma}_{\mathcal{B}^i})] = \mathrm{E}[\Delta(f'^{\gamma}_{\mathcal{B}^i})]$. Next, the independence of $\mathcal{B}^i$'s allows us to apply Piling-up Lemma [25] to continue as follows,

$$\frac{4L\log 2}{n} = \mathrm{E}\left[\Delta(h^{\bar{\gamma}}) \prod_{j=1}^{4} \Delta\left(h^{\gamma}_{\mathcal{B}^i_{t_j+1}}\right)\right] = \Delta(h^{\bar{\gamma}}) \prod_{j=1}^{4} \mathrm{E}\left[\Delta\left(h^{\gamma}_{\mathcal{B}^i_{t_j+1}}\right)\right].$$

Because we know from Section 5 that $\mathrm{E}[\Delta(h^{\gamma}_{\mathcal{B}^i_{t+1}})]$ does not depend on t and i, we finally have

$$\frac{4L\log 2}{n} = \Delta(h^{\bar{\gamma}}) \cdot \mathrm{E}^4\left[\Delta\left(h^{\gamma}_{\mathcal{B}_{t+1}}\right)\right]. \tag{20}$$

As we want to minimize n, according to Eq.(18), we would like to find some γ ($3 \leq |\gamma| < 6$) such that $\mathrm{E}[\Delta(f^{\gamma}_{\mathcal{B}^i})]$ is large, and above all, strictly positive. In order to have $\mathrm{E}[\Delta(f^{\gamma}_{\mathcal{B}^i})] > 0$, we must have $\Delta(h^{\bar{\gamma}}) > 0$ first, by Eq.(20). According to results of [16,17,12,23], only two aforementioned choices satisfy our predefined prerequisite about γ (i.e. both the first and last coordinates of γ are one): either $\gamma = (1,1,1,1,1)$ with $\Delta(h^{\bar{\gamma}}) = \lambda^2 \approx 2^{-6.71}$, or $\gamma = (1,1,0,1)$ with $\Delta(h^{\bar{\gamma}}) = \lambda'^2 = 2^{-8}$. For $\gamma = (1,1,1,1,1)$, from last section, we know that $\mathrm{E}[\Delta(h^{\gamma}_{\mathcal{B}_{t+1}})] \approx 2^{-2.9}$. So we conclude from Eq.(20) that $n \approx 2^{25.4}$ frames of keystreams generated by the same key $\mathcal{K}$ suffice to recover the $L = 49$-bit subkey K. Analogously, for $\gamma = (1,1,0,1)$, we have $\mathrm{E}[\Delta(h^{\gamma}_{\mathcal{B}_{t+1}})] = 2^{-3}$ from last section. And it results in $n \approx 2^{26.5}$ frames to recover $L = 33$-bit subkey.

Let us discuss the time complexity of the attack now. For all $J = (J_1, J_2) \in GF(2)^{L-1} \times GF(2)$, and let $J_1 = (J_{1,1}, \ldots, J_{1,4})$ where $J_{1,i} \in GF(2)^{4(\ell-2)}$, we define $\mathcal{H}, \mathcal{H}'$:

$$\mathcal{H}(J) = \sum_{i=1}^{n} 1_{\mathcal{G}'_{t_1}(\mathcal{P}^i),\ldots,\mathcal{G}'_{t_4}(\mathcal{P}^i)=J_1 \text{ and } \bar{\gamma}\cdot(Z^i_{t'} \oplus L'_{t'}(\mathcal{P}^i))=J_2},$$

$$\mathcal{H}'(J) = \begin{cases} 0, & \text{if } \prod_{i=1}^{4} \epsilon(h^{\gamma}_{J_{1,i}}) = 0 \\ \log 2^r \cdot D_{J_1}\left(J_2 \oplus \bigoplus_{i=1}^{4} g^{\gamma}(J_{1,i})\right), & \text{otherwise} \end{cases}$$

where $D_{J_1} = D_{h^{\gamma}_{J_{1,1}}} \otimes D_{h^{\gamma}_{J_{1,2}}} \otimes D_{h^{\gamma}_{J_{1,3}}} \otimes D_{h^{\gamma}_{J_{1,4}}}$. Let $\mathcal{H}''(K) = \widehat{\mathcal{H}}(K) \cdot \widehat{\mathcal{H}'}(K)$. By Theorem 1 in Section 3, we have $G(K) = \frac{1}{2L}\widehat{\mathcal{H}''}(K)$. This means that after precomputing $\widehat{\mathcal{H}'}$ in time $O(L \cdot 2^L)$, our partial key-recovery attack would be dominated by twice FWT, i.e. $\widehat{\mathcal{H}}, \widehat{\mathcal{H}}''$ with time $O(L \cdot 2^{L+1})$. Algorithm 2 illustrates the above basic partial key-recovery attack. Note that without the optimization technique of Theorem 1, the deterministic smart distinguisher has to perform $O(n \cdot 2^L)$ operations otherwise, which makes our attack impractical.

Algorithm 2. The basic partial key-recovery attack on two-level E0

Parameters:
1: $\gamma, t', t_1, t_2, t_3, t_4, L$
2: n set by Eq.(20)
Inputs:
3: $\mathcal{P}^i$ for $i = 1, 2, \ldots, n$
4: $Z_{t'}^i$ for $i = 1, 2, \ldots, n$
Preprocessing:
5: compute $H', \widehat{H}'$
Processing:
6: compute $H, \widehat{H}$
7: compute $H'' = \widehat{H} \cdot \widehat{H}'$ and $\widehat{H}''$
8: output K with the maximum $\widehat{H}''(K)$

Furthermore, by Table 2, we discovered a special property

$$\epsilon(h_{B_{t+1}B_{t+2}}^{\gamma}) \equiv \epsilon(h_{\overline{B}_{t+1}\overline{B}_{t+2}}^{\gamma}) \equiv -\epsilon(h_{\overline{B}_{t+1}B_{t+2}}^{\gamma}) \equiv -\epsilon(h_{B_{t+1}\overline{B}_{t+2}}^{\gamma}) \qquad (21)$$

for all $\mathcal{B}_{t+1} = B_{t+1}B_{t+2} \in GF(2)^8$ with $\gamma = (1, 1, 0, 1)$, where the bar operator denotes the bitwise complement of the 4-bit binary vector. This means that for our 33-bit partial key-recovery attack, we always have $4^4 = 256$ equivalent key candidates[16] (see Appendix for details), which helps to decrease the computation time on $\widehat{H}''$ (see [23]) from $33 \times 2^{33} \approx 2^{38}$ to $25 \times 2^{25} \approx 2^{30}$. In total we have the running time $2^{38} + 2^{30} \approx 2^{38}$ for Algorithm 2.

We have implemented the full Algorithm 2 with $\gamma = (1, 1, 0, 1), t' = 1, n = 2^{26}$ frames (slightly less than the theoretical estimate $2^{26.5}$) on the Linux platform, 2.4G CPU, 2G RAM, 128GB hard disk. It turned out that after one run of a 37-hour precomputation (i.e. Line **5** in Algorithm 2 which stores a 64GB table in the hard disk), of all the 30 runs tested so far, our attack *never fails to successfully* recover the right 25-bit key in about 19 hours. Computing $H, \widehat{H}, H'', \widehat{H}''$ takes time 27 minutes, 18 hours, 45 minutes and 20 seconds respectively. The running time is dominated by FWT[17] $\widehat{H}$, which only takes a negligible portion of CPU time and depends dominantly on the performance of the hardware, i.e. the external data transfer rate[18] between the hard disk and PC's main memory.

Inspired by the multi-bias analysis on the traditional distinguisher in [23], the advanced multi-bias analysis (see Appendix) which is an extension of this

---

[16] The term "equivalent key candidate" is exclusively used for our attack, which doesn't mean that they are equivalent keys for the Bluetooth encryption.

[17] The result is stored in a 32GB table in the hard disk.

[18] In our PC it is 32MB/s, which is common in the normal PC nowadays.

section allows us to reach the data complexity $n \approx 2^{23.8}$ frames with the same time complexity. Once we recover the first $(33 - 8) = 25$-bit subkey, we just increment (or decrement) t' by one and use the knowledge of those subkey bits to reiterate Algorithm 2 to recover more key bits similarly as was done in [24]. Since only 17 new key bits are involved, which reduce to the 13-bit equivalent key, it's much faster to recover those key bits. Finally, we perform an exhaustive search over the equivalent key candidates in negligible time, whose total number is upper bounded by $2^{\frac{8|\mathcal{K}|}{32}} = 2^{\frac{|\mathcal{K}|}{4}}$. The final complexity of the complete key-recovery attack is bounded by one run of Algorithm 2, i.e. $O(2^{38})$. Table 3 compares our attacks with the best known attacks [13,14,16,24] on two-level E0 for effective key size $|\mathcal{K}| = 128$. Note that with $|\mathcal{K}| = 64$, Bluetooth key loading at E0 level one makes the bits of the subkey K linearly independent for all $t' \in \bigcup_{k=0}^{2}\{8k + 1, \ldots, 8k + 5\}$. Therefore, the attack complexities remain to be on the same order.

Table 3. Comparison of our attacks with the best attacks on two-level E0 for $|\mathcal{K}| = 128$

Attack		Precomputation	Time	Frames	Data	Memory
Fluhrer-Lucks	[13]	-	2^{73}	-	2^{43}	2^{51}
Fluhrer	[14]	2^{80}	2^{65}	2	$2^{12.4}$	2^{80}
Golić et al.	[16]	2^{80}	2^{70}	45	2^{17}	2^{80}
Lu-Vaudenay	[24]	-	2^{40}	2^{35}	$2^{39.6}$	2^{35}
Our	basic	2^{38}	2^{38}	$2^{26.5}$	$2^{31.1}$	2^{33}
Attacks	advanced	2^{38}	2^{38}	$2^{23.8}$	$2^{28.4}$	2^{33}

7   Conclusion

In this paper, we have generalized the concept of conditional correlations in [1,21,22] to study conditional correlation attacks against stream ciphers and other cryptosystems, in case the computation of the output allows for side information related to correlations conditioned on the input. A general framework has been developed for smart distinguishers, which exploit those generalized conditional correlations. In particular, based on the theory of the traditional distinguisher [5] we derive the number of samples necessary for a smart distinguisher to succeed. It is demonstrated that the generalized conditional correlation is no smaller than the unconditional correlation. Consequently, the smart distinguisher improves on the traditional basic distinguisher (in the worst case the smart distinguisher degrades into the traditional one); the smart distinguisher could be efficient even if no ordinary correlations exist. As an application of our generalized conditional correlations, a conditional correlation attack on the two-level Bluetooth E0 is developed and optimized. Whereas the analysis in [24] was based on a traditional distinguishing attack using the strongest (unconditional) 5-bit correlation, we have successfully demonstrated the superiority of our attack over [24] by showing a best attack using 4-bit conditional correlations, which are *not* suitable for the attack in [24] as the corresponding ordinary correlations are all *zeros*. Our best attack fully recovers the original encryption key using the first 24 bits of $2^{23.8}$ frames and with 2^{38} computations. Compared with all existing attacks [13,14,16,20,24,29], this is clearly the fastest and only practical known-plaintext attack on Bluetooth encryption so far. It remains to be an interesting challenge to investigate the redundancy in the header of each frame for a practical ciphertext-only attack on Bluetooth encryption.

Acknowledgments

This work is supported in part by the National Competence Center in Research on Mobile Information and Communication Systems (NCCR-MICS), a center of the Swiss National Science Foundation under the grant number 5005-67322. The second author also receives his partial funding through Gebert Rüf Stiftung.

We owe many grateful thanks to Antoine Joux for his inspiring comments. We would also like to thank the anonymous reviewers for many helpful suggestions. The first author would like to thank LASEC staff Gildas Avoine, Thomas Baignères, Martine Corval, Matthieu Finiasz, Jean Monnerat and the former member Pascal Junod for their support during this project. The arduous experiment wouldn't succeed without great support of Thomas Baignères, Matthieu Finiasz and Pascal Junod as well as our robust computer.

References

1. Ross Anderson, *Searching for the Optimum Correlation Attack*, Fast Software Encryption 1994, Lecture Notes in Computer Science, vol.1008, B. Preneel Ed., Springer-Verlag, pp. 137-143, 1994
2. Frederik Armknecht, Matthias Krause, *Algebraic Attacks on Combiners with Memory*, Advances in Cryptology - CRYPTO 2003, Lecture Notes in Computer Science, vol.2729, D. Boneh Ed., Springer-Verlag, pp. 162-175, 2003
3. Frederik Armknecht, Joseph Lano, Bart Preneel, *Extending the Resynchronization Attack*, Selected Areas in Cryptography - SAC 2004, Lecture Notes in Computer Science, vol. 3357, H. Handschuh and A. Hasan Eds., Springer-Verlag, pp. 19-38, 2005 (extended version available at http://eprint.iacr.org/2004/232)
4. Frederik Armknecht, Willi Meier, *Fault Attacks on Combiners with Memory*, submitted
5. Thomas Baignères, Pascal Junod, Serge Vaudenay, *How Far Can We Go Beyond Linear Cryptanalysis?*, Advances in Cryptology - ASIACRYPT 2004, Lecture Notes in Computer Science, vol.3329, P. J. Lee Ed., Springer-Verlag, pp. 432-450, 2004
6. Bluetooth[TM], *Bluetooth Specification*, version 1.2, pp. 903-948, November, 2003, available at http://www.bluetooth.org
7. Anne Canteaut, Michael Trabbia, *Improved Fast Correlation Attacks Using Parity-check Equations of Weight 4 and 5*, Advances in Cryptology - EUROCRYPT 2000, Lecture Notes in Computer Science, vol.1807, B. Preneel Ed., Springer-Verlag, pp. 573-588, 2000

8. Vladimir V. Chepyzhov, Thomas Johansson, Ben Smeets, *A Simple Algorithm for Fast Correlation Attacks on Stream Ciphers*, Fast Software Encryption 2000, Lecture Notes in Computer Science, vol.1978, B. Schneier Ed., Springer-Verlag, pp. 181-195, 2000

9. Philippe Chose, Antoine Joux, Michel Mitton, *Fast Correlation Attacks: An Algorithmic Point of View*, Advances in Cryptology - EUROCRYPT 2002, Lecture Notes in Computer Science, vol.2332, L. R. Knudsen Ed., Springer-Verlag, pp. 209-221, 2002

10. Nicolas T. Courtois, *Fast Algebraic Attacks on Stream Ciphers with Linear Feedback*, Advances in Cryptology - CRYPTO 2003, Lecture Notes in Computer Science, vol.2729, D. Boneh Ed., Springer-Verlag, pp. 176-194, 2003

11. Thomas M. Cover, Joy A. Thomas, *Elements of Information Theory*, Wiley, 1991

12. Patrik Ekdahl, Thomas Johansson, *Some Results on Correlations in the Bluetooth Stream Cipher*, Proceedings of the 10th Joint Conference on Communications and Coding, Austria, 2000

13. Scott Fluhrer, Stefan Lucks, *Analysis of the E0 Encryption System*, Selected Areas in Cryptography - SAC 2001, Lecture Notes in Computer Science, vol. 2259, S. Vaudenay and A. Youssef Eds., Springer-Verlag, pp. 38-48, 2001

14. Scott Fluhrer, *Improved Key Recovery of Level 1 of the Bluetooth Encryption System*, available at `http://eprint.iacr.org/2002/068`

15. Jovan Dj. Golić, *Correlation Properties of a General Binary Combiner with Memory*, Journal of Cryptology, vol. 9, pp. 111-126, Nov. 1996

16. Jovan Dj. Golić, Vittorio Bagini, Guglielmo Morgari, *Linear Cryptanalysis of Bluetooth Stream Cipher*, Advances in Cryptology - EUROCRYPT 2002, Lecture Notes in Computer Science, vol. 2332, L. R. Knudsen Ed., Springer-Verlag, pp. 238-255, 2002

17. Miia Hermelin, Kaisa Nyberg, *Correlation Properties of the Bluetooth Combiner*, Information Security and Cryptology - ICISC'99, Lecture Notes in Computer Science, vol. 1787, JooSeok. Song Ed., Springer-Verlag, pp. 17-29, 2000

18. Thomas Johansson, Frederik Jönsson, *Improved Fast Correlation Attacks on Stream Ciphers via Convolutional Codes*, Advances in Cryptology - CRYPTO'99, Lecture Notes in Computer Science, vol.1666, M. Wiener Ed., Springer-Verlag, pp. 181-197, 1999

19. Thomas Johansson, Frederik Jönsson, *Fast Correlation Attacks through Reconstruction of Linear Polynomials*, Advances in Cryptology - CRYPTO 2000, Lecture Notes in Computer Science, vol.1880, M. Bellare Ed., Springer-Verlag, pp. 300-315, 2000

20. Matthias Krause, *BDD-Based Cryptanalysis of Keystream Generators*, Advances in Cryptology - EUROCRYPT 2002, Lecture Notes in Computer Science, vol. 2332, L. R. Knudsen Ed., Springer-Verlag, pp. 222-237, 2002

21. Sangjin Lee, Seongtaek Chee, Sangjoon Park, Sungmo Park, *Conditional Correlation Attack on Nonlinear Filter Generators*, Advances in Cryptology - ASIACRYPT 1996, Lecture Notes in Computer Science, vol.1163, Kwangjo Kim and Tsutomu Matsumoto Eds., Springer-Verlag, pp. 360-367, 1996

22. Bernhard Löhlein, *Attacks based on Conditional Correlations against the Nonlinear Filter Generator*, available at `http://eprint.iacr.org/2003/020`

23. Yi Lu, Serge Vaudenay, *Faster Correlation Attack on Bluetooth Keystream Generator E0*, Advances in Cryptology - CRYPTO 2004, Lecture Notes in Computer Science, vol.3152, M. Franklin Ed., Springer-Verlag, pp. 407-425, 2004

24. Yi Lu, Serge Vaudenay, *Cryptanalysis of Bluetooth Keystream Generator Two-level E0*, Advances in Cryptology - ASIACRYPT 2004, Lecture Notes in Computer Science, vol.3329, P. J. Lee Ed., Springer-Verlag, pp. 483-499, 2004
25. Mitsuru Matsui, *Linear Cryptanalysis Method for DES Cipher*, Advances in Cryptology - EUROCRYPT'93, Lecture Notes in Computer Science, vol.765, Springer-Verlag, pp. 386-397, 1993
26. Willi Meier, Othmar Staffelbach, *Fast Correlation Attacks on Certain Stream Ciphers*, Journal of Cryptology, vol. 1, pp. 159-176, Nov. 1989
27. Willi Meier, Othmar Staffelbach, *Correlation Properties of Combiners with Memory in Stream Ciphers*, Journal of Cryptology, vol. 5, pp. 67-86, Nov. 1992
28. Alfred J. Menezes, Paul C. van. Oorschot, Scott A. Vanstone, *Handbook of Applied Cryptography*, CRC, 1996
29. Markku Saarinen, *Re: Bluetooth and E0*, Posted at `sci.crypt.research`, 02/09/00
30. Thomas Siegenthaler, *Decrypting a class of Stream Ciphers using Ciphertext only*, IEEE Transactions on Computers, vol. C-34, pp. 81-85, Jan. 1985
31. Serge Vaudenay, *An Experiment on DES - Statistical Cryptanalysis*, Proceedings of the 3rd ACM Conferences on Computer Security, pp. 139-147, 1996

Appendix

Proof for $E[\Delta(f_\mathcal{B})] \geq \Delta(f)$

By Eq.(7), we have

$$E[\Delta(f_\mathcal{B})] = 2^r \sum_{A \in GF(2)^r} E\left[\left(D_{f_\mathcal{B}}(A) - \frac{1}{2^r}\right)^2\right], \tag{22}$$

where the expectation is taken over uniformly distributed $\mathcal{B}$ for the fixed A. On the other hand, since $D_f(A) = E[D_{f_\mathcal{B}}(A)]$ for any fixed A, we have

$$\Delta(f) = 2^r \sum_{A \in GF(2)^r} \left(D_f(A) - \frac{1}{2^r}\right)^2 \tag{23}$$

$$= 2^r \sum_{A \in GF(2)^r} \left(E[D_{f_\mathcal{B}}(A)] - \frac{1}{2^r}\right)^2 \tag{24}$$

$$= 2^r \sum_{A \in GF(2)^r} E^2\left[D_{f_\mathcal{B}}(A) - \frac{1}{2^r}\right], \tag{25}$$

by definition of Eq.(1), with all the expectation taken over uniformly distributed $\mathcal{B}$ for the fixed A. As we know from theory of statistics that for any fixed A,

$$0 \leq Var\left[D_{f_\mathcal{B}}(A) - \frac{1}{2^r}\right] = E\left[\left(D_{f_\mathcal{B}}(A) - \frac{1}{2^r}\right)^2\right] - E^2\left[D_{f_\mathcal{B}}(A) - \frac{1}{2^r}\right] \tag{26}$$

always holds, where equality holds iff $D_{f_\mathcal{B}}(A)$ is independent of $\mathcal{B}$. $\qquad\square$

Approximation of Distribution of $\mathcal{F}^\gamma_{\mathcal{B}^i}(\mathcal{X}^i)$ for Wrong Keys

Firstly, with $\widehat{K_1} \neq K_1$, the reason that we estimate $\mathcal{F}^\gamma_{\mathcal{B}^i}(\mathcal{X}^i)$ to be uniformly and independently distributed for all i can be explained as follows for the cases[19] when $\prod_{j=1}^4 \epsilon(h^\gamma_{\mathcal{B}^i_{t_j+1}}) \neq 0$. Assuming that $\mathcal{P}^i$'s are uniformly and independently distributed, we deduct by Eq.(12) that so are $\widehat{\mathcal{B}^i}$'s for every $\widehat{K}$, where $\widehat{\mathcal{B}^i} = (\widehat{\mathcal{B}^i_{t_1+1}}, \ldots, \widehat{\mathcal{B}^i_{t_4+1}})$. Hence, we estimate $g^\gamma(\widehat{\mathcal{B}^i_{t_j+1}})$ for $j = 1, \ldots, 4$ are also uniformly and independently distributed, which allows to conclude by Eq.(16) that $D_{\mathcal{F}^\gamma_{\mathcal{B}^i}}$ can be approximated by a uniformly distributed sequence.

Secondly, in the remaining one case of wrong guess such that $\widehat{K_1} = K_1$ and $\widehat{K_2} \neq K_2$, $\mathcal{F}^\gamma_{\mathcal{B}^i}(\mathcal{X}^i)$ is *no longer uniformly distributed*; but it is more favourable to us, because we have $\mathcal{F}^\gamma_{\mathcal{B}^i}(\mathcal{X}^i) = f^\gamma_{\mathcal{B}^i}(X^i) \oplus 1$ for all i such that $\prod_{j=1}^4 \epsilon(h^\gamma_{\mathcal{B}^i_{t_j+1}}) \neq 0$, whose distribution has larger Kullback-Leibler distance (see [11]) to $D_{f^\gamma_{\mathcal{B}^i}}$ than a uniform distribution does according to [5].

In all, we can pessimistically approximate $D_{\mathcal{F}^\gamma_{\mathcal{B}^i}}$ by a uniform distribution for each wrong guess $\widehat{K} \neq K$.

Advanced Application

Having studied how to apply Section 3 with $r = 1$ (namely the uni-bias-based approach) for an attack to E0 in Section 6, we wonder the possibility of improvement based on multi-biases in the same spirit as in [23], which are utilized by the traditional distinguisher.

For the reason of low time complexity of the attack, we still focus on analysis of 4-bit biases; additionally, we restrict ourselves to bi-biases analysis (i.e. $r = 2$) to simplify the presentation, which will be shown later to be optimal. Let $\Gamma = (\gamma_1, \gamma_2)$, where γ_1 is fixed to $(1, 1, 0, 1)$ and γ_2 with length $\ell_2 \overset{\text{def}}{=} |\gamma_2| = 4$ remains to be determined later such that the data complexity is lowered when we analyze the characteristics of bi-biases simultaneously for each frame instead of conducting the previous uni-bias-based analysis.

Recall that $g^{\gamma_1}(\mathcal{B}) : GF(2)^8 \to GF(2)$ in Section 6 was defined to be the most likely bit of $h^{\gamma_1}_{\mathcal{B}}(X)$ for a uniformly distributed $X \in GF(2)^4$ if it exists (i.e. $\epsilon(h^{\gamma_1}_{\mathcal{B}}) \neq 0$). We extend $g^{\gamma_1}(\mathcal{B}) : GF(2)^8 \to GF(2)$ to $g^\Gamma(\mathcal{B}) : GF(2)^8 \to GF(2)^2$ over all $\mathcal{B} \in GF(2)^8$ such that $\epsilon(h^{\gamma_1}_{\mathcal{B}}) \neq 0$, and let $g^\Gamma(\mathcal{B})$ be the most likely 2-bit binary vector $\beta = (\beta_1, \beta_2)$. Note that we can always easily determine the first bit β_1 because of the assumption $\epsilon(h^{\gamma_1}_{\mathcal{B}}) \neq 0$; with regards to determining the second bit β_2 in case that a tie occurs, we just let β_2 be a uniformly distributed bit. Let

$$h^\Gamma_{\mathcal{B}}(X) = (h^{\gamma_1}_{\mathcal{B}}(X), h^{\gamma_2}_{\mathcal{B}}(X)), \tag{27}$$

$$h^{\bar\Gamma}(\mathcal{B}, X) = (h^{\bar\gamma_1}(\mathcal{B}, X), h^{\bar\gamma_2}(\mathcal{B}, X)). \tag{28}$$

---

[19] By definition of $\mathcal{F}^\gamma_{\mathcal{B}^i}$, this is trivial for the cases when $\prod_{j=1}^4 \epsilon(h^\gamma_{\mathcal{B}^i_{t_j+1}}) = 0$.

Note that $h_{\mathcal{B}}^{\Gamma}(X)$ outputs the two bits which are generated by the same unknown X given $\mathcal{B}$; by contrast, $h^{\bar{\Gamma}}(\mathcal{B}, X)$ outputs the two bits which are generated by the unknown X and $\mathcal{B}$. We can extend $\mathcal{F}_{\mathcal{B}^i}^{\gamma_1}(\mathcal{X}^i)$ in Eq.(16) to $\mathcal{F}_{\mathcal{B}^i}^{\Gamma}(\mathcal{X}^i)$ by letting

$$
\mathcal{F}_{\mathcal{B}^i}^{\Gamma}(\mathcal{X}^i)
$$
$$
= \left(\bigoplus_{j=1}^{4} h_{\mathcal{B}_{t_j+1}^i}^{\gamma_1}(X_{t_j+1}^i) \oplus h^{\bar{\gamma}_1}(\mathcal{B}_{t'+1}^i, X_{t'+1}^i), \bigoplus_{j=1}^{4} h_{\mathcal{B}_{t_j+1}^i}^{\gamma_2}(X_{t_j+1}^i) \oplus h^{\bar{\gamma}_2}(\mathcal{B}_{t'+1}^i, X_{t'+1}^i) \right)
$$
$$
\oplus g^{\Gamma}(\widehat{\mathcal{B}_{t_j+1}^i}),
\tag{29}
$$

if $\prod_{j=1}^{4} \epsilon(h_{\mathcal{B}_{t_j+1}^i}^{\gamma_1}) \neq 0$; otherwise, we let it be a uniformly distributed two-bit vector. Similarly, we denote $\mathcal{F}_{\mathcal{B}^i}^{\Gamma}(\mathcal{X}^i)$ corresponding to the correct guess by $f_{\mathcal{B}^i}^{\Gamma}$.

It's easy to verify the assumption holds to apply Section 3 that says $D_{\mathcal{F}_{\mathcal{B}^i}^{\Gamma}}$ can still be approximated by a uniform distribution for each wrong guess on the key $\widehat{K} \neq K$. Moreover, by introducing the extended $f_{\mathcal{B}^i}^{\prime\Gamma}$ from $f_{\mathcal{B}^i}^{\prime\gamma_1}$ in Eq.(19) as

$$
f_{\mathcal{B}^i}^{\prime\Gamma}(\mathcal{X}^i) \overset{\text{def}}{=} (f_{\mathcal{B}^i}^{\prime\gamma_1}(\mathcal{X}^i), f_{\mathcal{B}^i}^{\prime\gamma_2}(\mathcal{X}^i))
\tag{30}
$$
$$
= \left(\bigoplus_{j=1}^{4} h_{\mathcal{B}_{t_j+1}^i}^{\gamma_1}(X_{t_j+1}^i) \oplus h^{\bar{\gamma}_1}(\mathcal{B}_{t'+1}^i, X_{t'+1}^i), \right.
$$
$$
\left. \bigoplus_{j=1}^{4} h_{\mathcal{B}_{t_j+1}^i}^{\gamma_2}(X_{t_j+1}^i) \oplus h^{\bar{\gamma}_2}(\mathcal{B}_{t'+1}^i, X_{t'+1}^i) \right).
$$

Theorem 5 can be extended to the generalized theorem below

Theorem 6. *For all* $\mathcal{B}^i = (\mathcal{B}_{t_1+1}^i, \mathcal{B}_{t_2+1}^i, \mathcal{B}_{t_3+1}^i, \mathcal{B}_{t_4+1}^i) \in GF(2)^{32}$, *we always have*

$$
\Delta(f_{\mathcal{B}^i}^{\Gamma}) = \Delta(f_{\mathcal{B}^i}^{\prime\Gamma}).
$$

Similar computation yields the same formula for data complexity we need as in Eq.(20)

$$
\frac{4L \log 2}{n} = \Delta(h^{\bar{\Gamma}}) \cdot \mathrm{E}^4 \left[\Delta \left(h_{\mathcal{B}_{t+1}}^{\Gamma} \right) \right].
\tag{31}
$$

Experimental result shows that with $\gamma_1 = (1, 1, 0, 1), \gamma_2 = (1, 0, 1, 1)$, we achieve optimum $\Delta(h_{\mathcal{B}_{t+1}}^{\Gamma}) \approx 2^{-2.415}$ (in comparison to $\Delta(h_{\mathcal{B}_{t+1}}^{\gamma_1}) = 2^{-3}$ previously), though $\Delta(h^{\bar{\Gamma}})$ always equals $\Delta(h^{\bar{\gamma}_1})$ regardless of the choice of γ_2; additionally, $\Delta(h^{\bar{\Gamma}}) \equiv 0$ if $\gamma_1, \gamma_2 \neq (1, 1, 0, 1)$. Therefore, we have the minimum data complexity $n \approx 2^{23.8}$ frames. And the time complexity remains the same according to Theorem 1 in Section 3.

Equivalent Keys

Recall that in Subsection 6.1 we have the 33-bit key $K = (K_1, K_2)$, with $K_1 = (\mathcal{G}_{t_1}(\mathcal{K}), \ldots, \mathcal{G}_{t_4}(\mathcal{K}))$. For simplicity, we let $K_{1,i} = \mathcal{G}_{t_i}(\mathcal{K})$. Define the following 8-bit masks (in hexadecimal):

$$\text{mask}_0 = 0x00, \ \text{mask}_1 = 0xff, \ \text{mask}_2 = 0x0f, \ \text{mask}_3 = 0xf0.$$

Then for any K, we can replace $K_{1,i}$ by $K_{1,i} \oplus \text{mask}_j$ for any $i = 1, 2, \ldots, 4$ and $j \in \{0, 1, 2, 3\}$ and replace K_2 by $K_2 \oplus \lceil \frac{j}{2} \rceil$. Denote this set containing $4^4 = 2^8$ elements by $\langle K \rangle$. We can easily verify that the Walsh coefficients $\widehat{\mathcal{H}''}$ of the element in the set equals by following the definition of convolution between $\mathcal{H}$ and $\mathcal{H}'$:

$$\mathcal{H} \otimes \mathcal{H}'(K) = \sum_{K'} \mathcal{H}(K') \mathcal{H}'(K \oplus K'). \tag{32}$$

Since if $R \in \langle K \rangle$ then $R \oplus K' \in \langle K \oplus K' \rangle$ for all K'. And $\mathcal{H}'$ maps all the elements of the same set to the same value from Section 6, we conclude the set defined above form an equivalent class of the candidate keys. Thus, we have 2^8 equivalent 33-bit keys.

Unconditional Characterizations of Non-interactive Zero-Knowledge

Rafael Pass[1,*] and Abhi Shelat[2,**]

[1] MIT CSAIL
pass@csail.mit.edu
[2] IBM Zurich Research
abhi@zurich.ibm.com

Abstract. Non-interactive zero-knowledge (NIZK) proofs have been investigated in two models: the *Public Parameter* model and the *Secret Parameter model*. In the former, a public string is "ideally" chosen according to some efficiently samplable distribution and made available to both the Prover and Verifier. In the latter, the parties instead obtain correlated (possibly different) private strings. To add further choice, the definition of zero-knowledge in these settings can either be *non-adaptive* or *adaptive*.

In this paper, we obtain several *unconditional* characterizations of computational, statistical and perfect NIZK for all combinations of these settings. Specifically, we show:

In the secret parameter model, **NIZK = NISZK = NIPZK = AM**. In the public parameter model,
▷ for the non-adaptive definition, **NISZK ⊆ AM ∩ coAM**,
▷ for the adaptive one, it also holds that **NISZK ⊂ BPP/1**,
▷ for computational NIZK for "hard" languages, one-way functions are both *necessary* and *sufficient*.

From our last result, we arrive at the following *unconditional* characterization of computational NIZK in the public parameter model (which complements well-known results for interactive zero-knowledge):

Either NIZK proofs exist only for "easy" languages (i.e., languages that are not hard-on-average), or they exist for all of **AM** (i.e., all languages which admit non-interactive proofs).

1   Introduction

A zero-knowledge proof system is a protocol between two parties, a Prover, and a Verifier, which guarantees two properties: a malicious Prover cannot convince the Verifier of a false theorem; a malicious Verifier cannot learn anything from an interaction beyond the validity of the theorem.

Non-interactive zero-knowledge (NIZK) was proposed by Blum, Feldman, and Micali [BFM88] to investigate the *minimal* interaction necessary for zero-

* Research supported by an Akamai Presidential Fellowship.
** This research was completed while at MIT CSAIL.

knowledge proofs. To achieve the absolute minimal amount of interaction —that is, a single message from the Prover to the Verifier— some *setup* assumptions are provably necessary [GO94]. These setup assumptions can be divided into two groups:

1. **Public Parameter Setup.** The originally proposed setup is the *Common Random String Model* in which a uniformly random string is made available to both the Prover and Verifier. Many NIZK schemes have been implemented in this model [SMP87, BFM88, FLS90, DMP88, BDMP91, KP98, DCO⁺01]. A slight relaxation of this model is the *Public Parameter model*, also known as the *Common Reference String Model*, in which a string is "ideally" chosen according to some polynomial-time samplable distribution and made available to both the Prover and Verifier. Such a setup can be used to select —say— safe primes, group parameters, or public keys for encryption schemes, etc. See for example [Dam00, CLOS02].

2. **Secret Parameter Setup.** Cramer and Damgård [CD04] explicitly introduce the Secret Parameter setup model in which the Prover and Verifier obtain correlated (possibly different) private information.
   More generally, the secret parameter model encompasses the *Pre-processing Model* in which the Prover and Verifier engage in an arbitrary interactive protocol, at the end of which, both Prover and Verifier receive a private output. (This follows because any arbitrary protocol for pre-processing can be viewed as a polynomial-time sampler from a well-defined distribution.) Such a setup model is studied in [KMO89, DMP88, Dam93].

The above setup models can be *implemented* in a variety of ways, which may or may not require their own independent assumptions (For example, secure two-party computations protocols can be used to pick a random string.) In this paper we defer the discussion of *how* trusted setups are implemented, and choose instead to focus on the relative *power* of the models.

We restrict our study to the simplest setting in which only a single theorem is proven. Also, we consider security against unbounded provers. (That is, we consider *proof systems* as opposed to *argument systems*.) Following similar studies in the interactive setting —see for example [Vad99, SV03, Vad04]— we allow the honest prover algorithm to be inefficient (although some of our constructions have efficient prover algorithm for languages in **NP**).

Our investigation also considers both *adaptive* and *non-adaptive* definitions of zero-knowledge for non-interactive proofs. Briefly, the difference between these two is that the adaptive variant guarantees that the zero-knowledge property holds even if the theorem statement is chosen *after* the trusted setup has finished, whereas the non-adaptive variant does not provide this guarantee.

1.1 Our Results

SECRET PARAMETER MODEL. One suspects that the secret-parameter setup is more powerful than its public-parameter counterpart. Indeed, in game theory, a well-known result due to Aumann [Aum74] states that players having access to *correlated* secret strings can achieve a larger class of equilibria, and in particular, better payoffs, than if they only share a common public string. As we shall see, this intuition carries over in a strong way to the cryptographic setting. But first, we show that,

> **Informal Theorem** [Upper bound]. In the secret parameter model, non-interactive *perfect* zero-knowledge proofs exist *unconditionally* for all languages in **AM**.

This result is obtained by combining the work of [FLS90] with an adaptation of Kilian's work on implementing commitments using oblivious transfer [Kil88].

Previously, for general **NP** languages, only *computational* NIZK proof systems were known in the secret-parameter setup model [DMP88, FLS90, KMO89, DFN05]. Furthermore, these systems relied on various computational assumptions, such as the existence of one-way permutations. Recently, Cramer and Damgård [CD04] constructed statistical NIZK proofs in this model for *specific* languages related to discrete logarithms. (On the other hand, their results apply to an unbounded number of proofs, whereas ours do not.)

As a corollary of our result, we obtain a complete characterization of computational, statistical and perfect NIZK in the secret parameter model. Namely, we show that **NIP** = **NIZK** = **NISZK** = **NIPZK** = **AM**, where **NIP** denotes the class of languages having non-interactive proofs, and **NIZK**, **NISZK** and **NIPZK** denotes the classes of languages having non-interactive computational, statistical and perfect zero-knowledge proofs.

PUBLIC PARAMETER MODEL: STATISTICAL NIZK. We next turn our attention to the public parameter model, and show that, in contrast to the Secret Parameter model, statistical NIZK proofs for **NP**-complete languages are unlikely to exist.[1]

> **Informal Theorem** [Lower bound]. In the public parameter model, non-interactive statistical (non-adaptive) zero-knowledge proof systems only exist for languages in **AM** ∩ **coAM**.

Previously, Aiello and Håstad [AH91] showed a similar type of lower bound for *interactive* zero-knowledge proofs. Although their results extend to the case of NIZK in the *common random string* model, they do not extend to the *general* public parameter model.[2] Indeed, our proof relies on different (and considerably simpler) techniques.

In the case of statistical *adaptive* NIZK, we present a stronger result.

---

[1] This follows because unless the polynomial hierarchy collapses, **NP** is not contained in **AM** ∩ **coAM** [BHZ87].

[2] This follows because the definition of zero-knowledge requires the simulator to output the random coins of the Verifier, and this is essential to the result in [AH91]. In contrast, the definition of NIZK in the Public Parameter model does not require the Simulator to output the random coins used by the trusted-party to generate the public parameter.

Informal Theorem [Lower bound]. Non-interactive statistical adaptive zero-knowledge proof systems only exist for languages in **BPP**/1 (i.e., the class of languages decidable in probabilistic polynomial time with *one* bit of advice, which depends only on the length of the instance).

By an argument of Adleman, this in particular means that all languages which have statistical adaptive NIZK in the public-parameter model can be decided by polynomial-sized circuits.

We note that a similar strengthening for the non-adaptive case is unlikely, as statistical non-interactive zero-knowledge proof systems for languages which are conjectured to be "hard" are known (e.g., see [GMR98]).

PUBLIC PARAMETER MODEL: COMPUTATIONAL NIZK. Due to the severe lower bounds for statistical NIZK, we continue our investigation by considering computational NIZK in the public parameter model. We first show that one-way functions are both necessary and sufficient in the public parameter model.

Informal Theorem [Upper bound]. If (non-uniform) one-way functions exist, then computational NIZK proof systems in the public parameter model exist for every language in **AM**.

Informal Theorem [Lower bound]. The existence of computational NIZK systems in the public parameter model for a hard-on-average language implies the existence of (non-uniform) one-way functions.

Our upper bound, which applies to the stronger adaptive definition, improves on the construction of Feige, Lapidot, and Shamir [FLS90] which uses one-way permutations (albeit in the common random string model, whereas our construction requires a public parameter). Our lower bound, which applies to the weaker non-adaptive definition, was only known for interactive zero-knowledge proofs [OW93]. We therefore present a (quite) different and relatively simple direct proof for the case of NIZK in the public parameter model.

As a final point, by combining our last two theorems, we obtain the following unconditional characterization of computational NIZK proofs in the public parameter model:

Either NIZK proofs exist only for "easy" languages (i.e., languages that are not hard-on-average), or NIZK proofs exist unconditionally for every language in **AM** *(i.e., for every language which admits a non-interactive proof).*

This type of "all-or-nothing" property was known for interactive zero-knowledge proofs, but not for NIZK since prior constructions of NIZK relied on one-way permutations.

ADDITIONAL CONTRIBUTIONS. As already mentioned, some proofs in this paper extend previously known results for interactive zero-knowledge proofs to the non-interactive setting. We emphasize that our proofs are not mere adaptations of prior results — indeed the results of Aiello and Håstad and of Ostrovsky and

Wigderson are complicated and technically challenging. In contrast, in the non-interactive setting, we obtain equivalent results in a much simpler way. This suggests the use of non-interactive zero-knowledge as a "test-bed" for understanding the (seemingly) more complicated setting of interactive zero-knowledge.

1.2   Other Related Work

In terms of understanding NIZK, two prior works, [DCPY98] and [GSV99], offer complete problems for non-interactive statistical zero-knowledge. Both of these works apply to the non-adaptive definition and only the common *random* string model. We emphasize that these results do not directly extend to the more general public parameter model. In particular, complete problems for **NISZK** in the public parameter model are not known (see the remarks following Thm. 4).

As mentioned earlier, many prior works, e.g. [AH91, Oka96, SV03, GV98, Vad99], address the problem of obtaining unconditional characterizations of statistical zero-knowledge in the interactive setting. More recently, Vadhan [Vad04] also obtains unconditional characterizations of computational zero-knowledge.

OPEN QUESTIONS. While our NIZK proof system in the secret parameter model has an efficient prover strategy, our proof system in the public parameter model does not. Indeed, resolving whether one-way functions suffice for *efficient-prover* NIZK systems is a long-standing open question with many important implications. A positive answer to this question would, for example, lead to the construction of CCA2-secure encryption schemes from any semantically-secure encryption scheme.

2   Definitions

We use standard notation for probabilistic experiments introduced in [GMR85], and abbreviate probabilistic polynomial time as p.p.t.

2.1   Non-interactive Proofs in the Trusted Setup Model

In the trusted setup model, every non-interactive proof system has an associated distribution $\mathcal{D}$ over binary strings of the form (s_V, s_P). During a setup phase, a trusted party samples from $\mathcal{D}$ and privately hands the Prover s_P and the Verifier s_V. The Prover and Verifier then use their respective values during the proof phase. We emphasize that our definition only models *single-theorem* proof systems (i.e., after setup, only one theorem of a fixed size can be proven).[3]

Definition 1 (Non-interactive Proofs in the Secret/Public Parameter Model). *A triple of algorithms, $(\mathcal{D}, P, V)$, is called a* non-interactive proof system *in the* secret parameter model *for a language L if the algorithm $\mathcal{D}$ is probabilistic polynomial-time, the algorithm V is a deterministic polynomial-time and there exists a negligible function μ such that the following two conditions hold:*

---

[3] While our definition only considers single-theorem proof systems, all of our results extend also to proof systems for an *a priori* bounded number of fixed-size statements.

– COMPLETENESS: *For every* $x \in L$

$$\Pr\left[(s_V, s_P) \leftarrow \mathcal{D}(1^x); \; \pi \leftarrow P(x, s_P) \; : \; V(x, s_V, \pi) = 1 \right] \geq 1 - \mu(|x|)$$

– SOUNDNESS: *For every* $x \notin L$, *every algorithm* B

$$\Pr\left[(s_V, s_P) \leftarrow \mathcal{D}(1^x); \; \pi' \leftarrow B(x, s_P) \; : \; V(x, s_V, \pi') = 1 \right] \leq \mu(|x|)$$

If $\mathcal{D}$ is such that s_V is always equal to s_P then we say that $(\mathcal{D}, P, V)$ is in the public parameter model.

Remark 1. In our definition, as with the original one in [BFM88], the Verifier is modeled by a deterministic polynomial time machine. By a standard argument due to Babai and Moran [BM88], this choice is without loss of generality since a probabilistic Verifier can be made to run deterministically through repetition and the embedding of the Verifier's random coins in the setup information.

Let **NIP** denote the class of languages having non-interactive proof systems. For the rest of this paper, we distinguish the secret parameter model from the public parameter model using the superscripts ^{SEC} and ^{PUB} respectively. We start by observing that $\mathbf{NIP}^{\mathrm{PUB}}$ and $\mathbf{NIP}^{\mathrm{SEC}}$ are equivalent. The proof appears in the full version.

Lemma 1. $\mathbf{AM} = \mathbf{NIP}^{\mathrm{PUB}} = \mathbf{NIP}^{\mathrm{SEC}}$

2.2 Zero Knowledge

We next introduce non-interactive zero-knowledge proofs. In the original *non-adaptive* definition of zero-knowledge from [BFM88], there is one simulator, which, after seeing the statement to be proven, generates both the public string and the proof at the same time. In a later *adaptive* definition from [FLS90], there are two simulators— the first of which must output a string before seeing any theorems. The stronger *adaptive* definition guarantees zero-knowledge *even* when the statements are chosen after the trusted setup has finished.[4] Here, we choose to present a weaker (and simpler) *adaptive* definition similar to the one used in [CD04]. The main reasons for this choice are that (a) a weaker definition only strengthens our lower bounds and (b) our definition is meaningful also for languages outside of **NP**, whereas the definitions of [FLS90, Gol04] only apply to languages in **NP**. Nevertheless, we mention that for languages in **NP**, our upper bounds (and of course the lower bounds) also hold for the stricter adaptive definitions of [FLS90, Gol04].

Definition 2 (Non-interactive Zero-Knowledge in the Secret/Public Parameter Model). *Let $(\mathcal{D}, P, V)$ be an non-interactive proof system in the secret (public) parameter model for the language L. We say that $(\mathcal{D}, P, V)$ is non-adaptively zero-knowledge in the secret (public) parameter model if there exists*

---

[4] One might also study an adaptive notion of soundness for non-interactive proofs. We do not pursue this line since every sound non-interactive proof system can be made adaptively sound via parallel repetition.

a p.p.t. simulator S such that the following two ensembles are computationally indistinguishable by polynomial-sized circuits (when the distinguishing gap is a function of $|x|$)

$$\{(s_V, s_P) \leftarrow \mathcal{D}(1^n);\ \pi \leftarrow P(s_P, x)\ :\ (s_V, \pi)\}_{x \in L}$$
$$\{((s'_V, \pi') \leftarrow S(x)\ :\ (s'_V, \pi')\}_{x \in L}$$

*We say that $(\mathcal{D}, P, V)$ is **adaptively zero-knowledge** in the secret (public) parameter model if there exists two p.p.t. simulators S_1, S_2 such that the following two ensembles are computationally indistinguishable by polynomial-sized circuits.*

$$\{(s_V, s_P) \leftarrow \mathcal{D}(1^n);\ \pi \leftarrow P(s_P, x)\ :\ (s_V, \pi)\}_{x \in L}$$
$$\{(s'_V, \mathsf{aux}) \leftarrow S_1(1^n);\ \pi' \leftarrow S_2(x, \mathsf{aux})\ :\ (s'_V, \pi')\}_{x \in L}$$

*We furthermore say that $(\mathcal{D}, P, V)$ is **perfect (statistical) zero-knowledge** if the above ensembles are identically distributed (statistically close).*

For notation purposes, we will use **NIZK**, **NISZK**, and **NIPZK** to denote the class of languages having computational, statistical, and perfect non-interactive zero-knowledge proof systems respectively.

3   The Hidden Bits Model

In order to prove our main theorems, we first review the "hidden bits" model described in [FLS90]. In this model, the Prover and Verifier share a hidden string R, which only the Prover can access. Additionally, the Prover can selectively reveal to the Verifier any portion of the string R. Informally, a proof in the hidden bits model consists of a triplet (π, R_I, I) where I is a sequence of indicies, $I \subseteq \{1, 2, ..., |R|\}$ representing the portion of R that the prover wishes to reveal to the verifier, R_I is the substring of R indexed by I, and π is a proof string. For a formal definition of this model, see Goldreich [Gol01] from which we borrow notation.

The following theorem is shown by Feige, Lapidot and Shamir.

Theorem 1 ([FLS90]). *There exists a non-interactive perfect zero-knowledge proof system in the hidden bits model for any language in* **NP**.

We extend their result to any language in **AM** by using the standard technique of transforming an **AM** proof into the **NP** statement that "there exists a short Prover message which convinces the polynomial-time Verifier."

Theorem 2. *There exists a non-interactive perfect zero-knowledge proof system in the hidden bits model for any language in* **AM**.

Looking ahead, in Sect. 4 we extend Thm. 2 to show that the class of non-interactive perfect zero-knowledge proofs in the hidden bits model is in fact *equivalent* to **AM**.

4   The Secret Parameter Model

Feige, Lapidot and Shamir show how to implement the hidden-bits model with a one-way permutation in the public parameter model. Their implementation, however, degrades the quality of zero-knowledge — in particular, the resulting protocol is only computational zero-knowledge. Below, we show how to avoid this degradation in the secret parameter model.

Lemma 2. *Let (P, V) be a non-interactive perfect zero-knowledge proof system for the language L in the hidden bits model. Then, there exists a non-interactive perfect adaptive zero-knowledge proof system (P', V') for the language L in the secret parameter model. Furthermore if, (P, V) has an efficient prover, then (P', V') has one as well.*

Proof Sketch. We implement the hidden bits model by providing the Prover and Verifier correlated information about each bit of the hidden string. In particular, each bit is split into shares using a simple secret sharing scheme. The Prover is given *all* of the shares, while the Verifier is only given a random subset of them (which is unknown to the Prover). This is done in such a way that the Verifier has no information about the bit, but nonetheless, the Prover cannot reveal the bit in two different ways except with exponentially small probability. We note that this technique is reminiscent to the one used in [Kil88] to obtain commitments from oblivious transfer and to the one in [KMO89] to obtain NIZK with pre-processing (we remark that their resulting NIZK still requires additional computational assumptions, even when ignoring the assumptions necessary for their pre-processing). Our protocol is described in Fig. 1 and a complete proof is given in the full version.   □

Armed with this Lemma, we can now prove our main theorem concerning non-interactive zero-knowledge in the secret parameter model.

Theorem 3. $\mathbf{NIP}^{\mathrm{SEC}} = \mathbf{NIZK}^{\mathrm{SEC}} = \mathbf{NISZK}^{\mathrm{SEC}} = \mathbf{NIPZK}^{\mathrm{SEC}} = \mathbf{AM}$

Proof. $\mathbf{NIPZK}^{\mathrm{SEC}} \subseteq \mathbf{NISZK}^{\mathrm{SEC}} \subseteq \mathbf{NIZK}^{\mathrm{SEC}} \subseteq \mathbf{NIP}^{\mathrm{SEC}}$ follows by definition. Lemma 1 shows that $\mathbf{NIP}^{\mathrm{SEC}} = \mathbf{AM}$, therefore, it suffices to show that $\mathbf{AM} \subseteq \mathbf{NIPZK}^{\mathrm{SEC}}$. This follows by combining Lemma 2 and Thm. 2.   □

RELATED CHARACTERIZATIONS. We note that Lemma 2 also gives an upper bound on the class of perfect zero-knowledge proofs in the hidden bits model. As a corollary, we obtain the following characterization.

Corollary 1. *The class of perfect zero-knowledge proofs in the hidden bits model equals* $\mathbf{AM}$.

5   The Public Parameter Model - Statistical NIZK

In this section we present severe lower bounds for the class of statistical NIZK in the public parameter model. (This stands in stark contrast to the secret

---

Proof System $(\mathcal{D}, P', V')$ – **NIZK in the Secret Parameter model**

Common Input: an instance x of a language L with witness relation R_L and 1^n: security parameter.

Private-output set-up: $\mathcal{D}(1^n) \to (s_P, s_V)$ proceeds as follows on input 1^n:

  1. **(Pick a random string)** Sample m random bits, $\sigma = \sigma_1, \ldots, \sigma_m$.
  2. **(Generate XOR shares)** For $i \in [1, m]$ and $j \in [1, n]$, sample a random bit τ_i^j. Let $\overline{\tau}_i^j = \sigma_i \oplus \tau_i^j$. (Notice that the n pairs $(\tau_i^j, \overline{\tau}_i^j)$ for $j \in [1, n]$ are n random "XOR shares" of the bit σ_i.)
  3. **(Select half of each share)** For $i \in [1, m]$ and $j \in [1, n]$, sample a random bit b_n^j. Let ρ_i^j as follows:

$$\rho_i^j = \begin{cases} \tau_i^j, & \text{if } b_i^j = 0 \\ \overline{\tau}_i^j & \text{otherwise} \end{cases}$$

  (In other words, the values $\{\rho_i^j\}$ are randomly selected "halves" from each of the n XOR shares for σ_i.)
  4. The private output s_P is the set of nm pairs $(\tau_i^j, \overline{\tau}_i^j)$ for $i, j \in [1, m] \times [1, n]$. Note that the string σ is easily derived from s_P.
  5. The private output s_V is the set of nm pairs $\{(\rho_i^j, b_i^j)\}$ for $i, j \in [1, m] \times [1, n]$.

Prover algorithm: On input (x, s_P),

  1. Compute $R = \sigma_1, \ldots, \sigma_m$ by setting $\sigma_i = \tau_i^1 \oplus \overline{\tau}_i^1$.
  2. Run the algorithm $(\pi, R_I, I) \leftarrow P(x, R)$. Recall that the set R_I consists of bits $\{r_i \mid i \in I\}$ and I consists of indices in $[1, m]$.
  3. Output $(\pi, R_I, I, \{o_i \mid i \in I\})$ where o_i denotes the opening of bit σ_i. That is, for all $i \in I$, o_i consists of all n shares $((\tau_i^1, \overline{\tau}_i^1), \ldots, (\tau_i^n, \overline{\tau}_i^n))$ of σ_i.

Verifier algorithm: On input $(x, s_V, \pi, R_I, I, \{o_i | i \in I\})$,

  1. Verify that each opening in R_I is consistent with o_i and with s_V. That is, for $i \in I$, inspect the n pairs, $(\tau_i^1, \overline{\tau}_i^1), \ldots, (\tau_i^n, \overline{\tau}_i^n)$ in o_i, and check that for all $j \in [1, n]$, ρ_i^j is equal to either τ_i^j or $\overline{\tau}_i^j$ (depending on whether $b_i^j = 0$ or 1 respectively). If any single check fails, then reject the proof. Finally, check that $r_i = \tau_i^1 \oplus \overline{\tau}_n^1$.
  2. Verify the proof by running $V(x, R_I, I, \pi)$ and accept if and only if V accepts.

Fig. 1. NIZK in the Secret Parameter model

parameter model, where statistical NIZK can be obtained for all of **AM**.) We first present a lower bound for statistical NIZK under the non-adaptive definition of zero-knowledge. We thereafter sharpen the bound under the more restrictive adaptive definition.

5.1   The Non-adaptive Case

In analogy with the result by [AH91] for interactive zero-knowledge, we show that only languages in the intersection of **AM** and **coAM** have statistical NIZK proof systems in the public parameter model.

Theorem 4. *If L has a statistical non-adaptive NIZK proof system in the public parameter model, then $L \subseteq \mathbf{AM} \cap \mathbf{coAM}$.*

Proof Sketch. Let $(\mathcal{D}, P, V)$ be a statistical NIZK proof system in the public parameter for the language L with simulator S. We show that $L \in \mathbf{AM}$ and that $L \in \mathbf{coAM}$. The former statement follows directly from Lemma 1. To prove the latter one, we present a two-round proof system for proving $x \notin L$. (Note that by the results of [GS86, BM88] it is sufficient to present a two-round *private* coin proof system.)

Verifier Challenge:
1. Run the simulator $(\sigma_0, \pi') \leftarrow S(x)$ and the sampling algorithm $\sigma_1 \leftarrow D(1^{|x|})$ to generate public parameter strings σ_0 and σ_1.
2. Run V on input (σ_0, π') to check if the honest verifier accepts the simulated proof. If V rejects, then output "accept" and halt.
3. Otherwise, flip a coin $b \in 0, 1$ and send $\alpha = \sigma_b$ to the prover.

The Prover response:
1. Upon receiving an input string α, check if there exists a proof π which the honest verifier V accepts (i.e., $V(x, \alpha, \pi) = 1$).
2. If so, output $\beta = 0$; otherwise, output $\beta = 1$.

The Verifier acceptance condition:
1. Upon receiving string β, output "accept" if $\beta = b$, and reject otherwise.

COMPLETENESS. We show that if $x \notin L$, then the Prover (almost) always convinces the Verifier. If the Verifier sent the string σ_0, the Prover always responds with $\beta = 0$, which makes the Verifier always accept. This follows since the Verifier only sends σ_0 if the simulated proof was accepting, which implies that there is at least one accepting proof of $x \in L$ for (P, V). If the Verifier sent the string σ_1, then by the soundness of (P, V), the probability (over the coins of the Verifier) that there exists a proof for $x \in L$ is negligible. Therefore, except with negligible probability, the Prover responds with $\beta = 1$ and the Verifier accepts.

SOUNDNESS. Intuitively, this protocol relies on the same logic as the graph non-isomorphism protocol. If $x \in L$, then the (exponential time) Prover cannot distinguish whether α was generated by the simulator or by the sampler $\mathcal{D}$, and therefore can only convince the Verifier with probability $1/2$. This follows from the statistical zero-knowledge property of (P, V). It only remains to show that the probability (over the random coins of the Verifier) that the Verifier accepts statements $x \in L$ in step (2), without further interaction, is negligible. This follows from the zero-knowledge (and completeness) property of (P, V). Otherwise, V would distinguish between simulated proofs and real ones (since by completeness, the honest prover P succeeds with high probability.)   □

Remark 2. Using techniques from the proof of Thm. 4, one can show that the class $\mathbf{NISZK}^{\mathrm{PUB}}$ reduces to the problem of Statistical Difference, which is complete for $\mathbf{SZK}$ [SV03][5]. Thus, an alternative way to prove this theorem would be to present such a reduction and then invoke the results of [AH91].

---

[5] This should be contrasted with Statistical Difference from Random and Image Density, which are the complete problems for $\mathbf{NISZK}$ in the Common Random String model. These problems are not known to be reducible to Statistical Difference.

5.2   The Adaptive Case

In this section we sharpen our results from the previous section when instead considering the adaptive variant of zero-knowledge.

Theorem 5. *If L has a non-interactive adaptive statistical zero-knowledge proof in the public parameter model, then $L \subset \mathbf{BPP}/1$.*

Proof Sketch. Let $(\mathcal{D}, P, V)$ be a non-interactive adaptive statistical zero-knowledge proof system for L with simulators S_1 and S_2.

We first observe that by the statistical zero-knowledge property, for every n for which L contains an instance of length n, the output of $S_1(1^n)$ must be statistically close to the output of $\mathcal{D}(1^n)$. This follows because the output of $S_1(1^n)$ is independent of the theorem statement.

This observation suggests the following probabilistic polynomial time decision procedure $D(x)$ for L, which obtains a one-bit non-uniform advice indicating whether L contains any instances of length $|x|$.

On input an instance x,
1. If the non-uniform advice indicates that L contains no instances of length $|x|$, directly reject.
2. Otherwise, run $(\sigma', \mathsf{aux}) \leftarrow S_1(1^{|x|})$ to generate a public parameter.
3. Run $\pi' \leftarrow S_2(x, \mathsf{aux})$ to produce a putative proof.
4. Run $V(x, \sigma', \pi')$ and accept iff V accepts.

Note that when $x \in L$, then D accepts with overwhelming probability due to the completeness and zero-knowledge property of $(\mathcal{D}, P, V)$. If $x \notin L$ and there are no instances of length $|x|$ in L, then D always rejects due to the non-uniform advice. It remains to show that when $x \notin L$, and there exists instances of length $|x|$ in L, then D rejects with high probability.

Assume, for sake of reaching contradiction, that there exists a polynomial $p(\cdot)$ such that for infinitely many lengths n, L contains instances of length n yet there exists an instance $x \notin L$ of length n, such that

$$\Pr\left[(\sigma', \mathsf{aux}) \leftarrow S_1(1^{|x|}); \; \pi' \leftarrow S_2(x, \mathsf{aux}) \; : \; V(x, \sigma', \pi') = 1\right] \geq \frac{1}{p(n)} \qquad (1)$$

We show how this contradicts the fact that the output of S_1 and $\mathcal{D}$ are statistically close (when L contains instances of length n). By the soundness of $(\mathcal{D}, P, V)$, there exists a negligible function μ such that for any unbounded prover P^*,

$$\Pr\left[\sigma \leftarrow \mathcal{D}(1^{|x|}); \; \pi' \leftarrow P^*(x, \sigma) \; : \; V(x, \sigma, \pi') = 1\right] \leq \mu(|x|) \qquad (2)$$

Consider an exponential time non-uniform distinguisher C, which on input σ'' (and advice x), enumerates all proof strings π' to determine if any of them convince V to accept x. If so, C outputs 0, and otherwise outputs 1.

If σ'' is generated by S_1, then by (1), such a proof string π' exists with noticeable probability. On the other hand, if σ'' comes from $\mathcal{D}$, then by (2), such a proof string only exists with negligible probability. We conclude that C distinguishes the output of S_1 from that of $\mathcal{D}$ with a non-negligible advantage. $\square$

6   The Public Parameter Model - Computational NIZK

In this section we show that one-way functions are sufficient and necessary for computational NIZK for languages that are hard-on-average. Combining these two results, we obtain the following unconditional characterization : *Either* **NIZK**[PUB] *only contains "easy" languages (i.e., languages that are not hard-on-average), or it "hits the roof", (i.e., contains all of* **AM***).*

PRELIMINARIES. Let us first define one-way functions and hard-on-average languages. As is standard in the context of zero-knowledge proofs, we define hardness in terms of infeasability for *non-uniform* p.p.t.

Definition 3 (One-way function). *A function* $f : \{0,1\}^* \to \{0,1\}^*$ *is called* one-way *if the following two conditions hold:*

- *Easy to compute: There exists a (deterministic) polynomial-time algorithm E such that on input x, E outputs $f(x)$.*
- *Hard to invert: For every non-uniform p.p.t. algorithm A, every sufficiently large integer n, and every polynomial $p(\cdot)$,*

$$\Pr\left[x \leftarrow \{0,1\}^n; y \leftarrow A(f(x)) \ : \ f(y) = f(x)\right] < \frac{1}{p(n)}$$

Definition 4 (Hard-on-average language). *A language L is hard-on-average if there exists a p.p.t. sampling algorithm G such that for every non-uniform p.p.t. algorithm A, every polynomial $p(\cdot)$, and every sufficiently large n,*

$$\Pr\left[x \leftarrow G(1^n) \ : \ A(x) \ correctly \ decides \ whether \ x \in L\right] < \frac{1}{2} + \frac{1}{p(n)}$$

6.1   OWFs Are Sufficient

We show how to implement the hidden bits model in the public-parameter model based on a one-way function. Recall that [FLS90] implements the hidden bits model using a one-way permutation and a hard-core predicate. The reason for using a one-way permutation is to give the Prover a short certificate for opening each bit *in only one way* (the certificate being the pre-image of the one-way permutation). A similar technique fails with one-way functions since a string may have either zero or many pre-images, and therefore a malicious Prover may be able to open some hidden bits as either zero or one.

Another approach would be to use a one-way function in order to construct a pseudo-random generator [HILL99], and then to represent a 0 value as a pseudo-random string and a 1 as a truly random string (in some sense, this technique is reminiscent of the one used by Naor for bit commitment schemes from pseudo-random generators [Nao91]). The Prover can thus open a 0 value by revealing a seed to the pseudo-random string. However, there is no way for the Prover to convince a Verifier that a string is truly random.

We overcome this problem by forming a reference string consisting of *pairs* of $2k$-bit strings, (α, β) in which exactly one of the two strings is pseudo-random while the other is truly random. More precisely, the 0-value is encoded as a pair in which α is generated pseudo-randomly by expanding a k bit seed into a $2k$ bit string, while β is chosen uniformly at random from $\{0,1\}^{2k}$. The 1-value is encoded the opposite way: α is chosen randomly, while β is generated pseudo-randomly. The Prover can now reveal a 0 or a 1 by revealing the seed for either α or β.

Lemma 3. *Assume the existence of one-way functions. Let (P, V) be a non-interactive (adaptive) zero-knowledge proof system for the language $L \in \mathbf{NP}$ in the hidden bits model. If P is an efficient prover, then, there exists a non-interactive (adaptive) zero-knowledge proof system (P', V') for the language L in the public parameter model.*

Proof Sketch. Let (P, V) be an NIZK proof system in the hidden bits model, let $G : \{0,1\}^k \rightarrow \{0,1\}^{2k}$ be a pseudo-random generator and let $L \in \mathbf{NP}$ be a language with witness relation R_L. Consider protocol (P', V') described in Fig. 2.

COMPLETENESS. Completeness follows from the corresponding completeness of (P, V) and the fact that P' aborts only with negligible probability.

SOUNDNESS. Assume for the moment that a cheating prover P'^* is only able to open R in one manner. In this case, the soundness of (P, V) carries over to (P', V') in the same way as in Lemma 2. All that remains is to show that R can only be opened in one way. Below, we argue that this happens with high probability.

Note that there are a maximum of 2^n pseudo-random strings in G's support. On the other hand, there are 2^{2n} strings of length $2n$. Therefore, a randomly sampled length-$2n$ string will be pseudo-random with probability at most 2^{-n}. Thus, for any pair (a_i, b_i), the probability that both values are pseudo-random is at most 2^{-n}. By the union bound, the probability that there is one such pair in s is upper-bounded by $n2^{-n}$.

ZERO-KNOWLEDGE. We present a simulator $S' = S'_1, S'_2$ for $(\mathcal{D}, P', V')$ which uses the simulator S for (P, V) as a subroutine. First, $(s, \mathsf{aux}) \leftarrow S'_1(1^n)$ generates s as a sequence of pairs (α'_i, β'_i) in which *both* α'_i and β'_i are pseudo-random. The aux value contains all of the seeds, u_i, w_i, for the pseudo-random values α'_i and β'_i respectively. The simulator S'_2 works by running simulator $S(x)$ to generate $(\pi', R'_I, I) \leftarrow S(x)$ and then outputting $(\pi', R'_I, I, \{v'_i \mid i \in I\})$ where v'_i equals u_i if $r_i = 0$ and w_i otherwise. In order to show the validity of the simulation, consider the following four hybrid distributions.

- Let H_1 denote the ensemble (s, π) in which the honest Prover runs on a string s generated according to $\mathcal{D}$.
- Let H_2 denote the output of the above experiment with the exception that $\mathcal{D}$ provides all pre-images $\{v_i\}$ to an efficient prover algorithm P_{eff}, which also

---

Proof System $(\mathcal{D}, P', V')$ – NIZK in the Public Parameter model

Common Input: an instance $x \in L$ and a security parameter 1^n

Public Parameter set-up: $\mathcal{D}(1^n) \to s$, where $\mathcal{D}$ proceeds as follows :

    1. Select m random bits $\sigma = \sigma_1, \ldots, \sigma_m$.

    2. For each $i \in [1, m]$, generate two strings α_i and β_i as follows:

        $\alpha_i \leftarrow G(v_i)$ where v_i is a uniformly chosen string of length k.

        $\beta_i \leftarrow_r \{0, 1\}^{2k}$

    3. Let $\tau_i = \begin{cases} (\alpha_i, \beta_i) & \text{if } \sigma_i = 1 \\ (\beta_i, \alpha_i) & \text{otherwise} \end{cases}$

    4. Output $s = \tau_1, \ldots, \tau_m$.

Prover's algorithm: On input x, s,

    1. Compute $R = \sigma_1, \ldots, \sigma_m$ from s by the following procedure. Parse s into m pairs $(a_1, b_1), \ldots, (a_m, b_m)$. For each pair (a_i, b_i), determine (in exponential time) which of either a_i or b_i are pseudo-random (i.e, in the range of G). In the former case, set $\sigma_i = 0$, and in the latter, $\sigma_i = 1$, and let v_i denote the seed used to generate the pseudo-random value. If both a_i and b_i are in the range of G, then output **abort**.

    2. Compute the lexographically first witness w satisfying $R_L(x, w)$.

    3. Run the Prover algorithm $(\pi, R_I, I) \leftarrow P(x, w, R)$. Recall that the set R_I consists of bits $\{r_i \mid i \in I\}$ and I consists of indices in $[1, m]$.

    4. Output $(\pi, R_I, I, \{v_i \mid i \in I\})$.

Verifier's algorithm: On input $(x, \pi, R_I, I, \{v_i \mid i \in I\})$

    1. Verify each opening in R_I is consistent with s and v_i. Parse s into m pairs $(a_1, b_1), \ldots, (a_m, b_m)$. For each $i \in I$, run $t \leftarrow G(v_i)$ and if $t = a_i$, set $\sigma_i = 1$, if $t = b_i$, then set $\sigma_i = 0$ (if neither or both conditions are met, then reject the proof). Finally, verify that $r_i = \sigma_i$.

    2. Run the Verifier algorithm $V(x, \pi, R_I, I)$ and accept iff V accepts.

Fig. 2. NIZK in the Public Parameter model

receives the lexographically first witness w for x and then only runs Step 3 and 4 of P''s algorithm.

– Let H_3 denote the output of the second experiment with the exception that s is generated by $S_1'(1^n)$, and that furthermore, $S_1'(1^n)$ gives either u_i or w_i (randomly chosen) to P_{eff} for all $i \in [1, m]$.

– Let H_4 denote the output of the third experiment with the exception that π is generated by $S_2'(x, \text{aux})$ and u_i, w_i in aux is given to P_{eff}. Notice that this distribution corresponds exactly to the output of S'.

In the full version we show that the above hybrid distributions are all indistinguishable, which concludes the proof. □

Remark 3. Note that we explicitly require two properties from the NIZK proof system (P, V) in the hidden bits model: first, that P is an efficient Prover, and secondly, that the zero-knowledge property is defined for non-uniform distinguishers. Both of these requirements stem from the fact that the Prover in our new protocol is unbounded, which creates complications in the hybrid arguments.

Theorem 6. *If (non-uniform) one-way functions exist, then for both adaptive and non-adaptive definitions of zero-knowledge,* $\mathbf{NIZK}^{\mathrm{PUB}} = \mathbf{NIP}^{\mathrm{PUB}} = \mathbf{AM}$.

Proof. By Thm. 1 and Lemma 3, $\mathbf{NP} \subseteq \mathbf{NIZK}^{\mathrm{PUB}}$. Using techniques from the proof of Thm. 2, we can extend this result to show that $\mathbf{AM} \subseteq \mathbf{NIZK}^{\mathrm{PUB}}$. By definition, $\mathbf{NIZK}^{\mathrm{PUB}} \subseteq \mathbf{NIP}^{\mathrm{PUB}}$. Finally, by Lemma 1, $\mathbf{NIP}^{\mathrm{PUB}} = \mathbf{AM}$. $\square$

6.2   OWFs Are Necessary

We proceed to show that (non-uniform) one-way functions are *necessary* for non-interactive zero-knowledge for "hard" languages. This stands in contrast to the secret parameter model where unconditional results are possible.

Theorem 7. *If there exists a non-adaptive NIZK proof system for a hard-on-average language L, then (non-uniform) one-way functions exist.*

Proof Sketch. Let $(\mathcal{D}, P, V)$ be a non-adaptive NIZK system for L in the public parameter model and let S be the simulator for (P, V). Furthermore, suppose that L is hard-on-average for the polynomial-time samplable distribution G. Now, consider the following two distributions:

$$\{(s_V, s_P) \leftarrow \mathcal{D}(1^n), x \leftarrow G(1^n) \ : \ x, s_V\} \tag{3}$$

$$\{(s'_V, \pi) \leftarrow S(x, 1^n), x \leftarrow G(1^n) \ : \ x, s'_V\} \tag{4}$$

We show that the above distributions are (non-uniformly) computationally indistinguishable, but statistically "far". By a result of Goldreich [Gol90] (relying on [HILL99]) the existence of such distributions implies the existence of (non-uniform) one-way functions. $\square$

Claim. The distributions (3) and (4) are computationally indistinguishable.

Proof Sketch. We start by noting that conditioned on x being a member of language L, the above distributions are computationally indistinguishable by the zero-knowledge property of (P, V). It then follows from the hardness of L that the above distributions must be computationally indistinguishable, even without this restriction. $\square$

Claim. The distributions (3) and (4) are *not* statistically indistinguishable.

Proof Sketch. We show that the distributions (3) and (4) are statistically "far" conditioned on instances $x \notin L$. It then follows from the fact that L is roughly balanced over G (due the hard-on-average property of L over G) that (3) and (4) are statistically "far" apart.

Note that on instances $x \notin L$, the soundness property of (P, V) guarantees that very few strings generated by $\mathcal{D}$ have proofs which are accepted by the

Verifier (otherwise, a cheating prover can, in exponential time, find such proofs and thereby violate the soundness condition). On the other hand, since L is hard-on-average, and since S runs in polynomial time, most of the strings s_V generated by S have proofs which are accepted by V (otherwise, S can be used to decide L). Therefore, the distributions (3) and (4) are statistically far apart, conditioned on instances $x \notin L$. □

Acknowledgments. We would like to thank Silvio Micali and the anonymous referees for their helpful suggestions.

References

[AH91]   W. Aiello and J. Håstad. Statistical zero-knowledge languages can be recognized in two rounds. *J. Comput. Syst. Sci*, 42:327–345, 1991.

[Aum74]   R. Aumann. Subjectivity and correlation in randomized strategies. *J. Math. Econ.*, 1:67–96, 1974.

[BM88]   L. Babai and S. Moran. Arthur-merlin games: A randomized proof system, and a hierarchy of complexity classes. *J. Comput. Syst. Sci*, 36(2):254–276, 1988.

[BDMP91]   M. Blum, A. De Santis, S. Micali, and G. Persiano. Noninteractive zero-knowledge. *SIAM J. Computing*, 20(6):1084–1118, 1991.

[BFM88]   M. Blum, P. Feldman, and S. Micali. Non-interactive zero-knowledge and its applications. In *STOC 88*, pages 103–112, 1988.

[BHZ87]   R. Boppana, J. Håstad, and S. Zachos. Does co-NP have short interactive proofs? *Inf. Process. Lett.*, 25(2):127–132, 1987.

[CLOS02]   R. Canetti, Y. Lindell, R. Ostrovsky, and A. Sahai. Universally composable two-party and multi-party secure computation. In *STOC 02*, pages 494–503, 2002.

[CD04]   R. Cramer and I. Damgård. Secret-key zero-knowledge and non-interactive verifiable exponentiation. In *TCC 04*, 2004.

[Dam93]   I. Damgård. Non-interactive circuit based proofs and non-interactive perfect zero knowledge with preprocessing. In *EUROCRYPT 92*, pages 341–355, 1993.

[Dam00]   I. Damgård. Efficient concurrent zero-knowledge in the auxiliary string model. In *EUROCRYPT 2000*, pages 418–430, 2000.

[DFN05]   I. Damgård, N. Fazio, and A. Nicolosi. Secret-key zero-knowledge protocols for NP and applications to threshold cryptography. Manuscript, 2005.

[DCO⁺01]   A. De Santis, G. Di Crescenzo, R. Ostrovsky, G. Persiano, and Amit Sahai. Robust non-interactive zero knowledge. *CRYPTO 01*, pages 566–598, 2001.

[DCPY98]   A. De Santis, G. Di Crescenzo, G. Persiano, and M. Yung. Image density is complete for non-interactive-SZK. In *ICALP 98*, pages 784–795, 1998.

[DMP88]   A. De Santis, S. Micali, and G. Persiano. Non-interactive zero-knowledge with preprocessing. In *CRYPTO 88*, pages 269–282, 1988.

[FLS90]   U. Feige, D. Lapidot, and A. Shamir. Multiple non-interactive zero knowledge proofs based on a single random string. In *FOCS 90*, pages 308–317, 1990.

[GMR98]   R. Gennaro, D. Micciancio, and T. Rabin. An efficient non-interactive statistical zero-knowledge proof system for quasi-safe prime products. In *CCS 98*, pages 67–72, 1998.

[Gol90]   O. Goldreich. A note on computational indistinguishability. *Inf. Process. Lett.*, 34(6):277–281, 1990.

[Gol01]   O. Goldreich. *Foundations of Cryptography: Vol I.* 2001.

[Gol04]   O. Goldreich. *Foundations of Cryptography: Vol II.* 2004.

[GO94]   O. Goldreich and Y. Oren. Definitions and properties of zero-knowledge proof systems. *J. Crypt.*, 7(1):1–32, 1994.

[GSV99]   O. Goldreich, A. Sahai, and S. Vadhan. Can statistical zero knowledge be made non-interactive? or on the relationship of SZK and NISZK. In *CRYPTO 99*, pages 467–484, 1999.

[GV98]   O. Goldreich and S. Vadhan. Comparing entropies in statistical zero-knowledge with applications to the structure of SZK. In *Computational Complexity*, 1998.

[GMR85]   S. Goldwasser, S. Micali, and C. Rackoff. The knowledge complexity of interactive proof-systems. In *STOC 85*, pages 291–304, 1985.

[GS86]   S. Goldwasser and M. Sipser. Private coins versus public coins in interactive proof systems. In *STOC 86*, pages 59–68, 1986.

[HILL99]   J. Håstad, R. Impagliazzo, L. A. Levin, and M. Luby. A pseudorandom generator from any one-way function. *SIAM J. Comput.*, 28(4):1364–1396, 1999.

[Kil88]   J. Kilian. Founding cryptography on oblivious transfer. In *STOC 88*, pages 20–31, 1988.

[KMO89]   J. Kilian, S. Micali, and R. Ostrovsky. Minimum resource zero-knowledge proofs. In *FOCS 89*, pages 474–479, 1989.

[KP98]   J. Kilian and E. Petrank. An efficient non-interactive zero-knowledge proof system for NP with general assumptions. *J. Crypt.*, 11(1):1–27, 1998.

[Nao91]   M. Naor. Bit commitment using pseudorandomness. *J. Crypt.*, 4(2):151–158, 1991.

[Oka96]   T. Okamoto. On relationships between statistical zero-knowledge proofs. In *STOC 96*, pages 649–658, 1996.

[OW93]   R. Ostrovsky and A. Wigderson. One-way fuctions are essential for non-trivial zero-knowledge. In *ISTCS*, pages 3–17, 1993.

[SV03]   A. Sahai and S. Vadhan. A complete problem for statistical zero knowledge. *J. ACM*, 50(2):196–249, 2003.

[SMP87]   A. De Santis, S. Micali, and G. Persiano. Non-interactive zero-knowledge proof systems. In *CRYPTO 87*, pages 52–72, 1987.

[Vad99]   S. Vadhan. *A Study of Statistical Zero-Knowledge Proofs.* PhD thesis, MIT, 1999.

[Vad04]   S. Vadhan. An unconditional study of computational zero knowledge. In *FOCS 04*, pages 176–185, 2004.

Impossibility and Feasibility Results
for Zero Knowledge with Public Keys*

Joël Alwen[1], Giuseppe Persiano[2], and Ivan Visconti[2]

[1] Technical University of Vienna, A-1010 Vienna, Austria
e9926980@stud3.tuwien.ac.at
[2] Dipartimento di Informatica ed Appl., Università di Salerno,
84081 Baronissi (SA), Italy
{giuper, visconti}@dia.unisa.it

Abstract. In this paper, we continue the study of the round complexity of black-box zero knowledge in the bare public-key (BPK, for short) model previously started by Micali and Reyzin in [11]. Specifically we show the impossibility of 3-round concurrent (and thus resettable) black-box zero-knowledge argument systems with sequential soundness for non-trivial languages. In light of the previous state-of-the-art, our result completes the analysis of the round complexity of black-box zero knowledge in the BPK model with respect to the notions of soundness and black-box zero knowledge.

Further we give sufficient conditions for the existence of a 3-round resettable zero-knowledge *proof* (in contrast to argument) system with concurrent soundness for $\mathcal{NP}$ in the upperbounded public-key model introduced in [14].

1 Introduction

The classical notion of a zero-knowledge proof system has been introduced in [1]. Roughly speaking, in a zero-knowledge proof system a prover can prove to a verifier the validity of a statement without releasing any additional information. Recently, starting with the work of Dwork, Naor, and Sahai [2], the concurrent and asynchronous execution of zero-knowledge protocols has been considered. In this setting, several concurrent executions of the same protocol take place and a malicious adversary controls the verifiers and the scheduling of the messages.

Motivated by considerations regarding smart cards, the notion of *resettable zero knowledge* (rZK, for short) was introduced in [3]. An rZK proof remains "secure" even if the verifier is allowed to tamper with the prover and to reset the prover in the middle of a proof to any previous state, then asking different questions. It is easy to see that concurrent zero knowledge is a special case of resettable zero knowledge and, currently, rZK is the strongest notion of

---

* The work presented in this paper has been supported in part by the European Commission through the IST Programme under contract IST-2002-507932 ECRYPT.

zero knowledge that has been studied when security against malicious verifiers is considered. Unfortunately, in the plain model, if we only consider black-box zero knowledge, constant-round concurrent (and therefore resettable) zero knowledge is possible only for trivial languages (see [4]). Moreover, the existence of a constant-round non-black-box concurrent zero-knowledge argument system is currently an open question (see [5] for the main results on non-black-box zero knowledge). An almost constant-round non-black-box concurrent zero-knowledge argument system has been recently given in [6] by assuming the existence of only one (stateful) prover.

Such negative results have motivated the introduction of the *bare public-key model* [3] (BPK, for short). Here each possible verifier deposits a public key pk in a public file and keeps private the associated secret information sk. From then on, all provers interacting with such a verifier will use pk and the verifier can not change pk from proof to proof. Note that in the BPK model there is no interactive preprocessing stage, no trusted third party or reference string and the public file can be completely under the control of the adversary. Consequently the BPK model is considered a very weak set-up assumption compared to some previously proposed models [2,7,8,9,10]. In this model, however, the notion of soundness is more subtle. This was first noted in [11], where the existence of four distinct and increasingly stronger notions of soundness: one-time, sequential, concurrent and resettable soundness is shown. Moreover it was pointed out that the constant-round rZK argument system in the BPK model of [3] did not seem to be *concurrently sound*.

In [11], a 3-round one-time sound black-box rZK argument system and a 4-round sequentially sound black-box rZK argument system for $\mathcal{NP}$ in the BPK model are given. Moreover it is shown that in the BPK model neither zero knowledge in less than 3 rounds nor black-box zero-knowledge with resettable soundness are possible for non-trivial languages. Two main problems were left open in [11] (see page 553 of [11] and page 13 of [12]).

The first open problem, namely the existence of a constant-round rZK argument system with concurrent soundness in the BPK model has been recently solved in [13] where an (optimal) 4-round protocol is presented. Before this result, a 3-round resettable zero-knowledge argument system with concurrent soundness has been presented by Micali and Reyzin in [14] in the upperbounded public-key (UPK, for short) model, where the verifier has a counter and his public key can be used only an a-priori fixed polynomial number of times.

The second open problem, namely the existence of a 3-round {resettable, concurrent, sequential} zero-knowledge argument with sequential soundness in the BPK, has been very partially addressed in [13] where a 3-round sequentially sound sequential zero-knowledge argument system has been presented.

The most interesting open problem in the BPK model is therefore the existence of a 3-round resettable (or even only concurrent) zero-knowledge argument system with sequential soundness for $\mathcal{NP}$.

Our Contribution. In this paper we show that 3-round black-box concurrent (and therefore resettable) zero-knowledge argument systems with sequential soundness in the BPK model exist only for trivial languages. As a consequence, we have that in the BPK model, sequential soundness in 3 rounds can be achieved for non-trivial languages only when no more than black-box sequential zero knowledge is required. This is in contrast to both one-time soundness and concurrent soundness where 3 rounds and 4 rounds respectively, have been shown to be both necessary and sufficient for sequential, concurrent and resettable zero knowledge. Our result closes the analysis on the round complexity of the BPK model with respect to notions of soundness and zero knowledge, see Fig. 1.

The intuition behind our impossibility result goes as follows. In the impossibility proof of 3-round black-box zero knowledge given in [15] by Goldreich and Krawczyk (on which our result is based) a deciding machine runs the simulator to determine if x is in L. In case the simulator outputs an accepting transcript even when $x \notin L$, the work of the simulator can be used by an adversarial prover that violates the soundness of the protocol. The translation in [11] of this proof to the BPK model works when the proof system is sound against *concurrent* malicious provers (because a rewind can be simulated with a new concurrent session). The proof in this paper observes that for 3-round black-box concurrent zero knowledge, there is at least one proof that the simulator does *not* rewind and therefore sequential soundness will suffice.

We also give some sufficient conditions for achieving 3-round resettable zero-knowledge proof (in contrast to argument) systems with concurrent soundness in the UPK model. Moreover, our construction does not use assumptions with respect to superpolynomial-time algorithms (i.e., complexity leveraging).

	3-Round OTS	3-Round SS	4-Round CS
sZK	[MR Crypto 01]	[DPV Crypto 04]	Folklore
cZK	[MR Crypto 01]	Impossible [This Paper]	[DPV Crypto 04]
rZK	[MR Crypto 01]	Impossible [This Paper]	[DPV Crypto 04]

Fig. 1. The round complexity of black-box zero knowledge in the BPK model

2   The Public-Key Models

Here we describe the BPK and the UPK models that we consider for our results. We give the definitions of zero-knowledge proof and argument systems with respect to the notions of soundness and zero knowledge that we use in the paper. For further details, see [11,14,12].

The BPK model assumes that: 1) there exists a polynomial-size collection of records associating identities with public keys in the public file F; 2) an (honest) prover is an interactive deterministic polynomial-time algorithm that takes as input a security parameter 1^k, F, an n-bit string x, such that $x \in L$ where L is an $\mathcal{NP}$-language, an auxiliary input y, a reference to an entry of F and a random tape; 3) an (honest) verifier V is an interactive deterministic polynomial-time algorithm that works in the following two stages: 1) in a first stage on input a security parameter 1^k and a random tape, V generates a key pair $(\mathtt{pk}, \mathtt{sk})$ and stores it's identity associated with $\mathtt{pk}$ in an entry of the file F; 2) in the second stage, V takes as input $\mathtt{sk}$, a statement $x \in L$ and a random string, V performs an interactive protocol with a prover, and outputs "accept" or "reject"; 4) the first interaction of a prover and a verifier starts after all verifiers have completed their first stage.

Definition 1. *Given an $\mathcal{NP}$-language L and its corresponding relation R_L, we say that a pair of probabilistic polynomial-time algorithms $\langle P, V \rangle$ is* **complete** *for L, if for all n-bit strings $x \in L$ and any witness y such that $(x, y) \in R_L$, the probability that V on input x when interacting with P on input x and y, outputs "reject" is negligible in n.*

Malicious Provers and Attacks in the BPK *Model.* Let s be a positive polynomial and $P^\star$ be a probabilistic polynomial-time algorithm that takes as first input 1^n.

$P^\star$ is an *s-sequential malicious prover* if it runs in at most $s(n)$ stages in the following way: in stage 1, $P^\star$ receives a public key $\mathtt{pk}$ and outputs an n-bit string x_1. In every even stage, $P^\star$ starts from the final configuration of the previous stage, sends and receives messages of a single interactive protocol on input $\mathtt{pk}$ and can decide to abort the stage in any moment and to start the next one. In every odd stage $i > 1$, $P^\star$ starts from the final configuration of the previous stage and outputs an n-bit string x_i.

$P^\star$ is an *s-concurrent malicious prover* if on input a public key $\mathtt{pk}$ of V, it can perform the following $s(n)$ interactive protocols with V: 1) if P^* is already running i protocols $0 \le i < s(n)$ it can start a new protocol with V choosing a new statement to be proved; 2) it can output a message for any running protocol, receive immediately the response from V and continue.

Given an s-sequential malicious prover $P^\star$ and an honest verifier V, a *sequential attack* is performed in the following way: 1) the first stage of V is run on input 1^n and a random string so that a pair $(\mathtt{pk}, \mathtt{sk})$ is obtained; 2) the first stage of $P^\star$ is run on input 1^n and $\mathtt{pk}$ and x_1 is obtained; 3) for $1 \le i \le s(n)/2$ the $2i$-th stage of $P^\star$ is run letting it interact with V which receives as input $\mathtt{sk}, x_i$ and a random string r_i, while the $(2i+1)$-th stage of $P^\star$ is run to obtain x_i.

Given an s-concurrent malicious prover P^* and an honest verifier V, a *concurrent attack* is performed in the following way: 1) the first stage of V is run on input 1^n and a random string so that a pair $(\mathtt{pk}, \mathtt{sk})$ is obtained; 2) P^* is run on input 1^n and $\mathtt{pk}$; 3) whenever P^* starts a new protocol choosing a statement, V is run on inputs the new statement, a new random string and $\mathtt{sk}$.

Definition 2. *Given a complete pair $\langle P, V \rangle$ for an $\mathcal{NP}$-language L in the* BPK *model, then $\langle P, V \rangle$ is a* **concurrently** *(resp.,* **sequentially***) sound interactive argument system for L if for all positive polynomials s, for all s-concurrent (resp., s-sequential) malicious provers $P^\star$ and for any false statement "$x \in L$" the probability that in an execution of a concurrent (resp. sequential) attack, V outputs "accept" for such a statement is negligible in n.*

In the definition above, if the malicious prover P^* is computationally unbounded, then $\langle P, V \rangle$ is a proof (and not only an argument) system.

Definition 3. *Let $\langle P, V \rangle$ be an interactive proof or argument system for a language L. We say that a probabilistic polynomial-time adversarial verifier V^* is a* **concurrent adversary in the** BPK **model** *if on input polynomially many values $\bar{x} = x_1, \ldots, x_{\text{POLY}(n)}$, it first generates the public file F with POLY(n) public keys and then concurrently interacts with POLY(n) number of independent copies of P (each with a valid witness for the statement), with common input $\bar{x}$ and without any restrictions over the scheduling of the messages in the different interactions with P. Moreover we say that the transcript of such a concurrent interaction consists of $\bar{x}$ and the sequence of prover and verifier messages exchanged during the interaction. We refer to* $\text{view}_{V^*}^P(\bar{x})$ *as the random variable describing the content of the random tape of V^* and the transcript of the concurrent interactions between P and V^*.*

Definition 4. *Let $\langle P, V \rangle$ be an interactive argument or proof system for a language L in the* BPK *model. We say that $\langle P, V \rangle$ is black-box* **concurrent zero knowledge** *if there exists a probabilistic polynomial-time algorithm S such that for each polynomial-time concurrent adversary V^*, let $S_{V^*}(\bar{x})$ be the output of S on input $\bar{x}$ and black-box access to V^*, then if $x_1, \ldots, x_{\text{POLY}(n)} \in L$, the ensembles $\{\text{view}_{V^*}^P(\bar{x})\}$ and $\{S_{V^*}(\bar{x})\}$ are computationally indistinguishable.*

Definition 5. *An interactive argument system $\langle P, V \rangle$ in the BPK model is black-box* **resettable zero knowledge** *if there exists a probabilistic polynomial-time algorithm S such that for all probabilistic polynomial time adversaries V^*, for all pairs of polynomials (s, t) and for all $x_i \in L$ where $|x_i| = n$ and $i = 1, \ldots, s(n)$, V^* runs in at most $t(n)$ steps and the following two distributions are indistinguishable:*

1. *the output of V^* that generates F with $s(n)$ entries and interacts (even concurrently) a polynomial number of times with each $P(x_i, y_i, j, r_k, F)$ where y_i is a witness for $x_i \in L$, $|x_i| = n$, j is the index (i.e. associated identity) of the public key of V^* in F and r_k is a random tape for $1 \le i, j, k \le s(n)$;*
2. *the output of S given black-box access to V^* on input $x_1, \ldots, x_{s(n)}$.*

*Moreover we define such an adversarial verifier V^* as an (s, t)-**resetting malicious verifier**.*

The definitions in the UPK model are very similar to the ones given for the BPK model. The only interesting difference is the attack of the malicious prover. Indeed, in the UPK model there is a bound on the number of sessions that the malicious prover can open in order to prove a false statement. Indeed, the verifier only uses his public key an a-priori fixed polynomial number of times and he uses a counter (a value that is persistent between the sessions) to verify that his key is not yet expired. For details see [12].

3   3-Round Sequentially Sound cZK in the BPK Model

In this section we will concentrate entirely on the proof of the following theorem.

Theorem 1. *Any black-box concurrent zero-knowledge argument system satisfying sequential soundness in the* BPK *model for a language L outside of $\mathcal{BPP}$ requires at least 4 rounds.*

3.1   Techniques for Achieving the Result

In [11,12], it has been proven that in the BPK model concurrent soundness can not be achieved in less than 4 rounds. That proof mainly follows the proof of the following theorem by Goldreich and Krawczyk in [15]: "In the plain model, any black-box zero-knowledge argument system for a language outside of $\mathcal{BPP}$ requires at least 4 rounds". Indeed, the proof given in [15], crucially uses the fact that if there exists a simulator M, then the work of M can be used either to decide the language or to violate the soundness of the protocol. Indeed, in case the simulator outputs an accepting transcript for a false statement, an adversarial prover can prove the same false statement by emulating the rewinds of the simulator by means of concurrent sessions. Also in [11,12], the same analysis is carefully repeated by using a concurrent malicious prover, thus proving that black-box zero knowledge with concurrent soundness needs at least 4 rounds for non-trivial languages in the BPK model.

However, when the prover can only open sequential sessions, the previously discussed approach does not work anymore. Actually, only sequential soundness and zero knowledge are not enough to prove the impossibility result. Indeed, in [13], it has been shown that in the BPK model a 3-round sequential black-box zero-knowledge argument system with sequential soundness exists. It is therefore vital to use the *concurrent* black-box zero-knowledge property along with sequential soundness in order to obtain the desired claim. Note that the black-box concurrent zero-knowledge property (i.e., the existence of a strong simulator that works in an hostile setting) has been previously used in [16,17,4] to show that black-box concurrent zero knowledge can not be achieved respectively in 4, 7 and finally in a constant number of rounds in the plain model. However these previous results do not help at all in the BPK model, since constant-round black-box concurrent (and even resettable) zero knowledge has been achieved in the BPK model. In particular only 3 rounds are necessary for one-time soundness [11] and 4 rounds for concurrent soundness [13].

Our proof of Theorem 1 therefore exploits a joined use of techniques for proving impossibility results for 3-round black-box zero knowledge as well as impossibility results for black-box concurrent zero knowledge in the plain model.

High-Level Overview. For the sake of simplifying the presentation, we now only consider conversation-based protocols, i.e., protocols where at the end of each proof the verifiers decides to accept or to not accept the proof without using private data. Therefore, in a conversation-based protocol, by simply looking at the transcript of the protocol it is possible to efficiently decide whether the verifier accepts or not.

We will show that for any language L with a 3-round sequentially sound black-box concurrent zero knowledge argument system $\Pi = \langle P, V \rangle$ in the BPK model , we can use the simulator M for concurrent zero knowledge as a black box to design an efficient deciding machine D for L.

More precisely, instead of only working with the honest verifier (in contrast to the proofs of [11,15]) we will first design a concurrent adversarial verifier $V^\star$ (as well as a useful variant $\tilde{V}^\star$) which D will let interact with M on input "$x \in L$" in order to decide $x \in L$. Specifically D runs M against $V^\star$ and decides the language based on whether M outputs an accepting transcript. To show the correctness of D we design a $V^\star$ which opens nested sessions (which we refer to as levels), each corresponding to a different public key. To be precise the behavior of $V^\star$ is the following:

1. Upon receiving the first message of a session at level i, $V^\star$ initiates a new session at level $i + 1$ by using the $(i + 1)$-th entry of the public file, until $i = \text{POLY}(k)$ (for some fixed polynomial $\text{POLY}()$) and only continues with the session at level i once level $i + 1$ has been successfully completed.
2. Before sending the second message in the 3-round protocol $V^\star$ uses a family of pseudorandom functions indexed by its random tape and evaluated upon $V^\star$'s entire view of the current interaction with M (including all levels) in order to generate a new value to be used as a source of randomness for computing its response message for a given session.

In order to keep things as simple as possible during the proof, we will also describe a variant of $V^\star$ called $\tilde{V}^\star$ which acts just as $V^\star$ except at level j (where the value of j is specified as input at $\tilde{V}^\star$'s startup). For this special level $\tilde{V}^\star$ outputs all messages received from the simulator (i.e., its complete view of the interaction at all levels with M thus far) to its output tape and responds with messages read on its input tape. In other words $\tilde{V}^\star$ acts as a proxy for an external algorithm at level j. We will use $\tilde{V}^\star$ to contradict the soundness of Π in case of failure of D with a polynomially related probability.

Next we will define two mutually exclusive categories of executions of a simulator such that any possible execution of a concurrent simulator M must fall into precisely one of the two categories. In the proof we will show that although D only works for one of the two categories, the other can be simply ignored as such executions would need exponential time (in the security parameter k) and therefore can happen only with negligible probability (since the expected running time of the simulator is polynomial in k).

D decides whether x is in the language as follows: if M outputs an accepting transcript, D will accept x otherwise D will reject x. Notice that if $x \in L$, then by the fact that M is a simulator for concurrent zero knowledge, it follows that D will accept x with overwhelming probability. The case $x \notin L$ is more complex. We will design a cheating sequential prover $P^\star$ which runs $\tilde{V}^\star$ against M, using $\tilde{V}^\star$ as a proxy for sessions with an honest verifier V. Then we show that for any $x \notin L$, the probability that $P^\star$ succeeds in cheating an honest verifier V is polynomially related to the probability of M outputting an accepting transcript for x. Here "convinces" refers to the prover convincing the verifier of the theorem $x \in L$ in some session and $\approx_{poly}$ stands for polynomially related. Thus by soundness, M will only successfully prove a false statement with a negligible probability. Therefore we can conclude that D is a deciding machine for L contradicting the assumption that $L \notin \mathcal{BPP}$.

Now we shall begin the detailed discussion by describing the aforementioned adversary $V^\star$.

3.2 The Adversarial Verifier $V^\star$

In general $V^\star$ acts exactly as the honest verifier V would except for a few special deviations.

Initialization Phase. Let POLY() be some fixed polynomial. $V^\star$ (honestly and independently) generates $p = \text{POLY}(k)$ public-private key pairs $\{(\text{pk}_i, \text{sk}_i)\}_i$ for $i \in \{1, 2, .., p\}$ placing all public keys $\text{pk} = \{\text{pk}_i\}_{i \le p}$ in the public file F. That is $V^\star$ simply runs V's initialization algorithm p times, each time with a new uniformly and independently chosen random string, publishing all (public) output.

Interactive Phase. $V^\star$ maintains an internal counter i which is initialized to 1. The counter is used to keep track of the current level, i.e. the index of the public key in F which is to be used for that level.

We denote with a triple (a_i, b_i, c_i) a 3-*round protocol played at level i*, where a_i denotes the first message of the prover to the verifier, b_i the second message of the verifier to the prover, and c_i denotes the last message of the prover to the verifier.

Upon receiving a_i for $i < p$, $V^\star$ initiates a new session in a concurrent fashion, at level $i + 1$ requesting a proof of the same statement "$x \in L$". Only once this new level has been successfully completed, does it continue execution at level i. (If $i = p$ then $V^\star$ continues execution of the current session without initiating any new levels.) See Fig. 2.

At level i, once $V^\star$ has received a_i but before initiating the next level (and thus before choosing and sending b_i), $V^\star$ sets the string $r_i' = f_{r_i}(view_{V^\star})$ where:

1. r_i is $V^\star$'s random tape for this session.
2. $view_{V^\star}$ is $V^\star$'s view of the entire interaction at all levels so far up to level i.
3. $f_{r_i} \in \{PRF_r\}_{r \in R}$ is the pseudorandom function with the index r_i where R is the set of all possible random tapes with which $V^\star$ can be initialized.

Now for the rest of this session, $V^\star$ uses r_i' as its random tape. Once the new random string has been defined, $V^\star$ continues as the honest verifier does until the end of the session.

The $\tilde{V}^\star$ Variation. For reasons of simplicity and overview we will now describe a related verifier algorithm $\tilde{V}^\star$. The only difference between $V^\star$ and $\tilde{V}^\star$ is that

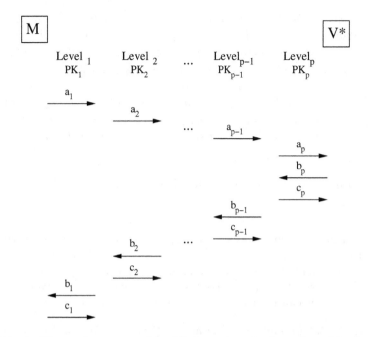

Fig. 2. The randomness used by V^* for each reply depends on his view

the latter, on initialization, reads an integer j along with a key $\mathbf{pk}$, on its input tape. $\tilde{V}^*$ generates the public file just as V^* does, except for substituting $(V, \mathbf{pk})$ for the identity j. Further whenever $\tilde{V}^*$ must send b_j, it writes its entire view $view_{\tilde{V}_j^*}$ to its output tape[1], and pauses execution until a value for b_j is written to its input tape which it forwards to the prover as its message. Finally, upon receiving c_j, it writes c_j and the same view $view_{\tilde{V}_j^*}$ it outputted when getting b_j at the beginning of this session. ($\tilde{V}^*$ will later be run as a subroutine by another algorithm which will tell it which messages to use for level j but let it act just as V^* would at all other levels. $view_{\tilde{V}_j^*}$ can be seen as an identifier for a given session.)

3.3   The Executions of the Simulator

Next we consider two (mutually exclusive) categories of executions of a simulator.

Definition 6. *Let M be the simulator that is guaranteed to exists for a black-box concurrent zero knowledge protocol $\Pi = \langle P, V, \rangle$ and let V^* be a probabilistic polynomial-time interactive algorithm. Then when M interacts with V^*, if there exists a level j, $j \leq p$ with the following properties:*

---

[1] This view (which includes the random tape assigned by M) is the exact same view which V^* uses as the argument to its pseudorandom function when creating r_j'.

1. Let $\{Views_{V_j^*}\}$ be the set of all of views seen so far by V^* during the entire interaction when at level j. Then $view_{V_j^*}$, the view of V^* just before playing b_j (at level j), is a new one, i.e. $view_{V_j^*} \notin \{Views_{V_j^*}\}$. Notice that the view at level j includes all messages a_i at levels i for $i \leq j$ (see Fig. 2).

2. From when b_j has been sent until when c_j has been sent, V^* never has a new view $view'_{V^*} \notin \{Views_{V_j^*}\}$. In other words after receiving b_j but before sending c_j for a given session at level j, M does not request a rewind such that V^* reaches a point where it sends a b'_j such that it has a view $view'_{V_j^*}$ which it has never had before.

then we call this execution of M a j-**deciding execution for** V^*.

Comments:

- If we consider a simulator M running against the adversarial verifier V^* specified above then the views considered in the definition are exactly those used as arguments to the pseudorandom function when setting r'_j.
- We use the term "j-deciding" because the simulator's j-th session will help to prove that it can be used to decide that language efficiently.

Definition 7. Let M and V^* be as above. If there exists no j as above (that is, in order to complete every session resulting in an accepting view of V^*, M requires at least one sequence of events (including a rewind) resulting in a new $View'_{V^*}$ at level j) then we call this execution of M a **hard execution for** V^*.

Comments and Examples. First we note that it is clear from the definition that given an execution of M, it must be either a j-deciding or hard for V^* but never both since either at least one such j exists or not. In order to clarify the two definitions we now give a simple example of a V^* with only one level of interaction with M.

Suppose $A = (a, b, c)$ and $B = (a', b', c')$ are the accepting transcripts of two different sessions (of the same protocol) and let R stand for "M rewinds V^*". Now suppose M and V^* interact as follows:

$$(a, b, R, a', b', R, a, b, c) = (m_1, m_2, m_3, m_4, m_5, m_6, m_7, m_8, m_9)$$

In this case the triple (a, b, c) does not fulfill the requirement to make this execution of M a 1-deciding execution for V^* because although (m_1, m_2, m_9) is an accepting transcript, between m_2 and m_9 being received by V^*, M rewinds V^* and causes it to have a new view (with messages m_3, m_4 and m_5) which violates the second point of the definition. Further although (m_7, m_8, m_9) is a transcript started and successfully completed without any rewind at all, this sequence violates the first point of the definition of a j-deciding execution since before playing m_8, V^* now has an old view: specifically that which it had before playing m_2. Therefore this is an hard execution for V^*. If, on the other hand, the interaction began only with m_4, or if the last message were a message $m_{6'} = c'$ instead of m_6 then M would be a 1-deciding simulator for V^*.

The intuition behind the crucial part of the proof is to show that if an execution of M when interacting with a specially designed adversarial verifier is a hard execution then M must perform an exponential number of rewinds in order to finish which implies that all hard executions with this verifier can only happen with negligible probability. If however they are j-deciding then in case the execution resulted in a false proof (i.e. if $x \notin L$) then the execution can be used by a specially designed malicious prover to break soundness. This is used to establish the correctness of a deciding machine which uses the simulator and an adversarial verifier as subroutines.

We now have all tools we need and can begin with the main proof.

3.4   The Proof of Theorem 1

Proof. Assume, by contradiction, there exists a language $L \notin \mathcal{BPP}$ with a 3-round black-box concurrent zero-knowledge argument system $\langle P, V \rangle$ in the BPK model enjoying sequential soundness, and let $V^\star$ and $\tilde{V}^\star$ be the adversarial verifiers as defined above. Then by the concurrent zero-knowledge property of $\langle P, V \rangle$, there exists an expected polynomial-time simulator M which, given oracle access to any concurrent verifier $V^\star$ will, for any true statement $x \in L$, output a transcript indistinguishable from that of $V^\star$ interacting with the honest prover P.

In order to reach the contradiction we construct an efficient deciding algorithm D which, on input x, decides membership in the language L which would imply $L \in \mathcal{BPP}$. To decide whether x is in L, D runs M while simulating $V^\star$ to it. If M outputs an accepting transcript, D will output $x \in L$, otherwise D will output $x \notin L$. Thus we will need to prove that with overwhelming probability M will output an accepting transcript if and only if $x \in L$.

Proposition 1. *If $x \in L$ then D will accept the proof with overwhelming probability.*

Proof. A session with a verifier using a pseudorandom tape looks the same as one with a verifier using a random tape, otherwise it is possible to break the randomness property of PRF. Since $V^\star$, after setting its random tape, uses the same algorithm as V for the rest of the interaction, $V^\star$ would, by completeness of $\langle P, V \rangle$, when interacting with P, accept for any given session at any level. Since there are only polynomially many sessions, but each has an overwhelming probability of resulting in an accepting view of $V^\star$, with overwhelming probability they will all result in an accepting view of $V^\star$. Thus with overwhelming probability $V^\star$ will accept when interacting with P. Therefore by the concurrent zero-knowledge property of $\langle P, V \rangle$, given a true statement as common input M will, with overwhelming probability, outputs an accepting view.

Proposition 2. *Given $x \notin L$ then D will reject x with overwhelming probability.*

Proof. To show this we will design a malicious prover $P^\star$ which can use M to prove any statement with a polynomially related probability to the probability of M outputting an accepting transcript for the same statement. We can then conclude that M outputs an accepting transcript for $x \notin L$ with negligible probability since otherwise the soundness of $\langle P, V \rangle$ is violated. However $P^\star$ will only work for j-deciding executions, so we will also show that an execution of M can be hard for $V^\star$ only with negligible probability.

We define $P^\star$ to be a polynomial-time algorithm with black-box access to M. Given x as input, $P^\star$ interacts with the honest verifier V with the goal of convincing V of the false statement "$x \in L$". It does this by running $\tilde{V}^\star$ with a random guess j' for j and the public key produced by V on its input, against M with the statement "$x \in L$". $P^\star$ then acts as a man-in-the-middle between $\tilde{V}^\star$ and V for all interaction at level j, and maintains a set of pairs: the views outputted by $\tilde{V}^\star$ and the corresponding response supplied by V. If $\tilde{V}^\star$ outputs a previously seen view then $P^\star$ does not forward the query to V, instead it simply answers by using the response previously stored in his memory. Further whenever $P^\star$ initiates a session with V (in order to get a response b_j for a session), $P^\star$ stores in a registry the view given in the output by $\tilde{V}^\star$. Thus when M outputs the message c_j of a session along with a view, $P^\star$ checks the value of the registry to see whether c_j corresponds to the currently open session with V. If this is the case then $P^\star$ forwards c_j to V thereby completing its current interaction with V. In other words the registry is used to store what is in essence, a unique ID for a session, namely the view which $V^\star$ would use as an argument when setting its pseudorandom tape r'. For a detailed description of $P^\star$ see Fig. 3.

Lemma 1. *When interacting with $P^\star$, M will act as when interacting with $V^\star$ with overwhelming probability.*

Proof. We must show that M can not tell the difference between interacting with $V^\star$ and with $P^\star$. Since $P^\star$ has oracle access to $\tilde{V}^\star$ it has complete control over its input including its random tape and can, in particular, perform all necessary rewinds and secret key operations. Thus it suffices to consider level j'. Because V and $V^\star$ follow the same algorithm in deciding the message b, apart from what they use as a random tape, the only concerns are rewinds and the fact that V's tapes are truly random and not chosen depending on any variables such as its view. Specifically:

1. M noticing that V is not being rewound when M requests a rewind to an old view.
2. M noticing that V is not being rewound when M requests a rewind to a new view (but in fact V simply uses a new random tape).
3. M noticing that V is not basing its randomness on the entire view at all levels.

The first concern is easily taken care of by pointing out in such a case $P^\star$ responds from memory, i.e. exactly as it did last time, as is expected, thus maintaining M's view. For the second concern, if M were able to tell the difference between $P^\star$ and $V^\star$ in such a situation, this would imply that it can tell the difference between a new random tape being used and a new pseudorandom tape. However such an M could then be used as a black box to create a distinguishing

Common input: security parameter k, public key pk and "$x \in L$".

P^* **Setup:**

1. P^* chooses a random $j' \leftarrow [1, p = \texttt{POLY}(k)]$.
2. P^* initializes $\tilde{V}^*$ with j' and pk on its input tape.
3. P^* initializes M with the public file generated by $\tilde{V}^*$. Further P^* connects the communication tapes of M and $\tilde{V}^*$ so that they interact with each other.
4. During the entire interaction P^* will maintain the set $S = \{View_{\tilde{V}^*}, b_{j'}\}$ of views given in output by $\tilde{V}^*$ at level j' and the corresponding responses written back by V. This set is initialized as empty.
5. During the entire interaction P^* will maintain the registry $View$ which is initialized as empty. (This will be used to keep track of the defining properties of the current open session with V.)
6. P^* initiates $\tilde{V}^*$ with the statement "$x \in L$" all the while proxying between V and $\tilde{V}^*$.

P^* **Interaction:**

- If $\tilde{V}^*$ outputs a view $View_{\tilde{V}^*}$ (we denote by $a_{j'}$ the most recent first round for level j' of $View_{\tilde{V}^*}$) then P^* checks if the view is already in S:
    - TRUE: Respond with the corresponding $b_{j'}$ from S.
    - FALSE: Close any open session with V and begin a new one. Set the registry $View = View_{\tilde{V}^*}$. Send $a_{j'}$ to V receiving $b_{j'}$ in response. Append $(View_{\tilde{V}^*}, b_{j'})$ to S. And finally write $b_{j'}$ to $\tilde{V}^*$'s input tape.
- If $\tilde{V}^*$ outputs $(c_{j'}, View_{\tilde{V}^*})$, then if $(a_{j'}, b_{j'}, c_{j'})$ is not an accepting transcript then P^* sends back an abort to $\tilde{V}^*$; otherwise P^* checks if $View = View_{\tilde{V}^*}$:
    - TRUE: Forward $c_{j'}$ to V thus completing the session with V.
    - FALSE: Drop $c_{j'}$.

Fig. 3. Using M as a black box to convince V of a false statement

algorithm for the *PRF* family which contradicts the families randomness property. This leaves the third concern which is dealt with by pointing out that a new session with V (where V will thus have a new random tape) resulting in a new response b is only requested by P^* upon receiving a new view. In other words the random tape used in generating the response b which M receives for a session, is created new only if the view at the beginning of a session is new. If the view is old then the old response (stored in memory by P^*) is used, which implies (from M's point of view) that the same random tape was used to generate b. Thus the random tape used to produce the message received by M depends only and completely upon the view before playing b.

As we will show below $P^\star$ will succeed in its goal if and only if the execution of M is j-deciding for $V^\star$ and if it guesses the value of j correctly.

Lemma 2. *An execution of M is j-deciding for $V^\star$ with overwhelming probability.*

Proof. Suppose, by contradiction, that an execution of M is hard for $V^\star$. In this case we will show through analysis of the interaction between M and $V^\star$ that this execution of M needs exponential time. Specifically we show, by induction on the level i, that M must perform exponentially (in k) many rewinds.

By definition of a hard execution for $V^\star$, M must perform at least one rewind for each session it solves when interacting with $V^\star$. In particular this means that for any new session at level $i = p$, M must rewind. The same holds true at level $i-1$. Now since this is a hard simulator, for every level at least one of the rewinds leads to a new view upon receiving a_{i-1}. This implies that when $V^\star$ initiates a session at level i it will also have a new view, compared to all previous sessions at level i, and thus will use a new random string r'_i. In other words this will be a new session and will therefore require M to rewind again since this is a hard execution for $V^\star$.

Again since this is a hard execution for $V^\star$, there must be a rewind after b_{i-1} has been received which leads to a new view at level $i - 1$, which means at least 2 new sessions at level i will need to be solved. Therefore at least 2 rewinds are required to solve all sessions at level i. The first rewind being used to solve the first session which must be successfully completed before b_{i-1} is sent, and the second rewind being used to solve the new session (with the new view) created after the rewind at level $i - 1$.

Thus by induction M must perform at least: $\sum_{i=1}^{p} 2^{i-1} = 2^p - 1$ rewinds. Since the expected polynomial-time of M is polynomial in k and $p = \texttt{POLY}(k)$, a hard execution for $V^\star$ can only happen with negligible probability.

Result. Since by Lemma 2, the execution of M is j-deciding for $V^\star$ and by lemma 1, M essentially acts the same when interacting with $P^\star$ in the setup described in $P^\star$'s definition, we can conclude that an execution of M is j-deciding for $P^\star$ as well with overwhelming probability.

Now $P^\star$ has a (non-negligible) probability $\frac{1}{p}$ of correctly guessing $j' = j$ and in such a case, by the definition of a j-deciding execution there will be at least one session where M causes no new view between beginning and end of (a_j, b_j, c_j) when it is solved for the first time. (Specifically there will be at least one session where $\tilde{V}^\star$ outputs $(c_{j'}, view_{\tilde{V}^\star})$ such that the registry $View = view_{\tilde{V}^\star}$.) This means that $P^\star$ will successfully complete at least one session with V. Therefore by soundness of $\langle P, V \rangle$ we have shown that on input a false statement, M will with overwhelming probability output a reject transcript.

To conclude that we have now a deciding algorithm D we need to first deal with the small matter of M running in *expected* polynomial-time rather then strict polynomial-time. This is dealt with in the same way as in [15] (remark 6.1), namely D simply terminates the execution of M after a fixed polynomial number of steps, chosen in such a way that M has a good chance of simulating the conversation.

Thus with proposition 1 and proposition 2 we have shown the efficient algorithm D will, with overwhelming probability, decide correctly contradicting $L \notin \mathcal{BPP}$.

Extension. As we have pointed out, the proof presented in this paper only applies to conversation-based argument systems. We remark that all argument systems known to us are indeed conversation-based. We remark though that our proof can be extended to cover all argument systems by considering an adversarial verifier $V^\star$ that imposes a more sophisticated schedule and rejects with some non-negligible probability the interactions.

4 3-Round RZK Proofs in the UPK Model

In this section we present the first 3-round rZK *proof* (in contrast to argument) system with concurrent soundness for all $\mathcal{NP}$ in the UPK model. Moreover, our construction does not need hardness assumptions with respect to superpolynomial-time algorithms (i.e., complexity leveraging). Similarly to the constant-round zero-knowledge proofs in the plain model of [18], we use at the beginning (i.e., in our case the construction of the public file) an unconditionally hiding commitment scheme. Obviously this can be implemented in one round (therefore in the UPK model) under the assumption that non-interactive unconditionally-hiding commitments exist. This is a strong assumption since so far no construction of non-interactive unconditionally-hiding commitments has been given in the plain model. Therefore, according to the current state-of-the-art, some variations of the model have to be considered. For instance a two-round interactive preprocessing allows to commit with unconditional hiding. This can be alternatively implemented by requiring first a non-interactive set-up stage performed by the provers, then a non-interactive set-up stage performed by the verifiers on input the output of the stage of the provers.

Notice that the hash-based commitment scheme used in the preprocessing stage in [14] actually is a non-interactive unconditionally hiding commitment scheme. However such a scheme can only achieve security with respect to uniform adversarial verifiers [19]. In our construction by using hash-based commitments we obtain the same security achieved in [14].

We stress that the 3-round rZK argument system with concurrent soundness presented in [14] is not a proof system since the verifier sends to the prover a non-interactive zero-knowledge proof of knowledge that is only computational zero knowledge. Moreover, if the proof system of [18] is implemented in 3 rounds in the UPK model, the adversarial verifier can play the second round by choosing one of the polynomially bounded number of public keys and therefore the simulator can not complete the proof with the same probability of the honest prover.

High-Level Overview. The main idea of the previous constructions of resettable zero-knowledge argument systems in the public-key models, is that the simulator obtains (by means of rewinds) the secret key (or at least the output of a computation that needs the secret key) of the verifier and then can perform a straight-line simulation. However, when the adversarial prover is computationally unbounded, on input the public key of the verifier, the prover can compute the secret key and then he could succeed in proving a false statement. We therefore use a different technique. We assume that the verifier generates a public key that corresponds to a super-polynomial number of secret keys each with the same probability. Moreover, we assume that no polynomial-time adversarial verifier can compute more than a bounded polynomial-number of secret keys. A good candidate for this component is a non-interactive unconditionally hiding commitment scheme. Indeed, in this case a commitment corresponds to any possible message of the corresponding message space. This guarantees that even an unbounded prover, by simply looking at the public key of the verifier, does not get any information about the secret key of the verifier. Moreover the binding property guarantees that the polynomial-time adversarial verifier does not find a pair of messages that correspond to the same commitment. This is important for the zero-knowledge property since we will use the fact that the simulator and the verifier will know the *same* secret key. Indeed, the mere possession of one valid secret key does not work since the unbounded prover can compute any secret and therefore could use it for proving false statements. It is therefore necessary to formalize that only knowledge of the specific secret key known by the verifier allows one to simulate the proof without knowledge of the witness for the statement on input.

We formalize this idea by requiring that the prover commits to the witness by means of an unconditionally binding commitment scheme, the verifier sends his secret key and then the prover uses a one-round resettable witness-indistinguishable proof system (e.g., using a ZAP [20], by assuming that the verifier puts in his public key also the first message of the ZAP) for proving that the committed witness, is either a witness for "$x \in L$" or is precisely the secret key sent by the verifier so far (here we use the FLS-paradigm [21]). Intuitively, the protocol is sound with respect to an unbounded prover since when the prover commits to the witness, he has only seen an unconditionally hiding commitment. Therefore he has no advantage for guessing the specific secret key known to the verifier in the set of all possible secret keys. However, notice that the verifier can not use more than once a public key. The protocol is zero-knowledge because the simulator extracts the only secret key that the adversarial verifier knows. Moreover, the fact that the simulator uses a different witness with respect to the one used by the prover in the last proof is not detected by a polynomial-time resetting adversarial verifier since the use of a different witness is plugged in a resettable witness indistinguishable proof system.

The formal protocol considers a bounded polynomial number of public keys since the protocol works in the upperbounded (UPK) public-key model. Since in each session the verifier uses a different secret key, we obtain also concurrent soundness with only 3 rounds.

References

1. Goldwasser, S., Micali, S., Rackoff, C.: The Knowledge Complexity of Interactive Proof Systems. SIAM J. on Computing **18** (1989) 186–208
2. Dwork, C., Naor, M., Sahai, A.: Concurrent Zero-Knowledge. Proc. of STOC '98, ACM (1998) 409–418
3. Canetti, R., Goldreich, O., Goldwasser, S., Micali, S.: Resettable Zero-Knowledge. Proc. of STOC '00, ACM (2000) 235–244
4. Canetti, R., Kilian, J., Petrank, E., Rosen, A.:   Black-Box Concurrent Zero-Knowledge Requires $\omega(\log n)$ Rounds. Proc. of STOC '01, ACM (2001) 570–579
5. Barak, B.: How to Go Beyond the Black-Box Simulation Barrier. Proc. of FOCS '01, (2001) 106–115
6. Persiano, G., Visconti, I.: Single-Prover Concurrent Zero Knowledge in Almost Constant Rounds. Proc. of ICALP '05. LNCS, Springer Verlag (2005)
7. Dwork, C., Sahai, A.: Concurrent Zero-Knowledge: Reducing the Need for Timing Constraints. Proc. of Crypto '98. Vol. 1462 of LNCS. (1998) 442–457
8. Goldreich, O.: Concurrent Zero-Knowledge with Timing, Revisited. Proc. of STOC '02, ACM (2002) 332–340
9. Damgard, I.: Efficient Concurrent Zero-Knowledge in the Auxiliary String Model. Proc. of Eurocrypt '00. Vol. 1807 of LNCS (2000) 418–430
10. Blum, M., De Santis, A., Micali, S., Persiano, G.: Non-Interactive Zero-Knowledge. SIAM J. on Computing **20** (1991) 1084–1118
11. Micali, S., Reyzin, L.: Soundness in the Public-Key Model. Proc. of Crypto '01. Vol. 2139 of LNCS (2001) 542–565
12. Reyzin, L.: Zero-Knowledge with Public Keys. PhD thesis, Massachusetts Institute of Technology (2001)
13. Di Crescenzo, G., Persiano, G., Visconti, I.:   Constant-Round Resettable Zero Knowledge with Concurrent Soundness in the Bare Public-Key Model. Proc. of Crypto '04. Vol. 3152 of LNCS (2004) 237–253
14. Micali, S., Reyzin, L.: Min-Round Resettable Zero-Knowledge in the Public-key Model. Proc. of Eurocrypt '01. Vol. 2045 of LNCS (2001) 373–393
15. Goldreich, O., Krawczyk, H.: On the composition of zero-knowledge proof systems. SIAM J. on Computing **25** (1996) 169–192
16. Kilian, J., Petrank, E., Rackoff, C.: Lower Bounds for Zero Knowledge on the Internet. Proc. of FOCS '98. (1998) 484–492
17. Rosen, A.: A Note on the Round-Complexity of Concurrent Zero-Knowledge. Proc. of Crypto '00. Vol. 1880 of LNCS (2000) 451–468
18. Goldreich, O., Kahan, A.: How to Construct Constant-Round Zero-Knowledge Proof Systems for NP. Journal of Cryptology **9** (1996) 167–190
19. Reyzin, L.: Personal communication (2005)
20. Dwork, C., Naor, M.: Zaps and their Applications. Proc. of FOCS '00. (2000) 283–293
21. Feige, U., Lapidot, D., Shamir, A.: Multiple Non-Interactive Zero Knowledge Proofs Under General Assumptions. SIAM J. on Computing **29** (1999) 1–28

Communication-Efficient Non-interactive Proofs of Knowledge with Online Extractors

Marc Fischlin*

Institute for Theoretical Computer Science,
ETH Zürich, Switzerland
marc.fischlin@inf.ethz.ch
http://www.fischlin.de/

Abstract. We show how to turn three-move proofs of knowledge into non-interactive ones in the random oracle model. Unlike the classical Fiat-Shamir transformation our solution supports an online extractor which outputs the witness from such a non-interactive proof instantaneously, without having to rewind or fork. Additionally, the communication complexity of our solution is significantly lower than for previous proofs with online extractors. We furthermore give a superlogarithmic lower bound on the number of hash function evaluations for such online extractable proofs, matching the number in our construction, and we also show how to enhance security of the group signature scheme suggested recently by Boneh, Boyen and Shacham with our construction.

1 Introduction

The Fiat-Shamir transformation [17] is a well-known technique to remove interaction from proofs of knowledge and to derive signature schemes from such proofs. The starting point is a three-move proof between a prover, holding a witness w to a public value x, and a verifier. The prover sends a commitment com, then receives a random challenge ch from the verifier and finally replies with resp. For the non-interactive version the prover computes the challenge himself by applying a hash function H to the commitment. The security of this transformation has later been analyzed under the idealized assumption that the hash function behaves as a random oracle [8,27], and has led to security proofs for related signature schemes.

Limitations. In the interactive case, all common knowledge extractors work by repeatedly rewinding the prover to the step after having sent com and completing the executions with independent random challenges. This eventually yields two valid executions (com, ch, resp), (com, ch', resp') for different challenges ch ≠ ch' from which the extractor can compute the witness w. The same technique is reflected in the security proofs of the non-interactive version: The extractor

---

* This work was supported by the Emmy Noether Programme Fi 940/1-1 of the German Research Foundation (DFG).

continuously rewinds to the point where the prover has asked the random oracle H about com and completes the executions with independent hash values to find two valid executions (called "forking" in [27]).

The notable fact above is that, although the proof is non-interactive, the extractor still works by rewinding. As pointed out by [29] for example, this causes problems for some cryptographic schemes. Consider for example the ElGamal encryption $(R, C) = (g^r, pk^r \cdot m)$ for messages m. One attempt to make this scheme secure against chosen-ciphertext attacks is to append a non-interactive proof of knowledge for $r = \log R$ to the ciphertext. The idea is that, giving such a proof, any party generating a ciphertext would already "know" r and therefore $m = C \cdot pk^{-r}$. In other words, decryption queries in a chosen-ciphertext attack should be simulatable with the help of the knowledge extractor. However, this intuition cannot be turned into a proof, at least not with the rewinding extractor. Consider for example an adversary which sequentially puts n hash queries for the proofs of knowledge and then asks a decryption oracle for ciphertexts involving these hash queries *in reverse order*. Then, to answer each decryption query the extractor would have to rewind to the corresponding hash query. By this, it destroys all previously simulated decryption queries and must redo them from scratch, and the overall running time would become exponential in n.

We remark that the rewinding strategy also causes a loose security reduction. The results in [27] show that, if the adversary makes Q queries to the random oracle and forges, say, Schnorr signatures in time T with probability ϵ, then we can compute discrete logarithms in expected time QT/ϵ with constant probability. Hence, the number of hash queries enters multiplicatively in the time/success ratio. In contrast, for RSA-PSS and similar schemes [11,12,21] tight reductions are known. For other schemes like discrete-logarithm signatures different approaches relying on potentially stronger assumptions have been taken to get tight security results [18].

Constructing Online Extractors. The solution to the problems above is to use extractors which output the witness immediately, i.e., without having to rewind. Following the terminology of [29], where this problem was discussed but circumvented differently, we call them *online extractors*.[1] Informally, such an extractor is given the value x, a valid proof π produced by a prover and all hash queries and answers the prover made for generating this proof (i.e., even queries which are later ignored in the proof). The extractor then computes the witness w without further communication with the prover. Note that here we use the fact that we work in the random oracle model, where the extractor sees the hash queries.

One known possibility to build such online extractors is to use cut-and-choose techniques combined with hash trees [26,22]. That is, one limits the challenge space to logarithmically many bits and repeats the following atomic protocol sufficiently often in parallel. The prover computes the initial commitment com of the interactive protocols and computes the answers for all possible challenges. Since there are only polynomially many challenges and answers, the prover can

---

[1] Sometimes such extractors are also called straight-line extractors in the literature.

build a binary hash tree with all answers at the leaves. Then he computes the actual challenge as the hash value over com and the root of the tree, and opens only the corresponding answer and all siblings on the path up to the root as a proof of correctness. For reasonable parameters these revealed hash values easily add about $10,000$ to $25,000$ bits to the non-interactive proof for all executions together, and thus cause a significant communication overhead.

Here we propose a different approach to build online extractors, producing much shorter proofs than the tree-based solution while having comparable extraction error and running time characteristics. In this introduction we provide a simplified description of our solution, omitting some necessary modifications. We also start with a polynomially bounded challenge space and a non-constant number of parallel executions. For each execution i the prover first computes com_i but now tries all polynomially many challenges $\mathsf{ch}_i = 0, 1, 2, \ldots$ and answers $\mathsf{resp}_i = \mathsf{resp}_i(\mathsf{ch}_i)$ till it finds one for which a predetermined (at most logarithmic) number of least significant bits of $H(x, \vec{\mathsf{com}}, i, \mathsf{ch}_i, \mathsf{resp}_i)$ are all zero. The prover outputs the vector $(\vec{\mathsf{com}}, \vec{\mathsf{ch}}, \vec{\mathsf{resp}})$, no further hash values need to be included, and the verifier now also checks in all executions that the lower bits of the hash values are zero.

The honest prover is able to find a convincing proof after a polynomial number of trials for each execution (except with negligible probability which can be adapted through parameters). It is also clear that any prover who probes at most one valid challenge-response pair for each execution most likely does not find a hash value with zero-bits.[2] If, on the other hand, the prover tries at least two samples, then the knowledge extractor can find them in the list of hash queries and compute the witness. It follows that the (online) extraction probability is negligibly close to the verifier's acceptance probability.

Our construction, outlined above, still requires a non-constant number of parallel repetitions in order to decrease the soundness error from polynomial to negligible. However, for proofs which are already based on small challenges, such as RSA with small exponents or "more quantum-resistant" alternatives like the recently proposed lattice-based proofs with bit challenges [24], several repetitions have to be carried out anyway, and our construction only yields an insignificant overhead in such cases. For other scenarios, like proofs of knowledge for discrete logarithms, the repetitions may still be acceptable, e.g., if the proof is only executed occasionally as for key registration. Alternatively, for the discrete logarithm for example, the prover can precompute the commitments $\mathsf{com}_i = g^{r_i}$ offline and the verifier is able to use batch verification techniques [6] to reduce the computational cost.

A Lower Bound. Both the hash-tree construction and our solution here require a non-constant number of repetitions of the atomic protocol. An interesting question is if one can reduce this number. As a step towards disproving this

---

[2] We presume that it is infeasible to find distinct responses to a single challenge. Indeed, this requirement is not necessary for the Fiat-Shamir transformation, yet all proofs we know of have this additional property.

we show that the number of hash function evaluations for the prover must be superlogarithmic in order to have an online extractor (unless finding witnesses is trivial, of course).[3] While this superlogarithmic bound would be straightforward if we restrict the hash function's output size to a few bits, our result holds independently of the length of hash values.

The proof of our lower bound requires that the knowledge extractor does not have the ability to choose the hash values. If we would allow the extractor to program the random oracle then we could apply the hash function to generate a common random string and run a non-interactive zero-knowledge proof of knowledge in the standard model (based on additional assumptions, though) [16]. For unrestricted (but polynomial) output length a single hash function evaluation for both the prover and verifier would then suffice. For non-programming extractors the number of hash function evaluations in our construction and the hash-tree solution are optimal with respect to general protocols.

A Word About Random Oracles. Our solution is given in the random oracle model, and a sequence of works [10,20,23,4] has shown that constructions in this model may not yield a secure scheme in the real world when the oracle is instantiated by some function. It is therefore worthwhile to take a look at the way we utilize the random oracle. In our transformation we essentially use the random oracle as a predicate with the following properties: The only way to evaluate this predicate is by computing it explicitly (thus "knowing" the input), that predicate outcomes are well distributed (i.e., random), the predicate values for related inputs are uncorrelated.

In comparison to the Fiat-Shamir transformation our construction somewhat "decouples" the hash function from the protocol flow. Indeed, the dependency of the answer and the hash function in the Fiat-Shamir transformation is exploited by Goldwasser and Tauman [20] to prove insecurity of the random oracle approach for the transformation. Because of the aforementioned separation of the protocol flow and the hash function in our solution, the counterexample in [20] does not seem to carry over (yet, similar results may hold here as well). The point is that our solution is provided as an alternative to the Fiat-Shamir transformation, given one accepts the random oracle model as a viable way to design efficient non-interactive proofs. Finding truly efficient non-interactive proofs of knowledge without random oracles is still open.

Applications. Clearly, proofs of knowledge with online extractors are especially suitable for settings with concurrent executions such as key registration steps. As another, more specific example, we show that our method can be used to enhance security of the Boneh et al. group signature scheme [5]. Roughly, a group signature scheme allows a set of users to sign messages such that a signature does not reveal the actual signer, yet a group manager holding a secret information

---

[3] To be more precise, we give a slightly stronger result relating the number of hash queries of the verifier and the prover. This stronger result shows for example that hard relations cannot have efficient provers if the verifier only makes a constant number of hash function queries to verify proofs.

can revoke this anonymity and identify the true signer. A stringent formalization of these two properties, called full anonymity and full traceability, has been put forth by Bellare et al. [7].

Although achieving strong traceability guarantees the protocol by Boneh et al. only realizes a slightly weaker anonymity notion. In the original definition [7] anonymity of a signer of a message should hold even if an adversary can request the group manager to open identities for other signatures (thus resembling chosen-ciphertext attacks on encryption schemes). In the weaker model such open queries are not allowed, and this property is consequently called CPA-full-anonymity in [5].

Without going into technical details we remark that the weaker anonymity property in [5] originates from the underlying (variation of the) ElGamal encryption scheme and its CPA-security. A promising step towards fully anonymous group signature is therefore to turn the ElGamal encryption into a CCA-secure scheme. As explained before, standard Fiat-Shamir proofs of knowledge for the randomness used to generate ciphertexts do not work because of the rewinding problems. And although there is a very efficient method to secure basic ElGamal against chosen-ciphertexts in the random oracle model [1], this technique inherently destroys the homomorphic properties of the ciphertexts. But this homomorphic property is crucial to the design of the group signature scheme as it allows to efficiently prove relations about the encrypted message.

Proofs of knowledge with online extractors provide a general solution. However, since one of the initial motivations of [5] was to design a scheme with *short* signatures of a couple of hundred bits only, the aforementioned hash-tree based constructions with their significant communication overhead, for example, are prohibitively expensive. We show that with our protocol we obtain a fully-anonymous scheme and for reasonable parameters the length of signatures increases from $1,500$ to about $5,000$ bits. In comparison, the RSA-based group signature scheme in [2], presumably one of the most outstanding group signature schemes, still requires more than $10,000$ bits. Based on implementation results about elliptic curves [15], and the fact that the signer in the scheme by Ataniese et al. [2] cannot apply Chinese-Remainder techniques to compute the exponentiations with $1,000$ and more bits, we estimate that our variation of the Boneh group signature is still more efficient, despite the repetitions for the proof of knowledge. This is especially true for the verifier who can apply batch verification techniques on top.

Organization. In Section 2 we give the basic definitions of three-move Fiat-Shamir proofs of knowledge and non-interactive ones with online extractors in the random oracle model. The main part of the paper, Section 3, presents our construction and the lower bound. Some of the proofs and the construction of secure signature schemes from our solution have been omitted due to lack of space. Section 4 finally presents our enhancement of the Boneh et al. group signature scheme.

2   Definitions

A security parameter k in our setting is an arbitrary string describing general parameters. In the most simple case $k = 1^\kappa$ describes the length κ of the cryptographic primitives in unary. More generally, k can for example consist of the description of a group $\mathcal{G}$ of prime order q and generator g of that group, i.e., $k = (\mathcal{G}, q, g)$. The security parameter also describes a sequence of efficiently verifiable relations $W = (W_k)_k$.

A (possibly negative) function $f(k)$ is called *negligible* if $f(k) \le 1/p(k)$ for any polynomial $p(k)$ and all sufficiently large k's. A function which is not negligible is called *noticeable*. For two functions f, g we denote by $f \gtrsim g$ the fact that $g - f$ is negligible. Accordingly, $f \approx g$ stands for $f \gtrsim g$ and $g \gtrsim f$. A function f with $f \gtrsim 1$ is called *overwhelming*.

We usually work in the random oracle model where parties have access to a random function H with some domain and range depending on k. We note that we do not let the relation W depend on the random oracle H in order to avoid "self-referencing" problems. We occasionally let an algorithm "output a random oracle", $H \leftarrow A$, meaning that A generates a description of a (pseudo)random function H.

We require some additional properties of the underlying Fiat-Shamir proof to make our transformation work. First, we need that the prover's initial commitment com has nontrivial entropy. This can be achieved easily by appending a superlogarithmic number of public random bits to com if necessary. Second, we need that the verifier sends a uniform bit string as the challenge ch; all common proofs have this property. Third, we require that the prover's response is quasi unique, i.e., it should be infeasible to find another valid response resp' to a proof (com, ch, resp), even if one knows the witness. This holds for example if resp is uniquely determined by x, com, ch, e.g., as for the protocols by Guillou-Quisquater [19] and Schnorr [28], but also for Okamoto's witness-indistinguishable variations these protocols [25] (if the parameter k contains the system parameters like the RSA modulus N with unknown factorization).

Definition 1. *A Fiat-Shamir proof of knowledge (with $\ell(k)$-bit challenges) for relation W is a pair (P, V) of probabilistic polynomial-time algorithms $P = (P_0, P_1)$, $V = (V_0, V_1)$ with the following properties.*

[Completeness.] For any parameter k, any $(x, w) \in W_k$, any $(\mathsf{com}, \mathsf{ch}, \mathsf{resp}) \leftarrow (P(x, w), V_0(x))$ it holds $V_1(x, \mathsf{com}, \mathsf{ch}, \mathsf{resp}) = 1$.

[Commitment Entropy.] For parameter k, for any $(x, w) \in W_k$, the min-entropy of $\mathsf{com} \leftarrow P_0(x, w)$ is superlogarithmic in k.

[Public Coin.] For any k, any $(x, w) \in W_k$ any $\mathsf{com} \leftarrow P_0(x, w)$ the challenge $\mathsf{ch} \leftarrow V_0(x, \mathsf{com})$ is uniform on $\{0, 1\}^{\ell(k)}$.

[Unique Responses.] For any probabilistic polynomial-time algorithm A, for parameter k and $(x, \mathsf{com}, \mathsf{ch}, \mathsf{resp}, \mathsf{resp}') \leftarrow A(k)$ we have, as a function of k,

$$\mathrm{Prob}\left[V_1(x, \mathsf{com}, \mathsf{ch}, \mathsf{resp}) = V_1(x, \mathsf{com}, \mathsf{ch}, \mathsf{resp}) = 1 \wedge \mathsf{resp} \ne \mathsf{resp}'\right] \approx 0.$$

[Special Soundness.] There exists a probabilistic polynomial-time algorithm K, the knowledge extractor, such that for any k, any $(x, w) \in W_k$, any pairs $(\mathsf{com}, \mathsf{ch}, \mathsf{resp})$, $(\mathsf{com}, \mathsf{ch}', \mathsf{resp}')$ with $V_1(x, \mathsf{com}, \mathsf{ch}, \mathsf{resp}) = V_1(x, \mathsf{com}, \mathsf{ch}', \mathsf{resp}') = 1$ and $\mathsf{ch} \neq \mathsf{ch}'$, for $w' \leftarrow K(x, \mathsf{com}, \mathsf{ch}, \mathsf{resp}, \mathsf{ch}', \mathsf{resp}')$ it holds $(x, w') \in W_k$.

[Honest-Verifier Zero-Knowledge.] There exists a probabilistic polynomial-time algorithm Z, the zero-knowledge simulator, such that for any pair of probabilistic polynomial-time algorithms $D = (D_0, D_1)$ the following distributions are computationally indistinguishable[4]:

- Let $(x, w, \delta) \leftarrow D_0(k)$, and $(\mathsf{com}, \mathsf{ch}, \mathsf{resp}) \leftarrow (P(x, w), V_0(x))$ if $(x, w) \in W_k$, and $(\mathsf{com}, \mathsf{ch}, \mathsf{resp}) \leftarrow \bot$ otherwise. Output $D_1(\mathsf{com}, \mathsf{ch}, \mathsf{resp}, \delta)$.
- Let $(x, w, \delta) \leftarrow D_0(k)$, and $(\mathsf{com}, \mathsf{ch}, \mathsf{resp}) \leftarrow Z(x, \mathrm{YES})$ if $(x, w) \in W_k$, and $(\mathsf{com}\,\mathsf{ch}, \mathsf{resp}) \leftarrow Z(x, \mathrm{NO})$ otherwise. Output $D_1(\mathsf{com}, \mathsf{ch}, \mathsf{resp}, \delta)$.

Below we sometimes use a stronger kind of zero-knowledge property which basically says that the simulator is able to generate proofs for a specific challenge, as long as this challenge is given in advance. To formalize this let V_0^{CH} be a verifier which on input x, ch merely outputs ch. Then a Fiat-Shamir proof of knowledge (with $\ell(k)$-bit challenges) is *special* zero-knowledge if the following holds:

[Special Zero-Knowledge.] There exists a probabilistic polynomial-time algorithm Z, the special zero-knowledge simulator, such that for any pair of probabilistic polynomial-time algorithms $D = (D_0, D_1)$ the following distributions are computationally indistinguishable:

- Let $(x, w, \mathsf{ch}, \delta) \leftarrow D_0(k)$, and $(\mathsf{com}, \mathsf{ch}, \mathsf{resp}) \leftarrow (P(x, w), V_0^{\mathrm{CH}}(x, \mathsf{ch}))$ if $(x, w) \in W_k$, and $(\mathsf{com}, \mathsf{ch}, \mathsf{resp}) \leftarrow \bot$ else. Output $D_1(\mathsf{com}, \mathsf{ch}, \mathsf{resp}, \delta)$.
- Let $(x, w, \mathsf{ch}, \delta) \leftarrow D_0(k)$, and $(\mathsf{com}, \mathsf{ch}, \mathsf{resp}) \leftarrow Z(x, \mathsf{ch}, \mathrm{YES})$ if $(x, w) \in W_k$, and $(\mathsf{com}, \mathsf{ch}, \mathsf{resp}) \leftarrow Z(x, \mathsf{ch}, \mathrm{NO})$ else. Output $D_1(\mathsf{com}, \mathsf{ch}, \mathsf{resp}, \delta)$.

We note that all common protocols obey this special zero-knowledge property. In fact, it is easy to see that *any* Fiat-Shamir proof of knowledge is special zero-knowledge if the challenge size $\ell(k) = O(\log k)$ is logarithmic (which holds for our transformation in the next section). The idea is to simply run many copies of the (regular) zero-knowledge simulator to find a transcript including a matching challenge.

We next define non-interactive proofs of knowledge with online extractors. We note that, in the random oracle model, it is easy to see that the verifier can be assumed wlog. to be deterministic.

Definition 2. *A pair (P, V) of probabilistic polynomial-time algorithms is called a non-interactive zero-knowledge proof of knowledge for relation W with an online extractor (in the random oracle model) if the following holds.*

[Completeness.] For any oracle H, any $(x, w) \in W_k$ and any $\pi \leftarrow P^H(x, w)$ we have $\mathrm{Prob}\left[V^H(x, \pi) = 1\right] \gtrsim 1$.

---

[4] Meaning that the probability that D_1 outputs 1 is the same in both experiments, up to a negligible difference.

[Zero-Knowledge.] *There exist a pair of probabilistic polynomial-time algorithms* $Z = (Z_0, Z_1)$, *the zero-knowledge simulator, such that for any pair of proba-bilistic polynomial-time algorithms* $D = (D_0, D_1)$ *the following distributions are computationally indistinguishable:*
  – *Let* H *be a random oracle*, $(x, w, \delta) \leftarrow D_0^H(k)$, *and* $\pi \leftarrow P^H(x, w)$ *if* $(x, w) \in W_k$, *and* $\pi \leftarrow \bot$ *otherwise. Output* $D_1^H(\pi, \delta)$.
  – *Let* $(H_0, \sigma) \leftarrow Z_0(k)$, $(x, w, \delta) \leftarrow D_0^{H_0}(k)$, *and* $(H_1, \pi) \leftarrow Z_1(\sigma, x, \text{YES})$ *if* $(x, w) \in W_k$, *and* $(H_1, \pi) \leftarrow Z_1(\sigma, x, \text{NO})$ *otherwise. Output* $D_1^{H_1}(\pi, \delta)$.
[Online Extractor.] *There exist a probabilistic polynomial-time algorithm* K, *the online extractor, such that the following holds for any algorithm* A. *Let* H *be a random oracle*, $(x, \pi) \leftarrow A^H(k)$ *and* $\mathcal{Q}_H(A)$ *be the sequence of queries of* A *to* H *and* H's *answers. Let* $w \leftarrow K(x, \pi, \mathcal{Q}_H(A))$. *Then, as a function of* k,

$$\text{Prob}\left[(x, w) \notin W_k \wedge V^H(x, \pi) = 1\right] \approx 0.$$

Note that we allow the zero-knowledge simulator to program the random oracle, but only in two stages. Namely, Z_0 generates H_0 for D_0 and then Z_1 selects H_1 for the find-stage of D_1. Since the adversary D_0 in the first stage can pass on all interactions with H_0 to D_1 through the state information δ, the simulator Z_1 must guarantee that H_1 is consistent with H_0. However, Z_1 now has the opportunity to adapt oracle H_1 with respect to the adversarial chosen theorem x. Simulator Z_1 also gets the information whether x is in the language or not (in which case the simulator can simply set $\pi \leftarrow \bot$).

3   Constructions

Our starting point are interactive Fiat-Shamir proofs with logarithmic challenge length ℓ. Note that such proofs can be easily constructed from proofs with smaller challenge length l by combining $\lceil \ell/l \rceil$ parallel executions. It is easy to verify that all required properties are preserved by these parallel executions, including unique responses and honest-verifier zero-knowledge. Analogously, we can go the other direction and limit the challenge size to at most ℓ bits while conserving the properties.

3.1   Generic Construction

Recall the idea of our construction explained in the introduction. In each of the r repetitions we let the prover search through challenges and responses to find a tuple $(\text{com}, \text{ch}, \text{resp})$ whose b least significant bits of the hash are 0^b for a small b. From now on we assume for simplicity that H only has b output bits; this can always be achieved by cutting off the leading bits.

Instead of demanding that all r hash values equal 0^b we give the honest prover more flexibility and let the verifier accept also proofs $(\text{com}_i, \text{ch}_i, \text{resp}_i)_{i=1,2,...,r}$ such that the sum of the r hash values $H(x, \overrightarrow{\text{com}}, i, \text{ch}_i, \text{resp}_i)$ (viewed as nat-ural numbers) does not exceed some parameter S. With this we can bound the prover's number of trials in each execution by 2^t for another parameter t, slightly larger than b, and guarantee that the prover terminates in strict polynomial time.

For sake of concreteness the reader may think of $b = 9$ (output length of the hash function), $t = 12$ (challenge size), $r = 10$ (number of repetitions) and $S = 10 = r$ (maximum sum). For these values the probability of the honest prover failing to find a valid proof is about 2^{-60}, and the knowledge extractor will obtain the witness whenever the proof is valid, except with probability approximately $Q \cdot 2^{-70}$ where Q denotes the number of hash queries the prover makes.

Construction 1. *Let $(P_{\mathsf{FS}}, V_{\mathsf{FS}})$ be an interactive Fiat-Shamir proof of knowledge with challenges of $\ell = \ell(k) = O(\log(k))$ bits for relation W. Define the parameters b, r, S, t (as functions of k) for the number of test bits, repetitions, maximum sum and trial bits such that $br = \omega(\log k)$, $2^{t-b} = \omega(\log k)$, $b, r, t = O(\log k)$, $S = O(r)$ and $b \leq t \leq \ell$. Define the following non-interactive proof system for relation W in the random oracle model, where the random oracle maps to b bits.*

[Prover.] The prover P^H on input (x, w) first runs the prover $P_{\mathsf{FS}}(x, w)$ in r independent repetitions to obtain r commitments $\mathsf{com}_1, \ldots, \mathsf{com}_r$. Let $\vec{\mathsf{com}} = (\mathsf{com}_1, \ldots, \mathsf{com}_r)$. Then P^H does the following, either sequentially or in parallel for each repetition i. For each $\mathsf{ch}_i = 0, 1, 2, \ldots, 2^t - 1$ (viewed as t-bit strings) it lets P_{FS} compute the final responses $\mathsf{resp}_i = \mathsf{resp}_i(\mathsf{ch}_i)$ by rewinding, until it finds the first one such that $H(x, \vec{\mathsf{com}}, i, \mathsf{ch}_i, \mathsf{resp}_i) = 0^b$; if no such tuple is found then P^H picks the first one for which the hash value is minimal among all 2^t hash values. The prover finally outputs $\pi = (\mathsf{com}_i, \mathsf{ch}_i, \mathsf{resp}_i)_{i=1,2,\ldots,r}$.

[Verifier.] The verifier V^H on input x and $\pi = (\mathsf{com}_i, \mathsf{ch}_i, \mathsf{resp}_i)_{i=1,2,\ldots,r}$ accepts if and only if $V_{1,\mathsf{FS}}(x, \mathsf{com}_i, \mathsf{ch}_i, \mathsf{resp}_i) = 1$ for each $i = 1, 2, \ldots, r$, and if $\sum_{i=1}^{r} H(x, \vec{\mathsf{com}}, i, \mathsf{ch}_i, \mathsf{resp}_i) \leq S$.

Note that for common iterated hash functions like SHA-1 the prover and the verifier can store the intermediate hash value of the prefix $(x, \vec{\mathsf{com}})$ and need not compute it from scratch for each of the r repetitions.

Our protocol has a small completeness error. For deterministic verifiers this error can be removed in principle by standard techniques, namely, by letting the prover check on behalf of the verifier that the proof is valid before outputting it; if not the prover simply sends the witness to the verifier. In practice, in case of this very unlikely event, the prover may just compute a proof from scratch.

Theorem 2. *Let $(P_{\mathsf{FS}}, V_{\mathsf{FS}})$ be an interactive Fiat-Shamir proof of knowledge for relation W. Then the scheme (P, V) in Construction 1 is a non-interactive zero-knowledge proof of knowledge for relation W (in the random oracle model) with an online extractor.*

Proof. (Sketch) We show that completeness, zero-knowledge and online extraction according to the definition are satisfied.

Completeness. For the completeness we show that the prover fails to convince the verifier with negligible probability only. For this let s_i be the random value $H(x, \vec{\mathsf{com}}, i, \mathsf{ch}_i, \mathsf{resp}_i)$ associated to the output of the i-th execution. Then,

$$\mathrm{Prob}[\exists i : \mathsf{s}_i > S] \leq r \cdot \left(1 - (S+1)2^{-b}\right)^{2^t} \leq r \cdot e^{-(S+1)2^{t-b}}$$

because in each of the at most 2^t tries the prover gets a random hash value of at most S with probability at least $(S+1)2^{-b}$, and all hash values are independent. The probability of having a value larger than S in one execution is thus negligible as r is logarithmic and 2^{t-b} is superlogarithmic. Hence, the sum of all r values exceeds rS with negligible probability only, and we from now on we can condition on the event that the sum of all s_i is at most rS. We also presume $r \geq 2$ in the sequel, else the claim already follows.

In order for the honest prover to fail the sum T of the r values $\mathsf{s}_1, \ldots, \mathsf{s}_r \geq 0$ must be larger than S. For any such $T = S+1, S+2, \ldots, rS$ there are at most $\binom{T+r-1}{r-1}$ ways to split the sum T into r non-negative integers $s_1, \ldots, s_r$. This is upper bounded by

$$\binom{T+r-1}{r-1} \leq \left(\frac{e(rS+r-1)}{r-1}\right)^{r-1} \leq (e(2S+1))^{r-1} \leq e^{r\ln(e(2S+1))}$$

On the other hand, the probability of obtaining such a sum for a given partition, $\mathsf{s}_1 = s_1, \ldots, \mathsf{s}_r = s_r$, is at most

$$\prod_{i=1}^{r} \mathrm{Prob}[\mathsf{s}_i = s_i] \leq \prod_{i=1}^{r} \mathrm{Prob}[\mathsf{s}_i \geq s_i] \leq \prod_{i=1}^{r} \left(1 - s_i 2^{-b}\right)^{2^t}$$

$$\leq \prod_{i=1}^{r} e^{-s_i 2^{t-b}} = e^{-(\sum s_i)2^{t-b}} = e^{-T2^{t-b}} \leq e^{-(S+1)2^{t-b}}$$

By choice of the parameters the probability of getting a sum T with $S < T \leq rS$ is therefore limited by $\exp(r\ln(e(2S+1)) - (S+1)2^{t-b})$. Since $\ln(2S+1) \leq S+1$, $r = O(\log k)$ and $2^{t-b} = \omega(\log k)$ this is negligible.

Zero-Knowledge. The zero-knowledge simulator $Z = (Z_0, Z_1)$ in the first stage simply lets H_0 be a (pseudo)random oracle. For the second stage, Z_1 defines H_1 to be consistent with H_0 on previous queries. For any other query to H_1 simulator Z_1, on input x (and YES, the case NO is trivial), first samples 2^t random b-bit strings for each i and assigns them to the t-bit challenges $\mathsf{ch}_i \in \{0,1\}^t$. Let $\tau_i : \{0,1\}^t \rightarrow \{0,1\}^b$ describe this assignment. Let ch_i be the first one (in lexicographic order) obtaining the minimum over all these 2^t values. Z_1 next runs the (wlog.) special zero-knowledge simulator Z_{FS} of the underlying Fiat-Shamir proof r times on x and each ch_i to obtain r tuples $(\mathsf{com}_i, \mathsf{ch}_i, \mathsf{resp}_i)$. It then defines the hash function H_1 for any query $(x, \vec{\mathsf{com}}, i, \mathsf{ch}_i^*, \mathsf{resp}_i^*)$ with $V_{1,\mathsf{FS}}(x, \mathsf{com}_i, \mathsf{ch}_i^*, \mathsf{resp}_i^*) = 1$ to be the value $\tau_i(\mathsf{ch}_i^*)$. All other values of H_1 are chosen (pseudo)randomly. The simulator outputs $\pi = (\mathsf{com}_i, \mathsf{ch}_i, \mathsf{resp}_i)_{i=1,2,\ldots,r}$ as the proof. Zero-knowledge of the simulator above follows from the special zero-knowledge property of the Fiat-Shamir protocol (together with a hybrid argument).

Online Extraction. We present a knowledge extractor $K(x, \pi, \mathcal{Q}_H(A))$ that, except with negligible probability over the choice of H, is able to output a witness w to x for an accepted proof $\pi = (\mathsf{com}_i, \mathsf{ch}_i, \mathsf{resp}_i)_{i=1,2,\ldots,r}$. Algorithm K browses through the list of queries and answers $\mathcal{Q}_H(A)$ and searches for a query $(x, \overrightarrow{\mathsf{com}}, i, \mathsf{ch}_i, \mathsf{resp}_i)$ as well another query $(x, \overrightarrow{\mathsf{com}}, i, \mathsf{ch}'_i, \mathsf{resp}'_i)$ for a different challenge $\mathsf{ch}_i \neq \mathsf{ch}'_i$ but such that $V_{\mathsf{FS}}(x, \mathsf{com}_i, \mathsf{ch}'_i, \mathsf{resp}'_i) = 1$. If it finds two such queries it runs the knowledge extractor K_{FS} of the Fiat-Shamir proof on these values and copies its output; if there are no such queries then K outputs $\perp$. It is clear that K succeeds every time it finds two valid queries for the same prefix $(x, \overrightarrow{\mathsf{com}}, i)$. On the other hand, by the choice of parameters the probability of making the verifier accept while probing at most one challenge-response pair is negligible.   □

We remark that the upper bounds derived on the number of representations of T with r integers, for completeness and extraction, have not been optimized. Moreover, we providently note that our knowledge extractor only needs the hash queries in $\mathcal{Q}_H(A)$ with prefix $(x, \overrightarrow{\mathsf{com}})$ to extract the witness for theorem x; all other queries are irrelevant to K.

Comparison to Hash-Tree Construction. We compare our construction with online extractors based on hash tress. Recall that, for the hash tree construction, in each of the r repetitions the prover computes the commitment com and all possible responses $\mathsf{resp}(\mathsf{ch})$ for challenges $\mathsf{ch} \in \{0,1\}^b$. Hash values of all 2^b responses are placed as leaves in a hash tree, and a binary tree of depth b is computed. This requires altogether $2^b + 2^b - 1 \approx 2^{b+1}$ hash function evaluations. The challenge is computed as the hash value over all commitments and tree roots, and in each tree the corresponding leaf is opened together with the siblings on the path.

To compare the efficiency of the two approaches, we set $b = 9$, $t = 12$, $r = 10$ and $S = 10$ for our construction and $b' = 8$ and $r' = 10$ for the hash-tree construction. Then the total number of hash function evaluations is roughly $r \cdot 2^9$ in both cases, and the number of executions of the underlying protocol are identical. In favor of the hash tree construction it must be said that our solution requires twice as many response computations on the average, though.

We have already remarked that the communication complexity of the hash-tree construction is significantly larger than for our construction, i.e., the partly disclosed hash trees add $br = 90$ hash values (typically of 160 or more bits) to the proof, while our solution does not add any communication overhead. As for the extraction error, the exact analysis for our construction with the given parameters shows that the extractor fails with probability at most $Q \cdot 2^{-72}$ where Q is the maximal number of hash queries (assuming that finding distinct responses is beyond feasibility). The extraction for the hash-tree construction basically fails only if one manages to guess all r challenges in advance and to put only one correct answer in each tree. This happens with probability approximately $Q/2^{br} = Q \cdot 2^{-80}$ and is only slightly smaller than for our construction. Yet, extraction in the hash-tree construction also requires that no collisions for the hash function are found. Finally, we note that the honest prover always manages to convince the honest verifier for the hash-tree construction whereas our protocol has a small completeness error.

Properties. Concerning the type of zero-knowledge, if there is a unique response for each x, com, ch, then our transformation converts an honest-verifier perfect zero-knowledge protocol into a statistical zero-knowledge one (against malicious verifiers). The small error is due to the negligible collision probability of commitments and applies to the standard Fiat-Shamir transformation as well.

As for proving logical combinations, given two interactive Fiat-Shamir protocols for two relations W^0, W^1 it is known [13,9,14] how to construct three-move proofs showing that one knows at least one of the witnesses to x^0, x^1 (i.e., prove OR), or one can also show that one knows both witnesses (i.e., prove AND). Since the derived protocols in both cases preserve the zero-knowledge and extraction property, and therefore constitute themselves Fiat-Shamir proofs of knowledge, our conversion can also be carried out for proving such logical statements.

Simulation Soundness and Secure Signatures. Our proof system even achieves the stronger notion of simulation soundness, i.e., even if the zero-knowledge simulator has simulated several proofs for adversarial chosen theorems, the online extractor can still extract the witness from the adversarial proof for a valid theorem (as long as either the theorem or the proof is new). It is then straightforward to construct a secure signature scheme from this simulation sound proof of knowledge, with a tight security reduction. The formal description is omitted.

3.2   Lower Bound for Hash Queries of Online Extractors

In this section we show our superlogarithmic lower bound on the number of hash function evaluations for non-programming online extractors. For notational convenience we let $f(k) = O_{\mathsf{K}}(\log k)$ or $f(k) = poly_{\mathsf{K}}(k)$ refer to a function $f(k)$ which grows only logarithmically or polynomially, restricted to all $k \in \mathsf{K}$, i.e., there is a constant c such that $f(k) \leq c \log k$ or $f(k) \leq k^c$ for all $k \in \mathsf{K}$. For any $k \notin \mathsf{K}$ the function f might exceed these bounds.

For our result we assume that the underlying relation W of the proof of knowledge is accompanied by an efficiently samplable, yet hard to invert procedure generating (x, w). For example, for the discrete-logarithm problem and parameter $k = (\mathcal{G}, q, g)$ this procedure picks $w \leftarrow \mathbb{Z}_q$ and computes $x \leftarrow g^w$. More formally, we say that the relation W has a *one-way instance generator* $\mathcal{I}$ if for any parameter k algorithm $\mathcal{I}$ returns in probabilistic polynomial-time $(x, w) \in W_k$, but such that for any probabilistic polynomial-time algorithm I, for $(x, w) \leftarrow \mathcal{I}(k)$ and $w' \leftarrow I(x)$ the probability $\text{Prob}[(x, w') \in W_k]$ is negligible (as a function of k).

Proposition 1. *Let (P, V) be a non-interactive zero-knowledge proof of knowledge for relation W with an online extractor K in the random oracle model. Let $\rho = \rho(k)$ and $\nu = \nu(k)$ be the maximum number of hash oracle queries the prover P resp. the verifier V makes to generate and to check a proof π. Then*

$\max_{v=0,1,\ldots,\nu} \binom{\rho}{v} = poly_K(k)$ *for an infinite set* K *implies that* W *does not have a one-way instance generator* $\mathcal{I}$.

Clearly, $\binom{\rho}{v}$ obtains its maximum at $\binom{\rho}{\lceil \rho/2 \rfloor}$, where $\lceil \rho/2 \rfloor$ is the rounded-off integer of $\rho/2$, and if $\rho = O_K(\log k)$ for an infinite set K, then $\binom{\rho}{\lceil \rho/2 \rfloor} \leq (2e)^{\lceil \rho/2 \rfloor} = poly_K(k)$ for the same set K, and the requirements of the proposition are satisfied. This implies that $\rho = \omega(\log k)$ must grow superlogarithmically for a one-way instance generator. Similarly, if the verifier only makes a constant number of hash function queries then the prover must perform a superpolynomial number of hash function evaluations, or else the instance generator cannot be one-way.

Proof. (Sketch) The high level idea of the proof is that, under the assumption that $\max_v \binom{\rho}{v} = poly_K(k)$ is polynomial, we can imagine that the prover tries to guess in advance the verifier's queries (among the ρ queries) and only makes those queries. This strategy will succeed with sufficiently large probability by assumption. Then, replacing the hash queries $\mathcal{Q}_H(P)$ the prover makes to generate the proof π by the queries $\mathcal{Q}_H(V)$ the verifier makes to verify the proof suffices to extract the witness. Specifically, we prove that $K(x, \pi, \mathcal{Q}_H(V))$ then returns a witness w with noticeable probability. Replacing the original proof by an indistinguishable one from the zero-knowledge simulator $Z(x)$ (without access to w) and running $K(x, \pi, \mathcal{Q}_H(V))$ on this proof implies that we can compute the witness w with noticeable probability from x alone. $\qquad\square$

Optimality of the Bound. Our lower bounds make essential use of the fact that the extractor cannot program the random oracle. In fact, if K was allowed to choose oracle values, then the oracle H (with unrestricted output length) could be defined to generate a sufficiently large common reference string and to run a non-interactive zero-knowledge proof of knowledge with online extractor in the standard model [16]. A single hash function evaluation would then suffice.

Also, the superlogarithmic bound cannot be improved for non-programming extractors. Namely, if we run the hash-tree construction or an easy modification of our solution for binary challenges and superlogarithmic r, then we get a negligible extraction error and make only $O(r)$ hash function queries.

4   Application to Group Signatures

In this section we show how to lift the CPA-anonymous group signature scheme by Boneh et al. [5] to a fully anonymous one. As explained in the introduction, the idea is to append a non-interactive proof of knowledge with online extractor for an ElGamal-like encryption. Although we give a brief introduction to group signatures we refer the reader to the work by Bellare et al. [7] for a comprehensive overview about (the security of) group signatures. Recall from the introduction that the two important security properties are full anonymity, the impossibility of identifying the author of a signature, and full traceability, the impossibility of generating a signature which cannot be traced to the right origin.

Very roughly, a group signature scheme consists of a (fixed) set of users, each user receiving a secret through an initial *key generation* phase carried out by a trusted third party. In addition, a public group key is established in this phase. Each user can run the *sign protocol* to generate a signature on behalf of the group. This signature is *verifiable* through the group's public key, yet outsiders remain oblivious about the actual signer. Only the group manager can revoke this anonymity and *open* the signature through an additional secret key.

The original scheme by Boneh works over bilinear group pairs (G_1, G_2) where deciding the Diffie-Hellman problem is easy. That is, for groups G_1, G_2 of prime order q generated by g_1, g_2 there is an efficiently computable isomorphism $\psi : G_2 \to G_1$ with $\psi(g_2) = g_1$, and an efficiently computable non-degenerated bilinear mapping e with $e(u^a, v^b) = e(u, v)^{ab}$ for all $u \in G_1, v \in G_2$ and $a, b \in \mathbb{Z}_q$.

For the security of the scheme it is assumed that the q-strong Diffie-Hellman problem —given $(g_1, g_2, g_2^{\gamma}, g_2^{(\gamma^2)}, \ldots, g_2^{(\gamma^q)})$ find $(g_1^{\gamma+x}, x)$ for *any* $x \in \mathbb{Z}_q^*$— is intractable. See [3] for more details. It is also presumed that the decision linear assumption in G_1 holds, namely that it is infeasible to distinguish tuples $(u, v, h, u^a, v^b, h^{a+b})$ and (u, v, h, u^a, v^b, h^c) for $u, v, h \leftarrow G_1$ and $a, b, c \in \mathbb{Z}_q$. This assumption implies that ElGamal-like encryptions $(u^a, v^b, h^{a+b} \cdot m)$ of messages m under public key (u, v, h) are semantically secure.

In the original scheme of Boneh et al. [5] the group's public key contains a value $w = g_2^{\gamma}$ and each user receives a pair (A_i, x_i) with $A_i = g_1^{1/(\gamma+x_i)}$ as the secret key. In addition, the group manager's public key consists of a public encryption key (u, v, h) such that $u^{\xi_1} = v^{\xi_2} = h$ for secret key ξ_1, ξ_2. To sign a message m the user encrypts A_i with the manager's public key as $T_1 \leftarrow u^a$, $T_2 \leftarrow v^b$ and $T_3 \leftarrow A_i h^{a+b}$ for random $a, b \leftarrow \mathbb{Z}_q$. In addition, the signer also computes a non-interactive proof τ (in the random oracle model) that (T_1, T_2, T_3) encrypts such an A_i with $e(A_i, wg_2^{x_i}) = e(g_1, g_2)$. The details of this zero-knowledge proof are irrelevant for our discussion here, we merely remark that the message m enters in this proof and that an independent random oracle is needed for this part. To verify a signature one verifies this proof τ. To revoke anonymity the group manager verifies the signatures and then decrypts $A_i = T_3/T_1^{\xi_1} T_2^{\xi_2}$ and recovers the user's identity through A_i.

We now augment the original scheme by our proof of knowledge for the ElGamal encryption:

Construction 3. *Define the following group signature scheme:*

[Key Generation.] Compute the public key as before by picking a bilinear group pair $\mathcal{G} = (G_1, G_2)$ and generators g_1, g_2, h. Sample $\xi_1, \xi_2, \gamma \leftarrow \mathbb{Z}_q^*$ and let $u^{\xi_1} = v^{\xi_2} = h$ and $w = g_2^{\gamma}$. The public key gpk consists of $(\mathcal{G}, g_1, g_2, h, u, v, w)$. Each of the n users obtains some $x_i \leftarrow \mathbb{Z}_q^*$ and $A_i = g_1^{1/(\gamma+x_i)}$ as the secret key. The group manager receives $(\xi_1, \xi_2, A_1, \ldots, A_n)$ as the secret key.

[Signing.] To sign a message $m \in \{0,1\}^*$ under a secret key (A_i, x_i) the user takes the group key $gpk = (\mathcal{G}, g_1, g_2, h, u, v, w)$ and does the following:

– As in the original scheme pick $a, b \leftarrow \mathbb{Z}_q$ and encrypt A_i under the group manager's public key, $T_1 \leftarrow u^a$, $T_2 \leftarrow v^b$ and $T_3 \leftarrow A_i h^{a+b}$.

- *Compute as before a non-interactive proof τ that $A_i = g_1^{1/(\gamma+x_i)}$ is encrypted in (T_1, T_2, T_3) for some $x_i \in \mathbb{Z}_q$, involving the message m.*
- *Additionally, compute a non-interactive zero-knowledge proof of knowledge π for α, β, i.e., run P^H on $(gpk, T_1, T_2, T_3, \tau, m, \alpha, \beta)$ for relation $W_k = \{((gpk, T_1, T_2, T_3, \tau, m), (\alpha, \beta)) \mid u^\alpha = T_1, v^\beta = T_2\}$ to obtain π.*
- *Output $(T_1, T_2, T_3, \pi, \tau)$ as the signature to m.*

[Verification.] *To verify a signature $(T_1, T_2, T_3, \pi, \tau)$ for a message m run the original verifier of the signature scheme and also run the verifier V^H of the non-interactive proof of knowledge on $(gpk, (T_1, T_2, T_3, \tau, m), \pi)$. Accept if both verifications succeed.*

[Open.] *To reveal the identity of a signer for a signature $(T_1, T_2, T_3, \tau, \pi)$ the group manager first verifies the validity of the signature (including the proof π). If correct, then the manager decrypts as in the original scheme to recover some $A = T_3/(T_1^{\xi_1} T_2^{\xi_2})$ and compares this value to the list of A_i's to find the user index i.*

For system parameters suggested in [5], namely, $|q| = 170$ bits and $|G_1| = 171$ bits, the original signature length is $1,533$ bits. If we use the same values $b = 9, r = S = 10, t = 12$ as in the previous section for our proof system, then our scheme adds about $2r \cdot 170 + rt = 3,520$ bits to signatures through the r repetitions of the atomic protocol for proving the AND of the two discrete logarithms. This proof requires $2r$ answers in $\mathbb{Z}_q$ (as usual in the discrete logarithm case, the commitments are not included in the proof π) and r challenges of t bits. Although the communication complexity of this new scheme is significantly larger, it is still superior to RSA-based group signatures where signatures easily exceed $10,000$ bits [2].

Interestingly, we still expect our version of the group signature scheme to be more efficient than the RSA-based scheme in [2], where half a dozen exponentiations with large exponents of more than thousand bits have to be carried out without Chinese Remainder. According to implementation results in [15] a single exponentiation for elliptic curves is estimated to be about ten times faster than such RSA exponentiations; the exact figures of course depend on implementation details. The proof of the following proposition is omitted for space reasons.

Proposition 2. *Under the Decision Linear Diffie-Hellman and the q-strong Diffie-Hellman assumption the group signature scheme in Construction 3 is a fully anonymous and fully traceable group signature scheme in the random oracle model.*

Acknowledgments

We thank the anonymous reviewers of Crypto 2005 for very comprehensive comments and suggestions.

References

1. Masayuki Abe. *Combining Encryption and Proof of Knowledge in the Random Oracle Model.* The Computer Journal, 47(1):58–70, 2004.
2. Giuseppe Ateniese, Jan Camenisch, Marc Joye, and Gene Tsudik. *A Practical and Provably Secure Coalition-Resistant Group Signature Scheme.* Advances in Cryptology — Crypto 2000, Volume 1880 of Lecture Notes in Computer Science. Springer-Verlag, 2000.
3. Dan Boneh and Xavier Boyen. *Short Signatures Without Random Oracles.* Advances in Cryptology — Eurocrypt 2004, Volume 3027 of Lecture Notes in Computer Science, pages 56–73. Springer-Verlag, 2004.
4. Mihir Bellare, Alexandra Boldyreva, and Adriana Palacio. *An Un-Instantiable Random-Oracle-Model Scheme for a Hybrid-Encryption Problem.* Advances in Cryptology — Eurocrypt 2004, Volume 3027 of Lecture Notes in Computer Science. Springer-Verlag, 2004.
5. Dan Boneh, Xavier Boyen, and Hovav Shacham. *Short Group Signatures.* Advances in Cryptology — Crypto 2004, Volume 3152 of Lecture Notes in Computer Science, pages 41–55. Springer-Verlag, 2004.
6. Mihir Bellare, Juan Garay, and Tal Rabin. *Fast Batch Verification for Modular Exponentiation and Digital Signatures.* Advances in Cryptology — Eurocrypt'98, Volume 1403 of Lecture Notes in Computer Science, pages 236–250. Springer-Verlag, 1998.
7. Mihir Bellare, Daniele Micciancio, and Bogdan Warinschi. *Foundations of Group Signatures: Formal Definitions, Simplified Requirements, and a Construction Based on General Assumptions.* Advances in Cryptology — Eurocrypt 2003, Volume 2656 of Lecture Notes in Computer Science, pages 614–629. Springer-Verlag, 2003.
8. M. Bellare and P. Rogaway. *Random Oracles are Practical: A Paradigm for Designing Efficient Protocols.* Proceedings of the Annual Conference on Computer and Communications Security (CCS). ACM Press, 1993.
9. Ronald Cramer, Ivan Damgård, and Berry Schoenmakers. *Proofs of Partial Knowledge and Simplified Desing of Witness Hiding Protocols.* Advances in Cryptology — Crypto'94, Volume 839 of Lecture Notes in Computer Science, pages 174–187. Springer-Verlag, 1995.
10. Ran Canetti, Oded Goldreich, and Shai Halevi. *The Random Oracle Methodology, Revisited.* Proceedings of the Annual Symposium on the Theory of Computing (STOC) 1998, pages 209–218. ACM Press, 1998.
11. Jean-Sebastien Coron. *On the Exact Security of Full Domain Hash.* Advances in Cryptology — Crypto 2000, Volume 1880 of Lecture Notes in Computer Science, pages 229–235. Springer-Verlag, 2000.
12. Jean-Sebastien Coron. *Optimal Security Proofs for PSS and Other Signature Schemes.* Advances in Cryptology — Eurocrypt 2002, Volume 2332 of Lecture Notes in Computer Science, pages 272–287. Springer-Verlag, 2002.
13. David Chaum and Torben Pedersen. *Wallet Databases with Observers.* Advances in Cryptology — Crypto'92, Volume 740 of Lecture Notes in Computer Science, pages 89–105. Springer-Verlag, 1992.
14. Alfredo De Santis, Giovanni Di Crescenzo, Giuseppe Persiano, and Moti Yung. *On Monotone Formula Closure of SZK.* Proceedings of the Annual Symposium on Foundations of Computer Science (FOCS)'94, pages 454–465. IEEE Computer Society Press, 1994.

15. Erik De Win, Serge Mister, Bart Preneel, and Michael Wiener. *On the Performance of Signature Schemes Based on Elliptic Curves*. Algorithmic Number Theory Symposium — ANTS–III, Volume 1423 of Lecture Notes in Computer Science, pages 252–266. Springer-Verlag, 1998.

16. Alfredo De Santis and Giuseppe Persiano. *Zero-Knowledge Proofs of Knowledge Without Interaction*. Proceedings of the Annual Symposium on Foundations of Computer Science (FOCS)'92, pages 427–436. IEEE Computer Society Press, 1992.

17. A. Fiat and A. Shamir. *How to Prove Yourself: Practical Solutions to Identification and Signature Schemes*. Advances in Cryptology — Crypto'86, Volume 263 of Lecture Notes in Computer Science, pages 186–194. Springer-Verlag, 1986.

18. Eu-Jin Goh and Stanislaw Jarecki. *Signature Scheme as Secure as the Diffie-Hellman Problem*. Advances in Cryptology — Eurocrypt 2003, Volume 2656 of Lecture Notes in Computer Science, pages 401–415. Springer-Verlag, 2003.

19. Louis Guillou and Jean-Jacques Quisquater. *A Practical Zero-Knowledge Protocol Fitted to Security Microprocessor Minimizing Both Trasmission and Memory*. Advances in Cryptology — Eurocrypt'88, Volume 330 of Lecture Notes in Computer Science, pages 123–128. Springer-Verlag, 1988.

20. Shafi Goldwasser and Yael Tauman. *On the (In)security of the Fiat-Shamir Paradigm*. Proceedings of the Annual Symposium on Foundations of Computer Science (FOCS) 2003, pages 102–113. IEEE Computer Society Press, 2003.

21. Jonothan Katz and Nan Wang. *Efficiency Improvement for Signature Schemes with Tight Security Reductions*. Proceedings of the Annual Conference on Computer and Communications Security (CCS). ACM Press, 2003.

22. R. Merkle. *A Digital Signature Based on a Conventional Encryption Function*. Advances in Cryptology — Crypto'87, Volume 293 of Lecture Notes in Computer Science, pages 369–378. Springer-Verlag, 1988.

23. Ueli Maurer, Renato Renner, and Clemens Holenstein. *Indifferentiability, Impossibility Results on Reductions, and Applications to the Random Oracle Methodology*. Theory of Cryptography Conference (TCC) 2004, Volume 2951 of Lecture Notes in Computer Science, pages 21–39. Springer-Verlag, 2004.

24. Daniele Micciancio and Salil Vadhan. *Statistical Zero-Knowledge Proofs with Efficient Provers: Lattice Problems and More*. Advances in Cryptology — Crypto 2003, Volume 2729 of Lecture Notes in Computer Science, pages 282–298. Springer-Verlag, 2003.

25. T. Okamoto. *Provable Secure and Practical Identification Schemes and Corresponding Signature Schemes*. Advances in Cryptology — Crypto'92, Volume 740 of Lecture Notes in Computer Science, pages 31–53. Springer-Verlag, 1992.

26. Rafael Pass. *On Deniability in the Common Reference String and Random Oracle Model*. Advances in Cryptology — Crypto 2003, Volume 2729 of Lecture Notes in Computer Science, pages 316–337. Springer-Verlag, 2003.

27. David Pointcheval and Jacques Stern. *Security Arguments for Digital Signatures and Blind Signatures*. Journal of Cryptology, 13(3):361–396, 2000.

28. C.P. Schnorr. *Efficient Signature Generation by Smart Cards*. Journal of Cryptology, 4:161–174, 1991.

29. Victor Shoup and Rosario Gennaro. *Securing Threshold Cryptosystems against Chosen Ciphertext Attack*. Journal of Cryptology, 15(2):75–96, 2002.

A Formal Treatment of Onion Routing

Jan Camenisch and Anna Lysyanskaya

[1] IBM Research, Zurich Research Laboratory,
CH–8803 Rüschlikon
jca@zurich.ibm.com
[2] Computer Science Department, Brown University,
Providence, RI 02912 USA
anna@cs.brown.edu

Abstract. Anonymous channels are necessary for a multitude of privacy-protecting protocols. Onion routing is probably the best known way to achieve anonymity in practice. However, the cryptographic aspects of onion routing have not been sufficiently explored: no satisfactory definitions of security have been given, and existing constructions have only had ad-hoc security analysis for the most part.

We provide a formal definition of onion-routing in the universally composable framework, and also discover a simpler definition (similar to CCA2 security for encryption) that implies security in the UC framework. We then exhibit an efficient and easy to implement construction of an onion routing scheme satisfying this definition.

1 Introduction

The ability to communicate anonymously is requisite for most privacy-preserving interactions. Many cryptographic protocols, and in particular, all the work on group signatures, blind signatures, electronic cash, anonymous credentials, etc., assume anonymous channels as a starting point.

One means to achieve anonymous communication are mix-networks [6]. Here, messages are wrapped in several layers of encryption and then routed through intermediate nodes each of which peels off a layer of encryption and then forwards them in random order to the next node. This process is repeated until all layers are removed. The way messages are wrapped (which determines their path through the network) can either be fixed or can be chosen by each sender for each message.

The former case is usually preferred in applications such as e-voting where one additionally want to ensure that no message is dropped in transit. In that case, each router is required to prove that it behaved correctly: that the messages it outputs are a permutation of the decryption of the messages it has received. The communication model suitable for such a protocol would have a broadcast channel or a public bulletin board; this is not considered efficient in a standard point-to-point network.

In the latter case, where the path is chosen on a message-by-message basis, one often calls the wrapped messages *onions* and speaks of *onion routing* [12,10].

V. Shoup (Ed.): Crypto 2005, LNCS 3621, pp. 169–187, 2005.

An onion router is simply responsible for removing a layer of encryption and sending the result to the next onion router. Although this means that onion routing cannot provide robustness (a router may drop an onion and no one will notice), the simplicity of this protocol makes it very attractive in practice. In fact, there are several implementations of onion routing available (see Dingledine et al. [10] and references therein). Unfortunately, these implementations use ad-hoc cryptography instead of provably secure schemes.

The only prior attempt to formalize and construct a provably secure onion routing scheme is due to Möller [16]. Contrary to his claimed goals, it is not hard to see that his definition of security does not guarantee that the onion's distance to destination is hidden from a malicious router. Additionally, his definition does not consider adaptive attacks aimed to break the anonymity properties of onion routing. Thus, although his work represents a first step in the right direction, it falls short of giving a satisfactory definition. His construction does not seem to meet our definition, but has some similarity to our construction.

Alternative means of achieving anonymous communications include Chaum's dining cryptographer networks [7,8] and Crowds [18].

Onion Routing: Definitional Issues. The state of the literature on anonymous channels today is comparable to that on secure encryption many years ago. While there is a good intuitive understanding of what functionality and security properties an anonymous channel must provide, and a multitude of constructions that seek to meet this intuition, there is a lack of satisfactory *definitions* and, as a result, of provably secure constructions. Indeed, realizing anonymous channels — and constructions aside, simply reasoning about the degree of anonymity a given routing algorithm in a network can provide — remains a question still largely open to rigorous study.

This paper does not actually give a definition of an anonymous channel. We do not know how to define it in such a way that it is, on the one hand, realizable, and, on the other hand, meets our intuitive understanding of what an anonymous channel must accomplish. The stumbling block is that, to realize anonymous channels, one must make non-cryptographic assumptions on the network model. The fact that a solution is proven secure under one set of assumptions on the network does not necessarily imply that it is secure under another set of assumptions.

For example, if one is trying to obtain anonymous channels by constructing a mix network [6], one must make the assumption that (1) there is a dedicated mix network where at least one server is honest; and, more severely, (2) everyone sends and receives about equal amount of traffic and so one cannot match senders to receivers by analyzing the amount of traffic sent and received. In fact, that second assumption on the network was experimentally shown to be crucial — it is known how to break security of mix networks using statistics on network usage where the amount of traffic sent and received by each party is not prescribed to be equal, but rather there is a continuous flow of traffic [14,9,23].

In cryptography, however, this is a classical situation. For example, semantic security [13] was introduced to capture what the adversary already knows about the plaintext (before the ciphertext is even formed) by requiring that a cryptosystem be secure for all a-priori distributions on the plaintext, even those chosen by the adversary. Thus, the cryptographic issue of secure encryption, was separated from the non-cryptographic modelling of the adversary's a-priori information. We take a similar approach here.

An onion routing scheme can provide some amount of anonymity when a message is sent through a sufficient number of honest onion routers and there is enough traffic on the network overall. However, nothing can really be inferred about how much anonymity an onion routing algorithm provides without a model that captures network traffic appropriately. As a result, security must be defined with the view of ensuring that the cryptographic aspects of a solution remain secure even in the worst-case network scenario.

Our Results. Armed with the definitional approach outlined above, we give a definition of security of an onion routing scheme in the universally composable framework [4]. We chose this approach not because we want onion routing to be universally composable with other protocols (we do, but that's a bonus side effect), but simply because we do not know how to do it in any other way! The beauty and versatility of the UC framework (as well as the related reactive security framework [17,1]) is that it guarantees that the network issues are orthogonal to the cryptographic ones — i.e., the cryptographic aspects remain secure under the worst-case assumptions on the network behavior. (Similarly to us, Wikström [22] gives a definition of security in the UC framework for general mix networks.)

Definitions based on the UC-framework, however, can be hard to work with. Thus we also give a *cryptographic* definition, similar to CCA2-security for encryption [11]. We show that in order to satisfy our UC-based definition, it is sufficient to give an onion routing scheme satisfying our cryptographic definition.

Finally, we give a construction that satisfies our cryptographic definition.

Overview of Our Definition and Solution. Our ideal functionality does not reveal to an adversarial router any information about onions apart from the prior and the next routers; in particular, the router does not learn how far a given message is from its destination. This property makes traffic analysis a lot harder to carry out, because now any message sent between two onion routers looks the same, even if one of the routers is controlled by the adversary, no matter how close it is to destination [2]. It is actually easy to see where this property comes in. Suppose that it were possible to tell by examining an onion, how far it is from destination. In order to ensure mixing, an onion router that receives an onion O that is h hops away from destination must buffer up several other onions that are also h hops away from destination before sending O to the next router. Overall, if onions can be up to N hops away from destination, each router will be buffering $\Theta(N)$ onions, a few for all possible values of h. This makes onion routing slow and expensive. In contrast, if an onion routing scheme hides distance to destination, then a router may just buffer a constant number of onions before sending them off.

However, achieving this in a cryptographic implementation seems challenging; let us explain why. In onion routing, each onion router P_i, upon receipt of an onion O_i, decrypts it ("peels off" a layer of encryption) to obtain the values P_{i+1} and O_{i+1}, where P_{i+1} is the identity of the next router in the chain, and O_{i+1} is the data that needs to be sent to P_{i+1}.

Suppose that the outgoing onion O_{i+1} is just the decryption of the incoming onion O_i. Semantic security under the CCA2 attack suggests that, even under active attack from the adversary, if P_i is honest, then the only thing that the incoming onion O_i reveals about the corresponding outgoing onion O_{i+1} is its length.

In the context of encryption, the fact that the length is revealed is a necessary evil that cannot be helped. In this case, however, the problem is not just that the length is revealed, but that, in a secure (i.e., probabilistic) cryptosystem, the length of a plaintext is *always* smaller than the length of a ciphertext.

One attempt to fix this problem is to require that P_i not only decrypt the onion, but also pad it so $|O_i| = |O_{i+1}|$. It is clear that just padding will not work: $|O_{i+1}|$ should be formed in such a way that even P_{i+1} (who can be malicious), upon decrypting O_{i+1} and obtaining the identity of P_{i+2} and the data O_{i+2}, still cannot tell that the onion O_{i+1} was padded, i.e., router P_{i+1} cannot tell that he is not the first router in the chain. At first glance, being able to pad the onion seems to contradict non-malleability: if you can pad it, then, it seems, you can form different onions with the same content and make the scheme vulnerable to adaptive attacks.

Our solution is to use CCA2 encryption *with tags* (or labels) [21,19,3], in combination with a pseudorandom permutation (block cipher). We make router P_i pad the onion is such a way that the next router P_{i+1} cannot tell that it was padded; and yet the fact this is possible does not contradict the non-malleability of the scheme because this padding is *deterministic*. The onion will only be processed correctly by P_{i+1} when the tag that P_{i+1} receives is correct, and the only way to make the tag correct is if P_i applied the appropriate deterministic padding. To see how it all fits together, see Section 4.1.

2 Onion Routing in the UC Framework

Setting. Let us assume that there is a network with J players $P_1, \ldots, P_J$. For simplicity, we do not distinguish players as senders, routers, and receivers; each player can assume any of these roles. In fact, making such a distinction would not affect our protocol at all and needs to be considered in its application only. We define onion routing in the public key model (i.e., in the hybrid model where a public-key infrastructure is already in place) where each player has an appropriately chosen identity P_i, a registered public key PK_i corresponding to this identity, and these values are known to each player.

In each instance of a message that should be sent, for some (s, r), we have a sender P_s (s stands for "sender") sending a message m of length ℓ_m (the length ℓ_m is a fixed parameter, all messages sent must be the same length) to recipient P_r (r stands for "recipient") through $n < N$ additional routers $P_{o_1}, \ldots, P_{o_n}$ (o stands for "onion router"), where the system parameter $N - 1$ is an upper bound on the number of routers that the sender can choose. How each sender selects his onion routers $P_{o_1}, \ldots, P_{o_n}$ is a non-cryptographic problem independent of the current exposition. The input to the onion sending procedure consists of the message m that P_s wishes to send to P_r, a list of onion routers $P_{o_1}, \ldots, P_{o_n}$, and the necessary public keys and parameters. The input to the onion routing procedure consists of an onion O, the routing party's secret key SK, and the necessary public keys and parameters. In case the routing party is in fact the recipient, the routing procedure will output the message m.

Definition of Security. The honest players are modelled by imagining that they obtain inputs (i.e., the data m they want to send, the identity of the recipient P_r, and the identities of the onion routers $P_{o_1}, \ldots, P_{o_n}$) from the environment $\mathcal{Z}$, and then follow the protocol (either the ideal or the cryptographic one). Similarly, the honest players' outputs are passed to the environment.

Following the standard universal composability approach (but dropping most of the formalism and subtleties to keep presentation compact), we say that an onion routing protocol is secure if there exists a simulator (ideal-world adversary) $\mathcal{S}$ such that no polynomial-time in λ (the security parameter) environment $\mathcal{Z}$ controlling the inputs and outputs of the honest players, and the behavior of malicious players, can distinguish between interacting with the honest parties in the ideal model through $\mathcal{S}$, or interacting with the honest parties using the protocol.

We note that the solution we present is secure in the public-key model, i.e., in the model where players publish the keys associated with their identities in some reliable manner. In the proof of security, we will allow the simulator $\mathcal{S}$ to generate the keys of all the honest players.

The Ideal Process. Let us define the ideal onion routing process. Let us assume that the adversary is static, i.e., each player is either honest or corrupted from the beginning, and the trusted party implementing the ideal process knows which parties are honest and which ones are corrupted.

Ideal Onion Routing Functionality: Internal Data Structure.

- The set *Bad* of parties controlled by the adversary.
- An onion O is stored in the form of $(sid, P_s, P_r, m, n, \mathcal{P}, i)$ where: sid is the identifier, P_s is the sender, P_r is the recipient, m is the message sent through the onion routers, $n < N$ is the length of the onion path, $\mathcal{P} = (P_{o_1}, \ldots, P_{o_n})$ is the path over which the message is sent (by convention, $P_{o_0} = P_s$, and $P_{o_{n+1}} = P_r$), i indicates how much of the path the message has already traversed (initially, $i = 0$). An onion has reached its destination when $i = n + 1$.

- A list L of onions that are being processed by the adversarial routers. Each entry of the list consists of $(temp, O, j)$, where $temp$ is the temporary id that the adversary needs to know to process the onion, while $O = (sid, P_s, P_r, m, n, \mathcal{P}, i)$ is the onion itself, and j is the entry in $\mathcal{P}$ where the onion should be sent next (the adversary does not get to see O and j). Remark: Note that entries are never removed from L. This models the replay attack: the ideal adversary is allowed to resend an onion.
- For each honest party P_i, a buffer B_i of onions that are currently being held by P_i. Each entry consists of $(temp', O)$, where $temp'$ is the temporary id that an honest party needs to know to process the onion and $O = (sid, P_s, P_r, m, n, \mathcal{P}, i)$ is the onion itself (the honest party does not get to see O). Entries from this buffer are removed if an honest party tells the functionality that she wants to send an onion to the next party.

Ideal Onion Routing Functionality: Instructions. The ideal process is activated by a message from router P, from the adversary $\mathcal{S}$, or from itself. There are four types of messages, as follows:

(Process_New_Onion, $P_r, m, n, \mathcal{P}$). Upon receiving such a message from P_s, where $m \in \{0, 1\}^{\ell_m} \cup \{\bot\}$, do:

1. If $|\mathcal{P}| \geq N$, reject.
2. Otherwise, create a new session id sid, and let $O = (sid, P, P_r, m, n, \mathcal{P}, 0)$. Send itself message (Process_Next_Step, O).

(Process_Next_Step, O). This is the core of the ideal protocol. Suppose $O = (sid, P_s, P_r, m, n, \mathcal{P}, i)$. The ideal functionality looks at the next part of the path. The router P_{o_i} just processed[1] the onion and now it is being passed to $P_{o_{i+1}}$. Corresponding to which routers are honest, and which ones are adversarial, there are two possibilities for the next part of the path:

I) Honest next. Suppose that the next node, $P_{o_{i+1}}$, is honest. Here, the ideal functionality makes up a random temporary id $temp$ for this onion and sends to $\mathcal{S}$ (recall that $\mathcal{S}$ controls the network so it decides which messages get delivered): "Onion $temp$ from P_{o_i} to $P_{o_{i+1}}$." It adds the entry $(temp, O, i+1)$ to list L. (See (Deliver_Message, $temp$) for what happens next.)

II) Adversary next. Suppose that $P_{o_{i+1}}$ is adversarial. Then there are two cases:

- There is an honest router remaining on the path to the recipient. Let P_{o_j} be the next honest router. (I.e., $j > i$ is the smallest integer such that P_{o_j} is honest.) In this case, the ideal functionality creates a random temporary id $temp$ for this onion, and sends the message "Onion $temp$ from P_{o_i}, routed through $(P_{o_{i+1}}, \ldots, P_{o_{j-1}})$ to P_{o_j}" to the ideal adversary $\mathcal{S}$, and stores $(temp, O, j)$ on the list L.

---

[1] In case $i = 0$, processed means having originated the onion and submitted it to the ideal process.

– P_{o_i} is the last honest router on the path; in particular, this means that P_r is adversarial as well. In that case, the ideal functionality sends the message "Onion from P_{o_i} with message m for P_r routed through $(P_{o_{i+1}}, \ldots, P_{o_n})$" to the adversary $\mathcal{S}$. (Note that if $P_{o_i+1} = P_r$, the list $(P_{o_{i+1}}, \ldots, P_{o_n})$ will be empty.)

(Deliver_Message, $temp$). This is a message that $\mathcal{S}$ sends to the ideal process to notify it that it agrees that the onion with temporary id $temp$ should be delivered to its current destination. To process this message, the functionality checks if the temporary identifier $temp$ corresponds to any onion O on the list L. If it does, it retrieves the corresponding record $(temp, O, j)$ and update the onion: if $O = (sid, P_s, P_r, m, n, \mathcal{P}, i)$, it replaces i with j to indicate that we have reached the j'th router on the path of this onion. If $j < n + 1$, it generates a temporary identifier $temp'$, sends "Onion $temp'$ received" to party P_{o_j}, and stores the resulting pair $(temp', O = (sid, P_s, P_r, m, n, \mathcal{P}, j))$ in the buffer B_{o_j} of party P_{o_j}. Otherwise, $j = n + 1$, so the onion has reached its destination: if $m \neq \perp$ it sends "Message m received" to router P_r; otherwise it does not deliver anything[2].

(Forward_Onion, $temp'$). This is a message from an honest ideal router P_i notifying the ideal process that it is ready to send the onion with id $temp'$ to the next hop. In response, the ideal functionality

– Checks if the temporary identifier $temp'$ corresponds to any entry in B_i. If it does, it retrieves the corresponding record $(temp', O)$.
– Sends itself the message (Process_Next_Step, O).
– Removes $(temp', O)$ from B_i.

This concludes the description of the ideal functionality. We must now explain how the ideal honest routers work. When an honest router receives a message of the form "Onion $temp'$ received" from the ideal functionality, it notifies environment $\mathcal{Z}$ about it and awaits instructions for when to forward the onion $temp'$ to its next destination. When instructed by $\mathcal{Z}$, it sends the message "Forward_Onion $temp'$" to the ideal functionality.

It's not hard to see that $\mathcal{Z}$ learns nothing else than pieces of paths of onions formed by honest senders (i.e., does not learn a sub-path's position or relations among different sub-paths). Moreover, if the sender and the receiver are both honest, the adversary does not learn the message.

2.1   Remarks and Extensions

Mixing Strategy. It may seem that, as defined in our ideal functionality, the adversary is too powerful because, for example, it is allowed to route just one onion at a time, and so can trace its entire route. In an onion routing

---

[2] This is needed to account for the fact that the adversary inserts onions into the network that at some point do not decrypt correctly.

implementation however, the instructions for which onion to send on will not come directly from the adversary, but rather from an honest player's mixing strategy. That is, each (honest) router is notified that an onion has arrived and is given a handle *temp* to that onion. Whenever the router decides (under her mixing strategy) that the onion *temp* should be sent on, she can notify the ideal functionality of this using the handle *temp*. A good mixing strategy will limit the power of the adversary to trace onions in the ideal world, which will translate into limited capability in the real world as well. What mixing strategy is a good one depends on the network. Additionally, there is a trade-off between providing more anonymity and minimizing latency of the network. We do not consider any of these issues in this paper but only point out that our scheme guarantees the maximum degree of security that any mixing strategy can inherently provide.

Replay Attacks. The definition as is allows replay attacks by the adversary. The adversary controls the network and can replay any message it wishes. In particular, it can take an onion that party P_i wants to send to P_j and deliver it to P_j as many times as it wishes. However, it is straightforward to modify our security definition and our scheme so as to prevent replay attacks. For instance, we could require that the sender inserts time stamps into all onions. I.e., a router P_i, in addition to the identity of the next router P_{i+1}, will also be given a time **time** and a random identifier oid_i (different for each onion and router). An onion router will drop the incoming onion when either the time **time** $+ t_\Delta$ (where t_Δ is a parameter) has passed or it finds oid_i in its database. If an onion is not dropped, the router will store oid_i until time **time** $+ t_\Delta$ has passed. It is not difficult to adapt our scheme and model to reflect this. We omit details to keep this exposition focused.

Forward Security. Forward secrecy is a desirable property in general, and in this context in particular [5,10]. Our scheme can be constructed from any CCA2-secure cryptosystem, and in particular, from a forward-secure one.

The Response Option. Another desirable property of an onion routing scheme is being able to respond to a message received anonymously. We address this after presenting our construction.

3   A Cryptographic Definition of Onion Routing

Here we give a cryptographic definition of an onion routing scheme and show why a scheme satisfying this definition is sufficient to realize the onion routing functionality described in the previous section.

Definition 1 (Onion routing scheme I/O). *A set of algorithms* $(G,$ FormOnion, ProcOnion) *satisfies the I/O spec for an onion routing scheme for message space* $M(1^\lambda)$ *and set of router names* $\mathcal{Q}$ *if:*

- *G is a key generation algorithm, possibly taking as input some public parameters* p, *and a router name* P: $(PK, SK) \leftarrow G(1^\lambda, p, P)$.

- FormOnion *is a probabilistic algorithm that on input a message* $m \in M(1^\lambda)$, *an upper bound on the number of layers* N, *a set of router names* $(P_1, \ldots, P_{n+1})$ *(each* $P_i \in Q$, $n \leq N$), *and a set of public keys corresponding to these routers* $(PK_1, \ldots, PK_{n+1})$, *outputs a set of onion layers* $(O_1, \ldots, O_{n+1})$. *(As* N *is typically a system-wide parameter, we usually omit to give it as input to this algorithm.)*

- ProcOnion *is a deterministic algorithm that, on input an onion* O, *identity* P, *and a secret key* SK, *peels off a layer of the onion to obtain a new onion* O' *and a destination* P' *for where to send it:* $(O', P') \leftarrow$ ProcOnion(SK, O, P).

Definition 2 (Onion evolution, path, and layering). *Let* $(G,$ FormOnion, ProcOnion$)$ *satisfy the onion routing I/O spec. Let* p *be the public parameters. Suppose that we have a set* Q, $\perp \notin Q$, *consisting of a polynomial number of (honest) router names. Suppose that we have a public-key infrastructure on* Q, *i.e., corresponding to each name* $P \in Q$ *there exists a key pair* $(PK(P), SK(P))$, *generated by running* $G(1^\lambda, p, P)$. *Let* O *be an onion received by router* $P \in Q$. *Let* $\mathcal{E}(O, P) = \{(O_i, P_i) \; : \; i \geq 1\}$ *be the maximal ordered list of pairs such that* $P_1 = P$, $O_1 = O$, *and for all* $i > 1$, $P_i \in Q$, *and* $(O_i, P_i) =$ ProcOnion$(SK(P_{i-1}), O_{i-1}, P_{i-1})$. *Then* $\mathcal{E}(O, P)$ *is the* evolution *of onion* O *starting at* P. *Moreover, if* $\mathcal{E}(O, P) = \{(O_i, P_i)\}$ *is the evolution of an onion, then* $\mathcal{P}(O, P) = \{P_i\}$ *is the* path *of the onion, while* $\mathcal{L}(O, P) = \{O_i\}$ *is the* layering *of the onion.*

Onion-correctness is the simple condition that if an onion is formed correctly and then the correct routers process it in the correct order, then the correct message is received by the last router P_{n+1}.

Definition 3 (Onion-correctness). *Let* $(G,$ FormOnion, ProcOnion$)$ *satisfy the I/O spec for an onion routing scheme. Then for all settings of the public parameters* p, *for all* $n < N$, *and for all* Q *with a public-key infrastructure as in Definition 2, for any path* $\mathcal{P} = (P_1, \ldots, P_{n+1})$, $\mathcal{P} \subseteq Q$, *for all messages* $m \in M(1^\lambda)$, *and for all onions* O_1 *formed as*

$$(O_1, \ldots, O_{n+1}) \leftarrow \mathsf{FormOnion}(m, N, (P_1, \ldots, P_{n+1}), (PK(P_1), \ldots, PK(P_{n+1})))$$

the following is true: (1) correct path: $\mathcal{P}(O_1, P_1) = (P_1, \ldots, P_{n+1})$; *(2)* correct layering: $\mathcal{L}(O_1, P_1) = (O_1, \ldots, O_{n+1})$; *(3)* correct decryption: $(m, \perp) =$ ProcOnion$(SK(P_{n+1}), O_{n+1}, P_{n+1})$.

Onion-integrity requires that even for an onion created by an adversary, the path is going to be of length at most N.

Definition 4 (Onion-integrity). *(Sketch) An onion routing scheme satisfies onion-integrity if for all probabilistic polynomial-time adversaries, the probability (taken over the choice of the public parameters* p, *the set of honest router names* Q *and the corresponding PKI as in Definition 2) that an adversary with adaptive access to* ProcOnion$(SK(P), \cdot, P)$ *procedures for all* $P \in Q$, *can produce and send to a router* $P_1 \in Q$ *an onion* O_1 *such that* $|\mathcal{P}(O_1, P_1)| > N$, *is negligible.*

Our definition of onion security is somewhat less intuitive. Here, an adversary is launching an adaptive attack against an onion router P. It gets to send onions to this router, and see how the router reacts, i.e., obtain the output of $\mathsf{ProcOnion}(SK(P), \cdot, P)$. The adversary's goal is to distinguish whether a given challenge onion corresponds to a particular message and route, or a random message and null route. The unintuitive part is that the adversary can also succeed by *re-wrapping* an onion, i.e., by adding a layer to its challenge onion.

Definition 5 (Onion-security). *(Sketch) Consider an adversary interacting with an onion routing challenger as follows:*

1. *The adversary receives as input a challenge public key PK, chosen by the challenger by letting $(PK, SK) \leftarrow G(1^\lambda, p)$, and the router name P.*
2. *The adversary may submit any number of onions O_i of his choice to the challenger, and obtain the output of $\mathsf{ProcOnion}(SK, O_i, P)$.*
3. *The adversary submits n, a message m, a set of names $(P_1, \ldots, P_{n+1})$, and index j, and n key pairs $1 \leq i \leq n+1$, $i \neq j$, (PK_i, SK_i). The challenger checks that the router names are valid[3], that the public keys correspond to the secret keys, and if so, sets $PK_j = PK$, sets bit b at random, and does the following:*
   − *If $b = 0$, let*

   $$(O_1, \ldots, O_j, \ldots, O_{n+1}) \leftarrow \mathsf{FormOnion}(m, (P_1, \ldots, P_{n+1}), (PK_1, \ldots, PK_{n+1}))$$

   − *Otherwise, choose $r \leftarrow M(1^\lambda)$, and let*

   $$(O_1, \ldots, O_j) \leftarrow \mathsf{FormOnion}(r, (P_1, \ldots, P_j), (PK_1, \ldots, PK_j))$$

4. *Now the adversary is allowed get responses for two types of queries:*
   − *Submit any onion $O_i \neq O_j$ of his choice and obtain $\mathsf{ProcOnion}(SK, O_i, P)$.*
   − *Submit a secret key SK', an identity $P' \neq P_{j-1}$, and an onion O' such that $O_j = \mathsf{ProcOnion}(SK', O', P')$; if P' is valid, and (SK', O', P') satisfy this condition, then the challenger responds by revealing the bit b.*
5. *The adversary then produces a guess b'.*

We say that a scheme with onion routing I/O satisfies onion security if for all probabilistic polynomial time adversaries $\mathcal{A}$ of the form described above, there is a negligible function ν such that the adversary's probability of outputting $b' = b$ is at most $1/2 + \nu(\lambda)$.

This definition of security is simple enough, much simpler than the UC-based definition described in the previous section. Yet, it turns out to be sufficient. Let us give an intuitive explanation why. A simulator that translates between a real-life adversary and an ideal functionality is responsible for two things: (1) creating some fake traffic in the real world that accounts for everything that happens in

---

[3] In our construction, router names are formed in a special way, hence this step is necessary for our construction to satisfy this definition.

the ideal world; and (2) translating the adversary's actions in the real world into instructions for the ideal functionality.

In particular, in its capacity (1), the simulator will sometimes receive a message from the ideal functionality telling it that an onion *temp* for honest router P_j is routed through adversarial routers $(P_1, \ldots, P_{j-1})$. The simulator is going to need to make up an onion O_1 to send to the adversarial party P_1. But the simulator is not going to know the message contained in the onion, or the rest of the route. So the simulator will instead make up a random message r and compute the onion so that it decrypts to r when it reaches the honest (real) router P_j. I.e, it will form O_1 by obtaining $(O_1, \ldots, O_j) \leftarrow \mathsf{FormOnion}(r, (P_1, \ldots, P_j), (PK_1, \ldots, PK_j))$. When the onion O_j arrives at P_j from the adversary, the simulator knows that it is time to tell the ideal functionality to deliver message *temp* to honest ideal P_j.

Now, there is a danger that this may cause errors in the simulation as far as capacity (2) is concerned: the adversary may manage to form another onion $\tilde{O}$, and send it to an honest router $\tilde{P}$, such that $(O_j, P) \in \mathcal{E}(\tilde{O}, \tilde{P})$. The simulator will be unable to handle this situation correctly, as the simulator relies on its ability to correctly decrypt and route all real-world onions, while in this case, the simulator does not know how to decrypt and route this "fake" onion past honest router P_j. A scheme satisfying the definition above would prevent this from happening: the adversary will not be able to form an onion $O' \neq O_{j-1}$ sent to an honest player P' such that $(P_j, O_j) = \mathsf{ProcOnion}(SK(P'), O', P')$.

In the full version of this paper, we give a formal proof of the following theorem:

Theorem 1. *An onion routing scheme* $(G, \mathsf{FormOnion}, \mathsf{ProcOnion})$ *satisfying onion-correctness, integrity and security, when combined with secure point-to-point channels, yields a UC-secure onion routing scheme.*

4   Onion Routing Scheme Construction

Tagged Encryption. The main tool in our construction is a CCA2-secure cryptosystem (Gen, E, D) that supports tags. Tags were introduced by Shoup and Gennaro [21]. The meaning of a tagged ciphertext is that the tag provides the context within which the ciphertext is to be decrypted. The point is that an adversary cannot attack the system by making the honest party under attack decrypt this ciphertext out of context. The input to E is (PK, m, T), where T is a tag, such that $D(SK, c, T')$ should fail if $c \leftarrow E(PK, m, T)$ and $T' \neq T$. In the definition of CCA2-security for tagged encryption, the adversary is, as usual, given adaptive access to the decryption oracle D throughout its attack; it chooses two messages (m_0, m_1) and a tag T and is given a challenge ciphertext $c \leftarrow E(PK, m_b, T)$ for a random bit b. The adversary is allowed to issue further queries $(c', T') \neq (c, T)$ to D. The definition of security stipulates that the adversary cannot guess b with probability non-negligibly higher than $1/2$. We omit the formal definition of CCA2-security with tags, and refer the reader to prior work.

Pseudorandom Permutations. We also use pseudorandom permutations (PRPs). Recall [15] that a polynomial-time algorithm $p_{(\cdot)}(\cdot)$ defines a pseudorandom permutation family if for every key $K \in \{0,1\}^*$, $p_K : \{0,1\}^{\ell(|K|)} \mapsto \{0,1\}^{\ell(|K|)}$ (where the function $\ell(\cdot)$ is upper-bounded by a polynomial, and is called the "block length" of p) is a permutation and is indistinguishable from a random permutation by any probabilistic poly-time adversary $\mathcal{A}$ with adaptive access to both p_K and p_K^{-1}. We have the same key K define a set of simultaneously pseudorandom permutations $\{p_K^i : 1 \leq i \leq \ell(|K|)\}$, where i is the block length for a permutation p_K^i. (This can be obtained from any standard pseudorandom permutation family by standard techniques. For example, let $K_i = F_K(i)$, where F is a pseudorandom function, and let $p_K^i = p_{K_i}^i$.)

Notation. In the sequel, we will denote p_K^i by p_K because the block length is always clear from the context. Let $\{m\}_K$ denote $p_K^{|m|}(m)$. Let $\{m\}_{K^{-1}}$ denote $(p^{-1})_K^{|m|}(m)$. By '$\circ$' we denote concatenation.

Parameters. Let λ be the security parameter. It guides the choice of ℓ_K which is the length of a PRP key, and of ℓ_C, which is the upper bound on the length of a ciphertext formed using the CCA2 secure cryptosystem (Gen, E, D) when the security parameter is λ. Let ℓ_m be the length of a message being sent. Let $\ell_H = \ell_K + \ell_C$.

Non-standard Assumption on the PRP. We assume that, if P_1 and P_2 are two strings of length $2\ell_K$ chosen uniformly at random, then it is hard to find N keys $K_1, \ldots, K_N$ and a string C of length ℓ_C such that

$$\{\{\ldots\{P_1 \circ 0^{\ell_C}\}_{K_1^{-1}} \ldots\}_{K_{N-1}^{-1}}\}_{K_N^{-1}} \in \{P_1 \circ C, P_2 \circ C\}$$

In the random-oracle model, it is easy to construct a PRP with this property: if p is a PRP, define p' as $p'_K = p_{\mathcal{H}(K)}$ where $\mathcal{H}$ is a random oracle. If this assumption can hold in the standard model, then our construction is secure in the plain model as well.

4.1   Construction of Onions

We begin with intuition for our construction. Suppose that the sender P_s would like to route a message m to recipient $P_r = P_{n+1}$ through intermediate routers $(P_1, \ldots, P_n)$. For a moment, imagine that the sender P_s has already established a common one-time secret key K_i with each router P_i, $1 \leq i \leq n+1$. In that setting, the following construction would work and guarantee some (although not the appropriate amount of) security:

Intuition: Construction 1. For simplicity, let $N = 4$, $n = 3$, so the sender is sending message m to P_4 via intermediate routers P_1, P_2 and P_3. Send to P_1 the onion O_1 formed as follows:

$$O_1 = (\{\{\{\{m\}_{K_4}\}_{K_3}\}_{K_2}\}_{K_1}, \{\{\{P_4\}_{K_3}\}_{K_2}\}_{K_1}, \{\{P_3\}_{K_2}\}_{K_1}, \{P_2\}_{K_1})$$

Upon receipt of this $O_1 = (M^{(1)}, H_3^{(1)}, H_2^{(1)}, H_1^{(1)})$, P_1 will remove a layer of encryption using key K_1, and obtain

$$(\{M^{(1)}\}_{K_1^{-1}}, \{H_3^{(1)}\}_{K_1^{-1}}, \{H_2^{(1)}\}_{K_1^{-1}}, \{H_1^{(1)}\}_{K_1^{-1}}) =$$
$$(\{\{\{m\}_{K_4}\}_{K_3}\}_{K_2}, \{\{P_4\}_{K_3}\}_{K_2}, \{P_3\}_{K_2}\}, P_2)$$

Now P_1 knows that P_2 is the next router. It could, therefore, send to P_2 the set of values $(\{M^{(1)}\}_{K_1^{-1}}, \{H_3^{(1)}\}_{K_1^{-1}}, \{H_2^{(1)}\}_{K_1^{-1}})$. But then the resulting onion O_2 will be shorter than O_1, which in this case would make it obvious to P_2 that he is only two hops from the recipient; while we want P_2 to think that he could be up to $N - 1$ hops away from the recipient. Thus, P_1 needs to pad the onion somehow. For example, P_1 picks a random string R_1 of length $|P_1|$ and sets:

$$(O_2, P_2) = \mathsf{ProcOnion}(K_1, O_1, P_1)$$
$$= ((\{M^{(1)}\}_{K_1^{-1}}, R_1, \{H_3^{(1)}\}_{K_1^{-1}}, \{H_2^{(1)}\}_{K_1^{-1}}), \{H_1^{(1)}\}_{K_1^{-1}})$$
$$= ((\{\{\{m\}_{K_4}\}_{K_3}\}_{K_2}, R_1, \{\{P_4\}_{K_3}\}_{K_2}, \{P_3\}_{K_2}\}), P_2)$$

Upon receipt of this $O_2 = (M^{(2)}, H_3^{(2)}, H_2^{(2)}, H_1^{(2)})$, P_2 will execute the same procedure as P_1, but using his key K_2, and will obtain onion O_3 and the identity of router P_3. Upon receipt of O_3, P_3 will also apply the same procedure and obtain O_4 and the identity of the router P_4. Finally, P_4 will obtain:

$$(O_5, P_5) = \mathsf{ProcOnion}(K_4, O_4, P_4)$$
$$= ((\{M^{(4)}\}_{K_4^{-1}}, R_4, \{H_3^{(4)}\}_{K_4^{-1}}, \{H_2^{(4)}\}_{K_4^{-1}}), \{H_1^{(4)}\}_{K_4^{-1}})$$
$$= ((m, R_4, \{R_3\}_{K_4^{-1}}, \{\{R_2\}_{K_3^{-1}}\}_{K_4^{-1}}), \{\{\{R_1\}_{K_2^{-1}}\}_{K_3^{-1}}\}_{K_4^{-1}})$$

How does P_4 know that he is the recipient? The probability over the choice of K_4 that P_5 obtained this way corresponds to a legal router name is negligible. Alternatively, P_4 may be able to tell if, by convention, a legal message m must begin with k 0's, where k is a security parameter.

Intuition: Construction 2. Let us now adapt Construction 1 to the public-key setting. It is clear that the symmetric keys K_i, $1 \leq i \leq n + 1$, need to be communicated to routers P_i using public-key encryption. In Construction 1, the only header information $H_1^{(i)}$ for router P_i was the identity of the next router, P_{i+1}. Now, the header information for router P_i must also include a public-key ciphertext $C_{i+1} = E(PK_{i+1}, K_{i+1}, T_{i+1})$, which will allow router P_{i+1} to obtain his symmetric key K_{i+1}. We need to explain how these ciphertexts are formed. Let us first consider C_1. Tag T_1 is used to provide the context within which router P_1 should decrypt C_1. C_1 exists in the context of the message part and the header of the onion, and therefore the intuitive thing to do is to set $T_1 = \mathcal{H}(M^{(1)}, H^{(1)})$, where $\mathcal{H}$ is a collision-resistant hash function. Similarly, $T_i = \mathcal{H}(M^{(i)}, H^{(i)})$, because router P_i uses the same ProcOnion procedure as router P_1. Therefore, to compute C_1, the sender first needs to generate the keys $(K_1, \ldots, K_{n+1})$, then compute $(C_2, \ldots, C_{n+1})$. Then the sender will have enough information to obtain the tag T_1 and to compute C_1.

So, let us figure out how to compute O_2. Consider how P_1 will process O_1 (adapting Construction 1):

$$
\begin{aligned}
(O_2, P_2) &= \mathsf{ProcOnion}(SK(P_1), O_1, P_1) \\
&= (M^{(2)}, H^{(2)}, C_2, P_2) \\
&= (\{M^{(1)}\}_{K_1^{-1}}, (R_1, \{H_3^{(1)}\}_{K_1^{-1}}, \{H_2^{(1)}\}_{K_1^{-1}}), \{H_1^{(1)}\}_{K_1^{-1}}) \\
&= (\{\{\{m\}_{K_4}\}_{K_3}\}_{K_2}, (R_1, \{\{C_4, P_4\}_{K_3}\}_{K_2}, \{C_3, P_3\}_{K_2}\}), C_2, P_2)
\end{aligned}
$$

We need to address how the value R_1 is formed. On the one hand, we have already established (in Construction 1) that it needs to be random-looking, as we need to make sure that P_2 does not realize that R_1 is a padding, rather than a meaningful header. On the other hand, consider the ciphertext $C_2 \leftarrow E(PK(P_2), K_2, T_2)$, where, as we have established $T_2 = \mathcal{H}(M^{(2)}, H^{(2)})$. So, as part of the header $H^{(2)}$, the value R_1 needs to be known to the sender at FormOnion time, to ensure that the ciphertext C_2 is formed using the correct tag T_2. Thus, let us set R_1 *pseudorandomly*, as follows: $R_1 = \{P_1 \circ 0^{\ell_C}\}_{K_1^{-1}}$, where recall that ℓ_C is the number of bits required to represent the ciphertext C_1. Similarly, $R_i = \{P_i \circ 0^{\ell_C}\}_{K_i^{-1}}$. (Why include the value P_i into the pad? This is something we need to make the proof of security go through. Perhaps it is possible to get rid of it somehow.)

Now we can explain how FormOnion works (still using $N = 4$, $n = 3$): pick symmetric keys (K_1, K_2, K_3, K_4). Let $R_i = \{P_i \circ 0^{\ell_C}\}_{K_i^{-1}}$ for $1 \leq i \leq 4$. First, form the innermost onion O_4, as follows:

$$
O_4 = (\{m\}_{K_4}, (R_3, \{R_2\}_{K_3^{-1}}, \{\{R_1\}_{K_3^{-1}}\}_{K_2^{-1}}), C_4 \leftarrow E(PK(P_4), K_4, T_4))
$$

where recall that $T_4 = \mathcal{H}(M^{(4)}, H^{(4)})$. Now, for $1 < i \leq 4$, to obtain O_{i-1} from $O_i = (M^{(i)}, (H_3^{(i)}, H_2^{(i)}, H_1^{(i)}), C_i)$, let

$$
\begin{aligned}
M^{(i-1)} &= \{M^{(i)}\}_{K_{i-1}} & H_3^{(i-1)} &= \{H_2^{(i)}\}_{K_{i-1}} \\
H_2^{(i-1)} &= \{H_1^{(i)}\}_{K_{i-1}} & H_1^{(i-1)} &= \{C_i, P_i\}_{K_{i-1}} \\
T_{i-1} &= \mathcal{H}(M^{(i-1)}, H^{(i-1)}) & C_{i-1} &\leftarrow E(PK(P_{i-1}), K_{i-1}, T_{i-1})
\end{aligned}
$$

It is easy to verify that the onions (O_1, O_2, O_3, O_4) formed this way will satisfy the correctness property (Definition 3).

We are now ready to describe our construction more formally. Note that without the intuition above, the more formal description of our construction may appear somewhat terse.

Setup. The key generation/setup algorithm G for a router is as follows: run $Gen(1^k)$ to obtain (PK, SK). Router name P must be a string of length $2\ell_K$, chosen uniformly at random by a trusted source of randomness; this needs to be done so that even for a PK chosen by an adversary, the name P of the corresponding router is still a random string. (In the random oracle model, this can be obtained by querying the random-oracle-like hash function on input PK.) Register (P, PK) with the PKI.

Forming an Onion. On input message $m \in \{0,1\}^{\ell_m}$, a set of router names $(P_1, \ldots, P_{n+1})$, and a set of corresponding public keys $(PK_1, \ldots, PK_{n+1})$, the algorithm FormOnion does:

1. (Normalize the input). If $n + 1 < N$, let $P_i = P_{n+1}$, and let $PK_i = PK_{n+1}$ for all $n + 1 < i \leq N$.

2. (Form inner layer). To obtain the inner onion O_N, choose symmetric keys $K_i \leftarrow \{0,1\}^{\ell_K}$, for $1 \leq i \leq N$. Let $R_i = \{P_i \circ 0^{\ell_C}\}_{K_i^{-1}}$. Let $M^{(N)} = \{m\}_{K_N}$. As for the header, $H_{N-1}^{(N)} = R_{N-1}$, $H_{N-2}^{(N)} = \{R_{N-2}\}_{K_{N-1}^{-1}}$, and, in general, $H_i^{(N)} = \{\ldots \{R_i\}_{K_{i+1}^{-1}} \ldots\}_{K_{N-1}^{-1}}$ for $1 \leq i < N - 1$. Let $T_N = \mathcal{H}(M^{(N)}, H_{N-1}^{(N)}, \ldots, H_1^{(N)})$. Finally, let $C_N \leftarrow E(PK_N, K_N, T_N)$. Let $O_N = (M^{(N)}, H_{N-1}^{(N)}, \ldots, H_1^{(N)}, C_N)$.

3. (Adding a layer). Once $O_i = (M^{(i)}, H_{N-1}^{(i)}, \ldots, H_1^{(i)}, C_i)$ is computed for any $1 < i \leq N$, compute O_{i-1} as follows: $M^{(i-1)} = \{M^{(i)}\}_{K_{i-1}}$; $H_j^{(i-1)} = \{H_{j-1}^{(i)}\}_{K_{i-1}}$ for $1 < j \leq N$; $H_1^{(i-1)} = \{P_i, C_i\}_{K_{i-1}}$. Let $T_{i-1} = \mathcal{H}(M^{(i-1)}, H_{N-1}^{(i-1)}, \ldots, H_1^{(i-1)})$. Finally, let $C_{i-1} \leftarrow E(PK_{i-1}, K_{i-1}, T_{i-1})$. The resulting onion is $O_{i-1} = (M^{(i-1)}, H_{N-1}^{(i-1)}, \ldots, H_1^{(i-1)}, C_{i-1})$.

Processing an Onion. On input a secret key SK, an onion $O = (M, H_N, \ldots, H_1, C)$, and the router name P, do: (1) compute tag $T = \mathcal{H}(M, H_N, \ldots, H_1)$; (2) let $K = D(SK, C, T)$; if $K = \perp$, reject; otherwise (3) let $(P', C') = \{H_1\}_{K^{-1}}$; (4) if P' does not correspond to a valid router name, output $(\{M\}_{K^{-1}}, \perp)$ (that means that P is the recipient of the message $m = \{M\}_{K^{-1}}$); otherwise (5) send to P' the onion $O' = (\{M\}_{K^{-1}}, \{P \circ 0^{\ell_C}\}_{K^{-1}}, \{H_N\}_{K^{-1}}, \ldots, \{H_2\}_{K^{-1}}, C')$

Theorem 2. *The construction described above is correct, achieves integrity, and is onion-secure in the PKI model where each router's name is chosen as a uniformly random string of length $2\ell_K$, and assuming that (1) (Gen, E, D) is a CCA-2 secure encryption with tags; (2) p is a PRP simultaneously secure for block lengths ℓ_M and ℓ_H for which the non-standard assumption holds, and (3) hash function $\mathcal{H}$ is collision-resistant.*

Proof. (Sketch) Correctness follows by inspection. Integrity is the consequence of our non-standard assumption: Suppose that our goal is to break the non-standard assumption. So we are given as input two strings P_1' and P_2'. We set up the the set of honest players $\mathcal{Q}$, together with their key pairs, as in Definition 2, giving each player a name chosen at random and assigning the strings P_1' and P_2' as names for two randomly chosen routers. Note that as our reduction was the one to set up all the keys for the honest routers, it is able to successfully answer all ProcOnion queries on their behalf, as required by Definition 4. Suppose the adversary is capable of producing an onion whose path is longer than N. With probability $1/|Q|$, this onion O_1 is sent to router $P_1 = P_1'$. Let $\{(P_1, O_1, K_1), \ldots, (P_i, O_i, K_i), \ldots\}$ be the evolution of this onion augmented by the symmetric keys $(K_1, \ldots, K_i, \ldots)$ that router P_i obtains while running ProcOnion$(SK(P_i), O_i, P_i)$. According to our ProcOnion construction, the value (if any) that router P_N obtains as a candidate for $(P_{N+1} \circ C_{N+1})$ is the string $\{H_1^{(N)}\}_{K_N^{-1}} = \{\{\ldots\{P_1 \circ 0^{\ell_C}\}_{K_1^{-1}}\ldots\}_{K_{N-1}^{-1}}\}_{K_N^{-1}} = P \circ C$. For this to be a valid onion O_{N+1}, P must be a valid router name. If $P = P_1$, then we have broken our assumption. Otherwise $P \neq P_1$, but then with probability at least $1/|Q|$, $P = P_2'$ and so we also break the non-standard assumption.

It remains to show onion-security. First, we use a counting argument to show that, with probability $1 - 2^{-\ell_K + \Theta(\log |Q|)}$ over the choice of router names, the adversary cannot re-wrap the challenge onion.

Suppose that the challenger produces the onion layers $(O_1, \ldots, O_j)$. Consider the header $H_{N-1}^{(j)}$ of the onion O_j. By construction, $H_{N-1}^{(j)} = R^{(j-1)} = \{P_{j-1} \circ 0^{\ell_C}\}_{K_{j-1}^{-1}}$. Also by construction, any SK, $O' = (M', H', C')$ and P' such that $O_j = $ ProcOnion(SK, O', P') must satisfy $\{P' \circ 0^{\ell_C}\}_{(K')^{-1}} = H_{N-1}^{(j)}$, where K' is the decryption of C' under key SK. Thus, to re-wrap the onion, the adversary must choose P_{j-1}, P' and K' such that $\{P_{j-1} \circ 0^{\ell_C}\}_{K_{j-1}^{-1}} = \{P' \circ 0^{\ell_C}\}_{(K')^{-1}}$.

Let P be a router name, and let K be a key for the PRP p. Let

$$Bad(P, K) = \{P' : \exists K' \text{ such that } P' \neq P \wedge \{P \circ 0^{\ell_C}\}_{K^{-1}} = \{P' \circ 0^{\ell_C}\}_{(K')^{-1}}\} .$$

As there are at most 2^{ℓ_K} choices for K', and p is a permutation, for all (P, K), $|Bad(P, K)| \leq 2^{\ell_K}$. Let $Bad(Q, K) = \{P' : \exists P \in Q \text{ such that } P' \in Bad(P, K)\}$. Then $|Bad(Q, K)| \leq |Q| \max_P |Bad(P, K)| \leq |Q| 2^{\ell_K}$.

Assume without loss of generality that the key K_{j-1} is fixed. Thus, for this onion to be "re-wrappable," it must be the case that there exists some $P' \in Bad(Q, K_{j-1})$ that corresponds to a valid router name, i.e. $Q \cap Bad(Q, K_{j-1}) \neq \emptyset$. As any $P' \in Q$ is chosen uniformly out of a set of size $2^{2\ell_K}$, while $|Bad(Q, K_{j-1})| \leq 2^{\ell_K + \log |Q|}$, it is easy to see that the probability over the choice of K_{j-1} and the router names for the set Q that the onion is "re-wrappable," is only $2^{-\ell_K + \Theta(\log |Q|)}$.

It remains to show that no adversary can guess the challenger's bit b, provided (as we have shown) that it cannot re-wrap the onion. This proof follows the standard "sequence of games" [20] argument. Suppose that we set up the following experiments. In experiment (1), the challenger interacts with the adversary as in Definition 5 when $b = 0$, using FormOnion. In experiment (2), the challenger departs from the first experiment in that it deviates from the usual FormOnion algorithm in forming the ciphertext C_j as $C_j \leftarrow E(PK, K', T_j)$, where $K' \neq K_j$ is an independently chosen key. It is easy to see that distinguishing experiments (1) and (2) is equivalent to breaking either the CCA2 security of the underlying cryptosystem, or the collision-resistance property of $\mathcal{H}$.

In experiment (3), the challenger forms O_j as follows: Choose keys $K_1, \ldots K_{j-1}$, and K'. Let $R_i = \{P_i \circ 0^{\ell_C}\}_{K_i^{-1}}$ for $1 \le i < j$. $M^{(j)} \leftarrow \{0,1\}^{\ell_m}$, $H_i^{(j)} = \{\ldots \{R_i\}_{K_{i+1}^{-1}} \ldots\}_{K_{j-1}^{-1}}$ for $1 \le i < j$, $H_i^{(j)} \leftarrow \{0,1\}^{\ell_H}$ for $j \le i \le N-1$. Finally, $C_j \leftarrow E(PK, K', T_j)$. The other onions, O_{j-1} through O_1, are formed using the "adding a layer" part of the FormOnion construction. It can be shown (omitted here for lack of space) that an adversary who can distinguish experiments (2) and (3) can distinguish p_{K_j} from a random permutation. The intuition here is that in experiment (3), everything that's supposed to be the output of p_{K_j} or $p_{K_j}^{-1}$ is random.

In experiment (4), the onion is formed by running FormOnion$(r, (P_1, \ldots, P_j),$ $(PK_1, \ldots, PK_j))$, except that C_j is formed as $C_j \leftarrow E(PK, K', T_j)$. Telling (3) and (4) apart is also equivalent to distinguishing p from a random permutation. The intuition here is that in experiment (4) the first $j-1$ parts of the header of onion O_j are formed as in experiment (3), while the rest are formed differently, and permuted using key K_j.

Finally, experiment (5) does what the challenger would do when $b = 1$. It is easy to see that distinguishing between (4) and (5) is equivalent to breaking CCA2 security of the cryptosystem or collision-resistant of $\mathcal{H}$.

4.2   Response Option

Suppose that P_s wants to send an anonymous message m to P_r and wants P_r to be able to respond. Our construction allows for that possibility (however we omit the definition and proof of security).

The sender chooses a path $(P'_1, \ldots, P'_n)$ for the return onion, (so $P'_0 = P_r$, and $P'_{n+1} = P_s$). Next, the sender forms $(O'_1, \ldots, O'_{n+1}) = $ FormOnion$(\varepsilon, (P'_1, \ldots, P'_{n+1}), (PK(P'_1), \ldots, PK(P'_{n+1})))$. It then chooses a symmetric authentication and encryption key a and remembers all the keys $(K'_1, \ldots, K'_{n+1})$ used during FormOnion. Finally, it forms its message as $m' = m \circ a \circ O'_1 \circ P'_1$, and forms its actual onion in the usual way, i.e., chooses intermediate routers $(P_1, \ldots, P_n)$ and sets $(O_1, \ldots, O_{n+1}) \leftarrow $ FormOnion$(m', (P_1, \ldots, P_n, P_r), (PK(P_1), \ldots, PK(P_n), PK(P_r)))$.

Upon receipt of $m' = (m, a, O'_1, P'_1)$, P_s responds as follows. Suppose his response is M. He encrypts and authenticates M using a, forming a ciphertext c_1. He then sends (c_1, O'_1) to P'_1, with the annotation that this is a response onion. A router P receiving a message (c, O') with the annotation that this is a response onion, applies ProcOnion to onion O' only, ignoring c. Recall that as a result of this, P' obtains (O'', P'') (what to send to the next router and who the next router is) and the key K'. It then sends the values $(\{c\}_{K'}, O'')$ to P'', also with the annotation that this is a response onion. Eventually, if all goes well, the tuple $(\{\ldots \{c_1\}_{K'_1} \ldots\}_{K'_n}, O'_{n+1})$ reaches P_s, who, upon processing O'_{n+1} recognizes that he is the recipient of this return onion, and is then able to obtain c_1 using the keys $K'_1, \ldots, K_n$ it stored, and to validate and decrypt c_1 using the key a. Note that, due to the symmetric authentication step using the key a, if P_r is honest, then no polynomial-time adversary can make P_s accept an invalid response.

Acknowledgments

We thank Ron Rivest for pointing out that our cryptographic definition must guarantee onion-integrity in addition to correctness and security. We are grateful to Leo Reyzin for valuable discussions. We thank the anonymous referees for their thoughtful comments. Jan Camenisch is supported by the IST NoE ECRYPT and by the IST Project PRIME, which receive research funding from the European Community's Sixth Framework Programme and the Swiss Federal Office for Education and Science. Anna Lysyanskaya is supported by NSF CAREER Grant CNS-0374661.

References

1. M. Backes, B. Pfitzmann, and M. Waidner. A general composition theorem for secure reactive systems. In *TCC 2004*, vol. 2951 of *LNCS*, pp. 336–354.
2. O. Berthold, A. Pfitzmann, and R. Standtke. The disadvantages of free MIX routes and how to overcome them. In *Proceedings of Designing Privacy Enhancing Technologies*, vol. 2009 of *LNCS*, pp. 30–45. Springer-Verlag, July 2000.
3. J. Camenisch and V. Shoup. Practical verifiable encryption and decryption of discrete logarithms. In *Advances in Cryptology — CRYPTO 2003*, LNCS, 2003.
4. R. Canetti. Universally composable security: A new paradigm for cryptographic protocols. In *Proc. 42nd IEEE Symposium on Foundations of Computer Science (FOCS)*, pp. 136–145, 2001.
5. R. Canetti, S. Halevi, and J. Katz. A forward-secure public-key encryption scheme. In *EUROCRYPT 2003*, vol. 2656 of *LNCS*, pp. 255–271. Springer Verlag, 2003.
6. D. Chaum. Untraceable electronic mail, return addresses, and digital pseudonyms. *Communications of the ACM*, 24(2):84–88, Feb. 1981.
7. D. Chaum. Security without identification: Transaction systems to make big brother obsolete. *Communications of the ACM*, 28(10):1030–1044, Oct. 1985.
8. D. Chaum. The dining cryptographers problem: Unconditional sender and recipient untraceability. *Journal of Cryptology*, 1:65–75, 1988.
9. G. Danezis. The traffic analysis of continuous-time mixes. In *Privacy Enhancing Technologies (PET)*, 2004.
10. R. Dingledine, N. Mathewson, and P. F. Syverson. Tor: The second-generation onion router. In *USENIX Security Symposium*, pp. 303–320. USENIX, 2004.
11. D. Dolev, C. Dwork, and M. Naor. Non-malleable cryptography. *SIAM Journal on Computing*, 2000.
12. D. M. Goldschlag, M. G. Reed, and P. F. Syverson. Onion routing for anonymous and private internet connections. *Comm. of the ACM*, 42(2):84–88, Feb. 1999.
13. S. Goldwasser and S. Micali. Probabilistic encryption. *Journal of Computer and System Sciences*, 28(2):270–299, Apr. 1984.
14. D. Kesdogan, D. Agrawal, and S. Penz. Limits of anonymity in open environments. In *Information Hiding 2003*, vol. 2578 of *LNCS*, pp. 53–69. Springer, 2003.
15. M. Luby and C. Rackoff. How to construct pseudorandom permutations and pseudorandom functions. *SIAM J. Computing*, 17(2):373–386, Apr. 1988.
16. B. Möller. Provably secure public-key encryption for length-preserving Chaumian mixes. In *Cryptographer's Track — RSA 2003*, pp. 244–262. Springer, 2003.

17. B. Pfitzmann and M. Waidner. A model for asynchronous reactive systems and its application to secure message transmission. In *IEEE Symposium on Research in Security and Privacy*, pp. 184–200. IEEE Computer Society Press, 2001.
18. M. K. Reiter and A. D. Rubin. Crowds: anonymity for Web transactions. *ACM Transactions on Information and System Security (TISSEC)*, 1(1):66–92, 1998.
19. V. Shoup. A proposal for an ISO standard for public key encryption. http://eprint.iacr.org/2001/112, 2001.
20. V. Shoup. Sequences of games: a tool for taming complexity in security proofs. http://eprint.iacr.org/2004/332, 2004.
21. V. Shoup and R. Gennaro. Securing threshold cryptosystems against chosen ciphertext attack. In *EUROCRYPT '98*, vol. 1403 of *LNCS*. Springer, 1998.
22. D. Wikström. A universally composable mix-net. In *Theory of Cryptography Conference*, vol. 2951 of *LNCS*, pp. 317–335. Springer, 2004.
23. Y. Zhu, X. Fu, B. Graham, R. Bettati, and W. Zhao. On flow correlation attacks and countermeasures in mix networks. In *PET*, 2004.

Simple and Efficient Shuffling with Provable Correctness and ZK Privacy

Kun Peng, Colin Boyd, and Ed Dawson

Information Security Institute,
Queensland University of Technology
{k.peng, c.boyd, e.dawson}@qut.edu.au
http://www.isrc.qut.edu.au

Abstract. A simple and efficient shuffling scheme containing two protocols is proposed. Firstly, a prototype, Protocol-1 is designed, which is based on the assumption that the shuffling party cannot find a linear relation of the shuffled messages in polynomial time. As application of Protocol-1 is limited, it is then optimised to Protocol-2, which does not need the assumption. Both protocols are simpler and more efficient than any other shuffling scheme with unlimited permutation. Moreover, they achieve provable correctness and ZK privacy.

Keywords: Shuffling, permutation, correctness, privacy, zero knowledge.

1 Introduction

Shuffling is a very important cryptographic primitive. In a shuffling, a party re-encrypts and shuffles a number of input ciphertexts to the same number of output ciphertexts and publicly proves the validity of his operation. Its most important application is to build up anonymous channels used in e-voting [13], anonymous email [4] and anonymous browsing [7] etc. It is also employed in other cryptographic applications like multiparty computation [17] and electronic auction [18]. Two properties must be satisfied in a shuffling. The first property is correctness, which requires the shuffling party's validity proof to guarantee that the plaintexts of the outputs are a permutation of the plaintexts of the inputs. The second property is privacy, which requires the validity proof of the shuffling to be zero knowledge.

Recently, several shuffling schemes [1,2,6,13,8,19,15] have been proposed. Among them, [2] is a slight modification of [1]; [15] is a Paillier-encryption-based version of [6]; a similar idea is used in [13] and [8]. Except [19], all of them employ complicated proof techniques to prove correctness of the shuffling. The shuffling in [1] and [2] employs a large and complex shuffling circuit; [6] and [15] explicitly deal with a $n \times n$ matrix (n is the number of inputs); [13] and [8] employ proof of equality of product of exponents. Complexity of the proof causes several drawbacks. Firstly, correctness of the shuffling is not always strict. More precisely, in [8], if an input is shuffled to its minus ($g^q = -1 \mod 2q + 1$

V. Shoup (Ed.): Crypto 2005, LNCS 3621, pp. 188–204, 2005.

where q and $2q + 1$ are primes and the order of g modulo $2q + 1$ is $2q$), the proof can be accepted with a probability no smaller than 0.5. Secondly, some details of the proof (for example, the efficiency optimisation mechanism in [8]) are too complex to be easily understood or strictly analysed. Thirdly, the proofs in [6], [13] and [15] are not honest-verifier zero knowledge as pointed out in [10], [15] and [14]. So their privacy cannot be strictly and formally guaranteed. Finally, the proof is inefficient in all of them except [19]. Especially, the computational cost in [1] and [2] are linear in $n \log n$ while [13] and [8] need seven rounds of communication.

Although [19] is simple and very efficient, it has two drawbacks. Firstly, only a fraction of all the possible permutations are permitted. Secondly, it needs an assumption called linear ignorance assumption in this paper.

Definition 1. *Let $D()$ be the decryption function for an encryption scheme with plaintext space $\{0, 1, \ldots, q - 1\}$. Suppose an adversary $\mathcal{A}$ is given a set of n valid ciphertexts $c_1, c_2, \ldots, c_n$. $\mathcal{A}$ succeeds if it outputs integers $l_1, l_2, \ldots, l_n$, not all zero, such that $\sum_{i=1}^{n} l_i D(c_i) = 0 \bmod q$. The* linear ignorance assumption *states that there is no efficient adversary that can succeed with non-negligible probability.*

In [19], linear ignorance assumption is used against the shuffling party, who receives some ciphertext to shuffle and acts as the adversary. It is assumed in [19] that given the ciphertexts to shuffle, the probability that the shuffling party can efficiently find a linear relation about the messages encrypted in them is negligible. When the encryption scheme is semantically secure and the distribution of $D(c_1), D(c_2), \ldots, D(c_n)$ is unknown, this assumption is reasonable. However, if some party with some information about $D(c_1), D(c_2), \ldots, D(c_n)$ collude with the shuffling party, this assumption fails.

In this paper, two correct and private shuffling protocols, denoted as Protocol-1 and Protocol-2, are proposed. Protocol-1 is a prototype and needs the linear ignorance assumption against the shuffling party. So the shuffling party's knowledge of the shuffled messages is strictly limited in Protocol-1. Therefore, Protocol-1 is not suitable for applications like e-voting, where the shuffling party (tallier) may get some information about the shuffled messages from some message providers (colluding voters). Protocol-2 is an optimization of Protocol-1. It requires slightly more computation than Protocol-1, but concretely realises linear ignorance of the shuffling party in regard to the ciphertexts to shuffle. Namely, in Protocol-2, linear ignorance of the shuffling party in regard to the ciphertexts is not an assumption but a provable fact, which is an advantage over [19] and Protocol-1. As a result, Protocol-2 does not need the linear ignorance assumption, so is suitable for a much wider range of applications than Protocol-1. Both the new shuffling protocols are honest-verifier zero knowledge and more efficient than [1,2,6,13,8,15]. Moreover, neither of them limits the permutation, which is an advantage over [19].

2   The Shuffling Protocol

Let n be the number of inputs. An additive homomorphic semantically-secure encryption scheme[1] like Paillier encryption [16] is employed where $E(m, r)$ stands for encryption of message m using random integer r, $RE(c, r)$ stands for re-encryption of ciphertext c using random integer r and $D(c)$ stands for decryption of ciphertext c. Additive homomorphism of the encryption scheme implies $RE(c, r) = cE(0, r)$. Let q be the modulus of the message space, which has no small factor. Any computation in any matrix or vector is modulo q in this paper. In encryption or re-encryption the random factor r is chosen from a set Q dependent on the encryption algorithm. $|m|$ stands for the bit length of an integer m. L is a security parameter, such that 2^L is no larger than the smallest factor of q.

M' stands for the transpose matrix of a matrix M. A matrix is called a permutation matrix if there is exactly one 1 in every row and exactly one 1 in every column in this matrix while the other elements in this matrix are zeros. ZP ($x_1, x_2, \ldots, x_k \mid f_1, f_2, \ldots, f_l$) stands for a ZK proof of knowledge of secret integers $x_1, x_2, \ldots, x_k$ satisfying conditions $f_1, f_2, \ldots, f_l$. $ExpCost(x)$ stands for the computational cost of an exponentiation computation with a x bit exponent. In this paper, it is assumed that $ExpCost(x)$ equals $1.5x$ multiplications. $ExpCost^n(x)$ stands the computational cost of the product of n exponentiations with x-bit exponents. Bellare $et\ al.$ [3] showed that $ExpCost^n(x)$ is no more than $n + 0.5nx$ multiplications.

In a shuffling, ciphertexts $c_1, c_2, \ldots, c_n$ encrypting messages $m_1, m_2, \ldots, m_n$ are sent to a shuffling party, who shuffles the ciphertexts into $c'_1, c'_2, \ldots, c'_n$ and has to prove that $D(c'_1), D(c'_2), \ldots, D(c'_n)$ is a permutation of $D(c_1), D(c_2), \ldots, D(c_n)$. Batch verification techniques in [17] indicate that if

$$\sum_{i=1}^{n} s_i D(c_i) = \sum_{i=1}^{n} s_{\pi(i)} D(c'_i) \bmod q \qquad (1)$$

can be satisfied with a non-negligible probability where $s_1, s_2, \ldots, s_n$ are randomly chosen and $\pi()$ is a permutation, the shuffling is correct and $D(c'_i) = D(c_{\pi(i)})$ for $i = 1, 2, \ldots, n$. However, direct verification of Equation (1) requires knowledge of $\pi()$. To protect privacy of the shuffling, $\pi()$ must not appear in the verification. Groth's shuffling scheme [8] shows that to prove Equation (1) without revealing $\pi()$ is complicated and inefficient. In the new shuffling scheme a much simpler method is employed. Firstly, it is proved that the shuffling party knows $t_1, t_2, \ldots, t_n$ such that

$$\sum_{i=1}^{n} s_i D(c_i) = \sum_{i=1}^{n} t_i D(c'_i) \bmod q \qquad (2)$$

---

[1] An encryption algorithm with encryption function $E()$ is additive homomorphic if $E(m_1)E(m_2) = E(m_1 + m_2)$ for any messages m_1 and m_2. An encryption algorithm is semantically-secure if given a ciphertext c and two messages m_1 and m_2, such that $c = E(m_i)$ where $i = 1$ or 2, there is no polynomial algorithm to find out i.

where it is not required to prove that $t_1, t_2, \ldots, t_n$ are a permutation of $s_1, s_2, \ldots, s_n$. This proof does not reveal the permutation, but is not strong enough to guarantee validity of the shuffling. Actually, Equation (2) only implies that under the linear ignorance assumption against the shuffling party there exists a matrix M such that $(D(c_1'), D(c_2'), \ldots, D(c_n')) \cdot M = (D(c_1), D(c_2), \ldots, D(c_n))$. As M need not be a permutation matrix, this proof only guarantees that $D(c_1), D(c_2), \ldots, D(c_n)$ is a linear combination of $D(c_1'), D(c_2'), \ldots, D(c_n')$ under the linear ignorance assumption against the shuffling party. However, repeating this proof in a non-linear manner can guarantee M is a permutation matrix under the linear ignorance assumption against the shuffling party. In Protocol-1, given random integers s_i and s_i' from $\{0, 1, \ldots, 2^L - 1\}$ for $i = 1, 2, \ldots n$, the shuffling party has to prove that he knows secret integers t_i and t_i' from Z_q for $i = 1, 2, \ldots n$, such that

$$\sum_{i=1}^{n} s_i D(c_i) = \sum_{i=1}^{n} t_i D(c_i') \bmod q$$

$$\sum_{i=1}^{n} s_i' D(c_i) = \sum_{i=1}^{n} t_i' D(c_i') \bmod q$$

$$\sum_{i=1}^{n} s_i s_i' D(c_i) = \sum_{i=1}^{n} t_i t_i' D(c_i') \bmod q$$

Note that $s_i s_i'$ and $t_i t_i'$ in the third equation breaks the linear relation among the three equations. Under the linear ignorance assumption against the shuffling party, the three equations above can guarantee correctness of the shuffling with an overwhelmingly large probability. In Protocol-2, every input to be shuffled is randomly distributed into two inputs, each in one of two input sets. Then the two sets of inputs are shuffled separately using the same permutation. As the distribution is random, the input messages in both shufflings are random and are unknown even to the original message providers. So it is impossible for the shuffling party to find any linear relation of the input messages in either shuffling as the employed encryption algorithm is semantically secure. As the two shufflings are identical, their outputs can be combined to be the final shuffled outputs.

2.1 Protocol-1

In Protocol-1, it is assumed that the shuffling party cannot find a linear relation of $m_1, m_2, \ldots, m_n$ in polynomial time. Protocol-1 is as follows.

1. The shuffling party randomly chooses $\pi()$, a permutation of $\{1, 2, \ldots, n\}$, and integers r_i from Q for $i = 1, 2, \ldots n$. He then outputs $c_i' = RE(c_{\pi(i)}, r_i)$ for $i = 1, 2, \ldots n$ while concealing $\pi()$.
2. A verifier randomly chooses and publishes s_i from $\{0, 1, \ldots, 2^L - 1\}$ for $i = 1, 2, \ldots n$. The shuffling party chooses r_i' from Q for $i = 1, 2, \ldots n$ and

publishes $c_i'' = c_i'^{t_i} E(0, r_i')$ for $i = 1, 2, \ldots n$ where $t_i = s_{\pi(i)}$. The shuffling party publishes ZK proof

$$ZP\ (\ t_i, r_i'\ |\ c_i'' = c_i'^{t_i} E(0, r_i')\ ) \qquad \text{for } i = 1, 2, \ldots n \tag{3}$$

and

$$ZP\ (\ r_i, t_i, r_i'\ \text{for } i = 1, 2, \ldots, n\ |\ \prod_{i=1}^{n} c_i^{s_i} \prod_{i=1}^{n} (E(0, r_i))^{t_i} E(0, r_i') = \prod_{i=1}^{n} c_i''\ ) \tag{4}$$

3. The verifier randomly chooses and publishes s_i' from $\{0, 1, \ldots, 2^L - 1\}$ for $i = 1, 2, \ldots n$. The shuffling party sets $t_i' = s_{\pi(i)}'$ for $i = 1, 2, \ldots n$ and publishes ZK proof

$$ZP\ (\ r_i, t_i, r_i', t_i'\ \text{for } i = 1, 2, \ldots n\ |\ \prod_{i=1}^{n} c_i^{s_i'} \prod_{i=1}^{n} (E(0, r_i))^{t_i'} = \prod_{i=1}^{n} c_i'^{t_i'},$$

$$\prod_{i=1}^{n} c_i^{s_i s_i'} \prod_{i=1}^{n} (E(0, r_i))^{t_i t_i'} (E(0, r_i'))^{t_i'} = \prod_{i=1}^{n} c_i''^{t_i'}\ ) \tag{5}$$

If the shuffling party is honest and sets $t_i = s_{\pi(i)}$ and $t_i' = s_{\pi(i)}'$, he can pass the verification as $\sum_{i=1}^{n} t_i D(c_i') = \sum_{i=1}^{n} s_{\pi(i)} D(c_{\pi(i)}) = \sum_{i=1}^{n} s_i D(c_i)$; $\sum_{i=1}^{n} t_i' D(c_i') = \sum_{i=1}^{n} s_{\pi(i)}' D(c_{\pi(i)}) = \sum_{i=1}^{n} s_i' D(c_i)$ and $\sum_{i=1}^{n} t_i t_i' D(c_i') = \sum_{i=1}^{n} s_{\pi(i)} s_{\pi(i)}' D(c_{\pi(i)}) = \sum_{i=1}^{n} s_i s_i' D(c_i)$. Theorem 1 shows that if the shuffling party can pass the verification with a non-negligible probability, his shuffling is correct.

Theorem 1. *If the verifier chooses his challenges s_i and s_i' randomly and the shuffling party in Protocol-1 can provide ZK proofs (3), (4) and (5) with a probability larger than 2^{-L}, there exists a $n \times n$ permutation matrix M such that $(m_1', m_2', \ldots, m_n')M = (m_1, m_2, \ldots, m_n)$ under the linear ignorance assumption against the shuffling party.*

To prove Theorem 1, the following lemmas are proved first.

Lemma 1. *If given random integers s_i from $\{0, 1, \ldots, 2^L - 1\}$ for $i = 1, 2, \ldots, n$, a party can find in polynomial time integers t_i from Z_q for $i = 1, 2, \ldots, n$ with a probability larger than 2^{-L}, such that $\sum_{i=1}^{n} s_i m_i = \sum_{i=1}^{n} t_i m_i' \bmod q$, then he can find in polynomial time a matrix M such that $(m_1', m_2', \ldots, m_n')M = (m_1, m_2, \ldots, m_n)$.*

Proof: Given any integer k in $\{1, 2, \ldots, n\}$ there must exist integers $s_1, s_2, \ldots, s_{k-1}, s_{k+1}, \ldots, s_n$ in $\{0, 1, \ldots, 2^L - 1\}$ and two different integers s_k and $\hat{s}_k$ in $\{0, 1, \ldots, 2^L - 1\}$ such that given $s_1, s_2, \ldots, s_n$ and $\hat{s}_k$, the party can find in polynomial time t_i and $\hat{t}_i$ from Z_q for $i = 1, 2, \ldots, n$ to satisfy the following two equations.

$$\sum_{i=1}^{n} s_i m_i = \sum_{i=1}^{n} t_i m_i' \bmod q \tag{6}$$

$$(\sum_{i=1}^{k-1} s_i m_i)\hat{s}_k m_k \sum_{i=k+1}^{n} s_i m_i = \sum_{i=1}^{n} \hat{t}_i m'_i \bmod q \tag{7}$$

Otherwise, for any $s_1, s_2, \ldots, s_{k-1}, s_{k+1}, \ldots, s_n$ there is at most one s_k to satisfy equation $\sum_{i=1}^{n} s_i m_i = \sum_{i=1}^{n} t_i m'_i \bmod q$. This deduction implies that among the 2^{nL} possible combinations of $s_1, s_2, \ldots, s_n$, the party can find in polynomial time t_i for $i = 1, 2, \ldots, n$ to satisfy $\sum_{i=1}^{n} s_i m_i = \sum_{i=1}^{n} t_i m'_i \bmod q$ for at most $2^{(n-1)L}$ combinations. This conclusion leads to a contradiction: given random integers s_i from $\{0, 1, \ldots, 2^L - 1\}$ for $i = 1, 2, \ldots, n$ the party can find in polynomial time t_i for $i = 1, 2, \ldots, n$ to satisfy $\sum_{i=1}^{n} s_i m_i = \sum_{i=1}^{n} t_i m'_i \bmod q$ with a probability no larger than 2^{-L}.

Subtracting (7) from (6) yields

$$(s_k - \hat{s}_k)m_k = \sum_{i=1}^{n} (t_i - \hat{t}_i)m'_i \bmod q$$

Note that $s_k \in \{0, 1, \ldots, 2^L - 1\}$, $\hat{s}_k \in \{0, 1, \ldots, 2^L - 1\}$, $s_k \neq \hat{s}_k$ and 2^L is no larger than the smallest factor of q. So $s_k - \hat{s}_k \neq 0 \bmod q$. Namely, given a non-zero integer $s_k - \hat{s}_k$, the party can find in polynomial time $t_i - \hat{t}_i$ for $i = 1, 2, \ldots, n$ such that $(s_k - \hat{s}_k)m_k = \sum_{i=1}^{n} (t_i - \hat{t}_i)m'_i \bmod q$. So, for any k in $\{1, 2, \ldots, n\}$ the party knows a vector $V_k = ((t_1 - \hat{t}_1)/(s_k - \hat{s}_k), (t_2 - \hat{t}_2)/(s_k - \hat{s}_k), \ldots, (t_n - \hat{t}_n)/(s_k - \hat{s}_k))'$ such that $m_k = (m'_1, m'_2, \ldots, m'_n)V_k$. Therefore, the party can find in polynomial time a matrix M such that $(m_1, m_2, \ldots, m_n) = (m'_1, m'_2, \ldots, m'_n)M$ where $M = (V_1, V_2, \ldots, V_n)$. □

Lemma 2. *If a party can find in polynomial time a $n \times n$ singular matrix M such that $(m'_1, m'_2, \ldots, m'_n)M = (m_1, m_2, \ldots, m_n)$ where $(m_1, m_2, \ldots, m_n)$ and $(m'_1, m'_2, \ldots, m'_n)$ are two vectors, then he can find in polynomial time a linear relation about $m_1, m_2, \ldots, m_n$.*

Proof: Suppose $M = (V_1, V_2, \ldots, V_n)$. Then $m_i = (m'_1, m'_2, \ldots, m'_n)V_i$.

As M is singular and the party can find in polynomial time M, he can find in polynomial time integers $l_1, l_2, \ldots, l_n$ and k such that $\sum_{i=1}^{n} l_i V_i = (0, 0, \ldots, 0)$ where $1 \leq k \leq n$ and $l_k \neq 0 \bmod q$. So

$$\sum_{i=1}^{n} l_i m_i = \sum_{i=1}^{n} l_i (m'_1, m'_2, \ldots, m'_n)V_i = (m'_1, m'_2, \ldots, m'_n) \sum_{i=1}^{n} l_i V_i = 0$$

Namely, the party can find in polynomial time $l_1, l_2, \ldots, l_n$ to satisfy $\sum_{i=1}^{n} l_i m_i = 0$ where $1 \leq k \leq n$ and $l_k \neq 0 \bmod q$. □

Lemma 3. *If a party can find a $n \times n$ non-singular matrix M and integers $l_1, l_2, \ldots, l_n$ and k in polynomial time such that $(m'_1, m'_2, \ldots, m'_n) = (m_1, m_2, \ldots, m_n)M$, $\sum_{i=1}^{n} l_i m'_i = 0$, $1 \leq k \leq n$ and $l_k \neq 0 \bmod q$ where $(m_1, m_2, \ldots, m_n)$ and $(m'_1, m'_2, \ldots, m'_n)$ are two vectors, then he can find a linear relation about $m_1, m_2, \ldots, m_n$ in polynomial time.*

Proof: As $(m'_1, m'_2, \ldots, m'_n) = (m_1, m_2, \ldots, m_n)M$ and $\sum_{i=1}^{n} l_i m'_i = 0$,

$$\sum_{i=1}^{n} l_i (m_1, m_2, \ldots, m_n) V_i = 0 \quad \text{where } M = (V_1, V_2, \ldots, V_n)$$

So

$$(m_1, m_2, \ldots, m_n) \sum_{i=1}^{n} l_i V_i = 0$$

Note that $\sum_{i=1}^{n} l_i V_i \neq (0, 0, \ldots, 0)$ as M is non-singular, $1 \leq k \leq n$ and $l_k \neq 0 \bmod q$. Therefore, the party can find a linear relation about $m_1, m_2, \ldots, m_n$ in polynomial time.   □

Lemma 4. *If given random integers s_i from $\{0, 1, \ldots, 2^L - 1\}$ for $i = 1, 2, \ldots, n$, a party can find a $n \times n$ non-singular matrix M and integers t_i from Z_q for $i = 1, 2, \ldots, n$ in polynomial time such that $(m_1, m_2, \ldots, m_n) = (m'_1, m'_2, \ldots, m'_n)M$ and $\sum_{i=1}^{n} s_i m_i = \sum_{i=1}^{n} t_i m'_i \bmod q$ where $(m_1, m_2, \ldots, m_n)$ and $(m'_1, m'_2, \ldots, m'_n)$ are two vectors, then $(s_1, s_2, \ldots, s_n)M = (t_1, t_2, \ldots, t_n)$ under the linear ignorance assumption against the shuffling party.*

Proof:

$$(m_1, m_2, \ldots, m_n) = (m'_1, m'_2, \ldots, m'_n)M$$

implies

$$m_i = (m'_1, m'_2, \ldots, m'_n)V_i \qquad \text{for } i = 1, 2, \ldots, n$$

where $M = (V_1, V_2, \ldots, V_n)$.

So

$$\sum_{i=1}^{n} s_i m_i = \sum_{i=1}^{n} t_i m'_i \bmod q$$

implies

$$(m'_1, m'_2, \ldots, m'_n) \sum_{i=1}^{n} s_i V_i = (m'_1, m'_2, \ldots, m'_n) \begin{pmatrix} t_1 \\ t_2 \\ \vdots \\ t_n \end{pmatrix}$$

So given random integers s_i from $\{0, 1, \ldots, 2^L - 1\}$ for $i = 1, 2, \ldots, n$, the party can find matrix $M = (V_1, V_2, \ldots, V_n)$ and integers t_i from Z_q for $i = 1, 2, \ldots, n$ in polynomial time such that

$$(m'_1, m'_2, \ldots, m'_n)\left(\sum_{i=1}^{n} s_i V_i - \begin{pmatrix} t_1 \\ t_2 \\ \vdots \\ t_n \end{pmatrix}\right) = 0 \qquad (8)$$

As M is non-singular,

$$(m_1', m_2', \ldots, m_n') = (m_1, m_2, \ldots, m_n)M^{-1}$$

So

$$\sum_{i=1}^n s_i V_i - \begin{pmatrix} t_1 \\ t_2 \\ \vdots \\ t_n \end{pmatrix} = \begin{pmatrix} 0 \\ 0 \\ \vdots \\ 0 \end{pmatrix}$$

otherwise according to Lemma 3 the party can find a linear relation about $m_1, m_2, \ldots, m_n$ in polynomial time, which is contradictory to the linear ignorance assumption against the shuffling party. So

$$\sum_{i=1}^n s_i V_i = \begin{pmatrix} t_1 \\ t_2 \\ \vdots \\ t_n \end{pmatrix} \quad \text{and thus } M' \begin{pmatrix} s_1 \\ s_2 \\ \vdots \\ s_n \end{pmatrix} = \begin{pmatrix} t_1 \\ t_2 \\ \vdots \\ t_n \end{pmatrix}$$

Namely,

$$(s_1, s_2, \ldots, s_n)M = (t_1, t_2, \ldots, t_n) \qquad \square$$

Lemma 5. *If $\sum_{i=1}^n y_i s_i = 0 \bmod q$ with a probability larger than 2^{-L} for random integers $s_1, s_2, \ldots, s_n$ from $\{0, 1, 2, \ldots, 2^L - 1\}$, then $y_i = 0 \bmod q$ for $i = 1, 2, \ldots, n$.*

Proof: Given any integer k in $\{1, 2, \ldots, n\}$, there must exist integers $s_1, s_2, \ldots, s_{k-1}, s_{k+1}, \ldots, s_n$ in $\{0, 1, \ldots, 2^L - 1\}$ and two different integers s_k and $\hat{s}_k$ in $\{0, 1, \ldots, 2^L - 1\}$ such that the following two equations are correct.

$$\sum_{i=1}^n y_i s_i = 0 \bmod q \qquad (9)$$

$$(\sum_{i=1}^{k-1} y_i s_i) + y_k \hat{s}_k + \sum_{i=k+1}^n y_i s_i = 0 \bmod q \qquad (10)$$

Otherwise, for any $s_1, s_2, \ldots, s_{k-1}, s_{k+1}, \ldots, s_n$ there is at most one s_k to satisfy equation $\sum_{i=1}^n y_i s_i = 0 \bmod q$. This deduction implies among the 2^{nL} possible combinations of $s_1, s_2, \ldots, s_n$, equation $\sum_{i=1}^n y_i s_i = 0 \bmod q$ is correct for at most $2^{(n-1)L}$ combinations. This conclusion leads to a contradiction: given random integers s_i from $\{0, 1, \ldots, 2^L - 1\}$ for $i = 1, 2, \ldots, n$, equation $\sum_{i=1}^n y_i s_i = 0 \bmod q$ is correct with a probability no larger than 2^{-L}.

Subtracting (10) from (9) yields

$$y_k(s_k - \hat{s}_k) = 0 \bmod q$$

Note that $GCD(s_k - \hat{s}_k, q) = 1$ as 2^L is no larger than the smallest factor of q, $s_k \neq \hat{s}_k$ and $s_k, \hat{s}_k$ are L-bit integers. So, $y_k = 0 \bmod q$. Note that k can be any integer in $\{1, 2, \ldots, n\}$. Therefore $y_i = 0 \bmod q$ for $i = 1, 2, \ldots, n$. $\square$

Proof of Theorem 1:
According to additive homomorphism of the employed encryption algorithm, ZK proofs (3), (4) and (5) guarantee that the shuffling party can find integers t_i and t'_i for $i = 1, 2, \ldots, n$ to satisfy

$$\sum_{i=1}^{n} s_i m_i = \sum_{i=1}^{n} t_i m'_i \bmod q \tag{11}$$

$$\sum_{i=1}^{n} s'_i m_i = \sum_{i=1}^{n} t'_i m'_i \bmod q \tag{12}$$

$$\sum_{i=1}^{n} s_i s'_i m_i = \sum_{i=1}^{n} t_i t'_i m'_i \bmod q \tag{13}$$

where $m'_i = D(c'_i)$ and s_i and s'_i for $i = 1, 2, \ldots, n$ are randomly chosen by the verifier.

According to Lemma 1, the shuffling party knows a matrix M such that

$$(m'_1, m'_2, \ldots, m'_n)M = (m_1, m_2, \ldots, m_n) \tag{14}$$

According to Lemma 2, M is non-singular under the linear ignorance assumption against the shuffling party.

According to Lemma 4, Equations (14) together with Equations (11), (12) and (13) implies

$$(s_1, s_2, \ldots, s_n)M = (t_1, t_2, \ldots, t_n) \tag{15}$$
$$(s'_1, s'_2, \ldots, s'_n)M = (t'_1, t'_2, \ldots, t'_n) \tag{16}$$
$$(s_1 s'_1, s_2 s'_2, \ldots, s_n s'_n)M = (t_1 t'_1, t_2 t'_2, \ldots, t_n t'_n) \tag{17}$$

under the linear ignorance assumption against the shuffling party.

Equation (15), Equation (16) and Equation (17) respectively imply

$$(s_1, s_2, \ldots, s_n)V_1 = t_1 \tag{18}$$
$$(s'_1, s'_2, \ldots, s'_n)V_1 = t'_1 \tag{19}$$
$$(s_1 s'_1, s_2 s'_2, \ldots, s_n s'_n)V_1 = t_1 t'_1 \tag{20}$$

where $M = (V_1, V_2, \ldots, V_n)$.

Equation (18) and Equation (19) imply

$$(s'_1, s'_2, \ldots, s'_n)V_1(s_1, s_2, \ldots, s_n)V_1 = t_1 t'_1 \tag{21}$$

Equation (20) and Equation (21) imply

$$(s'_1, s'_2, \ldots, s'_n)V_1(s_1, s_2, \ldots, s_n)V_1 = (s_1 s'_1, s_2 s'_2, \ldots, s_n s'_n)V_1$$

So

$$(s_1', s_2', \ldots, s_n') V_1 (s_1, s_2, \ldots, s_n) V_1 = (s_1', s_2', \ldots, s_n') \begin{pmatrix} v_{1,1} s_1 \\ v_{1,2} s_2 \\ \vdots \\ v_{1,n} s_n \end{pmatrix}$$

$$\text{where } V_1 = \begin{pmatrix} v_{1,1} \\ v_{1,2} \\ \vdots \\ v_{1,n} \end{pmatrix}$$

under the linear ignorance assumption against the shuffling party.

Note that $s_1', s_2', \ldots, s_n'$ are randomly chosen by the verifier. So according to Lemma 5,

$$V_1 (s_1, s_2, \ldots, s_n) V_1 = \begin{pmatrix} v_{1,1} s_1 \\ v_{1,2} s_2 \\ \vdots \\ v_{1,n} s_n \end{pmatrix}$$

under the linear ignorance assumption against the shuffling party. So

$$(s_1, s_2, \ldots, s_n) V_1 v_{1,i} = v_{1,i} s_i \text{ for } i = 1, 2, \ldots, n$$

under the linear ignorance assumption against the shuffling party.

Note that $V_1 \neq (0, 0, \ldots, 0)$ as M is non-singular. So there must exist integer k such that $1 \leq k \leq n$ and $v_{i,k} \neq 0 \bmod q$. So

$$(s_1, s_2, \ldots, s_n) V_1 = s_k$$

under the linear ignorance assumption against the shuffling party. Namely,

$$s_1 v_{1,1} + s_2 v_{1,2} + \ldots + s_n v_{1,n} = s_k \bmod q$$

and thus

$$s_1 v_{1,1} + s_2 v_{1,2} + \ldots + s_{k-1} v_{1,k-1} + (s_k - 1) v_{1,k} + s_{k+1} v_{1,k+1} + \ldots + s_n v_{1,n} = 0 \bmod q$$

under the linear ignorance assumption against the shuffling party.

Note that $s_1, s_2, \ldots, s_n$ are randomly chosen by the verifier. So according to Lemma 5, $v_{1,1} = v_{1,2} = \ldots = v_{1,k-1} = v_{1,k+1} = \ldots = v_{1,n} = 0$ and $v_{1,k} = 1$ under the linear ignorance assumption against the shuffling party. Namely, V_1 contains one 1 and $n - 1$ 0s under the linear ignorance assumption against the shuffling party.

For the same reason, V_i contains one 1 and $n - 1$ 0s for $i = 2, 3, \ldots, n$ under the linear ignorance assumption against the shuffling party. Note that M is non-singular. Therefore, M is a permutation matrix under the linear ignorance assumption against the shuffling party.   □

In some applications of shuffling like [17], only semantically encrypted ciphertexts $c_1, c_2, \ldots, c_n$ are given to the shuffling party while no information about $m_1, m_2, \ldots, m_n$ is known. So the linear ignorance assumption against the shuffling party (the shuffling party cannot find a linear relation about $m_1, m_2, \ldots, m_n$ in polynomial time) is satisfied. Therefore, the shuffling by Protocol-1 is correct in these applications according to Theorem 1.

2.2   Protocol-2

In Protocol-1, the linear ignorance assumption is necessary. That means Protocol-1 cannot guarantee correctness of the shuffling if someone with knowledge of any shuffled message colludes with the shuffling party. For example, when the shuffling is used to shuffle the votes in e-voting, some voters may collude with the shuffling party and reveal their votes. Then the shuffling party can tamper with some votes without being detected. So Protocol-1 is upgraded to Protocol-2, which can guarantee the linear ignorance and thus correctness of the shuffling without any assumption. The upgrade is simple. The input ciphertexts $c_1, c_2, \ldots, c_n$ are divided into two groups of random ciphertexts $d_1, d_2, \ldots, d_n$ and $e_1, e_2, \ldots, e_n$ such that $c_i = e_i d_i$ for $i = 1, 2, \ldots, n$. Then Protocol-1 can be applied to shuffle $d_1, d_2, \ldots, d_n$ and $e_1, e_2, \ldots, e_n$ using an identical permutation. After the shuffling, the two groups of outputs are combined to recover the re-encrypted permutation of $c_1, c_2, \ldots, c_n$. Protocol-2 is as follows.

1. The shuffling party calculates $d_i = h(c_i)$ for $i = 1, 2, \ldots, n$ where $h()$ is a random oracle query implemented by a hash function from the ciphertext space of the employed encryption algorithm to the same ciphertext space. Thus two groups of ciphertexts d_i for $i = 1, 2, \ldots, n$ and $e_i = c_i/d_i$ for $i = 1, 2, \ldots, n$ are obtained.

2. The shuffling party randomly chooses $\pi()$, a permutation of $\{0, 1, \ldots, n\}$ and integers r_i and u_i from Q for $i = 1, 2, \ldots n$. He then outputs $d'_i = RE(d_{\pi(i)}, r_i)$ and $e'_i = RE(e_{\pi(i)}, u_i)$ for $i = 1, 2, \ldots n$ while concealing $\pi()$.

3. The verifier randomly chooses and publishes s_i from $\{0, 1, \ldots, 2^L - 1\}$ for $i = 1, 2, \ldots n$. The shuffling party chooses r'_i from Q for $i = 1, 2, \ldots n$ and publishes $d''_i = d'^{t_i}_i E(0, r'_i)$ for $i = 1, 2, \ldots n$ where $t_i = s_{\pi(i)}$. The shuffling party publishes ZK proof

$$ZP\left(\, t_i, r'_i \mid d''_i = d'^{t_i}_i E(0, r'_i)\,\right) \qquad \text{for } i = 1, 2, \ldots n \qquad (22)$$

and

$$ZP\left(\, r_i, u_i, t_i, r'_i \text{ for } i = 1, 2, \ldots, n \mid \right.$$

$$\prod_{i=1}^{n} d_i^{s_i} \prod_{i=1}^{n} (E(0, r_i))^{t_i} E(0, r'_i) = \prod_{i=1}^{n} d''_i, \qquad (23)$$

$$\left. \prod_{i=1}^{n} e_i^{s_i} \prod_{i=1}^{n} (E(0, u_i))^{t_i} = \prod_{i=1}^{n} e'^{t_i}_i \,\right)$$

4. The verifier randomly chooses and publishes s'_i from $\{0, 1, \ldots, 2^L - 1\}$ for $i = 1, 2, \ldots n$. The shuffling party sets $t'_i = s'_{\pi(i)}$ for $i = 1, 2, \ldots n$ and publishes ZK proof

$$ZP\ (\ r_i, t_i, r'_i, t'_i \text{ for } i = 1, 2, \ldots n \mid \prod_{i=1}^{n} d_i^{s'_i} \prod_{i=1}^{n} (E(0, r_i))^{t'_i} = \prod_{i=1}^{n} d'^{t'_i}_i,$$

$$\prod_{i=1}^{n} d_i^{s_i s'_i} \prod_{i=1}^{n} (E(0, r_i))^{t_i t'_i} (E(0, r'_i))^{t_i} = \prod_{i=1}^{n} d''^{t'_i}_i\ ) \quad (24)$$

5. If the proofs above are verified to be valid, the outputs of the shuffling are $c'_i = d'_i e'_i$ for $i = 1, 2, \ldots n$.

Just like in Protocol-1, if the shuffling party is honest and sets $t_i = s_{\pi(i)}$ and $t'_i = s'_{\pi(i)}$, he can pass the verification in Protocol-2. Theorem 2 shows that if the shuffling party can pass the verification in Protocol-2 with a non-negligible probability, his shuffling is correct even without the linear ignorance assumption.

Theorem 2. *If the verifier chooses his challenges s_i and s'_i randomly and the shuffling party in Protocol-2 can provide ZK proofs (22), (23) and (24) with a probability larger than 2^{-L}, then there is an identical permutation from $D(d_1), D(d_2), \ldots, D(d_n)$ to $D(d'_1), D(d'_2), \ldots, D(d'_n)$ and from $D(e_1), D(e_2), \ldots, D(e_n)$ to $D(e'_1), D(e'_2), \ldots, D(e'_n)$.*

Proof: According to additive homomorphism of the employed encryption, ZK proofs (22), (23) and (24) guarantee that the shuffling party can find integers t_i and t'_i for $i = 1, 2, \ldots, n$ to satisfy

$$\sum_{i=1}^{n} s_i D(d_i) = \sum_{i=1}^{n} t_i D(d'_i) \bmod q \quad (25)$$

$$\sum_{i=1}^{n} s_i D(e_i) = \sum_{i=1}^{n} t_i D(e'_i) \bmod q \quad (26)$$

$$\sum_{i=1}^{n} s'_i D(d_i) = \sum_{i=1}^{n} t'_i D(d'_i) \bmod q \quad (27)$$

$$\sum_{i=1}^{n} s_i s'_i D(d_i) = \sum_{i=1}^{n} t_i t'_i D(d'_i) \bmod q \quad (28)$$

where s_i and s'_i for $i = 1, 2, \ldots, n$ are randomly chosen by the verifier.

Note that $d_1, d_2, \ldots, d_n$ are produced by the hash function $h()$, which is regarded as a random oracle. So to find a linear relation about $D(d_1), D(d_2), \ldots, D(d_n)$ is equivalent to repeatedly querying a random oracle for a vector of n random ciphertexts and then finding a linear relation on the plaintexts corrresponding to one of these vectors. This is infeasible as the employed encryption algorithm is semantically secure. So the probability that the shuffling party can find any linear relation about $D(d_1), D(d_2), \ldots, D(d_n)$ is negligible. For the same reason, the probability that the shuffling party can find any linear relation about $D(e_1), D(e_2), \ldots, D(e_n)$ is negligible.

According to Theorem 1, Equations (25), (27) and (28) imply that there exists a permutation matrix M such that

$$(D(d_1'), D(d_2'), \ldots, D(d_n'))M = (D(d_1), D(d_2), \ldots, D(d_n))$$

So according to Lemma 4,

$$(s_1, s_2, \ldots, s_n)M = (t_1, t_2, \ldots, t_n) \tag{29}$$

According to Lemma 1 and Lemma 4, Equation (26) implies that there exists a matrix $\hat{M}$ such that

$$(D(e_1'), D(e_2'), \ldots, D(e_n'))\hat{M} = (D(e_1), D(e_2), \ldots, D(e_n))$$

and

$$(s_1, s_2, \ldots, s_n)\hat{M} = (t_1, t_2, \ldots, t_n) \tag{30}$$

Subtracting (30) from (29) yields

$$(s_1, s_2, \ldots, s_n)(M - \hat{M}) = (0, 0, \ldots, 0)$$

According to Lemma 5, every column vector in matrix $M - \hat{M}$ contains n zeros. So $M = \hat{M}$. Therefore there is an identical permutation (matrix) from $D(d_1), D(d_2), \ldots, D(d_n)$ to $D(d_1'), D(d_2'), \ldots, D(d_n')$ and from $D(e_1), D(e_2), \ldots, D(e_n)$ to $D(e_1'), D(e_2'), \ldots, D(e_n')$. $\square$

According to Theorem 2, $D(d_1)D(e_1), D(d_2)D(e_2), \ldots, D(d_n)D(e_n)$ is permuted to $D(d_1')D(e_1'), D(d_2')D(e_2'), \ldots, D(d_n')D(e_n')$. Namely, $D(c_1'), D(c_2'), \ldots, D(c_n')$ is a permutation of $D(c_1), D(c_2), \ldots, D(c_n)$ even in the absence of the linear ignorance assumption.

3   Implementation and Cost

The additive homomorphic semantically secure encryption employed in Protocol-1 may be the modified ElGamal encryption [11,12] or Paillier encryption [16]. The implementation details and computational cost are slightly different with different encryption schemes. For example, the following Paillier encryption algorithm can be employed. $N = p_1 p_2$, $p_1 = 2p_1' + 1$, $p_2 = 2p_2' + 1$ where p_1, p_2, p_1' and p_2' are large primes and $GCD(N, p_1' p_2') = 1$. Integers a, b are randomly chosen from Z_N^* and $g = (1 + N)^a + b^N \bmod N$. The public key consists of N and g. The private key is $\beta p_1' p_2'$ where β is randomly chosen from Z_N^*. A message $m \in Z_N$ is encrypted to $c = g^m r^N \bmod N^2$ where r is randomly chosen from Z_N^*. The modulus of the message space is N. If Paillier encryption is employed, Protocol-1 can be implemented as follows.

1. The shuffling party randomly chooses integers r_i from Z_N^* for $i = 1, 2, \ldots n$. He then outputs $c_i' = c_{\pi(i)} r_i^N \bmod N^2$ for $i = 1, 2, \ldots n$.
2. After the verifier publishes s_i from $\{0, 1, \ldots, 2^L - 1\}$ for $i = 1, 2, \ldots n$, the shuffling party chooses r_i' from Z_N^* for $i = 1, 2, \ldots n$ and publishes $c_i'' = c_i'^{t_i} r_i'^N \bmod N^2$ for $i = 1, 2, \ldots n$ where $t_i = s_{\pi(i)}$. The shuffling party publishes ZK proof

$$ZP\ (\ t_i, r_i' \mid c_i'' = c_i'^{t_i} r_i'^N \bmod N^2\ ) \qquad \text{for } i = 1, 2, \ldots n \qquad (31)$$

and

$$ZP\ (\ R_1 \mid R_1^N = C_1 \bmod N^2\ ) \qquad (32)$$

where $R_1 = \prod_{i=1}^n r_i^{t_i} r_i' \bmod N^2$ and $C_1 = \prod_{i=1}^n c_i'' / \prod_{i=1}^n c_i^{s_i} \bmod N^2$.
3. After the verifier publishes s_i' from $\{0, 1, \ldots, 2^L - 1\}$ for $i = 1, 2, \ldots n$, the shuffling party sets $t_i' = s_{\pi(i)}'$ for $i = 1, 2, \ldots n$ and publishes ZK proof

$$ZP\ (\ R_2, R_3, t_i' \text{ for } i = 1, 2, \ldots n \mid C_2 R_2^N = \prod_{i=1}^n c_i'^{t_i'} \bmod N^2,$$
$$C_3 R_3^N = \prod_{i=1}^n c_i''^{t_i'} \bmod N^2\ ) \qquad (33)$$

where $R_2 = \prod_{i=1}^n r_i^{t_i'} \bmod N^2$, $R_3 = \prod_{i=1}^n r_i^{t_i t_i'} r_i'^{t_i'} \bmod N^2$, $C_2 = \prod_{i=1}^n c_i^{s_i'} \bmod N^2$ and $C_3 = \prod_{i=1}^n c_i^{s_i s_i'} \bmod N^2$.

Non-interactive implementation of ZK proof (31), (32) and (33) can be implemented as follows.

1. The shuffling party randomly chooses $W_1 \in Z_N^*$, $W_2 \in Z_N^*$, $W_3 \in Z_N^*$, $v_i \in Z_N$ for $i = 1, 2, \ldots, n$, $v_i' \in Z_N$ for $i = 1, 2, \ldots, n$ and $x_i \in Z_N^*$ for $i = 1, 2, \ldots, n$. He calculates $a_i = c_i'^{v_i} x_i^N \bmod N^2$ for $i = 1, 2, \ldots, n$, $f = W_1^N \bmod N^2$, $a = (\prod_{i=1}^n c_i'^{v_i'})/W_2^N \bmod N^2$ and $b = (\prod_{i=1}^n (c_i''^{v_i'})/W_3^N \bmod N^2$.
2. The shuffling party calculates $c = H(f, a, b, a_1, a_2, \ldots, a_n)$ where $H()$ is a random oracle query implemented by a hash function with a 128-bit output.
3. The shuffling party calculates $z_1 = W_1 R_1^c \bmod N^2$, $z_2 = W_2/R_2^c \bmod N^2$, $z_3 = W_3/R_3^c \bmod N^2$, $\alpha_i = x_i r_i^c \bmod N^2$ for $i = 1, 2, \ldots, n$, $\gamma_i = v_i + ct_i \bmod N$ for $i = 1, 2, \ldots, n$ and $\gamma_i' = ct_i' - v_i' \bmod N$ for $i = 1, 2, \ldots, n$.
4. The shuffling party publishes $z_1, z_2, z_3, \alpha_1, \alpha_2, \ldots, \alpha_n, \gamma_1, \gamma_2, \ldots, \gamma_n, \gamma_1', \gamma_2', \ldots, \gamma_n'$. Anyone can verify that

$$c = H(\ z_1^N/C_1^c,\ C_2^c/(z_2^N \prod_{i=1}^n c_i'^{\gamma_i'}),\ C_3^c/(bz_3^N \prod_{i=1}^n c_i''^{\gamma_i'}),$$
$$c_i'^{\gamma_i} \alpha_i^N/c_i''^c \text{ for } i = 1, 2, \ldots, n\ ) \qquad (34)$$

This implementation is a combination of ZK proof of knowledge of logarithm [20], ZK proof of equality of logarithms [5] and ZK proof of knowledge of root [9]. All the three proof techniques are correct and specially sound, so this implementation guarantees Equations (3), (4) and (5). All of the three proof techniques are honest-verifier zero knowledge. So if the hash function can be regarded as a random oracle query, this implementation is zero knowledge. Therefore, ZK privacy is achieved in Protocol-1. In this implementation, the computational cost of shuffling is n full length exponentiations[2]; the cost of proof is $3nExpCost(|N|) + 2ExpCost^n(|N|) + nExpCost(L) + 3ExpCost^n(L) + ExpCost^n(2L) + (n+3)ExpCost(128) + 3$, which is approximately equal to $11n/3 + 8nL/(3|N|) + 128(n+3)/|N| + 3$ full length exponentiations.

ZK proofs (22), (23) and (24) in Protocol-2 can be implemented similarly. When Paillier encryption is employed, the computational cost of shuffling is $2n$ full length exponentiations; the cost of proof is approximately equal to $11n/3 + 11nL/(3|N|) + 128(n+4)/|N| + 3$ full length exponentiations. It is well known [11,12] that ElGamal encryption can be modified to be additive homomorphic. If the additional DL search in the decryption function caused by the modification is not an efficiency concern (e.g. when the messages are in a known small set), the modified ElGamal encryption can also be applied to our shuffling. An ElGamal-based shuffling only uses ZK proof of knowledge of logarithm [20] and ZK proof of equality of logarithms [5]. Note that in the ElGamal-based shuffling each output ciphertext must be verified to be in the ciphertext space. When a prime p is the multiplication modulus, the ciphertext space is the cyclic subgroup G with order q where q is a prime and $p = 2q + 1$. If an output is in $Z_p^* - G$, Proofs (3), (4), (5) cannot guarantee correctness of the shuffling. The implementation and cost of the ElGamal-based shuffling are similar to those of Paillier-based shuffling in both Protocol-1 and Protocol-2.

In summary, both protocols can be efficiently implemented with either Paillier encryption or ElGamal encryption to achieve correctness and privacy in the shuffling.

4   Conclusion

Two new shuffling protocols are proposed in this paper. The first protocol is a prototype and based on an assumption. The second one removes the assumption and can be applied to more applications. Both protocols are simple and efficient, and achieve all the desired properties of shuffling. In Tables 1, the new shuffling protocols based on Paillier encryption are compared against the existing shuffling protocols. It is demonstrated in Table 1 that Protocol-2 is the only shuffling scheme with strict correctness, unlimited permutation, ZK privacy and without the linear ignorance assumption. In Table 1 the computational cost is counted in terms of full-length exponentiations (with 1024-bit exponent) where $L = 20$. It is demonstrated that the new shuffling protocols are more efficient than the existing shuffling schemes except [19], which is not a complete shuffling.

---

[2] An exponentiation is called full length if the exponent can be as long as the order of the base.

Table 1. Comparison of computation cost in full-length exponentiations

	Correctness	Permutation	Privacy	Linear ignor-ance assumption	Computation cost (shuffling and proof)	Communication Rounds
[1,2]	strict	unlimited	ZK	unnecessary	$\geq 16(n \log_2 n - 2n + 2)$	3
[6,15]	strict	unlimited	not ZK	unnecessary	$10n$	3
[13]	strict	unlimited	not ZK	unnecessary	$12n$	7
[8][a]	not strict	unlimited	ZK	unnecessary	$8n + 3n/\kappa + 3$	7
[19][b]	strict	limited	ZK	necessary	$2n + k(4k - 2)$	3
Protocol-1	strict	unlimited	ZK	necessary	$n + \frac{369}{96}n + \frac{27}{8} < 5n$	3
Protocol-2	strict	unlimited	ZK	unnecessary	$2n + \frac{3077}{768}n + 3.5 \approx 6n$	3

[a] κ is a chosen parameter.
[b] k is a small parameter determined by the flexibility of permutation and strength of privacy.

Acknowledgements

We acknowledge the support of the Australian Research Council through ARC Discovery Grant No. DP0345458.

References

1. M Abe. Mix-networks on permutation net-works. In *ASIACRYPT '98*, volume 1716 of *Lecture Notes in Computer Science*, pages 258–273, Berlin, 1999. Springer-Verlag.
2. Masayuki Abe and Fumitaka Hoshino. Remarks on mix-network based on permutation networks. In *Public Key Cryptography 2001*, volume 1992 of *Lecture Notes in Computer Science*, pages 317–324, Berlin, 2001. Springer-Verlag.
3. M Bellare, J A Garay, and T Rabin. Fast batch verification for modular exponentiation and digital signatures. In *EUROCRYPT '98*, volume 1403 of *Lecture Notes in Computer Science*, pages 236–250, Berlin, 1998. Springer-Verlag.
4. D Chaum. Untraceable electronic mail, return address and digital pseudonym. *Communications of the ACM, 24(2)*, pages 84–88, 1981.
5. D. Chaum and T. P. Pedersen. Wallet databases with observers. In *CRYPTO '92*, volume 740 of *Lecture Notes in Computer Science*, pages 89–105, Berlin, 1992. Springer-Verlag.
6. Jun Furukawa and Kazue Sako. An efficient scheme for proving a shuffle. In *CRYPTO '01*, volume 2139 of *Lecture Notes in Computer Science*, pages 368–387, Berlin, 2001. Springer.
7. Eran Gabber, Phillip B. Gibbons, Yossi Matias, and Alain Mayer. How to make personalized web browsing simple, secure, and anonymous. In *Proceedings of Financial Cryptography 1997*, volume 1318 of *Lecture Notes in Computer Science*, pages 17–31, Berlin, 1997. Springer.
8. Jens Groth. A verifiable secret shuffle of homomorphic encryptions. In *Public Key Cryptography 2003*, volume 2567 of *Lecture Notes in Computer Science*, pages 145–160, Berlin, 2003. Springer-Verlag.

9. L. C. Guillou and J. J. Quisquater. A "paradoxical" identity-based signature scheme resulting from zero-knowledge. In Shafi Goldwasser, editor, *CRYPTO '88*, volume 403 of *Lecture Notes in Computer Science*, pages 216–231, Berlin, 1989. Springer-Verlag.

10. J.Furukawa, H.Miyauchi, K.Mori, S.Obana, and K.Sako. An implementation of a universally verifiable electronic voting scheme based on shuffling. In *Proceedings of Financial Cryptography 2002*, volume 2357 of *Lecture Notes in Computer Science*, pages 16–30, Berlin, 2002. Springer.

11. Byoungcheon Lee and Kwangjo Kim. Receipt-free electronic voting through collaboration of voter and honest verifier. In *JW-ISC 2000*, pages 101–108, 2000.

12. Byoungcheon Lee and Kwangjo Kim. Receipt-free electronic voting scheme with a tamper-resistant randomizer. In *Information Security and Cryptology, ICISC 2002*, volume 2587 of *Lecture Notes in Computer Science*, pages 389–406, Berlin, 2002. Springer-Verlag.

13. C. Andrew Neff. A verifiable secret shuffle and its application to e-voting. In *ACM Conference on Computer and Communications Security 2001*, pages 116–125, 2001.

14. Lan Nguyen and Rei Safavi-Naini. An efficient verifiable shuffle with perfect zero-knowledge proof system. In *Cryptographic Algorithms and their Uses 2004*, pages 40–56, 2004.

15. Lan Nguyen, Rei Safavi-Naini, and Kaoru Kurosawa. Verifiable shuffles: A formal model and a paillier-based efficient construction with provable security. In *Applied Cryptography and Network Security, ACNS 2004*, volume 3089 of *Lecture Notes in Computer Science*, pages 61–75, Berlin, 2004. Springer-Verlag.

16. P Paillier. Public key cryptosystem based on composite degree residuosity classes. In *EUROCRYPT '99*, volume 1592 of *Lecture Notes in Computer Science*, pages 223–238, Berlin, 1999. Springer-Verlag.

17. Kun Peng, Colin Boyd, Ed Dawson, and Byoungcheon Lee. An efficient and verifiable solution to the millionaire problem. In *Pre-Proceedings of ICISC 2004*, pages 315–330, 2004.

18. Kun Peng, Colin Boyd, Edward Dawson, and Kapali Viswanathan. Efficient implementation of relative bid privacy in sealed-bid auction. In *The 4th International Workshop on Information Security Applications, WISA2003*, volume 2908 of *Lecture Notes in Computer Science*, pages 244–256, Berlin, 2003. Springer-Verlag.

19. Kun Peng, Colin Boyd, Edward Dawson, and Kapali Viswanathan. A correct, private and efficient mix network. In *2004 International Workshop on Practice and Theory in Public Key Cryptography*, pages 439–454, Berlin, 2004. Springer-Verlag.

20. C Schnorr. Efficient signature generation by smart cards. *Journal of Cryptology, 4, 1991*, pages 161–174, 1991.

Searchable Encryption Revisited: Consistency Properties, Relation to Anonymous IBE, and Extensions

Michel Abdalla[1], Mihir Bellare[2], Dario Catalano[1], Eike Kiltz[2],
Tadayoshi Kohno[2], Tanja Lange[3], John Malone-Lee[4],
Gregory Neven[5], Pascal Paillier[6], and Haixia Shi[2]

[1] Departement d'Informatique, École normale supérieure
{Michel.Abdalla, Dario.Catalano}@ens.fr
http://www.di.ens.fr/users/{mabdalla,catalano}
[2] Department of Computer Science & Engineering,
University of California San Diego
{mihir, ekiltz, tkohno, hashi}@cs.ucsd.edu
http://www.cs.ucsd.edu/users/{mihir,ekiltz,tkohno,hashi}
[3] Department of Mathematics, Technical University of Denmark
t.lange@mat.dtu.dk
http://www.ruhr-uni-bochum.de/itsc/tanja
[4] Department of Computer Science, University of Bristol
malone@compsci.bristol.ac.uk
http://www.cs.bris.ac.uk/~malone
[5] Department of Electrical Engineering, Katholieke Universiteit Leuven
Gregory.Neven@esat.kuleuven.be
http://www.neven.org
[6] Applied Research and Security Center, Gemplus Card International
Pascal.Paillier@gemplus.com

Abstract. We identify and fill some gaps with regard to consistency (the extent to which false positives are produced) for public-key encryption with keyword search (PEKS). We define computational and statistical relaxations of the existing notion of perfect consistency, show that the scheme of [7] is computationally consistent, and provide a new scheme that is statistically consistent. We also provide a transform of an anonymous IBE scheme to a secure PEKS scheme that, unlike the previous one, guarantees consistency. Finally we suggest three extensions of the basic notions considered here, namely anonymous HIBE, public-key encryption with temporary keyword search, and identity-based encryption with keyword search.

1 Introduction

There has recently been interest in various forms of "searchable encryption" [18,7,12,14,20]. In this paper, we further explore one of the variants of this goal, namely public-key encryption with keyword search (PEKS) as introduced by

V. Shoup (Ed.): Crypto 2005, LNCS 3621, pp. 205–222, 2005.

Boneh, Di Crescenzo, Ostrovsky and Persiano [7]. We begin by discussing consistency-related issues and results, then consider the connection to anonymous identity-based encryption (IBE) and finally discuss some extensions.

1.1 Consistency in PEKS

Any cryptographic primitive must meet two conditions. One is of course a security condition. The other, which we will here call a *consistency* condition, ensures that the primitive fulfills its function. For example, for public-key encryption, the security condition is privacy. (This could be formalized in many ways, eg. IND-CPA or IND-CCA.) The consistency condition is that decryption reverses encryption, meaning that if M is encrypted under public key pk to result in ciphertext C, then decrypting C under the secret key corresponding to pk results in M being returned.

PEKS. In a PEKS scheme, Alice can provide a *gateway* with a trapdoor t_w (computed as a function of her secret key) for any keyword w of her choice. A sender encrypts a keyword w' under Alice's public key pk to obtain a ciphertext C that is sent to the gateway. The latter can apply a test function Test to t_w, C to get back 0 or 1. The consistency condition as per [7] is that if $w = w'$ then Test(t_w, C) returns 1 and if $w \neq w'$ it returns 0. The security condition is that the gateway learn nothing about w' beyond whether or not it equals w. (The corresponding formal notion will be denoted PEKS-IND-CPA.) The application setting is that C can be attached to an email (ordinarily encrypted for Alice under a different public key), allowing the gateway to route the email to different locations (eg. Alice's desktop, laptop or pager) based on w while preserving privacy of the latter to the largest extent possible.

CONSISTENCY OF $\mathcal{BDOP}$-$\mathcal{PEKS}$. It is easy to see (cf. Proposition 1) that the main construction of [7] (a random oracle model, BDH-based PEKS-IND-CPA secure PEKS scheme that we call $\mathcal{BDOP}$-$\mathcal{PEKS}$) fails to meet the consistency condition defined in [7] and stated above. (Specifically, there are distinct keywords w, w' such that Test$(t_w, C) = 1$ for any C that encrypts w'.) The potential problem this raises in practice is that email will be incorrectly routed.

NEW NOTIONS OF CONSISTENCY. It is natural to ask if $\mathcal{BDOP}$-$\mathcal{PEKS}$ meets some consistency condition that is weaker than theirs but still adequate in practice. To answer this, we provide some new definitions. Somewhat unusually for a consistency condition, we formulate consistency more like a security condition, via an experiment involving an adversary. The difference is that this adversary is not very "adversarial": it is supposed to reflect some kind of worst case but not malicious behavior. However this turns out to be a difficult line to draw, definitionally, so that some subtle issues arise. One advantage of this approach is that it naturally gives rise to a hierarchy of notions of consistency, namely perfect, statistical and computational. The first asks that the advantage of any (even computationally unbounded) adversary be zero; the second that the advantage of any (even computationally unbounded) adversary be negligible; the third that the advantage of any polynomial-time adversary be negligible. We note that perfect consistency as per our definition coincides with consistency as per [7], and so our notions can be viewed as natural weakenings of theirs.

AN ANALOGY. There is a natural notion of *decryption error* for encryption schemes [13, Section 5.1.2]. A perfectly consistent PEKS is the analog of an encryption scheme with zero decryption error (the usual requirement). A statistically consistent PEKS is the analog of an encryption scheme with negligible decryption error (a less common but still often used condition [2,10]). However, computational consistency is a non-standard relaxation, for consistency conditions are typically not computational. This is not because one cannot define them that way (one could certainly define a computational consistency requirement for encryption) but rather because there has never been any motivation to do so. What makes PEKS different, as emerges from the results below, is that computational consistency is relevant and arises naturally.

CONSISTENCY OF $\mathcal{BDOP}$-$\mathcal{PEKS}$, REVISITED. The counter-example showing that $\mathcal{BDOP}$-$\mathcal{PEKS}$ is not perfectly consistent extends to show that it is not statistically consistent either. However, we show (cf. Theorem 4) that $\mathcal{BDOP}$-$\mathcal{PEKS}$ is computationally consistent. In the random-oracle model, this is not under any computational assumption: the limitation on the running time of the adversary is relevant because it limits the number of queries the adversary can make to the random oracle. When the random oracle is instantiated via a hash function, we would need to assume collision-resistance of the hash function. The implication of this result is that $\mathcal{BDOP}$-$\mathcal{PEKS}$ is probably fine to use in practice, in that incorrect routing of email, while possible in principle, is unlikely to actually happen.

A STATISTICALLY CONSISTENT PEKS SCHEME. We provide the first construction of a PEKS scheme that is *statistically* consistent. The scheme is in the RO model, and is also PEKS-IND-CPA secure assuming the BDH problem is hard.

The motivation here was largely theoretical. From a foundational perspective, we wanted to know whether PEKS was an anomaly in the sense that only computational consistency is possible, or whether, like other primitives, statistical consistency could be achieved. However, it is also true that while computational consistency is arguably enough in an application, statistical might be preferable because the guarantee is unconditional.

1.2 PEKS and Anonymous IBE

$\mathcal{BDOP}$-$\mathcal{PEKS}$ is based on the Boneh-Franklin IBE ($\mathcal{BF}$-$\mathcal{IBE}$) scheme [8]. It is natural to ask whether one might, more generally, build PEKS schemes from IBE schemes in some blackbox way. To this end, a transform of an IBE scheme into a PEKS scheme is presented in [7]. Interestingly, they note that the property of the IBE scheme that appears necessary to provide PEKS-IND-CPA of the PEKS scheme is not the usual IBE-IND-CPA but rather anonymity. (An IBE scheme is anonymous if a ciphertext does not reveal the identity of the recipient [3].) While

[7] stops short of stating and proving a formal result here, it is not hard to verify that their intuition is correct. Namely one can show that if the starting IBE scheme $\mathcal{IBE}$ meets an appropriate formal notion of anonymity (IBE-ANO-CPA, cf. Sect. 3) then $\mathcal{PEKS}$ = bdop-ibe-2-peks($\mathcal{IBE}$) is PEKS-IND-CPA.

CONSISTENCY IN bdop-ibe-2-peks. Unfortunately, we show (cf. Theorem 6) that there are IBE schemes for which the PEKS scheme output by bdop-ibe-2-peks is not even computationally consistent. This means that bdop-ibe-2-peks is not in general a suitable way to turn an IBE scheme into a PEKS scheme. (Although it might be in some cases, and in particular is when the starting IBE scheme is $\mathcal{BF}\text{-}\mathcal{IBE}$, for in that case the resulting PEKS scheme is $\mathcal{BDOP}\text{-}\mathcal{PEKS}$.)

new-ibe-2-peks. We propose a randomized variant of the bdop-ibe-2-peks transform that we call new-ibe-2-peks, and prove that if an IBE scheme $\mathcal{IBE}$ is IBE-ANO-CPA and IBE-IND-CPA then the PEKS scheme new-ibe-2-peks($\mathcal{IBE}$) is PEKS-IND-CPA and computationally consistent. We do not know of a transform where the resulting PEKS scheme is statistically or perfectly consistent.

ANONYMOUS IBE SCHEMES. The above motivates finding anonymous IBE schemes. Towards this, we begin by extending Halevi's condition for anonymity [15] to the IBE setting. Based on this, we are able to give a simple proof that the (random-oracle model) $\mathcal{BF}\text{-}\mathcal{IBE}$ scheme [8] is IBE-ANO-CPA assuming the BDH problem is hard (cf. Theorem 8). (We clarify that a proof of this result is implicit in the proof of security of the $\mathcal{BF}\text{-}\mathcal{IBE}$ based $\mathcal{BDOP}\text{-}\mathcal{PEKS}$ scheme given in [7]. Our contribution is to have stated the formal definition of anonymity and provided a simpler proof via the extension of Halevi's condition.) Towards answering the question of whether there exist anonymous IBE schemes in the standard (as opposed to random oracle) model, we present in [1] an attack to show that Water's IBE scheme [19] is not IBE-ANO-CPA.

1.3   Extensions

ANONYMOUS HIBE. We provide definitions of anonymity for hierarchical IBE (HIBE) schemes. Our definition can be parameterized by a level, so that we can talk of a HIBE that is anonymous at level l. We note that the HIBE schemes of [11,6] are not anonymous, even at level 1. (That of [16] appears to be anonymous at both levels 1 and 2 but is very limited in nature and thus turns out not to be useful for our applications.) We modify the construction of Gentry and Silverberg [11] to obtain a HIBE that is (HIBE-IND-CPA and) anonymous at level 1. The construction is in the random oracle model and assumes BDH is hard.

PETKS. In a PEKS scheme, once the gateway has the trapdoor for a certain period, it can test whether this keyword was present in any past ciphertexts or future ciphertexts. It may be useful to limit the period in which the trapdoor can be used. Here we propose an extension of PEKS that we call public-key encryption with temporary keyword search (PETKS) that allows this. A trapdoor here is created for a time interval $[s..e]$ and will only allow the gateway to test whether ciphertexts created in this time interval contain the keyword. We provide definitions of privacy and consistency for PETKS, and then show how to implement it with overhead that is only logarithmic in the total number of time periods. Our construction can use any HIBE that is anonymous at level 1. Using the above-mentioned HIBE we get a particular instantiation that is secure in the random-oracle model if BDH is hard.

IBEKS. We define the notion of an identity-based encryption with keyword search scheme. This is just like a PEKS scheme except that encryption is performed given only the identity of the receiver and a master public-key, just like in an IBE scheme. We show how to implement IBEKS given any level-2 anonymous HIBE scheme. However no suitable implementation of the latter is known, so we have no concrete implementation of IBEKS.

1.4   Remarks

LIMITED PEKS SCHEMES. Boneh et. al. [7] also present a couple of PEKS schemes that are what they call *limited*. In the first scheme, the public key has size polynomial in the number of keywords that can be used. In the second scheme, the key and ciphertext have size polynomial in the number of trapdoors that can be securely issued to the gateway. Although these schemes are not very interesting due to their limited nature, one could ask about their consistency. In the full version of this paper [1] we extend our definitions of consistency to this limited setting and then show that the first scheme is statistically consistent while the second scheme is computationally consistent, and statistically consistent under some conditions.

CONSISTENCY OF OTHER SEARCHABLE ENCRYPTION SCHEMES. Of the other papers on searchable encryption of which we are aware [18,12,14,20], none formally define or rigorously address the notion of consistency for their respective types of searchable encryption schemes. Goh [12] and Golle, Staddon, and Waters [14] define consistency conditions analogous to BDOP's "perfect consistency" condition, but none of the constructions in [12,14] satisfy their respective perfect consistency condition. Song, Wagner, and Perrig [18] and Waters et al. [20] do not formally state and prove consistency conditions for their respective searchable encryption schemes, but they, as well as Goh [12], do acknowledge and informally bound the non-zero probability of a false positive.

2   Consistency in PEKS

PEKS. A *public key encryption with keyword search* (PEKS) scheme [7] $\mathcal{PEKS} =$ (KG, PEKS, Td, Test) consists of four polynomial-time algorithms. Via $(pk, sk) \xleftarrow{\$}$ $\mathsf{KG}(1^k)$, where $k \in \mathbb{N}$ is the security parameter, the randomized key-generation algorithm produces keys for the receiver; via $C \xleftarrow{\$} \mathsf{PEKS}^H(pk, w)$ a sender encrypts a keyword w to get a ciphertext; via $t_w \xleftarrow{\$} \mathsf{Td}^H(sk, w)$ the receiver computes a trapdoor t_w for keyword w and provides it to the gateway; via $b \leftarrow \mathsf{Test}^H(t_w, C)$ the gateway tests whether C encrypts w, where b is a bit with 1 meaning "accept" or "yes" and 0 meaning "reject" or "no". Here H is a random oracle whose domain and/or range might depend on k and pk. (In constructs we might use multiple random oracles, but since one can always obtain these from a single one [5], definitions will assume just one.)

CONSISTENCY. The requirement of [7] can be divided into two parts. First, $\mathsf{Test}(t_w, C)$ always accept when C encrypts w. More formally, for all $k \in \mathbb{N}$ and all $w \in \{0,1\}^*$ we ask that $\Pr[\mathsf{Test}^H(\mathsf{Td}^H(sk, w), \mathsf{PEKS}^H(pk, w)) = 1] = 1$, where the probability is taken over the choice of $(pk, sk) \xleftarrow{\$} \mathsf{KG}(1^k)$, the random choice of H, and the coins of all the algorithms in the expression above. Since we will always require this too, it is convenient henceforth to take it as an integral part of the PEKS notion and not mention it again, reserving the term "consistency" to only refer to what happens when the ciphertext encrypts a keyword different from the one for which the gateway is testing. In this regard, the requirement of [7], which we will call *perfect consistency*, is that $\mathsf{Test}(t_{w'}, C)$ always reject when C doesn't encrypt w'. More formally, for all $k \in \mathbb{N}$ and all distinct $w, w' \in \{0,1\}^*$, we ask that $\Pr[\mathsf{Test}^H(\mathsf{Td}^H(sk, w'), \mathsf{PEKS}^H(pk, w)) = 1] = 0$, where the probability is taken over the choice of $(pk, sk) \xleftarrow{\$} \mathsf{KG}(1^k)$, the random choice of H, and the coins of all the algorithms in the expression above. (We note that [7] provide informal rather than formal statements, but it is hard to interpret them in any way other than what we have done.)

PRIVACY. Privacy for a PEKS scheme [7] asks that an adversary should not be able to distinguish between the encryption of two challenge keywords of its choice, even if it is allowed to obtain trapdoors for any non-challenge keywords. Formally, let $\mathcal{A}$ be an adversary and let k be the security parameter. Below, we depict an experiment, where $b \in \{0,1\}$ is a bit, and also the oracle provided to the adversary in this experiment.

Experiment $\mathbf{Exp}_{\mathcal{PEKS},\mathcal{A}}^{\text{peks-ind-cpa-b}}(k)$

> $WSet \leftarrow \emptyset$; $(pk, sk) \xleftarrow{\$} \mathsf{KG}(1^k)$
> Pick random oracle H
> $(w_0, w_1, st) \xleftarrow{\$} \mathcal{A}^{\text{TRAPD}(\cdot), H}(\mathbf{find}, pk)$
> $C \xleftarrow{\$} \mathsf{PEKS}^H(pk, w_b)$
> $b' \xleftarrow{\$} \mathcal{A}^{\text{TRAPD}(\cdot), H}(\mathbf{guess}, C, st)$
> If $\{w_0, w_1\} \cap WSet = \emptyset$ then return b' else return 0

Oracle $\text{TRAPD}(w)$

> $WSet \leftarrow WSet \cup \{w\}$
> $t_w \xleftarrow{\$} \mathsf{Td}^H(sk, w)$
> Return t_w

The PEKS-IND-CPA-*advantage* $\mathbf{Adv}_{\mathcal{PEKS},\mathcal{A}}^{\text{peks-ind-cpa}}(k)$ of $\mathcal{A}$ is defined as

$$\Pr\left[\mathbf{Exp}_{\mathcal{PEKS},\mathcal{A}}^{\text{peks-ind-cpa-1}}(k) = 1\right] - \Pr\left[\mathbf{Exp}_{\mathcal{PEKS},\mathcal{A}}^{\text{peks-ind-cpa-0}}(k) = 1\right].$$

KG(1^k)
  $(\mathbb{G}_1, \mathbb{G}_2, p, e) \xleftarrow{\$} \mathcal{G}(1^k)$; $P \xleftarrow{\$} \mathbb{G}_1^*$; $s \xleftarrow{\$} \mathbb{Z}_p^*$
  $pk \leftarrow (\mathbb{G}_1, \mathbb{G}_2, p, e, P, sP)$; $sk \leftarrow (pk, s)$
  Return (pk, sk)

PEKS$^{H_1, H_2}(pk, w)$
  Parse pk as $(\mathbb{G}_1, \mathbb{G}_2, p, e, P, sP)$
  $r \xleftarrow{\$} \mathbb{Z}_p^*$; $T \leftarrow e(H_1(w), sP)^r$
  $C \leftarrow (rP, H_2(T))$; Return C

Td$^{H_1}(sk, w)$
  Parse sk as $(pk = (\mathbb{G}_1, \mathbb{G}_2, p, e, P, sP), s)$
  $t_w \leftarrow (pk, sH_1(w))$; Return t_w

Test$^{H_1, H_2}(t_w, C)$
  Parse t_w as $((\mathbb{G}_1, \mathbb{G}_2, p, e, P, sP), X)$
  Parse C as (U, V) ; $T \leftarrow e(X, U)$
  If $V = H_2(T)$ then return 1
  Else return 0

Fig. 1. Algorithms constituting the PEKS scheme $\mathcal{BDOP\text{-}PEKS}$. $\mathcal{G}$ is a pairing parameter generator and $H_1 \colon \{0,1\}^* \to \mathbb{G}_1$ and $H_2 \colon \mathbb{G}_2 \to \{0,1\}^k$ are random oracles.

A scheme $\mathcal{PEKS}$ is said to be PEKS-IND-CPA-*secure* if the advantage is a negligible function in k for all polynomial-time adversaries $\mathcal{A}$.

PARAMETER GENERATION ALGORITHMS AND THE BDH PROBLEM. All pairing based schemes will be parameterized by a *pairing parameter generator*. This is a randomized polynomial-time algorithm $\mathcal{G}$ that on input 1^k returns the description of an additive cyclic group $\mathbb{G}_1$ of prime order p, where $2^k < p < 2^{k+1}$, the description of a multiplicative cyclic group $\mathbb{G}_2$ of the same order, and a non-degenerate bilinear pairing $e \colon \mathbb{G}_1 \times \mathbb{G}_1 \to \mathbb{G}_2$. See [8] for a description of the properties of such pairings. We use $\mathbb{G}_1^*$ to denote $\mathbb{G}_1 \setminus \{0\}$, i.e. the set of all group elements except the neutral element. We define the advantage $\mathbf{Adv}_{\mathcal{G}, \mathcal{A}}^{bdh}(k)$ of an adversary $\mathcal{A}$ in solving the BDH problem relative to a pairing parameter generator $\mathcal{G}$ as the probability that, on input $(1^k, (\mathbb{G}_1, \mathbb{G}_2, p, e), P, aP, bP, cP)$ for randomly chosen $P \xleftarrow{\$} \mathbb{G}_1^*$ and $a, b, c \xleftarrow{\$} \mathbb{Z}_p^*$, adversary $\mathcal{A}$ outputs $e(P, P)^{abc}$. We say that the BDH problem is hard relative to this generator if $\mathbf{Adv}_{\mathcal{G}, \mathcal{A}}^{bdh}$ is a negligible function in k for all polynomial-time adversaries $\mathcal{A}$.

CONSISTENCY OF $\mathcal{BDOP\text{-}PEKS}$. Figure 1 presents the $\mathcal{BDOP\text{-}PEKS}$ scheme. It is based on a pairing parameter generator $\mathcal{G}$.

Proposition 1. *The $\mathcal{BDOP\text{-}PEKS}$ scheme is not perfectly consistent.*

Proof. Since the number of possible keywords is infinite, there will certainly exist distinct keywords $w, w' \in \{0,1\}^*$ such that $H_1(w) = H_1(w')$. The trapdoors for such keywords will be the same, and so Test$^{H_1, H_2}$(Td(sk, w), PEKS$^{H_1, H_2}(pk, w')$) will always return 1. □

It is tempting to say that, since H_1 is a random oracle, the probability of a collision is small, and thus the above really does not matter. Whether or not this is true depends on how one wants to define consistency, which is the issue we explore next.

NEW NOTIONS OF CONSISTENCY. The full version of this paper [1] considers various possible relaxations of perfect consistency and argues that the obvious ones are inadequate either because $\mathcal{BDOP}\text{-}\mathcal{PEKS}$ continues to fail them or because they are too weak. It then motivates and explains our approach and definitions. In this extended abstract we simply state our definitions, referring the reader to [1] for more information.

Definition 2. *Let* $\mathcal{PEKS} = (\mathsf{KG}, \mathsf{PEKS}, \mathsf{Td}, \mathsf{Test})$ *be a PEKS scheme. Let* $\mathcal{U}$ *be an adversary and let* k *be the security parameter. Consider the following experiment:*

> **Experiment** $\mathbf{Exp}_{\mathcal{PEKS},\mathcal{U}}^{\text{peks-consist}}(k)$
>
> $(pk, sk) \xleftarrow{\$} \mathsf{KG}(1^k)$; Pick random oracle H
>
> $(w, w') \xleftarrow{\$} \mathcal{U}^H(pk)$; $C \xleftarrow{\$} \mathsf{PEKS}^H(pk, w)$; $t_{w'} \xleftarrow{\$} \mathsf{Td}^H(sk, w')$
>
> If $w \neq w'$ and $\mathsf{Test}^H(t_{w'}, C) = 1$ then return 1 else return 0

We define the advantage of $\mathcal{U}$ *as*

$$\mathbf{Adv}_{\mathcal{PEKS},\mathcal{U}}^{\text{peks-consist}}(k) = \Pr\left[\mathbf{Exp}_{\mathcal{PEKS},\mathcal{U}}^{\text{peks-consist}}(k) = 1\right].$$

The scheme is said to be perfectly consistent *if this advantage is 0 for all (computationally unrestricted) adversaries* $\mathcal{U}$, statistically consistent *if it is negligible for all (computationally unrestricted) adversaries, and* computationally consistent *if it is negligible for all polynomial-time adversaries.*

We have purposely re-used the term perfect consistency, for in fact the above notion of perfect consistency coincides with the one from [7] recalled above.

CONSISTENCY OF $\mathcal{BDOP}\text{-}\mathcal{PEKS}$, REVISITED. Having formally defined the statistical and computational consistency requirements for PEKS schemes, we return to evaluating the consistency of $\mathcal{BDOP}\text{-}\mathcal{PEKS}$. We first observe that Proposition 1 extends to show:

Proposition 3. *The* $\mathcal{BDOP}\text{-}\mathcal{PEKS}$ *scheme is not statistically consistent.*

The proof is in [1]. On the positive side, the following, proved in [1], means that $\mathcal{BDOP}\text{-}\mathcal{PEKS}$ is probably just fine in practice:

Theorem 4. *The* $\mathcal{BDOP}\text{-}\mathcal{PEKS}$ *scheme is computationally consistent in the random oracle model.*

A STATISTICALLY CONSISTENT PEKS SCHEME. We present the first PEKS scheme that is (PEKS-IND-CPA and) *statistically* consistent. To define the scheme, we first introduce the function $f(k) = k^{\lg(k)}$. (Any function that is super-polynomial but sub-exponential would suffice. This choice is made for concreteness.) The algorithms constituting our scheme $\mathcal{PEKS}\text{-}\mathcal{STAT}$ are then depicted in Fig. 2. We are denoting by $|x|$ the length of a string x. The scheme uses ideas from the $\mathcal{BDOP}\text{-}\mathcal{PEKS}$ scheme [7] as well as from the $\mathcal{BF}\text{-}\mathcal{IBE}$ scheme [8], but adds some new elements. In particular the random choice of "session" key K, and the fact that the random oracle H_2 is length-increasing, are important. The first thing we stress about the scheme is that the algorithms are

KG(1^k)

$(\mathbb{G}_1, \mathbb{G}_2, p, e) \xleftarrow{\$} \mathcal{G}(1^k)$; $P \xleftarrow{\$} \mathbb{G}_1^*$

$s \xleftarrow{\$} \mathbb{Z}_p^*$; $pk \leftarrow (1^k, P, sP, \mathbb{G}_1, \mathbb{G}_2, p, e)$

$sk \leftarrow (pk, s)$; Return (pk, sk)

PEKS$^{H_1, H_2, H_3, H_4}(pk, w)$

Parse pk as $(1^k, P, sP, \mathbb{G}_1, \mathbb{G}_2, p, e)$

If $|w| \geq f(k)$ then return w

$r \xleftarrow{\$} \mathbb{Z}_p^*$; $T \leftarrow e(sP, H_1(w))^r$

$K_1 \leftarrow H_4(T)$; $K_2 \leftarrow H_2(T)$

$K \xleftarrow{\$} \{0,1\}^k$; $c \leftarrow K_1 \oplus K$

$t \leftarrow H_3(K \| w)$

Return (rP, c, t, K_2)

Td$^{H_1}(sk, w)$

Parse sk as $(pk = (1^k, P, sP, \mathbb{G}_1, \mathbb{G}_2, p, e), s)$

$t_w \leftarrow (pk, sH_1(w), w)$

Return t_w

Test$^{H_1, H_2, H_3, H_4}(t_w, C)$

Parse t_w as

$\quad ((1^k, P, sP, \mathbb{G}_1, \mathbb{G}_2, p, e), sH_1(w), w)$

If $|w| \geq f(k)$ then

$\quad$ If $C = w$ then return 1 else return 0

If C cannot be parsed as (rP, c, t, K_2)

$\quad$ then return 0

$T \leftarrow e(rP, sH_1(w))$

$K \leftarrow c \oplus H_4(T)$

If $K_2 \neq H_2(T)$ then return 0

If $t = H_3(K \| w)$ then return 1 else return 0

Fig. 2. Algorithms constituting the PEKS scheme $\mathcal{PEKS\text{-}STAT}$. Here $f(k) = k^{\lg(k)}$, $\mathcal{G}$ is a pairing parameter generator and $H_1 : \{0,1\}^* \to \mathbb{G}_1$, $H_2 : \mathbb{G}_2 \to \{0,1\}^{3k}$, $H_3 : \{0,1\}^* \to \{0,1\}^k$, and $H_4 : \{0,1\}^* \to \{0,1\}^k$ are random oracles. In implementations we require that the lengths of the encodings of (rP, c, t, K_2) be polynomial in k.

---

polynomial-time. This is because polynomial time means in the length of the inputs, and the input of (say) the encryption algorithm includes w as well as 1^k, so it can test whether $|w| \geq f(k)$ in polynomial time. Now the formal statement of our result is the following:

Theorem 5. *The $\mathcal{PEKS\text{-}STAT}$ scheme is statistically consistent in the random oracle model. Further, $\mathcal{PEKS\text{-}STAT}$ is PEKS-IND-CPA-secure in the random oracle model assuming that the BDH problem is hard relative to generator $\mathcal{G}$.*

We refer to [1] for the proof, and provide a little intuition here. Privacy when the adversary is restricted to attacking the scheme only on keywords of size at most $f(k)$ can be shown based on techniques used to prove IBE-IND-CPA of the $\mathcal{BF\text{-}IBE}$ scheme [8] and to prove anonymity of the same scheme (cf. Theorem 8). When the keyword has length at least $f(k)$ it is sent in the clear, but intuitively the reason this does not violate privacy is that the adversary is poly(k) time and thus cannot even write down such a keyword in order to query it to its challenge oracle. More interesting is the proof of statistical consistency. The main issue is that the computationally unbounded consistency adversary $\mathcal{U}$ can easily find any collisions that exist for the random-oracle hash functions. The scheme is designed so that the adversary effectively has to find a large number of collisions to win. It uses the fact that H_2 is with high probability injective, and then uses a counting argument based on an occupancy problem bound.

3   PEKS and Anonymous IBE

IBE. An *identity-based encryption* (IBE) scheme [17,8] $I\mathcal{BE} = (\mathsf{Setup}, \mathsf{KeyDer}, \mathsf{Enc}, \mathsf{Dec})$ consists of four polynomial-time algorithms. Via $(pk, msk) \xleftarrow{\$} \mathsf{Setup}(1^k)$ the randomized key-generation algorithm produces master keys for security parameter $k \in \mathbb{N}$; via $usk[id] \xleftarrow{\$} \mathsf{KeyDer}^H(msk, id)$ the master computes the secret key for identity id; via $C \xleftarrow{\$} \mathsf{Enc}^H(pk, id, M)$ a sender encrypts a message M to identity id to get a ciphertext; via $M \leftarrow \mathsf{Dec}^H(usk, C)$ the possessor of secret key usk decrypts ciphertext C to get back a message. Here H is a random oracle with domain and range possibly depending on k and pk. (In constructs we might use multiple random oracles, but since one can always obtain these from a single one [5], definitions will assume just one.) Associated to the scheme is a message space MsgSp where for $\mathsf{MsgSp}(k) \subseteq \{0,1\}^*$ for every $k \in \mathbb{N}$. For consistency, we require that for all $k \in \mathbb{N}$, all identities id and messages $M \in \mathsf{MsgSp}(k)$ we have $\Pr[\mathsf{Dec}^H(\mathsf{KeyDer}^H(msk, id), \mathsf{Enc}^H(pk, id, M)) = M] = 1$, where the probability is taken over the choice of $(pk, msk) \xleftarrow{\$} \mathsf{Setup}(1^k)$, the random choice of H, and the coins of all the algorithms in the expression above. Unless otherwise stated, it is assumed that $\{0,1\}^k \subseteq \mathsf{MsgSp}(k)$ for all $k \in \mathbb{N}$.

PRIVACY AND ANONYMITY. Privacy (IBE-IND-CPA) follows [8] while anonymity (IBE-ANO-CPA) is a straightforward adaptation of [3] to IBE schemes. Let $I\mathcal{BE} = (\mathsf{Setup}, \mathsf{KeyDer}, \mathsf{Enc}, \mathsf{Dec})$ be an IBE scheme. Let $\mathcal{A}$ be an adversary and let k be the security parameter. Now consider the following experiments, where $b \in \{0,1\}$ is a bit:

Experiment $\mathbf{Exp}^{\text{ibe-ind-cpa-}b}_{I\mathcal{BE},\mathcal{A}}(k)$	**Experiment $\mathbf{Exp}^{\text{ibe-ano-cpa-}b}_{I\mathcal{BE},\mathcal{A}}(k)$**
$IDSet \leftarrow \emptyset$; $(pk, msk) \xleftarrow{\$} \mathsf{Setup}(1^k)$	$IDSet \leftarrow \emptyset$; $(pk, msk) \xleftarrow{\$} \mathsf{Setup}(1^k)$
Pick random oracle H	Pick random oracle H
$(id, M_0, M_1, st) \xleftarrow{\$} \mathcal{A}^{\text{KEYDER},H}(\texttt{find}, pk)$	$(id_0, id_1, M, st) \xleftarrow{\$} \mathcal{A}^{\text{KEYDER},H}(\texttt{find}, pk)$
$C \xleftarrow{\$} \mathsf{Enc}^H(pk, id, M_b)$	$C \xleftarrow{\$} \mathsf{Enc}^H(pk, id_b, M)$
$b' \xleftarrow{\$} \mathcal{A}^{\text{KEYDER},H}(\texttt{guess}, C, st)$	$b' \xleftarrow{\$} \mathcal{A}^{\text{KEYDER},H}(\texttt{guess}, C, st)$
If $\{M_0, M_1\} \not\subseteq \mathsf{MsgSp}(k)$ then return 0	If $M \notin \mathsf{MsgSp}(k)$ then return 0
If $id \notin IDSet$ and $|M_0| = |M_1|$	If $\{id_0, id_1\} \cap IDSet = \emptyset$
Then return b' else return 0	Then return b' else return 0

where the oracle $\text{KEYDER}(id)$ is defined as

$$IDSet \leftarrow IDSet \cup \{id\} \; ; \; usk[id] \xleftarrow{\$} \mathsf{KeyDer}^H(msk, id) \; ; \; \text{Return } usk[id]$$

For prop $\in \{\text{ind}, \text{ano}\}$, we define the advantage $\mathbf{Adv}^{\text{ibe-prop-cpa}}_{I\mathcal{BE},\mathcal{A}}(k)$ of $\mathcal{A}$ in the corresponding experiment as

$$\Pr\left[\mathbf{Exp}^{\text{ibe-prop-cpa-1}}_{I\mathcal{BE},\mathcal{A}}(k) = 1\right] - \Pr\left[\mathbf{Exp}^{\text{ibe-prop-cpa-0}}_{I\mathcal{BE},\mathcal{A}}(k) = 1\right].$$

IBE scheme $\mathcal{IBE}$ is said to be IBE-IND-CPA-*secure* (resp., IBE-ANO-CPA-*secure*) if $\mathbf{Adv}_{\mathcal{IBE},\mathcal{A}}^{\text{ibe-ind-cpa}}$ (resp., $\mathbf{Adv}_{\mathcal{IBE},\mathcal{A}}^{\text{ibe-ano-cpa}}$) is a negligible function in k for all polynomial-time adversaries $\mathcal{A}$.

bdop-ibe-2-peks. The bdop-ibe-2-peks transform [7] takes input an IBE scheme $\mathcal{IBE}$ = (Setup, KeyDer, Enc, Dec) and returns the PEKS scheme $\mathcal{PEKS}$ = (KG, Td, PEKS, Test) where KG(1^k) = Setup(1^k), Td(sk, w) = KeyDer(sk, w), PEKS(pk, w) = Enc($pk, w, 0^k$), and Test(t_w, C) returns 1 iff Dec(t_w, C) = 0^k. Since $\mathcal{BF\text{-}IBE}$ is anonymous (Theorem 8), since $\mathcal{BDOP\text{-}PEKS}$ is exactly $\mathcal{BF\text{-}IBE}$ transformed via bdop-ibe-2-peks, and since $\mathcal{BDOP\text{-}PEKS}$ is not statistically consistent (Proposition 3), we can conclude that the bdop-ibe-2-peks transformation does not necessarily yield a statistically consistent PEKS scheme. The following theorem strengthens this result, showing that, under the minimal assumption of the existence of an IBE-IND-CPA- and IBE-ANO-CPA-secure IBE scheme, there exists an IBE scheme such that the resulting PEKS scheme via bdop-ibe-2-peks fails to be even computationally consistent.

Theorem 6. *For any* IBE-ANO-CPA-*secure and* IBE-IND-CPA-*secure IBE scheme* $\mathcal{IBE}$, *there exists another* IBE-ANO-CPA-*secure and* IBE-IND-CPA-*secure IBE scheme* $\overline{\mathcal{IBE}}$ *such that the PEKS scheme* $\mathcal{PEKS}$ *derived from* $\overline{\mathcal{IBE}}$ *via* bdop-ibe-2-peks *is not computationally consistent.*

The full proof can be found in the full version [1]; here we sketch the main idea. Given IBE scheme $\mathcal{IBE}$, consider the following IBE scheme $\overline{\mathcal{IBE}}$. The public key includes a normal public key for $\mathcal{IBE}$ and a random string R of length k. A message M is encrypted by encrypting $M \| R$ under $\mathcal{IBE}$. When decrypting a ciphertext C, the result is parsed as $M \| R'$. If $R' = R$, then M is returned as the plaintext, otherwise the decryption algorithm returns 0^k. The resulting Test algorithm will return 1, except with negligible probability, regardless of what trapdoor is being used. This is because the decryption algorithm returns 0^k whenever the last k bits of the plaintext are not equal to R. Intuitively, this should happen with all but negligible probability since, if $\mathcal{IBE}$ is IBE-IND-CPA-secure, a portion of a string encrypted to one identity should not correctly decrypt with the secret key for a different identity.

FIXING THE bdop-ibe-2-peks TRANSFORMATION. The negative result in Theorem 6 raises the question: Does the existence of IBE schemes imply the existence of computationally consistent PEKS schemes? We answer that in the affirmative by presenting a revision to the BDOP transformation, called new-ibe-2-peks, that transforms any IBE-IND-CPA- and IBE-ANO-CPA-secure IBE scheme into a PEKS-IND-CPA-secure and computationally consistent PEKS scheme. Our new new-ibe-2-peks transform, instead of always encrypting the same message 0^k, chooses and encrypts a random message R and appends R in the clear to the ciphertext. Thus, given IBE scheme $\mathcal{IBE}$ = (Setup, KeyDer, Enc, Dec), the PEKS scheme $\mathcal{PEKS}$ = new-ibe-2-peks($\mathcal{IBE}$) = (KG, Td, PEKS, Test) is such that KG(1^k) = Setup(1^k), Td(sk, w) = KeyDer(sk, w), PEKS(pk, w) = (Enc(pk, w, R), R) where $R \overset{\$}{\leftarrow} \{0, 1\}^k$, and Test($t_w, C = (C_1, C_2)$) returns 1 iff Dec($t_w, C_1$)

returns C_2. Intuitively, this construction avoids the problem of oddly-behaving
Dec algorithms by making sure that the *only* way to ruin the consistency of
the PEKS scheme is by correctly guessing the value encrypted by a ciphertext,
using the secret key of a different identity, which should not be possible for an
IBE-IND-CPA-secure IBE scheme. Hence, the consistency of the resulting PEKS
scheme is due to the data privacy property of the IBE scheme, while the data
privacy property of the PEKS scheme comes from the anonymity of the IBE
scheme. We prove the theorem statement below in the full version [1].

Theorem 7. *Let* $I\mathcal{BE}$ *be an IBE scheme and let* $P\mathcal{EKS}$ *be the PEKS scheme
built from* $I\mathcal{BE}$ *via* new-ibe-2-peks. *If* $I\mathcal{BE}$ *is* IBE-IND-CPA-*secure, then* $P\mathcal{EKS}$
is computationally consistent. Further, if $I\mathcal{BE}$ *is* IBE-ANO-CPA-*secure, then*
$P\mathcal{EKS}$ *is* PEKS-IND-CPA-*secure.*

ANONYMOUS IBE SCHEMES. Theorem 7 motivates a search for IBE-ANO-CPA-
secure IBE schemes. The following shows that the Boneh-Franklin BasicIdent
IBE scheme, $B\mathcal{F}\text{-}I\mathcal{BE}$, is anonymous. The proof is simple due to our use of an
extension to Halevi's technique for proving the anonymity of public key encryp-
tion schemes [15]. This extension and the proof can be found in [1].

Theorem 8. *The* $B\mathcal{F}\text{-}I\mathcal{BE}$ *scheme is* IBE-ANO-CPA-*secure in the random oracle model assuming that BDH is hard relative to the underlying pairing generator.*

4   Extensions

We propose three extensions of concepts seen above, namely anonymous HIBEs,
public-key encryption with temporary keyword search, and identity-based PEKS.

4.1   Anonymous HIBE

HIBEs. A *hierarchical identity-based encryption* (HIBE) scheme [16,11,6] is a
generalization of an IBE scheme in which an identity is a vector of strings $id =
(id_1, \ldots, id_l)$ with the understanding that when $l = 0$ this is the empty string ε.
The number of components in this vector is called the level of the identity and is
denoted $|id|$. If $0 \leq i \leq l$ then $id|_i = (id_1, \ldots, id_i)$ denotes the vector containing
the first i components of id (this is ε if $i = 0$). If $|id'| \geq l+1$ $(l \geq 0)$ and $id'|_l = id$
then we say that id is an ancestor of id', or equivalently, that id' is a descendant
of id. If the level of id' is $l + 1$ then id is a parent of id', or, equivalently, id' is a
child of id. For any id with $|id| \geq 1$ we let $par(id) = id|_{|id|-1}$ denote its parent.
Two nodes $id = (id_1, \ldots, id_l)$ and $id' = (id'_1, \ldots, id'_l)$ at level l are said to be
siblings iff $id|_{l-1} = id'|_{l-1}$. Moreover, if $id_l < id'_l$ in lexicographic order, then id
is a left sibling of id' and id' is a right sibling of id. An identity at level one or
more can be issued a secret key by its parent. (And thus an identity can issue
keys for any of its descendants if necessary.)

Formally a HIBE scheme $H\mathcal{IBE} = (\mathsf{Setup}, \mathsf{KeyDer}, \mathsf{Enc}, \mathsf{Dec})$ consists of four
polynomial-time algorithms. Via $(pk, msk = usk[\varepsilon]) \xleftarrow{\$} \mathsf{Setup}(1^k)$, where $k \in \mathbb{N}$ is

a security parameter, the randomized key-generation algorithm produces master keys, with the secret key being associated to the (unique) identity at level 0; via $usk[id] \xleftarrow{\$} \mathsf{KeyDer}^H(usk[\mathrm{par}(id)], id)$ the parent of an identity id with $|id| \geq 1$ can compute a secret key for id; via $C \xleftarrow{\$} \mathsf{Enc}^H(pk, id, M)$ a sender encrypts a message M to identity id to get a ciphertext; via $M \leftarrow \mathsf{Dec}^H(usk[id], C)$ the identity id decrypts ciphertext C to get back a message. Here H is a random oracle with domain and range possibly depending on k and pk. (In constructs we might use multiple random oracles, but since one can always obtain these from a single one [5], definitions will assume just one.) Associated to the scheme is a message space MsgSp where for $\mathsf{MsgSp}(k) \subseteq \{0,1\}^*$ for every $k \in \mathbb{N}$. For consistency, we require that for all $k \in \mathbb{N}$, all identities id with $|id| \geq 1$ and all messages $M \in \mathsf{MsgSp}(k)$, $\Pr[\mathsf{Dec}^H(\mathsf{KeyDer}^H(usk[\mathrm{par}(id)], id), \mathsf{Enc}^H(pk, id, M)) = M] = 1$, where the probability is taken over the choice of $(pk, usk[\varepsilon]) \xleftarrow{\$} \mathsf{Setup}(1^k)$, the random choice of H, and the coins of all the algorithms in the expression above. Unless otherwise stated, it is assumed that $\{0,1\}^k \subseteq \mathsf{MsgSp}(k)$ for all $k \in \mathbb{N}$.

PRIVACY AND ANONYMITY. Let $d \colon \mathbb{N} \to \mathbb{N}$ be a maximum depth parameter. The notion of privacy, denoted HIBE-IND-CPA$[d(k)]$, is analogous to that for IBE schemes (IBE-IND-CPA) but using identity vectors rather than identity strings and where the adversary is not allowed to query the KEYDER oracle for the secret key of any ancestor of the identity under attack, and all identities id (in queries or challenges) must have $|id| \leq d(k)$. For anonymity, we define the notion of the scheme being anonymous at level $l \geq 1$, denoted HIBE-ANO-CPA$[l, d(k)]$. (Stronger notions are possible, but not needed here.) It is analogous to IBE-ANO-CPA except that the identities returned by the adversary must differ *only* in the l-th component. The adversary can ask the key derivation oracle KEYDER for the secret keys of all identities except for those of the challenge identities or any of their ancestors, and all identities id (in queries or challenges) must have $|id| \leq d(k)$. The definitions are provided in full in [1].

CONSTRUCTION. The HIBE scheme of [16] appears to be anonymous, but supports only two levels of identities, and is only resistant against limited collusions at the second level, and hence is not usable for our constructions that follow. Since the HIBE of [11] (here denoted $\mathcal{GS}\text{-}\mathcal{HIBE}$) is equivalent to the (provably anonymous as per Theorem 8) Boneh-Franklin IBE scheme [8] when restricted to the first level, one could hope that $\mathcal{GS}\text{-}\mathcal{HIBE}$ is level-1 anonymous, but this turns out not to be true, and the HIBE of [6] is not level-1 anonymous either. We suggest a modified version of $\mathcal{GS}\text{-}\mathcal{HIBE}$ called $m\mathcal{GS}\text{-}\mathcal{HIBE}$. It is depicted in Fig. 3. The following, proved in [1], implies in particular that $m\mathcal{GS}\text{-}\mathcal{HIBE}$ is the first full HIBE scheme providing any anonymity at all. The restriction on d is inherited from [11].

Theorem 9. *The $m\mathcal{GS}\text{-}\mathcal{HIBE}$ scheme is* HIBE-ANO-CPA$[1, d(k)]$-*secure and* HIBE-IND-CPA$[d(k)]$-*secure for any* $d(k) = O(\log(k))$ *in the random oracle model assuming the BDH problem is hard relative to the generator* $\mathcal{G}$.

Setup(1^k)

$(\mathbb{G}_1, \mathbb{G}_2, p, e) \xleftarrow{\$} \mathcal{G}(1^k)$; $P \xleftarrow{\$} \mathbb{G}_1^*$

$s_0 \xleftarrow{\$} \mathbb{Z}_p^*$; $S_0 \leftarrow 0$; $Q_0 \leftarrow s_0 P$

$pk \leftarrow (\mathbb{G}_1, \mathbb{G}_2, p, e, P, Q_0)$

$msk \leftarrow (pk, \varepsilon, S_0, s_0)$

Return (pk, msk)

KeyDer$^{H_{1,1},\ldots,H_{1,l}}(usk, id)$

Parse id as $(id_1, \ldots, id_{l+1})$

Parse usk as $(pk, id|_l, S_l, Q_1, \ldots, Q_{l-1}, s_l)$

Parse pk as $(\mathbb{G}_1, \mathbb{G}_2, p, e, P, Q_0)$

$S_{l+1} \leftarrow S_l + s_l H_{1,l+1}(id_{l+1})$

$Q_l \leftarrow s_l P$; $s_{l+1} \xleftarrow{\$} \mathbb{Z}_p^*$

Return $(pk, id, S_{l+1}, Q_1, \ldots, Q_l, s_{l+1})$

Enc$^{H_{1,1},\ldots,H_{1,l},H_2}(pk, id, m)$

Parse pk as $(\mathbb{G}_1, \mathbb{G}_2, p, e, P, Q_0)$

Parse id as $(id_1, \ldots, id_l)$

$r \xleftarrow{\$} \mathbb{Z}_p^*$; $C_1 \leftarrow rP$

For $i = 2, \ldots, l$ do $C_i \leftarrow r H_{1,i}(id_i)$

$C_{l+1} \leftarrow m \oplus H_2(e(rH_{1,1}(id_1), Q_0))$

Return $(C_1, \ldots, C_{l+1})$

Dec$^{H_2}(usk, C)$

Parse usk as $(pk, id, S_l, Q_1, \ldots, Q_{l-1}, s_l)$

Parse id as $(id_1, \ldots, id_l)$

Parse pk as $(\mathbb{G}_1, \mathbb{G}_2, p, e, P, Q_0)$

Parse C as $(C_1, \ldots, C_{l+1})$

$\kappa \leftarrow e(S_l, C_1) \cdot \prod_{i=2}^{l} e(Q_{i-1}, C_i)^{-1}$

Return $C_{l+1} \oplus H_2(\kappa)$

Fig. 3. Algorithms of the $m\mathcal{GS}\text{-}\mathcal{HIBE}$ scheme $m\mathcal{GS}\text{-}\mathcal{HIBE}$. $\mathcal{G}$ is a pairing parameter generator and $H_{1,i}: \{0,1\}^* \rightarrow \mathbb{G}_1^*$ and $H_2: \mathbb{G}_2 \rightarrow \{0,1\}^k$ are random oracles.

4.2 Temporarily Searchable Encryption

PETKS. *Public-key encryption with temporary keyword search* (PETKS) is a generalization of PEKS in which a trapdoor can be issued for any desired window of time rather than forever. Formally, the scheme $\mathcal{PETKS} = (\mathsf{KG}, \mathsf{Td}, \mathsf{PETKS}, \mathsf{Test}, N)$ consists of four polynomial-time algorithms and a polynomially bounded function $N: \mathbb{N} \rightarrow \mathbb{N}$. Via $(pk, sk) \xleftarrow{\$} \mathsf{KG}(1^k)$, the randomized key-generation algorithm produces keys for the receiver; via $C \xleftarrow{\$} \mathsf{PETKS}^H(pk, w, i)$ a sender encrypts a keyword w in time period $i \in [0..N(k) - 1]$ to get a ciphertext; via $t_w \xleftarrow{\$} \mathsf{Td}^H(sk, w, s, e)$ the receiver computes a trapdoor t_w for keyword w in period $[s..e]$ where $0 \leq s \leq e \leq N(k) - 1$, and provides it to the gateway; via $b \leftarrow \mathsf{Test}^H(t_w, C)$ the gateway tests whether C encrypts w, where b is a bit with 1 meaning "accept" or "yes" and 0 meaning "reject" or "no". Here H is a random oracle whose domain and/or range might depend on k and pk. (In constructs we might use multiple random oracles, but since one can always obtain these from a single one [5], definitions will assume just one.) We require that for all $k \in \mathbb{N}$, all s, e, i with $0 \leq s \leq i \leq e \leq N(k) - 1$, and all $w \in \{0,1\}^*$, we have $\Pr[\mathsf{Test}^H(\mathsf{Td}^H(sk, w, s, e), \mathsf{PETKS}^H(pk, w, i)) = 1] = 1$, where the probability is taken over the choice of $(pk, sk) \xleftarrow{\$} \mathsf{KG}(1^k)$, the random choice of H, and the coins of all the algorithms in the expression above.

CONSISTENCY. Computational, statistical and perfect consistency can be defined analogously to the way they were defined for PEKS. For details, see [1].

PRIVACY. Privacy for a PETKS scheme asks that an adversary should not be able to distinguish between the encryption of two challenge keywords of its choice in a time period $i \in [0..N(k) - 1]$ of its choice, even if it is allowed not only to obtain trapdoors for non-challenge keywords issued for any time interval, but also is allowed to obtain trapdoors for *any* keywords (even the challenge ones), issued for time intervals not containing i. A formal definition of the notion of privacy, which we denote PETKS-IND-CPA, is in [1].

CONSTRUCTIONS WITH LINEAR COMPLEXITY. PETKS is reminiscent of forward-security [4,9], and, as in these works, there are straightforward solutions with keys of length linear in N. One such solution is to use a standard PEKS scheme and generate a different key pair (pk_i, sk_i) for each time period $i \in [0..N(k)-1]$. Let $pk = (pk_0, \ldots, pk_{N(k)-1})$ be the PETKS public key and $sk = (sk_0, \ldots, sk_{N(k)-1})$ be the PETKS secret key. During time period i, the sender encrypts a keyword by encrypting under pk_i using the PEKS scheme. The trapdoor for interval $[s..e]$ consists of all PEKS trapdoors of periods $s, \ldots, e$. A somewhat more efficient solution is to let the PETKS master key pair be a single key pair for the standard PEKS scheme, and append the time period to the keyword when encrypting or computing trapdoors. This scheme achieves $O(N)$ public and secret key length, but still has linear trapdoor length, because the PETKS trapdoor still contains PEKS trapdoors for all time periods $s, \ldots, e$.

A CONSTRUCTION WITH $O(\log(N))$ COMPLEXITY. We now present a transformation hibe-2-petks of a HIBE scheme into a PETKS scheme that yields a PETKS scheme with complexity logarithmic in N for all parameters. The construction is very similar to the generic construction of forward-secure encryption from binary-tree encryption [9]. The number of time periods is $N(k) = 2^{t(k)}$ for some $t(k) = O(\log(k))$. If $i \in [0..N(k)-1]$, then let $i_1 \ldots i_{t(k)}$ denote its binary representation as a $t(k)$-bit string. Intuitively, our construction instantiates a HIBE of depth $t(k)+1$ with keywords as the first level of the identity tree and the time structure on the lower levels. The trapdoor for keyword w and interval of time periods $[s..e]$ consists of the user secret keys of all identities from $(w, s_1, \ldots, s_{t(k)})$ to $(w, e_1, \ldots, e_{t(k)})$, but taking advantage of the hierarchical structure to include entire subtrees of keys.

More precisely, let $\mathcal{HIBE} = $ (Setup, KeyDer, Enc, Dec) be a HIBE scheme. Then we associate to it a PETKS scheme $\mathcal{PETKS} = $ hibe-2-petks$(\mathcal{HIBE}, t(k)) = $ (KG, Td, PETKS, Test, N) such that $N(k) = 2^{t(k)}$, KG$(1^k) = $ Setup(1^k) and PETKS$(pk, w, i) = (i, C_1, C_2)$ where $C_1 \stackrel{\$}{\leftarrow} \{0,1\}^k$ and $C_2 \leftarrow $ Enc$(pk, (w, i_1, \ldots, i_{t(k)}), C_1)$. The trapdoor algorithm Td(sk, w, s, e) first constructs a set T of identities as follows. Let j be the smallest index so that $s_j \neq e_j$. Then T is the set containing $(w, s_1, \ldots, s_{t(k)})$, $(w, e_1, \ldots, e_{t(k)})$, the right siblings of all nodes on the path from $(w, s_1, \ldots, s_{j+1})$ to $(w, s_1, \ldots, s_{t(k)})$, and the left siblings of all nodes on the path from $(w, e_1, \ldots, e_{j+1})$ to $(w, e_1, \ldots, e_{t(k)})$. If j does not exist, meaning $s = e$, then $T \leftarrow \{(w, s_1, \ldots, s_{t(k)})\}$. The trapdoor t_w is the set of tuples $((w, i_1, \ldots, i_r), $ KeyDer$(sk, (w, i_1, \ldots, i_r)))$ for all $(i_1, \ldots, i_r) \in T$. To test a ciphertext (i, C_1, C_2), the Test algorithm looks up a tuple $((w, i_1, \ldots, i_r), usk[(w, i_1, \ldots, i_r)])$ in t_w, derives $usk[(w, i_1, \ldots, i_{t(k)})]$ using repetitive calls to the KeyDer algorithm, and returns 1 iff Dec$(usk[(w, i_1, \ldots, i_{t(k)})], C_2) = C_1$. The proof of the following is in [1]:

Theorem 10. *Let* $\mathcal{HIBE}$ *be a HIBE scheme, and let* $\mathcal{PETKS}$ = hibe-2-petks $(\mathcal{HIBE}, t(k))$ *where* $t(k) = O(\log(k))$. *If* $\mathcal{HIBE}$ *is* HIBE-ANO-CPA$[1, t(k) + 1]$-*secure, then* $\mathcal{PETKS}$ *is* PETKS-IND-CPA-*secure. Furthermore, if* $\mathcal{HIBE}$ *is* HIBE-IND-CPA$[t(k) + 1]$-*secure, then* $\mathcal{PETKS}$ *is computationally consistent.*

Since the $m\mathcal{GS}$-$\mathcal{HIBE}$ has user secret keys and ciphertexts of size linear in the depth of the tree, our resulting PETKS scheme has public and secret keys of size $O(1)$, ciphertexts of size $O(\log N)$ and trapdoors of size $O(\log^2 N)$. We note that in this case a user can decrypt ciphertexts intended for any of its descendants directly, without needing to derive the corresponding secret key first. This makes the call to the KeyDer algorithm in the Test algorithm superfluous, thereby improving the efficiency of Test.

4.3   ID-Based Searchable Encryption

In this section, we show how to combine the concepts of identity-based encryption and PEKS to obtain identity-based encryption with keyword search (IBEKS). Like in IBE schemes, this allows to use any string as a recipient's public key for the PEKS scheme.

IBEKS. An identity-based encryption with keyword search scheme $\mathcal{IBEKS}$ = (Setup, KeyDer, Td, Enc, Test) is made up of five algorithms. Via $(pk, msk) \xleftarrow{\$}$ Setup(1^k), where $k \in \mathbb{N}$ is the security parameter, the randomized setup algorithm produces master keys; via $usk[id] \xleftarrow{\$}$ KeyDer$^H(msk, id)$, the master computes the secret key for identity id; via $C \xleftarrow{\$}$ Enc$^H(pk, id, w)$, a sender encrypts a keyword w to identity id to get a ciphertext; via $t_w \xleftarrow{\$}$ Td$^H(usk[id], w)$, the receiver computes a trapdoor t_w for keyword w and identity id and provides it to the gateway; via $b \leftarrow$ Test$^H(t_w, C)$, the gateway tests whether C encrypts w, where b is a bit with 1 meaning "accept" or "yes" and 0 meaning "reject" or "no". As usual H is a random oracle whose domain and/or range might depend on k and pk. For consistency, we require that for all $k \in \mathbb{N}$, all identities id, and all $w \in \{0, 1\}^*$, we have Pr[Test$^H($Td$^H($KeyDer$^H(msk, id), w), Enc^H(pk, id, w)) = 1] = 1$, where the probability is taken over the choice of $(pk, msk) \xleftarrow{\$}$ Setup(1^k), the random choice of H, and the coins of all algorithms in the expression above.

CONSISTENCY AND PRIVACY. Computational, statistical and perfect consistency can be defined analogously to the way they were defined for PEKS. A privacy notion (denoted IBEKS-IND-CPA) can be obtained by appropriately combining ideas of the definitions of privacy for PEKS and IBE. For details, see [1].

CONSTRUCTION. We now propose a generic transformation, called hibe-2-ibeks, to convert any HIBE scheme with two levels into an IBEKS scheme. Given a HIBE scheme $\mathcal{HIBE}$ = (Setup, KeyDer, Enc, Dec) with two levels, hibe-2-ibeks returns the IBEKS scheme $\mathcal{IBEKS}$ = (Setup, $\overline{\text{KeyDer}}$, $\overline{\text{Enc}}$, Td, Test) such that $\overline{\text{KeyDer}}(msk, id) = (usk, id)$ where $usk \xleftarrow{\$}$ KeyDer(msk, id), $\overline{\text{Enc}}(pk, id, w) = (C_1, C_2)$ where $C_1 \xleftarrow{\$} \{0, 1\}^k$ and $C_2 = $ Enc$(pk, (id, w), C_1)$, Td$(\overline{usk} = (usk, id)$,

$w) = \mathsf{KeyDer}(usk, (id, w))$ and $\mathsf{Test}(t_w, (C_1, C_2))$ returns 1 iff $\mathsf{Dec}(t_w, C_2) = C_1$. The proof of the following is in [1]:

Theorem 11. *Let $\mathcal{HIBE}$ be a HIBE scheme and $\mathcal{IBEKS} = \mathsf{hibe\text{-}2\text{-}ibeks}(\mathcal{HIBE})$. If $\mathcal{HIBE}$ is HIBE-IND-CPA[2]-secure, then $\mathcal{IBEKS}$ is computationally consistent. Furthermore, if $\mathcal{HIBE}$ is HIBE-ANO-CPA[2, 2]-secure, then $\mathcal{IBEKS}$ is IBEKS-IND-CPA-secure.*

We know of no HIBE scheme that is anonymous at level 2, and thus we have no concrete instantiations of the above. (We exclude the scheme of [16] because it is not secure against a polynomial number of level-2 key extractions, as required for HIBE-ANO-CPA[2,2]-security and in particular for our construction.)

Acknowledgements

We thank Nigel Smart for suggesting the concept of temporarily searchable encryption. Second and tenth authors were supported in part by NSF grants ANR-0129617 and CCR-0208842 and by an IBM Faculty Partnership Development Award. Fourth author was supported by a DAAD postdoc fellowship. Fifth author was supported by an IBM Ph.D Fellowship. Eighth author was supported in part by the Flemish Government under GOA Mefisto 2006/06 and Ambiorix 2005/11, and by the European Commission through the IST Project PRIME. The rest of the authors were supported in part by the European Commission through the IST Program under Contract IST-2002-507932 ECRYPT.

References

1. M. Abdalla, M. Bellare, D. Catalano, E. Kiltz, T. Kohno, T. Lange, J. Malone-Lee, G. Neven, P. Paillier, and H. Shi. Searchable encryption revisited: Consistency properties, relation to anonymous ibe, and extensions. Full version of current paper. Available at IACR Cryptology ePrint Archive, http://eprint.iacr.org.
2. M. Ajtai and C. Dwork. A public-key cryptosystem with worst-case/average-case equivalence. In *29th ACM STOC*. ACM Press, 1997.
3. M. Bellare, A. Boldyreva, A. Desai, and D. Pointcheval. Key-privacy in public-key encryption. In *ASIACRYPT 2001, LNCS* 2248. Springer, 2001.
4. M. Bellare and S. K. Miner. A forward-secure digital signature scheme. In *CRYPTO'99, LNCS* 1666. Springer, 1999.
5. M. Bellare and P. Rogaway. Random oracles are practical: A paradigm for designing efficient protocols. In *ACM CCS 93*. ACM Press, 1993.
6. D. Boneh, X. Boyen, and E.-J. Goh. Hierarchical identity based encryption with constant size ciphertext. In *EUROCRYPT 2005, LNCS* 3494. Springer, 2005.
7. D. Boneh, G. Di Crescenzo, R. Ostrovsky, and G. Persiano. Public key encryption with keyword search. In *EUROCRYPT 2004, LNCS* 3027. Springer, 2004.
8. D. Boneh and M. K. Franklin. Identity based encryption from the Weil pairing. *SIAM Journal on Computing*, 32(3), 2003.
9. R. Canetti, S. Halevi, and J. Katz. A forward-secure public-key encryption scheme. In *EUROCRYPT 2003, LNCS* 2656. Springer, 2003.

10. C. Dwork, M. Naor, and O. Reingold. Immunizing encryption schemes from de-cryption errors. In *EUROCRYPT 2004, LNCS* 3027. Springer, 2004.
11. C. Gentry and A. Silverberg. Hierarchical ID-based cryptography. In *ASIACRYPT 2002, LNCS* 2501. Springer, 2002.
12. E.-J. Goh. Secure indexes. Cryptology ePrint Archive, Report 2003/216, 2003.
13. O. Goldreich. *Foundations of Cryptography: Basic Applications*, volume 2. Cambridge University Press, Cambridge, UK, 2004.
14. P. Golle, J. Staddon, and B. R. Waters. Secure conjunctive keyword search over encrypted data. In *ACNS 2004, LNCS* 3089. Springer, 2004.
15. S. Halevi. A sufficient condition for key-privacy. Cryptology ePrint Archive, Report 2005/005, 2005.
16. J. Horwitz and B. Lynn. Toward hierarchical identity-based encryption. In *EUROCRYPT 2002, LNCS* 2332. Springer, 2002.
17. A. Shamir. Identity-based cryptosystems and signature schemes. In *CRYPTO'84, LNCS* 196. Springer, 1985.
18. D. X. Song, D. Wagner, and A. Perrig. Practical techniques for searches on encrypted data. In *2000 IEEE Symposium on Security and Privacy*, 2000.
19. B. R. Waters. Efficient identity-based encryption without random oracles. In *EUROCRYPT 2005, LNCS* 3494. Springer, 2005.
20. B. R. Waters, D. Balfanz, G. Durfee, and D. K. Smetters. Building an encrypted and searchable audit log. In *ISOC NDSS 2004*, 2004.

Private Searching on Streaming Data*

Rafail Ostrovsky[1],[**] and William E. Skeith III[2]

[1] UCLA Computer Science Department
rafail@cs.ucla.edu
[2] UCLA Department of Mathematics
wskeith@math.ucla.edu

Abstract. In this paper, we consider the problem of private searching on streaming data. We show that in this model we can efficiently implement searching for documents under a secret criteria (such as presence or absence of a hidden combination of hidden keywords) under various cryptographic assumptions. Our results can be viewed in a variety of ways: as a generalization of the notion of a Private Information Retrieval (to the more general queries and to a streaming environment as well as to public-key program obfuscation); as positive results on privacy-preserving datamining; and as a delegation of hidden program computation to other machines.

Keywords: Code Obfuscation, Crypto-computing, Software security, Database security, Public-key Encryption with special properties, Private Information Retrieval, Privacy-Preserving Keyword Search, Secure Algorithms for Streaming Data, Privacy-Preserving Datamining, Secure Delegation of Computation, Searching with Privacy, Mobile code.

1 Introduction

DATA FILTERING FOR THE INTELLIGENCE COMMUNITY. As our motivating example, we examine one of the tasks of the intelligence community: to collect "potentially useful" information from huge streaming sources of data. The data sources are vast, and it is often impractical to keep all the data. Thus, streaming data is typically sieved from multiple data streams in an on-line fashion, one document/message/packet at a time, where most of the data is immediately dismissed and dropped to the ground, and only some small fraction of "potentially useful" data is retained. These streaming data sources, just to give a few examples, include things like packet traffic on some network routers, on-line news feeds (such as Reuters.com), some internet chat-rooms, or potentially terrorist-related blogs or web-cites. Of course, most of the data is totally innocent and is immediately dismissed except for some data that raises "red flags" is collected for later analysis "on the inside".

---

* For the full version of this paper, see the IACR E-Print Archive.

** Supported in part by Intel equipment grant, Teradata Corporation, NSF Cybertrust grant, OKAWA research award, B. John Garrick Foundation and Xerox Innovation group Award.

V. Shoup (Ed.): Crypto 2005, LNCS 3621, pp. 223–240, 2005.

In almost all cases, what's "potentially useful" and raises a "red flag" is classified, and satisfies a secret criteria (i.e., a boolean decision whether to keep this document or throw it away). The classified "sieving" algorithm is typically written by various intelligence community analysts. Keeping this "sieving" algorithm classified is clearly essential, since otherwise adversaries could easily avoid their messages from being collected by simply avoiding criteria that is used to collect such documents in the first place. In order to keep the selection criteria classified, one possible solution (and in fact the one that is used in practice) is to collect *all* streaming data "on the inside" —in a secure environment— and then filter the information, according to classified rules, throwing away most of it and keeping only a small fraction of data-items that are interesting according to the secret criteria, such as a set of keywords that raise a red-flag. While this certainly keeps the sieving information private, this solution requires **transferring** all streaming data to a classified environment, adding considerable cost, both in terms of communication cost and a potential delay or even loss of data, if the transfer to the classified environment is interrupted or dropped in transit. Furthermore, it requires considerable cost of **storage** of this (un-sieved) data in case the transfer to the classified setting is delayed.

Clearly, a far more preferable solution, is to sieve all these data-streams at their sources (even on the same computers or routers where the stream is generated or arrives in the first place). The issue, of course, is how can we do it while keeping sieving criteria classified, even if the computer where the sieving program executes falls into enemy's hands? Perhaps somewhat surprisingly, we show how to do just that while keeping the sieving criteria provably hidden from the adversary, even if the adversary gets to experiment with the sieving executable code and/or tries to reverse-engineer it. Put differently, we construct a "compiler" (i.e. of how to compile sieving rules) so that it is provably impossible to reverse-engineer the classified rules from the executable complied sieving code. Now, we state our results in a more general terms, that we believe are of independent interest:

PUBLIC-KEY PROGRAM OBFUSCATION: Very informally, given a program f from a complexity class $\mathcal{C}$, and a security parameter k, a **public-key program obfuscator** compiles f into (F, Dec), where F on any input computes an encryption of what f would compute on the same input. The decryption algorithm Dec decrypts the output of F. That is, for any input x, $Dec(F(x)) = f(x)$, but given code for F it is impossible to distinguish for any polynomial time adversary which f from complexity class $\mathcal{C}$ was used to produce F. We stress that in our definition, the program encoding length $|F|$ must polynomially depend only on $|f|$ and k, and the output length of $|F(x)|$ must polynomially depend only on $|f(x)|$ and k. It is easy to see that Single-Database Private Information Retrieval (including keyword search) can be viewed as a special case of public-key program obfuscator.

OBFUSCATING SEARCHING ON STREAMING DATA: We consider how to public-key program obfuscate Keyword Search algorithms on streaming data, where the size of the query (i.e. complied executable) must be *independent* of the size of stream (i.e., database), and that can be executed in an on-line environment, one document at a time. Our results also can be viewed as improvement and a speedup of the best previous results of single-round PIR with keyword search of Freedman, Ishai, Pinkas and Reingold [10]. In addition to the introduction of the streaming model, this paper also improves the previous work on keyword PIR by allowing for the simultaneous return of multiple documents that match a set of keywords, and also the ability to more efficiently perform different types of queries beyond just searching for a single keyword. For example, we show how to search for the disjunction of a set of keywords and several other functions.

OUR RESULTS: We consider a dictionary of finite size (e.g., an English dictionary) D that serves as the universe for our keywords. Additionally, we can also have keywords that must be absent from the document in order to match it. We describe the various properties of such filtering software below. A filtering program F stores up to some maximum number m of matching documents in an encrypted buffer B. We provide several methods for constructing such software F that saves up to m matching documents with overwhelming probability and saves non-matching documents with negligible probability (in most cases, this probability will be identically 0), all without F or its encrypted buffer B revealing any information about the query that F performs. The requirement that non-matching documents are not saved (or at worst with negligible probability) is motivated by the streaming model: in general the number of non-matching documents will be vast in comparison to those that do match, and hence, to use only small storage, we must guarantee that non-matching documents from the stream do not collect in our buffer. Among our results, we show how to execute queries that search for documents that match keywords in a disjunctive manner, i.e., queries that search for documents containing one or more keywords from a keyword set. Based on the Paillier cryptosystem, [18], we provide a construction where the filtering software F runs in $O(l \cdot k^3)$ time to process a document, where k is a security parameter, and l is the length of a document, and stores results in a buffer bounded by $\mathcal{O}(m \cdot l \cdot k^2)$. We stress again that F processes documents one at a time in an online, streaming environment. The size of F in this case will be $O(k \cdot |D|)$ where $|D|$ is the size of the dictionary in words. Note that in the above construction, the program size is proportional to the dictionary size. We can in fact improve this as well: we have constructed a reduced program size model that depends on the *Φ-Hiding Assumption* [5]. The running time of the filtering software in this implementation is linear in the document size and is $O(k^3)$ in the security parameter k. The program size for this model is only $O(polylog(|D|))$. We also have an abstract construction based on any group homomorphic, semantically secure encryption scheme. Its performance depends on the performance of the underlying encryption scheme, but will generally perform similarly to the above constructions. As mentioned above, all of these constructions have size that is independent of the size of the data stream. Also, using the results of Boneh, Goh, and Nissim [2], we show how to execute queries that search for an "AND" of two sets of keywords (i.e., the query searches for documents that contain at least one word from K_1 *and* at least one word from K_2 for sets of keywords K_1, K_2), without asymptotically increasing the program size.

Our contributions can be divided into three major areas: Introduction of the streaming model; having queries simultaneously return multiple results; and the ability to extend the semantics of queries beyond just matching a single keyword.

COMPARISON WITH PREVIOUS WORK: A superficially related topic is that of "searching on encrypted data" (e.g., see [3] and the references therein). We note that this body of work is in fact not directly relevant, as the data (i.e. input stream) that is being searched is not encrypted in our setting.

The notion of obfuscation was considered by [1], but we stress that our setting is different, since our notion of public-key obfuscation allows the output to be encrypted, whereas the definition of [1] demands the output of the obfuscated code is given in the clear, making the original notion of obfuscation much more demanding.

Our notion is also superficially related to the notion of "crypto-computing" [19]. However, in this work we are concerned with programs that contain loops, and where we cannot afford to expand this program into circuits, as this will blow-up the program size.

Our work is most closely related to the notion of Single-database Private Information Retrieval (PIR), that was introduced by Kushilevitz and Ostrovsky [13] and has received a lot of subsequent attention in the literature [13,5,8,17,14,4,20,15,10]. (In the setting of multiple, non-communicating databases, the PIR notion was introduced in [7].) In particular, the first PIR with poly-logarithmic overhead was shown by Cachin, Micali and Stadler [5], and their construction can be executed in the streaming environment. Thus the results of this paper can be viewed as a generalization of their work as well. The setting of single-database PIR keyword search was considered in [13,6,12] and more recently by Freedman, Ishai, Pinkas and Reingold [10]. The issue of multiple matches of a single keyword (in a somewhat different setting) was considered by Kurosawa and Ogata [12].

There are important differences between previous works and our work on single-database PIR keyword search: in the streaming model, the program size must be *independent* of the size of the stream, as the stream is assumed to be an arbitrarily polynomially-large source of data and the complier does not need to know the size of the stream when creating the obfuscated query. In contrast, in all previous non-trivial PIR protocols, when creating the query, the user of the PIR protocol must know the upper bound on the database size while creating the PIR query. Also, as is necessary in the streaming model, the memory needed for our scheme is bounded by a value proportional to the size of a document as well as an upper bound on the total number of documents we wish to collect, but is independent of the size of the stream of documents. Finally, we have also extended the types of queries that can be performed. In previous work on keyword PIR, a single keyword was searched for in a database and a single result returned. If one wanted to query an "OR" of several keywords, this would require creating several PIR queries, and then sending each to the database. We however show how to intrinsically extend the types of queries that can be performed, without loss of efficiency or with multiple queries. In particular, all of our systems can efficiently perform an "OR" on a set of keywords and its negation (i.e. a document matches if certain keyword is absent from the document). In addition, we show how to privately execute a query that searches for an "AND" of two sets of keywords, meaning that a document will match if and only if it contains at least one word from each of the keyword sets without the increase in program (or dictionary) size.

2   Definitions and Preliminaries

2.1   Basic Definitions

For a set V we denote the power set of V by $\mathcal{P}(V)$.

Definition 1. *Recall that a function $g : \mathbb{N} \to \mathbb{R}^+$ is said to be* negligible *if for any $c \in \mathbb{N}$ there exists $N_c \in \mathbb{Z}$ such that $n \geq N_c \Rightarrow g(n) \leq \frac{1}{n^c}$.*

Definition 2. *Let $\mathcal{C}$ be a class of programs, and let $f \in \mathcal{C}$. We define a* public key program obfuscator in the weak sense *to be an algorithm*

$$\mathsf{Compile}(f, r, 1^k) \mapsto \{F(\cdot, \cdot), \mathsf{Decrypt}(\cdot)\}$$

where r is randomness, k is a security parameter, and F and Decrypt *are algorithms with the following properties:*

- *(Correctness) F is a probabilistic function such that $\forall x, \Pr_{R,R'}\Big[\mathsf{Decrypt}$*
  $(F(x, R')) = f(x)\Big] \geq 1 - neg(k)$
- *(Compiled Program Conciseness) There exists a constant c such that $|f| \geq \frac{|F(\cdot,\cdot)|}{(|f|+k)^c}$*
- *(Output Conciseness) There exists a constant c such that For all x, R $|f(x)| \geq \frac{|F(x,R)|}{k^c}$*
- *(Privacy) Consider the following game between an adversary A and a challenger C:*
  1. *On input of a security parameter k, A outputs two functions $f_1, f_2 \in \mathcal{C}$.*
  2. *C chooses a $b \in \{0, 1\}$ at random and computes $\mathsf{Compile}(f_b, r, k) = \{F, \mathsf{Decrypt}\}$ and sends F back to A.*
  3. *A outputs a guess b'.*
  We say that the adversary wins if $b' = b$, and we define the adversary's advantage to be $\mathrm{Adv}_A(k) = |\Pr(b = b') - \frac{1}{2}|$. Finally we say that the system is secure if for all $A \in PPT$, $\mathrm{Adv}_A(k)$ is a negligible function in k.

We also define a stronger notion of this functionality, in which the decryption algorithm does not give any information about f beyond what can be learned from the output of the function alone.

Definition 3. *Let $\mathcal{C}$ be a class of programs, and let $f \in \mathcal{C}$. We define a public key program obfuscator in the strong sense to be a triple of algorithms* (Key-Gen, Compile, Decrypt) *defined as follows:*

1. Key-Gen(k): *Takes a security parameter k and outputs a public key and a secret key $A_{public}, A_{private}$.*
2. Compile$(f, r, A_{public}, A_{private})$: *Takes a program $f \in \mathcal{C}$, randomness r and the public and private keys, and outputs a program $F(\cdot, \cdot)$ that is subject to the same Correctness and conciseness properties as in Definition 2.*
3. Decrypt$(F(x), A_{private})$: *Takes output of F and the private key and recovers $f(x)$, just as in the correctness of Definition 2.*

Privacy is also defined as in Definition 2, however in the first step the adversary A receives as an additional input A_{public} and we also require that Decrypt *reveals no information about f beyond what could be computed from $f(x)$: Formally, for all adversaries $A \in PPT$ and for all history functions h there exists a simulating program $B \in PPT$ that on input $f(x)$ and $h(x)$ is computationally indistinguishable from A on input* (Decrypt$, F(x), h(x))$.

Now, we give instantiations of these general definitions to the class of programs $\mathcal{C}$ that we show how to handle: We consider a universe of words $W = \{0,1\}^*$, and a dictionary $D \subset W$ with $|D| = \alpha < \infty$. We think of a document just to be an ordered, finite sequence of words in W, however, it will often be convenient to look at the set of distinct words in a document, and also to look at some representation of a document as a single string in $\{0,1\}^*$. So, the term *document* will often have several meanings, depending on the context: if M is said to be a *document*, generally this will mean M is an ordered sequence in W, but at times, (e.g., when M appears in set theoretic formulas) *document* will mean (finite) element of $\mathcal{P}(W)$ (or possibly $\mathcal{P}(D)$), and at other times still, (say when one is talking of bit-wise encrypting a document) we'll view M as $M \in \{0,1\}^*$. We define a *set of keywords* to be any subset $K \subset D$. Finally, we define a *stream* of documents S just to be any sequence of documents.

We will consider only a few types of queries in this work, however would like to state our definitions in generality. We think of a *query type*, $\mathcal{Q}$ as a class of logical expressions in $\wedge$, $\vee$, and $\neg$. For example, $\mathcal{Q}$ could be the class of expressions using only the operation $\wedge$. Given a query type, one can plug in the number of variables, call it α for an expression, and possibly other parameters as well, to select a specific boolean expression, Q in α variables from the class $\mathcal{Q}$. Then, given this logical expression, one can input $K \subset D$ where $K = \{k_i\}_{i=1}^{\alpha}$ and create a function, call it $Q_K : \mathcal{P}(D) \rightarrow \{0,1\}$ that takes documents, and returns 1 if and only if a document matches the criteria. $Q_K(M)$ is computed simply by evaluating Q on inputs of the form $(k_i \in M)$. We will call Q_K a *query over keywords K*.

We note again that our work does not show how to privately execute arbitrary queries, despite the generality of these definitions. In fact, extending the query semantics is an interesting open problem.

Definition 4. *For a query Q_K on a set of keywords K, and for a document M, we say that M matches query Q_K if and only if $Q_K(M) = 1$.*

Definition 5. *For a fixed query type $\mathcal{Q}$, a private filter generator consists of the following probabilistic polynomial time algorithms:*

1. Key-Gen(k): *Takes a security parameter k and generates public key A_{public}, and a private key $A_{private}$.*
2. Filter-Gen$(D, Q_K, A_{public}, A_{private}, m, \gamma)$: *Takes a dictionary D, a query $Q_K \in \mathcal{Q}$ for the set of keywords K, along with the private key and generates a search program F. F searches any document stream S (processing one document at a time and updating B) collects up to m documents that match Q_K in B, outputting an encrypted buffer B that contains the query results, where $|B| = \mathcal{O}(\gamma)$ throughout the execution.*
3. Filter-Decrypt$(B, A_{private})$: *Decrypts an encrypted buffer B, produced by F as above, using the private key and produces output B^*, a collection of the matching documents from S.*

Definition 6. *(Correctness of a Private Filter Generator)*
Let $F = $ Filter-Gen$(D, Q_K, A_{public}, A_{private}, m, \gamma)$, where D is a dictionary, Q_K is a query for keywords K, $m, \gamma \in \mathbb{Z}^+$ and $(A_{public}, A_{private}) = $ Key-Gen(k). We say that a private filter generator is correct *if the following holds:*
Let F run on any document stream S, and set $B = F(S)$.
Let $B^* = $ Filter-Decrypt$(B, A_{private})$. Then,

- *If $|\{M \in S \mid Q_K(M) = 1\}| \leq m$ then*

$$\Pr\Big[B^* = \{M \in S \mid Q_K(M) = 1\}\Big] > 1 - neg(\gamma)$$

- *If $|\{M \in S \mid Q_K(M) = 1\}| > m$ then*

$$\Pr\Big[(B^* \subset \{M \in S \mid Q_K(M) = 1\}) \vee (B^* = \bot)\Big] > 1 - neg(\gamma)$$

where $\bot$ is a special symbol denoting buffer overflow, and the probabilities are taken over all coin-tosses of F, Filter-Gen and of Key-Gen.

Definition 7. *(Privacy) Fix a dictionary D. Consider the following game between an adversary A, and a challenger C. The game consists of the following steps:*

1. *C first runs Key-Gen(k) to obtain $A_{public}, A_{private}$, and then sends A_{public} to A.*
2. *A chooses two queries for two sets of keywords, $Q_{0 K_0}, Q_{1 K_1}$, with $K_0, K_1 \subset D$ and sends them to C.*
3. *C chooses a random bit $b \in \{0, 1\}$ and executes Filter-Gen$(D, Q_{b K_b}, A_{public}, A_{private}, m, \gamma)$ to create F_b, the filtering program for the query $Q_{b K_b}$, and then sends F_b back to A.*

4. $A(F_b)$ can experiment with code of F_b in an arbitrary non-black-box way finally output $b' \in \{0, 1\}$.

The adversary wins the game if $b = b'$ and loses otherwise. We define the adversary A's advantage in this game to be $\mathrm{Adv}_A(k) = \left| \Pr(b = b') - \frac{1}{2} \right|$ We say that a private filter generator is semantically secure if for any adversary $A \in PPT$ we have that $\mathrm{Adv}_A(k)$ is a negligible function, where the probability is taken over coin-tosses of the challenger and the adversary.

2.2   Combinatorial Lemmas

We require in our definitions that matching documents are saved with overwhelming probability in the buffer B (in terms of the size of B), while nonmatching documents are not saved at all (at worst, with negligible probability). We accomplish this by the following method: If the document is of interest to us, we throw it at random γ times into the buffer. What we are able to guarantee is that if only one document lands in a certain location, and no other document lands in this location, we will be able to recover it. If there is a collision of one or more documents, we assume that all documents at this location are lost (and furthermore, we guarantee that we will detect such collisions with overwhelming probability). To amplify the probability that matching documents survive, we throw each γ times, and we make the total buffer size proportional to $2\gamma m$, where m is the upper bound on the number of documents we wish to save. Thus, we need to analyze the following combinatorial game, where each document corresponds to a ball of different color.

Color-Survival Game: Let $m, \gamma \in \mathbb{Z}^+$, and suppose we have m different colors, call them $\{color_i\}_{i=1}^m$, and γ balls of each color. We throw the γm balls uniformly at random into $2\gamma m$ bins, call them $\{bin_j\}_{j=1}^{2\gamma m}$. We say that a ball "survives" in bin_j, if no other ball (of any color) lands in bin_j. We say that $color_i$ "survives" if at least one ball of color $color_i$ survives. We say that the game succeeds if all m colors survive, otherwise we say that it fails.

Lemma 1. The probability that the color-survival game fails is negligible in γ.

Proof. See full version.

Another issue is how to distinguish valid documents in the buffer from collisions of two or more matching documents in the buffer. (In general it is unlikely that the sum of two messages in some language will look like another message in the same language, but we need to guarantee this fact.) This can also be accomplished by means of a simple probabilistic construction. We will append to each document k bits, where exactly $k/3$ randomly chosen bits from this string are set to 1. When reading the buffer results, we will consider a document to be good if exactly $k/3$ of the appended bits are 1's. If a buffer collision occurs between two matching documents, the buffer at this location will store the sum of the messages, and the sum of 2 or more of the k-bit strings. [1] We need to analyze the probability that after adding the k-bit strings, the resulting string *still* has exactly $k/3$ bits set to 1, and show that this probability is negligible in k. We will phrase the problem in terms of "balls in bins" as before: let $Bins = \{bin_j\}_{j=1}^k$ be distinguishable bins, and let $\mathcal{T}(Bins)$ denote the process of selecting $k/3$ bins uniformly at random from all $\binom{k}{k/3}$ choices, and putting one ball in each of these bins. For a fixed randomness, we can formalize $\mathcal{T}$ as a map $\mathcal{T} : \mathbb{Z}^k \to \mathbb{Z}^k$ such that for any $v = (v_1, ..., v_k) \in \mathbb{Z}^k$, $0 \le (\mathcal{T}(v)_j - v_j) \le 1$ for all $j \in \{1, ..., k\}$, and $\sum_{j=1}^k (\mathcal{T}(v)_j - v_j) = k/3$. Let $N(bin_j)$ be the number of balls in bin_j. Now, after independently repeating this experiment with the same bins, which were initially empty, we will be interested in the probability that for exactly $2k/3$ bins, the number of balls inside is 0 mod n, for $n > 1$. I.e., after applying $\mathcal{B}$ twice, what is $\Pr[\ |\{j \mid N(bin_j) \equiv 0 \bmod n\}| = 2k/3]$?

Lemma 2. *Let $Bins = \{bin_j\}_{j=1}^k$ be distinguishable bins, all of which are empty. Let $Bins = \mathcal{T}^2(Bins)$. Then for all $n > 1$,*

$$\Pr\Big[|\{j \in \{1, ..., k\} \mid N(bin_j) \equiv 0 \ mod \ n\}| = 2k/3\Big]$$

is negligible in k.

Proof. See full version.

2.3 Organization of the Rest of This Paper

In what follows, we will give several constructions of private filter generators, beginning with our most efficient construction using a variant of the Paillier Cryptosystem [18],[9]. We also show a construction with reduced program size using the Cachin-Micali-Stadler PIR protocol [5], then we give a construction based on any group homomorphic semantically secure encryption scheme, and finally a construction based on the work of Boneh, Goh, and Nissim [2] that extends the query semantics to include a single "AND" operation without increasing the program size.

3 Paillier-Based Construction

Definition 8. *Let $(G_1, \cdot), (G_2, *)$ be groups. Let $\mathcal{E}$ be the probabilistic encryption algorithm and $\mathcal{D}$ be the decryption algorithm of an encryption scheme with plaintext set G_1 and ciphertext set G_2. The encryption scheme is said to be* group homomorphic *if the encryption map $\mathcal{E} : G_1 \to G_2$ has the following property:*

$$\forall \ a, b \in G_1, \ \mathcal{D}(\mathcal{E}(a \cdot b)) = \mathcal{D}(\mathcal{E}(a) * \mathcal{E}(b))$$

---

[1] If a document does not match, it will be encrypted as the 0 message, as will its appended string of k bits, so nothing will ever be marked as a collision with a non-matching document.

Note that since the encryption is probabilistic, we have to phrase the homomorphic property using $\mathcal{D}$, instead of simply saying that $\mathcal{E}$ is a homomorphism. Also, as standard notation when working with homomorphic encryption as just defined, we will use id_{G_1}, id_{G_2} to be the identity elements of G_1, G_2, respectively.

3.1   Private Filter Generator Construction

We now formally present the Key-Gen, Filter-Gen, and Buffer-Decrypt algorithms. The class $\mathcal{Q}$ of queries that can be executed is the class of all boolean expressions using only $\vee$. By doubling the program size, it is easy to handle an $\vee$ of both presence and absence of keywords. For simplicity of exposition, we describe how to detect collisions separately from the main algorithm.

Key-Gen(k) Execute the key generation algorithm for the Paillier cryptosystem to find an appropriate RSA number, n and its factorization $n = pq$. We will make one additional assumption on $n = pq$: we require that $|D| < \min\{p, q\}$. (We need to guarantee that any number $\leq |D|$ is a unit mod n^s.) Save n as A_{public}, and save the factorization as $A_{private}$.

Filter-Gen($D, Q_K, A_{public}, A_{private}, m, \gamma$) This algorithm outputs a search program F for the query $Q_K \in \mathcal{Q}$. So, $Q_K(M) = \bigvee_{w \in K}(w \in M)$.
We will use the Damgård-Jurik extension [9] to construct F as follows. Choose an integer $s > 0$ based upon the size of documents that you wish to store so that each document can be represented as a group element in $\mathbb{Z}_{n^s}$. Then F contains the following data:

- A buffer B consisting of $2\gamma m$ blocks with each the size of two elements of $\mathbb{Z}^*_{n^{s+1}}$ (so, we view each block of B as an ordered pair $(v_1, v_2) \in \mathbb{Z}^*_{n^{s+1}} \times \mathbb{Z}^*_{n^{s+1}}$). Furthermore, we will initialize every position to $(1, 1)$, two copies of the identity element.
- An array $\widehat{D} = \{\widehat{d_i}\}_{i=1}^{|D|}$ where each $\widehat{d_i} \in \mathbb{Z}^*_{n^{s+1}}$ such that $\widehat{d_i}$ is an encryption of $1 \in \mathbb{Z}_{n^s}$ if $d_i \in K$ and is an encryption of 0 otherwise. (Note: We of course use re-randomized encryptions of these values for each entry in the array.)

F then proceeds with the following steps upon receiving an input document M from the stream:

1. Construct a temporary collection $\widehat{M} = \{\widehat{d_i} \in \widehat{D} \mid d_i \in M\}$.
2. Compute

$$v = \prod_{\widehat{d_i} \in \widehat{M}} \widehat{d_i}$$

3. Compute v^M and multiply (v, v^M) into γ random locations in the buffer B, just as in our combinatorial game from section 2.2.

Note that the private key actually is not needed. The public key alone will suffice for the creation of F.

Buffer-Decrypt($B, A_{private}$) First, this algorithm simply decrypts B one block at a time using the decryption algorithm for the Paillier system. Each decrypted block will contain the 0 message (i.e., $(0,0)$) or a non-zero message, $(w_1, w_2) \in \mathbb{Z}_{n^s} \times \mathbb{Z}_{n^s}$. Blocks with the 0 message are discarded. A non-zero message (w_1, w_2) will be of the form (c, cM') if no collisions have occurred at this location, where c is the number of distinct keywords from K that appear in M'. So to recover M', simply compute w_2/w_1 and add this to the array B^*. Finally, output B^*.

In general, the filter generation and buffer decryption algorithms will make use of Lemma 2, having the filtering software append an extra r bits to each document, with exactly $r/3$ bits equal to 1, and then having the buffer decryption algorithm save documents to the output B^* if and only if exactly $r/3$ of the extra bits are 1. In any of our constructions, this can be accomplished by adding r extra blocks the size of the security parameter to an entry in the buffer to represent the bits. However, this is undesirable in our Paillier-based construction, as this would cause an increase (by a factor of $r/2$) to the size of the buffer.

Lemma 3. *With $\mathcal{O}(k)$ additional bits added to each block of B, we can detect all collisions of matching documents with probability $> 1 - neg(k)$.*

Proof. Since $log(|D|)$ is much smaller than the security parameter k, we can encode the bits from Lemma 2 using $\mathcal{O}(k)$ bits. For further details, see the full version of this paper.

3.2 Correctness

We need to show that if the number of matching documents is less than m, then

$$Pr\left[B^* = \{M \in S \mid Q_K(M) = 1\}\right] > 1 - neg(\gamma)$$

and otherwise, we have that B^* is a subset of the matching documents (or contains the overflow symbol, $\perp$). Provided that the buffer decryption algorithm can distinguish collisions in the buffer from valid documents (see above remark) this equates to showing that non-matching documents are saved with negligible probability in B and that matching documents are saved with overwhelming probability in B. These two facts are easy to show.

Theorem 1. *Assuming that the Paillier (and Damgård-Jurik) cryptosystems are semantically secure, then the private filter generator from the preceding construction is semantically secure according to Definition 7.*

Proof. Denote by $\mathcal{E}$ the encryption algorithm of the Paillier/Damgård-Jurik cryptosystem. Suppose that there exists an adversary A that can gain a non-negligible advantage ϵ in our semantic security game from Definition 7. Then A could be used to gain an advantage in breaking the semantic security of the Paillier encryption scheme as follows: Initiate the semantic security game for the Paillier encryption scheme with some challenger C. C will send us an integer n for the Paillier challenge. For messages m_0, m_1, we choose $m_0 = 0 \in \mathbb{Z}_{n^s}$ and choose $m_1 = 1 \in \mathbb{Z}_{n^s}$. After sending m_0, m_1 back to C, we will receive $e_b = \mathcal{E}(m_b)$, an encryption of one of these two values. Next we initiate the private filter generator semantic security game with A. A will give us two queries

Q_0, Q_1 in $\mathcal{Q}$ for some sets of keywords K_0, K_1, respectively. We use the public key n to compute an encryption of 0, call it $e_0 = \mathcal{E}(0)$. Now we pick a random bit q, and construct filtering software for Q_q as follows: we proceed as described above, constructing the array $\widehat{D}$ by using re-randomized encryptions, $\mathcal{E}(0)$ of 0 for all words in $D \setminus K_q$, and for the elements of K_q, we use $\mathcal{E}(0)e_b$, which are randomized encryptions of m_b. Now we give this program back to A, and A returns a guess q'. With probability $1/2$, e_b is an encryption of 0, and hence the program that we gave A does not search for anything at all, and in this event clearly A's guess is independent of q, and hence the probability that $q' = q$ is $1/2$. However, with probability $1/2$, $e_b = \mathcal{E}(1)$, hence the program we've sent A is filtering software that searches for Q_q, constructed exactly as in the Filter-Gen algorithm, and hence in this case with probability $1/2 + \epsilon$, A will guess q correctly, as our behavior here was indistinguishable from an actual challenger. We determine our guess b' as follows: if A guesses $q' = q$ correctly, then we will set $b' = 1$, and otherwise we will set $b' = 0$. Putting it all together, we can now compute the probability that our guess is correct: $\Pr(b' = b) = \frac{1}{2}\left(\frac{1}{2}\right) + \frac{1}{2}\left(\frac{1}{2} + \epsilon\right) = \frac{1}{2} + \frac{\epsilon}{2}$ and hence we have obtained a non-negligible advantage in the semantic security game for the Paillier system, a contradiction to our assumption. Therefore, our system is secure according to Definition 7.

4   Reducing Program Size Below Dictionary Size

In our other constructions, the program size is proportional to the size of the dictionary. By relaxing our definition slightly, we are able to provide a new construction using Cachin-Micali-Stadler PIR [5] which reduces the program size. Security of this system depends on the security of [5] which uses the Φ-Hiding Assumption.[2]

The basic idea is to have a standard dictionary D agreed upon ahead of time by all users, and then to replace the array $\widehat{M}$ in the filtering software with PIR queries that execute on a database consisting of the characteristic function of M with respect to D to determine if keywords are present or not. The return of the queries is then used to modify the buffer. This will reduce the size of the distributed filtering software. However, as mentioned above, we will need to relax our definition slightly and publish an upper bound U for $|K|$, the number of keywords used in a search.

4.1   Private Filter Generation

We now formally present the Key-Gen, Filter-Gen, and Buffer-Decrypt algorithms of our construction. The class $\mathcal{Q}$ of queries that can be executed by this protocol is again just the set of boolean expressions in only the operator $\vee$ over presence or absence of keywords, as discussed above. Also, an important note: for this

---

[2] It is an interesting open question how to reduce the program size under other cryptographic assumptions.

construction, it is necessary to know the set of keywords being used during key generation, and hence what we achieve here is only weak public key program obfuscation, as in Definition 2. For consistency of notation, we still present this as 3 algorithms, even though the key generation could be combined with the filter generation algorithm. For brevity, we omit the handling of collision detection, which is handled using Lemma 2.

Key-Gen(k, K, D) The CMS algorithms are run to generate PIR queries, $\{q_j\}$ for the keywords K, and the resulting factorizations of the corresponding composite numbers $\{m_j\}$ are saved as the key, $A_{private}$, while A_{public} is set to $\{m_j\}$.

Filter-Gen($D, Q_K, A_{public}, A_{private}, m, \gamma$) This algorithm constructs and outputs a private filter F for the query Q_K, using the PIR queries q_j that were generated in the Key-Gen(k, K, D) algorithm. It operates as follows.
F contains the following data:

- The array of CMS PIR queries, $\{q_j\}_{j=1}^{U}$ from the first algorithm, which are designed to retrieve a bit from a database having size equal to the number of words in the agreed upon dictionary, D. Only $|K|$ of these queries will be meaningful. For each $w \in K$, there will be a meaningful query that retrieves the bit at index corresponding to w's index in the dictionary. Let $\{p_{j,l}\}_{l=1}^{|D|}$ be the primes generated by the information in q_j, and let m_j be composite number part of q_j. The leftover $U - |K|$ queries are set to retrieve random bits.
- An array of buffers $\{B_j\}_{j=1}^{U}$, each indexed by blocks the size of elements of $\mathbb{Z}_{m_j}^*$, with every position initialized to the identity element.

The program then proceeds with the following steps upon receiving an input document M:

1. Construct the complement of the characteristic vector for the words of M relative to the dictionary D. I.e., create an array of bits $\bar{D} = \{\bar{d}_i\}$ of size $|D|$, such that $\bar{d}_i = 0 \Leftrightarrow d_i \in M$. We'll use this array as our database for the PIR protocols.

   Next, for each $j \in \{1, 2, ..., U\}$, do the following steps:
2. Execute query q_j on $\bar{D}$ and store the result in r_j.
3. Bitwise encrypt M, using r_j to encrypt a 1 and using the identity of $\mathbb{Z}_{m_j}^*$ to encrypt a 0.
4. Take the jth encryption from step 3 and position-wise multiply it into a random location in buffer B_j γ-times, as described in our color-survival game from section 2.

Buffer-Decrypt($B, A_{private}$) Simply decrypts each buffer B_j one block at a time by interpreting each group element with $p_{j,i}$th roots as a 0 and other elements as 1's, where i represents the index of the bit that is searched for by query q_j. All valid non-zero decryptions are stored in the output B^*.

4.2   Correctness of Private Filter

Since CMS PIR is not deterministic, it is possible that our queries will have the wrong answer at times. However, this probability is negligible in the security parameter. Again, as we've seen before, provided that the decryption algorithm can distinguish valid documents from collisions in buffer, correctness equates to storing non-matching documents in B with negligible probability and matching documents with overwhelming probability. These facts are easy to verify.

Theorem 2. *Assume the Φ-Assumption holds. Then the Private Filter Generator from the preceding construction is semantically secure according to Definition 2.*

Proof. (Sketch) If an adversary can distinguish any two keyword sets, then the adversary can also distinguish between two fixed keywords, by a standard hybrid argument. This is precisely what it means to violate the privacy definition of [5].

5   Eliminating the Probability of Error

In this section, we present ways to eliminate the probability of collisions in the buffer by using perfect hash functions. Recall the definition of perfect hash function. For a set $S \subset \{1, ..., m\}$, if a function $h : \{1, ..., m\} \rightarrow \{1, ..., n\}$ is such that $h|_S$ (the restriction of h to S) is injective, then h is called a *perfect hash function* for S. We will be concerned with families of such functions. We say that H is an (m, n, k)-*family of perfect hash functions* if $\forall S \subset \{1, ..., m\}$ with $|S| = k$, $\exists h \in H$ such that h is perfect for S.

We will apply these families in a very straightforward way. Namely, we define m to be the number of documents in the stream and k to be the number of documents we expect to save. Then, since there exist polynomial size (m, n, k)-families of perfect hash functions H, e.g., $|H| \leq \left\lceil \frac{\log \binom{m}{k}}{\log(n^k) - \log(n^k - k! \binom{n}{k})} \right\rceil$ [16], then our system could consist of $|H|$ buffers, each of size n documents, and our protocol would just write each (potential) encryption of a document to each of the $|H|$ buffers once, using the corresponding hash function from H to determine the index in the buffer. Then, no matter which of the $\binom{m}{k}$ documents were of interest, at least one of the functions in H would be injective on that set of indexes, and hence at least one of our buffers would be free of collisions.

6   Construction Based on Any Homomorphic Encryption

We provide here an abstract construction based upon an arbitrary homomorphic, semantically secure public key encryption scheme. The class of queries $\mathcal{Q}$ that are considered here is again, all boolean expressions in only the operation $\vee$, over presence or absence of keywords, as discussed above. This construction is similar to the Paillier-based construction, except that since we encrypt bitwise, we incur an extra multiplicative factor of the security parameter k in the buffer size. However, both the proof and the construction are somewhat simpler and can be based on any homomorphic encryption.

6.1   Construction of Abstract Private Filter Generator

Throughout this section, let $\mathcal{PKE} = \{\mathcal{KG}(k), \mathcal{E}(p), \mathcal{D}(c)\}$ be a public key encryption scheme. Here, $\mathcal{KG}, \mathcal{E}, \mathcal{D}$ are key generation, encryption, and decryption algorithms, respectively for any group homomorphic, semantically secure encryption scheme, satisfying Definition 8. We describe the Key-Gen, Filter-Gen, and Buffer-Decrypt algorithms. We will write the group operations of G_1 and G_2 multiplicatively. (As usual, G_1, G_2 come from a distribution of groups in some class depending on the security parameter, but to avoid confusion and unnecessary notation, we will always refer to them simply as G_1, G_2 where it is understood that they are actually sampled from some distribution based on k.)

Key-Gen(k) Execute $\mathcal{KG}(k)$ and save the private key as $A_{private}$, and save the public parameters of $\mathcal{PKE}$ as A_{public}.

Filter-Gen($D, Q_K, A_{public}, A_{private}, m, \gamma$) This algorithm constructs and outputs a filtering program F for Q_K, constructed as follows.
F contains the following data:

  - A buffer $B(\gamma)$ of size $2\gamma m$, indexed by blocks the size of an element of G_2 times the document size, with every position initialized to id_{G_2}.
  - Fix an element $g \in G_1$ with $g \neq id_{G_1}$. The program contains an array $\widehat{D} = \{\widehat{d}_i\}_{i=1}^{|D|}$ where each $\widehat{d}_i \in G_2$ such that $\widehat{d}_i$ is set to $\mathcal{E}(g) \in G_1$ if $d_i \in K$ and it is set to $\mathcal{E}(id_{G_1})$ otherwise. (Note: we are of course re-applying $\mathcal{E}$ to compute each encryption, and not re-using the same encryption with the same randomness over and over.)

F then proceeds with the following steps upon receiving an input document M:

1. Construct a temporary collection $\widehat{M} = \{\widehat{d}_i \in \widehat{D} \mid d_i \in M\}$.
2. Choose a random subset $S \subset \widehat{M}$ of size $\lceil |\widehat{M}|/2 \rceil$ and compute

$$v = \prod_{s \in S} s$$

3. Bitwise encrypt M using encryptions of id_{G_1} for 0's and using v to encrypt 1's to create a vector of G_2 elements.
4. Choose a random location in B, take the encryption of step 3, and position-wise multiply these two vectors storing the result back in B at the same location.
5. Repeat steps 2-4 $\left(\frac{c}{c-1}\right)\gamma$ times, where in general, c will be a constant approximately the size of G_1.

Buffer-Decrypt($B, A_{private}$) Decrypts B one block at a time using the decryption algorithm $\mathcal{D}$ to decrypt the elements of G_2, and then interpreting non-identity elements of G_1 as 1's and id_{G_1} as 0, storing the non-zero, valid messages in the output B^*.

7   Construction for a Single AND

7.1   Handling Several AND Operations by Increasing Program Size

We note that there are several simple (and unsatisfactory) modifications that can be made to our basic system to compute an AND. For example a query consisting of at most a c AND operations can be performed simply by changing the dictionary D to a dictionary D' containing all $|D|^c$ c-tuples of words in D, which of course comes at a polynomial blow-up[3] of program size.[4] So, only constant, or logarithmic size keyword sets can be used in order to keep the program size polynomial.

7.2   Executing a Single AND Without Increasing Program Size

Using the results of Boneh, Goh, and Nissim [2], we can extend the types of queries that can be privately executed to include queries involving a single AND of an OR of two sets of keywords without increasing the program size. This construction is very similar to the abstract construction, and hence several details that would be redundant will be omitted from this section. The authors of [2] build an additively homomorphic public key cryptosystem that is semantically secure under this subgroup decision problem. The plaintext set of the system is $\mathbb{Z}_{q_2}$, and the ciphertext set can be either $\mathbb{G}$ or $\mathbb{G}_1$ (which are both isomorphic to $\mathbb{Z}_n$). However, the decryption algorithm requires one to compute discrete logs. Since there are no known algorithms for efficiently computing discrete logs in general, this system can only be used to encrypt small messages. Using the bilinear map e, this system has the following homomorphic property. Let $F \in \mathbb{Z}_{q_2}[X_1, ..., X_u]$ be a multivariate polynomial *of total degree 2* and let $\{c_i\}_{i=1}^u$ be encryptions of $\{x_i\}_{i=1}^u$, $x_i \in \mathbb{Z}_{q_2}$. Then, one can compute an encryption c_F of the evaluation $F(x_1, ..., x_u)$ of F on the x_i with only the public key. This is done simply by using the bilinear map e in place of any multiplications in F, and then multiplying ciphertexts in the place of additions occurring in F. And once again, note that decryption is feasible only when the x_i are small, so one must restrict the message space to be a small subset of $\mathbb{Z}_{q_2}$. (In our application, we will always have $x_i \in \{0, 1\}$.) Using this cryptosystem in our abstract construction, we can easily extend the types of queries that can be performed.

7.3   Construction of Private Filter Generator

More precisely, we can now perform queries of the following form, where M is a document and $K_1, K_2 \subset D$ are sets of keywords:

$$(M \cap K_1 \neq \varnothing) \wedge (M \cap K_2 \neq \varnothing)$$

---

[3] Asymptotically, if we treat $|D|$ as a constant, the above observation allows a logarithmic number of AND operations with polynomial blow-up of program size. It is an interesting open problem to handle more than a logarithmic number of AND operations, keeping the program size polynomial.

[4] A naive suggestion that we received for an implementation of "AND" is to keep track of several buffers, one for each keyword or set of keywords, and then look for documents that appear in each buffer after the buffers are retrieved, however this will put many non-matching documents in the buffers, and hence is inappropriate for the streaming model. Furthermore, it really just amounts to searching for an OR and doing local processing to filter out the difference.

We describe the Key-Gen, Filter-Gen, and Buffer-Decrypt algorithms below.

Key-Gen(k) Execute the key generation algorithm of the BGN system to produce $A_{public} = (n, \mathbb{G}, \mathbb{G}_1, e, g, h)$ where g is a generator, $n = q_1 q_2$, and h is a random element of order q_1. The private key, $A_{private}$ is the factorization of n. We make the additional assumption that $|D| < q_2$.

Filter-Gen($D, Q_{K_1, K_2}, A_{public}, A_{private}, m, \gamma$) This algorithm constructs and outputs a private filter F for the query Q_{K_1, K_2}, constructed as follows, where this query searches for all documents M such that $(M \cap K_1 \neq \varnothing) \wedge (M \cap K_2 \neq \varnothing)$. F contains the following data:

- A buffer $B(\gamma)$ of size $2\gamma m$, indexed by blocks the size of an element of $\mathbb{G}_1$ times the document size, with every position initialized to the identity element of $\mathbb{G}_1$.
- Two arrays $\widehat{D_l} = \{\widehat{d_i^l}\}_{i=1}^{|D|}$ where each $\widehat{d_i^l} \in \mathbb{G}$, such that $\widehat{d_i^l}$ is an encryption of $1 \in \mathbb{Z}_n$ if $d_i \in K_l$ and an encryption of 0 otherwise.

F then proceeds with the following steps upon receiving an input document M:

1. Construct temporary collections $\widehat{M_l} = \{\widehat{d_i^l} \in \widehat{D_l} \mid d_i \in M\}$.
2. For $l = 1, 2$, compute

$$v_l = \prod_{\widehat{d_i^l} \in \widehat{M_l}} \widehat{d_i^l}$$

and

$$v = e(v_1, v_2) \in \mathbb{G}_1$$

3. Bitwise encrypt M using encryptions of 0 in $\mathbb{G}_1$ for 0's and using v to encrypt 1's to create a vector of $\mathbb{G}_1$ elements.
4. Choose γ random locations in B, take the encryption of step 3, and position-wise multiply these two vectors storing the result back in B at the same location.

Buffer-Decrypt($B, A_{private}$) Decrypts B one block at a time using the decryption algorithm from the BGN system, interpreting non-identity elements of $\mathbb{Z}_{q_2}$ as 1's and 0 as 0, storing the non-zero, valid messages in the output B^*.[5]

Theorem 3. *Assuming that the subgroup decision problem of [2] is hard, then the Private Filter Generator from the preceding construction is semantically secure according to Definition 7.*

References

1. B. Barak, O. Goldreich, R. Impagliazzo, S. Rudich, A. Sahai, S. Vadhan, and K. Yang. On the (im)possibility of software obfuscation. In *Crypto 2001*, pages 1–18, 2001. LNCS 2139.
2. D. Boneh, E. Goh, K. Nissim. Evaluating 2-DNF Formulas on Ciphertexts. TCC 2005: 325-341

---

[5] See footnote 3.

3. D. Boneh, G. Crescenzo, R. Ostrovsky, G. Persiano. Public Key Encryption with Keyword Search. EUROCRYPT 2004: 506-522

4. Y. C. Chang. Single Database Private Information Retrieval with Logarithmic Communication. ACISP 2004

5. C. Cachin, S. Micali, and M. Stadler. Computationally private information retrieval with polylogarithmic communication. In J. Stern, editor, *Advances in Cryptology – EUROCRYPT '99*, volume 1592 of *Lecture Notes in Computer Science*, pages 402–414. Springer, 1999.

6. B. Chor, N. Gilboa, M. Naor Private Information Retrieval by Keywords in Technical Report TR CS0917, Department of Computer Science, Technion, 1998.

7. B. Chor, O. Goldreich, E. Kushilevitz, and M. Sudan. Private information retrieval. In *Proc. of the 36th Annu. IEEE Symp. on Foundations of Computer Science*, pages 41–51, 1995. Journal version: *J. of the ACM*, 45:965–981, 1998.

8. G. Di Crescenzo, T. Malkin, and R. Ostrovsky. Single-database private information retrieval implies oblivious transfer. In *Advances in Cryptology - EUROCRYPT 2000*, 2000.

9. I. Damgård, M. Jurik. A Generalisation, a Simplification and some Applications of Paillier's Probabilistic Public-Key System. In Public Key Cryptography (PKC 2001)

10. M. Freedman, Y. Ishai, B. Pinkas and O. Reingold. Keyword Search and Oblivious Pseudorandom Functions. To appear in 2nd Theory of Cryptography Conference (TCC '05) Cambridge, MA, Feb 2005.

11. S. Goldwasser and S. Micali. Probabilistic encryption. In J. Comp. Sys. Sci, 28(1):270–299, 1984.

12. K. Kurosawa, W. Ogata. Oblivious Keyword Search. Journal of Complexity, Volume 20 , Issue 2-3 April/June 2004 Special issue on coding and cryptography Pages: 356 - 371

13. E. Kushilevitz and R. Ostrovsky. Replication is not needed: Single database, computationally-private information retrieval. In *Proc. of the 38th Annu. IEEE Symp. on Foundations of Computer Science*, pages 364–373, 1997.

14. E. Kushilevitz and R. Ostrovsky. One-way Trapdoor Permutations are Sufficient for Non-Trivial Single-Database Computationally-Private Information Retrieval. In *Proc. of EUROCRYPT '00*, 2000.

15. H. Lipmaa. An Oblivious Transfer Protocol with Log-Squared Communication. IACR ePrint Cryptology Archive 2004/063

16. K. Mehlhorn. On the Program Size of Perfect and Universal Hash Functions. In Proc. 23rd annual IEEE Symposium on Foundations of Computer Science, 1982, pp. 170-175.

17. M. Naor and B. Pinkas. Oblivious transfer and polynomial evaluation. *Proc. 31st STOC*, pp. 245–254, 1999.

18. P. Paillier. Public Key Cryptosystems based on Composite Degree Residue Classes. Advances in Cryptology - EUROCRYPT '99, LNCS volume 1592, pp. 223-238. Springer Verlag, 1999.

19. T. Sander, A. Young, M.Yung. Non-Interactive CryptoComputing For NC1 FOCS 1999: 554-567

20. J.P. Stern, A New and Efficient All or Nothing Disclosure of Secrets Protocol Asiacrypt 1998 Proceedings, Springer Verlag.

Privacy-Preserving Set Operations

Lea Kissner and Dawn Song

Carnegie Mellon University, Pittsburgh PA 15213
leak@cs.cmu.edu, dawnsong@cmu.edu

Abstract. In many important applications, a collection of mutually distrustful parties must perform private computation over multisets. Each party's input to the function is his private input multiset. In order to protect these private sets, the players perform *privacy-preserving* computation; that is, no party learns more information about other parties' private input sets than what can be deduced from the result. In this paper, we propose efficient techniques for privacy-preserving operations on multisets. By building a framework of multiset operations, employing the mathematical properties of polynomials, we design efficient, secure, and composable methods to enable privacy-preserving computation of the *union, intersection,* and *element reduction* operations. We apply these techniques to a wide range of practical problems, achieving more efficient results than those of previous work.

1 Introduction

Private computation over sets and multisets is required in many important applications. In the real world, parties often resort to use of a trusted third party, who computes a fixed function on all parties' private input multisets, or forgo the application altogether. This unconditional trust is fraught with security risks; the trusted party may be dishonest or compromised, as it is an attractive target. We design efficient *privacy-preserving* techniques and protocols for computation over multisets by mutually distrustful parties: no party learns more information about other parties' private input sets than what can be deduced from the result of the computation.

For example, to determine which airline passengers appear on a 'do-not-fly' list, the airline must perform a set-intersection operation between its private passenger list and the government's list. This is an example of the Set-Intersection problem. If a social services organization needs to determine the list of people on welfare who have cancer, the union of each hospital's lists of cancer patients must be calculated (but not revealed), then an intersection operation between the unrevealed list of cancer patients and the welfare rolls must be performed. This problem may be efficiently solved by composition of our private union and set-intersection techniques. Another example is privacy-preserving distributed network monitoring. In this scenario, each node monitors anomalous local traffic, and a distributed group of nodes collectively identify popular anomalous behaviors: behaviors that are identified by at least a threshold t number of monitors. This is an example of the Over-Threshold Set-Union problem.

V. Shoup (Ed.): Crypto 2005, LNCS 3621, pp. 241–257, 2005.

Contributions. In this paper, we propose efficient techniques for privacy-preserving operations on multisets. By building a framework of set operations using polynomial representations and employing the mathematical properties of polynomials, we design efficient methods to enable privacy-preserving computation of the *union, intersection, and element reduction*[1] multiset operations.

An important feature of our privacy-preserving multiset operations is that they can be composed, and thus enable a wide range of applications. To demonstrate the power of our techniques, we apply our operations to solve specific problems, including Set-Intersection (Section 5) and Over-Threshold Set-Union (Section 6). We also discuss a number of other applications in Section 7, such as constructing protocols for the composition of multiset operations, computing Threshold Set-Union and Cardinality Set-Intersection, and determining the Subset relation. Due to space constraints, we describe utilization of our techniques for efficiently and privately evaluating CNF boolean functions in [18].

Our protocols are more efficient than the results obtained from previous work. General multiparty computation is the best previous result for most of the problems that we address in this paper. Only the private Set-Intersection problem and two-party Cardinality Set-Intersection problem have been previously studied [12]. However, previous work only provides protocols for 3-or-more-party Set-Intersection secure only against honest-but-curious players; it is not obvious how to extend this work to achieve security against malicious players. Also, previous work focuses on achieving results for the Set-Intersection problem in isolation – these techniques cannot be used to compose set operations. In contrast, we provide efficient solutions for private multi-party Set-Intersection secure against malicious players [18], and our multiset intersection operator can be easily composed with other operations to enable a wide range of efficient private computation over multisets.

Our protocols are provably secure in the PPT-bounded adversary model. We consider both standard adversary models: honest-but-curious adversaries (HBC) and malicious adversaries. We prove the security of each of our protocols in the full version of this paper [18].

We discuss related work in Section 2. In Section 3, we introduce our adversary models, as well as our cryptographic tools. We describe our privacy-preserving set operation techniques in in Section 4. Section 5 gives a protocol and security analysis for the Set-Intersection problem, and Section 6 gives a protocol and security analysis for the Over-Threshold Set-Union problem. We discuss several additional applications of our techniques in Section 7.

2   Related Work

For most of the privacy-preserving set function problems we address in this paper (except for the Set-Intersection problem), the best previously known results

---

[1] The *element reduction* by d, $\mathrm{Rd}_d(A)$, of a multiset A is the multiset composed of the elements of A such that for every element a that appears in A at least $d' > d$ times, a is included $d' - d$ times in $\mathrm{Rd}_d(A)$.

are through general multiparty computation. General two-party computation was introduced by Yao [25], and general computation for multiple parties was introduced in [1]. In general multiparty computation, the players share the values of each input, and cooperatively evaluate the circuit. For each multiplication gate, the players must cooperate to securely multiply their inputs and re-share the result, requiring $O(n)$ communication for honest-but-curious players and $O(n^2)$ communication for malicious players [15]. Recent results that allow non-interactive private multiplication of shares [7] do not extend to our adversary model, in which any $c < n$ players may collude. Our results are more efficient than the general MPC approach; we compare communication complexity in [18].

The most relevant work to our paper is by Freedman, Nissim, and Pinkas (FNP) [12]. They proposed protocols for the problems related to Set-Intersection, based on the representation of sets as roots of a polynomial [12]. Their work does not utilize properties of polynomials beyond evaluation at given points. We explore the power of polynomial representation of multisets, using operations on polynomials to obtain composable privacy-preserving multisets operations.

Much work has been done in designing solutions for privacy-preserving computation of different functions. For example, private equality testing is the problem of set-intersection for the case in which the size of the private input sets is 1. Protocols for this problem are proposed in [9,21,19], and fairness is added in [2]. We do not enumerate the works of privacy-preserving computation of other functions here, as they address drastically different problems and cannot be applied to our setting.

3   Preliminaries

In this section, we describe our adversary models and the cryptographic tools used in this paper.

3.1   Adversary Models

In this paper, we consider two standard adversary models: honest-but-curious adversaries and malicious adversaries. Due to space constraints, we only provide intuition and informal definitions of these models. Formal definitions of these models can be found in [15].

Honest-But-Curious Adversaries. In this model, all parties act according to their prescribed actions in the protocol. Security in this model is straightforward: no player or coalition of $c < n$ players (who cheat by sharing their private information) gains information about other players' private input sets, other than what can be deduced from the result of the protocol. This is formalized by considering an ideal implementation where a trusted third party (TTP) receives the inputs of the parties and outputs the result of the defined function. We require that in the real implementation of the protocol—that is, one without a TTP—each party does not learn more information than in the ideal implementation.

Malicious Adversaries. In this model, an adversary may behave arbitrarily. In particular, we cannot hope to prevent malicious parties from refusing to participate in the protocol, choosing arbitrary values for its private input set, or aborting the protocol prematurely. Instead, we focus on the standard security definition (see, e.g., [15]) which captures the correctness and the privacy issues of the protocol. Informally, the security definition is based on a comparison between the ideal model and a TTP, where a malicious party may give arbitrary input to the TTP. The security definition is also limited to the case where at least one of the parties is honest. Let Γ be the set of colluding malicious parties; for any strategy Γ can follow in the real protocol, there is a translated strategy that it could follow in the ideal model, such that, to Γ, the real execution is computationally indistinguishable from execution in the ideal model.

3.2   Additively Homomorphic Cryptosystem

In this paper we utilize a semantically secure [16], additively homomorphic public-key cryptosystem. Let $E_{pk}(\cdot)$ denote the encryption function with public key pk. The cryptosystem supports the following operations, which can be performed without knowledge of the private key: (1) Given the encryptions of a and b, $E_{pk}(a)$ and $E_{pk}(b)$, we can efficiently compute the encryption of $a + b$, denoted $E_{pk}(a + b) := E_{pk}(a) +_h E_{pk}(b)$; (2) Given a constant c and the encryption of a, $E_{pk}(a)$, we can efficiently compute the encryption of ca, denoted $E_{pk}(c \cdot a) := c \times_h E_{pk}(a)$. When such operations are performed, we require that the resulting ciphertexts be re-randomized for security. In re-randomization, a ciphertext is transformed so as to form an encryption of the same plaintext, under a different random string than the one originally used. We also require that the homomorphic public-key cryptosystem support secure (n, n)-threshold decryption, i.e., the corresponding private key is shared by a group of n players, and decryption must be performed by *all* players acting together.

In our protocols for the malicious case, we require: (1) the decryption protocol be secure against malicious players, typically, this is done by requiring each player to prove in zero-knowledge that he has followed the threshold decryption protocol correctly [14]; (2) efficient construction of zero-knowledge proofs of plaintext knowledge; (3) optionally, efficient construction of certain zero-knowledge proofs, as detailed in [18].

Note that Paillier's cryptosystem [23] satisfies each of our requirements: it is additively homomorphic, supports ciphertexts re-randomization and threshold decryption (secure in the malicious case) [10,11], and allows certain efficient zero-knowledge proofs (standard constructions from [5,3], and proof of plaintext knowledge [6]).

In the rest of this paper, we simply use $E_{pk}(\cdot)$ to denote the encryption function of the homomorphic cryptosystem which satisfies all the aforementioned properties.

4   Techniques and Mathematical Intuition

In this section, we introduce our techniques for privacy-preserving computation of operations on multisets.

Problem Setting. Let there be n players. We denote the private input set of player i as S_i, and $|S_i| = k$ ($1 \leq i \leq n$). We denote the jth element of set i as $(S_i)_j$. We denote the domain of the elements in these sets as P, ($\forall_{i \in [n], j \in [k]} (S_i)_j \in P$).

Let R denote the plaintext domain $\mathrm{Dom}(E_{pk}(\cdot))$ (in Paillier's cryptosystem, R is Z_N). We require that R be sufficiently large that an element a drawn uniformly from R has only negligible probability of *representing an element of P*, denoted $a \in P$. For example, we could require that only elements of the form $b = a \mathbin{||} h(a)$ could represent an element in P, where $h(\cdot)$ denotes a cryptographic hash function [20]. That is, there exists an a of proper length such that $b = a \mathbin{||} h(a)$. If $|h(\cdot)| = \lg\left(\frac{1}{\epsilon}\right)$, then there is only ϵ probability that $a' \leftarrow R$ represents an element in P.

In this section, we first give background on polynomial representation of multisets, as well as the mathematical properties of polynomials that we use in this paper. We then introduce our privacy-preserving (TTP model) set operations using polynomial representations, then show how to achieve privacy in the real setting by calculating them using encrypted polynomials. Finally, we overview the applications of these techniques explored in the rest of the paper.

4.1   Background: Polynomial Rings and Polynomial Representation of Sets

The polynomial ring $R[x]$ consists of all polynomials with coefficients from R. Let $f, g \in R[x]$, such that $f(x) = \sum_{i=0}^{\deg(f)} f[i]x^i$, where $f[i]$ denotes the coefficient of x^i in the polynomial f. Let $f + g$ denote the addition of f and g, $f * g$ denote the multiplication of f and g, and $f^{(d)}$ denote the dth formal derivative of f. Note that the *formal derivative* of f is $\sum_{i=0}^{\deg(f)-1}(i+1)f[i+1]x^i$.

Polynomial Representation of Sets. In this paper, we use polynomials to represent multisets. Given a multiset $S = \{S_j\}_{1 \leq j \leq k}$, we construct a polynomial representation of S, $f \in R[x]$, as $f(x) = \prod_{1 \leq j \leq k}(x - S_j)$. On the other hand, given a polynomial $f \in R[x]$, we define the multiset S represented by the polynomial f as follows: an element $a \in S$ if and only if (1) $f(a) = 0$ and (2) a represents an element from P. Note that our polynomial representation naturally handles multisets: The element a appears in the multiset b times if $(x-a)^b \mid f \wedge (x-a)^{b+1} \nmid f$.

Note that previous work has proposed to use polynomials to represent sets [12] (as opposed to multisets). However, to the best of our knowledge, previous work has only utilized the technique of polynomial evaluation for privacy-preserving operations. As a result, previous work is limited to set intersection and cannot be composed with other set operators. In this paper, we propose a framework to perform various set operations using polynomial representations and construct efficient privacy-preserving set operations using the mathematical properties of polynomials. By utilizing polynomial representations as the intermediate form of representations of sets, our framework allows arbitrary composition of set operators as outlined in our grammar.

4.2   Our Techniques: Privacy-Preserving Set Operations

In this section, we construct algorithms for computing the polynomial representation of operations on sets, including union, intersection, and element reduction. We design these algorithms to be privacy-preserving in the following sense: the polynomial representation of any operation result reveals no more information than the set representation of the result. First, we introduce our algorithms for computing the polynomial representation of set operations union, intersection, and element reduction (with a trusted third party). We then extend these techniques to encrypted polynomials, allowing secure implementation of our techniques without a trusted third party. Note that the privacy-preserving set operations defined in this section may be *arbitrarily composed* (see Section 7), and constitute truly general techniques.

Set Operations Using Polynomial Representations. In this section, we introduce efficient techniques for set operations using polynomial representations. In particular, let f, g be polynomial representations of the multisets S, T. We describe techniques to compute the polynomial representation of their union, intersection, and element reduction by d. We design our techniques so that the polynomial representation of any operation result reveals no more information than the set representation of the result. This privacy property is formally stated in Theorems 1, 2, and 3, by comparing to the ideal model.

Union. We define the union of multisets $S \cup T$ as the multiset where each element a that appears in S $b_S \geq 0$ times and T $b_T \geq 0$ times appears in the resulting multiset $b_S + b_T$ times. We compute the polynomial representation of $S \cup T$ as follows, where f and g are the polynomial representation of S and T respectively:

$$f * g.$$

Note that $f * g$ is a polynomial representation of $S \cup T$ because (1) all elements that appear in either set S or T are preserved: $(f(a) = 0) \wedge (g(b) = 0) \rightarrow ((f*g)(a) = 0) \wedge ((f*g)(b) = 0)$; (2) as $f(a) = 0 \Leftrightarrow (x-a) \mid f$, duplicate elements from each multiset are preserved: $(f(a) = 0) \wedge (g(a) = 0) \rightarrow (x-a)^2 \mid (f*g)$. In addition, we prove that, given $f * g$, one cannot learn more information about S and T than what can be deduced from $S \cup T$, as formally stated in the following theorem:

Theorem 1. *Let* TTP1 *be a trusted third party which receives the private input multiset* S_i *from player* i *for* $1 \leq i \leq n$, *and then returns to every player the union multiset* $S_1 \cup \cdots \cup S_n$ *directly. Let* TTP2 *be another trusted third party, which receives the private input multiset* S_i *from player* i *for* $1 \leq i \leq n$, *andthen:*

(1) calculates the polynomial representation f_i for each S_i; (2) computes and returns to every player $\prod_{i=1}^{n} f_i$.

There exists a PPT translation algorithm such that, to each player, the results of the following two scenarios are distributed identically: (1) applying translation to the output of TTP1; *(2) returning the output of* TTP2 *directly.*

Proof. Theorem 1 is trivially true. (This theorem is included for completeness.)

Intersection. We define the intersection of multisets $S \cap T$ as the multiset where each element a that appears in S $b_S > 0$ times and T $b_T > 0$ times appears in the resulting multiset $\min\{b_S, b_T\}$ times. Let S and T be two multisets of equal size, and f and g be their polynomial representations respectively. We compute the polynomial representation of $S \cap T$ as:

$$f * r + g * s$$

where $r, s \leftarrow R^{\deg(f)}[x]$, where $R^b[x]$ is the set of all polynomials of degree $0, \ldots, b$ with coefficients chosen independently and uniformly from R: $r = \sum_{i=0}^{\beta} r[i]x^i$ and $s = \sum_{i=0}^{\beta} s[i]x^i$, where $\forall_{0 \leq i \leq \beta} r[i] \leftarrow R$, $\forall_{0 \leq i \leq \beta} s[i] \leftarrow R$.

We show below that $f * r + g * s$ is a polynomial representation of $S \cap T$. In addition, we prove that, given $f * r + g * s$, one cannot learn more information about S and T than what can be deduced from $S \cap T$, as formally stated in Theorem 2.

First, we must prove the following lemma:

Lemma 1. *Let $\hat{f}, \hat{g}$ be polynomials in $R[x]$ where R is a ring, $\deg(\hat{f}) = \deg(\hat{g}) = \alpha$, and $\gcd(\hat{f}, \hat{g}) = 1$. Let $r = \sum_{i=0}^{\beta} r[i]x^i$, and $s = \sum_{i=0}^{\beta} s[i]x^i$, where $\forall_{0 \leq i \leq \beta} r[i] \leftarrow R$, $\forall_{0 \leq i \leq \beta} s[i] \leftarrow R$ (independently) and $\beta \geq \alpha$.*

Let $\hat{u} = \hat{f} * r + \hat{g} * s = \sum_{i=0}^{\alpha+\beta} u[i]x^i$. Then $\forall_{0 \leq i \leq \alpha+\beta} \hat{u}[i]$ are distributed uniformly and independently over R.

We give a proof of Lemma 1 in Appendix A.

By this lemma, $f * r + g * s = \gcd(f, g) * u$, where u is distributed uniformly in $R^{\gamma}[x]$ for $\gamma = 2 \deg(f) - |S \cap T|$. Note that a is a root of $\gcd(f, g)$ and $(x - a)^{\ell_a} \mid \gcd(f, g)$ if and only if a appears ℓ_a times in $S \cap T$. Moreover, because u is distributed uniformly in $R^{\gamma}[x]$, with overwhelming probability the roots of u do not represent any element from P (as explained in the beginning of Section 4). Thus, the computed polynomial $f*r+g*s$ is a polynomial representation of $S \cap T$. Note that this technique for computing the intersection of two multisets can be extended to simultaneously compute the intersection of an arbitrary number of multisets in a similar manner. Also, given $f * r + g * s$, one cannot learn more information about S and T than what can be deduced from $S \cap T$, as formally stated in the following theorem:

Theorem 2. *Let* TTP1 *be a trusted third party which receives the private input multiset S_i from player i for $1 \leq i \leq n$, and then returns to every player the intersection multiset $S_1 \cap \cdots \cap S_n$ directly. Let* TTP2 *be another trusted third* *party, which receives the private input multiset S_i from player i for $1 \leq i \leq n$, and then: (1) calculates the polynomial representation f_i for each S_i; (2) chooses $r_i \leftarrow R^k[x]$; (3) computes and returns to each player $\sum_{i=1}^{n} f_i * r_i$.*

There exists a PPT translation algorithm such that, to each player, the results of the following two scenarios are distributed identically: (1) applying translation to the output of TTP1*; (2) returning the output of* TTP2 *directly.*

Proof (Proof sketch). Let the output of TTP1 be denoted T. The translation algorithm operates as follows: (1) calculates the polynomial representation g of T; (2) chooses the random polynomial $u \leftarrow R^{2k-|T|}[x]$; (3) computes and returns $g * u$.

Element Reduction. We define the operation of element reduction (by d) of multiset S (denoted $\mathrm{Rd}_d(S)$) as follows: for each element a that appears b times in S, it appears $\max\{b - d, 0\}$ times in the resulting multiset. We compute the polynomial representation of $\mathrm{Rd}_d(S)$ as:

$$f^{(d)} * F * r + f * s$$

where $r, s \leftarrow R^{\deg(f)}[x]$ and F is any polynomial of degree d, such that $\forall_{a \in P} \ F(a) \neq 0$. Note that a random polynomial of degree d in $R[x]$ has this property with overwhelming probability.

To show that formal derivative operation allows element reduction, we require the following lemma:

Lemma 2. *Let $f \in R[x]$, where R is a ring, $d \geq 1$.*

1. *If $(x - a)^{d+1} \mid f$, then $(x - a) \mid f^{(d)}$.*
2. *If $(x - a) \mid f$ and $(x - a)^{d+1} \nmid f$, then $(x - a) \nmid f^{(d)}$.*

Lemma 2 is a standard result [24]. By this lemma and $\gcd(F, f) = 1$, an element a is a root of $\gcd(f^{(d)}, f)$ and $(x - a)^{\ell_a} \mid \gcd(f^{(d)}, f)$ if and only if a appears ℓ_a times in $\mathrm{Rd}_d(S)$. By Lemma 1, $f^{(d)} * F * r + f * s = \gcd(f^{(d)}, f) * u$, where u is distributed uniformly in $R^\gamma[x]$ for $\gamma = 2k - |\mathrm{Rd}_d(S)|$. Thus, with overwhelming probability, any root of u does not represent any element from P. Therefore, $f^{(d)} * F * r + f * s$ is a polynomial representation of $\mathrm{Rd}_d(S)$, and moreover, given $f^{(d)} * F * r + f * s$, one cannot learn more information about S than what can be deduced from $\mathrm{Rd}_d(S)$, as formally stated in the following theorem:

Theorem 3. *Let F be a publicly known polynomial of degree d such that $\forall_{a \in P}$ $F(a) \neq 0$. Let* TTP1 *be a trusted third party which receives a private input multiset S, and then returns the reduction multiset $\mathrm{Rd}_d(S)$ directly. Let* TTP2 *be another trusted third party, which receives a private input multiset S, and then: (1) calculates the polynomial representation f of S; (2) chooses $r, s \leftarrow R^k[x]$; (3) computes and returns $f^{(d)} * F * r + f * s$.*

There exists a PPT translation algorithm such that the results of the following two scenarios are distributed identically: (1) applying translation to the output of TTP1*; (2) returning the output of* TTP2 *directly.*

Proof (Proof sketch). Let the output of TTP1 be denoted T. The translation algorithm operates as follows: (1) calculates the polynomial representation g of T; (2) chooses the random polynomial $u \leftarrow R^{2k-|T|}[x]$; (3) computes and returns $g * u$.

Operations with Encrypted Polynomials. In the previous section, we prove the security of our polynomial-based multiset operators when the polynomial representation of the result is computed by a trusted third party (TTP2). By using additively homomorphic encryption, we allow these results to be implemented as protocols in the real world without a trusted third party (i.e., the polynomial representation of the set operations is computed by the parties collectively without a trusted third party). In the algorithms given above, there are three basic polynomial operations that are used: addition, multiplication, and the formal derivative. We give algorithms in this section for computation of these operations with encrypted polynomials.

For $f \in R[x]$, we represent the *encryption of polynomial* f, $E_{pk}(f)$, as the ordered list of the encryptions of its coefficients under the additively homomorphic cryptosystem: $E_{pk}(f[0]), \ldots, E_{pk}(f[\deg(f)])$. Let f_1, f_2, and g be polynomials in $R[x]$ such that $f_1(x) = \sum_{i=0}^{\deg(f_1)} f_1[i]x^i$, $f_2(x) = \sum_{i=0}^{\deg(f_2)} f_2[i]x^i$, and $g(x) = \sum_{i=0}^{\deg(g)} g[i]x^i$. Let $a, b \in R$. Using the homomorphic properties of the homomorphic cryptosystem, we can efficiently perform the following operations on encrypted polynomials without knowledge of the private key:

- Sum of encrypted polynomials: given the encryptions of the polynomial f_1 and f_2, we can efficiently compute the encryption of the polynomial $g := f_1 + f_2$, by calculating $E_{pk}(g[i]) := E_{pk}(f_1[i]) +_h E_{pk}(f_2[i])$ $(0 \leq i \leq \max\{\deg(f_1), \deg(f_2)\})$
- Product of an unencrypted polynomial and an encrypted polynomial: given a polynomial f_2 and the encryption of polynomial f_1, we can efficiently compute the encryption of polynomial $g := f_1 * f_2$, (also denoted $f_2 *_h E_{pk}(f_1)$) by calculating the encryption of each coefficient $E_{pk}(g[i]) := (f_2[0] \times_h E_{pk}(f_1[i])) +_h \cdots +_h (f_2[i] \times_h E_{pk}(f_1[0]))$ $(0 \leq i \leq \deg(f_1) + \deg(f_2))$.
- Derivative of an encrypted polynomial: given the encryption of polynomial f_1, we can efficiently compute the encryption of polynomial $g := \frac{d}{dx} f_1$, by calculating the encryption of each coefficient $E_{pk}(g[i]) := (i+1) \times_h E_{pk}(f_1[i+1])$ $(0 \leq i \leq \deg(f_1) - 1)$.
- Evaluation of an encrypted polynomial at an unencrypted point: given the encryption of polynomial f_1, we can efficiently compute the encryption of $a := f_1(b)$, by calculating $E_{pk}(a) := (b^0 \times_h E_{pk}(f_1[0])) +_h \cdots +_h (b^{\deg(f)} \times_h E_{pk}(f_1[\deg(f_1)]))$.

It is easy to see that with the above operations on encrypted polynomials, we can allow the computation of the polynomial representations of set operations described in Section 4.2 without the trusted third party (TTP2) while enjoying the same security. As an example, we design in Section 5, a protocol for the Set-Intersection problem, and discuss several other selected applications in Section 7.

5   Application I: Private Set-Intersection

Problem Definition. Let there be n parties; each has a private input set S_i ($1 \leq i \leq n$) of size k. We define the *Set-Intersection* problem as follows: all players learn the intersection of all private input multisets without gaining any other information; that is, each player learns $S_1 \cap S_2 \cap \cdots \cap S_n$.

We design a protocol, secure against a coalition of honest-but-curious adversaries, in Section 5.1. We then describe variations of the problem and how to extend this protocol to be secure against malicious adversaries in Section 7.

5.1   Set-Intersection Protocol

Our protocol for the honest-but-curious case is given in Fig. 1. In this protocol, each player i ($1 \leq i \leq n$) first calculates a polynomial representation $f_i \in R[x]$ of his input multiset S_i. He then encrypts this polynomial f_i, and sends it to

---

Protocol: SET-INTERSECTION-HBC
Input: There are $n \geq 2$ honest-but-curious players, $c < n$ dishonestly colluding, each with a private input set S_i, such that $|S_i| = k$. The players share the secret key sk, to which pk is the corresponding public key to a homomorpic cryptosystem.

1. Each player $i = 1, \ldots, n$
   (a) calculates the polynomial $f_i = (x - (S_i)_1) \ldots (x - (S_i)_k)$
   (b) sends the encryption of the polynomial f_i to players $i + 1, \ldots, i + c$
   (c) chooses $c + 1$ polynomials $r_{i,0}, \ldots, r_{i,c} \leftarrow R^k[x]$
   (d) calculates the encryption of the polynomial $\phi_i = f_{i-c} * r_{i,i-c} + \cdots + f_{i-1} * r_{i,i-1} + f_i * r_{i,0}$, utilizing the algorithms given in Sec. 4.2.
2. Player 1 sends the encryption of the polynomial $\lambda_1 = \phi_1$, to player 2
3. Each player $i = 2, \ldots, n$ in turn
   (a) receives the encryption of the polynomial λ_{i-1} from player $i - 1$
   (b) calculates the encryption of the polynomial $\lambda_i = \lambda_{i-1} + \phi_i$ by utilizing the algorithms given in Sec. 4.2.
   (c) sends the encryption of the polynomial λ_i to player $i + 1 \mod n$
4. Player 1 distributes the encryption of the polynomial $p = \lambda_n = \sum_{i=1}^{n} f_i * \left(\sum_{j=0}^{c} r_{i+j,j} \right)$ to all other players.
5. All players perform a group decryption to obtain the polynomial p.
6. Each player $i = 1, \ldots, n$ determines the intersection multiset: for each $a \in S_i$, he calculates b such that $(x - a)^b | p \ \wedge \ (x - a)^{b+1} \nmid p$. The element a appears b times in the intersection multiset.

Fig. 1. Set-Intersection protocol for the honest-but-curious case

c other players $i + 1, \ldots, i + c$. For each encrypted polynomial $E_{pk}(f_i)$, each player $i + j$ $(0 \leq j \leq c)$ chooses a random polynomial $r_{i+j,j} \in R^k[x]$. Note that at most c players may collude, thus $\sum_{j=0}^{c} r_{i+j,j}$ is both uniformly distributed and known to no player. They then compute the encrypted polynomial $\left(\sum_{j=0}^{c} r_{i+j,j} \right) *_h E_{pk}(f_i)$. From these encrypted polynomials, the players compute the encryption of $p = \sum_{i=1}^{n} f_i * \left(\sum_{j=0}^{c} r_{i+j,j} \right)$. All players engage in group decryption to obtain the polynomial p. Thus, by Theorem 2, the players have privately computed p, a polynomial representing the intersection of their private input multisets. Finally, to reconstruct the multiset represented by polynomial p, the player i, for each $a \in S_i$, calculates b such that $(x - a)^b | p \wedge (x - a)^{b+1} \nmid p$. The element a appears b times in the intersection multiset.

Security Analysis. We show that our protocol is correct, as each player learns the appropriate answer set at its termination, and secure in the honest-but-curious model, as no player gains information that it would not gain when using its input in the ideal model. A formal statement of these properties is as follows:

Theorem 4. *In the Set-Intersection protocol of Fig. 1, every player learns the intersection of all players' private inputs, $S_1 \cap S_2 \cap \cdots \cap S_n$, with overwhelming probability.*

Theorem 5. *Assuming that the additively homomorphic, threshold cryptosystem $E_{pk}(\cdot)$ is semantically secure, with overwhelming probability, in the Set-Intersection protocol of Fig. 1, any coalition of fewer than n PPT honest-but-curious players learns no more information than would be gained by using the same private inputs in the ideal model with a trusted third party.*

We provide proof sketches for Theorems 4 and 5 in [18].

6   Application II: Private Over-Threshold Set-Union

Problem Definition. Let there be n players; each has a private input set S_i $(1 \leq i \leq n)$ of size k. We define the *Over-Threshold Set-Union* problem as follows: all players learn which elements appear in the union of the players' private input multisets at least a threshold number t times, and the number of times these elements appeared in the union of players' private inputs, without gaining any other information. For example, assume that a appears in the combined private input of the players 15 times. If $t = 10$, then all players learn a has appeared 15 times. However, if $t = 16$, then no player learns a appears in any player's private input. This problem can be computed as $\mathrm{Rd}_{t-1}(S_1 \cup \cdots \cup S_n)$.

In Section 6.1, we design a protocol for the Over-Threshold Set-Union problem, secure against honest-but-curious adversaries. This protocol is significantly more efficient than utilizing general multiparty computation (the best previous solution for this problem). We describe a variation of the problem and security against malicious adversaries in Section 7.

6.1   Over-Threshold Set-Union Protocol

We describe our protocol secure against honest-but-curious players for the Over-Threshold Set-Union problem in Fig. 2. In this protocol, each player i $(1 \leq i \leq n)$ first calculates f_i, the polynomial representation of its input multiset S_i. All players then compute the encryption of polynomial $p = \prod_{i=1}^{n} f_i$, the polynomial representation of $S_1 \cup \cdots \cup S_n$. Players $i = 1, \ldots, c+1$ then each chooses random polynomials r_i, s_i, and calculates the encryption of the polynomial $F * p^{(t-1)} * r_i + p * s_i$ as shown in Fig. 2. All players then calculate the encryption of the polynomial $\Phi = F * p^{(t-1)} * \left(\sum_{i=1}^{c+1} r_i \right) + p * \left(\sum_{i=1}^{c+1} s_i \right)$ and perform a group decryption to obtain Φ. As at most c players may dishonestly collude, the polynomials $\sum_{i=1}^{c+1} r_i$,

---

Protocol: OVER-THRESHOLD SET-UNION-HBC
Input: There are $n \geq 2$ honest-but-curious players, $c < n$ dishonestly colluding, each with a private input set S_i, such that $|S_i| = k$. The players share the secret key sk, to which pk is the corresponding public key for a homomorphic cryptosystem. The threshold number of repetitions at which an element appears in the output is t. F is a fixed polynomial of degree $t - 1$ which has no roots representing elements of P.

1. Each player $i = 1, \ldots, n$ calculates the polynomial $f_i = (x - (S_i)_1) \ldots (x - (S_i)_k)$
2. Player 1 sends the encryption of the polynomial $\lambda_1 = f_1$ to player 2
3. Each player $i = 2, \ldots, n$
   (a) receives the encryption of the polynomial λ_{i-1} from player $i - 1$
   (b) calculates the encryption of the polynomial $\lambda_i = \lambda_{i-1} * f_i$ by utilizing the algorithm given in Sec. 4.2.
   (c) sends the encryption of the polynomial λ_i to player $i + 1 \mod n$
4. Player 1 distributes the encryption of the polynomial $p = \lambda_n = \prod_{i=1}^{n} f_i$ to players $2, \ldots, c + 1$
5. Each player $i = 1, \ldots, c + 1$
   (a) calculates the encryption of the $t - 1$th derivative of p, denoted $p^{(t-1)}$, by repeating the algorithm given in Sec. 4.2.
   (b) chooses random polynomials $r_i, s_i \leftarrow R^{nk}[x]$
   (c) calculates the encryption of the polynomial $p * s_i + F * p^{(t-1)} * r_i$ and sends it to all other players.
6. All players perform a group decryption to obtain the polynomial $\Phi = F * p^{(t-1)} * \left(\sum_{i=1}^{c+1} r_i \right) + p * \left(\sum_{i=1}^{c+1} s_i \right)$.
7. Each player $i = 1, \ldots, n$, for each $j = 1, \ldots, k$
   (a) chooses a random element $b_{i,j} \leftarrow R$
   (b) calculates $u_{i,j} = b_{i,j} \times \Phi((S_i)_j) + (S_i)_j$
8. All players distribute/shuffle the elements $u_{i,j}$ $(1 \leq i \leq n, 1 \leq j \leq k)$ such that each player learns all of the elements, but does not learn their origin. Each element $a \in P$ that appears b times in the shuffled elements is an element in the threshold set that appears b times in the players' private inputs.

Fig. 2. Over-Threshold Set-Union protocol for the honest-but-curious case

$\sum_{i=1}^{c+1} s_i$ are uniformly distributed and known to no player. By Theorem 3, Φ is a polynomial representation of $\mathrm{Rd}_{t-1}(S_1 \cup \cdots \cup S_n)$.

Each player $i = 1, \ldots, n$ then chooses $b_{i,j} \leftarrow R$ and computes $u_{i,j} = b_{i,j} \times \Phi((S_i)_j) + (S_i)_j$ $(1 \leq j \leq k)$. Each element $u_{i,j}$ equals $(S_i)_j$ if $(S_i)_j \in \mathrm{Rd}_{t-1}(S_1 \cup \cdots \cup S_n)$, and is otherwise uniformly distributed over R. The players then shuffle these elements $u_{i,j}$, such that each player learns all of the elements, but does not learn which player's set they came from. The shuffle can be easily accomplished with standard techniques [4,17,8,13,22], with communication complexity at most $O(n^2 k)$. The multiset formed by those shuffled elements that represent elements of P is $\mathrm{Rd}_{t-1}(S_1 \cup \cdots \cup S_n)$.

Security Analysis We show that our protocol is correct, as each player learns the appropriate answer set at its termination, and secure in the honest-but-curious model, as no player gains information that it would not gain when using its input in the ideal model with a trusted third party. A formal statement of these properties is as follows:

Theorem 6. *In the Over-Threshold Set-Union protocol of Fig. 2, with overwhelming probability, every honest-but-curious player learns each element a which appears at least t times in the union of the n players' private inputs, as well as the number of times it so appears.*

Theorem 7. *Assuming that the additively homomorphic, threshold cryptosystem $E_{pk}(\cdot)$ is semantically secure, with overwhelming probability, in the Over-Threshold Set-Union protocol of Fig. 2, any coalition of fewer than n PPT honest-but-curious players learns no more information than would be gained by using the same private inputs in the ideal model with a trusted third party.*

We provide proof sketches for Theorems 6 and 7 in [18].

7   Other Applications

Using the encrypted polynomial techniques of Section 4, we may construct efficient protocols for functions composed of multiset intersection, union, and element reduction. These functions are described by the following grammar:

$$\Upsilon ::= s \mid \mathrm{Rd}_d(\Upsilon) \mid \Upsilon \cap \Upsilon \mid s \cup \Upsilon \mid \Upsilon \cup s,$$

where s represents any multiset held by some player, and $d \geq 1$. Note that any monotone function on multisets can be expressed using the grammar above, and thus our techniques for privacy-preserving set operations are truly general. Additional techniques allow arbitrary composition of multiset operations are described in [18].

We design a protocol for Cardinality Set-Intersection, using polynomial evaluation. We describe a protocol for the Threshold Set-Union problem, a variant of Over-Threshold Set-Union. We also design protocols for several more variations on the Over-Threshold Set-Union problem, determining the subset relation, and for evaluation of boolean CNF formulas using our techniques; constructions and proofs are given in [18].

7.1 Cardinality Set-Intersection

We may use the technique of polynomial evaluation to design protocols for variants on the multiset functions previously described; this is closely related to techniques utilized in [12]. In the Cardinality Set-Intersection problem, each player learns $|S_1 \cap \cdots \cap S_n|$, without learning any other information. Our protocol for Cardinality Set-Intersection on sets proceeds as our protocol for Set-Intersection, given in Section 5, until the point where all players learn the encryption of p, the polynomial representation of $S_1 \cap \cdots \cap S_n$. Each player $i = 1, \ldots, n$ then evaluates this encrypted polynomial at each unique element $a \in S_i$, obtaining β_a, an encryption of $p(a)$. He then blinds each encrypted evaluation $p(a)$ by calculating $\beta'_a = b_a \times_h \beta_a$. All players then distribute and shuffle the ciphertexts β'_a constructed by each player, such that all players receive all ciphertexts, without learning their source. The players then decrypt these ciphertexts, finding that nb of the decryptions are 0, implying that there are b unique elements in $S_1 \cap \cdots \cap S_n$. Due to space constraints, we describe the details of our protocols for this problem in [18].

7.2 Threshold Set-Union

Using our techniques, we design efficient solutions to variations of the Over-Threshold Set-Union problem, including the *Threshold Set-Union* problem, where each player learns which elements appear in $\mathrm{Rd}_{t-1}(S_1 \cup \cdots \cup S_n)$ without gaining any other information. Note that this differs from the Over-Threshold Set-Union problem in that the players do not learn the number of times any element appears in $\mathrm{Rd}_{t-1}(S_1 \cup \cdots \cup S_n)$. Our protocol for the Threshold Set-Union problem follows our protocol for Over-Threshold Set-Union until all players have learned the encryption of the polynomial Φ, the polynomial representation of $\mathrm{Rd}_{t-1}(S_1 \cup \cdots \cup S_n)$. Each player $i = 1, \ldots, n$ then evaluates this encrypted polynomial at each element $a = (S_i)_j$, obtaining $U_{i,j}$, an encryption of $\Phi(a)$. He then chooses $b_{i,j} \leftarrow R$, and calculates $U'_{i,j} = b_{i,j} \times_h U_{i,j} +_h (S_i)_j$. Each element $U'_{i,j}$ is an encryption of $(S_i)_j$ if $(S_i)_j \in \mathrm{Rd}_{t-1}(S_1 \cup \cdots \cup S_n)$, and is otherwise uniformly distributed over R. All players then shuffle and distribute $U'_{i,j}$ $(1 \leq i \leq n, 1 \leq j \leq k)$, such that all players receive all ciphertexts, without learning their source. Shuffling can be easily accomplished with standard techniques [4,17,8,13,22], with communication complexity at most $O(n^2 k)$. The players then perform a calculation so that if any two shuffled ciphertexts are encryptions of the same plaintext, one will reveal the correct plaintext element, and the other will yield a uniformly distributed element of R. Thus each element of $\mathrm{Rd}_{t-1}(S_1 \cup \cdots \cup S_n)$ is revealed exactly once. Due to space constraints, we describe the details of our protocols for the Threshold Set-Union problem and several other variants in [18].

7.3 Private Subset Relation

Problem Statement. Let the set A be held by Alice. The set B may be the result of an arbitrary function over multiple players' input sets (for example as calculated using the grammar above). The *Subset* problem is to determine whether $A \subseteq B$ without revealing any additional information.

Let λ be the encryption of the polynomial p representing B. Note that $A \subseteq B \Leftrightarrow \forall_{a \in A} \, p(a) = 0$. Alice thus evaluates the encrypted polynomial λ at each element $a \in A$, homomorphically multiplies a random element by each encrypted evaluation, and adds these blinded ciphertexts to obtain β'. If β' is an encryption of 0, then $A \subseteq B$. More formally:

1. For each element $a = A_j$ $(1 \leq j \leq |A|)$, the player holding A:
   (a) calculates $\beta_j = \lambda(a)$
   (b) chooses a random element $b_j \leftarrow R$, and calculates $\beta'_j = b_j \times_h \beta_j$
2. The player holding A calculates $\beta' = \beta'_1 +_h \cdots +_h \beta'_{|A|}$
3. All players together decrypt β' to obtain y. If $y = 0$, then $A \subseteq B$.

This protocol can be easily extended to allow the set A to be held by multiple players, such that $A = A_1 \cup \cdots \cup A_\nu$, where each set A_i is held by a single player.

7.4   Security Against Malicious Parties

We can extend our Set-Intersection protocol in Figure 1, secure against honest-but-curious players, to one secure against malicious adversaries by adding zero-knowledge proofs or using cut-and-choose to ensure security. By adding zero-knowledge proofs to our Over-Threshold Set-Union and Cardinality Set-Intersection protocols secure against honest-but-curious adversaries, we extend our results to enable security against malicious adversaries. Due to space constraints, we provide details of these protocols, as well as security analysis, in [18].

Acknowledgments. We extend our thanks to Dan Boneh, Benny Pinkas, David Molnar, David Brumley, Alina Oprea, Luis von Ahn, and anonymous reviewers for their invaluable comments. We especially thank Benny Pinkas for suggestions on the presentation of this paper.

References

1. M. Ben-Or, S. Goldwasser, and A. Widgerson. Completeness theorems for non-cryptographic fault-tolerant distributed computation. In *Proc. of STOC*, 1988.
2. Fabirce Boudot, Berry Schoenmakers, and Jacques Traore. A fair and efficient solution to the socialist millionaires' problem. *Discrete Applied Mathematics*, 111:77–85, 2001.
3. Jan Camenisch. Proof systems for general statements about discrete logarithms. Technical Report 260, Dept. of Computer Science, ETH Zurich, Mar 1997.
4. David Chaum. Untraceable electronic mail, return addresses, and digital pseudonyms. *Communications of the ACM*, 24:84–8, 1981.
5. David Chaum, Jan-Hendrick Evertse, Jeroen van de Graaf, and Rene Peralta. Demonstrating possession of a discrete log without revealing it. In A.M. Odlyzko, editor, *Proc. of Crypto*, pages 200–212. Springer-Verlag, 1986.

6. R. Cramer, I. Damgård, and J. Buus Nielsen. Multiparty computation from threshold homomorphic encryption. In *Proc. of Eurocrypt*, pages 280–99. Springer-Verlag, 2001.
7. Ronald Cramer, Ivan Damgård, and Ueli Maurer. General secure multi-party computation from any linear secret sharing scheme. In *Proc. of Eurocrypt*. Springer-Verlag, May 2000.
8. Y. Desmedt and K. Kurosawa. How to break a practical mix and design a new one. In *Proc. of Eurocrypt*, pages 557–72. Springer-Verlag, 2000.
9. Ronald Fagin, Moni Naor, and Peter Winkler. Comparing information without leaking it. *Communications of the ACM*, 39:77–85, 1996.
10. P. Fouque, G. Poupard, and J. Stern. Sharing decryption in the context of voting of lotteries. In *Proc. of Financial Cryptography*, 2000.
11. Pierre-Alain Fouque and David Pointcheval. Threshold cryptosystems secure against chosen-ciphertext attacks. In *Proc. of Asiacrypt*, pages 573–84, 2000.
12. Michael Freedman, Kobi Nissim, and Benny Pinkas. Efficient private matching and set intersection. In *Proc. of Eurocrypt*, volume LNCS 3027, pages 1–19. Springer-Verlag, May 2004.
13. J. Furukawa and K. Sako. An efficient scheme for proving a shuffle. In *Proc. of Crypto*, pages 368–87. Springer-Verlag, 2001.
14. Rosario Gennaro and Victor Shoup. Securing threshold cryptosystems against chosen ciphertext attack. *Journal of Cryptology*, 15:75–96, 2002.
15. Oded Goldreich. The foundations of cryptography – volume 2. http://www.wisdom.weizmann.ac.il/ oded/foc-vol2.html.
16. Shafi Goldwasser and Silvio Micali. Probabilistic encryption. *Journal of Computer and Systems Science*, 28:270–99, 1984.
17. M. Jakobsson. A practical mix. In *Proc. of Eurocrypt*, pages 448–61. Springer-Verlag, 1998.
18. Lea Kissner and Dawn Song. Private and threshold set-intersection. Technical Report CMU-CS-05-113, Carnegie Mellon University, February 2005.
19. Helger Lipmaa. Verifiable homomorphic oblivious transfer and private equality test. In *Proc. of Asiacrypt*, pages 416–33, 2003.
20. Alfred J. Menezes, Paul C. van Oorschot, and Scott A. Vanstone. *Handbook of Applied Cryptography*. CRC Press, October 1996.
21. Moni Naor and Benny Pinkas. Oblivious transfer and polynomial evaluation. In *Proc. ACM Symposium on Theory of Computing*, pages 245–54, 1999.
22. A. Neff. A verifiable secret shuffle and its application to e-voting. In *ACM CCS*, pages 116–25, 2001.
23. Pascal Paillier. Public-key cryptosystems based on composite degree residuosity classes. In *Proc. of Asiacrypt*, pages 573–84, 2000.
24. Victor Shoup. A computational introduction to number theory and algebra. http://shoup.net/ntb/.
25. Andrew C-C Yao. Protocols for secure computations. In *Proc. of FOCS*, 1982.

A   Proof of Lemma

Theorem 1. *Let f, g be polynomials in $R[x]$ where R is a ring, $\deg(f) = \deg(g) = \alpha$, and $\gcd(f, g) = 1$. Let $r = \sum_{i=0}^{\beta} r[i]x^i$ and $s = \sum_{i=0}^{\beta} s[i]x^i$, where $\forall_{0 \le i \le \beta}\ r[i] \leftarrow R$, $\forall_{0 \le i \le \beta}\ s[i] \leftarrow R$ (independently) and $\beta \ge \alpha$.*

Let $u = f * r + g * s = \sum_{i=0}^{\alpha+\beta} u[i]x^i$. Then $\forall_{0 \le i \le \alpha+\beta}\ u[i]$ are distributed uniformly and independently over R.

Proof. For clarity, we give a brief outline of the proof before proceeding to the details. Given any fixed polynomials f, g, u, we calculate the number z of r, s pairs such that $f * r + g * s = u$. We may then check that, given any fixed polynomials f, g, the total number of possible r, s pairs, divided by z, is equal to the number of possible result polynomials u. This implies that, if $\gcd(f, g) = 1$ and we choose the coefficients of r, s uniformly and independently from R, the coefficients of the result polynomial u are distributed uniformly and independently over R.

We now determine the value of z, the number of r, s pairs such that $f * r + g * s = u$. Let us assume that there exists at least one pair $\hat{r}, \hat{s}$ such that $f * \hat{r} + g * \hat{s} = u$. For any pair $\hat{r}', \hat{s}'$ such that $f * \hat{r}' + g * \hat{s}' = u$, then

$$f * \hat{r} + g * \hat{s} = f * \hat{r}' + g * \hat{s}'$$
$$f * (\hat{r} - \hat{r}') = g * (\hat{s}' - \hat{s})$$

As $\gcd(f, g) = 1$, we may conclude that $g | (\hat{r} - \hat{r}')$ and $f | (\hat{s}' - \hat{s})$. Let $p * g = \hat{r} - \hat{r}'$ and $p * f = \hat{s}' - \hat{s}$. We now show that each polynomial p, of degree at most $\beta - \alpha$, determines exactly one unique pair $\hat{r}', \hat{s}'$ such that $f * \hat{r}' + g * \hat{s}' = u$. Note that $\hat{r}' = \hat{r} - g * p$, $\hat{s}' = \hat{s} + f * p$; as we have fixed $f, g, \hat{r}, \hat{s}$, a choice of p determines both $\hat{r}', \hat{s}'$. If these assignments were not unique, there would exist polynomials p, p' such that either $\hat{r}' = \hat{r} - g * p = \hat{r} - g * p'$ or $\hat{s}' = \hat{s} + f * p = \hat{s} + f * p'$; either condition implies that $p = p'$, giving a contradiction. Thus the number of polynomials p, of degree at most $\beta - \alpha$, is exactly equivalent to the number of r, s pairs such that $f * r + g * s = u$. As there are $|R|^{\beta - \alpha + 1}$ such polynomials p, $z = |R|^{\beta - \alpha + 1}$.

We now show that the total number of r, s pairs, divided by z, is equal to the number of result polynomials u. There are $|R|^{2\beta + 2}$ r, s pairs. As $\frac{|R|^{2\beta + 2}}{z} = \frac{|R|^{2\beta + 2}}{|R|^{\beta - \alpha + 1}} = |R|^{\alpha + \beta + 1}$, and there are $|R|^{\alpha + \beta + 1}$ possible result polynomials, we have proved the theorem true.

Collusion Resistant Broadcast Encryption with Short Ciphertexts and Private Keys

Dan Boneh[1,*], Craig Gentry[2], and Brent Waters[1]

[1] Stanford University
{dabo, bwaters}@cs.stanford.edu
[2] DoCoMo USA Labs
cgentry@docomolabs-usa.com

Abstract. We describe two new public key broadcast encryption systems for stateless receivers. Both systems are fully secure against any number of colluders. In our first construction both ciphertexts and private keys are of constant size (only two group elements), for any subset of receivers. The public key size in this system is linear in the total number of receivers. Our second system is a generalization of the first that provides a tradeoff between ciphertext size and public key size. For example, we achieve a collusion resistant broadcast system for n users where both ciphertexts and public keys are of size $O(\sqrt{n})$ for any subset of receivers. We discuss several applications of these systems.

1 Introduction

In a broadcast encryption scheme [FN93] a broadcaster encrypts a message for some subset S of users who are listening on a broadcast channel. Any user in S can use his private key to decrypt the broadcast. However, even if all users outside of S collude they can obtain no information about the contents of the broadcast. Such systems are said to be collusion resistant. The broadcaster can encrypt to any subset S of his choice. We use n to denote the total number of users.

Broadcast encryption has several applications including access control in encrypted file systems, satellite TV subscription services, and DVD content protection. As we will see in Section 4 we distinguish between two types of applications:

- Applications where we broadcast to large sets, namely sets of size $n - r$ for $r \ll n$. The best systems [NNL01, HS02, GST04] achieve a broadcast message containing $O(r)$ ciphertexts where each user's private key is of size $O(\log n)$.
- Applications where we broadcast to small sets, namely sets of size t for $t \ll n$. Until now, the best known solution was trivial, namely encrypt the broadcast message under each recipient's key. This broadcast message contains t ciphertexts and each user's private key is of size $O(1)$.

---

[*] Supported by NSF.

V. Shoup (Ed.): Crypto 2005, LNCS 3621, pp. 258–275, 2005.

In this paper we construct fully collusion secure broadcast encryption systems with short ciphertexts and private keys for *arbitrary* receiver sets. Our constructions use groups with an efficiently computable bilinear map. Our first construction provides a system in which both the broadcast message and user private keys are of constant size (a precise statement is given in the next section). No matter what the receiver set is, our broadcast ciphertext contains only two group elements. Each user's private key is just a single group element. Thus, when broadcasting to small sets our system generates far shorter ciphertexts than the trivial solution discussed above. However, the public key size in this system is linear in the number of recipients. This is not a large problem in applications such as encrypted file systems where the receivers have access to a large shared storage medium in which the public key can be stored. For other applications, such as content protection, we need to minimize both public key and ciphertext size.

Our second system is a generalization of the first that enables us to tradeoff public key size for ciphertext size. One interesting parametrization of our scheme gives a system where both the public key and the ciphertext are of size $O(\sqrt{n})$. This means that we can attach the public key to the encrypted broadcast and still achieve ciphertext size of $O(\sqrt{n})$. Consequently, we obtain a fully collusion secure broadcast encryption scheme with $O(\sqrt{n})$ ciphertext size (for any subset of users) where the users have a constant size private key.

In Section 2 we define our security model and the complexity assumption we use. In Section 3 we describe our systems and prove their semantic security. In Section 4 we discuss in detail several applications for these systems. Finally, in Section 5 we describe how to make our systems chosen-ciphertext secure.

1.1 Related Work

Fiat and Naor [FN93] were the first to formally explore broadcast encryption. They presented a solution for n users that is secure against a collusion of t users and has ciphertext size of $O(t \log^2 t \log n)$.

Naor et al. [NNL01] presented a fully collusion secure broadcast encryption system that is efficient for broadcasting to all but a small set of revoked users. Their scheme is useful for content protection where broadcasts will be sent to all but a small set of receivers whose keys have been compromised. Their scheme can be used to encrypt to $n - r$ users with a header size of $O(r)$ elements and private keys of size $O(\log^2 n)$. Further improvements [HS02, GST04] reduce the private key size to $O(\log n)$. Dodis and Fazio [DF02] extend the NNL (subtree difference) method into a public key broadcast system for a small size public key.

Other broadcast encryption methods for large sets include Naor and Pinkas [NP00] and Dodis and Fazio [DF03] as well as [AMM99, TT01]. For some fixed t all these systems can revoke any $r < t$ users where ciphertexts are always of size $O(t)$ and private keys are constant size. By running $\log n$ of these systems in parallel, where the revocation bound of the i'th system is $t_i = 2^i$ (as in [YJCK04]), one obtains a broadcast encryption system with the same parameters as [GST04]. Private key size is $O(\log n)$ and, when revoking r users, ciphertext size is proportional to $2^{\lceil \log_2 r \rceil} = O(r)$. This simple extension to the Naor and Pinkas system gives a broadcast system with similar parameters as the latest NNL derivative.

Wallner et al. [WHA97] and Wong [WGL98] independently discovered the logical-key-hierarchy scheme (LKH) for multicast group key management. Using these methods receivers maintain state and remain connected to receive key-update messages. The parameters of these schemes are improved in later work [CGI+99, CMN99, SM03]. Our broadcast system also gives a group key management method with short key update messages.

The security of our broadcast encryption relies on computational assumptions. Several other works [Sti97, ST98, SW98, GSY99, GSW00] explore broadcast encryption and tracing from an information theoretic perspective.

Boneh and Silverberg [BS03] show that n-linear maps give the ultimate fully collusion secure scheme with constant public key, private key, and ciphertext size. However, there are currently no known implementations of cryptographically useful n-linear maps for $n > 2$. Our results show that we can come fairly close using bilinear maps alone.

2 Preliminaries

We begin by formally defining public-key broadcast encryption systems. For simplicity we define broadcast encryption as a key encapsulation mechanism. We then state the complexity assumption needed for our proof of security.

2.1 Broadcast Encryption Systems

A broadcast encryption system is made up of three randomized algorithms:

Setup(n). Takes as input the number of receivers n. It outputs n private keys $d_1, \ldots, d_n$ and a public key PK.

Encrypt(S, PK). Takes as input a subset $S \subseteq \{1, \ldots, n\}$, and a public key PK. It outputs a pair (Hdr, K) where Hdr is called the header and $K \in \mathcal{K}$ is a message encryption key chosen from a finite key set $\mathcal{K}$. We will often refer to Hdr as the broadcast ciphertext.

Let M be a message to be broadcast that should be decipherable precisely by the receivers in S. Let C_M be the encryption of M under the symmetric key K. The broadcast consists of (S, Hdr, C_M). The pair (S, Hdr) is often called the full header and C_M is often called the broadcast body.

Decrypt$(S, i, d_i, \text{Hdr}, PK)$. Takes as input a subset $S \subseteq \{1, \ldots, n\}$, a user id $i \in \{1, \ldots, n\}$ and the private key d_i for user i, a header Hdr, and the public key PK. If $i \in S$, then the algorithm outputs a message encryption key $K \in \mathcal{K}$. Intuitively, user i can then use K to decrypt the broadcast body C_M and obtain the message body M.

As usual, we require that the system be correct, namely that for all subsets $S \subseteq \{1, \ldots, n\}$ and all $i \in S$,

if $(PK, (d_1, \ldots, d_n)) \overset{\text{R}}{\leftarrow} Setup(n)$ and $(\text{Hdr}, K) \overset{\text{R}}{\leftarrow} Encrypt(S, PK)$
then $Decrypt(S, i, d_i, \text{Hdr}, PK) = K$.

We define chosen ciphertext security of a broadcast encryption system against a static adversary. Security is defined using the following game between an attack algorithm $\mathcal{A}$ and a challenger. Both the challenger and $\mathcal{A}$ are given n, the total number of users, as input.

Init. Algorithm $\mathcal{A}$ begins by outputting a set $S^* \subseteq \{1, \ldots, n\}$ of receivers that it wants to attack.

Setup. The challenger runs $Setup(n)$ to obtain a public key PK and private keys $d_1, \ldots, d_n$. It gives $\mathcal{A}$ the public key PK and all private keys d_j for which $j \notin S^*$.

Query phase 1. Algorithm $\mathcal{A}$ issues decryption queries $q_1, \ldots, q_m$ adaptively where a decryption query consists of (u, S, Hdr) where $S \subseteq S^*$ and $u \in S$. The challenger responds with $Decrypt(S, u, d_u, \text{Hdr}, PK)$.

Challenge. The challenger runs algorithm $Encrypt$ to obtain $(\text{Hdr}^*, K) \overset{\text{R}}{\leftarrow} Encrypt(S, PK)$ where $K \in \mathcal{K}$. Next, the challenger picks a random $b \in \{0, 1\}$. It sets $K_b = K$ and picks a random $K_{1-b} \in \mathcal{K}$. It then gives (Hdr^*, K_0, K_1) to algorithm $\mathcal{A}$.

Query phase 2. Algorithm $\mathcal{A}$ adaptively issues more decryption queries $q_{m+1}, \ldots, q_{q_D}$ where $q_i = (u, S, \text{Hdr})$ with $S \subseteq S^*$ and $u \in S$. The only constraint is that $\text{Hdr} \neq \text{Hdr}^*$. The challenger responds as in phase 1.

Guess. Algorithm $\mathcal{A}$ outputs its guess $b' \in \{0, 1\}$ for b and wins the game if $b = b'$.

Let $\mathsf{AdvBr}_{\mathcal{A}, n}$ denote the probability that $\mathcal{A}$ wins the game when the challenger and $\mathcal{A}$ are given n as input.

Definition 1. *We say that a broadcast encryption system is (t, ϵ, n, q_D) CCA secure if for all t-time algorithms $\mathcal{A}$ that make a total of q_D decryption queries, we have that $|\mathsf{AdvBr}_{\mathcal{A}, n} - \frac{1}{2}| < \epsilon$.*

The game above models an attack where all users not in the set S^* collude to try and expose a broadcast intended for users in S^* only. The set S^* is chosen by the adversary. Note that the adversary is non-adaptive; it chooses S^*, and obtains the keys for users outside of S^*, before it even sees the public key PK. An adaptive adversary could request user keys adaptively. We only prove security of our system in the non-adaptive settings described above. It is an open problem to build a broadcast encryption system with the performance of our system which is secure against adaptive adversaries. We note that similar formal definitions for broadcast encryption security were given in [BS03, DF03].

As usual, we define semantic security for a broadcast encryption scheme by preventing the attacker from issuing decryption queries.

Definition 2. *We say that a broadcast encryption system is (t, ϵ, n) semantically secure if it is $(t, \epsilon, n, 0)$ CCA secure.*

In Section 3 we first construct *semantically secure* systems with constant ciphertext and private key size. We come back to chosen ciphertext security in Section 5.

2.2   Bilinear Maps

We briefly review the necessary facts about bilinear maps and bilinear map groups. We use the following standard notation [Jou00, JN03, BF01]:

1. $\mathbb{G}$ and $\mathbb{G}_1$ are two (multiplicative) cyclic groups of prime order p;
2. g is a generator of $\mathbb{G}$.
3. $e : \mathbb{G} \times \mathbb{G} \to \mathbb{G}_1$ is a bilinear map.

Let $\mathbb{G}$ and $\mathbb{G}_1$ be two groups as above. A bilinear map is a map $e : \mathbb{G} \times \mathbb{G} \to \mathbb{G}_1$ with the following properties:

1. For all $u, v \in \mathbb{G}$ and $a, b \in \mathbb{Z}$, we have $e(u^a, v^b) = e(u, v)^{ab}$, and
2. The map is not degenerate, i.e., $e(g, g) \neq 1$.

We say that $\mathbb{G}$ is a bilinear group if the group action in $\mathbb{G}$ can be computed efficiently and there exists a group $\mathbb{G}_1$ and an efficiently computable bilinear map $e : \mathbb{G} \times \mathbb{G} \to \mathbb{G}_1$ as above. Note that $e(,)$ is symmetric since $e(g^a, g^b) = e(g, g)^{ab} = e(g^b, g^a)$.

2.3   Complexity Assumptions

Security of our system is based on a complexity assumption called the bilinear Diffie-Hellman Exponent assumption (BDHE). This assumption was previously introduced in [BBG05].

Let $\mathbb{G}$ be a bilinear group of prime order p. The ℓ-BDHE problem in $\mathbb{G}$ is stated as follows: given a vector of $2\ell + 1$ elements

$$\left(h, g, g^\alpha, g^{(\alpha^2)}, \ldots, g^{(\alpha^\ell)}, g^{(\alpha^{\ell+2})}, \ldots, g^{(\alpha^{2\ell})} \right) \in \mathbb{G}^{2\ell+1}$$

as input, output $e(g, h)^{(\alpha^{\ell+1})} \in \mathbb{G}_1$. Note that the input vector is missing the term $g^{(\alpha^{\ell+1})}$ so that the bilinear map seems to be of little help in computing the required $e(g, h)^{(\alpha^{\ell+1})}$.

As shorthand, once g and α are specified, we use g_i to denote $g_i = g^{(\alpha^i)} \in \mathbb{G}$. An algorithm $\mathcal{A}$ has advantage ϵ in solving ℓ-BDHE in $\mathbb{G}$ if

$$\Pr\left[\mathcal{A}\left(h, g, g_1, \ldots, g_\ell, g_{\ell+2}, \ldots, g_{2\ell} \right) = e(g_{\ell+1}, h) \right] \geq \epsilon$$

where the probability is over the random choice of generator g in $\mathbb{G}$, the random choice of h in $\mathbb{G}$, the random choice of α in $\mathbb{Z}_p$, and the random bits used by $\mathcal{A}$.

The decisional version of the ℓ-BDHE problem in $\mathbb{G}$ is defined analogously. Let $\boldsymbol{y}_{g,\alpha,\ell} = (g_1, \ldots, g_\ell, g_{\ell+2}, \ldots, g_{2\ell})$. An algorithm $\mathcal{B}$ that outputs $b \in \{0,1\}$ has advantage ϵ in solving decision ℓ-BDHE in $\mathbb{G}$ if

$$\left| \Pr\left[\mathcal{B}\big(g, h, \boldsymbol{y}_{g,\alpha,\ell}, e(g_{\ell+1}, h)\big) = 0\right] - \Pr\left[\mathcal{B}\big(g, h, \boldsymbol{y}_{g,\alpha,\ell}, T\big) = 0\right] \right| \geq \epsilon$$

where the probability is over the random choice of generators g, h in $\mathbb{G}$, the random choice of α in $\mathbb{Z}_p$, the random choice of $T \in \mathbb{G}_1$, and the random bits consumed by $\mathcal{B}$. We refer to the distribution on the left as $\mathcal{P}_{BDHE}$ and the distribution on the right as $\mathcal{R}_{BDHE}$.

Definition 3. *We say that the (decision) (t, ϵ, ℓ)-BDHE assumption holds in $\mathbb{G}$ if no t-time algorithm has advantage at least ϵ in solving the (decision) ℓ-BDHE problem in $\mathbb{G}$.*

Occasionally we drop the t and ϵ and refer to the (decision) ℓ-BDHE in $\mathbb{G}$. We note that the ℓ-BDHE assumption is a natural extension of the bilinear-DHI assumption previously used in [BB04, DY05]. Furthermore, Boneh et al. [BBG05] show that the ℓ-BDHE assumption holds in generic bilinear groups [Sho97].

3 Construction

We are now ready to present our broadcast encryption system. We first present a special case system where ciphertexts and private keys are always constant size. The public key grows linearly with the number of users. We then present a generalization that allows us to balance the public key size and the ciphertext size. Private keys are still constant size. We prove security of this general system.

3.1 A Special Case

We begin by describing a broadcast encryption system for n users where the ciphertexts and private keys are constant size. The public key grows linearly in the number of users.

***Setup*(n):** Let $\mathbb{G}$ be a bilinear group of prime order p. The algorithm first picks a random generator $g \in \mathbb{G}$ and a random $\alpha \in \mathbb{Z}_p$. It computes $g_i = g^{(\alpha^i)} \in \mathbb{G}$ for $i = 1, 2, \ldots, n, n+2, \ldots, 2n$. Next, it picks a random $\gamma \in \mathbb{Z}_p$ and sets $v = g^\gamma \in \mathbb{G}$. The public key is:

$$PK = (g, g_1, \ldots, g_n, g_{n+2}, \ldots, g_{2n}, v) \in \mathbb{G}^{2n+1}$$

The private key for user $i \in \{1, \ldots, n\}$ is set as: $d_i = g_i^\gamma \in \mathbb{G}$. Note that $d_i = v^{(\alpha^i)}$. The algorithm outputs the public key PK and the n private keys $d_1, \ldots, d_n$.

Encrypt(S, PK)**:** Pick a random t in $\mathbb{Z}_p$ and set $K = e(g_{n+1}, g)^t \in \mathbb{G}_1$. The value $e(g_{n+1}, g)$ can be computed as $e(g_n, g_1)$. Next, set

$$\text{Hdr} = \left(g^t, \ \left(v \cdot \prod_{j \in S} g_{n+1-j}\right)^t \right) \in \mathbb{G}^2$$

and output the pair (Hdr, K).

Decrypt$(S, i, d_i, \text{Hdr}, PK)$**:** Let $\text{Hdr} = (C_0, C_1)$ and recall that $d_i \in \mathbb{G}$. Then, output

$$K = e(g_i, C_1) \ / \ e(d_i \cdot \prod_{\substack{j \in S \\ j \neq i}} g_{n+1-j+i}, \ C_0)$$

Note that a private key is only one group element in $\mathbb{G}$ and the ciphertext, Hdr, is only two group elements. Furthermore, since $e(g_{n+1}, g)$ can be precomputed, encryption requires no pairings. Nevertheless, the system is able to broadcast to any subset of users and is fully collusion resistant. We prove security in Section 3.3 where we discuss a more general system.

We verify that the system is correct — i.e., that the decryption algorithm works correctly — by observing that, for any $i \in S$, the quotient of the terms

$$e(g_i, C_1) = e(g, g)^{\alpha^i \cdot t(\gamma + \sum_{j \in S} \alpha^{n+1-j})} = e(g, g)^{t(\gamma \alpha^i + \sum_{j \in S} \alpha^{n+1-j+i})} \quad \text{and}$$

$$e(C_0, \ d_i \cdot \prod_{\substack{j \in S \\ j \neq i}} g_{n+1-j+i}) = e(g, g)^{t \cdot (\gamma \alpha^i + \sum_{\substack{j \in S \\ j \neq i}}^{j \in S} \alpha^{n+1-j+i})}$$

is $K = e(g_{n+1}, g)^t = e(g, g)^{t \alpha^{n+1}}$, as required.

Efficient Implementation. For any large number of receivers, decryption time will be dominated by the $|S| - 2$ group operations needed to compute $\prod_{\substack{j \neq i \\ j \neq i}}^{j \in S} g_{n+1-j+i}$. However, we observe that if the receiver had previously computed the value $w = \prod_{j \neq i}^{j \in S'} g_{n+1-j+i}$ for some receiver set S' that is similar to S then, the receiver can compute $\prod_{j \neq i}^{j \in S} g_{n+1-j+i}$ with just δ group operations using the cached value w, where δ is the size of the set difference between S and S'.

This observation is especially useful when the broadcast system is intended to broadcast to large sets, i.e. sets of size $n - r$ for $r \ll n$. The private key d_i could include the value $\prod_{j \neq i}^{j \in [1,n]} g_{n+1-j+i} \in \mathbb{G}$ which would enable the receiver to decrypt using only r group operations. Furthermore, user i would only need r elements from the public key PK.

We note that computation time for encryption will similarly be dominated by the $|S| - 1$ group operations to compute $\prod_{j \in S} g_{n+1-j}^t$ and similar performance optimizations (e.g. precomputing $\prod_{j=1}^{n} g_{n+1-j}$) apply. We also note that in the secret key settings (where the encryptor is allowed to keep secret information) the encryptor need only store (g, v, α) as opposed to storing the entire public key vector.

3.2   A General Construction

Next, we present a more general broadcast encryption system. The idea is to run A parallel instances of the system in the previous section where each instance can broadcast to at most $B < n$ users. As a result we can handle as many as $n = AB$ users. However, we substantially improve performance by sharing information between the A instances. In particular, all instances will share the same public key values $g, g_1, \ldots, g_B, g_{B+2}, \ldots, g_{2B}$.

This generalized system enables us to tradeoff the public key size for ciphertext size. Setting $B = n$ gives the system of the previous section. However, setting $B = \lfloor \sqrt{n} \rfloor$ gives a system where both public key and ciphertext size are about $\sqrt{n}$ elements. Note that either way, the private key is always just one group element.

For fixed positive integer B, the B-broadcast encryption system works as follows:

Setup$_B(n)$: The algorithm will set up $A = \lceil \frac{n}{B} \rceil$ instances of the scheme. Let $\mathbb{G}$ be a bilinear group of prime order p. The algorithm first picks a random generator $g \in \mathbb{G}$ and a random $\alpha \in \mathbb{Z}_p$. It computes $g_i = g^{(\alpha^i)} \in \mathbb{G}$ for $i = 1, 2, \ldots, B, B+2, \ldots, 2B$. Next, it picks random $\gamma_1, \ldots, \gamma_A \in \mathbb{Z}_p$ and sets $v_1 = g^{\gamma_1}, \ldots, v_A = g^{\gamma_A} \in \mathbb{G}$. The public key is:

$$PK = (g, g_1, \ldots, g_B, g_{B+2}, \ldots, g_{2B}, v_1, \ldots, v_A) \in \mathbb{G}^{2B+A}$$

The private key for user $i \in \{1, \ldots, n\}$ is defined as follows: write i as $i = (a-1)B + b$ for some $1 \le a \le A$ and $1 \le b \le B$ (i.e. $a = \lceil i/B \rceil$ and $b = i \bmod B$). Set the private key for user i as:

$$d_i = g_b^{\gamma_a} \in \mathbb{G} \qquad \text{(note that} \quad d_i = v_a^{(\alpha^b)} \text{)}$$

The algorithm outputs the public key PK and the n private keys $d_1, \ldots, d_n$.

Encrypt(S, PK): For each $\ell = 1, \ldots, A$ define the subsets $\hat{S}_\ell$ and S_ℓ as

$$\hat{S}_\ell = S \cap \{(\ell-1)B+1, \ldots, \ell B\}, \qquad S_\ell = \{x - \ell B + B \mid x \in \hat{S}_\ell\} \subseteq \{1, \ldots, B\}$$

In other words, $\hat{S}_\ell$ contains all users in S that fall in the ℓ'th interval of length B and S_ℓ contains the indices of those users relative to the beginning of the interval. Pick a random t in $\mathbb{Z}_p$ and set $K = e(g_{B+1}, g)^t \in \mathbb{G}_1$. Set

$$\text{Hdr} = \left(g^t, \; (v_1 \cdot \prod_{j \in S_1} g_{B+1-j})^t, \; \ldots, \; (v_A \cdot \prod_{j \in S_A} g_{B+1-j})^t \right) \in \mathbb{G}^{A+1}$$

Output the pair (Hdr, K). Note that Hdr contains $A + 1$ elements.

Decrypt$(S, i, d_i, \text{Hdr}, PK)$: Let $\text{Hdr} = (C_0, C_1, \ldots, C_A)$ and recall that $d_i \in \mathbb{G}$. Write i as $i = (a-1)B + b$ for some $1 \le a \le A$ and $1 \le b \le B$. Then

$$K = e(g_b, C_a) \; / \; e(d_i \cdot \prod_{\substack{j \in S_a \\ j \neq b}} g_{B+1-j+b}, \; C_0)$$

Verifying that the decryption algorithm works correctly is analogous to the calculation in the previous section. We note that when $B = n$ then $A = 1$ and we obtain the system of the previous section.

Efficiency. A user's private key size again will only consist of one group element. The ciphertext consists of $A + 1$ group elements and the public key is $2B + A$ elements. Our choice of B depends on the application. As we will see, in some cases we want $B = n$ to obtain the smallest possible ciphertext. In other cases we want $B = \sqrt{n}$ to minimize the concatenation of the ciphertext and public key.

The decryption time for user $i = (a-1)B+b$ will be dominated by $|S_a|-2 < B$ group operations. Similar caching techniques to those described in the end of Section 3.1 can be used to improve performance.

3.3   Security

We now prove the semantic security of the general system of Section 3.2.

Theorem 1. *Let $\mathbb{G}$ be a bilinear group of prime order p. For any positive integers B, n ($n > B$) our B-broadcast encryption system is (t, ϵ, n) semantically secure assuming the decision (t, ϵ, B)-BDHE assumption holds in $\mathbb{G}$.*

Proof. Suppose there exists a t-time adversary, $\mathcal{A}$, such that $\mathsf{AdvBr}_{\mathcal{A}, n} > \epsilon$ for a system parameterized with a given B. We build an algorithm, $\mathcal{B}$, that has advantage ϵ in solving the decision B-BDHE problem in $\mathbb{G}$. Algorithm $\mathcal{B}$ takes as input a random decision B-BDHE challenge $(g, h, \boldsymbol{y}_{g, \alpha, B}, Z)$, where $\boldsymbol{y}_{g, \alpha, B} = (g_1, \ldots, g_B, g_{B+2}, \ldots, g_{2B})$ and Z is either $e(g_{B+1}, h)$ or a random element of $\mathbb{G}_1$ (recall that $g_i = g^{(\alpha^i)}$ for all i). Algorithm $\mathcal{B}$ proceeds as follows.

Init. Algorithm $\mathcal{B}$ runs $\mathcal{A}$ and receives the set S of users that $\mathcal{A}$ wishes to be challenged on.

Setup. $\mathcal{B}$ needs to generate a public key PK and private keys d_i for $i \notin S$. The crux of the proof is in the choice of $v_1, \ldots, v_A$. Algorithm $\mathcal{B}$ chooses random $u_i \in \mathbb{Z}_p$ for $1 \le i \le A$. We again define the subsets $\hat{S}_i$ and S_i as

$$\hat{S}_i = S \cap \{(i-1)B+1, \ldots, iB\} \quad \text{and} \quad S_i = \{x - iB + B \mid x \in \hat{S}_i\} \subseteq \{1, \ldots, B\}$$

For $i = 1, \ldots, A$ algorithm $\mathcal{B}$ sets $\quad v_i = g^{u_i} \left(\prod_{j \in S_i} g_{B+1-j} \right)^{-1}$.   It gives $\mathcal{A}$ the public key

$$PK = (g, g_1, \ldots, g_B, g_{B+2}, \ldots, g_{2B}, v_1, \ldots, v_A) \in \mathbb{G}^{2B+A}$$

Note that since g, α and the u_i values are chosen uniformly at random, this public key has an identical distribution to that in the actual construction.

Next, the adversary needs all private keys that are not in the target set S. For all $i \notin S$ we write i as $i = (a - 1)B + b$ for some $1 \le a \le A$ and $1 \le b \le B$. Algorithm $\mathcal{B}$ computes d_i as

$$d_i = g_b^{u_i} \cdot \prod_{j \in S_a} (g_{B+1-j+b})^{-1}$$

Indeed, we have that $\quad d_i = (g^{u_i} \prod_{j \in S_a} (g_{B+1-j})^{-1})^{(\alpha^b)} = v_a^{(\alpha^b)} \quad$ as required. The main point is that since $i \notin S$ we know that $b \notin S_a$ and hence the product defining d_i does not include the term g_{B+1}. It follows that algorithm $\mathcal{B}$ has all the necessary values to compute d_i.

Challenge. To generate the challenge, $\mathcal{B}$ computes Hdr as $(h, h^{u_1}, \ldots, h^{u_A})$. It then randomly chooses a bit $b \in \{0, 1\}$ and sets $K_b = Z$ and picks a random K_{1-b} in $\mathbb{G}_1$. It gives (Hdr, K_0, K_1) as the challenge to $\mathcal{A}$.

We claim that when $Z = e(g_{B+1}, h)$ (i.e. the input to $\mathcal{B}$ is a B-BDHE tuple sampled from $\mathcal{P}_{BDHE}$) then (Hdr, K_0, K_1) is a valid challenge to $\mathcal{A}$ as in a real attack. To see this, write $h = g^t$ for some (unknown) $t \in \mathbb{Z}_p$. Then, for all $i = 1, \ldots, A$ we have

$$h^{u_i} = (g^{u_i})^t = (g^{u_i} (\prod_{j \in S_i} g_{B+1-j})^{-1} (\prod_{j \in S_i} g_{B+1-j}))^t = (v_i \prod_{j \in S_i} g_{B+1-j})^t$$

Therefore, by definition, $(h, h^{u_1}, \ldots, h^{u_A})$ is a valid encryption of key $e(g_{B+1}, g)^t$. Furthermore, $e(g_{B+1}, g)^t = e(g_{B+1}, h) = Z = K_b$ and hence (Hdr, K_0, K_1) is a valid challenge to $\mathcal{A}$.

On the other hand, when Z is random in $\mathbb{G}_1$ (i.e. the input to $\mathcal{B}$ is sampled from $\mathcal{R}_{BDHE}$) then K_0, K_1 are just random independent elements of $\mathbb{G}_1$.

Guess. The adversary, $\mathcal{A}$ outputs a guess b' of b. If $b' = b$ the algorithm $\mathcal{B}$ outputs 0 (indicating that $Z = e(g_{B+1}, h)$). Otherwise, it outputs 1 (indicating that Z is random in $\mathbb{G}_1$).

We see that $\Pr[\mathcal{B}(g, h, \mathbf{y}_{g,\alpha,B}, Z) = 0] = \frac{1}{2}$ if $(g, h, \mathbf{y}_{g,\alpha,B}, Z)$ is sampled from $\mathcal{R}_{BDHE}$. If $(g, h, \mathbf{y}_{g,\alpha,B}, Z)$ is sampled from $\mathcal{P}_{BDHE}$ then $| \Pr[\mathcal{B}(g, h, \mathbf{y}_{g,\alpha,B}, Z) = 0] - \frac{1}{2}| = \text{AdvBr}_{A,n} \ge \epsilon$. It follows that $\mathcal{B}$ has advantage at least ϵ in solving decision B-BDHE in $\mathbb{G}$. This concludes the proof of Theorem 1.   □

Note that the proof of Theorem 1 does not use the random oracle model. The system can be proved secure using the weaker *computational B-BDHE* assumption (as opposed to decision B-BDHE), using the random oracle model. In that case the advantage of the simulator is at least ϵ/q, where q is the maximum number of random oracle queries made by the adversary.

4   Applications

We describe how our system can be used for a number of specific applications. The first application, file sharing in encrypted file systems, is an example of broadcasts to small sets. The second application, encrypted email for large mailing lists, shows that the majority of the public key can be shared by many broadcast systems so that the public key for a new broadcast system is constant size. The third application, DVD content protection, is an example of broadcasts to large sets.

4.1   File Sharing in Encrypted File Systems

Encrypted file systems let users encrypt files on disk. For example, Windows EFS encrypts the file contents using a file encryption key K_F and then places an encryption of K_F in the file header. If n users have access to the file, EFS encrypts K_F under the public keys of all n users and places the resulting n ciphertexts in the file header. Related designs can be found in the SiRiUS [GSMB03] and Plutus [KRS$^+$03] file systems.

Abstractly, access control in an encrypted file system can be viewed as a broadcast encryption problem. The file system is the broadcast channel and the key K_F is broadcast (via the file header) to the subset of users that can access file F. Many encrypted file systems implement the straightforward broadcast system where the number of ciphertexts in the file header grows linearly in the number of users that can access the file. As a result, there is often a hard limit on the number of users that can access a file. For example, the following quote is from Microsoft's knowledge base:

> "EFS has a limit of 256KB in the file header for the EFS metadata. This limits the number of individual entries for file sharing that may be added. On average, a maximum of 800 individual users may be added to an encrypted file."

A natural question is whether we can implement file sharing in an encrypted file system without resorting to large headers. Remarkably, there is no known combinatorial solution that performs better than the straightforward solution used in EFS. The broadcast system of Section 3.1 performs far better and provides a system with the following parameters:

- The public key (whose size is linear in n) is stored on the file system. Even for a large organization of 100,000 users this file is only 4MB long (using a standard security parameter where each group element is 20 bytes).
- Each user is given a private key that contains only one group element.
- Each file header contains $([S], C)$ where $[S]$ is a description of the set S of users who can access F and C is a fixed size ciphertext consisting of only two group elements.

Since S tends to be small relative to n, its shortest description is simply an enumeration of the users in S. Assuming 32-bit user ID's, a description of a set S of size r takes $4r$ bytes. Hence, the file header grows with the size of S, but only at a rate of 4 bytes per user. In EFS the header grows by one public key ciphertext per user. For comparison, we can accommodate sharing among 800 users using a header of size $4 \times 800 + 40 = 3240$ bytes which is far less than

EFS's header size. Even if EFS were using short ElGamal ciphertexts on elliptic curves, headers would grow by 44 bytes per user which would result in headers that are 11 times bigger than headers in our system.

Our system has a few more properties that make it especially useful for cryptographic access control. We describe these next.

1. Incremental sharing. Suppose a file header contains a ciphertext $C = (C_0, C_1)$ which is the encryption of K_F for a certain set S of users. Let $C_0 = g^t$ and $C_1 = (v \cdot \prod_{j \in S} g_{n+1-j})^t$. Suppose the file owner wishes to add access rights for some user $u \in \{1, \ldots, n\}$. This is easy to do given t. Simply set $C_1 \leftarrow C_1 \cdot g_{n+1-u}^t$. Similarly, to revoke access rights for user u set $C_1 \leftarrow C_1/g_{n+1-u}^t$.

This incremental sharing mechanism requires the file owner to remember the random value $t \in \mathbb{Z}_p$ for every file. Alternatively, the file owner can embed a short nonce T_F in every file header and derive the value t for that file by setting $t \leftarrow PRF_k(T_F)$ where k is a secret key known only to the file owner. Hence, changing access permissions can be done efficiently with the file owner only having to remember a single key k. Note that when access rights to a file F change it is sometimes desirable to re-encrypt the file using a new key K_F^{new}. Modifying the existing header to encrypt a new K_F^{new} for the updated access list is just as easy.

2. Incremental growth of the number of users. In many cases a broadcast encryption system must be able to handle the incremental addition of new users. It is desirable to have a system that does not a-priori restrict the total number of users it can handle. Our system supports this by slowly expanding the public key as the number of users in the system grows. To do so, at system initialization the setup algorithm picks a large value of n (say $n = 2^{64}$) that is much larger than the maximum number of users that will ever use the system. At any one time if there are j users in the system the public key will be $g_{n-j+1}, \ldots, g_n, g_{n+2}, \ldots, g_{n+j}$. Whenever a new user joins the system we simply add two more elements to the public key. Note that user i must also be given g_i as part of the private key and everything else remains the same.

4.2   Sending Encrypted Email to Mailing Lists

One interesting aspect of our broadcast encryption system is that the public values $\boldsymbol{y}_{g,\alpha,n} = (g_1, \ldots g_n, g_{n+2}, \ldots, g_{2n})$ can be shared among many broadcast systems and α can be erased. Suppose this $\boldsymbol{y}_{g,\alpha,n}$ is distributed globally to a large group of users (for example, imagine $\boldsymbol{y}_{g,\alpha,n}$ is pre-installed on every computer). Then creating a new broadcast system is done by simply choosing a random $\gamma \in \mathbb{Z}_p$, setting $v = g^\gamma$, and assigning private keys as $d_i = g_i^\gamma$. Since all broadcast systems use the same pre-distributed $\boldsymbol{y}_{g,\alpha,n}$, the actual public key for this new broadcast system is just one element, v. Theorem 1 shows that using the same $\boldsymbol{y}_{g,\alpha,n}$ for many broadcast systems is secure.

We illustrate this property with an example of secure mailing lists. Sending out encrypted email to all members of a mailing list is an example of a broadcast encryption system. Suppose the global public vector $\boldsymbol{y}_{g,\alpha,n}$ is shipped with the operating system and installed on every computer. We arbitrarily set $n = 50,000$ in which case the size of $\boldsymbol{y}_{g,\alpha,n}$ is about 2MB.

For every secure mailing list, the administrator creates a separate broadcast encryption system. Thus, every mailing list is identified by its public key $v = g^\gamma$. We assume the maximum number of members in a mailing list is less than $n = 50,000$ (larger lists can be partitioned into multiple smaller lists). Each time a new user is added onto the list, the user is assigned a previously unused index $i \in \{1, \ldots, n\}$ and given the secret key $d_i = g_i^\gamma$. The broadcast set S is updated to include i and all mailing list members are notified of the change in S. Similarly, if a user with index j is removed from the list, then j is removed from the set S and all members are notified. We note that any member, i, does not need to actually store S. Instead, member i need only store the value $\prod_{j \in S j \neq i} g_{n+1-j+i}$ needed for decryption. The member updates this value every time a membership update message is sent. To send email to a mailing list with public key v the server simply does a broadcast encryption to the current set of members S using $(\boldsymbol{y}_{g,\alpha,n}, v)$ as the public key.

In this mail system, the header of an email message sent to the list is constant size. Similarly, membership update messages are constant size. A mailing list member only needs to store two group elements for each list he belongs to (although we have to keep in mind the cost of storing $\boldsymbol{y}_{g,\alpha,n}$ which is amortized over all mailing lists). It is interesting to compare our solution to one using an LKH scheme [WHA97, WGL98]. In LKH email messages are encrypted under a group key. Using this type of a system each update message contains $O(\log(m))$ ciphertexts, and private keys are of size $O(\log(m))$ (per system), where m is the current group size. In our system, update messages and private user storage are much smaller.

4.3   Content Protection

Broadcast encryption applies naturally to protecting DVD content, where the goal is to revoke compromised DVD players. Recall that the public key in our system is needed for decryption and hence it must be embedded in the header of every DVD disk. Consequently, we are interested in minimizing the total length of the header and public key, namely minimize $|\text{Hdr}| + |PK|$.

Let n be the total number of DVD players (e.g. $n = 2^{32}$) and let r be the number of revoked players. Let $\ell_{\text{id}} = \log_2 n$ (e.g. $\ell_{\text{id}} = 32$) and let $\bar{k}$ be the size of a group element (e.g. $\bar{k} = 160$ bits). Then using our $\sqrt{n}$-broadcast system $(B = \sqrt{n})$ we can broadcast to sets of size $n - r$ using the following parameters:

$$\text{priv-key-size} = 4\bar{k}, \quad \text{and} \quad |\text{Hdr}| + |PK| + |S| = 4\bar{k}\lceil\sqrt{n}\rceil + r\ell_{\text{id}}$$

In comparison, the NNL system [NNL01] and its derivatives [HS02, GST04] can broadcast to sets of size $n - r$ using:

$$\text{priv-key-size} = O(k \log n), \quad \text{and} \quad \text{header-size} = O((k + \ell_{\text{id}}) \cdot r)$$

where k is the length of a symmetric key (e.g. $k = 128$ bits). Note that the broadcast header grows by $O(k + \ell_{\mathrm{id}})$ bits per revoked player. With our system the broadcast header only grows by ℓ_{id} bits per revoked player.

Example. Plugging in real numbers we obtain the following. When $n = 2^{32}, \bar{k} = 20$ bytes, and $\ell_{\mathrm{id}} = 4$ bytes, header size in our system is 5.12MB and each revocation adds 4 bytes to the header. In NNL-like systems, using $k = 128$-bit symmetric keys, each revocation adds about 40 bytes to the header, but there is no upfront 5MB fixed cost.

The best system is obtained by combining NNL with our system (using NNL when $r < \sqrt{n}$ and our system when $r > \sqrt{n}$). Thus, as long as things are stable, DVD disk distributors use NNL. In case of a disaster where, say, a DVD player manufacturer loses a large number of player keys, DVD disk distributors can switch to our system where the header size grows slowly beyond $O(\sqrt{n})$.

5   Chosen Ciphertext Secure Broadcast Encryption

We show how to extend the system of Section 3.1 to obtain chosen ciphertext security. The basic idea is to compose the system with the IBE system of [BB04] and then apply the ideas of [CHK04]. The resulting system is chosen ciphertext secure without using random oracles.

We need a signature scheme (*SigKeyGen, Sign, Verify*). We also need a collision resistant hash function that maps verification keys to $\mathbb{Z}_p$. Alternatively, we can simply assume (as we do below) that signature verification keys are encoded as elements of $\mathbb{Z}_p$. This greatly simplifies the notation.

As we will see, security of the CCA-secure broadcast system for n users is based on the $(n+1)$-BDHE assumption (as opposed to the n-BDHE assumption for the system of Section 3.1). Hence, to keep the notation consistent with Section 3.1 we will describe the CCA-secure system for $n - 1$ users so that security will depend on the n-BDHE assumption as before. The system works as follows:

Setup$(n - 1)$: Public key PK is generated as in Section 3.1. The private key for user $i \in \{1, \ldots, n-1\}$ is set as: $d_i = g_i^\gamma \in \mathbb{G}$.   The algorithm outputs the public key PK and the $n - 1$ private keys $d_1, \ldots, d_{n-1}$.

Encrypt(S, PK): Run the *SigKeyGen* algorithm to obtain a signature signing key K_{SIG} and a verification key V_{SIG}. Recall that for simplicity we assume $V_{\mathrm{SIG}} \in \mathbb{Z}_p$. Next, pick a random t in $\mathbb{Z}_p$ and set $K = e(g_{n+1}, g)^t \in \mathbb{G}_1$. Set

$$C = \left(g^t, \ \left(v \cdot g_1^{V_{\mathrm{SIG}}} \cdot \prod_{j \in S} g_{n+1-j} \right)^t \right) \in \mathbb{G}^2, \quad \mathrm{Hdr} = \left(C, Sign(C, K_{\mathrm{SIG}}), V_{\mathrm{SIG}} \right)$$

and output the pair (Hdr, K). Note that the only change to the ciphertext from Section 3.1 is the term $g_1^{V_{\mathrm{SIG}}}$ and the additional signature data.

Decrypt$(S, i, d_i, \mathrm{Hdr}, PK)$: Let $\mathrm{Hdr} = ((C_0, C_1), \ \sigma, \ V_{\mathrm{SIG}})$.
   1. Verify that σ is a valid signature of (C_0, C_1) under the key V_{SIG}. If invalid, output '?'.

2. Otherwise, pick a random $w \in \mathbb{Z}_p$ and compute

$$\hat{d}_0 = \left(d_i \cdot g_{i+1}^{V_{\text{SIG}}} \cdot \prod_{\substack{j \in S \\ j \neq i}} g_{n+1-j+i} \right) \cdot \left(v \cdot g_1^{V_{\text{SIG}}} \cdot \prod_{j \in S} g_{n+1-j} \right)^w \quad \text{and} \quad \hat{d}_1 = g_i g^w$$

3. Output $K = e(\hat{d}_1, C_1)/e(\hat{d}_0, C_0)$.

Correctness can be shown with a similar calculation to the one in Section 3.1. Note that private key size and ciphertext size are unchanged.

Unlike the system of Section 3.1, decryption requires a randomization value $w \in \mathbb{Z}_p$. This randomization ensures that for any $i \in S$ the pair $(\hat{d}_0, \hat{d}_1)$ is chosen from the following distribution

$$\left(g_{n+1}^{-1} \cdot \left(v \cdot g_1^{V_{\text{SIG}}} \cdot \prod_{j \in S} g_{n+1-j} \right)^r , \quad g^r \right)$$

where r is uniform in $\mathbb{Z}_p$. Note that this distribution is independent of i implying that all members of S generate a sample from the same distribution. Although this randomization slows down decryption by a factor of two, it is necessary for the proof of security.

We briefly recall that a signature scheme (*SigKeyGen, Sign, Verify*) is (t, ϵ, q_s) strongly existentially unforgeable if no t-time adversary who makes at most q_s signature queries is able to produce some new (message,signature) pair with probability at least ϵ. A complete definition is given in, e.g., [CHK04]. The following theorem proves chosen ciphertext security.

Theorem 2. *Let $\mathbb{G}$ be a bilinear group of prime order p. For any positive integer n, the broadcast encryption system above is $(t, \epsilon_1 + \epsilon_2, n - 1, q_D)$ CCA-secure assuming the decision (t, ϵ_1, n)-BDHE assumption holds in $\mathbb{G}$ and the signature scheme is $(t, \epsilon_2, 1)$ strongly existentially unforgeable.*

We give the proof of Theorem 2 in the full version of the paper [BGW05]. The proof does not use the random oracle model implying that the system is chosen-ciphertext secure in the standard model.

We also note that instead of the signature-based method of [CHK04] we could have used the more efficient MAC-based method of [BK05]. We chose to present the construction using the signature method to simplify the proof. The MAC-based method would also work.

6   Conclusions and Open Problems

We presented a fully collusion resistant broadcast encryption scheme with constant size ciphertexts and private keys for arbitrary receiver sets. In Section 5 we built a chosen-ciphertext secure broadcast system with the same parameters. A generalization of our basic scheme gave us a tradeoff between public key size and ciphertext size. With the appropriate parametrization we achieve a broadcast encryption scheme with $O(\sqrt{n})$ ciphertext and public key size. We discussed several applications such as encrypted file systems and content protection.

We leave as an open problem the question of building a public-key broadcast encryption system with the same parameters as ours which is secure against adaptive adversaries. We note that any non-adaptive scheme that is (t, ϵ, n) secure is also $(t, \epsilon/2^n, n)$ secure against adaptive adversaries. However, in practice this reduction is only meaningful for small values of n.

Another problem is to build a tracing traitors system [CFN94] with the same parameters as our system. Ideally, one could combine the two systems to obtain an efficient trace-and-revoke system. Finally, it is interesting to explore alternative systems with similar performance that can be proved secure under a weaker assumption.

References

[AMM99]   J. Anzai, N. Matsuzaki, and T. Matsumoto. A quick key distribution scheme with entity revocation. In *Proc. of Asiacrypt '99*, pages 333–347. Springer-Verlag, 1999.

[BB04]   D. Boneh and X. Boyen. Efficient selective-ID identity based encryption without random oracles. In *Proc. of Eurocrypt 2004*, volume 3027 of *LNCS*, pages 223–238. Springer-Verlag, 2004.

[BBG05]   D. Boneh, X. Boyen, and E. Goh. Hierarchical identity based encryption with constant size ciphertext. In *Proc. of Eurocrypt 2005*. Springer-Verlag, 2005.

[BF01]   D. Boneh and M. Franklin. Identity-based encryption from the Weil pairing. In *Proceedings of Crypto 2001*, volume 2139 of *LNCS*, pages 213–29. Springer-Verlag, 2001.

[BGW05]   D. Boneh, C. Gentry, and B. Waters. Collusion resistant broadcast encryption with short ciphertexts and private keys. Cryptology ePrint Archive, Report 2005/018, 2005. Full version of current paper.

[BK05]   D. Boneh and J. Katz. Improved efficiency for CCA-secure cryptosystems built using identity based encryption. In *Proc. of RSA-CT 2005*, volume 3376 of *LNCS*, pages 87–103. Springer-Verlag, 2005.

[BS03]   D. Boneh and A. Silverberg. Applications of multilinear forms to cryptography. *Contemporary Mathematics*, 324:71–90, 2003.

[CFN94]   B. Chor, A. Fiat, and M. Naor. Tracing traitors. In *Proc. of Crypto 1994*, volume 839 of *LNCS*, pages 257–270. Springer-Verlag, 1994.

[CGI+99]   R. Canetti, J. Garay, G. Itkis, D. Micciancio, M. Naor, and B. Pinkas. Multicast security: A taxonomy and some efficient constructions. In *Proc. IEEE INFOCOM 1999*, volume 2, pages 708–716. IEEE, 1999.

[CHK04]   R. Canetti, S. Halevi, and J. Katz. Chosen-ciphertext security from identity-based encryption. In *Proc. of Eurocrypt 2004*, volume 3027 of *LNCS*, pages 207–222. Springer-Verlag, 2004.

[CMN99]   R. Canetti, T. Malkin, and K. Nissim. Efficient communication-storage tradeoffs for multicast encryption. In *Proc. of Eurocrypt 1999*, pages 459–474. Springer-Verlag, 1999.

[DF02]   Y. Dodis and N. Fazio. Public key broadcast encryption for stateless receivers. In *Proc. of DRM 2002*, volume 2696 of *LNCS*, pages 61–80. Springer-Verlag, 2002.

[DF03]   Y. Dodis and N. Fazio. Public key broadcast encryption secure against adaptive chosen ciphertext attack. In *Proc. of PKC 2003*, pages 100–115. Springer-Verlag, 2003.

[DY05]   Y. Dodis and A. Yampolskiy. A verifiable random function with short proofs and keys. In *Proc. of PKC 2005*, pages 416–431. Springer-Verlag, 2005.

[FN93]   A. Fiat and M. Naor. Broadcast encryption. In *Proc. of Crypto 1993*, volume 773 of *LNCS*, pages 480–491. Springer-Verlag, 1993.

[GSMB03]   E. Goh, H. Shacham, N. Modadugu, and D. Boneh. Sirius: Securing remote untrusted storage. In *Proc. of NDSS 2003*, pages 131–145, 2003.

[GST04]   M.T. Goodrich, J.Z. Sun, and R. Tamassia. Efficient tree-based revocation in groups of low-state devices. In *Proc. of Crypto 2004*, volume 3152 of *LNCS*, pages 511–527. Springer-Verlag, 2004.

[GSW00]   J. Garay, J. Staddon, and A. Wool. Long-lived broadcast encryption. In *Proc. of Crypto 2000*, volume 1880 of *LNCS*, pages 333–352. Springer-Verlag, 2000.

[GSY99]   E. Gafni, J. Staddon, and Y.L. Yin. Efficient methods for integrating traceability and broadcast encryption. In *Proc. of Crypto 1999*, volume 1666 of *LNCS*, pages 372–387. Springer-Verlag, 1999.

[HS02]   D. Halevy and A. Shamir. The lsd broadcast encryption scheme. In *Proc. of Crypto 2002*, volume 2442 of *LNCS*, pages 47–60. Springer-Verlag, 2002.

[JN03]   A. Joux and K. Nguyen. Separating decision Diffie-Hellman from Diffie-Hellman in cryptographic groups. *J. of Cryptology*, 16(4):239–247, 2003. Early version in Cryptology ePrint Archive, Report 2001/003.

[Jou00]   A. Joux. A one round protocol for tripartite Diffie-Hellman. In *Proc. of ANTS IV*, volume 1838 of *LNCS*, pages 385–94. Springer-Verlag, 2000.

[KRS+03]   M. Kallahalla, E. Riedel, R. Swaminathan, Q. Wang, and K. Fu. Plutus: Scalable secure file sharing on untrusted storage. In *Proc. of USENIX Conf. on File and Storage Technologies (FAST)*, 2003.

[NNL01]   D. Naor, M. Naor, and J. Lotspiech. Revocation and tracing schemes for stateless receivers. In *Proc. of Crypto 2001*, volume 2139 of *LNCS*, pages 41–62. Springer-Verlag, 2001.

[NP00]   M. Naor and B. Pinkas. Efficient trace and revoke schemes. In *Proc. of Financial cryptography 2000*, volume 1962 of *LNCS*, pages 1–20. Springer-Verlag, 2000.

[Sho97]   V. Shoup. Lower bounds for discrete logarithms and related problems. In *Proc. of Eurocrypt 1997*, volume 1233 of *LNCS*, pages 256–266. Springer-Verlag, 1997.

[SM03]   A.T. Sherman and D.A. McGrew. Key establishment in large dynamic groups using one-way function trees. *IEEE Trans. Softw. Eng.*, 29(5):444–458, 2003.

[ST98]   D.R. Stinson and T.V. Trung. Some new results on key distribution patterns and broadcast encryption. *Des. Codes Cryptography*, 14(3):261–279, 1998.

[Sti97]   D.R. Stinson. On some methods for unconditionally secure key distribution and broadcast encryption. *Des. Codes Cryptography*, 12(3):215–243, 1997.

[SW98]   D.R. Stinson and R. Wei. Combinatorial properties and constructions of traceability schemes and frameproof codes. *SIAM J. Discret. Math.*, 11(1):41–53, 1998.

[TT01]   W. Tzeng and Z. Tzeng. A public-key traitor tracing scheme with revocation using dynamic shares. In *Proc. of PKC 2001*, pages 207–224. Springer-Verlag, 2001.

[WGL98]   C.K. Wong, M. Gouda, and S. Lam. Secure group communications using key graphs. In *Proc. of SIGCOMM 1998*, 1998.

[WHA97]   D.M. Wallner, E.J. Harder, and R.C. Agee. Key management for multicast: Issues and architectures. IETF draft wallner-key, 1997.

[YJCK04]   E. Yoo, N. Jho, J. Cheon, and M. Kim. Efficient broadcast encryption using multiple interpolation methods. In *Proc. of ICISC 2004*, LNCS. Springer-Verlag, 2004.

Generic Transformation for Scalable Broadcast Encryption Schemes*

Jung Yeon Hwang, Dong Hoon Lee, and Jongin Lim

Graduate School of Information Security CIST,
Korea University, Seoul 136-701, Korea
{videmot, donghlee, jilim}@korea.ac.kr

Abstract. Broadcast encryption schemes allow a message sender to broadcast an encrypted data so that only legitimate receivers decrypt it. Because of the intrinsic nature of one-to-many communication in broadcasting, transmission length may be of major concern. Several broadcast encryption schemes with good transmission overhead have been proposed. But, these broadcast encryption schemes are not practical since they are greatly sacrificing performance of other efficiency parameters to achieve good performance in transmission length.

In this paper we study a generic transformation method which transforms any broadcast encryption scheme to one suited to desired application environments while preserving security. Our transformation reduces computation overhead and/or user storage by slightly increasing transmission overhead of a given broadcast encryption scheme. We provide two transformed instances. The first instance is comparable to the results of the "stratified subset difference (SSD)" technique by Goodrich et al. and firstly achieves $\mathcal{O}(\log n)$ storage, $\mathcal{O}(\log n)$ computation, and $\mathcal{O}(\frac{\log n}{\log \log n} r)$ transmission, at the same time, where n is the number of users and r is the number of revoked users. The second instance outperforms the "one-way chain based broadcast encryption" of Jho et al., which is the best known scheme achieving less than r transmission length with reasonable communication and storage overhead.

1 Intoduction

In recent years broadcast encryption schemes have been intensively studied for lots of applications such as satellite-based commerce, multicast communication, secure distribution of copyright-protected material and DRM(Digital Rights Management), etc. Broadcast encryption (BE) schemes are one-to-many communication methods in which a message sender can broadcast an encrypted data to a group of users over an insecure channel so that only legitimate receivers decrypt it. Especially, a *stateless* BE scheme has a useful property that

---

* This research was supported by the MIC(Ministry of Information and Communication), Korea, under the ITRC(Information Technology Research Center) support program supervised by the IITA(Institute of Information Technology Assessment).

V. Shoup (Ed.): Crypto 2005, LNCS 3621, pp. 276–292, 2005.

any legitimate receiver with its initial set-up can obtain the current group session key only from the current transmission without the history of past transmissions. One of main security concerns in the stateless broadcast encryption schemes is how to efficiently exclude illegal (revoked) users from a privileged set, that is, how to ensure that only legal users decrypt a encrypted broadcast message.

Various BE schemes have been designed to improve efficiency. Efficiency of BE schemes is mainly measured by three parameters: the length of transmission messages, user storage, and computational overhead at a user device. The ultimate goal would be to achieve the best efficiency of all three parameters simultaneously. But it seems, to date, that there exists no BE scheme achieving this goal. As an alternative treatment, a trade-off between the parameters has been considered. In fact, schemes with a various efficiency trade-off fit into many real applications and moreover support the creation of potential application scenarios. Since a message sender in BE schemes broadcasts a message to possible huge number of users, efficiency in transmission overhead has been considered as a critical measure by service providers. Therefore, reducing storage or computation overhead without greatly sacrificing transmission overhead is important.

In most practical applications of BE, a group of users may be quite huge and BE schemes should basically provide scalability, i.e., suitability for a large number of group users. But, unfortunately, most of transmission-efficient BE schemes are not scalable since they requires large storage or computation at a user device. Especially, these schemes are not suitable to wireless networks where users are holding strictly resource-restricted mobile devices.

OUR CONTRIBUTIONS. In the paper we study a modular approach to transform an arbitrary BE scheme to a scalable one efficiently while preserving the security of the underlying scheme. We construct a compiler of which resulting scheme, for a large number of group users, maintains transmission overhead of the original scheme asymptotically but gains reduction in users storage and/or computation overhead. Hence, by applying our compiler to a known transmission-efficient BE scheme which is impractical due to large computation or user storage for keys, we can inexpensively construct an efficient and scalable solution regardless of the structure of the underlying BE scheme.

To illustrate our transformation, we concretely present two compiled instances which provide a good performance in various aspects, in fact, outperform the previously known schemes:

- Goodrich et al. [9] proposed the stratified subset difference (SSD) method, which achieves $\mathcal{O}(r)$ transmission and $\mathcal{O}(n^{\frac{1}{d}})$ computation and $\mathcal{O}(\log n)$ storage overhead per user, where n is the number of users, r is the number of revoked users, and d is a predefined constant. This is the best scheme achieving $\mathcal{O}(r)$ transmission and $\mathcal{O}(\log n)$ storage overhead simultaneously. But under $\mathcal{O}(\log n)$ computation restriction, the scheme needs $\mathcal{O}(\frac{1}{\log(\log n)} \log^2 n)$ storage, which is closer to $\mathcal{O}(\log^2 n)$ storage overhead per user. This should be undesirable in memory-constrained environments. Our first example is a

BE scheme which achieves $\mathcal{O}(\frac{\log n}{\log(\log n)}r)$ transmission, $\mathcal{O}(\log n)$ computation overhead, and $\mathcal{O}(\log n)$ (precisely $\log n + 1$) user storage, at the same time.
- Very recently, Jho et al. [14] proposed the "one-way chain based broadcast encryption schemes" of which one is the best scheme achieving less than r transmission messages with user computation overhead proportional to n at the worst case. But their scheme is still considered non-scalable because of excessive storage requirement, i.e., for a predetermined constant k, $\binom{n-1}{k}$ keys storage at a user device. The second example is a BE scheme in which the number of transmission messages is less than r only except for a small number of revoked users, i.e., 0.75% of n, while user storage and computation overhead are in a reasonable bound.

RELATED WORK. Since the first formal work of BE by Fiat and Naor [8], many researches [12] have been done to improve the efficiency in various aspects by using various trade-off methods and design approaches, i.e., combinatorial designs, logical key trees, algebraic approaches such as secret sharing, multi-linear mapping, and cryptographic tools such one-way accumulator.

Some BE schemes based on combinatorial design are suggested to provide information-theoretical security [10,11,17,18].

Based on a logical key tree structure, a number of broadcast encryption schemes [20,19,1,2,16,13,9] have been suggested. Significant works among them are the Subset Difference (SD) scheme [16] by Naor et al. and its improvement, the layered SD scheme [13] by Halevi and Shamir. These schemes achieve $\mathcal{O}(r)$ transmission complexity while maintaining $\mathcal{O}(\log n)$ computation overhead and $\mathcal{O}(\log^2 n)$ key storage per user. Recently Goodrich et al. [9] firstly proposed the stratified subset difference (SSD) method which satisfies $\mathcal{O}(\log n)$ keys storage per user (this is called the *log-key* restriction) and $\mathcal{O}(r)$ transmission overhead simultaneously but requires $\mathcal{O}(n^{\frac{1}{d}})$ computation overhead where d is a predetermined constant. Their security depends on the existence of pseudo-random sequence number generator.

To achieve more efficient transmission overhead, some schemes have used algebraic properties such as secret-sharing [15,3]. But these schemes have to broadcast at least r transmission messages in order to expose the shares of revoked users. Recently, a notable work based on a one-way accumulator was suggested by Attrapadung et al. to achieve $\mathcal{O}(1)$ transmission complexity [2]. Their method uses a trade-off between security against collusion and non-secret storage size. However, despite of constant transmission complexity, their scheme is considered as impractical in the case of large number of users because of massive requirement in non-secret keys and computation cost at user side. Boneh and Silverberg [6] proposed a zero-message BE scheme which requires only constant amount of non-secret storage by using n-linear maps of which construction seems to be very difficult for $n > 2$. Very recently, Boneh et. al. [5] proposed a (public-key) BE scheme using bilinear maps where transmission length is $\mathcal{O}(\sqrt{n})$, user key storage is a constant size and computation overhead is $\mathcal{O}(\sqrt{n})$. Security of their scheme is based on the so-called *Bilinear Diffie-Hellman Exponent* assumption.

ORGANIZATION. The rest of this paper is organized as follows. We review and define some notions of broadcast encryption in Section 2 and construct our compiler and analyze its efficiency in Section 3. We illustrate two compiled instances of our compiler in Section 4. We compare the resulting schemes with the SD [16], LSD [13], SSD [9], one-way chain based BE [14] schemes in Section 5. Finally, we conclude with some remarks on other issues in Section 6.

2   Broadcast Encryption

In this section we briefly review and define the notion of broadcast encryption. Generally BE schemes are classified into two types: symmetric key and public key based BE schemes. In the symmetric key setting, the only trusted group center GC can generate a broadcast message to users while, in the public key setting, any users are allowed to broadcast a message. We denote by $\mathcal{U}$ the set of users and by $\mathcal{R} \subset \mathcal{U}$ the set of revoked users. The following is a formal definition of a symmetric key based BE scheme.

BROADCAST ENCRYPTION SCHEME. A BE scheme B is a triple of polynomial-time algorithms (SetUp, BEnc, Dec), i.e., setup, broadcast encryption, and decryption:

- SetUp, the randomized algorithm takes as input a security parameter 1^λ and user set $\mathcal{U}$. It generates private information SKEY_u for user $u \in \mathcal{U}$. Private information of group center GC is defined as the set $\mathsf{SKEY}_\mathcal{U}$ of private information of all users.
- BEnc, the randomized algorithm takes as input a security parameter 1^λ, private information $\mathsf{SKEY}_\mathcal{U}$ of GC, a set $\mathcal{R}$ of revoked users, and a message M to be broadcast. It first generates a session key GSK and outputs $(Hdr_\mathcal{R}, C_{\mathsf{GSK},M})$ where a header Hdr is information for a privileged user to compute GSK and $C_{\mathsf{GSK},M}$ is a ciphertext of M encrypted under the symmetric key GSK.

    Broadcast message consists of $[\mathcal{R}, Hdr_\mathcal{R}, C_{\mathsf{GSK},M}]$. The pair $(\mathcal{R}, Hdr_\mathcal{R})$ and $C_{\mathsf{GSK},M}$ are often called the full header and the body, respectively.
- Dec, the (deterministic) algorithm takes as input a user index ind_u, private information SKEY_u of u, the set of revoked users $\mathcal{R}$, and a header $Hdr_\mathcal{R}$. If $u \in \mathcal{U} \backslash \mathcal{R}$ then it outputs the session key GSK.

In public key broadcast encryption, the setup algorithm additionally generates the public keys $\mathsf{PK}_\mathcal{U}$ of users and $\mathsf{PK}_\mathcal{U}$ instead of the private information $\mathsf{SKEY}_\mathcal{U}$ of GC is taken as input in the algorithms BEnc and Dec.

Input terms in the above description may be extended by allowing additional input terms such as a revocation threshold value, i.e., the maximum number of users that can be revoked.

In [16] Naor et. al. presented the so-called *Subset-Cover* framework. The idea of this abstract method is to define a specific subset and associate each subset with a (subset) key SK, which is made available only to the users of the given subset. To cover the set $\mathcal{U} \backslash \mathcal{R}$ of privileged users, $\mathcal{U} \backslash \mathcal{R}$ are partitioned into collection of such pre-defined subsets and the (subset) keys SK_i associated to the subsets are used to encrypt a session key GSK. In this case the header consists of ciphertexts of GSK, i.e., $Hdr_\mathcal{R} = [\mathcal{E}_{SK_1}(\mathsf{GSK}), ..., \mathcal{E}_{SK_t}(\mathsf{GSK})]$ where $\mathcal{E}$ is a symmetric encryption scheme.

EFFICIENCY. Let n and r be the numbers of total users and revoked users for a given BE scheme B, respectively. Efficiency of BE schemes is mainly measured by three parameters: transmission overhead, user storage, and computational overhead.

- $TO_\mathsf{B}(r, n)$: Transmission overhead is defined as the total length (number of bits) of a header in a broadcast message transmitted. We exclude the information of indices of revoked users and the body from the transmission overhead since the information are equivalently needed for all BE schemes.
- $SO_\mathsf{B}(n)$: User storage overhead is defined as the maximum number of private keys initially given to a user.
- $CO_\mathsf{B}(n)$: Computational overhead is defined as the maximum number of basic computation done by a user device.

SECURITY. Basically a BE scheme should provide resiliency to collusion of any set of revoked users. According to the capabilities of an adversary and security goal, we can formally define several types of the security notion of broadcast encryption. Here we briefly present the so-called CCA1-security [4] (chosen ciphertext security in the pre-processing mode [7]) of broadcast encryption, which is believed to be sufficient for most applications. Especially we note that the Subset-Cover framework of [16] in which *computationally independent* keys are used as a message encryption key, is suitable to this notion.

To measure the CCA1-security of a BE scheme B we consider the following game between an adversary $\mathcal{A}$ and a challenger which models adaptive adversarial actions, user corruption and chosen ciphertext attack, etc:

- **Setup.** The challenger runs $\mathsf{SetUp}(1^\lambda, \mathcal{U})$ algorithm and generates private information of users $u \in \mathcal{U}$.
- **Adversarial Action.** $\mathcal{A}$ corrupts any user u' to obtain private information $\mathsf{SKEY}_{u'}$ and asks to any (non-corrupted) user to decrypt a ciphertext C created by $\mathcal{A}$. $\mathcal{A}$ also gets the encryption of a message M selected by itself when it chooses a set $\mathcal{R}$ of revoked users.
- **Challenge.** As a challenge, $\mathcal{A}$ outputs a message CM and a set $\mathcal{R}'$ of revoked users including all ones corrupted by $\mathcal{A}$. The challenger selects a random bit $b \in \{0, 1\}$. If $b=1$ the challenger runs BEnc with $\mathcal{R}'$ to obtain $C=(Hdr_{\mathcal{R}'}, C_{\mathsf{GSK},CM})$. Otherwise it computes $C=(Hdr'_{\mathcal{R}'}, C_{\mathsf{GSK}',RM})$ where RM is a random message whose length is similar to that of the message CM. Then it gives C to $\mathcal{A}$.
- **Guess.** $\mathcal{A}$ outputs its guess $b' \in \{0, 1\}$.

Let CGues denote the event that the adversary correctly guesses the bit b in the above game. The advantage of an adversary $\mathcal{A}$ is defined as $\mathsf{Adv}_{\mathcal{A},\mathsf{B}}(\lambda) = |2 \cdot Pr[\mathsf{CGues}] - 1|$ where $Pr[\mathsf{CGues}]$ is the probability of CGues. We say that a BE scheme B is *CCA1-secure* if for any probabilistic polynomial time adversary $\mathcal{A}$, the advantage $\mathsf{Adv}_{\mathcal{A},\mathsf{B}}(\lambda)$ is negligible.

3   Generic Transformation for Scalable Broadcast Encryption

In this section we present a compiler transforming a broadcast encryption scheme impractical due to computation overhead or user storage for huge number of users to a scalable one. We assume that the number of group users is denoted by $n=w^s$. The variables w and s are to be defined to reduce user storage or computation overhead in advance.

We first provide an overview of our construction intuitively. The main idea of our method is to apply a given broadcast encryption scheme B to a relatively small subset in a hierarchical and independent manner. To implement such a concept, we use a complete w-ary tree with height s, where each user is associated with a leaf. In the tree the root is labeled with a special symbol $b_0=e$. If a node at depth less than s is labeled with β then its b_i-th child is labeled with βb_i where $b_i \in \{1,\ldots,w\}$. That is, $v_{b_0 b_1 \cdots b_{k-1}}$ is a node in level k where $b_0 b_1 \cdots b_{k-1}$ is the concatenation of all indices on the path from the root to the node. Let *sibling set* $S_{b_0 b_1 \cdots b_j}$ be a set of nodes with a same parent $v_{b_0 b_1 \cdots b_j}$ in the tree. The BE scheme B is applied to each sibling set $S_{b_0 b_1 \cdots b_j}$ independently, as if nodes in $S_{b_0 b_1 \cdots b_j}$ are users for an independent BE scheme. To revoke a user, by considering all nodes on the path from the revoked leaf (i.e., user) to the root as revoked nodes, we independently apply the revocation method of B to each sibling set including a node along in the path from the root to the revoked leaf.

3.1   Our Compiler

Given any BE scheme $\mathsf{B} = (\mathsf{SetUp}, \mathsf{BEnc}, \mathsf{Dec})$, our compiler constructs a BE scheme $\overline{\mathsf{B}} = (\overline{\mathsf{SetUp}}, \overline{\mathsf{BEnc}}, \overline{\mathsf{Dec}})$ as follows:

- $\overline{\mathsf{SetUp}}$: For given security parameter 1^λ and a set $\mathcal{U}$ of group users, the algorithm performs the following:
  - First $\overline{\mathsf{SetUp}}$ makes a complete w-ary tree $\mathcal{T}_{|w|}$ in which each leaf is associated to each user. Next, (if necessary) $\overline{\mathsf{SetUp}}$ constructs a user structure for each sibling set in $\mathcal{T}_{|w|}$ according to B.
  - Independently running SetUp of B on each sibling set $S_{b_0 b_1 \cdots b_j}$, $(0 \le j \le s-1)$, $\overline{\mathsf{SetUp}}$ assigns keys to each node (including an interior node). For distinction we denote the BE scheme B and its SetUp applied to $S_{b_0 b_1 \cdots b_j}$ by $\mathsf{B}_{b_0 b_1 \cdots b_j}$ and $\mathsf{B}.\mathsf{SetUp}_{b_0 b_1 \cdots b_j}$, respectively. That is, each node (which is not actually a user in the tree) in $S_{b_0 b_1 \cdots b_j}$ is assigned user keys by $\mathsf{B}_{b_0 b_1 \cdots b_j}$. Let $K_{b_0 b_1 \cdots b_j b_{j+1}}$ be the set of keys assigned to a node $v_{b_0 b_1 \cdots b_j b_{j+1}}$ in $S_{b_0 b_1 \cdots b_j}$. $\overline{\mathsf{SetUp}}$ then provides each leaf $v_{b_0 b_1 \cdots b_s}$ (i.e., user) with a set

$$UK_{v_{b_0 b_1 \cdots b_s}} = K_{b_0} \cup K_{b_0 b_1} \cup \cdots \cup K_{b_0 b_1 \cdots b_s},$$

where K_{b_0} is a singleton set of an initial session key.

- $\overline{\text{BEnc}}$: For given message M and a set $\mathcal{R}$ of r revoke users, it performs the followings to generate a broadcast message: it first makes the Steiner Tree $\mathcal{ST}$ induced by $\mathcal{R}$, that is, the minimal subtree of $\mathcal{T}_{|w|}$ which connects the root of $\mathcal{T}_{|w|}$ to all leaves in $\mathcal{R}$. Starting from $\mathcal{ST}$ as an initial tree, it recursively removes leaves from $\mathcal{ST}$ until $\mathcal{ST}$ becomes a single node.

  1. Find a sibling set S consisting of leaves of $\mathcal{ST}$.
  2. If $|S|=w$, then it removes from $\mathcal{ST}$ all leaves in S and makes their parent node a leaf.
  3. Otherwise, it applies revocation method of BEnc to S and generates ciphertexts of a group session key. Then it removes all leaves in S from $\mathcal{ST}$ and makes their parent node a leaf.

- $\overline{\text{Dec}}$: For given legal user $v_{b_0 b_1 \cdots b_s} \in \mathcal{U} \backslash \mathcal{R}$, it first finds the user's ancestor $v_{b_0 b_1 \cdots b_t}$ in the lowest level such that $v_{b_0 b_1 \cdots b_t c_{t+1} \cdots c_s}$ is a revoked user. To decrypt a group session key, it uses a key assinged to revoke a node $v_{b_0 b_1 \cdots b_t c_{t+1} \cdots c_s}$ from $S_{b_0 b_1 \cdots b_t}$.

As an example shown in Figure 1, we consider a complete 5-ary tree with height 3 for a set of 125 users $\mathcal{U}=\{u_1, \cdots, u_{125}\}$. A leaf v_{e235}, which is associated with user u_{40}, receives a set of keys $UK_{v_{e235}}=K_e \cup K_{e2} \cup K_{e23} \cup K_{e235}$ where K_e is a singleton set of an initial group session key, K_{e2} is a set of keys assigned to a node v_{e2} in sibling set S_{e2} by B.SetUp$_{e2}$, K_{e23} is a set of keys assigned to a node v_{e23} in sibling set S_{e23} by B.SetUp$_{e23}$ and K_{e235} is a set of keys assigned to a node v_{e235} in the sibling set S_{e235} by B.SetUp$_{e235}$, as in Figure 1.

To revoke $\{v_{e125}, v_{e434}\}$, as in Figure 2, consider the minimal subtree $\mathcal{ST}$ which connects the root to the leaves v_{e125} and v_{e434}. Taking all nodes with a same parent in $\mathcal{ST}$ revoked in their sibling set S_α, we apply revocation method of B$_\alpha$ to the sibling set S_α. Revocation methods of B$_{e12}$, B$_{e43}$, B$_{e1}$, B$_{e4}$, B$_e$ are sequentially applied to the sibling sets S_{e12}, S_{e43}, S_{e1}, S_{e4}, S_e in a bottom-up manner, respectively.

In the construction of our compiler, a single broadcast encryption scheme are independently applied to each sibling set in $\mathcal{T}_{|w|}$. But the construction allows

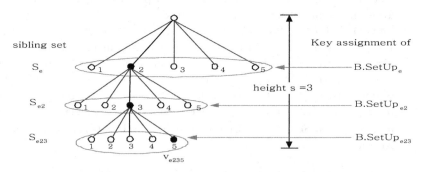

Fig. 1. Key assignment in our compiler : a complete 5-ary tree

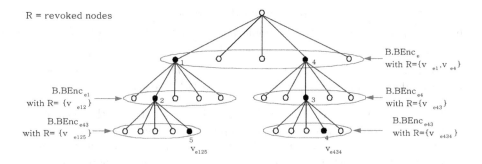

Fig. 2. Revocation in our compiler

that different broadcast encryption schemes are applied to different sibling sets, in order to provide flexibility depending on the resource restriction of client devices. We observe that nodes in the higher level (i.e., closer to the root) become useless more quickly as revoked users are uniformly distributed. Utilizing this observation, we can use a BE scheme assigning less keys per node at a higher level, which will increase the number of transmission messages slightly during initial period. This must be a good trade-off because the initial transmission overhead is relatively small.

Basically the security of our modular method is based on the security of a given BE scheme and the independence usage of the scheme. By using a standard hybrid argument, we can prove the following lemma. The proof will appear in the full version of the paper.

Lemma 1. *The compiled scheme preserves the security of the underlying broadcast encryption scheme.*

3.2   Performance Analysis

We analyze efficiency of the presented compiler with respect to three efficiency parameters: transmission overhead, user storage overhead, computational overhead at a user device.

USER STORAGE OVERHEAD. In a compiled BE scheme, the number of keys that a user should store is $|UK_{v_{b_0 b_1 b_2 \cdots b_s}}| = |K_{b_0}| + |K_{b_0 b_1}| + \cdots + |K_{b_0 b_1 \cdots b_s}| = 1 + s \cdot SO_B(n^{1/s})$. BE schemes satisfying $\mathcal{O}(\log n)$ storage restriction have been considered important [9] since they are well suited to low-memory devices in wireless mobile networks. We note that the compiled BE scheme $\bar{B}$ preserves $\mathcal{O}(\log n)$ key restriction of the underlying BE scheme B. Concretely, $SO_{\bar{B}}(n)$ is $\mathcal{O}(\log n)$ since $1 + s \cdot SO_B(n^{1/s}) \le 1 + s \cdot (c \cdot \log_w n^{\frac{1}{s}} + 1) = 1 + (c+1) \cdot \log_w n \le 1 + (c+1) \cdot \log_2 n$ where c is a constant factor. If storage size in the underlying scheme is less than $\log_w n$ such as a constant then storage size in the compiled scheme increases up to $\log_w n$ which is still satisfying $\mathcal{O}(\log n)$ storage restriction.

COMPUTATION OVERHEAD. In a compiled BE scheme, the maximum number of the basic operations which a user should perform is $CO_B(n^{1/s})(=CO_B(n^{\frac{1}{\log_w n}}))$ since the size of each sibling set at each level is $n^{1/s}$. If $s=\frac{\log_w n}{\log_w(\log_w n)}$ then $n^{1/s}=\log_w n$. If t different BE schemes B_i $(1 \leq i \leq t)$ are used for sibling sets in the setup algorithm then $CO_{\overline{B}}(n)= \text{Max } \{CO_{B_i}(n^{1/s})|1 \leq i \leq t\}$.

TRANSMISSION OVERHEAD. Generally it is not easy to analyze the asymptotic behavior of transmission overhead in compiled BE schemes since BE schemes show various transmission overhead. However we assume that transmission overhead in a given BE scheme is monotone increasing (possibly non-decreasing) as the number of revoked users increases. In this case, transmission overhead $TO_{\overline{B}}(r, n)$ in a compiled BE scheme is upper-bounded by $s \cdot TO_B(r, n^{1/s})$.

In particular, if a given BE scheme satisfies the Subset-Cover framework we can concretely show that $TO_{\overline{B}}(r, n)$ is recursively described as follows:

$$\begin{aligned}
&r(s-i-1)TO_B(1,\omega)+(r \bmod \omega^i)TO_B(1+\lceil\tfrac{r-\omega^i}{\omega^i}\rceil,\omega) && \text{if } \omega^i \leq r < \omega^i\gamma,\\
&\quad + (\omega^i - (r \bmod \omega^i))TO_B(\lceil\tfrac{r-\omega^i}{\omega^i}\rceil,\omega) && \\
&r(s-i-1)TO_B(1,\omega) + (\omega^{i+1}-r) && \text{if } \omega^i\gamma \leq r < \omega^{i+1}.
\end{aligned}$$

where $\omega=n^{1/s}$ and γ is a number such that the maximum number of transmission ciphertexts in B for γ revoked users is $n-\gamma$. The concrete analysis appears in Appendix.

4   Compiled Instances

We apply our compiler to several transmission-efficient schemes, which have inefficiency in computational overhead or user keys storage for huge number of users, to gain scalable and efficient BE schemes. The transformation provides reduction in user storage and/or computation overhead by slightly increasing transmission overhead of a given BE scheme.

4.1   Broadcast Encryption Scheme for User Devices with Low-Resource

In this section we present a BE scheme which achieves $\mathcal{O}(\log n)$ user storage, $\mathcal{O}(\log n)$ computation overhead, and $\mathcal{O}(\frac{\log n}{\log(\log n)}r)$ transmission overhead at the same time. To achieve this goal, we first construct a BE scheme B1 which requires $2r$ transmission messages and only $1+\log_2 n$ key storage per user, but n operations per user. Next, by applying the compiler to B1, we gain the desired scheme.

Broadcast Encryption Scheme B1. As a structure of B1 scheme we consider a segment of the number line $\mathcal{L}$ where numbers are linearly ordered by their magnitude. For any points i and j ($\geq i$), we denote the set $\{k|i \leq k \leq j\}$, called as a closed interval, by $S_{[i,j]}$. For example, $S_{[2,6]}=\{2,3,4,5,6\}$.

We define two *one-way chains*, $\mathcal{C}^+_{[i,j]}$ and $\mathcal{C}^-_{[i,j]}$ associated with $S_{[i,j]}$, and, for a given function $F \colon \{0,1\}^\ell \rightarrow \{0,1\}^\ell$, *chain-values* corresponding to them as follows:

- $\mathcal{C}^+_{[i,j]}$ is a one-way chain such that starts from i and positively goes through $i+1, \cdots, j-1$ and then ends at j. For a given $\mathsf{sd}_i \in \{0,1\}^\ell$, the chain-value of $\mathcal{C}^+_{[i,j]}$ is $F^{|j-i|}(\mathsf{sd}_i)$.
- $\mathcal{C}^-_{[i,j]}$ is a one-way chain such that starts from j and negatively goes through $j-1, \cdots, i+1$ and then ends at i. For a given $\mathsf{sd}_j \in \{0,1\}^\ell$, the chain-value of $\mathcal{C}^-_{[i,j]}$ is $F^{|j-i|}(\mathsf{sd}_j)$.

$F^d(\mathsf{sd})$ is computed by repeatedly applying the function F to sd d times.

SetUp. For a given security parameter 1^λ and a set $\mathcal{U}$ of users, the algorithm SetUp performs the following: First it arranges all users in $\mathcal{U}$ on a segment of the number line $\mathcal{L}$ linearly by the magnitude. A point i in $\mathcal{L}$ is associated with a user u_i. Next, to give a user a set of private keys, it executes the following key assignment.

Starting from $S_{[1,n]}$ as an initial closed interval SetUp performs the following recursively: For a given closed interval $S_{[i,j]}$ for $1 \leq i < j \leq n$, SetUp selects random and independent labels sd_i and sd_j, and assigns these to users u_i and u_j. SetUp computes chain-values by consecutively applying F to labels sd_i and sd_j, respectively. Then SetUp assigns $F^{k-i}(\mathsf{sd}_i)$ and $F^{j-k}(\mathsf{sd}_j)$ to a user u_k. Next SetUp divides the closed interval $S_{[i,j]}$ to get two sub-intervals $S_{[i,m]}$ and $S_{[m+1,j]}$ where $m = \frac{i+j-1}{2}$. While a sub-interval is not a singleton, SetUp applies the above assignment method to the sub-intervals repeatedly. The label sd_i (sd_j), which is assigned to the previous closed interval, is reused in a sub-interval $S_{[i,m]}$ ($S_{[m+1,j]}$) and label sd_m (sd_{m+1}) is newly selected and assigned to a user u_m (u_{m+1}, respectively).

By using the above method, SetUp provides a user with $1+\log_2 n$ keys since $1+\log_2 n$ closed intervals including the user are gained from the above binary division and, for each interval, only one key value is newly assigned to the user.

For an example, for $\mathcal{U} = \{u_1, \cdots, u_{32}\}$, SetUp provides a user u_6 with 6 ($= 1 + \log_2 2^5$) keys, i.e., chain-values $F^5(\mathsf{sd}_1)$, $F^{26}(\mathsf{sd}_{32})$, $F^{10}(\mathsf{sd}_{16})$, $F^2(\mathsf{sd}_8)$, $F(\mathsf{sd}_5)$, and sd_6 associated to 4 closed intervals, $S_{[1,32]}$, $S_{[1,16]}$, $S_{[1,8]}$, $S_{[5,8]}$ and $S_{[5,6]}$, as in Figure 3.

Broadcast Encryption. The revocation method of B1 is based on the following singleton revocation: For a given closed interval $S_{[i,j]}$ of $\mathcal{L}$, to revoke a user u_t, that is, a point $t \in S_{[i,j]}$, the remaining users are covered by two one-way chains $\mathcal{C}^+_{[i,t-1]}$ and $\mathcal{C}^-_{[t+1,j]}$, which proceed from each end point toward opposite directions. The use of these two chains obviously excludes a point t in a subset $S_{[i,j]}$. The keys associated with $\mathcal{C}^+_{[i,t-1]}$ and $\mathcal{C}^-_{[t+1,j]}$ are $F^{t-i}(\mathsf{sd}_i)$ and $F^{j-t}(\mathsf{sd}_j)$, respectively.

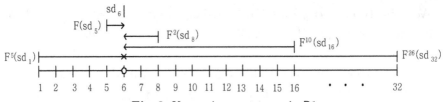

Fig. 3. Key assignment to u_6 in B1

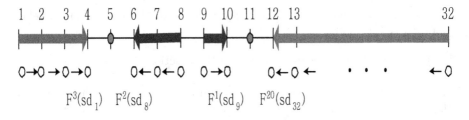

Fig. 4. Revocation in B1

For given r revoked users, BEnc applies the above single revocation method to each disjoint sub-interval including one with one revoked user. In order to apply the method systematically BEnc uses a binary division. That is, for a given set of revoked points $\mathcal{R}=\{u_{i_1},...,u_{i_r}\}$, BEnc finds a division point d_i firstly separating each pair of consecutive revoked nodes u_{i_j} and u_{i_j+1} by performing a binary search on $\mathcal{L}$. BEnc then partitions $\mathcal{L}$ so that $\mathcal{L} = S_{[d_0,d_1]} \cup S_{[d_1+1,d_2]} \cup \cdots \cup S_{[d_{r-1}+1,d_r]}$ where $i_j \in S_{[d_{j-1},d_j]}$, $d_0=1$ and $d_r=n$. Finally BEnc covers each subset by using the above single revocation method.

For example, as shown in Figure 4, for $\mathcal{U}=\{u_1,\cdots,u_{32}\}$ and $\mathcal{R}=\{u_5,u_{11}\}$, the set $\mathcal{U}\backslash\mathcal{R}$ of remaining users is partitioned as follows:

$$\mathcal{L}\backslash\{5,11\}=S_{[1,8]}\cup S_{[9,32]}= (\mathcal{C}^+_{[1,4]}\cup\mathcal{C}^-_{[6,8]}) \cup(\mathcal{C}^+_{[9,10]}\cup\mathcal{C}^-_{[12,32]}).$$

Then four keys $F^3(\mathsf{sd}_1)$, $F^2(\mathsf{sd}_6)$, $F^1(\mathsf{sd}_9)$, and $F^{20}(\mathsf{sd}_{32})$ are assigned to four one-way chains, $\mathcal{C}^+_{[1,4]}$, $\mathcal{C}^-_{[6,8]}$, $\mathcal{C}^+_{[9,10]}$, and $\mathcal{C}^-_{[12,32]}$, respectively.

After construction of cover sets, BEnc applies another one-way function F' to the chain-values and then uses the resulting values as keys to encrypt a group session key.

Decryption. For given legal user $u_k \in \mathcal{U}\backslash\mathcal{R}$, the decryption algorithm Dec first finds two consecutive revoked users u_{i_j} and $u_{i_{j+1}}$ such that $k\in S_{[i_j,i_{j+1}]}$. Next, by using a binary search, Dec finds the division point d which firstly separates two points i_j and i_{j+1}. If $d \geq k$ then it computes $F^{k-i_j}(F^{d-k}(\mathsf{sd}_d))=F^{d-i_j}(\mathsf{sd}_d)$. Otherwise, it computes $F^{i_{j+1}-k}(F^{k-d-1}(\mathsf{sd}_{d+1})) =F^{i_{j+1}-d-1}(\mathsf{sd}_{d+1})$.

SECURITY. We can easily show the correctness of B1 that every privileged user can decrypt an encrypted group session key. Revoked users are excluded by one-wayness of one-way chain and so cannot obtain useful information to decrypt an encrypted group session key. Formally, we show that B1 scheme is resilient to collusion of any set of revoked users by using the following lemma and the similar idea in [16]. In the lemma we assume that F and F' are pseudo random permutations in the sense that no probabilistic polynomial-time adversary can distinguish the output of F (and F') on a randomly selected input from a truly random string of similar length with non-negligible probability.

Lemma 2. *The above key assignment satisfies the key-indistinguishability condition [16] under the pseudo-randomness of given functions F and F'.*

We can prove the lemma by using a hybrid argument on the length of one-way chains, i.e., showing that the gap between true randomness and pseudo-randomness is negligible.

EFFICIENCY. In the presented scheme, at most two ciphertexts of a group session key per revoked user are generated. Hence the number of total ciphertexts consisting of a header Hdr is at most $2 \cdot r$ for r revoked users. But computation overhead is proportional to n.

When we apply the compiler to B1 scheme, in the resulting scheme $\overline{B1}$, the compiled BE scheme $\overline{B1}$ satisfies $\mathcal{O}(\log n)$ key restriction since user keys storage in the original BE scheme B1 is $1 + \log_2 n$. However, we can show that user storage overhead does not change, i.e., $1 + \log_2 n$ since one private node key assigned to the parent node of a given node can be deleted and so $1 - s + s \cdot (\log_2 n^{\frac{1}{s}} + 1) = 1 + \log_2 n$.

Computation overhead is reduced to $\mathcal{O}(n^{\frac{1}{s}})$ for $s = \log_w n$. If we choose the variables $w = n^{\frac{\log_2(\log_2 n)}{\log_2 n}}$ and $s = \frac{\log_2 n}{\log_2(\log_2 n)}$ then we also reduce $\mathcal{O}(n^{\frac{1}{s}})$ computation overhead to $\mathcal{O}(\log n)$. However transmission overhead slightly increases by at most a factor of s from $2r$. More precisely, transmission overhead is described by the recursive formula in Session 3.2 since B1 satisfies the Subset-Cover framework.

REMARK. Based on a similar approach using one-way chains, Goodrich et al. [9] presented the SSD (stratified subset difference) scheme for low-memory devices. But, unlike the work in [9], our method does not use a tree structure. This eliminates the cost for traversing internal nodes in the tree, which causes increase in computation overhead. In addition, with respect to efficiency, the SSD scheme achieves $\mathcal{O}(\log n)$, more precisely $2d \log_2 n$, user storage overhead and $\mathcal{O}(r)$ transmission overhead, but $\mathcal{O}(n^{\frac{1}{d}})$ computation overhead where d is a predetermined constant. When $\mathcal{O}(\log n)$ computation restriction is strictly required, the constant d should be as large as $\frac{\log_2 n}{\log_2(\log_2 n)}$ and user storage overhead also becomes, rather than $\mathcal{O}(\log n)$, closer to $\mathcal{O}(\log^2 n)$, which is relatively heavy and so undesirable in memory-constrained environments.

4.2 Transmission-Efficient Broadcast Encryption Schemes

In this section, to construct a scalable transmission-efficient BE schemes, we further apply our compiler to a previously known transmission-efficient BE scheme, but inefficient in computation cost and user storage size for a huge number of users.

Recently, Jho et al. [14] have presented BE schemes where the number of transmission messages is less than the number of revoked users r, i.e., $\frac{1}{k} \cdot r$ for a predetermined constant k. To bring the number of transmission messages down, they used a fine strategy to cover several subsets of privileged users by using only one key. Their basic scheme requires $\mathcal{O}(n^k)$ user storage overhead and $\mathcal{O}(n)$ computation overhead. To reduce storage and computation overhead further, they presented interval or partition-based construction to deal with relatively small number of users. Unfortunately, in their methods, user storage overhead is still heavy or initial transmission length is relatively large.

By applying our compiler to their schemes, we construct a scalable BE scheme $\overline{B2}$, which has $\frac{1}{2} \cdot r$ transmission messages (except only for a small number of revoked users) with a reasonable user storage and computation overhead. As the underlying schemes for our compiler, we apply two different BE schemes in [14] at different depth of a w-ary tree. One is the BE scheme using simple one-way ring where the number of transmission messages is r. This scheme is applied to every sibling set not in the bottom level. The other is the BE scheme based on a so-called $\mathsf{HOC}(2,[m,2])$, which is a combination of $\mathsf{HOC}(2{:}m)$ with simple hierarchical ring with depth 2 and $\mathsf{OFBE}(m{:}2)$ using 1-jump one-way chain. $\mathsf{HOC}(2,[m,2])$ has $\lfloor \frac{1}{2}r \rfloor + 1$ transmission ciphertexts and relatively low user storage compared to that of the one-way chain-based scheme. This scheme is applied to every sibling set in the bottom level, i.e., sibling set consisting of leaves in the tree. The efficiency for $\overline{B2}$ is as shown in Table 1.

Table 1. Efficiency of $\overline{B2}$

	$TO_{\overline{B2}}(r, n)$	$SO_{\overline{B2}}(n)$	$CO_{\overline{B2}}(n)$
$\overline{B2}$	$\leq \frac{1}{2}r + n^{\frac{s-1}{s}}$	$(s-1)n^{1/s} + \frac{(n^{1/s})^2 - 2n^{1/s} + 24}{8} - s$	$\mathcal{O}(n^{\frac{1}{s}})$

We note that, for $w(=n^{1/s})=100$, $n^{(s-1)/s}$ is less than 1% of n. The compiled scheme $\overline{B2}$ provides similar (or less) transmission overhead, compared to the schemes in [14] while gains reasonably low user storage and computation overhead. For comparison between the schemes, refer to Session 5.

Similarly applying our compiler to other BE schemes such as BE schemes based on a secret sharing [3,15], one-way accmulators [2], or complicated operations etc., gives scalable transformations of these BE schemes under different security assumptions in information theoretical or computational aspects.

5   Efficiency Comparison Between Proposed Schemes

In this section we compare the efficiency between our compiled BE schemes with SD [16], LSD [13], SSD [9], $(1,100)$-π_1 [14] schemes. In the following we assume that the size of a key is 128 bits, which is considered reasonably secure currently.

Table 2. Comparison between $\overline{B1}$, $\overline{B2}$, SD[16], LSD[13], and SSD[9] for $n=10^8$

Scheme	Transmission Overhead	User Storage Overhead	Computation Overhead
SD [16]	$\leq 2r - 1$	368 ($5.74Kbyte$)	27
LSD [13]	$\leq 4r$	143 ($2.24Kbyte$)	27
SSD [9]	$\leq 2sr(s=4)$	213 ($3.33Kbyte$)	100
(1,100)-π_1 [14]	$\leq 2r+0.01n$	5274 ($82.4Kbyte$)	100
$\overline{B1}$	$\leq 2r+0.01n$	27 ($0.422Kbyte$)	100
$\overline{B2}$	$\leq 0.5r+0.01n$	1528 ($23.875Kbyte$)	100

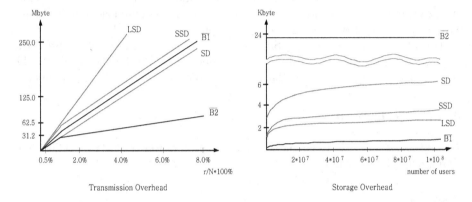

Fig. 5. Transmission and storage overhead for $n=10^8$ for the worst case

The number of computations means the number of basic operations needed to compute a key encrypting a group session key.

For a specific example, we consider the case of $n=10^8$ users and $w=100$. As we show in Figure 5, the number of transmission ciphertexts of $\overline{B2}$ is similar to that of the SD scheme at initial interval where the number r of revoked users is smaller than 0.75 % of the total users. But, except this interval, the number of transmission messages of $\overline{B2}$ becomes, at worst case, about $\frac{1}{4}$ of the number of transmission messages of the SD scheme. The number of keys stored by a user in $\overline{B2}$ is about 4 times as many as that of the SD scheme. But this difference is acceptable in many applications.

In particular, $\overline{B1}$ satisfies log-key restriction strictly, and suitable to low-memory applications where the memory is less 1 Kbyte such as a smart card. This allows a message sender to revoke any r users with transmission overhead being similar to that of the SD scheme [16].

In Table 2. "$\leq$" in the first column means upper-bound of the number of transmission ciphertexts of a group session key. Since the original BE schemes B1 and B2 are defined in Subset-Cover framework transmission overheads in the compiled schemes $\overline{B1}$ and $\overline{B2}$ are described by the recursive formula in Section 3.2. More concretely, if $10^{-4}n \leq r \leq 10^{-2}n$ then $TO_{\overline{B1}}(r,n) \leq 4r+10^{-4}n$ and

$TO_{\overline{B2}}(r, n) \leq r + 10^{-4}n$. Else $10^{-6}n \leq r \leq 10^{-4}n$ then $TO_{\overline{B1}}(r, n) \leq 6r + 10^{-4}n$ and $TO_{\overline{B2}}(r, n) \leq 1.5r + 10^{-4}n$.

6   Conclusion

We have presented a modular method transforming broadcast encryption schemes, which are impractical due to computation complexity or user keys storage for huge number of users, to scalable ones. As concrete examples, we have presented some compiled instances: The first is a BE scheme achieving $\mathcal{O}(\log n)$ user storage, $\mathcal{O}(\log n)$ computation overhead, and $\mathcal{O}(\frac{\log n}{\log \log n} \cdot r)$ transmission overhead at the same time. The second is a transmission-efficient BE scheme with a reasonably low user storage and computation overhead.

For all schemes based on the Subset-Cover framework, our compiler provides a traitor tracing method by using a similar method in [16]. Further study would be a method to apply our modular approach to other traitor tracing methods.

Acknowledgments

The authors would like to thank the anonymous reviewers of CRYPTO 2005 for giving helpful comments.

References

1. T. Asano, *A Revocation Scheme with Minimal Storage at Receivers*, In Advances in Cryptology-Asiacrypt 2002, Springer-Verlag, LNCS vol. 2501, pp.433-450, 2002.
2. N. Attrapadung, K. Kobara and H. Imai, *Broadcast Encryption with Short Keys and Transmissions*, ACM Workshop On Digital Rights Management 2003, pp.55-66, 2003.
3. J. Anzai, N. Matsuzaki and T. Matsumoto, *Quick Group Key Distribution Scheme with Entity Revocation*, In Advances in Cryptology-Asiacrypt 1999, Springer-Verlag, LNCS vol. 1716, pp.333-347, 1999.
4. M. Bellare, A. Desai, E.Jokipii and P. Rogaway, *A Concrete Security Treatment of Symmetric Encryption: Analysis of the DES Modes of Operation*, In Proceedings of the 38th Annual Symposium on Foundations of Computer Science - FOCS'97, pp.394-403, 1997.
5. D. Boneh, C. Gentry and B. Waters, *Collusion Resistant Broadcst Encryption With Short Ciphertexts and Private Keys*, Available from http://eprint.iacr.org 2005. To appear in CRYPTO 2005.
6. D. Boneh and A. Silverberg, *Applications of Multilinear Forms to Cryptography*, Available from http://eprint.iacr.org 2002.
7. D. Dolev, C. Dwork, M. Naor, Pinkas, *Nonmalleable Cryptography,*, SIAM Journal on Discrete Mathematics, 30(2), pp.391-437, 2000.
8. A. Fiat and M. Naor, *Broadcast Encryption*, In Advances in Cryptology-CRYPTO 1994, Springer-Verlag, LNCS vol. 773, pp.480-491, 1994.
9. M. T. Goodrich, J. Z. Sun, and R. Tamassia, *Efficient Tree-Based Revocation in Groups of Low-State Devices*, In Advances in Cryptology-CRYPTO 2004, Springer-Verlag, LNCS vol. 3152, pp.511-527, 2004.

10. J. A. Garay, J. Staddon, and A. Wool, *Long-Lived Broadcast Encryption*, In Advances in Cryptology-CRYPTO 2000, Springer-Verlag, LNCS vol. 1880, pp.333-352, 2000.
11. E. Gafni, J. Staddon, and Y. L. Yin, *Efficient Methods for Integrating Traceability and Broadcast Encryption*, In Advances in Cryptology-CRYPTO 1999, Springer-Verlag, LNCS vol. 1666, pp.372-387, 1999.
12. J. Horwitz, *A Survey of Broadcast Encryption*, Manuscript, 2003.
13. D. Halevy and A. Shamir, *The LSD Broadcast Encryption Scheme*, In Advances in Cryptology-CRYPTO 2002, Springer-Verlag, LNCS vol. 2442, pp.41-60, 2002.
14. N.-S. Jho, J. Y. Hwang, J. H. Cheon, M. Kim, D. H. Lee and E. S. Yoo, *One-way chain Based Broadcast Encryption Scheme*, In Advances in Cryptology-Eurocrypt 2005, Springer, LNCS vol. 3494, pp.559-574, 2005.
15. M. Naor and B. Pinkas, *Efficient Trace and Revoke Scheme*, Financial Cryptography FC 2000, Springer-Verlag, LNCS vol. 1962, pp.1-20, 2000.
16. D. Naor, M. Naor, and J. Lotspiech, *Revocation and Tracing Schemes for Stateless Receivers*, In Advances in Cryptology-CRYPTO 2001, Springer-Verlag, LNCS vol. 2139, pp.41-62, 2001.
17. D. R. Stinson and T. V. Trung, *Some New Results on Key Distribution Patterns and Broadcast Encryption*, Designs, Codes and Cryptography, vol 14., no. 3, pp.261-279, 1998.
18. D. R. Stinson and R. Wei, *Combinatorial Properties and Constructions of Traceability Schemes and Frameproof Codes*, SIAM Journal on Discrete Mathematics, vol 11., no. 1, pp.41-53, 1998.
19. D. M. Wallner, E. G. Harder, and R. C. Agee, *Key Agreement for Multicast :Issues and Architecture*, In internet draft draft-waller-key-arch-01.txt, Sep, 1998.
20. C. K. Wong and S. S. Lam, *Digital Signatures for Flows and Multicasts*, IEEE/ACM Transactions on Networking, vol. 7, no. 4: pp. 502-513, 1999.

A   Analysis of Transmission Efficiency of Our Compiler

Let $\omega=n^{1/s}$ and γ be a number satisfying $TO_B(\gamma,\omega)=\omega-\gamma$. To analyze transmission efficiency, we use the following observations: The worst case occurs when revoked users have the least number of common ancestors. If there is no revoked user, then GC uses an initial group session key to cover all users and hence there is no transmission messages (ciphertext of the group session key). If $r = 1$, we obtain the formula (1) since there is one revoked node in each level and so total s sibling sets to be covered, and $TO_B(1,\omega)$ transmission messages for each sibling set are required. If $2 \leq r < \omega$, then there is r revoked nodes in each level and total $r(s-1)$ sibling sets should be covered. In this case, if $\gamma < r$, then $\omega - r$ transmission messages are transmitted for the first level. Therefore, we obtain the formula (2) and (3) for $2 \leq r < \omega$. If $\omega \leq r < \omega\gamma$, then we do not need to consider nodes in the first level since all nodes in the first level are revoked. In level 2, $(r \mod \omega)$ sibling sets have $1 + \lceil \frac{r-\omega}{\omega} \rceil$ revoked nodes and $\omega - (r \mod \omega)$ sibling sets have $\lceil \frac{r-\omega}{\omega} \rceil$ revoked node. In level j ($3 \leq j \leq s$), $r(s-2)$ messages should be transmitted to cover $r(s-2)$ sibling sets since one revoked node exists in r sibling sets in level j ($1 \leq j \leq s-2$). Hence we obtain the formula (4).

Now we can easily generalize the formula (3) and (4) to the formula (5) and (6), and again the formula (5) and (6) to get the formula (7), (8), (9) inductively.

$$r = 1, \quad sTO_B(1, \omega) \tag{1}$$

$$2 \leq r < \gamma, \quad r(s-1)TO_B(1, \omega) + TO_B(r, \omega) \tag{2}$$

$$\gamma \leq r < \omega, \quad r(s-1)TO_B(1, \omega) + (\omega - r) \tag{3}$$

$$\omega \leq r < \omega\gamma, \quad r(s-2)TO_B(1, \omega) + (r \bmod \omega)TO_B(1 + \lceil \tfrac{r-\omega}{\omega} \rceil, \omega)$$
$$+ (\omega - (r \bmod \omega))TO_B(\lceil \tfrac{r-\omega}{\omega} \rceil, \omega) \tag{4}$$

$$\vdots \qquad\qquad \vdots$$

$$\omega^{i-1}\gamma \leq r < \omega^i, \quad r(s-i)TO_B(1, \omega) + (\omega^i - r) \tag{5}$$

$$\omega^i \leq r < \omega^i\gamma, \quad r(s-i-1)TO_B(1, \omega)$$
$$+ (r \bmod \omega^i)TO_B(1 + \lceil \tfrac{r-\omega^i}{\omega^i} \rceil, \omega)$$
$$+ (\omega^i - (r \bmod \omega^i))TO_B(\lceil \tfrac{r-\omega^i}{\omega^i} \rceil, \omega) \tag{6}$$

$$\vdots \qquad\qquad \vdots$$

$$\omega^{s-2}\gamma \leq r < \omega^{s-1}, \quad rTO_B(1, \omega) + (\omega^{s-1} - r) \tag{7}$$

$$\omega^{s-1} \leq r < \omega^{s-1}\gamma, \quad (r \bmod \omega^{s-1})TO_B(1 + \lceil \tfrac{r-\omega^{s-1}}{\omega^{s-1}} \rceil, \omega)$$
$$+ (\omega^{s-1} - (r \bmod \omega^{s-1}))TO_B(\lceil \tfrac{r-\omega^{s-1}}{\omega^{s-1}} \rceil, \omega) \tag{8}$$

$$\omega^{s-1}\gamma \leq r < n, \quad n - r \tag{9}$$

Authenticating Pervasive Devices with Human Protocols

Ari Juels[1] and Stephen A. Weis[2]

[1] RSA Laboratories, Bedford, MA, USA
ajuels@rsasecurity.com
[2] Massachusetts Institute of Technology, Cambridge, MA, USA
sweis@mit.edu

Abstract. Forgery and counterfeiting are emerging as serious security risks in low-cost pervasive computing devices. These devices lack the computational, storage, power, and communication resources necessary for most cryptographic authentication schemes. Surprisingly, low-cost pervasive devices like Radio Frequency Identification (RFID) tags share similar capabilities with another weak computing device: people.

These similarities motivate the adoption of techniques from human-computer security to the pervasive computing setting. This paper analyzes a particular human-to-computer authentication protocol designed by Hopper and Blum (HB), and shows it to be practical for low-cost pervasive devices. We offer an improved, concrete proof of security for the HB protocol against passive adversaries.

This paper also offers a new, augmented version of the HB protocol, named HB$^+$, that is secure against active adversaries. The HB$^+$ protocol is a novel, symmetric authentication protocol with a simple, low-cost implementation. We prove the security of the HB$^+$ protocol against active adversaries based on the hardness of the Learning Parity with Noise (LPN) problem.

Keywords: Authentication, HumanAut, Learning Parity with Noise (LPN), pervasive computing, RFID.

1   Introduction

As low-cost computing devices become more pervasive, counterfeiting may become a more serious security threat. For example, the security of access control or payment systems will rely on the authenticity of low-cost devices. Yet in many settings, low-cost pervasive devices lack the resources to implement standard cryptographic authentication protocols. Low-cost Radio Frequency Identification (RFID) tags exemplify such resource-constrained devices. Viewing them as possible beneficiaries of our work, we use RFID tags as a basis for our discussions of the issues surrounding low-cost authentication.

Low-cost RFID tags in the form of Electronic Product Codes (EPC) are poised to become the most pervasive device in history [10]. Already, there are

V. Shoup (Ed.): Crypto 2005, LNCS 3621, pp. 293–308, 2005.

billions of RFID tags on the market, used for applications like supply-chain management, inventory monitoring, access control, and payment systems. Proposed as a replacement for the Universal Product Code (UPC) (the barcode found on most consumer items), EPC tags are likely one day to be affixed to everyday consumer products.

Today's generation of basic EPC tags lack the computational resources for strong cryptographic authentication. These tags may only devote hundreds of gates to security operations. Typically, EPC tags will passively harvest power from radio signals emitted by tag readers. This means they have no internal clock, nor can perform any operations independent of a reader.

In principle, standard cryptographic algorithms – asymmetric or symmetric – can support authentication protocols. But implementing an asymmetric cryptosystem like RSA in EPC tags is entirely infeasible. RSA implementations require tens of thousands of gate equivalents. Even the storage for RSA keys would dwarf the memory available on most EPC tags.

Standard symmetric encryption algorithms, like DES or AES, are also too costly for EPC tags. While current EPC tags may have at most 2,000 gate equivalents available for security (and generally much less), common DES implementations require tens of thousands of gates. Although recent light-weight AES implementations require approximately 5,000 gates [11], this is still too expensive for today's EPC tags.

It is easy to brush aside consideration of these resource constraints. One might assume that Moore's Law will eventually enable RFID tags and similar devices to implement standard cryptographic primitives like AES. But there is a countervailing force: Many in the RFID industry believe that pricing pressure and the spread of RFID tags into ever more cost-competitive domains will mean little effective change in tag resources for some time to come, and thus a pressing need for new lightweight primitives.

Contribution. This paper's contribution is a novel, lightweight, symmetric-key authentication protocol that we call HB^+. HB^+ may be appropriate for use in today's generation of EPC tags or other low-cost pervasive devices. We prove the security of this protocol against both passive eavesdroppers and adversaries able to adaptively query legitimate tags. We also offer an improved, concrete security reduction of a prior authentication protocol HB that is based on the same underlying hardness problem.

Organization. In Section 2, we describe the basic "human authentication" or "HumanAut" protocol, due to Hopper and Blum (HB), from which we build an authentication protocol appropriate for RFID tags that is secure against passive eavesdroppers. We discuss the underlying hardness assumption, the "Learning Parity with Noise" (LPN) problem, in Section 3. Section 4 offers our new, enhanced variant of the HB protocol, HB^+, that is secure against adversaries able to query legitimate tags *actively*. Section 5 presents an improved, concrete reduction of the LPN problem to the security of the HB protocol, and shows a concrete reduction of security from the HB protocol to the HB^+ protocol. Finally, Section 6 states several open problems related to this work.

Many details – and our proofs in particular – are omitted from this version of the paper due to lack of space. A more complete version of the paper is available at www.ari-juels.com or crypto.csail.mit.edu/~sweis.

1.1   The Problem of Authentication

It seems inevitable that many applications will come to rely on basic RFID tags or other low-cost devices as authenticators. For example, the United States Food and Drug Administration (FDA) proposed attaching RFID tags to prescription drug containers in an attempt to combat counterfeiting and theft [13].

Other RFID early-adopters include public transit systems and casinos. Several cities around the world use RFID bus and subway fare cards, and casinos are beginning to deploy RFID-tagged gambling chips and integrated gaming tables. Some people have even had basic RFID tags with static identifiers implanted in their bodies as payment devices or medical-record locators [37].

Most RFID devices today promiscuously broadcast a static identifier with no explicit authentication procedure. This allows an attacker to surreptitiously scan identifying data in what is called a *skimming* attack. Besides the implicit threat to privacy, skimmed data may be used to produce cloned tags, exposing several lines of attack.

For example, in a *swapping* attack, a thief skims valid RFID tags attached to products inside a sealed container. The thief then manufactures cloned tags, seals them inside a decoy container (containing, e.g., fraudulent pharmaceuticals), and swaps the decoy container with the original. Thanks to the ability to clone a tag and prepare the decoy in advance, the thief can execute the physical swap very quickly. In the past, corrupt officials have sought to rig elections by conducting this type of attack against sealed ballot boxes [34].

Clones also create denial-of-service issues. If multiple, valid-looking clones appear in a system like a casino, must they be honored as legitimate? Or must they all be rejected as frauds? Cloned tags could be intentionally designed to corrupt supply-chain databases or to interfere with retail shopping systems. Denial of service is especially critical in RFID-based military logistics systems.

Researchers have recently remonstrated practical cloning attacks against real-world RFID devices. Mandel, Roach, and Winstein demonstrated how to read access control proximity card data from a range of several feet and produce low-cost clones [24] (despite the fact that these particular proximity cards only had a legitimate read range of several inches). A team of researchers from Johns Hopkins University and RSA Laboratories recently elaborated attacks against a cryptographically-enabled RFID transponder that is present in millions of payment and automobile immobilization systems [6]. Their attacks involved extraction of secret keys and simulation of target transponders; they demonstrated an existing risk of automobile theft or payment fraud from compromise of RFID systems.

Example EPC Specifications

Storage: 128-512 bits of read-only storage.
Memory: 32-128 bits of volatile read-write memory.
Gate Count: 1000-10000 gates.
Security Gate Count Budget: 200-2000 gates.
Operating Frequency: 868-956 MHz (UHF).
Scanning Range: 3 meters.
Performance: 100 read operations per second.
Clock Cycles per Read: 10,000 clock cycles.
Tag Power Source: Passively powered by Reader via RF signal.
Power Consumption: 10 microwatts.
Features: Anti-Collision Protocol Support
Random Number Generator

Fig. 1. Example specification for a 5-10¢ low-cost RFID tag

1.2   Previous Work on RFID Security

As explained above, securing RFID tags is challenging because of their limited resources and small physical form. Figure 1 offers specifications that might be realistic for a current EPC tag. Such limited power, storage, and circuitry, make it difficult to implement traditional authentication protocols. This problem has been the topic of a growing body of literature.

A number of proposals for authentication protocols in RFID tags rely on the use of symmetric-key primitives. The authors often resort to a hope for enhanced RFID tag functionality in the future, and do not propose use of any particular primitive. We do not survey this literature in any detail here, but refer the reader to, e.g., [15, 29, 32, 33, 36, 39].

Other authors have sought to enforce privacy or authentication in RFID systems while avoiding the need for implementing standard cryptographic primitives on tags, e.g., [12, 18, 19, 21, 20, 29].

Feldhofer, Dominikus, and Wolkerstorfer [11] propose a low-cost AES implementation, potentially useful for higher-cost RFID tags, but still out of reach for basic tags in the foreseeable future.

1.3   Humans vs. RFID Tags

Low-cost RFID tags and other pervasive devices share many limitations with another weak computing device: human beings. The target cost for a EPC-type RFID tag is in the US$0.05-0.10 (5-10¢) range. The limitations imposed at these costs are approximated in Figure 1. We will see that in many ways, the computational capacities of people are similar.

Like people, tags can neither remember long passwords nor keep long calculations in their working memory. Tags may only be able to store a short secret of perhaps 32-128 bits, and be able to persistently store 128-512 bits overall. A working capacity of 32-128 bits of volatile memory is plausible in a low-cost tag, similar to how most human beings can maintain about seven decimal digits in their immediate memory [28].

Neither tags nor humans can efficiently perform lengthy computations. A basic RFID tag may have a total of anywhere from 1000-10000 gates, with only 200-2000 budgeted specifically for security. (Low-cost tags achieve only the lower range of these figures.) As explained above, performing modular arithmetic over large fields or evaluating standardized cryptographic functions like AES is currently not feasible in a low-cost device or for many human beings.

Tags and people each have comparative advantages and disadvantages. Tags are better at performing logical operations like ANDs, ORs and XORs. Tags are also better at picking random values than people – a key property we rely on for the protocols presented here. However, tag secrets can be completely revealed through physical attacks, such as electron microscope probing [1]. In contrast, physically attacking people tends to yield unreliable results.

Because of their similar sets of capabilities, this paper considers adopting human authentication protocols in low-cost pervasive computing devices. The motivating human-computer authentication protocols we consider were designed to allow a person to log onto an untrusted terminal while someone spies over his/her shoulder, without the use of any scratch paper or computational devices. Clearly, a simple password would be immediately revealed to an eavesdropper.

Such protocols are the subject of Carnegie Mellon University's HumanAut project. Earlier work by Matsumoto and Imai [26] and Matsumoto [25] propose human authentication protocols that are good for a small number of authentications [38]. Naor and Pinkas describe a human authentication scheme based on "visual cryptography" [30]. However, this paper will focus primarily on the the human authentication protocols of Hopper and Blum [16, 17].

2   The HB Protocol

We begin by reviewing Hopper and Blum's secure human authentication protocol [16, 17], which we will refer to as the HB protocol. We then place it in the RFID setting. The HB protocol is only secure against passive eavesdroppers – not active attackers. In Section 4, we augment the HB protocol against active adversaries that may initiate their own tag queries.

Suppose Alice and a computing device C share an k-bit secret x, and Alice would like to authenticate herself to C. C selects a random challenge $a \in \{0,1\}^k$ and sends it to Alice. Alice computes the binary inner-product $a \cdot x$, then sends the result back to C. C computes $a \cdot x$, and accepts if Alice's parity bit is correct.

In a single round, someone imitating Alice who does not know the secret x will guess the correct value $a \cdot x$ half the time. By repeating for r rounds, Alice can lower the probability of naïvely guessing the correct parity bits for all r rounds to 2^{-r}.

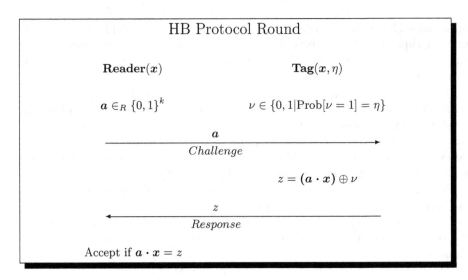

Fig. 2. A single round of the HB authentication protocol

Of course, an eavesdropper capturing $O(k)$ valid challenge-response pairs between Alice and C can quickly calculate the value of x through Gaussian elimination. To prevent revealing x to passive eavesdroppers, Alice can inject noise into her response. Alice intentionally sends the wrong response with constant probability $\eta \in (0, \frac{1}{2})$. C then authenticates Alice's identity if fewer than ηr of her responses are incorrect.

Figure 2 illustrates a round of the HB protocol in the RFID setting. Here, the tag plays the role of the prover (Alice) and the reader of the authenticating device C. Each authentication consists of r rounds, where r is a security parameter.

The HB protocol is very simple to implement in hardware. Computing the binary inner product $a \cdot x$ only requires bitwise AND and XOR operations that can be computed on the fly as each bit of a is received. There is no need to buffer the entire value a. The noise bit ν can be cheaply generated from physical properties like thermal noise, shot noise, diode breakdown noise, metastability, oscillation jitter, or any of a slew of other methods. Only a single random bit value is needed in each round. This can help avoid localized correlation in the random bit stream, as occurs in chaos-based or diode breakdown random number generators.

Remark: The HB protocol can be also deployed as a *privacy-preserving* identification scheme. A reader may initiate queries to a tag without actually knowing whom that tag belongs to. Based on the responses, a reader can check its database of known tag values and see if there are any likely matches. This preserves the privacy of a tag's identity, since an eavesdropper only captures an instance of the LPN problem, which is discussed in the Section 3.

3   Learning Parity in the Presence of Noise

Suppose that an eavesdropper, i.e., a passive adversary, captures q rounds of the HB protocol over several authentications and wishes to impersonate Alice. Consider each challenge a as a row in a matrix A; similarly, let us view Alice's set of responses as a vector z. Given the challenge set A sent to Alice, a natural attack for the adversary is to try to find a vector x' that is functionally close to Alice's secret x. In other words, the adversary might try to compute a x' which, given challenge set A in the HB protocol, yields a set of responses that is close to z. (Ideally, the adversary would like to figure out x itself.)

The goal of the adversary in this case is akin to the core problem on which we base our investigations in this paper. This problem is known as the *Learning Parity in the Presence of Noise*, or LPN problem. The LPN problem involves finding a vector x' such that $|(A \cdot x') \oplus z| \leq \eta q$, where $|v|$ represents the Hamming weight of vector v. Formally, it is as follows:

Definition 1 (LPN Problem). *Let A be a random $q \times k$ binary matrix, let x be a random k-bit vector, let $\eta \in (0, \frac{1}{2})$ be a constant noise parameter, and let ν be a random q-bit vector such that $|\nu| \leq \eta q$. Given A, η, and $z = (A \cdot x) \oplus \nu$, find a k-bit vector x' such that $|(A \cdot x') \oplus z| \leq \eta q$.*

The LPN problem may also be formulated and referred to as as the *Minimum Disagreement Problem* [9], or the problem of finding the closest vector to a random linear error-correcting code; also known as the syndrome decoding problem [2, 23]. Syndrome decoding is the basis of the McEliece public-key cryptosystem [27] and other cryptosystems, e.g., [8, 31]. Algebraic coding theory is also central to Stern's public-key identification scheme [35]. Chabaud offers attacks that, although infeasible, help to establish practical security parameters for error-correcting-code based cryptosystems [7].

The LPN problem is known to be NP-Hard [2], and is hard even within an approximation ratio of two [14]. A longstanding open question is whether this problem is difficult for random instances. A result by Kearns proves that the LPN is not efficiently solvable in the statistical query model [22]. An earlier result by Blum, Furst, Kearns, and Lipton [3] shows that given a random k-bit vector a, an adversary who could weakly predict the value $a \cdot x$ with advantage $\frac{1}{k^c}$ could solve the LPN problem. Hopper and Blum [16, 17] show that the LPN problem is both pseudo-randomizable and log-uniform.

The best known algorithm to solve random LPN instances is due to Blum, Kalai, and Wasserman, and has a subexponential runtime of $2^{O(\frac{k}{\log k})}$ [4]. Based on a concrete analysis of this algorithm, we discuss estimates for lower-bounds on key sizes for the HB and HB$^+$ protocols in the full version of the paper.

As mentioned above, the basic HB protocol is only secure against passive eavesdroppers. It is not secure against an active adversary with the ability to query tags. If the same challenge a is repeated $\Omega((1 - 2\eta)^{-2})$ times, an adversary can learn the error-free value of $a \cdot x$ with very high probability. Given $\Omega(k)$ error-free values, an adversary can quickly compute x through Gaussian elimination.

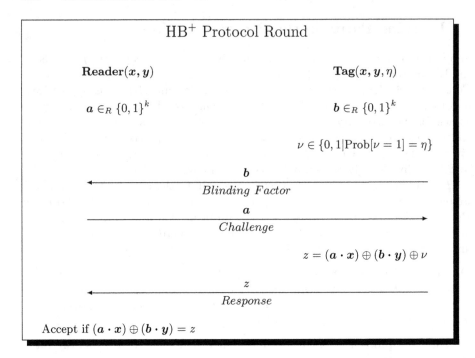

Fig. 3. A single round of the HB$^+$ protocol

4   Authentication Against Active Adversaries

In this section, we show how to strengthen the HB protocol against active adversaries. We refer to the improved protocol as HB$^+$. HB$^+$ prevents corrupt readers from extracting tag secrets through adaptive (non-random) challenges, and thus prevents counterfeit tags from successfully authenticating themselves. Happily, HB$^+$ requires marginally more resources than the "passive" HB protocol in the previous section.

4.1   Defending Against Active Attacks: The HB$^+$ Protocol

The HB$^+$ protocol is quite simple, and shares a familiar "commit, challenge, respond" format with classic protocols like Fiat-Shamir identification. Rather than sharing a single k-bit random secret x, the tag and reader now share an additional k-bit random secret y.

Unlike the case in the HB protocol, the tag in the HB$^+$ protocol first generates random k-bit "blinding" vector b and sends it to the reader. As before, the reader challenges the tag with an k-bit random vector a.

The tag then computes $z = (a \cdot x) \oplus (b \cdot y) \oplus \nu$, and sends the response z to the reader. The reader accepts the round if $z = (a \cdot x) \oplus (b \cdot y)$. As before, the reader authenticates a tag after r rounds if the tag's response is incorrect in less than ηr rounds. This protocol is illustrated in Figure 3.

One reason that Hopper and Blum may not have originally proposed this protocol improvement is that it is inappropriate for use by humans. It requires the tag (playing the role of the human), to generate a random k-bit string b on each query. If the tag (or human) does not generate uniformly distributed b values, it may be possible to extract information on x or y.

To convert HB^+ into a two-round protocol, an intuitive idea would be to have the tag transmit its b vector along with its response bit z. Being able to choose b *after* receiving a, however, may give too much power to an adversarial tag. In particular, our security reduction in Section 5.4 relies on the tag transmitting its b value first. It's an open question whether there exists a secure two-round version of HB^+. Another open question is whether security is preserved if a and b are transmitted simultaneously on a duplex channel.

Beyond the requirements for the HB protocol, HB^+ only requires the generation of k random bits for b and additional storage for an k-bit secret y. As before, computations can be performed bitwise; there is no need for the tag to store the entire vectors a or b. Overall, this protocol is still quite efficient to implement in hardware, software, or perhaps even by a human being with a decent randomness source.

4.2   Security Intuition

As explained above, an active adversary can defeat the basic HB protocol and extract x by making adaptive, non-random a challenges to the tag. In our augmented protocol HB^+, an adversary can still, of course, select a challenges to mount an active attack.

By selecting its own random blinding factor b, however, the tag effectively prevents an adversary from actively extracting x or y with non-random a challenges. Since the secret y is independent of x, we may think of the tag as initiating an independent, interleaved HB protocol with the roles of the participants reversed. In other words, an adversary observing b and $(b \cdot y) \oplus \nu$ should not be able to extract significant information on y.

Recall that the value $(b \cdot y) \oplus \nu$ is XORed with the the output of the original, reader-initiated HB protocol, $a \cdot x$. This prevents an adversary from extracting information through non-random a challenges. Thus, the value $(b \cdot y) \oplus \nu$ effectively "blinds" the value $a \cdot x$ from both passive and active adversaries.

This observation underlies our proof strategy for the security of HB^+. We argue that an adversary able to efficiently learn y can efficiently solve the LPN problem. In particular, an adversary that does not know y cannot guess $b \cdot y$, and therefore cannot learn information about x from a tag response z.

The blinding therefore protects against leaking the secret x in the face of active attacks. Without knowledge of x or y, an adversary cannot create a fake tag that will respond correctly to a challenge a. In other words, cloning will be infeasible. In Section 5, we will present a concrete reduction from the LPN problem to the security of the HB^+ protocol. In other words, an adversary with some significant advantage of impersonating a tag in the HB^+ protocol can be used to solve the LPN problem with some significant advantage.

5   Security Proofs

We will first present concrete security notation in Section 5.1. Section 5.2 reviews key aspects of the Blum et al. proof strategy that reduces the LPN problem to the security of the HB protocol [3]. We offer a more thorough and concrete version of the Blum et al. reduction in Section 5.3. In Section 5.4, we present a concrete reduction from the HB protocol to the HB$^+$ protocol. Finally, in Section 5.5, we combine these results to offer a concrete reduction of the LPN problem to the security of the HB$^+$ protocol.

5.1   Notation and Definitions

We define a tag-authentication system in terms of a pair of probabilistic functions $(\mathcal{R}, \mathcal{T})$, namely a reader function $\mathcal{R}$ and a tag function $\mathcal{T}$.

The tag function $\mathcal{T}$ is defined in terms of a noise parameter η, a k-bit secret $\boldsymbol{x}$, and a set of q random k-bit vectors $\{\boldsymbol{a}^{(i)}\}_{i=1}^{q}$ that we view for convenience as a matrix $\boldsymbol{A}$. Additionally, $\mathcal{T}$ includes a k-bit secret y for protocol HB$^+$. We let q be the maximum number of protocol invocations on $\mathcal{T}$ in our experiments.

For protocol HB, we denote the fully parameterized tag function by $\mathcal{T}_{x,\boldsymbol{A},\eta}$. On the ith invocation of this protocol, $\mathcal{T}$ is presumed to output $(\boldsymbol{a}^{(i)}, (\boldsymbol{a}^{(i)} \cdot \boldsymbol{x}) \oplus \nu)$. Here ν is a bit of noise parameterized by η. This models a passive eavesdropper observing a round of the HB protocol. Note that the oracle $\mathcal{T}_{x,\boldsymbol{A},\eta}$ takes no input and essentially acts as an interface to a flat transcript. For this protocol, the reader $\mathcal{R}_x$ takes as input a pair $(\boldsymbol{a}, z)$. It outputs either "accept" or "reject".

For protocol HB$^+$, we denote a fully parameterized tag function as $\mathcal{T}_{x,y,\eta}$. This oracle internally generates random blinding vectors $\boldsymbol{b}$. On the ith invocation of $\mathcal{T}$ for this protocol, the tag outputs some random $\boldsymbol{b}^{(i)}$, takes a challenge vector $\boldsymbol{a}^{(i)}$ (that could depend on $\boldsymbol{b}^{(i)}$) as input, and outputs $z = (\boldsymbol{a}^{(i)} \cdot \boldsymbol{x}) \oplus (\boldsymbol{b}^{(i)} \cdot \boldsymbol{y}) \oplus \nu$. This models an active adversary querying a tag in a round of the HB$^+$ protocol. For this protocol, the reader $\mathcal{R}_{x,y}$ takes as input a triple $(\boldsymbol{a}, \boldsymbol{b}, z)$ and outputs either "accept" or "reject".

For both protocols HB and HB$^+$, we consider a two-phase attack model involving an adversary comprising a pair of functions $\mathcal{A} = (\mathcal{A}_{query}, \mathcal{A}_{clone})$, a reader $\mathcal{R}$, and a tag $\mathcal{T}$. In the first, "query" phase, the adversarial function $\mathcal{A}_{query}$ has oracle access to $\mathcal{T}$ and outputs some state σ.

The second, "cloning" phase involves the adversarial function $\mathcal{A}_{clone}$. The function $\mathcal{A}_{clone}$ takes as input a state value σ. In HB$^+$, it outputs a blinding factor $\boldsymbol{b}'$ (when given the input command "initiate"). In both HB and HB$^+$, when given the input command "guess", $\mathcal{A}_{clone}$ takes the full experimental state as input, and outputs a response bit z'.

We presume that a protocol invocation takes some fixed amount of time (as would be the case, for example, in an RFID system). We characterize the total protocol time by three parameters: the number of queries to a $\mathcal{T}$ oracle, q; the computational runtime t_1 of $\mathcal{A}_{query}$; and the computational runtime t_2 of $\mathcal{A}_{clone}$. Let D be some distribution of $q \times k$ matrices. We let $\xleftarrow{R}$ denote uniform random assignment. Other notation should be clear from context.

Experiment $\mathbf{Exp}_{A,D}^{HB-attack}[k,\eta,q]$

$\quad x \xleftarrow{R} \{0,1\}^k;$

$\quad A \xleftarrow{R} D$

$\quad \sigma \leftarrow A_{query}^{Tx,A,\eta};$

$\quad a' \xleftarrow{R} \{0,1\}^k;$

$\quad z' \leftarrow A_{clone}(\sigma,a',\text{"guess"});$

$\quad$ Output $\mathcal{R}_x(a',z').$

Experiment $\mathbf{Exp}_A^{HB^+-attack}[k,\eta,q]$

$\quad x,y \xleftarrow{R} \{0,1\}^k;$

$\quad \sigma \leftarrow A_{query}^{Tx,y,\eta};$

$\quad b' \leftarrow A_{clone}(\sigma,\text{"initiate"});$

$\quad a' \xleftarrow{R} \{0,1\}^k;$

$\quad z' \leftarrow A_{clone}(\sigma,a',b',\text{"guess"});$

$\quad$ Output $\mathcal{R}_{x,y}(a',b',z').$

Consider $\mathcal{A}$'s advantage for key-length k, noise parameter η, over q rounds. In the case of the HB-attack experiment, this advantage will be over matrices A drawn from the distribution D:

$$\mathsf{Adv}_{A,D}^{HB-attack}(k,\eta,q) = \left| Pr\left[\mathbf{Exp}_{A,D}^{HB-attack}[k,\eta,q] = \text{"accept"}\right] - \frac{1}{2} \right|$$

Let $Time(t_1,t_2)$ represent the set of all adversaries $\mathcal{A}$ with runtimes t_1 and t_2, respectively. Denote the maximum advantage over $Time(t_1,t_2)$:

$$\mathsf{Adv}_D^{HB-attack}(k,\eta,q,t_1,t_2) = \max_{A \in Time(t_1,t_2)} \{\mathsf{Adv}_{A,D}^{HB-attack}(k,\eta,q)\}$$

The definitions for Adv are exactly analogous for HB^+-attack, except that there is no input distribution D, as adversarial queries are active.

5.2 Blum et al. Proof Strategy Outline

Given an adversary $\mathcal{A}$ that achieves the advantage $\mathsf{Adv}_{A,U}^{HB-attack}(k,q,\eta,t_1,t_2) = \epsilon$, Blum et al. [3] offer a proof strategy to extract bits of x, and thus solve the LPN problem. If ϵ is a non-negligible function of k, then x can be extracted by their reduction in polynomial time.

To extract the ith bit of the secret x, the Blum et al. reduction works as follows. The reduction takes a given LPN instance (A,z) and randomly modifies it to produce a new instance (A',z').

The modification involves two steps. First, a vector x' is chosen uniformly at random and $z' = (z \oplus A) \cdot x' = (A \cdot (x \oplus x')) \oplus \nu$ is computed. Note that thanks to the random selection of x', the vector $(x \oplus x')$ is uniformly distributed. Second, the ith column of A is replaced with random bits. To view this another way, denote the subspace of matrices obtained by uniformly randomizing the ith column of A as R_i^A. The second step of the modification involves setting $A' \xleftarrow{R} R_i^A$. Once computed as described, the modified problem instance (A',z') is fed to an HB adversary $\mathcal{A}_{query}$.

Suppose that the ith bit of $(x \oplus x')$, which we denote $(x \oplus x')_i$, is a binary '1'. In this case, since A is a randomly distributed matrix (because HB challenges are random), and the secret x is also randomly distributed, the bits of z' are random. In other words, thanks to the '1' bit, the randomized ith row of A'

"counts" in the computation of z', which therefore comes out random. Hence z' contains no information about the correct value of $A \cdot (x \oplus x')$ or about the secret x. Since $\mathcal{A}_{query}$ cannot pass any meaningful information in σ to $\mathcal{A}_{clone}$ in this case, $\mathcal{A}_{clone}$ can do no better than random guessing of parity bits, and enjoys no advantage.

In contrast, suppose that $(x \oplus x')_i$, is a binary '0'. In this case, the ith row of A' does not "count" in the computation of z', and does not have a randomizing effect. Hence z' may contain meaningful information about the secret x in this case. As a result, when $\mathcal{A}_{clone}$ shows an advantage over modified problem instances (A', z') for a particular fixed choice of x', it is clear for those instances that $(x \oplus x')_i = 0$, i.e. $x_i = x_i'$.

In summary then, the Blum et al. reduction involves presentation of suitably modified problem instances (A', z') to HB adversary $\mathcal{A}$. By noting choices of x' for which $\mathcal{A}$ demonstrates an advantage, it is possible in principle to learn individual bits of the secret x. With presentation of enough modified problem instances to $\mathcal{A}$, it is possible to learn x completely with high probability.

5.3   Reduction from LPN to HB-Attack

We will show a concrete reduction from the LPN problem to the HB-attack experiment. This is essentially a concrete version of Blum et al.'s asymptotic reduction strategy from [3] and is an important step in proving Theorem 1.

Unfortunately, the original Blum et al. proof strategy does not account for the fact that while $\mathcal{A}$'s advantage may be non-negligible over random matrices, it may actually be negligible over modified (A', z') values, i.e., over the distribution R_i^A. Matrices are not independent over this distribution: Any two sample matrices are identical in all but one column. Thus, it is possible in principle that $\mathcal{A}$ loses its advantage over this distribution of matrices and that the reduction fails to work. This is a problem that we must remedy here.

We address the problem by modifying a given sample matrix only once. A modified matrix A' in our reduction is uniformly distributed. This is because it is chosen uniformly from a random R_i^A subspace associated with a random matrix A. Additionally, since we use a fresh sample for each trial, our modified matrices are necessarily independent of each other. The trade-off is that kL times as many sample matrices are needed for our reduction, where L is the number of trials per bit.

This is an inefficient solution in terms of samples. It is entirely possible that the adversary's advantage is preserved when, for each column j, samples are drawn from the $R_i^{A_j}$ subspace for a matrix A_j. It might even be possible to devise a rigorous reduction that uses a single matrix A for all columns. We leave these as open questions.

Lemma 1. *Let* $\mathsf{Adv}_U^{HB-Attack}(k, \eta, q, t_1, t_2) = \epsilon$, *where U is a uniform distribution over binary matrices $\mathbb{Z}_2^{q \times k}$, and let $\mathcal{A}$ be an adversary that achieves this ϵ-advantage. Then there is an algorithm $\mathcal{A}'$ with running time $t_1' \leq kLt_1$ and $t_2' \leq kLt_2$, where $L = \frac{8(\ln k - \ln \ln k)}{(1 - 2\eta)^2 \epsilon^2}$, that makes $q' \leq kLq + 1$ queries that can correctly extract all k bits of x with probability $\epsilon' \geq \frac{1}{k}$.*

5.4   Reduction from HB-Attack to HB$^+$-Attack

We show that an HB$^+$-attack adversary with ζ-advantage can be used to build an HB-attack adversary with advantage $\frac{\zeta^3}{4} - \frac{\zeta^3+1}{2k}$. We provide concrete costs of this reduction that will be used for Theorem 1.

Lemma 3. *If* $\mathsf{Adv}_U^{HB^+-Attack}(k, \eta, q, t_1, t_2) = \zeta$, *then*

$$\mathsf{Adv}_U^{HB-Attack}(k, \eta, q', t_1', t_2') \geq \frac{\zeta^3}{4} - \frac{\zeta^3+1}{2k},$$

where $q' \leq q(2 + \log_2 q)$, $t_1' \leq kq't_1$, $t_2' \leq 2kt_2$, *and* $k \geq 9$.

(Lemma 2, a technical lemma, is omitted here. For consistency, however, we retain the numbering of lemmas from the full version of this paper.)

Lemma 3 is the main technical core of the paper, but its proof must be omitted here due to lack of space. It is worth briefly explaining the proof intuition. The proof naturally involves a simulation where the HB-attack adversary $\mathcal{A}$ makes calls to the furnished HB$^+$-attack adversary, which we call $\mathcal{A}^+$. In other words, $\mathcal{A}$ simulates the environment for $\mathbf{Exp}_{\mathcal{A}^+}^{HB^+-attack}$. The goal of $\mathcal{A}$ is to use $\mathcal{A}^+$ to compute a correct target response w to an HB challenge vector $\boldsymbol{a}$ that $\mathcal{A}$ itself receives in an experiment $\mathbf{Exp}_{\mathcal{A}}^{HB-attack}$.

$\mathcal{A}$ makes its calls to $\mathcal{A}^+$ in a special way: It "cooks" transcripts obtained from its own HB oracle before passing them to $\mathcal{A}^+$ during its simulation of the query phase of $\mathbf{Exp}^{HB^+-attack}$. The "cooked" transcripts are such that the target value w is embedded implicitly in a secret bit of the simulated HB$^+$ oracle.

In its simulation of the cloning phase of $\mathbf{Exp}^{HB^+-attack}$, the adversary $\mathcal{A}$ extracts the embedded secret bit using a standard cryptographic trick. After $\mathcal{A}^+$ has committed a blinding value $\boldsymbol{b}$, $\mathcal{A}$ rewinds $\mathcal{A}^+$ to as to make two different challenges $\boldsymbol{a}^{(0)}$ and $\boldsymbol{a}^{(1)}$ relative to $\boldsymbol{b}$. By looking at the difference in the responses, $\mathcal{A}$ can extract the embedded secret bit and compute its own target response w.

There are two main technical challenges in the proof. The first is finding the right embedding of w in a secret bit of the simulated HB$^+$-oracle. Indeed, our approach is somewhat surprising. One might intuitively expect $\mathcal{A}$ instead to cause $\mathcal{A}^+$ to emit a *response* equal to w during the simulation; after all, w itself is intended to be a tag response furnished by $\mathcal{A}$, rather than a secret bit. (We could not determine a good way to have w returned as a response.) The second challenge comes in the rewinding and extraction. There is the possibility of a non-uniformity in the responses of $\mathcal{A}^+$. An important technical lemma (Lemma 2) is necessary to bound this non-uniformity. The statement and proof of Lemma 2 are omitted here, but provided in the full version of the paper.

5.5   Reduction of LPN to HB$^+$-Attack

By combining Lemmas 1 and 3, we obtain a concrete reduction of the LPN problem to the HB$^+$-attack experiment. Given an adversary that has an ϵ-advantage against the HB$^+$-attack experiment within a specific amount of time and queries, we can construct an adversary that solves the LPN problem within a concrete upper bound of time and queries. The following theorem follows directly from Lemmas 1 and 3.

Theorem 1. *Let* $\mathsf{Adv}^{HB^+-Attack}(k, \eta, q, t_1, t_2) = \zeta$, *where* U *is a uniform distribution over binary matrices* $\mathbb{Z}_2^{q \times k}$, *and let* $\mathcal{A}$ *be an adversary that achieves this* ζ-*advantage. Then there is an algorithm that can solve a random* $q' \times k$ *instance of the LPN problem in time* (t'_1, t'_2) *with probability* $\frac{1}{k}$, *where* $t'_1 \leq k^2 Lq(2 + \log_2 q)t_1$, $t'_2 \leq 2k^2 Lt_2$, $q' \leq kLq(2 + \log_2 q)$, *and* $L = \frac{128k^4(\ln k - \ln \ln k)}{(1-2\eta)^2(\zeta^3(k-2)+2)^2}$.

To put this in asymptotic terms, the LPN problem may be solved by an adversary where $\mathsf{Adv}^{HB^+-Attack}(k, \eta, q, t_1, t_2) = \zeta$ in time $O(\frac{(k^5 \log k)(q \log q) \ t}{(1-2\eta^2)\zeta^6})$, where $t = t_1 + t_2$.

6   Conclusion and Open Questions

In summary, this paper presents a new authentication protocol named HB^+ that is appropriate for low-cost pervasive computing devices. The HB^+ protocol is secure in the presence of both passive and active adversaries and should be implementable within the tight resource constraints of today's EPC-type RFID tags. The security of the HB^+ protocol is based on the LPN problem, whose hardness over random instances remains an open question.

It is also open question whether the two-round variant of HB^+ that was briefly mentioned in Section 4.1 is secure. The security of concurrent executions of the HB^+ protocol is also unknown. Our security proof uses a rewinding technique that would be take time exponential in the number of concurrent rounds.

Our results here do not offer direct practical guidance for parameterization in real RFID tags. It would be desirable to see a much tighter concrete reduction than we give here. One avenue might be improvement to the Blum et al. reduction. As mentioned in Section 5.3, the efficiency of the modified concrete version of Blum et al. reduction [3] may be improved. Our version uses sample values only once. It may be possible to use a single sample to generate several trials per column, or perhaps to generate trials for every column. This lowers the concrete query costs. It is unclear, however, whether the reduction holds over non-uniform input distributions.

Finally, there is second human authentication protocol by Hopper and Blum, based on the "Sum of k Mins" problem and error-correcting challenges [5, 17]. Unlike the HB protocol, this protocol is already supposed to be secure against active adversaries. However, the hardness of the "Sum of k Mins" has not been studied as much as the LPN problem, nor is it clear whether this protocol can efficiently be adapted for low-cost devices. These remain open avenues of research.

References

[1] ANDERSON, R., AND KUHN, M. Low Cost Attacks on Tamper Resistant Devices. In *International Workshop on Security Protocols* (1997), vol. 1361 of *Lecture Notes in Computer Science*, pp. 125–136.

[2] BERLEKAMP, E. R., MCELIECE, R. J., AND TILBORG, V. On the Inherent Intractability of Certain Coding Problems. *IEEE Transactions on Information Theory 24* (1978), 384–386.

[3] BLUM, A., FURST, M., KEARNS, M., AND LIPTON, R. J. Cryptographic Primitives Based on Hard Learning Problems. In *Advances in Cryptology – CRYPTO'93* (1994), vol. 773 of *Lecture Notes in Computer Science*, pp. 278–291.

[4] BLUM, A., KALAI, A., AND WASSERMAN, H. Noise-Tolerant Learning, the Parity Problem, and the Statistical Query Model. *Journal of the ACM 50*, 4 (July 2003), 506–519.

[5] BLUM, M., LUBY, M., AND RUBINFELD, R. Self-Testing/Correcting with Applications to Numerical Problems. In *Symposium on Theory of Computation* (1990), pp. 73–83.

[6] BONO, S., GREEN, M., STUBBLEFIELD, A., JUELS, A., RUBIN, A., AND SZYDLO, M. Security Analysis of a Cryptographically-Enabled RFID Device. In *USENIX Security* (2005). To appear. Available at http://rfidanalysis.org/.

[7] CHABAUD, F. On the Security of Some Cryptosystems Based on Error-Correcting Codes. In *Advances in Cryptology - EUROCRYPT* (1995), vol. 950 of *Lecture Notes in Computer Science*, pp. 131–139.

[8] COURTOIS, N., FINIASZ, M., AND SENDRIER, N. How to Achieve a McEliece-based Digital Signature Scheme. In *Advances in Cryptology - ASIACRYPT* (2001), vol. 2248 of *Lecture Notes in Computer Science*, pp. 157–174.

[9] CRAWFORD, J. M., KEARNS, M. J., AND SHAPIRE, R. E. The Minimal Disagreement Parity Problem as a Hard Satisfiability Problem. Tech. rep., Computational Intelligence Research Laboratory and AT&T Bell Labs, February 1994.

[10] EPCGLOBAL. Website. http://www.epcglobalinc.org/, 2005.

[11] FELDHOFER, M., DOMINIKUS, S., AND WOLKERSTORFER, J. Strong Authentication for RFID Systems using the AES Algorithm. In *Cryptographic Hardware in Embedded Systems (CHES)* (2004).

[12] FLOERKEMEIER, C., AND LAMPE, M. Issues with RFID Usage in Ubiquitous Computing Applications. In *Pervasive Computing (PERVASIVE)* (2004), vol. 3001 of *Lecture Notes in Computer Science*, pp. 188–193.

[13] FOOD AND DRUG ADMINISTRATION. Combating counterfeit drugs. Tech. rep., US Department of Health and Human Services, Rockville, Maryland, Februrary 2004.

[14] HÅSTAD, J. Some Optimal Inapproximability Results. In *Symposium on Theory of Computing* (1997), pp. 1–10.

[15] HENRICI, D., AND MÜLLER, P. Hash-based Enhancement of Location Privacy for Radio-Frequency Identification Devices using Varying Identifiers. In *Pervasive Computing and Communications (PerCom)* (2004), IEEE Computer Society, pp. 149–153.

[16] HOPPER, N., AND BLUM, M. A Secure Human-Computer Authentication Scheme. Tech. Rep. CMU-CS-00-139, Carnegie Mellon University, 2000.

[17] HOPPER, N. J., AND BLUM, M. Secure Human Identification Protocols. In *Advances in Cryptology - ASIACRYPT* (2001), vol. 2248 of *Lecture Notes in Computer Science*, pp. 52–66.

[18] JUELS, A. Minimalist Cryptography for RFID Tags. In *Security in Communication Networks* (2004), C. Blundo and S. Cimato, Eds., vol. 3352 of *Lecture Notes in Computer Science*, Springer-Verlag, pp. 149–164.

[19] JUELS, A. "Yoking Proofs" for RFID Tags. In *Pervasive Computing and Communications Workshop* (2004), IEEE Press.

[20] JUELS, A., AND PAPPU, R. Squealing Euros: Privacy Protection in RFID-Enabled Banknotes. In *Financial Cryptography* (2003), vol. 2742 of *Lecture Notes in Computer Science*, pp. 103–121.

[21] JUELS, A., RIVEST, R. L., AND SZYDLO, M. The blocker tag: selective blocking of RFID tags for consumer privacy. In *Proceedings of the 10th ACM conference on Computer and communication security* (2003), ACM Press, pp. 103–111.

[22] KEARNS, M. Efficient Noise-Tolerant Learning from Statistical Queries. *Journal of the ACM 45*, 6 (November 1998), 983–1006.

[23] MACWILLIAMS, F., AND SLOANE, N. *The Theory of Error-Correcting Codes*. North-Holland, Amsterdam, 1977.

[24] MANDEL, J., ROACH, A., AND WINSTEIN, K. MIT Proximity Card Vulnerabilities. Tech. rep., Massachusetts Institute of Technology, March 2004.

[25] MATSUMOTO, T. Human-computer Cryptography: An Attempt. In *Computer and Communications Security* (1996), ACM Press, pp. 68–75.

[26] MATSUMOTO, T., AND IMAI, H. Human Identification through Insecure Channel. In *Advances in Cryptology - EUROCRYPT* (1991), vol. 547 of *Lecture Notes in Computer Science*, pp. 409–421.

[27] MCELIECE, R. J. DSN Progress Report. Tech. Rep. 42–44, JPL-Caltech, 1978.

[28] MILLER, G. A. The Magical Number Seven, Plus or Minus Two: Some Limits on Our Capacity for Processing Information. *Psychological Review 63* (1956), 81–97.

[29] MOLNAR, D., AND WAGNER, D. Privacy and Security in Library RFID : Issues, Practices, and Architectures. In *Computer and Communications Security* (2004), B. Pfitzmann and P. McDaniel, Eds., ACM, pp. 210 – 219.

[30] NAOR, M., AND PINKAS, B. Visual Authentication and Identification. In *Advances in Cryptology - CRYPTO* (1997), vol. 1294 of *Lecture Notes in Computer Science*, pp. 322–336.

[31] NIEDERREITER, H. Knapsack-Type Cryptosystems and Algebraic Coding Theory. *Problems of Control and Information Theory 15*, 2 (1986), 159–166.

[32] OHKUBO, M., SUZUKI, K., AND KINOSHITA, S. Efficient Hash-Chain Based RFID Privacy Protection Scheme. In *Ubiquitous Computing (UBICOMP)* (September 2004).

[33] SARMA, S. E., WEIS, S. A., AND ENGELS, D. W. RFID Systems and Security and Privacy Implications. In *Workshop on Cryptographic Hardware and Embedded Systems* (2002), vol. 2523, Lecture Notes in Computer Science, pp. 454–470.

[34] SHAMOS, M. I. Paper v. Electronic Voting Records - An Assessment, 2004. Paper written to accompany panel presentation at Computers, Freedom, and Privacy Conference '04. Available at http://euro.ecom.cmu.edu/people/faculty/mshamos/paper.htm.

[35] STERN, J. A New Paradigm for Public Key Identification. *IEEE Transactions on Information Theory 42*, 6 (1996), 1757–1768.

[36] VAJDA, I., AND BUTTYAN, L. Lightweight Authentication Protocols for Low-Cost RFID Tags. In *Ubiquitious Computing (UBICOMP)* (2003).

[37] VERICHIP. Website. http://www.4verichip.com/, 2005.

[38] WANG, C.-H., HWANG, T., AND TSAI, J.-J. On the Matsumoto and Imai's Human Identification Scheme. In *EuroCrypt '95* (1995), vol. 921 of *Lecture Notes in Computer Science*, pp. 382–392.

[39] WEIS, S. A., SARMA, S. E., RIVEST, R. L., AND ENGELS, D. W. Security and Privacy Aspects of Low-Cost Radio Frequency Identification Systems. In *Security in Pervasive Computing* (2004), vol. 2802 of *Lecture Notes in Computer Science*, pp. 201–212.

Secure Communications over Insecure Channels Based on Short Authenticated Strings

Serge Vaudenay

EPFL CH-1015 Lausanne, Switzerland
http://lasecwww.epfl.ch

Abstract. We propose a way to establish peer-to-peer authenticated communications over an insecure channel by using an extra channel which can authenticate very short strings, e.g. 15 bits. We call this SAS-based authentication as for authentication based on Short Authenticated Strings. The extra channel uses a weak notion of authentication in which strings cannot be forged nor modified, but whose delivery can be maliciously stalled, canceled, or replayed. Our protocol is optimal and relies on an extractable or equivocable commitment scheme.

This approach offers an alternative (or complement) to public-key infrastructures, since we no longer need any central authority, and to password-based authenticated key exchange, since we no longer need to establish a confidential password. It can be used to establish secure associations in ad-hoc networks. Applications could be the authentication of a public key (e.g. for SSH or PGP) by users over the telephone, the user-aided pairing of wireless (e.g. Bluetooth) devices, or the restore of secure associations in a disaster case, namely when one remote peer had his long-term keys corrupted.

1 On Building Secure Communications

One of the key issue of modern cryptography is the problem of establishing a secure peer-to-peer communication over an insecure channel. Assuming that we can establish a private and authenticated key, standard tunneling techniques can achieve it. In the seminal work of Merkle [32] and Diffie and Hellman [18], the private and authenticated key establishment problem was reduced to establishing a communication in which messages are authenticated. Public key cryptosystems such as RSA [39] further reduce to the establishment of an authenticated public key. Note that the seed authentication is also a limiting factor for quantum cryptography [10].

Another major step was the notion of password-based authenticated key agreement which was first proposed by Bellovin and Merritt [8,9] and whose security was proven by Bellare, Pointcheval, and Rogaway [5] in the random oracle model. Another protocol, provably secure in the standard model, was proposed by Katz, Ostrovsky, and Yung [29]. Here, we assume that a private and authenticated short password was set up prior to the protocol. The key agreement protocol is such that no *offline* dictionary attack is feasible against the password so that the threat model restricts to *online* password-guessing attacks which are easily detectable.[1] When compared to the above approach, we thus reduce the size of the initial key, but we require its confidentiality again.

---

[1] See Chapter 7 of [12] for a survey on password-based authenticated key agreement.

V. Shoup (Ed.): Crypto 2005, LNCS 3621, pp. 309–326, 2005.

3-party models offer other solutions. The Needham-Schroeder model [34] assumes that everyone has a private authenticated key with a Trusted Third Party (TTP). Kerberos [30] is a popular application. The authenticated (only) key model is achieved with the notion of certificate by a Certificate Authority (CA). TLS [19] typically uses X.509 [27] certificates. Note that TLS authenticates the server to the client (which is enough to open a secure tunnel), but that the client authentication is typically based on a (short) password through the tunnel. Finally, fully password-based 3-party authenticated key agreement was studied by Abdalla, Fouque, and Pointcheval [3].

Ah-hoc networks cannot assume the availability of a central third party and setting up a secure network is a real challenge. Networks which are not attended by a human operator (e.g. sensor networks) can use a pragmatic solution such as the "resurrecting duckling" paradigm of Stajano and Anderson [40]. Smaller networks which are attended by a human operator such as networks of personal mobile devices (laptops, cell phones, PDAs, headsets, ...) can use the human operator as a third party, but must minimize his job. A familiar example is the Bluetooth [2] pairing: the operator picks a random PIN code and types it on devices to be associated, and a pairing protocol is run through a wireless link to establish a 128-bit private authenticated key. Operator-to-device transmissions is assumed to be secure (i.e. confidential and authenticated). However, as shown by Jakobsson and Wetzel [28], the standard Bluetooth pairing protocol is insecure unless we assume that either the radio communications in the pairing protocol are confidential as well, or the PIN code is long enough.

	long key	short key
A + C channel	symmetric-key cryptography	password-based authenticated key agreement
A channel	public-key cryptography	*SAS-based authentication*

Fig. 1. Two-Party Private and Authenticated Key Establishment Paradigms

Solutions to the secure communications over insecure channels therefore seem to go to two opposite directions (which further translate in a 3-party model): remove the confidential channel (and use public keys) or use short passwords rather than long secret keys. A natural additional step consists of combining the two approaches: using an extra channel which only provides authentication and which is limited to the transmission of short bitstrings. A straightforward solution consists of authenticating every message of a regular key agreement protocol such as the Diffie-Hellman protocol [18] as suggested by Balfanz et al. [4]. The size of messages is typically pretty high, but can be reduced by authenticating only the hashed values of the messages. By using a collision resistant hash function, the number of bits to authenticate typically reduces to 160 bits, but a 160-bit string is still pretty long: by using the encoding rules of the RFC 1760 [23] standard we can represent 160 bits in a human friendly way by using 16 small English words. A second solution by Hoepman [25,26] can significantly reduce this number. It is based on special purpose hash functions. However, the security proof is incomplete and no hash functions with the required properties happen to exist. Another approach by Gehrmann, Mitchell, and Nyberg [21] (dedicated to the Bluetooth pairing problem) called MANA I (as for Manual Authentication), based on a universal hash function family, can perform a message authentication. They however require a stronger notion of authentication channel. Interestingly, those protocols were proposed to replace the current insecure Bluetooth pairing protocol and are now suggesting new solutions in more traditional secure communication standards, e.g. IKEv2 (see Nyberg [35]).

In this paper, we study solutions which can achieve message authentication by using the (weak) authentication of a short bitstring. We call them SAS-based schemes as for "Short Authenticated Strings". A typical application is the pairing problem in wireless networks such as Bluetooth. Another application is secure peer-to-peer communication: if two persons who know each other want to set up a secure communication they can exchange SAS on a postcard, by fax, over a phone call, a voice message, or when they physically encounter.

The other MANA protocols [21,22], as well as the extension of the Hoepman protocols by Peyrin and Vaudenay [38], can be seen as a 3-party translation called the "User-Aided Key Exchange (UAKE)". The user becomes a real participant in the protocol who does simple computations like comparing strings or picking a random one. The security proof in the present paper could equally apply to these cases.

2 Preliminaries

2.1 Communication and Adversarial Models

We consider a communication network with (insecure but cheap) broadband communication channels and narrowband channels which can be used to authenticate short messages. Authenticated channels are related to a node identity ID. An illustration for this model is the location-limited channel of Balfanz et al. [4]. For instance, a user A working on his laptop in an airport lounge would like to print a confidential document on a laser printer B through a wireless link. The user reading a message on the LCD screen of B and typing it on the laptop keyboard is an authenticated channel from the identified printer to the laptop. A SAS-based authentication protocol can be used to transmit and authenticate the public key of the printer by keeping small the transmission over the authenticated channel. Another example is when Bob would like to authenticate the PGP public key of Alice in his key ring. If he can recognize her voice, she can spell a SAS on his voice mail. If he can recognize her signature, she can send a signed SAS to him by fax or even on a postcard.

Adversarial Model. Except for the authentication channels, we assume that the adversary has full control on the communication channels. In particular, she can prevent a message from being delivered, she can delay it, replay it, modify it, change the recipient address, and of course, read it. We adopt here the stronger security model of Bellare-Rogaway [6,7] which even assumes that the adversary has full control on which node launches a new instance of a protocol, and on which protocol instance runs a new step of the protocol. Bellare-Rogaway [6,7] considered protocols for access control or key agreement which basically have no input. Here, protocols do have inputs and we assume that the adversary can choose it. Namely, we assume that the adversary has access to a launch(n, r, x) oracle in which n is a node of the network, r is a character (i.e. a role to play in the protocol), and x is the input of the protocol for this character. This oracle returns a unique instance tag π_n^i. Since a node can a priori run concurrent protocols, there may be several instances related to the same node n. To simplify we restrict ourselves to 2-party protocols so that there are only two characters Alice and Bob in the protocol. Any node can play any of these characters. A q-shot adversary is an adversary limited to q launch$(\cdot, \text{Alice}, \cdot)$ queries and q launch$(\cdot, \text{Bob}, \cdot)$ queries. The adversary also has access to the oracle send(π_n^i, m) which sends a message m to a given instance and returns an m' message which is meant to be sent to the other participant. For example, a protocol with input x and y respectively can be run on node A and B by

1. $\pi_a \leftarrow$ launch(A, Alice, x)
2. $\pi_b \leftarrow$ launch(B, Bob, y)
3. $m_1 \leftarrow$ send$(\pi_a, \emptyset)$
4. $m_2 \leftarrow$ send(π_b, m_1)
5. $m_3 \leftarrow$ send(π_a, m_2)
6. ...

until a message is a termination message. Note that the Bellare-Rogaway model [6,7] considers additional oracles reveal(π_n^i) (which reveals the output from a protocol instance), corrupt(n, x) (which corrupts the collection of instances related to the node n and forces their private states to become x), and test (which is specific to the semantic security of key agreement protocols). These oracles are not relevant here since we never use long-term secrets and the output of the protocols is not secret.

Authentication Channel. The authentication channels provide to the recipient of a message the insurance on whom sent it as is. In particular the adversary cannot modify it (i.e. integrity is implicitly protected). On the other hand she can stall it, remove it, or replay it. We stress that those channels are not assumed to provide confidentiality. Formally, an authentication channel from a node n refers to the identifier ID_n. The send oracle maintains an unordered set of authenticated messages in all authenticated channels from the node n. Only send oracles with a π_n^i instance can insert a new message x in this set. Later, when a send oracle is queried with any instance tag and a message of form authenticate$_{\text{ID}_n}(x)$, it is accepted by the oracle only if x is in the set. Note that concurrent or successive instances related to the same node write in the same set: messages from the node are authenticated, but the connection to the right instance is not. Authenticated channels can typically be implemented, e.g. by human telephone conversations, voice mail messages, handwritten postcards, etc.

Stronger Authentication Channel. We can also consider *strong authentication channels*, namely authenticated channels which provide an additional security property. We list here a few possible properties in an informal way.

Stall-free transmission: from the time an authenticated message is released by a send oracle to the time it is given as input to a send oracle query, no other oracle query can be made. Hence, either the message is treated by the immediately following oracle query, or it is never used.

Transmission with acknowledgment: messages are released together with a destination node identifier, and the sending instance is given a way to check whether at least one instance related to this node has received the message or not.

Listener-ready transmission: similarly, the sending instance can check if an instance related to the destination node is currently listening to the authenticated channel.

Face-to-face conversations achieve all properties. Telephone conversations achieve the last two properties: Alice starts talking to Bob when she is aware that Bob is listening, and subtle human senses assure her that Bob has heard her message. Less interactive communications such as voice mail messages do not provide these properties: Bob may not even be aware that Alice wants to send him a message, and Alice has neither way to know when Bob has received it, nor insurance that her message was recorded.

Message Authentication Protocol. Our message authentication protocols have input m on the side of the claimant Alice and output $I||\hat{m}$ on the side of the verifier Bob. Intuitively, they should be such that $I = \mathsf{ID}_A$ and $\hat{m} = m$, meaning that $\hat{m}$ coming to node B was authenticated as sent by ID_A, the identifier of Alice.

On a global perspective, several $\mathsf{launch}(A_k, \mathrm{Alice}, m_k)$ and $\mathsf{launch}(B_\ell, \mathrm{Bob}, \emptyset)$ are queried, which creates several $\pi_{A_k}^{i_k}$ instances of Alice (authentication claims) and several $\pi_{B_\ell}^{j_\ell}$ instances of Bob (authentication verifications). If no attack occurs then we have a perfect matching between the k's and ℓ's, related instances have matching conversations which fully follow the protocol specifications, and the $\pi_{B_\ell}^{j_\ell}$ ends with output $\mathsf{ID}_{A_k}||m_k$ for the matching k. In any other case we say that an attack occurred. We say that an attack is successful if there exists at least an instance $\pi_{B_\ell}^{j_\ell}$ which terminated and output $I||\hat{m}$ such that there is no k for which $I = \mathsf{ID}_{A_k}$ and $\hat{m} = m_k$. Note that many protocol instances can endlessly stay in an unterminated state or turn in an abort state. In particular, we do not consider denial-of-services attacks.

2.2 Commitment Schemes

Our protocols are based on commitment schemes. They are used to commit on an arbitrary non-hidden message m together with a hidden k-bit string r. We formalize them by three algorithms.

setup which generates a random parameter K_P (which is used by all other algorithms and omitted from notations for simplicity reasons) and a secret key K_S.

commit(m, r) which takes a message $x = m||r$ and produces two strings: a *commit* value c and a *decommit* value d. Here, we consider that x includes a part m which is not meant to be hidden and a part r which is a hidden k-bit string. We can call m a tag for the commitment so that we have a *tag-based commitment* to r. Note that this algorithm is typically non deterministic.

open(m, c, d) which takes m, c, and d and yields a message r or an error signal. We require this algorithm to be *deterministic* and to be such that whenever there exists r such that (c, d) is a possible output for commit(m, r), open(m, c, d) yields r.

Note that the setup plays no real role so far. It is used in extensions of commitment schemes. We keep it anyway to have definitions well suited to all kinds of commitment schemes that will be used. Commitment schemes have two security properties.

- (T, ε)-*hiding*: no algorithm $\mathcal{A}$ bounded by a time complexity T can win the following game by interacting with a challenger C with a probability higher than $2^{-k} + \varepsilon$.
    1. C runs setup and sends K_P to $\mathcal{A}$.
    2. $\mathcal{A}$ selects a tag m and sends it to C.
    3. C picks a random r, runs commit on (m, r), gets (c, d), and sends c to $\mathcal{A}$.
    4. $\mathcal{A}$ yields r' and wins if $r = r'$.
    When $T = +\infty$ and $\varepsilon = 0$, we say that the scheme is *perfectly hiding*.
- (T, ε)-*binding*: no algorithm $\mathcal{A}$ bounded by a time complexity T can win the following game by interacting with a challenger C with a probability higher than $2^{-k} + \varepsilon$.
    1. C runs setup and sends K_P to $\mathcal{A}$.
    2. $\mathcal{A}$ selects a tag m and sends it to C.
    3. $\mathcal{A}$ selects a c and sends it to C.
    4. C picks a random r and sends it to $\mathcal{A}$.
    5. $\mathcal{A}$ computes a d and wins if (m, c, d) opens to r.
    When $T = +\infty$ and $\varepsilon = 0$, we say that the scheme is *perfectly binding*.

Commitment schemes can be relative to an oracle, in which case all algorithms and adversaries have access to the oracle. However, they have no access to the complete history of oracle calls. Extensions of commitment schemes have extra algorithms which do have access to this history.

Extractable Commitment. In this extension of commitment schemes, there is an additional *deterministic* algorithm $\text{extract}_{K_S}(m, c)$ which yields r when there exists d such that (m, c, d) opens to r. When using oracles, this algorithm is given the history of oracle queries. Clearly, extractable commitments are perfectly binding. Adversaries playing the hiding game can make oracle calls to extract, except on the committed m tag.

Equivocable Commitment. In this extension of commitment schemes, there are two algorithms $\text{simcommit}_{K_S}(m)$ and $\text{equivocate}_{K_S}(m, c, r, \xi)$. simcommit returns a fake commit value c and an information ξ, and equivocate returns a decommit value d such that (m, c, d) opens to an arbitrary r for (c, ξ) obtained from simcommit. For any $K_P \| K_S$ and any m, the distribution of fake commit values is assumed to be identical to the distribution of real commit values to any r with tag m. From this we deduce that the commitment is perfectly hiding. Adversaries playing the binding game can make oracle calls to simcommit and equivocate, except on the committed m tag, and are assumed not to see ξ. Namely, the equivocate oracle works only if there was a matching oracle call to simcommit before, and gets ξ directly from the history. In our paper, we further assume that adversaries are limited to a single query to simcommit and equivocate. This is a quite restrictive assumption, but it will be enough for our purpose.

Example 1 (Ideal commitment model). A first commitment scheme model which can be used is the *ideal commitment model*. Here, we assume that the network includes a trusted third party (TTP) with whom anyone can communicate in a perfectly secure way. The commit algorithm consists of securely sending a message x to the TTP. The TTP attaches it to a nonce value c which is returned to the sender and inserts (c, x) in an archive with a protection flag set. There is no decommit value, but the same sender can replace the disclosure of it by a decommit call to the TTP. Then, the TTP clears the protection flag of the (c,x) entry which becomes readable by anyone. Obviously we obtain a perfectly binding and hiding scheme. Note that it is extractable and equivocable.

Example 2 (Extractable commitment based on a random oracle). We take an easy commitment scheme which was already mentioned in Pass [37]. Here, we use a random oracle H which upon a query d returns a random value $c \leftarrow H(d)$ in $\{0,1\}^{\ell_c}$. The commit(m,r) algorithm simply picks a random value e in $\{0,1\}^{\ell_e}$, takes $d = r||e$, and calls $c \leftarrow H(m||d)$. The open(m,c,d) algorithm simply extracts r from d and checks that $c = H(m||d)$. When all oracle queries to H produce no collision on $m||c$, commitments can trivially be extracted form the history. When the number of queries is q, this is the case, except with probability less than $q^2.2^{-\ell_c-1}$. We prove that the best strategy for an adversary to play the hiding game is to look for $r||e$ exhaustively by trying it with H. Actually, if for each r the adversary has queried q_r values $H(m||r||e)$ on the committed tag m, the probability of success when answering r is one if the right $r||e$ was found, which happens with probability $2^{-\ell_e-k} \sum_s q_s$, and $(2^{\ell_e} - q_r)/(2^{\ell_e+k} - \sum_s q_s)$ in the other case. Hence, the overall probability of success is $2^{-k} + 2^{-\ell_e-k}(\sum_s q_s - q_r)$ which is at most $2^{-k} + q.2^{-\ell_e-k}$. Hence, when H is limited to q accesses, the commitment scheme is $(+\infty, q.2^{-\ell_e-k})$-extractable with probability at least $1 - q^2.2^{-\ell_c-1}$. So with $\ell_c = 2\ell_e$ and $\ell_e = 80$, the scheme is pretty safe until the complexity reaches a number of oracle calls within the order of magnitude of 2^{80-k}.

Example 3 (Equivocable commitment based on a signature scheme). One can easily prove (see Appendix B) that the notion $\mathsf{SSTC}(2)$ of *simulation-sound trapdoor commitment* as defined in MacKenzie-Yang [31], where the adversary is given two equivocate oracle calls, provides equivocable commitments following our definition. Hence, from [31] we get a nice equivocable commitment scheme based on DSA [1] and another one based on the Cramer-Shoup signature scheme [14]. We can also use the stronger notion of non-malleable commitments [15,16,20], and in particular the Damgård-Groth commitment scheme [17] based on the strong RSA assumption in the common reference string (CRS) model.

Introducing a public key may look paradoxical since the purpose of our work is to get rid of any a priori authenticated public key. This is the puzzling aspect of the CRS model: we assume usage of a public key for which a secret key exists, but that nobody can use it. We can even rely on the uniform random string (URS) model (see [17]) in which the public key is a uniformly distributed reference bitstring.

2.3 Previous Work

The Gehrmann-Mitchell-Nyberg MANA I [21] protocol is depicted in Fig. 2.[2] By convention we put a hat on received messages which are not authenticated since they can differ from sent messages in the case of an active attack. Other MANA protocols are designed for two devices attended by a user who can do simple operations. MANA I is the only protocol in which the user is a passive messenger who only forwards messages.

---

[2] Note that the original MANA I protocol is followed by an authenticated acknowledgment.

MANA I uses a universal hash function family H. Proposed constructions lead to 16–20 bit long SAS values. Although MANA I is essentially non-interactive, the security requires a stronger authentication channel. Otherwise, one can run the following attack.

1. $\pi_a \leftarrow \text{launch}(A, \text{Alice}, m)$
2. $\pi_b \leftarrow \text{launch}(B, \text{Bob}, \emptyset)$
3. $m \leftarrow \text{send}(\pi_a, \emptyset)$
4. $\text{authenticate}_{\text{Alice}}(K\|\mu) \leftarrow \text{send}(\pi_a, \emptyset)$
5. find $\hat{m} \neq m$ such that $H_K(\hat{m}) = \mu$ by random search
6. $\text{send}(\pi_b, \hat{m})$
7. $\text{send}(\pi_b, \text{authenticate}_{\text{Alice}}(K\|\mu))$

MANA I is nevertheless secure when using an authentication channel which provides stall-free transmission or listener-ready transmission as defined in Section 2.1.

Fig. 2. The MANA I Protocol

The Hoepman authenticated key agreement protocol [25] is depicted in Fig. 3. It consists of a commitment exchange and an authentication exchange, followed by a regular Diffie-Hellman protocol [18].[3] The protocol is based on the decisional Diffie-Hellman problem in a group G. It works with the hypothesis that H_1 and H_2 are two hash functions such that H_2 is balanced from G to $\{0,1\}^k$, and given a uniformly distributed X in G, the two random variables $H_1(X)$ and $H_2(X)$ are independent. Although no example which meet these criteria is provided in [25][4], Hoepman provided a sketch of security proof for the complete protocol[5] in the Bellare-Rogaway model.

3   Non-interactive Message Authentication

We first present a solution based on a collision resistant hash function inspired by Balfanz et al. [4]. Since the result is quite straightforward, the proof is omitted.

---

[3] Note that first committing to the Diffie-Hellman values was already suggested by Mitchell-Ward-Wilson [33].

[4] One can note that the criterion on H_2 seemingly suggests that the order of G should be a multiple of 2^k which is not the case in classical Diffie-Hellman groups so (H_1, H_2) instances may not exist at all.

[5] Fig. 3 presents a simplified version of the protocol. The complete protocol is followed by a key confirmation and a key derivation based on the leftover hash lemma [24] (see Boneh [11]).

Alice   Bob
pick $x_A \in_U \{0, \ldots, |G| - 1\}$ pick $x_B \in_U \{0, \ldots, |G| - 1\}$
$\quad y_A \leftarrow g^{x_A}$ $y_B \leftarrow g^{x_B}$

$h_A \leftarrow H_1(y_A)$ $\xrightarrow{\quad h_A \quad}$

$\xleftarrow{\quad h_B \quad}$ $h_B \leftarrow H_1(y_B)$

$\text{SAS}_A \leftarrow H_2(y_A)$ $\xrightarrow{\text{authenticate}_{\text{Alice}}(\text{SAS}_A)}$

$\xleftarrow{\text{authenticate}_{\text{Bob}}(\text{SAS}_B)}$ $\text{SAS}_B \leftarrow H_2(y_B)$

$\xrightarrow{\quad y_A \quad}$ check $\hat{h}_A, \text{SAS}_A$

check $\hat{h}_B, \text{SAS}_B$ $\xleftarrow{\quad y_B \quad}$

$\quad k_A \leftarrow (\hat{y}_B)^{x_A}$ $k_B \leftarrow (\hat{y}_A)^{x_B}$

output: Bob, k_A **output: Alice**, k_B

Fig. 3. The Hoepman Authenticated Key Agreement Protocol

Fig. 4. Message Authentication using a Collision Resistant Hash Function

Theorem 4. *Let μ be the overall time complexity of the message authentication proto-col in Fig. 4. Given an adversary of time complexity T, number of launch oracle queries Q, and probability of success p, we can find collisions on H within a complexity $T + \mu Q$ and same probability.*

One advantage of this protocol is that it is non-interactive. Collision resistance requires the number of authenticated bits to be at least 160 which is still quite large. We can actually half this number by using the Pasini-Vaudenay protocol [36] based on a hash function resisting to second preimage attacks (a.k.a. weakly collision resistant hash function).[6] Note that the MANA I protocol requires less bits, but a stronger authentica-tion channel which renders the protocol "less non-interactive".

4   SAS-Based Message Authentication

In Fig. 5 is depicted a SAS-based message authentication protocol. Basically, Alice first commits on her (non-hidden) input message m together with a hidden random string R_A. After reception of m and the commit value c, Bob picks a random string R_B and gives it to Alice. Alice then opens her commitment by sending a d and sends $\text{SAS} = R_A \oplus R_B$ through her authenticated channel. Bob can finally check the consistency of this string

---

[6] This protocol consists of sending $m||c||d||\text{authenticate}(H(c))$ where (c, d) is obtained from m by using a trapdoor commitment.

with the commitment. This protocol can be used to authenticate in both directions and henceforth for authenticated key agreement as detailed in Appendix A.

With weak authentication channels, the adversary can run as follows:

- impersonate Bob and start the protocol with Alice with $m, c, \hat{R}_B, d$,
- stall the SAS message,
- launch several Bob's and impersonate Alice with $\hat{m}, \hat{c}, \hat{d}$ until Alice's SAS matches,
- deliver the SAS and complete the protocol.

This attack works within a number of trials around 2^k. Note that the attack is not discrete on Bob's side since many protocols abort. Similarly, the adversary can launch many instances of Alice and make a catalog of Alice's SAS messages. After Alice has performed quite a lot of protocols, the catalog can be close to the complete catalog of 2^k messages. With this collection, the adversary can impersonate Alice. Note that the attack is pretty discrete here. We can further trade the number of instances of Bob against the number of instances of Alice and have a birthday paradox effect. Namely, with $2^{\frac{k}{2}}$ concurrent runs of Alice and Bob we have fair chances of success.

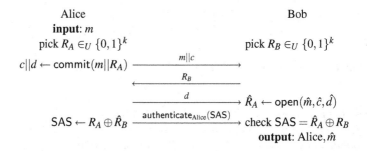

Fig. 5. SAS-based Message Authentication

Theorem 5. *We consider one-shot adversaries against the message authentication protocol in Fig. 5. We denote by T and p their time complexity and probability of success, respectively. We assume that the commitment scheme is either (T_C, ε)-extractable or (T_C, ε)-equivocable. There exists a (small) constant μ such that for any adversary, we have either $p \leq 2^{-k} + \varepsilon$ or $T \geq T_C - \mu$.*

Our results seemingly suggest that for any secure commitment scheme the success probability of practical one-shot attacks is bounded by $2^{-k} + \varepsilon$ where ε is negligible. Our results are pretty tight since adversaries with a probability of success 2^{-k} clearly exist.

Proof. Due to the protocol specifications, a successful adversary must perform the following sequences of steps to interact with Alice and Bob. The way the adversary interleaves the two sequences is free.

1. select m, $\pi_a \leftarrow \text{launch}(\cdot, \text{Alice}, m)$ 1. $\pi_b \leftarrow \text{launch}(\cdot, \text{Bob}, \emptyset)$
2. $c \leftarrow \text{send}(\pi_a, \emptyset)$ 2. select $\hat{m} || \hat{c}$, $R_B \leftarrow \text{send}(\pi_b, \hat{m} || \hat{c})$
3. select $\hat{R}_B$, $d \leftarrow \text{send}(\pi_a, \hat{R}_B)$ 3. select $\hat{d}$, $\text{send}(\pi_b, \hat{d})$

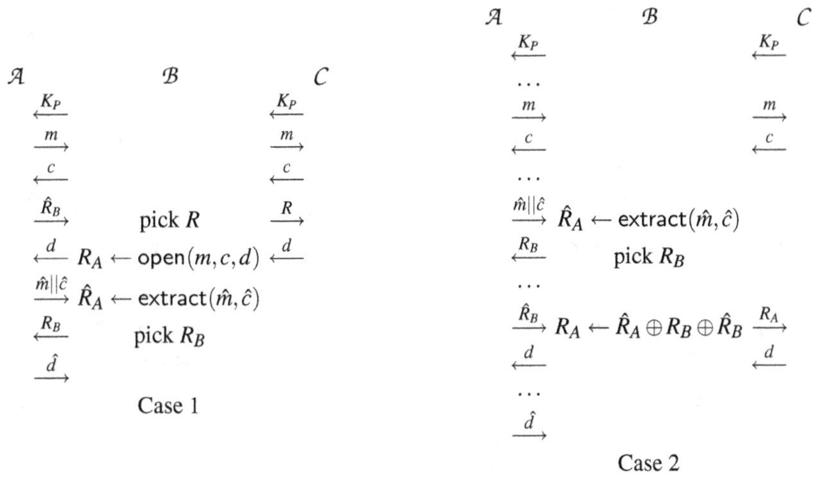

Fig. 6. Reduction with Extractable Commitments

The attack runs like a game for the adversary $\mathcal{A}$ who wins if (m,c,d) (resp. $(\hat{m},\hat{c},\hat{d})$) opens to R_A (resp. $\hat{R}_A$) such that $R_A \oplus \hat{R}_B = \hat{R}_A \oplus R_B$. The game starts by receiving the selected public parameter K_P (if any) for the commitment scheme. We can make $\mathcal{A}$ play with a simulator for Alice and Bob. Note that an attack implies that $m \neq \hat{m}$.

Extractable Commitments. We construct from $\mathcal{A}$ an adversary $\mathcal{B}$ who plays an augmented hiding game with the help of one query to the extract oracle. The augmented hiding game consists of the regular one followed by C sending the right decommit value d. Obviously, an adversary playing this augmented game can be transformed into an adversary playing the regular game with the same winning probability. As depicted in Fig. 6, $\mathcal{B}$ first receives K_P and sends it to $\mathcal{A}$. Then, he simulates Alice and Bob to $\mathcal{A}$. Whenever $\mathcal{A}$ sends m to Alice, $\mathcal{B}$ submits it in the augmented hiding game. Then, $\mathcal{B}$ receives a challenge c which is sent back to $\mathcal{A}$ (i.e., Alice does not compute any commitment but rather uses the challenge). Whenever $\mathcal{A}$ sends $\hat{m}\|\hat{c}$ to Bob, $\mathcal{B}$ can extract $\hat{R}_A$ by calling extract$(\hat{m},\hat{c})$. When $\mathcal{A}$ sends $\hat{R}_B$ to Alice, we distinguish two cases.

Case 1. If $\mathcal{A}$ did not send $\hat{m}\|\hat{c}$ to Bob yet, there is essentially one way to interleave the two sequences which consists in first playing with Alice, then playing with Bob. Here, $\mathcal{B}$ answers a random R to the challenge and wins with probability 2^{-k}. $\mathcal{B}$ continues the simulation and plays with $\mathcal{A}$ by receiving d and sending it from Alice to $\mathcal{A}$. Then, $\mathcal{A}$ sends $\hat{m}\|\hat{c}$ to Bob from which $\mathcal{B}$ extracts $\hat{R}_A$. Bob's simulation picks a random R_B, but R_A, $\hat{R}_A$, and $\hat{R}_B$ are fixed. So, $\mathcal{A}$ wins with probability 2^{-k}. Hence, $\mathcal{A}$ and $\mathcal{B}$ win their respective game with the same probability in this case.

Case 2. If now $\mathcal{A}$ has sent $\hat{m}\|\hat{c}$ to Bob, $\mathcal{B}$ can compute $R_A = \hat{R}_A \oplus R_B \oplus \hat{R}_B$ and answer R_A. $\mathcal{B}$ receives d and sends it to $\mathcal{A}$. A typical example is depicted on Fig. 6. Here, $\mathcal{A}$ and $\mathcal{B}$ win or loose at the same time, thus win with the same probability.

We observe that the simulation by $\mathcal{B}$ is perfect and that the extraction is legitimate since $m \neq \hat{m}$. We deduce that we can win the hiding game with the same probability as the

$$
\begin{array}{lll}
\mathcal{A} & \mathcal{B} & \mathcal{C} \\
\xleftarrow{K_P} & & \xleftarrow{K_P} \\
\xrightarrow{m} & & \\
\xleftarrow{c} & c \leftarrow \mathsf{simcommit}(m) & \\
\xrightarrow{\hat{R}_B} & \text{pick } R_A & \\
\xleftarrow{d} & d \leftarrow \mathsf{equivocate}(m,c,R_A) & \\
\xrightarrow{\hat{m}||\hat{c}} & & \xrightarrow{\hat{m}||\hat{c}} \\
\xrightarrow{R_B} & R_B \leftarrow R_A \oplus \hat{R}_A \oplus \hat{R}_B & \xleftarrow{\hat{R}_A} \\
\xrightarrow{\hat{d}} & & \xrightarrow{\hat{d}} \\
& \text{Case 1} &
\end{array}
$$

$$
\begin{array}{lll}
\mathcal{A} & \mathcal{B} & \mathcal{C} \\
\xleftarrow{K_P} & & \xleftarrow{K_P} \\
\dots & & \\
\xrightarrow{m} & & \\
\xleftarrow{c} & c \leftarrow \mathsf{simcommit}(m) & \\
\dots & & \\
\xrightarrow{\hat{m}||\hat{c}} & & \xrightarrow{\hat{m}||\hat{c}} \\
\xleftarrow{R_B} & \text{pick } R_B & \xleftarrow{\hat{R}_A} \\
\dots & & \\
\xrightarrow{\hat{R}_B} & R_A \leftarrow \hat{R}_A \oplus R_B \oplus \hat{R}_B & \\
\xleftarrow{d} & d \leftarrow \mathsf{equivocate}(m,c,R_A) & \\
\dots & & \\
\xrightarrow{\hat{d}} & & \xrightarrow{\hat{d}} \\
& \text{Case 2} &
\end{array}
$$

Fig. 7. Reduction with Equivocable Commitments

attack. Let μ be the complexity of the protocol plus one oracle call. The complexity of $\mathcal{B}$ is essentially the complexity of $\mathcal{A}$ plus μ. The success probability of the attack is thus at most $2^{-k} + \varepsilon$.

Equivocable Commitments. From $\mathcal{A}$ we construct an adversary $\mathcal{B}$ who plays the binding game with the help of one query to the simcommit and equivocate oracles. (See Fig. 7.) Namely, $\mathcal{B}$ runs $\mathcal{A}$ and first forwards K_P. Whenever $\mathcal{B}$ receives $\hat{m}$ and $\hat{c}$ from $\mathcal{A}$, he submits it in the binding game. Whenever $\mathcal{B}$ must send c to $\mathcal{A}$, he launches the simcommit oracle to get c. When $\mathcal{A}$ sends $\hat{R}_B$ to Bob, we distinguish two cases.

Case 1. If $\mathcal{A}$ did not send $\hat{m}||\hat{c}$ to Bob yet, there is essentially one way to interleave the two sequences. Here, $\mathcal{B}$ picks a random R_A and equivocate his commitment by calling equivocate(m,c,R_A) so that he can send d to $\mathcal{A}$. When receiving the challenge $\hat{R}_A$, $\mathcal{B}$ chooses $R_B = R_A \oplus \hat{R}_A \oplus \hat{R}_B$ and send it to $\mathcal{A}$.

Case 2. If now $\mathcal{A}$ has sent $\hat{m}||\hat{c}$ to Bob, $\mathcal{B}$ has already answered some random R_B to $\mathcal{A}$ and received some challenge $\hat{R}_A$. He can thus compute $R_A = \hat{R}_A \oplus R_B \oplus \hat{R}_B$ and equivocate his commitment by calling equivocate(m,c,R_A). A typical example is depicted on Fig. 6.

Here, $\mathcal{A}$ and $\mathcal{B}$ win or loose at the same time, thus win with the same probability so we conclude as for the extractable commitments. □

5   On the Selection of the SAS Length

We now study the security in a multiparty and concurrent setting.

Lemma 6. *We consider a message authentication protocol with claimant Alice and verifier Bob in which a single SAS is sent. We denote by μ_A (resp. μ_B) the complexity of*

Alice's (resp. Bob's) part. We consider adversaries such that the number of instances of Alice (resp. Bob) is at most Q_A (resp. Q_B). We further denote by T_0 and p_0 their time complexity and probability of success, respectively. There is a generic transformation which, for any Q_A, Q_B, and any adversary, transforms it into a one-shot adversary with complexity $T \leq T_0 + \mu_A Q_A + \mu_B Q_B$ and probability of success $p \geq p_0/Q_A Q_B$.

The lemma tells us that once we proved that a protocol resists one-shot adversaries up to a probability of success of p, then it resists to adversaries up to a probability of success which is close to $Q_A Q_B p$. With the protocol in Fig. 5, this probability is basically $Q_A Q_B.2^{-k}$. This bound is tight as shown by the attacks in Section 4.

Proof. Let us consider an adversary $\mathcal{A}$. We number every instance of Alice and every instance of Bob by using two separate counters. We say that an instance π_a of Alice is compatible with an instance π_b of Bob if π_b succeeded and received an authenticated message which was sent by π_a. The number of possible compatible pairs of instances is upper bounded by $K = Q_A Q_B$. When an attack is successful, it yields a random pair (I, J) of compatible instances of Alice and Bob.

We transform $\mathcal{A}$ into a one-shot adversary $\mathcal{B}$ as follows: we run $\mathcal{A}$ and simulate launch and send oracle calls. We pick a random pair (I^*, J^*) with uniformly distributed $I^* \in \{1, \ldots, Q_A\}$ and $J^* \in \{1, \ldots, Q_B\}$. When $\mathcal{A}$ queries launch($\cdot$, Alice, $\cdot$) for the I^*th time, $\mathcal{B}$ forwards the query to the real launch oracle. send queries to the related instance are also forwarded to the real send oracle. The same holds for the J^*th query launch($\cdot$, Bob, $\cdot$). Clearly, the attack succeeds with probability $p_0/Q_A Q_B$ on the only non-simulated instances. It runs with complexity $T_0 + \mu_A Q_A + \mu_B Q_B$.   □

For applications, we assume that the number of network nodes is $N \approx 2^{20}$, and that the number of protocol runs per node is limited to $R \approx 2^{10}$. Actually, the protocols are not meant to be run so many times: only for seed authentication. Indeed, they can be used to authenticate a public key, and authentication can later be done using the public key itself so the protocol is no longer useful. We target a probability of success limited to $p \approx 2^{-10}$. Using Th. 5 and the previous lemma tells us that we can take $k \geq \log_2 \frac{Q_A Q_B}{p}$.

When considering the probability of success *at large* over the network, i.e. the probability that an attack occurs somewhere in the network, we have the constraint $Q_A + Q_B \leq NR$. Thus we have $Q_A Q_B \leq N^2 R^2/4$ for our message authentication protocol. Thus we can take $k = \log_2 \frac{N^2 R^2}{4p} = 68$ which is already shorter than the solution based on hash functions.

When considering the probability of success *against a target verifier* node, i.e. the probability that a given user will accept a forged message, we take $Q_B \leq R$ as an additional constraint. Thus we have $Q_A Q_B \leq NR^2$ which leads us to $k = \log_2 \frac{NR^2}{p} = 50$. By using the encoding rules of the RFC 1760 [23] standard, this represents five 4-character human-friendly (or at least English) words.

Credit cards ATM use 4-digit PIN codes which are confidential and quite strongly authenticated. Protocols are also limited to three trials. In our settings, this translates into a 3-shot 2-party model: $N = 2$, $R = 3$, and $p = 3 \cdot 10^{-4}$. To reach the same security level with weak authentication and no confidentiality, we need a SAS of size $k = \log_2 \frac{NR^2}{4p} \approx 15$ bits, i.e. a 5-digit PIN code.

6   Conclusion

We have shown how to achieve authentication over an insecure channel by using a narrowband authentication channel. The later channel is used to authenticate a short string: the SAS. Using weak authentication, we can obtain high security level in a multiparty setting by using 50-bit SAS. Note that in a 3-shot 2-party adversarial model, a 15-bit SAS (i.e. a 5-digit PIN code) is enough. This is similar to the MANA I protocol, except that we no longer require a strong notion of authentication. SAS channels are widely available for human beings: they can transmit SAS by fax, voice mail, type them on mobile devices, etc.

Our protocol is well suited to ad-hoc message authentication. It can be used for PKI-less public key transmission or to run a key agreement protocol. It can also be used to restore a secure association in disaster cases when two remote peers have compromised their secret keys or a PKI is badly broken. Another application could be a Bluetooth-like pairing between physically identified wireless devices with higher security: we no longer rely on the secrecy of a PIN code but on the authentication through a human user of a short string.

Our protocol relies on a commitment scheme and is provably secure in the strongest security model so far, namely the Bellare-Rogaway model, by using extractable or equivocable commitment schemes. They can be constructed in the ideal commitment model, in the random oracle commitment model, and in the CRS model.

Acknowledgments

I wish to thank Anna Lysyanskaya, Ivan Damgård, Kaisa Nyberg, Moti Yung, Phil MacKenzie, and the anonymous reviewers for insightful references and comments. This paper was highly inspired by [32] which was originally submitted (and rejected!) in 1975, so I would like to thank Ralph Merkle for 30 years of research fun.

This work was supported in part by the National Competence Center in Research on Mobile Information and Communication Systems (NCCR-MICS), a center supported by the Swiss National Science Foundation under grant number 5005-67322.

References

1. Digital Signature Standard (DSS). *Federal Information Processing Standards* publication #186-2. U.S. Department of Commerce, National Institute of Standards and Technology, 2000.
2. Specification of the Bluetooth System. Vol. 2: Core System Package. Bluetooth Specification version 1.2, 2003.
3. M. Abdalla, P.-A. Fouque, D. Pointcheval. Password-Based Authenticated Key Exchange in the Three-Party Setting. In *Public Key Cryptography'05*, Les Diablerets, Switzerland, Lecture Notes in Computer Science 3386, pp. 65–84, Springer-Verlag, 2005.
4. D. Balfanz, D. K. Smeeters, P. Stewart, H. Chi Wong. Talking to Strangers: Authentication in Ad-Hoc Wireless Networks. In *Network and Distributed System Security Symposium Conference (NDSS 02)*, San Diego, California, U.S.A., The Internet Society, 2002.

5. M. Bellare, D. Pointcheval, P. Rogaway. Authenticated Key Exchange Secure against Dictionary Attacks. In *Advances in Cryptology EUROCRYPT'00*, Brugge, Belgium, Lecture Notes in Computer Science 1807, pp. 139–155, Springer-Verlag, 2000.

6. M. Bellare, P. Rogaway. Entity Authentication and Key Distribution. In *Advances in Cryptology CRYPTO'93*, Santa Barbara, California, U.S.A., Lecture Notes in Computer Science 773, pp. 232–249, Springer-Verlag, 1994.

7. M. Bellare, P. Rogaway. Provably Secure Session Key Distribution: the Three Party Case. In *Proceedings of the 27th ACM Symposium on Theory of Computing*, Las Vegas, Nevada, U.S.A., pp. 57–66, ACM Press, 1995.

8. S. M. Bellovin, M. Merritt. Encrypted Key Exchange: Password-Based Protocols Secure Against Dictionary Attacks. In *IEEE symposium on Research in Security and Privacy*, Oakland, California, USA, pp. IEEE Computer Society Press, 72–84, 1992.

9. S. M. Bellovin, M. Merritt. Augmented Encrypted Key Exchange. In *1st ACM Conference on Computer and Communications Security*, Fairfax, Virginia, U.S.A., pp. 244–250, ACM Press, 1993.

10. C. H. Bennett, G. Brassard. Quantum Cryptography: Public Key Distribution and Coin Tossing. In *Proc. IEEE International Conference on Computers, Systems, and Signal Processing*, Bangalore, India, pp. 175–179, IEEE Press, 1984.

11. D. Boneh. The Decision Diffie-Hellman Problem. In *Proceedings of the 3rd Algorithmic Number Theory Symposium*, Portland, Oregon, U.S.A., Lecture Notes in Computer Science 1423, pp. 48–63, Springer-Verlag, 1998.

12. C. Boyd, A. Mathuria. *Protocols for Authentication and Key Establishment*, Information Security and Cryptography, Springer Verlag, 2003.

13. M. Čagalj, S. Čapkun, J.-P. Hubaux. Key Agreement in Peer-to-Peer Wireless Networks. To appear in the Proceedings of the IEEE, fall 2005.

14. R. Cramer, V. Shoup. Signature Schemes based on the Strong RSA Assumption. *ACM Transactions on Information and System Security*, vol. 3, pp. 161–185, 2000.

15. G. Di Crescenzo, Y. Ishai, R. Ostrovsky. Non-Interactive and Non-Malleable Commitment. In *Proceedings of the 30th ACM Symposium on Theory of Computing*, Dallas, Texas, U.S.A., pp. 141–150, ACM Press, 1998.

16. G. Di Crescenzo, J. Katz, R. Ostrovsky, A. Smith. Efficient and Non-Interactive Non-Malleable Commitments. In *Advances in Cryptology EUROCRYPT'01*, Innsbruck, Austria, Lecture Notes in Computer Science 2045, pp. 40–59, Springer-Verlag, 2001.

17. I. Damgård, J. Groth. Non-interactive and Reusable Non-malleable Commitment Schemes. In *Proceedings of the 35th ACM Symposium on Theory of Computing*, San Diego, California, U.S.A., pp. 426–437, ACM Press, 2003.

18. W. Diffie, M. E. Hellman. New Directions in Cryptography. *IEEE Transactions on Information Theory*, vol. IT-22, pp. 644–654, 1976.

19. T. Dierks, C. Allen. The TLS Protocol Version 1.0. RFC 2246, standard tracks, the Internet Society, 1999.

20. D. Dolev, C. Dwork, M. Naor. Non-Malleable Cryptography. *SIAM Journal of Computing*, vol. 30, pp. 391–437, 2000.

21. C. Gehrmann, C. Mitchell, K. Nyberg. Manual Authentication for Wireless Devices. *RSA Cryptobytes*, vol. 7, pp. 29–37, 2004.

22. C. Gehrmann, K. Nyberg. Security in Personal Area Networks. In *Security for Mobility*, C. Mitchell (Ed.), pp. 191–230, IEE, 2004.

23. N. Haller. The S/KEY One-Time Password System. RFC 1760, 1995.

24. J. Håstad, R. Impagliazzo, L. Levin, M. Luby. A Pseudorandom Generator from any One-way Function. *SIAM Journal on Computing*, vol. 4, pp. 1364–1396, 1999.

25. J.-H. Hoepman. The Ephemeral Pairing Problem. In *Financial Cryptography*, Key West, Florida, USA, Lecture Notes in Computer Science 3110, pp. 212–226, Springer-Verlag, 2004.

26. J.-H. Hoepman. Ephemeral Pairing on Anonymous Networks. In *Proceedings of the Second International Conference on Security in Pervasive Computing (SPC'05)*, Boppard, Germany, Lecture Notes in Computer Science 3450, pp. 101–116, Springer-Verlag, 2005.

27. R. Housley, W. Ford, W. Polk, D. Solo. Internet X.509 Public Key Infrastructure Certificate and CRL Profile. Internet Standard. RFC 2459, The Internet Society, 1999.

28. M. Jakobsson, S. Wetzel. Security Weaknesses in Bluetooth. In *Topics in Cryptology (CT–RSA'01)*, San Francisco, California, USA, Lecture Notes in Computer Science 2020, pp. 176–191, Springer-Verlag, 2001.

29. J. Katz, R. Ostrovsky, M. Yung. Efficient Password-Authenticated Key Exchange using Human-Memorable Passwords. In *Advances in Cryptology EUROCRYPT'01*, Innsbruck, Austria, Lecture Notes in Computer Science 2045, pp. 475–494, Springer-Verlag, 2001.

30. J. Kohl, C. Neuman. The Kerberos Network Authentication Service (V5). Internet standard. RFC 1510, 1993.

31. P. MacKenzie, K. Yang. On Simulation-Sound Trapdoor Commitments. In *Advances in Cryptology EUROCRYPT'04*, Interlaken, Switzerland, Lecture Notes in Computer Science 3027, pp. 382–400, Springer-Verlag, 2004.

32. R. C. Merkle. Secure Communications over Insecure Channels. *Communications of the ACM*, vol. 21, pp. 294–299, 1978.

33. C. Mitchell, M. Ward, P. Wilson. On Key Control in Key Agreement Protocols. *Electronics Letters*, vol. 34, pp. 980–981, 1998.

34. R. M. Needham, M. D. Schroeder. Using Encryption for Authentication in Large Networks of Computers. *Communications of the ACM*, vol. 21, pp. 993–999, 1978.

35. K. Nyberg. IKE in Ad-hoc IP Networking. In *Security in Ad-hoc and Sensor Networks, 1st European Workshop (ESAS'04)*, Heidelberg, Germany, Lecture Notes in Computer Science 3313, pp. 139–151, Springer-Verlag, 2005.

36. S. Pasini, S. Vaudenay. Optimized Message Authentication Protocols. Unpublished.

37. R. Pass. On Deniability in the Common Reference String and Random Oracle Model. In *Advances in Cryptology CRYPTO'03*, Santa Barbara, California, U.S.A., Lecture Notes in Computer Science 2729, pp. 316–337, Springer-Verlag, 2003.

38. T. Peyrin, S. Vaudenay. The Pairing Problem with User Interaction. In *Security and Privacy in the Age of Ubiquitous Computing IFIP TC11 20th International Information Security Conference (SEC'05)*, Chiba, Japan, pp. 251–265, Springer-Verlag, 2005.

39. R. L. Rivest, A. Shamir and L. M. Adleman. A Method for Obtaining Digital Signatures and Public-key Cryptosystem. *Communications of the ACM*, vol. 21, pp. 120–126, 1978.

40. F. Stajano, R. Anderson. The Resurrecting Duckling: Security Issues for Ad-hoc Wireless Networks. In *Proceedings of the 7th International Workshop on Security Protocols*, Cambridge, United Kingdom, Lecture Notes in Computer Science 1796, pp. 172–194, Springer-Verlag, 1999.

A   Message Cross-Authentication

We consider protocols which perform message authentication in both directions at the same time. These protocols have inputs m_A on the side of Alice and m_B on the side of Bob, and outputs $I_B||\hat{m}_B$ on the side of Alice and $I_A||\hat{m}_A$ on the side of Bob. They should be such that they achieve message authentication in both directions. In Fig. 8 is a message cross-authentication protocol. It requires k authenticated bits in both ways.

Obviously, an adversary against this protocol transforms into an adversary against the message authentication protocol in Fig. 5. So Theorem 5 holds for this new protocol as well. This protocol can be used e.g. to run the Diffie-Hellman authenticated key agreement protocol [18] with $m_A = g^{x_A}$ and $m_B = g^{x_B}$. This is essentially the protocol called DH-SC in [13].

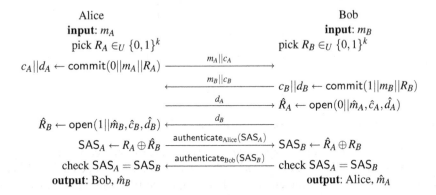

Fig. 8. SAS-based Cross Authentication

B   Using Simulation-Sound Trapdoor Commitments

MacKenzie-Yang [31] defines SSTC commitments by five algorithms setup′, commit′, verify′, simcommit′, and equivocate′. The only syntaxic difference with our definition is in the verify′ algorithm which replaces open, but without message recovery. Namely, commit′(m,r) yields a pair (c,e) and verify′$(m,c,r,e) = \text{true}$ whenever (c,e) is a possible output of commit′(m,r). Obviously, by letting $d = r||e$, we define a commitment scheme in our sense.

In [31], the hiding game is restricted to a 2-fold game. Namely, the adversary yields r_0 and r_1 together with c, and the challenger picks r equal to one of these two values. The adversary should have a probability of success less than $\frac{1}{2} + \varepsilon$. The following lemma proves that such a commitment scheme is 2ε-hiding in our sense.

Lemma 7. *There exists a (small) constant* ν *such that for any T and* ε, *a* $(T + \nu, \varepsilon)$-2-*fold-hiding commitment scheme is a* $(T, 2\varepsilon)$-*hiding commitment scheme.*

Proof. Let $\mathcal{A}$ be an adversary of complexity at most T which plays our 2^k-fold hiding game. We construct an adversary $\mathcal{B}$ for the 2-fold hiding game as follows.

1. $\mathcal{B}$ receives K_P and forwards it to $\mathcal{A}$.
2. $\mathcal{A}$ sends m to $\mathcal{B}$.
3. $\mathcal{B}$ picks two random different r_0 and r_1 and plays m, r_0, r_1.
4. $\mathcal{B}$ receives a challenge c which commits to either r_0 or r_1 and forwards it to $\mathcal{A}$.
5. $\mathcal{A}$ answers to the challenge by a string r.
6. If $r = r_b$, $\mathcal{B}$ answers b. Otherwise, $\mathcal{B}$ picks a random bit b and answers b.

We let ν be the complexity of $\mathcal{B}$ without $\mathcal{A}$. Obviously, $\mathcal{B}$ perfectly simulates a challenger for the 2^k-fold game to $\mathcal{A}$. Let $p' = 2^{-k} + \varepsilon'$ be the probability of success of $\mathcal{A}$. When $\mathcal{A}$ is successful, so is $\mathcal{B}$. When $\mathcal{A}$ is not successful and $r \notin \{r_0, r_1\}$, $\mathcal{B}$ succeeds with probability $\frac{1}{2}$. When $\mathcal{A}$ is not successful and $r \in \{r_0, r_1\}$, $\mathcal{B}$ fails. Hence, the probability that $\mathcal{B}$ answers correctly to the 2-fold game is

$$
\begin{aligned}
p &= p' + \frac{1-p'}{2}\left(1 - \frac{1}{2^k - 1}\right) \\
&= \frac{1}{2} + p'\frac{2^k}{2(2^k - 1)} - \frac{1}{2(2^k - 1)} \\
&= \frac{1}{2} + \varepsilon'\frac{2^k}{2(2^k - 1)}
\end{aligned}
$$

Since we must have $p - \frac{1}{2} \le \varepsilon$, we deduce $\varepsilon' \le 2\varepsilon(1 - 2^{-k}) \le 2\varepsilon$.   □

In [31], our binding game is replaced by the ability to produce a collision, namely a (m, c, d, d') quadruplet such that (m, c, d) and (m, c, d') successfully open to two different values. Following the simulation-sound binding property definition, the adversary has access to simcommit$'$ and equivocate$'$ like in our definition. Namely, they do not see ξ and cannot decommit values which are not issued by simcommit$'$. By usual rewinding techniques, we show that from an adversary who wins our binding property with probability $2^{-k} + \varepsilon$ in time T and one simcommit query we can make an adversary who finds collisions with probability $2^{-2k}\varepsilon$ in time $2T$ and two simcommit queries.

On Codes, Matroids and Secure Multi-party Computation from Linear Secret Sharing Schemes

Ronald Cramer[1], Vanesa Daza[2,*], Ignacio Gracia[2,*], Jorge Jiménez Urroz[2,*], Gregor Leander[3], Jaume Martí-Farré[2,*], and Carles Padró[2,*]

[1] CWI, Amsterdam & Mathematical Institute, Leiden University
cramer@cwi.nl
[2] Dept. of Applied Maths. IV, Technical University of Catalonia, Barcelona
{vdaza, ignacio, jjimenez, jaumem, matcpl}@ma4.upc.edu
[3] CITS Research Group, Ruhr-University Bochum
leander@cits.ruhr-uni-bochum.de

Abstract. Error correcting codes and matroids have been widely used in the study of ordinary secret sharing schemes. In this paper, we study the connections between codes, matroids and a special class of secret sharing schemes, namely multiplicative linear secret sharing schemes. Such schemes are known to enable multi-party computation protocols secure against general (non-threshold) adversaries.

Two open problems related to the complexity of multiplicative LSSSs are considered in this paper.

The first one deals with strongly multiplicative LSSSs. As opposed to the case of multiplicative LSSSs, it is not known whether there is an efficient method to transform an LSSS into a strongly multiplicative LSSS for the same access structure with a polynomial increase of the complexity. We prove a property of strongly multiplicative LSSSs that could be useful in solving this problem. Namely, using a suitable generalization of the well-known Berlekamp-Welch decoder, we show that all strongly multiplicative LSSSs enable efficient reconstruction of a shared secret in the presence of malicious faults.

The second one is to characterize the access structures of ideal multiplicative LSSSs. Specifically, we wonder whether all self-dual vector space access structures are in this situation. By the aforementioned connection, this in fact constitutes an open problem about matroid theory, since it can be re-stated in terms of representability of identically self-dual matroids by self-dual codes. We introduce a new concept, the flat-partition, that provides a useful classification of identically self-dual matroids. Uniform identically self-dual matroids, which are known to be representable by self-dual codes, form one of the classes. We prove that this property also holds for the family of matroids that, in a natural way, is the next class in the above classification: the identically self-dual bipartite matroids.

* These authors' research is partially supported by the Spanish Ministry of Education and Science under project TIC 2003-00866.

V. Shoup (Ed.): Crypto 2005, LNCS 3621, pp. 327–343, 2005.

Keywords: multi-party computation, multiplicative linear secret sharing schemes, identically self-dual matroids, self-dual codes, efficient error correction.

1   Introduction

Two open problems on multiplicative linear secret sharing schemes are studied in this paper. Our results deal with the connections between linear codes, representable matroids and linear secret sharing schemes. Some facts about these connections are recalled in Section 1.2. The reader is referred to [22] for a general reference on Matroid Theory and to [5,19,28,29] for more information about the relation between secret sharing schemes and matroids.

1.1   Multiplicative Linear Secret Sharing Schemes and General Secure Multi-party Computation

In a $\mathbb{K}$ -*linear secret sharing scheme* ($\mathbb{K}$-LSSS) on the set of players $P = \{1, \dots, n\}$, the share of every player $i \in P$ is a vector in E_i, a vector space of finite dimension over the finite field $\mathbb{K}$, and is computed as a fixed linear function of the secret value $k \in \mathbb{K}$ and some other randomly chosen elements in $\mathbb{K}$.

More formally, any sequence $\Pi = (\pi_1, \dots, \pi_n, \pi_{n+1})$ of surjective linear mappings $\pi_i \colon E \to E_i$, where E and E_i are vector spaces of finite dimension over $\mathbb{K}$ and $E_{n+1} = \mathbb{K}$, defines a $\mathbb{K}$-linear secret sharing scheme $\Sigma_{n+1}(\Pi)$ on the set of players $P = \{1, \dots, n\}$. For any vector $\mathbf{x} \in E$, the values $(\pi_i(\mathbf{x}))_{1 \le i \le n}$ are shares of the secret value $k = \pi_{n+1}(\mathbf{x}) \in \mathbb{K}$. The access structure, $\Gamma_{n+1}(\Pi)$ of this scheme, that is, the family of qualified subsets, consists of all subsets $A \subset P_{n+1}$ such that $\bigcap_{i \in A} \ker \pi_i \subset \ker \pi_{n+1}$.

Linear secret sharing schemes are usually defined in a more general way by considering that the vector space E_{n+1} corresponding to the secret value is not necessarily equal to $\mathbb{K}$. We are not going to consider such LSSSs in this paper.

The *complexity* of a LSSS Σ is defined as $\lambda(\Sigma) = \sum_{i=1}^{n} \dim E_i \ge n$, which corresponds to the total number of field elements that are distributed. The schemes with complexity $\lambda(\Sigma) = n$ are called *ideal*. For any finite field $\mathbb{K}$ and for any access structure Γ, there exists a $\mathbb{K}$-LSSS for Γ [14]. We notate $\lambda_{\mathbb{K}}(\Gamma)$ for the minimum complexity of the $\mathbb{K}$-LSSSs with access structure Γ. If there exists an ideal $\mathbb{K}$-LSSS for Γ, that is, if $\lambda_{\mathbb{K}}(\Gamma) = n$, we say that Γ is a $\mathbb{K}$-*vector space access structure*.

Linear secret sharing schemes were first considered, only in the ideal case, in [4]. General Linear secret sharing schemes were introduced by Simmons [27], Jackson and Martin [15] and Karchmer and Wigderson [16] under other names such as geometric secret sharing schemes or monotone span programs.

In a LSSS, any linear combination of the shares of different secrets results in shares for the same linear combination of the secret values. Because of that, LSSSs are used as a building block of multi-party computation protocols. Nevertheless, if we require protocols computing any arithmetic circuit, a similar property is needed for the multiplication of two secrets, that is, the LSSS must be *multiplicative*.

We illustrate the multiplicative property of LSSSs by analyzing the Shamir's (d, n)-threshold scheme [26]. In this scheme, the secret $k \in \mathbb{K}$ and the shares $k_i \in \mathbb{K}$, where $i = 1, \ldots, n$, are the values of a random polynomial with degree at most $d - 1$ in some given points. The secret is recovered by Lagrange interpolation. If $n \geq 2d - 1$, the product kk' of the two secret values is a linear combination of any $2d - 1$ values $c_i = k_i k_i'$. This linear combination is obtained by interpolating the product of the two random polynomials that were used to distribute the shares. This multiplicative property of the Shamir's scheme is used in [3,8,9,11] to construct multi-party computation protocols that are secure against a threshold-based adversary.

In order to obtain efficient multi-party computation protocols for a general adversary structure, a generalization of the multiplicative property of the Shamir's scheme to any linear secret sharing scheme is proposed in [10].

Specifically, a linear secret sharing scheme over the finite field $\mathbb{K}$ is said to be *multiplicative* if every player $i \in P$ can compute, from his shares k_i, k_i' of two shared secrets $k, k' \in \mathbb{K}$, a value $c_i \in \mathbb{K}$ such that the product kk' is a linear combination of all the values $c_1, \ldots, c_n$. We say that a secret sharing scheme is *strongly multiplicative* if, for any subset $A \subset P$ such that $P - A$ is not qualified, the product kk' can be computed using only values from the players in A.

Observe that the Shamir's (d, n)-secret sharing scheme is multiplicative if and only if $n \geq 2d - 1$, and it is strongly multiplicative if and only if $n \geq 3d - 2$. An access structure is said to be $\mathcal{Q}_2$, or $\mathcal{Q}_3$, if the set of players is not the union of any two, or, respectively, three, unqualified subsets. In general, as a consequence of the results in [10,13], an access structure Γ can be realized by a multiplicative LSSS if and only if it is $\mathcal{Q}_2$, and Γ admits an strongly multiplicative LSSS if and only if it is $\mathcal{Q}_3$.

Cramer, Damgård and Maurer [10] presented a method to construct, from any $\mathbb{K}$-MLSSS Σ with $\mathcal{Q}_2$ access structure Γ, an error-free multi-party computation protocol secure against a passive adversary which is able to corrupt any set of players $B \notin \Gamma$ and computing any arithmetic circuit C over $\mathbb{K}$. The complexity of this protocol is polynomial in the size of C, $\log |\mathbb{K}|$ and $\lambda(\Sigma)$. They prove a similar result for an active adversary. In this case, the resulting protocol is perfect with zero error probability if the LSSS is strongly multiplicative, with a $\mathcal{Q}_3$ access structure Γ.

One of the key results in [10] is a method to construct, from any $\mathbb{K}$-LSSS Σ with $\mathcal{Q}_2$ access structure Γ, a multiplicative $\mathbb{K}$-LSSS Σ' with the same access structure and complexity $\lambda(\Sigma') = 2\lambda(\Sigma)$. That is, if $\mu_{\mathbb{K}}(\Gamma)$ denotes the minimum complexity of all $\mathbb{K}$-MLSSSs with access structure Γ, the above result means that $\mu_{\mathbb{K}}(\Gamma) \leq 2\lambda_{\mathbb{K}}(\Gamma)$ for any finite field $\mathbb{K}$ and for any $\mathcal{Q}_2$ access structure Γ.

Therefore, in the passive adversary case, the construction of efficient multi-party computation protocols can be reduced to the search of efficient linear secret sharing schemes. Specifically, a multi-party computation protocol computing any arithmetic circuit C over $\mathbb{K}$ and secure against a passive adversary which is able to corrupt any set of players $B \notin \Gamma$ can be efficiently constructed from any LSSS whose access structure Γ' is $\mathcal{Q}_2$ and $\Gamma' \subset \Gamma$.

This is not the situation when an active adversary is considered, because it is not known whether it is possible to construct, for any $\mathcal{Q}_3$ access structure Γ, a strongly multiplicative LSSS whose complexity is polynomial on the complexity of the best LSSS for Γ.

Nevertheless, the active adversary case is also solved in [10] if an exponentially small error probability is allowed. A construction is given in [10] for the active adversary case that efficiently provides, from any LSSS with $\mathcal{Q}_3$ access structure Γ, a multiparty computation protocol with exponentially small error probability, secure against an active adversary which is able to corrupt any set of players not in Γ.

1.2   Codes, Matroids and Secret Sharing Schemes

Let us take $Q = \{1, \ldots, n, n+1\}$ and $P_i = Q - \{i\}$ for any $i \in Q$. This notation will be used all through the paper. From now on, vectors appearing in matrix operations will be considered as one-row matrices.

Let $\Pi = (\pi_1, \ldots, \pi_n, \pi_{n+1})$ be a sequence of surjective linear mappings $\pi_i \colon E \to \mathbb{K}$, that is, non-zero vectors in the dual space E^*. We are going to suppose always that those vectors span E^*. Observe that Π can be seen as a linear mapping $\Pi \colon E \to \mathbb{K}^{n+1}$ and, once a basis of E is fixed, it can be represented by the $d \times (n+1)$ matrix $M = M(\Pi)$ such that $\Pi(\mathbf{x}) = \mathbf{x}M$ for all $\mathbf{x} \in E$. Observe that $\mathrm{rank}(M) = d$ and that the i-th column of M corresponds to the linear form π_i.

The matrix M is a generator matrix of a $[n+1, d]$-linear code $\mathcal{C} = \mathcal{C}(\Pi)$. The columns of M define a $\mathbb{K}$-representable matroid $\mathcal{M} = \mathcal{M}(\Pi)$ on the set of points Q. This matroid depends only on the code $\mathcal{C}$, that is, it does not depend on the choice of the generator matrix M. In this situation, we say that $\mathcal{M}$ is the matroid associated to the code $\mathcal{C}$ and also that the code $\mathcal{C}$ is a $\mathbb{K}$-representation of the matroid $\mathcal{M}$. Observe that different codes can represent the same matroid. Important properties about the weight distribution of a linear code can be studied from its associated matroid. Several results on this relation between matroids and codes are given in [1,6,7,12] and other works.

Besides, the code $\mathcal{C}$ defines an ideal linear secret sharing scheme $\Sigma_i(\Pi)$ for every $i \in Q$. Every codeword of $\mathcal{C}$ is in the form $(\pi_1(\mathbf{x}), \ldots, \pi_i(\mathbf{x}), \ldots, \pi_{n+1}(\mathbf{x}))$ and can be seen as a distribution of shares for the secret value $\pi_i(\mathbf{x}) \in \mathbb{K}$ among the players in $P_i = Q - \{i\}$. Observe that the access structure $\Gamma_i(\Pi)$ of the scheme $\Sigma_i(\Pi)$, which is a $\mathbb{K}$-vector space access structure, consists of all subsets $A \subset P_i$ such that $\pi_i \in \langle \pi_j : j \in A \rangle$. Therefore, $A \subset P_i$ is a minimal qualified subset in that structure if and only if $A \cup \{i\}$ is a circuit of the matroid $\mathcal{M}(\Pi)$. As a consequence, the access structures $\Gamma_i(\Pi)$ are determined by the matroid $\mathcal{M}(\Pi)$. This connection between ideal secret sharing schemes and matroids, which applies to non-linear schemes as well, was discovered by Brickell and Davenport [5] and has been studied afterwards by several authors [2,19,20,21,28,29,30]. It plays a key role in one of the main open problems in secret sharing: the characterization of the access structures of ideal secret sharing schemes.

Actually, non-ideal linear secret sharing schemes can also be represented as linear codes. In the general case, several columns of the generator matrix are assigned to every player.

Error correction in linear codes is related to an important property of secret sharing schemes: the possibility of reconstructing the shared secret value even if some shares are corrupted.

The different notions of *duality* that are defined for codes, for matroids and for access structures are closely related.

Let N be a parity check matrix for the code $\mathcal{C} = \mathcal{C}(\Pi)$. That is, N is a $(n-d+1) \times (n+1)$ matrix with $\mathrm{rank}(N) = n-d+1$ and $MN^{\top} = 0$, where $N^{\top}$ denotes the transpose of N. The matrix N is a generator matrix of a $[n+1, n-d+1]$-linear code $\mathcal{C}^{\perp}$, which is called the *dual code* of the code $\mathcal{C}$. The code $\mathcal{C}$ is said to be *self-dual* if $\mathcal{C}^{\perp} = \mathcal{C}$. In this case, $2d = n+1$ and $MM^{\top} = 0$ for every generator matrix M.

If the linear code $\mathcal{C}$ defines a (not necessarily ideal) LSSS with access structure Γ, then the dual code $\mathcal{C}^{\perp}$ defines a LSSS for the *dual access structure* $\Gamma^* = \{A \subset P : P - A \notin \Gamma\}$. As a consequence of this fact, $\lambda_{\mathbb{K}}(\Gamma^*) = \lambda_{\mathbb{K}}(\Gamma)$ for every access structure Γ and for every finite field $\mathbb{K}$.

The matroid $\mathcal{N}$ associated to the dual code $\mathcal{C}^{\perp}$ is the *dual matroid* of the matroid $\mathcal{M}$ corresponding to $\mathcal{C}$, that is, the family of bases of $\mathcal{N} = \mathcal{M}^*$ is $\mathcal{B}(\mathcal{M}^*) = \{B \subset Q : Q - B \in \mathcal{B}(\mathcal{M})\}$, where $\mathcal{B}(\mathcal{M})$ is the family of bases of $\mathcal{M}$.

Moreover, for every $i \in Q$, if Γ_i and Γ_i' are the access structures on the set P_i that are determined, respectively, by the matroids $\mathcal{M}$ and $\mathcal{M}^*$, then $\Gamma_i' = \Gamma_i^*$. Therefore, the dual of a $\mathbb{K}$-representable matroid is also $\mathbb{K}$-representable and the same applies to $\mathbb{K}$-vector space access structures.

Observe that the matroid $\mathcal{M}$ associated to a self-dual code is identically self-dual, that is, $\mathcal{M} = \mathcal{M}^*$. Nevertheless, it is not known whether every representable identically self-dual matroid can be represented by a self-dual code.

Duality plays an important role in the study of the multiplicative property of LSSSs. First of all, an access structure Γ is $\mathcal{Q}_2$ if and only if $\Gamma^* \subset \Gamma$. This fact and the aforementioned relation between duality in codes and LSSSs are the key points in the proof of the bound $\mu_{\mathbb{K}}(\Gamma) \leq 2\lambda_{\mathbb{K}}(\Gamma)$ given in [10]. Besides, the ideal LSSS defined by a self-dual code is multiplicative and, hence, its access structure is such that $\mu_{\mathbb{K}}(\Gamma) = \lambda_{\mathbb{K}}(\Gamma)$.

2   Our Results

2.1   On Strongly Multiplicative Linear Secret Sharing Schemes

The first open problem we consider in this paper deals with the efficient construction of strongly multiplicative LSSSs. As we said before, no efficient general reductions are known for it at all, except for some upper bounds on the minimal complexity of strongly multiplicative LSSSs in terms of certain threshold circuits. That is, the existence of a transformation that renders an LSSS strongly multiplicative at the cost of increasing its complexity at most polynomially is an unsolved question.

We shed some light on that problem by proving a new property of strongly multiplicative LSSSs. Using a suitable generalization of the well-known Berle-kamp-Welch decoder for Reed-Solomon codes, we show Theorem 1, which is proved in Section 4, that all strongly multiplicative LSSSs allow for efficient reconstruction of a shared secret in the presence of malicious faults. In this way, we find an interesting connection between the problem of the strong multiplication in LSSSs and the existence of codes with efficient decoding algorithms.

Theorem 1. *Let* $\mathbf{s} = (s_1, \ldots, s_n)$ *be a full vector of shares for a secret* $s \in \mathbb{K}$, *computed according to a strongly multiplicative* $\mathbb{K}$-*LSSS with access structure* Γ *on* n *players. Let* $\mathbf{e}$ *denote the all zero vector, except where it states the errors that a set of players* $A \notin \Gamma$ *have introduced in their respective shares. Define* $\mathbf{c} = \mathbf{s} + \mathbf{e}$. *Then the secret* s *can be recovered from* $\mathbf{c}$ *in time* poly$(n, \log |\mathbb{K}|)$.

2.2   On Ideal Multiplicative Linear Secret Sharing Schemes

The characterization of the access structures of ideal MLSSSs is the second open problem that is studied in this work. That is, we are interested in determining which $\mathcal{Q}_2$ vector space access structures can be realized by an ideal MLSSS or, equivalently, for which $\mathcal{Q}_2$ access structures there exists a finite field $\mathbb{K}$ with $\mu_{\mathbb{K}}(\Gamma) = \lambda_{\mathbb{K}}(\Gamma) = n$.

This is a case of the more general problem of determining the cases in which the factor 2 loss in the construction of MLSSSs given in [10] is necessary. That is, to find out in which situations the bound $\mu_{\mathbb{K}}(\Gamma) \leq 2\lambda_{\mathbb{K}}(\Gamma)$ can be improved.

The (d, n)-threshold structures with $n \geq 2d - 1$ are examples of access structures that can be realized by an ideal LSSS. Other examples are obtained from self-dual codes. If the linear code $\mathcal{C}(\Pi)$ is self-dual, then, the ideal LSSSs $\Sigma_i(\Pi)$, where $i \in Q$, are multiplicative. Therefore, for every $i \in Q$, the vector space access structure $\Gamma_i = \Gamma_i(\Pi)$ is such that $\mu_{\mathbb{K}}(\Gamma_i) = \lambda_{\mathbb{K}}(\Gamma_i) = n$. Observe that those access structures are self-dual, that is, $\Gamma_i^* = \Gamma_i$.

On the other hand, there exist examples of $\mathcal{Q}_2$ access structures Γ such that $\lambda_{\mathbb{K}}(\Gamma) = n$ for some finite field $\mathbb{K}$ but do not admit any ideal MLSSS over any finite field. The arguments that are used to prove this fact do not apply if a self-dual vector space access structure is considered. An infinite family of such examples will be given in the full version of the paper.

Self-dual access structures coincide with the *minimally* $\mathcal{Q}_2$ access structures, that is, with the $\mathcal{Q}_2$ access structures Γ such that any substructure $\Gamma' \subsetneq \Gamma$ is not $\mathcal{Q}_2$. The results in this paper lead us to believe that any self-dual vector space access structure can be realized by an ideal multiplicative linear secret sharing scheme and, hence, to state the following open problem. One of the goals of this paper is to move forward in the search of its solution.

Problem 1. To determine whether there exists, for any self-dual $\mathbb{K}$-vector space access structure Γ, an ideal multiplicative $\mathbb{L}$-LSSS, being the finite field $\mathbb{L}$ an algebraic extension of $\mathbb{K}$.

Since $\mu_{\mathbb{K}}(\Gamma) \leq 2\lambda_{\mathbb{K}}(\Gamma)$ for any $\mathcal{Q}_2$ access structure Γ, to study this open problem seems to have a limited practical interest. Nevertheless, its theoretical interest can be justified by several reasons.

First, due to the minimality of the $\mathcal{Q}_2$ property, self-dual access structures are an extremal case in the theory of MLSSSs. Moreover, self-duality seems to be in the core of the multiplicative property. For instance, the construction in [10] providing the bound $\mu_{\mathbb{K}}(\Gamma) \leq 2\lambda_{\mathbb{K}}(\Gamma)$ is related to self-dual codes and, hence, to ideal MLSSSs for self-dual access structures.

Besides, the interest of Problem 1 is increased by the fact that, as we pointed out before, it can be stated in terms of an interesting open problem about the relation between Matroid Theory and Code Theory. Namely, by studying how the connection between codes, matroids and LSSSs applies to multiplicative LSSSs, we prove in Section 5.1 that Problem 1 is equivalent to the following one.

Problem 2. To determine whether every identically self-dual $\mathbb{K}$-representable matroid can be represented by a self-dual linear code over some finite field $\mathbb{L}$, an algebraic extension of $\mathbb{K}$.

Finally, we think that the results and techniques in this paper, and the ones that possibly will be found in future research on that problem, can provide a better understanding of the multiplicative property and may be useful to find new results on the existence of efficient strongly multiplicative LSSSs. In particular, the study of the characterization of the access structures of ideal strongly multiplicative LSSSs, which should be also attacked by using Matroid Theory, may lead to interesting advances on that problem. For instance, one can observe a remarkable difference in the strong multiplicative case: the minimality of the $\mathcal{Q}_3$ property does not imply any important matroid property comparable to self-duality.

We say that a matroid is *self-dually $\mathbb{K}$-representable* if it can be represented by a self-dual code over the finite field $\mathbb{K}$. Any self-dually representable matroid is identically self-dual and representable. The open problem we consider here is to decide whether the reciprocal of this fact is true.

The uniform matroids $U_{d,n}$ and the $\mathbb{Z}_2$-representable matroids are the only families of matroids for which it is known that all identically self-dual matroids are self-dually representable.

There exist several methods to combine some given matroids into a new one. The *sum*, which is defined in Section 5.3, is one of them. We show in Section 5.3 that the the sum of two self-dually representable matroids is equally self-dually representable and that Problem 2 can be restricted to indecomposable matroids, that is, matroids that are not a non-trivial sum of two other matroids.

In order to take the first steps in solving Problem 2, we introduce the concept of *flat-partition* of a matroid, which is defined in Section 5.3. On one hand, we use the flat-partitions to characterize in Proposition 4 the indecomposable identically self-dual matroids. On the other hand, the number of flat-partitions provide a useful classification of identically self-dual matroids. The identically self-dual matroids that do not admit any flat-partition are exactly the uniform matroids $U_{d,2d}$, which, as we said before, are self-dually representable.

We prove in Theorem 2 that the identically self-dual matroids with exactly one flat-partition are self-dually representable as well. These matroids are precisely the identically self-dual bipartite matroids. In a *bipartite matroid*, the set of points is divided in two parts and all points in each part are symmetrical. The access structures defined by these matroids are among the *bipartite access structures*, which were introduced in [23]. As a consequence of the results in [23], bipartite matroids are representable. Bipartite matroids have been independently studied in [20,21], where they are called *matroids with two uniform components*.

Bipartite access structures are also interesting for their applications because they appear in a natural way in situations in which the players are divided into two different classes. They are closely related to other families of access structures that have practical interest as well: the hierarchical access structures [30] and the weighted threshold access structures [2,26].

Theorem 2. *Let $\mathcal{M}$ be an identically self-dual bipartite matroid. Then, $\mathcal{M}$ can be represented by a self-dual linear code over some finite field $\mathbb{K}$. Equivalently, every self-dual bipartite vector space access structure can be realized by an ideal MLSSS over some finite field $\mathbb{K}$.*

Therefore, the bipartite matroids form another family of matroids for which all identically self-dual matroids are self-dually representable. Most of the identically self-dual matroids in this family are indecomposable. So, the existence of self-dual codes that represent them could not be derived from other matroids that were known to be self-dually representable.

3   Multiplicative Linear Secret Sharing Schemes

Some definitions and basic results about multiplicative linear secret sharing schemes are given in the following.

We begin by recalling some notation and elementary facts about bilinear forms. If $\alpha, \beta \colon E \to \mathbb{K}$ are linear forms, $\alpha \otimes \beta$ denotes the bilinear form $\alpha \otimes \beta \colon E \times E \to \mathbb{K}$ defined by $(\alpha \otimes \beta)(\mathbf{x}, \mathbf{y}) = \alpha(\mathbf{x})\beta(\mathbf{y})$. These bilinear forms span the vector space of all bilinear forms on E, which is denoted by $E^* \otimes E^*$ and has dimension d^2, where $d = \dim E$. Actually, if $\{\mathbf{e}^1, \ldots, \mathbf{e}^d\}$ is a basis of E^*, then $\{\mathbf{e}^i \otimes \mathbf{e}^j : 1 \leq i, j \leq d\}$ is a basis of $E^* \otimes E^*$. Since $E^{**} = E$, the vector space of the bilinear forms on E^* is $E \otimes E$, which is spanned by $\{\mathbf{x} \otimes \mathbf{y} : \mathbf{x}, \mathbf{y} \in E\}$. Finally, observe that $(E \otimes E)^* = E^* \otimes E^*$. This is due to the fact that any bilinear form $\alpha \otimes \beta \in E^* \otimes E^*$ induces a linear form $\alpha \otimes \beta \colon E \otimes E \to \mathbb{K}$, determined by $(\alpha \otimes \beta)(\mathbf{x} \otimes \mathbf{y}) = \alpha(\mathbf{x})\beta(\mathbf{y})$.

If $\Sigma = \Sigma_{n+1}(\pi_1, \ldots, \pi_n, \pi_{n+1})$ is an LSSS and $A \subset P_{n+1}$, we notate Σ_A for the natural restriction of Σ to the players in A, that is, the scheme defined by the linear mappings $((\pi_i)_{i \in A}, \pi_{n+1})$. The next definition deals with general (not necessarily ideal) LSSSs.

Definition 1. *Let* $\Sigma = \Sigma_{n+1}(\pi_1, \ldots, \pi_n, \pi_{n+1})$ *be a* $\mathbb{K}$-*LSSS with access structure* Γ. *The scheme* Σ *is said to be* multiplicative *if, for every* $i \in P_{n+1} = \{1, \ldots, n\}$, *there exists a bilinear form* $\phi_i \colon E_i \times E_i \to \mathbb{K}$ *such that* $(\pi_{n+1} \otimes \pi_{n+1})(\mathbf{x}_1, \mathbf{x}_2) = \sum_{i=1}^{n} \phi_i(\pi_i(\mathbf{x}_1), \pi_i(\mathbf{x}_2))$ *for any pair of vectors* $\mathbf{x}_1, \mathbf{x}_2 \in E$. *We say that* Σ *is* strongly multiplicative *if the scheme* $\Sigma_{P_{n+1}-A}$ *is multiplicative for every* $A \subset P_{n+1}$ *with* $A \notin \Gamma$.

It is not difficult to check that the access structure of a multiplicative LSSS must be $\mathcal{Q}_2$. Equally, strongly multiplicative LSSSs only exist for $\mathcal{Q}_3$ access structures.

Let $\Sigma = \Sigma_{n+1}(\Pi)$ be an ideal LSSS. Every bilinear form $\phi \colon \mathbb{K} \times \mathbb{K} \to \mathbb{K}$ can be defined by $\phi(x, y) = \lambda xy$ for some $\lambda \in \mathbb{K}$. Therefore, Σ is multiplicative if and only if there exist values $\lambda_i \in \mathbb{K}$ such that $\pi_{n+1} \otimes \pi_{n+1} = \sum_{i=1}^{n} \lambda_i (\pi_i \otimes \pi_i)$. Equally, Σ is strongly multiplicative if and only if, for every $A \notin \Gamma_{n+1}(\Pi)$, there exist values $\lambda_{i,A} \in \mathbb{K}$ such that $\pi_{n+1} \otimes \pi_{n+1} = \sum_{i \in P_{n+1}-A} \lambda_{i,A}(\pi_i \otimes \pi_i)$. The values λ_i or $\lambda_{i,A}$ form the *recombination vector* introduced in [10].

Since the bilinear forms $\pi_i \otimes \pi_i$ can be seen as vectors in $(E \otimes E)^*$, we can consider the LSSS $\Sigma_{n+1}^{\mu}(\Pi) = \Sigma_{n+1}(\pi_1 \otimes \pi_1, \ldots, \pi_n \otimes \pi_n, \pi_{n+1} \otimes \pi_{n+1})$, which has access structure $\Gamma_{n+1}^{\mu}(\Pi) = \Gamma_{n+1}(\pi_1 \otimes \pi_1, \ldots, \pi_n \otimes \pi_n, \pi_{n+1} \otimes \pi_{n+1})$. That is, $A \in \Gamma_{n+1}^{\mu}(\Pi)$ if and only if $\pi_{n+1} \otimes \pi_{n+1}$ is a linear combination of the vectors $\{\pi_i \otimes \pi_i : i \in A\}$.

Lemma 1. *Let* $\Sigma = \Sigma_{n+1}(\Pi)$ *be an ideal LSSS. Then, the following properties hold.*

1. $\Gamma_{n+1}^{\mu}(\Pi) \subset \Gamma_{n+1}(\Pi)$.
2. Σ *is multiplicative if and only if* $\Gamma_{n+1}^{\mu}(\Pi) \neq \emptyset$.
3. Σ *is strongly multiplicative if and only if* $(\Gamma_{n+1}(\Pi))^* \subset \Gamma_{n+1}^{\mu}(\Pi)$.

4   Reconstruction of a Secret in the Presence of Errors

In any LSSS with a $\mathcal{Q}_3$ access structure Γ, unique reconstruction of the secret from the full set of n shares is possible, even if the shares corresponding to an unqualified set $A \notin \Gamma$ are corrupted. Nevertheless, it is not known how to do that efficiently. In this section we prove Theorem 1, which implies that, if the LSSS is strongly multiplicative, there exists an efficient reconstruction algorithm.

We only consider here the *ideal* LSSS case. Proofs extend easily to the general case, at the cost of some notational headaches.

First we review the familiar case of Shamir's secret sharing scheme, where $t+1$ or more shares jointly determine the secret, and at most t shares do not even jointly contain any information about the secret. Exactly when $t < \frac{n}{3}$, unique reconstruction of the secret from the full set of n shares is possible, even if at most t shares are corrupted. This can be done efficiently, for instance by the Berlekamp-Welch decoding algorithm for Reed-Solomon codes.

Let p be a polynomial of degree at most t, and define $p(0) = s$. Let $\mathbf{s}$ be the vector with $s_i = p(i)$, $i = 1, \ldots, n$, and let $\mathbf{e}$ be a vector of Hamming-weight at most t. Write $\mathbf{c} = \mathbf{s} + \mathbf{e}$. Given $\mathbf{c}$ only, compute non-zero polynomials F and E

with $\deg(F) \le 2t$ and $\deg(E) \le t$, such that $F(i) = c_i \cdot E(i)$, for $i = 1, \ldots, n$. This is in fact a system of linear equations in the coefficients of F and E, and it has a non-trivial solution. Actually, for every polynomial E such that $E(i) = 0$ whenever the i-th share is corrupted, that is, $c_i \ne e_i$, the polynomials $F = pE$ and E are a solution to the system. Moreover, from Lagrange's Interpolation Theorem, all solutions are in this form. Therefore, for all F, E that satisfy the system, it holds that $E(i) = 0$ if the i-th share is corrupted. The corrupted shares are then deleted by removing all c_i with $E(i) = 0$ from $\mathbf{c}$. All that remains are incorrupted shares, that is, $c_j = s_j$, and there will be more than t of those left.

Below we present an efficient reconstruction algorithm for the more general situation where the secret is shared according to a strongly multiplicative LSSS with a $\mathcal{Q}_3$ access structure Γ. We do this by appropriately generalizing the Berlekamp-Welch algorithm. Note that such generalizations cannot generally rely on Lagrange's Interpolation Theorem, since LSSSs are not in general based on evaluation of polynomials.

Pellikaan [24] has previously generalized the Berlekamp-Welch algorithm and has shown that his generalized decoding algorithm applies to a much wider class of error correcting codes. Technically, our generalization bears some similarity to that of [24].

Strong multiplication was first considered in [10] and was used to construct efficient multi-party computation protocols with zero error in the active adversary model. More precisely it is used in the *Commitment Multiplication Protocol* to ensure that commitments for a, b and c are consistent in the sense that $ab = c$ with zero probability to cheat.

We now prove Theorem 1. Let $\Pi = (\pi_1, \ldots, \pi_n, \pi_{n+1})$ be a sequence of linear forms $\pi_i \colon E \to \mathbb{K}$ such that $\Sigma = \Sigma_{n+1}(\Pi)$ is a strongly multiplicative LSSS with $\mathcal{Q}_3$ access structure $\Gamma = \Gamma_{n+1}(\Pi)$. Let us consider also the scheme $\Sigma^\mu = \Sigma^\mu_{n+1}(\Pi) = \Sigma_{n+1}(\pi_1 \otimes \pi_1, \ldots, \pi_n \otimes \pi_n, \pi_{n+1} \otimes \pi_{n+1})$. From Lemma 1, the access structure of this scheme, $\Gamma^\mu = \Gamma^\mu_{n+1}(\Pi)$, is such that $\Gamma^* \subset \Gamma^\mu$.

Let us fix a basis for E and the induced basis of $E \otimes E$. Let M and $\widehat{M}$ be the matrices associated, respectively, to the schemes Σ and Σ^μ. Observe that, if $d = \dim E$, the matrix M has d rows and $n + 1$ columns while $\widehat{M}$ has d^2 rows and $n + 1$ columns.

If $\mathbf{u}, \mathbf{v} \in \mathbb{K}^k$, then $\mathbf{u} * \mathbf{v}$ denotes the vector $(u_1 v_1, \ldots, u_k v_k)$. Observe that

$$(\mathbf{x} \otimes \mathbf{y})\widehat{M} = ((\pi_i \otimes \pi_i)(\mathbf{x} \otimes \mathbf{y}))_{1 \le i \le n+1} = (\pi_i(\mathbf{x})\pi_i(\mathbf{y}))_{1 \le i \le n+1} = (\mathbf{x}M) * (\mathbf{y}M)$$

for every pair of vectors $\mathbf{x}, \mathbf{y} \in E$.

Let us consider $\mathbf{s}' = (s_1, \ldots, s_n, s_{n+1}) = \mathbf{x}M$. Then, $\mathbf{s} = (s_1, \ldots, s_n)$ is a full set of shares for the secret $s_{n+1} = \pi_{n+1}(\mathbf{x})$. Let $A \subset P_{n+1}$ be a non-qualified subset, that is, $A \notin \Gamma$. Let $\mathbf{e} = (e_1, \ldots, e_n)$ be a vector with $e_i = 0$ for every $i \notin A$. Write $\mathbf{c} = (c_1, \ldots, c_n) = \mathbf{s} + \mathbf{e}$. Given only $\mathbf{c}$, the secret s_{n+1} is recovered efficiently as follows.

Let $\widehat{N}$ and N be the matrices that are obtained, respectively, from $\widehat{M}$ and M by removing the last column. Observe that $\mathbf{c} = \mathbf{x}N + \mathbf{e}$. Let us consider the system of linear equations

$$\begin{cases} \widehat{\mathbf{y}}\widehat{N} = \mathbf{c} * (\mathbf{y}N) \\ \pi_{n+1}(\mathbf{y}) = 1 \end{cases}$$

where the unknowns are the d^2 coordinates of the vector $\widehat{\mathbf{y}} \in E \otimes E$ and the d coordinates of the vector $\mathbf{y} \in E$. We claim that this system of linear equations always has a solution and that $s_{n+1} = (\pi_{n+1} \otimes \pi_{n+1})(\widehat{\mathbf{y}})$ for every solution $(\widehat{\mathbf{y}}, \mathbf{y})$. Therefore, the secret s_{n+1} can be obtained from $\mathbf{c}$ by solving that system of linear equations.

This is argued as follows. Note that $(\widehat{\mathbf{y}}, \mathbf{y})$ is a solution if and only if $(\widehat{\mathbf{y}} - \mathbf{x} \otimes \mathbf{y})\widehat{N} = \mathbf{e} * (\mathbf{y}N)$. Since $A \notin \Gamma$, there exists a vector $\mathbf{z} \in E$ such that $\pi_{n+1}(\mathbf{z}) = 1$ while $\pi_i(\mathbf{z}) = 0$ for every $i \in A$. Observe that $(\mathbf{x} \otimes \mathbf{z}, \mathbf{z})$ is a solution for every vector $\mathbf{z} \in E$ in that situation. Indeed, $\mathbf{e} * (\mathbf{z}N) = 0$, because $\mathbf{z}N$ is zero where $\mathbf{e}$ is non-zero. Let $(\widehat{\mathbf{y}}, \mathbf{y})$ be an arbitrary solution and consider $(\widehat{\mathbf{y}} - \mathbf{x} \otimes \mathbf{y})\widehat{M} = (t_1, \ldots, t_n, t_{n+1})$. Then, $(t_1, \ldots, t_n)$ are shares of the secret t_{n+1} according to the LSSS Σ^μ. Since $(t_1, \ldots, t_n) = \mathbf{e} * (\mathbf{y}N)$, we get that $t_i = 0$ for every $i \in P_{n+1} - A$ and, hence, $t_{n+1} = 0$ because $P_{n+1} - A \in \Gamma^* \subset \Gamma^\mu$. Finally, $(\pi_{n+1} \otimes \pi_{n+1})(\widehat{\mathbf{y}} - \mathbf{x} \otimes \mathbf{y}) = t_{n+1} = 0$ and $(\pi_{n+1} \otimes \pi_{n+1})(\widehat{\mathbf{y}}) = (\pi_{n+1} \otimes \pi_{n+1})(\mathbf{x} \otimes \mathbf{y}) = \pi_{n+1}(\mathbf{x})\pi_{n+1}(\mathbf{y}) = s_{n+1}$. $\square$

A positive application of Theorem 1 is as follows. Using a strongly multiplicative LSSS, the Commitment Multiplication Protocol (CMP) from [10] is directly a Verifiable Secret Sharing scheme (VSS). This saves a multiplicative factor n in the volume of communication needed, since the general reduction from VSS to CMP is not needed in this case.

5 Ideal Multiplicative Linear Secret Sharing Schemes, Self-dual Linear Codes and Identically Self-dual Matroids

5.1 Equivalence Between the Two Problems

A matroid $\mathcal{M}$ is said to be *connected* if, for every two different points $i, j \in Q$, there exists a circuit C with $i, j \in C$. In a *connected access structure*, every participant is at least in a minimal qualified subset. If $\mathcal{M}(\Pi)$ is a connected matroid, all access structures $\Gamma_i(\Pi)$ are connected. Moreover, as a consequence of [22, Proposition 4.1.2], if one of the access structures $\Gamma_i(\Pi)$ is connected, then $\mathcal{M}(\Pi)$ is connected and, hence, all the other access structures $\Gamma_j(\Pi)$ are connected.

We say that a linear code $\mathcal{C}$ with generator matrix M is *almost self-dual* if there exists a non-singular diagonal matrix $D = \text{diag}(\lambda_1, \ldots, \lambda_n, \lambda_{n+1})$ such that MD is a parity check matrix.

Lemma 2. Let $\Pi = (\pi_1, \ldots \pi_{2d})$ be a sequence of linear forms in $E^* = (\mathbb{K}^d)^*$ such that the matroid $\mathcal{M}(\Pi)$ is identically self-dual and connected. In the space $\mathcal{S}(E)$ of the symmetric bilinear forms on E, the vectors $\{\pi_j \otimes \pi_j : j \in Q - \{i\}\}$

are linearly independent for any $i \in Q$. Besides, the code $C(\Pi)$ is almost self-dual if and only if the vectors $\{\pi_j \otimes \pi_j : j \in Q\}$ are linearly dependent.

Proof. Let us suppose that the vectors $\{\pi_j \otimes \pi_j : 1 \leq j \leq 2d - 1\}$ are linearly dependent. Then, we can suppose that $\pi_1 \otimes \pi_1 = \sum_{i=2}^{2d-1} \lambda_i(\pi_i \otimes \pi_i)$. The access structure $\Gamma_1(\Pi)$ is self-dual and connected. Then, there exists a minimal qualified subset $A \subset P_1$ such that $2d \in A$. We can suppose that $A = \{r+1, \ldots, 2d-1, 2d\}$. Since $\Gamma_1(\Pi)$ is self-dual, $P_1 - A = \{2, \ldots, r\}$ is not qualified. Then, there exists a vector $\mathbf{x} \in E$ such that $\pi_1(\mathbf{x}) = 1$ and $\pi_i(\mathbf{x}) = 0$ for every $i = 2, \ldots, r$. Therefore, $\pi_1 = \sum_{i=r+1}^{2d-1}(\lambda_i \pi_i(\mathbf{x}))\pi_i$, a contradiction with the fact that $A = \{r+1, \ldots, 2d-1, 2d\}$ is a *minimal* qualified subset of the access structure $\Gamma_1(\Pi)$.

Observe that $\sum_{i=1}^{2d} \lambda_i(\pi_i \otimes \pi_i) = 0$ if and only if the diagonal matrix $D = \operatorname{diag}(\lambda_1, \ldots, \lambda_{2d-1}, \lambda_{2d})$ is such that $MDM^\top = 0$. $\square$

By taking into account that a non-connected matroid can be divided into connected components [22, Proposition 4.1.2], the equivalence between Problems 1 and 2 is an immediate consequence of the following two propositions. We skip the proof of the first one.

Proposition 1. *Let $\mathcal{M}$ be an identically self-dual representable connected matroid on the set of points $Q = \{1, \ldots, 2d\}$ and let $\Gamma_{2d}(\mathcal{M})$ be the access structure induced by $\mathcal{M}$ on the set P_{2d}. Then $\Gamma_{2d}(\mathcal{M})$ can be realized by an ideal multiplicative $\mathbb{K}$-LSSS if and only if $\mathcal{M}$ can be represented by an almost self-dual code C over the field $\mathbb{K}$.*

Proposition 2. *Let $\mathcal{M}$ be an identically self-dual matroid that is represented, over the finite field $\mathbb{K}$, by an almost self-dual code. Then, there exists a finite field $\mathbb{L}$, which is an algebraic extension of $\mathbb{K}$, such that $\mathcal{M}$ is represented by a self-dual code over $\mathbb{L}$.*

Proof. Let C be an almost self-dual code over a finite field $\mathbb{K}$. Let M be a generator matrix and $D = \operatorname{diag}(\lambda_1, \ldots, \lambda_{2d-1}, \lambda_{2d})$ the non-singular diagonal matrix such that MD is a parity check matrix. Let us consider, in an extension field $\mathbb{L} \supset \mathbb{K}$, the diagonal matrix $D_1 = \operatorname{diag}(\sqrt{\lambda_1}, \ldots, \sqrt{\lambda_{2d-1}}, \sqrt{\lambda_{2d}})$. Then, the matrix $M_1 = MD_1$ is a generator matrix of a self-dual code C_1. The matroids associated to C and to C_1 are equal. $\square$

5.2 Known Families of Self-dually Representable Matroids

There are two families of matroids for which it is known that all identically self-dual matroids are self-dually representable.

The uniform matroids are the first example. A uniform matroid $U_{d,n}$ is identically self-dual if and only if $n = 2d$. The access structure $\Gamma_{2d}(U_{d,2d})$ is the threshold structure $\Gamma_{d,2d-1}$, which can be realized by an ideal multiplicative $\mathbb{K}$-LSSS for any finite field $\mathbb{K}$ with $|\mathbb{K}| \geq 2d$. Namely, the Shamir's polynomial scheme. Therefore, the matroid $U_{d,2d}$ can be represented by an almost self-dual code over any finite field $\mathbb{K}$ with $|\mathbb{K}| \geq 2d$.

The second family is formed by the $\mathbb{Z}_2$-representable matroids. For any of these matroids $\mathcal{M}$, there exists a unique $\mathbb{Z}_2$-representation. That is, there exists a unique linear code $\mathcal{C}$ over $\mathbb{Z}_2$ whose associated matroid is $\mathcal{M}$. If $\mathcal{M}$ is an identically self-dual $\mathbb{Z}_2$-representable matroid, the codes $\mathcal{C}$ and $\mathcal{C}^\perp$ are $\mathbb{Z}_2$-representations of $\mathcal{M}$ and, hence, $\mathcal{C} = \mathcal{C}^\perp$. Therefore, all identically self-dual $\mathbb{Z}_2$-representable matroids are self-dually $\mathbb{Z}_2$-representable. For instance, an identically self-dual binary matroid $\mathcal{M}$ on the set $Q = \{1, \ldots, 8\}$ is obtained from the eight vectors in the set $\{(v_1, v_2, v_3, v_4) \in \mathbb{Z}_2^4 : v_1 = 1\}$. All access structures that are obtained from $\mathcal{M}$ are isomorphic to the access structure defined by the Fano Plane by considering the points in the plane as the players and the lines as the minimal qualified subsets [18]. Therefore, this access structure can be realized by an ideal multiplicative $\mathbb{Z}_2$-LSSS.

5.3 Flat-Partitions and Sum of Matroids

We recall next the definition and some properties of the sum of two matroids. More information on that topic can be found in [22, Chapter 7].

Let $\mathcal{M}_1$ and $\mathcal{M}_2$ be connected matroids on the sets Q_1 and Q_2, respectively. Let $\mathcal{B}_1$ and $\mathcal{B}_2$ be their families of bases. Let us suppose that $Q_1 \cap Q_2 = \emptyset$ and let us take two points $q_1 \in Q_1$ and $q_2 \in Q_2$. The *sum of $\mathcal{M}_1$ and $\mathcal{M}_2$ at the points q_1 and q_2*, which will be denoted by $\mathcal{M} = \mathcal{M}_1 \oplus_{(q_1, q_2)} \mathcal{M}_2$, is the matroid on the set of points $Q = (Q_1 \cup Q_2) \setminus \{q_1, q_2\}$ whose family of bases is $\mathcal{B} = \mathcal{B}_1' \cup \mathcal{B}_2'$, where $\mathcal{B}_1' = \{B_1 \cup C_2 \subset Q : B_1 \in \mathcal{B}_1, C_2 \cup \{q_2\} \in \mathcal{B}_2\}$ and $\mathcal{B}_2' = \{C_1 \cup B_2 \subset Q : C_1 \cup \{q_1\} \in \mathcal{B}_1, B_2 \in \mathcal{B}_2\}$.

It is not difficult to check that $\mathcal{B}$ is the family of bases of a matroid and that $\mathcal{M}$ is a connected matroid with $\dim \mathcal{M} = \dim \mathcal{M}_1 + \dim \mathcal{M}_2 - 1$. The proof of the following proposition will be given in the full version of the paper.

Proposition 3. *The matroid $\mathcal{M} = \mathcal{M}_1 \oplus_{(q_1, q_2)} \mathcal{M}_2$ is identically self-dual if and only if both $\mathcal{M}_1$ and $\mathcal{M}_2$ are identically self-dual.*

We say that a sum of matroids $\mathcal{M}_1 \oplus \mathcal{M}_2$ is *trivial* if one of the matroids $\mathcal{M}_i$ is the uniform matroid $U_{1,2}$. In this case, $\mathcal{M}_1 \oplus U_{1,2} \cong \mathcal{M}_1$. A matroid $\mathcal{M}$ is *indecomposable* if it is not isomorphic to any non-trivial sum of matroids.

Let $\mathcal{M}$ be a matroid on a set of points Q and let (X_1, X_2) be a partition of Q. We say that (X_1, X_2) is a *flat-partition* of $\mathcal{M}$ if X_1 and X_2 are flats of $\mathcal{M}$. The next proposition, which is a consequence of the results in [22, Chapter 7], provides a characterization of indecomposable identically self-dual matroids in terms of their flat-partitions

Proposition 4. *Let $\mathcal{M}$ be a connected identically self-dual matroid. Then $\mathcal{M}$ is indecomposable if and only if there is no flat-partition (X_1, X_2) of $\mathcal{M}$ with $\mathrm{rank}(X_1) + \mathrm{rank}(X_2) = \dim(\mathcal{M}) + 1$.*

As a consequence of Proposition 3 and the next two propositions, whose proofs will be given in the full version of the paper, the search for an answer to Problem 2 can be restricted to indecomposable matroids.

Proposition 5. *Let $\mathcal{M} = \mathcal{M}_1 \oplus_{(q_1, q_2)} \mathcal{M}_2$ be a non-trivial sum of two identically self-dual matroids. Then $\mathcal{M}$ is $\mathbb{K}$-representable if and only if both $\mathcal{M}_1$ and $\mathcal{M}_2$ are $\mathbb{K}$-representable.*

Proposition 6. *Let $\mathcal{M}_1$ and $\mathcal{M}_2$ be two matroids that are represented over a finite field $\mathbb{K}$ by almost self-dual codes. Then, the sum $\mathcal{M} = \mathcal{M}_1 \oplus_{(q_1, q_2)} \mathcal{M}_2$ can be represented over $\mathbb{K}$ by an almost self-dual code. Besides, if $\mathcal{M}_1$ and $\mathcal{M}_2$ are self-dually $\mathbb{K}$-representable, the sum $\mathcal{M}$ is self-dually $\mathbb{L}$-representable, where $\mathbb{L}$ is an algebraic extension of $\mathbb{K}$ with $(\mathbb{K} : \mathbb{L}) \leq 2$.*

5.4   Identically Self-dual Bipartite Matroids

It is not hard to see that the uniform matroid $U_{d,2d}$ on the set $Q = \{1, \ldots, 2d\}$ does not admit any flat-partition. As a direct cosequence of the next lemma, any non-uniform identically self-dual matroid admits at least one flat partition.

Lemma 3. *Let $\mathcal{M}$ be an identically self-dual matroid and let $C \subset Q$ be a circuit of $\mathcal{M}$ with $\text{rank}(C) < \dim(\mathcal{M})$. Let us consider the flat $X_1 = \langle C \rangle$ and $X_2 = Q \setminus X_1$. Then, (X_1, X_2) is a flat-partition of $\mathcal{M}$.*

Proof. We have to prove that X_2 is a flat. Otherwise, there exists $x \in X_1 \cap \langle X_2 \rangle$. Since C is a circuit, there exists a basis B_1 of X_1 with $x \notin B_1$. Besides, there exists $C_2 \subset X_2$ such that $B = B_1 \cup C_2$ is a basis of $\mathcal{M}$. Let us consider the basis $B' = Q \setminus B$ and let us take $B_2 = B' \cap X_2$.

We are going to prove that $\langle B_2 \rangle = X_2$. If not, there exists $y \in X_2 \setminus \langle B_2 \rangle$. Observe that $y \in C_2$ and that $B_2 \cup \{y\}$ is an independent set. Therefore, $Q \setminus (B_2 \cup \{y\}) = X_1 \cup (C_2 \setminus \{y\})$ is a spanning set. Since $\langle B_1 \rangle = X_1$, we have that $B'' = B_1 \cup (C_2 \setminus \{y\})$ is equally a spanning set, a contradiction with $B'' \subsetneq B$. Therefore, $x \in \langle B_2 \rangle$, a contradiction with $B_2 \cup \{x\} \subset B'$. $\square$

As said before, any identically self-dual uniform matroid $U_{d,2d}$ can be represented by a self-dual code $\mathcal{C}$ over some finite field $\mathbb{K}$. By the above observation, this means that the identically self-dual matroids that do not admit any flat-partition are self-dually representable.

A natural question arising at this point is whether the same occurs with the identically self-dual matroids that admit exactly *one* flat-partition. Proposition 8 shows that these matroids coincide with the identically self-dual bipartite matroids.

Definition 2. *Let d, r_1 and r_2 be any integers such that $1 < r_i < d < r_1 + r_2$. Let us take $Q = \{1, \ldots, n, n+1\}$ and a partition (X_1, X_2) of Q with $|X_i| \geq r_i$. We define the matroid $\mathcal{M} = \mathcal{M}(X_1, X_2, r_1, r_2, d)$ by determining its bases: $B \subset Q$ is a basis of $\mathcal{M}$ if and only if $|B| = d$ and $d - r_j \leq |B \cap X_i| \leq r_i$, where $\{i, j\} = \{1, 2\}$. Observe that (X_1, X_2) is a flat-partition of Q with $\text{rank}(X_i) = r_i$. Any matroid in this form is said to be bipartite.*

We skip the proof of the next proposition, which determines which bipartite matroids are identically self-dual.

Proposition 7. *Let $\mathcal{M} = \mathcal{M}(X_1, X_2, r_1, r_2, d)$ be a bipartite matroid. Then, $\mathcal{M}$ is identically self-dual if and only if $|Q| = 2d$ and $|X_1| = d + r_1 - r_2$.*

Proposition 8. *Let $\mathcal{M}$ be a connected identically self-dual matroid. Then, $\mathcal{M}$ is bipartite if and only if it admits exactly one flat-partition.*

Proof. Let us suppose that $\mathcal{M}$ is bipartite, that is, $\mathcal{M} = \mathcal{M}(X_1, X_2, r_1, r_2, d)$. We have to prove that (X_1, X_2) is the only flat-partition of $\mathcal{M}$. Let (Y_1, Y_2) be a flat-partition of $\mathcal{M}$. We can suppose that $|Y_1| \geq d = \dim(\mathcal{M})$. If $|Y_1 \cap X_i| \geq d - r_j$ for all $\{i, j\} = \{1, 2\}$, there exists $B \subset Y_1$ such that $|B| = d$ and $d - r_j \leq |B \cap X_i| \leq r_i$. Since Y_1 does not contain any basis of $\mathcal{M}$, we get $|Y_1 \cap X_1| < d - r_2$ or $|Y_1 \cap X_2| < d - r_1$. Without loss of generality, we assume that $|Y_1 \cap X_2| < d - r_1$. Then, $|Y_1 \cap X_1| > r_1$ because $|Y_1 \cap X_1| + |Y_1 \cap X_2| \geq d$. Besides, since $d + r_2 - r_1 = |Y_1 \cap X_2| + |Y_2 \cap X_2|$, we have that $|Y_2 \cap X_2| > r_2$. Observe that, for $i = 1, 2$, any subset of r_i points in X_i is independent and, hence, $X_i \subset Y_i$ because Y_i is a flat and contains a basis of X_i. Therefore, $(X_1, X_2) = (Y_1, Y_2)$.

Let us suppose now that (X_1, X_2) is the only flat-partition of $\mathcal{M}$. We are going to prove that $\mathcal{M}$ is the bipartite matroid $\mathcal{M}(X_1, X_2, r_1, r_2, d)$, where $r_i = \mathrm{rank}(X_i)$ and $d = \dim(\mathcal{M})$. It is not difficult to check that $1 < r_i < d < r_1 + r_2$ and that $d - r_j \leq |B \cap X_i| \leq r_i$ if B is a basis of $\mathcal{M}$ and $\{i, j\} = \{1, 2\}$. We only have to prove that any set $B \subset Q$ such that $|B| = d$ and $d - r_j \leq |B \cap X_i| \leq r_i$ for $\{i, j\} = \{1, 2\}$ is a basis of $\mathcal{M}$. Let us suppose that, on the contrary, there exists such a subset B that is not a basis. Then, there exists a circuit $C \subset B$. Let us consider $Y_1 = \langle C \rangle$ and $Y_2 = Q \setminus Y_1$. From Lemma 3, (Y_1, Y_2) is a flat-partition of $\mathcal{M}$. The proof is concluded by showing that this flat-partition is different from (X_1, X_2). If $Y_1 = X_i$ for some $i = 1, 2$, we have $C \subset X_i$. Since $|C| \leq r_i$ and C is a circuit, $\mathrm{rank}(Y_1) < r_i$, a contradiction.   □

The access structures defined by bipartite matroids were first considered in [23], where the authors proved that they are vector space access structures, that is, they admit an ideal LSSS. As a direct consequence of this fact, any bipartite matroid is representable.

Theorem 2 extends this result of [23] by showing that, additionally, the identically self-dual bipartite matroids are self-dually representable. This is done by a refinement of the approach of [23] based on techniques from Algebraic Geometry.

From Propositions 4, 7 and 8, if $r_1 + r_2 - d > 1$, the identically self-dual bipartite matroid $\mathcal{M} = \mathcal{M}(X_1, X_2, r_1, r_2, d)$ is indecomposable. Therefore, we found a new large family of identically self-dual matroids giving an affirmative answer to Problem 2 and, hence, a new large family of self-dual vector space access structures for which Problem 1 has a positive answer.

The proof of Theorem 2, which is quite long and involved, will be given in the full version of the paper. In the following, we present a brief sketch of it. Given an identically self-dual bipartite matroid $\mathcal{M} = \mathcal{M}(X_1, X_2, r_1, r_2, d)$, one has to prove the existence of a finite field $\mathbb{K}$ and a set of $\mathbb{K}$-linear forms $\{\pi_1, \ldots, \pi_{2d}\}$ satisfying two requirements: first, they must be a $\mathbb{K}$-representation of the matroid $\mathcal{M}$ and, second, the vectors $\{\pi_i \otimes \pi_i : 1 \leq i \leq 2d\}$ must be linearly dependent.

In order to prove the existence of those linear forms, we conveniently choose some fixed vectors $\{\pi_1, \ldots, \pi_{n_1}\}$ corresponding to the points in the flat X_1 and a family of vectors $\{\mathbf{w}(x) \ : \ x \in \mathbb{K}\} \subset (\mathbb{K}^d)^*$ depending on one parameter. Afterwards, we use some Algebraic Geometry to prove that there exist vectors $\pi_{n_1+i} = \mathbf{w}(\beta_i^{-1})$, where $i = 1, \ldots, n_2 = |X_2|$, such that the vectors $\{\pi_1, \ldots, \pi_{2d}\}$ have the required properties. Specifically, the second requirement above is satisfied if the point $(\beta_1, \ldots, \beta_{n_2})$ is a zero of a system of polynomial equations on n_2 variables. These equations define an algebraic variety M in $\overline{\mathbb{Z}}_p^{n_2}$, where $\overline{\mathbb{Z}}_p$ is the algebraic closure of the finite field $\mathbb{Z}_p$. If p is large enough, the variety M is irreducible [25]. The first requirement is verified if other polynomials on the same variables are not zero in the point $(\beta_1, \ldots, \beta_{n_2})$. Every one of these equations defines and algebraic variety V_j in $\overline{\mathbb{Z}}_p^{n_2}$. We prove that M is not a subset of any of the varieties V_j and, since M is irreducible, this implies $M \not\subset \bigcup V_j$ [17]. Therefore there exists a point $(\beta_1, \ldots, \beta_{n_2}) \in M - (\bigcup V_j) \subset \overline{\mathbb{Z}}_p^{n_2}$. Finally, we take a finite field $\mathbb{K}$, an algebraic extension of $\mathbb{Z}_p$ containing all values β_i, and over that field, the linear forms $\pi_{n_1+i} = \mathbf{w}(\beta_i^{-1})$.

References

1. A. Barg. On some polynomials related to weight enumerators of linear codes. *SIAM J. Discrete Math.* **15** (2002) 155–164.
2. A. Beimel, T. Tassa, E. Weinreb. Characterizing Ideal Weighted Threshold Secret Sharing. *Theory of Cryptography, Second Theory of Cryptography Conference, TCC 2005. Lecture Notes in Comput. Sci.* **3378** (2005) 600–619.
3. M. Ben-Or, S. Goldwasser, A. Wigderson. Completeness theorems for non-cryptographic fault-tolerant distributed computation. *Proc. ACM STOC'88* (1988) 1–10.
4. E.F. Brickell. Some ideal secret sharing schemes. *J. Combin. Math. and Combin. Comput.* **9** (1989) 105–113.
5. E.F. Brickell, D.M. Davenport. On the classification of ideal secret sharing schemes. *J. Cryptology.* **4** (1991) 123–134.
6. T. Britz. MacWilliams identities and matroid polynomials. *Electron. J. Combin.* **9** (2002), Research Paper 19, 16 pp.
7. P.J. Cameron. Cycle index, weight enumerator, and Tutte polynomial. *Electron. J. Combin.* **9** (2002), Note 2, 10 pp.
8. R. Canetti, U. Feige, O. Goldreich, M. Naor. Adaptively secure multi-party computation. *Proc. ACM STOC'96* (1996) 639–648.
9. D. Chaum, C. Crépeau, I. Damgård. Multi-party unconditionally secure protocols. *Proc. ACM STOC'88* (1988) 11–19.
10. R. Cramer, I. Damgård, U. Maurer. General Secure Multi-Party Computation from any Linear Secret-Sharing Scheme. *Advances in Cryptology - EUROCRYPT 2000, Lecture Notes in Comput. Sci.* **1807** (2000) 316–334.
11. O. Goldreich, M. Micali, A. Wigderson. How to play any mental game or a completeness theorem for protocols with honest majority. *Proc.19th ACM Symposium on the Theory of Computing STOC'87* (1987) 218–229.
12. C. Greene. Weight enumeration and the geometry of linear codes. *Studies in Appl. Math.* **55** (1976) 119–128.

13. M. Hirt, U. Maurer. Complete characterization of adversaries tolerable in secure multi-party computation. *Proc. 16th Symposium on Principles of Distributed Computing PODC '97* (1997) 25–34.

14. M. Ito, A. Saito, T. Nishizeki. Secret sharing scheme realizing any access structure. *Proc. IEEE Globecom'87.* (1987) 99–102.

15. W.-A. Jackson, K.M. Martin. Geometric secret sharing schemes and their duals. *Des. Codes Cryptogr.* **4** (1994) 83–95.

16. M. Karchmer, A. Wigderson. On span programs. *Proceedings of the Eighth Annual Structure in Complexity Theory Conference* (San Diego, CA, 1993), 102–111, 1993.

17. E. Kunz. *Introduction to Commutative Algebra and Algebraic Geometry.* Birkhäuser, Boston, 1985.

18. J. Martí-Farré, C. Padró. Secret sharing schemes on access structures with intersection number equal to one. In *Proceedings of the Third Conference on Security in Communication Networks '02, Lecture Notes in Comput. Sci.* **2576** (2003) 354–363. Amalfi, Italy, 2002.

19. F. Matúš. Matroid representations by partitions. *Discrete Mathematics* **203** (1999) 169–194.

20. S.-L. Ng. A Representation of a Family of Secret Sharing Matroids. *Des. Codes Cryptogr.* **30** (2003) 5-19.

21. S.-L. Ng, M. Walker. On the composition of matroids and ideal secret sharing schemes. *Des. Codes Cryptogr.* **24** (2001) 49-67.

22. J.G. Oxley. *Matroid theory.* Oxford Science Publications. The Clarendon Press, Oxford University Press, New York, 1992.

23. C. Padró, G. Sáez. Secret sharing schemes with bipartite access structure. *IEEE Transactions on Information Theory.* **46** (2000) 2596–2604. A previous version appeared in *Advances in Cryptology - EUROCRYPT'98, Lecture Notes in Comput. Sci.* **1403** (1998) 500-511.

24. R. Pellikaan. On decoding by error location and dependent sets of error positions. *Discrete Math.* **106/107** (1992) 369–381.

25. Z. Reichstein, B. Youssin. Essential dimensions of algebraic groups and a resolution theorem for *G*-varieties. With an appendix by János Kollár and Endre Szabó. *Canad. J. Math.* **52** (2000) 1018–1056.

26. A. Shamir. How to share a secret. *Commun. of the ACM.* **22** (1979) 612–613.

27. G.J. Simmons. An introduction to shared secret and/or shared control schemes and their application. *Contemporary Cryptology. The Science of Information Integrity.* IEEE Press (1991), 441-497.

28. J. Simonis, A. Ashikhmin. Almost affine codes. *Des. Codes Cryptogr.* **14** (1998) 179–197.

29. D.R. Stinson. An explication of secret sharing schemes. *Des. Codes Cryptogr.* **2** (1992) 357–390.

30. T. Tassa. Hierarchical Threshold Secret Sharing. *Theory of Cryptography, First Theory of Cryptography Conference, TCC 2004. Lecture Notes in Comput. Sci.* **2951** (2004) 473–490.

Black-Box Secret Sharing from Primitive Sets in Algebraic Number Fields

Ronald Cramer[1,2], Serge Fehr[1], and Martijn Stam[3,*]

[1] CWI, Amsterdam, The Netherlands
{cramer, fehr}@cwi.nl
[2] Mathematical Institute, Leiden University, The Netherlands
[3] Department of Computer Science, University of Bristol, United Kingdom
stam@cs.bris.ac.uk

Abstract. A *black-box* secret sharing scheme (BBSSS) for a given access structure works in exactly the same way over any finite Abelian group, as it only requires black-box access to group operations and to random group elements. In particular, there is no dependence on e.g. the structure of the group or its order. The expansion factor of a BBSSS is the length of a vector of shares (the number of group elements in it) divided by the number of players n.

At CRYPTS 2002 Cramer and Fehr proposed a threshold BBSSS with an asymptotically minimal expansion factor $\Theta(\log n)$.

In this paper we propose a BBSSS that is based on a new paradigm, namely, *primitive sets in algebraic number fields*. This leads to a new BB-SSS with an expansion factor that is absolutely minimal up to an additive term of at most 2, which is an improvement by a constant additive factor.

We provide good evidence that our scheme is considerably more efficient in terms of the computational resources it requires. Indeed, the number of group operations to be performed is $\tilde{O}(n^2)$ instead of $\tilde{O}(n^3)$ for sharing and $\tilde{O}(n^{1.6})$ instead of $\tilde{O}(n^{2.6})$ for reconstruction.

Finally, we show that our scheme, as well as that of Cramer and Fehr, has asymptotically optimal randomness efficiency.

1 Introduction

The concept of *secret sharing* was introduced independently by Shamir [12] and by Blakley [1] as a means to protect a secret simultaneously from exposure and from being lost. It allows to share the secret among a set of n participants, in such a way that any coalitions of at least $t+1$ participants can reconstruct the secret (completeness) while any t or fewer participants have no information about it (privacy). The work of Shamir and Blakley spawned a tremendous amount of research [15].

Of particular interest to us is *black-box* secret sharing, introduced by Desmedt and Frankel [4]. A black-box secret sharing scheme is distinguished in that it

---

* This author is partially supported by the EPSRC and by the Commission of European Communities through the IST program under contract IST-2002-507932 ECRYPT.

V. Shoup (Ed.): Crypto 2005, LNCS 3621, pp. 344–360, 2005.

works over any finite Abelian group and requires only black-box access to the group operations and to random group elements. The distribution matrix and the reconstruction vectors are defined independently of the group from which the secret is sampled, and completeness and privacy are guaranteed to hold regardless of which group is chosen. Simple cases are $t = 0$ where each participant is given a copy of the secret, and $t = n-1$ where a straightforward additive sharing suffices. We will henceforth assume that $0 < t < n - 1$ to exclude these trivial cases.

The original motivation for looking at black-box secret sharing was their applicability to threshold RSA. Although threshold RSA can nowadays be much more conveniently dealt with using Shoup's threshold RSA technique [13] (or in the proactive case using the techniques of Frankel et al. [7]), black-box secret sharing still remains a useful primitive with several applications such as black-box ring multiparty computation [3] or threshold RSA with small public exponent (in which case Shoup's technique fails), and it may very well be relevant to new distributed cryptographic schemes, for instance based on class groups. Furthermore, this problem has turned into an interesting cryptographic problem in its own right.

The average number of group elements handed out to a participant in order to share a single group element is known as the *expansion factor*. The expansion factor expresses the bandwidth taken up by the scheme and is therefore an important property of the scheme. Desmedt and Frankel [4] proposed a scheme with expansion factor linear in the number n of participants. Their scheme is based on finding an invertible Vandermonde determinant over a cyclotomic number field. In subsequent works some improvements to the expansion factor were made, but all within a constant factor [5,10].

At Crypto 2002, Cramer and Fehr [2] used a new approach based on finding pairs of co-prime Vandermonde determinants over low degree integral extensions of $\mathbb{Z}$. This results in black-box secret sharing schemes with logarithmic expansion factor. They also show that this is asymptotically optimal by proving a tight lower bound. In fact, they prove that the expansion factor of their scheme is minimal up to an additive term of at most three.[1]

We improve these results on black-box secret sharing in several ways. We describe a novel technique for constructing black-box secret sharing schemes, by in a way combining the advantages of both approaches. Briefly, our approach requires to find *one* primitive Vandermonde determinant over a *low degree* integral extension of $\mathbb{Z}$. A Vandermonde determinant is primitive in an integral extension if its only rational integer divisors are $-1, +1$. This allows us to further reduce the gap between the expansion factor and the lower bound by one. By using a slight tweak which applies very generally to Shamir-like schemes, the expansion factor drops one more in case the number of participants is a power of two.[2]

---

[1] Note that Fehr [6, Corollary 4.1] incorrectly claims an additive term of at most two.

[2] This tweak is interesting in its own right: it allows one to do a Shamir-like secret sharing over a field F of size $|F| \geq n$, rather than $|F| > n$. Yet—although we are not aware of it being mentioned elsewhere in the literature—we dare not claim its novelty.

We also give evidence that the new approach not only leads to a scheme with (slightly) improved bandwidth, but also with significantly improved computational complexity. Indeed, it appears, and we can confirm this for all practical values of n, that sharing a group element with our scheme requires $\tilde{O}(n^2)$ and reconstructing the secret from the shares $\tilde{O}(n^{1.6})$ group operations, in contrast to $\tilde{O}(n^3)$ respectively $\tilde{O}(n^{2.6})$ or more for previous schemes. At present there is no such proof for general n.

Finally, we address the *randomness complexity* of black-box secret sharing, i.e., the number of random group elements that need to be sampled to share a group element. We prove the lower bound $\Omega(t \cdot \lg n)$, which meets the randomness complexity of our (as well as the scheme from [2]) black-box secret sharing scheme and hence shows that these schemes are also optimal with respect to their randomness complexity. We would like to point out that recently a similar lower bound has been proven in [9]. However, the proof given seems a bit vague as it makes use of a better lower bound result on the expansion factor of black-box secret sharing schemes than what can be proven.

The paper is organized as follows. In the following Section 2, we give some definitions and known results regarding black-box secret sharing, and in Section 3 we describe a framework on which previous as well as our new black-box secret sharing scheme are based. In Section 4, we then briefly describe the schemes from [4] and [2], before we discuss our new approach in Section 5. Section 6 is dedicated to the lower bound on the randomness complexity before we conclude in Section 7.

2   Definitions and Known Results

Throughout this section let n and $t < n$ be non-negative integers. Informally, in a black-box secret sharing scheme the shares are computed from the secret and from random group elements by solely using the group operations addition and subtraction (considering the group to be additive), i.e., by taking $\mathbb{Z}$-linear combinations of the secret and random group elements. Similarly, the secret is reconstructed by taking an appropriate $\mathbb{Z}$-linear combination of the shares. Additionally, the coefficients for these linear combinations are designed independently of and correctness and privacy hold regardless of the group to which the scheme is applied. This leads to the following formal definition due to [2].

We first introduce the notion of a labeled matrix. A *labeled matrix* consists of a matrix $M \in R^{d \times e}$ over some given ring R, together with a surjective function $\psi : \{1, \ldots, d\} \to \{1, \ldots, n\}$. We say that the j-th row of M is *labeled* by $\psi(j)$. For $\emptyset \neq A \subseteq \{1, \ldots, n\}$, $M_A \in R^{d_A \times e}$ denotes the restriction of M to those rows that are labeled by an $i \in A$. Similarly, for an arbitrary d-vector $\mathbf{x} = (x_1, \ldots, x_d)$ (over a possibly different domain), $\mathbf{x}_A$ denotes the restriction of $\mathbf{x}$ to those coordinates x_j with $\psi(j) \in A$. In order to simplify notation, we write M_i and $\mathbf{x}_i$ instead of M_A and $\mathbf{x}_A$ in case $A = \{i\}$, and we typically do not make ψ explicit.

Definition 1 (Black-Box Secret Sharing). *A labeled matrix $M \in \mathbb{Z}^{d \times e}$ over the integers is a* black-box secret sharing scheme *for n and t if the following holds.*

For an arbitrary finite Abelian group G *and an arbitrarily distributed* $s \in \mathsf{G}$ *let* $\mathbf{g} = (g_1, \ldots, g_e)^T \in \mathsf{G}^e$ *be drawn uniformly at random subject to* $g_1 = s$ *only. Define the share vector as* $\mathbf{s} = M\mathbf{g}$ *(where* $\mathbf{s}_i$ *is given to the i-th participant). Then, for any nonempty subset* $A \subseteq \{1, \ldots, n\}$:

i.   *(Completeness) If* $|A| > t$ *then there exists* $\boldsymbol{\lambda}(A) \in \mathbb{Z}^{d_A}$, *only depending on* M *and* A, *such that* $\mathbf{s}_A^T \cdot \boldsymbol{\lambda}(A) = s$ *with probability 1.*

ii. *(Privacy) If* $|A| \leq t$ *then* $\mathbf{s}_A$ *contains no Shannon information on* s.

Note that a black-box secret sharing scheme is linear by definition (essentially because a black-box group allows only linear operations).

Definition 2 (Expansion Factor and Randomness Complexity). *The ex-pansion factor* η *of a black-box secret sharing scheme* $M \in \mathbb{Z}^{d \times e}$ *for* n *and* t *is defined by* $\eta = d/n$, *and the randomness complexity* ρ *by* $\rho = e - 1$.

The expansion factor of a black-box secret sharing scheme measures the average number of group elements each participant receives (and need not be integral). For the trivial cases $t = 0$ and $t = n - 1$ the expansion factor 1 can be achieved. The randomness complexity determines the number of random group elements that need to be sampled to share a secret. The number of group operations during dealing and reconstructing depends both on d and e and on the size of the elements in the matrix M (optimizing the number of group operations given M is essentially an addition chain problem).

Theorem 1 ([2]). *Let* $M \in \mathbb{Z}^{d \times e}$ *be a labeled matrix. Define* $\boldsymbol{\varepsilon} = (1, 0, \ldots, 0) \in \mathbb{Z}^e$. *Then* M *is a black-box secret sharing scheme for* n *and* t *if and only if for every nonempty* $A \subseteq \{1, \ldots, n\}$ *the following holds.*

i.   *(Completeness) If* $|A| > t$ *then* $\boldsymbol{\varepsilon} \in \mathrm{im}(M_A^T)$.

ii. *(Privacy) If* $|A| \leq t$ *then there exists* $\boldsymbol{\kappa} = (\kappa_1, \ldots, \kappa_e)^T \in \ker(M_A)$ *with* $\kappa_1 = 1$.

Note the difference between this definition and that of monotone span programs (which is equivalent with linear secret sharing over finite fields). Whereas in the latter case the completeness condition and the privacy condition are character-ized by "to span or not to span", over $\mathbb{Z}$ and in the context of blackbox secret sharing this is slightly more subtle. See [2].

In [2] the above theorem is used to prove a lower bound on the expansion factor by looking at an instantiation of any given black-box secret sharing scheme over the group $\mathbb{F}_2$ and borrowing arguments from Karchmer and Wigderson [8]. The upper bound in the theorem below follows from the explicit construction of a black-box secret sharing scheme in [2].

Theorem 2 ([2]). *The minimal expansion factor* η *of a black-box secret sharing scheme for* n *and* t *with* $0 < t < n - 1$ *satisfies*

$$\lfloor \lg n \rfloor - 1 < \lg \frac{n+3}{2} \leq \eta \leq \lceil \lg(n+1) \rceil + 1 = \lfloor \lg n \rfloor + 2 .$$

If $t = 1$ *then it even holds that* $\eta \geq \lg n$.

3   Integral Extensions and *Weak* Black-Box Secret Sharing

Let R be a ring of the form $R = \mathbb{Z}[X]/(f)$ where f is a monic irreducible polynomial in $\mathbb{Z}[X]$ of degree m. We call such a ring an *integral extension* (of degree m).[3] Note that R is a free $\mathbb{Z}$-module[4] with basis $\bar{1}, \bar{X}, \ldots, \bar{X}^{m-1}$ (the residue classes of $1, X, \ldots, X^{m-1}$ modulo $f(X)$). Furthermore, let G be a finite Abelian (additive) group. Such a group is naturally a $\mathbb{Z}$-module. The fact is that the m-fold direct sum $\mathsf{G}^m = \mathsf{G} \oplus \cdots \oplus \mathsf{G}$ can be regarded as an R-module. Indeed, as a group, respectively as $\mathbb{Z}$-module, G^m is isomorphic to the tensor product $R \otimes_{\mathbb{Z}} \mathsf{G}$ (with isomorphism $(g_1, g_2, \ldots, g_m) \mapsto \bar{1} \otimes g_1 + \bar{X} \otimes g_2 + \cdots + \bar{X}^{m-1} \otimes g_m$); the latter though, sometimes referred to as the *extension of* G *over* R [11], is an R-module by "multiplication into the R-component".

Now, since G^m is an R-module, polynomials with coefficients in G^m can be evaluated over R. This allows us to perform a version of Shamir secret sharing [12]: Given the parameters n and t as well as the secret $s \in \mathsf{G}$, the dealer picks uniformly at random a *sharing polynomial*

$$g(x) = r_0 + \cdots + r_{t-1} x^{t-1} + \hat{s} x^t \in \mathsf{G}^m[x]$$

of degree t with coefficients in G^m such that its *leading* coefficient equals $\hat{s} = (s, 0, \ldots, 0) \in \mathsf{G}^m$ (we need to embed the secret s into G^m). Given n pairwise different evaluation points $\alpha_i \in R$, known to everyone, the dealer hands out share $s_i = g(\alpha_i) \in \mathsf{G}^m$ to participant i for $i = 1, \ldots, n$.

We would like to point out that by fixing the basis $\bar{1}, \bar{X}, \ldots, \bar{X}^{m-1}$ for R over $\mathbb{Z}$ and using standard techniques this candidate black-box secret sharing scheme can be described by a labeled integer matrix M and thus fits into the framework of our formal Definition 1; although, as discussed below, correctness holds only in a weak sense. The expansion factor is obviously $\eta = m$: each share is an element in G^m, and the randomness complexity is $\rho = t \cdot m$: the randomness is enclosed by the t non-leading coefficients of $g \in \mathsf{G}^m[x]$.

Jointly, any $t + 1$ participants know $t + 1$ points on a polynomial of degree t. Normally, when working over a field, this would allow them to reconstruct the entire polynomial using Lagrange interpolation. In our setting, where divisions cannot necessarily be done (in R), we will have to settle with a multiple $\Delta \cdot \hat{s} \in \mathsf{G}^m$ of the secret, where $\Delta \in R$ is some common multiple of the denominators of the Lagrange coefficients. A possible generic choice for Δ is the Vandermonde determinant

$$\Delta(\alpha_1, \ldots, \alpha_n) = \prod_{1 \le i < j \le n} (\alpha_i - \alpha_j) .$$

---

[3] Using standard terminology from algebraic number theory, R is an example of an *order*.

[4] Loosely speaking, a *module* is a vector space over a ring rather than over a field, and it is called *free* if it allows for a basis (which is not granted for general modules).

Reconstruction by a set A of $t+1$ participants can be expressed in the following formula:

$$\Delta \cdot \hat{s} = \sum_{i \in A} \left(\Delta \cdot \prod_{\substack{j \in A \\ j \neq i}} \frac{1}{\alpha_i - \alpha_j} \right) s_i \ .$$

Putting the secret into the leading coefficient as we do (rather than into the constant coefficient) of the sharing polynomial immediately leads to privacy. Essentially, for any $A \subset \{1, \ldots, n\}$ with $|A| \leq t$, privacy follows from the existence of the polynomial $\kappa = \prod_{i \in A}(x - \alpha_i) \in R[x]$ of degree at most t with leading coefficient 1 and with $\kappa(\alpha_i) = 0$ for all $i \in A$. Indeed, for any secrets $s, s' \in \mathsf{G}$ and any sharing polynomial $g \in \mathsf{G}^m[x]$ for s, the participants in A cannot distinguish between a sharing of s with sharing polynomial g and a sharing of s' with sharing polynomial $g' = g + (s' - s)\kappa$.

Introducing the notion of a δ-*weak* black-box secret sharing scheme for $\delta \in R$, to be understood in that the correctness condition of a black-box secret sharing scheme (Definition 1) only holds in that $\delta \cdot \hat{s}$ (rather than s) can be reconstructed while the privacy condition holds fully, we can summarize the observations of this section as follows.

Theorem 3. *Let R be an integral extension of degree m, and let $\alpha_1, \ldots, \alpha_n \in R$ be pairwise different. Then there exists a $\Delta(\alpha_1, \ldots, \alpha_n)$-weak black-box secret sharing scheme for n and t with expansion factor $\eta = m$ and randomness complexity $\rho = t \cdot m$.*

Note that an additional advantage of putting the secret into the leading coefficient of the sharing polynomial (rather than into the constant coefficient) is that 0 may be used as evaluation point. This extra evaluation point is relevant for the expansion factor if the number of participants is a power of 2. This "swapping" trick, putting the secret in the leading coefficient instead of in the constant term, is not exploited in [4] nor in [2], but it applies to their schemes as well.

4   Previous Schemes

Based on the common framework just described, we can summarize previous research. It all boils down to reconstructing s given $\Delta \cdot \hat{s}$ and the restriction the scheme poses on Δ for the reconstruction to be possible.

4.1   Using an Invertible Δ

Desmedt and Frankel [4] provide a solution for black-box secret sharing with expansion factor $O(n)$. They achieve this by selecting the polynomial f in such a way that Δ can be chosen to be a unit in $R = \mathbb{Z}[X]/(f)$. A necessary and sufficient condition for this is that there exist n evaluation points in the ring whose differences are all units in the ring.[5] In this case, all divisions required for Lagrange interpolation can in fact take place in the ring R, so Δ can be forgotten altogether.

The maximal cardinality of a subset of R such that all differences are units is called the Lenstra constant of the ring R. If we set $R = \mathbb{Z}[X]/(f(X))$, where $f(X) \in \mathbb{Z}[X]$ is the p-th cyclotomic polynomial, we have a ring of degree $p-1$ and with Lenstra constant p. So if we take p as the smallest prime greater than n, we have black-box secret sharing scheme with expansion factor $O(n)$ for n players. Finding integral extensions for which the Lenstra constant is exponential (or super-linear) in the degree of the ring is part of an open problem in number theory, as far as we know

4.2 Using Two Relatively Co-prime Δ's

Cramer and Fehr [2] propose scheme which has expansion factor $\lfloor \lg n \rfloor + 2$. In a nutshell, it shares the secret *twice* using weak secret sharing schemes with two different sets, say $\boldsymbol{\alpha} = (\alpha_1, \ldots, \alpha_n)$ and $\boldsymbol{\beta} = (\beta_1, \ldots, \beta_n)$, of evaluation points. This allows to reconstruct two different multiples of the secret: $\Delta(\boldsymbol{\alpha}) \cdot \hat{s}$ and $\Delta(\boldsymbol{\beta}) \cdot \hat{s}$. By ensuring that $\Delta(\boldsymbol{\alpha})$ and $\Delta(\boldsymbol{\beta})$ are co-prime, standard Euclidean techniques can be used to recover $\hat{s}$ and the real secret s: let a and b be such that $a \cdot \Delta(\boldsymbol{\alpha}) + b \cdot \Delta(\boldsymbol{\beta}) = 1$, then $a \cdot \Delta(\boldsymbol{\alpha})\hat{s} + b \cdot \Delta(\boldsymbol{\beta})\hat{s} = \hat{s} = (s, 0, \ldots, 0)$.

A small expansion factor can then be obtained by picking the α_i's in the integers and the β_i's in a suitable integral extension $R = \mathbb{Z}[X]/(f)$. A necessary and sufficient condition for the existence of $\boldsymbol{\beta}$ such that $\Delta(\boldsymbol{\alpha})$ and $\Delta(\boldsymbol{\beta})$ are co-prime is that for all primes p the *lowest* irreducible polynomial dividing f modulo p has degree $\underline{d_p}$ such that $n \leq p^{d_p}$. This can be satisfied by certain polynomials f of degree $m = \lceil \lg(n+1) \rceil = \lfloor \lg n \rfloor + 1$, and thus results in a total expansion factor $\lfloor \lg n \rfloor + 2$.

In [2] polynomials f of degree $\lfloor \lg n \rfloor + 1$ are considered that are irreducible modulo all the primes $p \leq n$. It is then possible to set $\alpha_i = i$ and β_i to the (residue class modulo f of the) unique polynomial of degree less than m with coefficients in $\{0, 1\}$ that evaluates to i in the point 2, in other words, whose coefficient vector is the binary representation of i. (Note that the "swapping" trick allows us to use a polynomial of degree $m = \lceil \lg n \rceil$ instead of $\lfloor \lg n \rfloor + 1$, although there is only a difference if n is a power of two.)

5 The New Black-Box Secret Sharing Scheme

5.1 The New Scheme: Using a Primitive Δ

As an example, consider the case $t = 1$, where only two participants are needed to reconstruct the secret (for any number of participants n). If we use $R = \mathbb{Z}[X]/(f)$ with *any* (monic and irreducible) f of degree at least $\lceil \lg n \rceil$, and

---

[5] If the secret is embedded in the constant term, the evaluation points need to be units themselves as well and evaluation in zero is prohibited.

the same $\{0,1\}$-polynomial evaluation points β_i as described in Section 4.2, then any pair $(i,j), i \neq j$ can reconstruct $(\beta_i - \beta_j)\hat{s}$. For concreteness, suppose $\deg f = 4$ and that $\beta_i - \beta_j$ equals $\bar{1} - \bar{X} + \bar{X}^3$. In this case we know that $(\beta_i - \beta_j)\hat{s} = (s, -s, 0, s)$. Indeed, as discussed in Section 3, $(\beta_i - \beta_j)\hat{s}$ is computed by associating $\hat{s} = (s, 0, \dots, 0)$ with $\bar{1} \otimes s$, computing $(\beta_i - \beta_j) \cdot (\bar{1} \otimes s) = (\beta_i - \beta_j) \otimes s = \bar{1} \otimes s + \bar{X} \otimes (-s) + \bar{X}^3 \otimes s$, and reading out the "coefficients". So, $(\beta_i - \beta_j)\hat{s}$ already contains s as a coordinate! There is no need for a second sharing or $\beta_i - \beta_j$ being a unit in R. Our choice of $\beta_i - \beta_j$ is inconsequential in this argument. If the β_i's are defined as they are as non-zero polynomials of degree smaller than m with coefficients in $\{0,1\}$ (regardless of f), then $(\beta_i - \beta_j)\hat{s}$ contains at least one copy of the secret or its negative.

In general, using the weak black-box secret sharing scheme the participants can reconstruct $\Delta \cdot \hat{s}$. Since $\hat{s} = (s, 0, \dots, 0)$, this module scalar-multiplication equals $\Delta \cdot \hat{s} = (\Delta_0 \cdot s, \dots, \Delta_{m-1} \cdot s)$ with integer Δ_i's such that $\Delta = \sum_{i=0}^{m-1} \Delta_i \cdot \bar{X}^i$. The secret s can be reconstructed from the Δ_i's if and only if the Δ_i's are co-prime, by using the extended Euclidean algorithm. In essence, the ideas of [4] (using a single weak black-box secret sharing scheme) and of [2] (recovering s from co-prime multiples) are combined. Contrary to the scheme from [2], we do not need a second sharing. This is where our improvement and lower expansion factor stem from.

A prime $p \in \mathbb{Z}$ is a divisor of all Δ_i's if and only if $\Delta \equiv 0 \bmod pR$. A sufficient and necessary condition on the set of interpolation points is that it is a primitive set in the integral extension R, as defined below.

Definition 3 (Primitive Elements and Sets). *Let R be an integral extension. Then $\delta \in R$ is primitive if its only rational integer divisors are 1 and -1, i.e., if $\delta \not\equiv 0 \bmod pR$ for all primes $p \in \mathbb{Z}$. A set $\{\alpha_1, \dots, \alpha_n\}$ in R is called primitive if its Vandermonde determinant $\Delta(\alpha_1, \dots, \alpha_n)$ is primitive.*

For an arbitrary $\mathbb{Z}$-basis of R, $p \in \mathbb{Z}$ dividing $\delta \in R$ is equivalent to p dividing all the rational integer coordinates of δ with respect to that basis. Therefore, $\delta \in R$ is primitive if and only if its coordinates have no non-trivial common factor in $\mathbb{Z}$. Note also that the required property is stronger than requiring the α_i's to be pairwise different modulo every prime p, since not every prime $p \in \mathbb{Z}$ is necessarily also prime in R.

For $f(X) \in \mathbb{Z}[X]$ and for a prime $p \in \mathbb{Z}$, define $f_p(X) \in \mathbb{F}_p[X]$ as f taken $\bmod p$, and write $f_p = f_{p,1}^{\epsilon_{p,1}} \cdots f_{p,\ell_p}^{\epsilon_{p,\ell_p}}$ for its factorization into powers of distinct irreducible polynomials in $\mathbb{F}_p[X]$. The degree of such $f_{p,i}$ is denoted $d_{p,i}$. Also define $\bar{d}_p = \max_{1 \leq i \leq \ell_p} d_{p,i}$.

Theorem 4. *Let $R = \mathbb{Z}[X]/(f)$ be an integral extension of degree $m > 1$. If $n \leq p^{\bar{d}_p}$ for every prime $p \in \mathbb{Z}$, then there exists a primitive set in R with cardinality n.*

This implies the existence of an integral extension of degree $\lceil \lg(n) \rceil$ with a primitive set of size n, by taking f such that f_p is irreducible for all primes p with $2 \leq p \leq n$. Such f can for instance be constructed using the Chinese Remainder Theorem, see also [2].

Corollary 1. *For any $t, n \in \mathbb{Z}$ with $0 < t < n - 1$ there exists a black-box secret sharing scheme M with expansion factor $\eta = \lceil \lg(n) \rceil$ and randomness complexity $\rho = t \cdot \lceil \lg(n) \rceil$.*

The *computational* efficiency of the scheme is discussed in Section 5.4.

5.2   Proof of Theorem 4

Let $p \in \mathbb{Z}$ be a prime. Then we have

$$R/pR \simeq \mathbb{F}_p[X]/(f_{p,1}^{\epsilon_{p,1}} \cdots f_{p,\ell_p}^{\epsilon_{p,\ell_p}}) \simeq \mathbb{F}_p[X]/(f_{p,1}^{\epsilon_{p,1}}) \times \cdots \times \mathbb{F}_p[X]/(f_{p,\ell_p}^{\epsilon_{p,\ell_p}})$$

and thus we have the canonical projection

$$R/pR \to \mathbb{F}_p[X]/(f_{p,1}) \times \cdots \times \mathbb{F}_p[X]/(f_{p,\ell_p}) \simeq \mathbb{F}_{p^{d_{p,1}}} \times \cdots \times \mathbb{F}_{p^{d_{p,\ell_p}}}$$

where (for any prime power q) $\mathbb{F}_q$ denotes the field with q elements. Hence, if $n \leq p^{d_p}$, then there clearly exist $\alpha_1, \ldots, \alpha_n \in R$ such that $\Delta(\alpha_1, \ldots, \alpha_n) \not\equiv 0 \bmod pR$: choose n distinct elements from $\mathbb{F}_{p^{d_p}}$ and lift them arbitrarily to elements in R. Furthermore, different solutions modulo a *finite* set of different primes p can be combined to a solution modulo all primes from that set by the Chinese Remainder Theorem. However, we are after a solution that holds modulo *all* primes simultaneously.

Instead, we construct a primitive set of size n by induction: as long as the upperbound on n as stated in the theorem is satisfied, then, given a primitive set $\{\alpha_1, \ldots, \alpha_{n-1}\} \subset R$, we can construct $\alpha_n \in R$ such that $\{\alpha_1, \ldots, \alpha_{n-1}, \alpha_n\} \subset R$ is a primitive set as well. For technical reasons to become clear later on, the actual induction hypothesis corresponds to a slightly stronger claim, but we suppress this at this point in the exposition.

Assume we are given a primitive set $\{\alpha_1, \ldots, \alpha_{n-1}\} \subset R$. Consider the polynomial

$$\Delta(\alpha_1, \ldots, \alpha_{n-1}, X) = \Delta(\alpha_1, \ldots, \alpha_{n-1}) \cdot \prod_{i<n} (\alpha_i - X) \in R[X].$$

Let $e_1 \ldots, e_m$ be some fixed $\mathbb{Z}$-basis of R, where m is the degree of $f(X)$. Clearly, there exist polynomials $F_1, \ldots, F_m \in \mathbb{Z}[X_1, \ldots, X_m]$ such that

$$\mathbf{d} = (F_1(x_1, \ldots, x_m), \ldots, F_m(x_1, \ldots, x_m)) \in \mathbb{Z}^m$$

represents the coordinate-vector (w.r.t. the chosen basis) of $\Delta(\alpha_1, \ldots, \alpha_{n-1}, x) \in R$ for an arbitrary $x = x_1 e_1 + \cdots + x_m e_m \in R$ $(x_1, \ldots, x_m \in \mathbb{Z})$.

Now suppose that the intersection between the ideal $\hat{I} = (F_1, \ldots, F_m) \cdot \mathbb{Z}[X_1, \ldots, X_m]$ and $\mathbb{Z}[X_1]$ contains a non-zero polynomial $g(X_1)$. In other words, there exist polynomials $\mu_1, \ldots, \mu_m \in \mathbb{Z}[X_1, \ldots, X_m]$ such that

$$g(X_1) = \sum_{i=1}^{m} \mu_i(X_1, \ldots, X_m) \cdot F_i(X_1, \ldots, X_m).$$

Therefore, if we choose $x_1 \in \mathbb{Z}$ such that $g(x_1) \neq 0$, then, *no matter how* $x_2, \ldots, x_m \in \mathbb{Z}$ *are chosen*, it will be the case that a given prime $p \in \mathbb{Z}$ does not divide all $F_i(x_1, \ldots, x_m) \in \mathbb{Z}$, or equivalently, $\Delta(\alpha_1, \ldots, \alpha_{n-1}, x) \not\equiv 0 \bmod pR$, unless perhaps when p divides $g(x_1)$.

Based on these observations the proof of the theorem essentially consists of two main steps. First, we show the existence of $g(X_1)$. Second, with a proper choice of $x_1 \in \mathbb{Z}$ such that $g(x_1) \neq 0$, we select for each prime $p \in \mathbb{Z}$ that divides $g(x_1)$ an element $a_{n,p} \in R$ such that its first coordinate is equal to $x_1 \bmod p$ and such that $\Delta(\alpha_1, \ldots, \alpha_{n-1}, a_{n,p}) \not\equiv 0 \bmod pR$.

The proof is then easily completed by constructing the desired $\alpha_n \in R$ such that its first coordinate is x_1 and such that $\alpha_n \equiv a_{n,p} \bmod pR$ for each of those finitely many primes p. This is by simple coordinate-wise application of the CRT. More precisely, let $\alpha_n = x_1 e_1 + x_2 e_2 + \cdots + x_m e_m$, where x_1 is as above, and for each $i \geq 2$, $x_i \in \mathbb{Z}$ is such that x_i is equivalent to the i-coordinate of $a_{n,p}$ modulo each those primes p.

We now start with the existence of $g(X_1)$. The argument utilizes the following well-known theorem from algebraic geometry (see e.g. [11]), which we state for convenience below.

Theorem 5 (Hilbert's Nullstellensatz). *Let K be an algebraically closed field, let $I \subset K[X_1, \ldots, X_r]$ be an ideal, and let $\mathcal{Z}(I) \subset K^r$ denote the algebraic variety $\{(z_1, \ldots, z_r) \in K^r \,|\, g(z_1, \ldots, z_r) = 0 \; \forall g \in I\}$. If $h \in K[X_1, \ldots, X_r]$ satisfies $h(z_1, \ldots, z_r) = 0$ for every $(z_1, \ldots, z_r) \in \mathcal{Z}(I)$, i.e., it vanishes on the variety, then there exists a positive integer k such that $h^k \in I$.*

Let $\bar{\mathbb{Q}}$ denote the algebraic closure of $\mathbb{Q}$, i.e., the field of all algebraic numbers, and let I denote the ideal $(F_1, \ldots, F_m) \cdot \bar{\mathbb{Q}}[X_1, \ldots, X_m]$. We claim that the algebraic variety $\mathcal{Z}(I)$ is finite. This is argued in two steps. Consider the tensor-product $\bar{\mathbb{Q}} \otimes_{\mathbb{Z}} R$, which has a natural ring structure. First, $\mathcal{Z}(I)$ is in one-to-one correspondence with the solutions to the univariate polynomial equation $\Delta(\alpha_1, \ldots, \alpha_{n-1}, x) = 0$ with $x \in \bar{\mathbb{Q}} \otimes_{\mathbb{Z}} R$, which we show below. Second, as a ring, $\bar{\mathbb{Q}} \otimes_{\mathbb{Z}} R$ is isomorphic to a finite product of fields.[6] Therefore, the univariate polynomial equation has at most a finite number of solutions, and the claim follows. One-to-one correspondence is argued as follows. The elements of $\bar{\mathbb{Q}} \otimes_{\mathbb{Z}} R$ uniquely correspond to the expressions of the form $\sum_{i=1}^m q_i \otimes e_i$ with the $q_i \in \bar{\mathbb{Q}}$. Using simple rewriting properties of tensor-product, it follows that $\Delta(\alpha_1, \ldots, \alpha_{n-1}, x) = 0$ for $x \in \bar{\mathbb{Q}} \otimes_{\mathbb{Z}} R$ if and only if $\sum_{i=1}^m F_i(q_1, \ldots, q_m) \otimes e_i = 0$. This happens if and only if all $F_i(q_1, \ldots, q_m)$ are 0, or equivalently, $(q_1, \ldots, q_m) \in \mathcal{Z}(I)$. Note that some of the properties of tensor product we have used above rely on the fact that R has a $\mathbb{Z}$-basis.

Finiteness of $\mathcal{Z}(I)$ implies the existence of a non-zero polynomial $\tilde{g}(X_1)$ in the intersection of I and $\bar{\mathbb{Q}}[X_1]$. Indeed, the polynomial $\prod_{z \in \mathcal{Z}(I)}(X_1 - z_1) \in \bar{\mathbb{Q}}[X_1]$ (where z_1 denotes the first coordinate of z) clearly vanishes on $\mathcal{Z}(I)$, and by the

---

[6] Indeed, $\bar{\mathbb{Q}} \otimes_{\mathbb{Z}} R \simeq \bar{\mathbb{Q}}[X]/(f) \simeq \prod_{i=1}^m \bar{\mathbb{Q}}$. The first isomorphism is by a standard fact that can be found e.g. in [11], and the second follows since f factors into distinct linear polynomials.

Nullstellensatz some power of this polynomial is in I. In turn this implies the existence of a non-zero polynomial $g(X_1)$ in the intersection of $\hat{I}$ and $\mathbb{Z}[X_1]$, as desired. This is an immediate consequence of basic field theory.[7]

With the existence of $g(X_1)$ settled, we proceed with the remainder of the proof. As a matter of terminology, for $\beta, \gamma \in R$, we will say that $\beta = \gamma$ *within* $\mathbb{F}_p^{d_{p,i}}$ if the canonical projections of β and γ coincide in that component. Similar for $\mathbb{F}_p[X]/(f_{p,i}^{\epsilon_{p,i}})$.

First we make the actual induction hypothesis precise. We assume there exists a primitive set $\alpha_1, \ldots, \alpha_{n-1} \in R$ such that *additionally* for every prime $p \in \mathbb{Z}$ with $1 < p < n$ it holds that $\Delta(\alpha_1, \ldots, \alpha_{n-1})$ is non-zero within the *largest* field $\mathbb{F}_{p^{\bar{d}_p}}$. The induction hypothesis is clearly satisfied in case of a single element set. If $n \leq p^{\bar{d}_p}$ for every prime $p \in \mathbb{Z}$, then we construct $\alpha_n \in R$ such that $\{\alpha_1, \ldots, \alpha_{n-1}, \alpha_n\}$ is a primitive set *and* such that the additional requirement is satisfied.

Instead of selecting $x_1 \in \mathbb{Z}$ arbitrarily such that $g(x_1) \neq 0$, we have to give a special treatment to the primes $p \in \mathbb{Z}$ with $1 < p < n$ first, for reasons to become clear later on. We start by choosing for every such prime p an $a_{n,p} \in R$ such that $\Delta(\alpha_1, \ldots, \alpha_{n-1}, a_{n,p})$ is non-zero within (the largest field) $\mathbb{F}_{p^{\bar{d}_p}}$. This can be done by virtue of the induction hypothesis and using arguments as in the beginning of the section. Then we choose $x_1 \in \mathbb{Z}$ such that modulo every prime $p \in \mathbb{Z}$ with $1 < p < n$, x_1 is congruent to the first coordinate of $a_{n,p}$, and such that $g(x_1)$ is non-zero. Such x_1 exists as g has only a finite number of zeroes.

Now fix any prime $p \in \mathbb{Z}$ with $p \geq n$ and p divides $g(x_1)$. We now select $a_{n,p}$ as required. We have $\Delta(\alpha_1, \ldots, \alpha_{n-1}) \neq 0$ within at least one of the $\mathbb{F}_p[X]/(f_{p,i}^{\epsilon_{p,i}})$ into which R/pR splits, by the induction hypothesis. Fix an index k for which this is the case.

We first treat the case when $f_p(X) \in \mathbb{F}_p[X]$ is irreducible (so $\ell_p = k = 1$). In this case $R/pR \simeq \mathbb{F}_p[X]/(f_p) \simeq \mathbb{F}_{p^m}$. Since $p \geq n$, there are $p^{m-1} \geq n$ elements in $\mathbb{F}_{p^m}$ with first coordinate x_1, and it is clearly possible to select $a_{n,p} \in R$ as required, i.e., its first coordinate is x_1 and $\Delta(\alpha_1, \ldots, \alpha_{n-1}, a_{n,p}) \not\equiv 0 \bmod pR$.

Second, suppose that the polynomial $f_p(X) \in \mathbb{F}_p[X]$ is reducible. Since $p \geq n$, $\mathbb{F}_{p^{d_{p,k}}} (\simeq \mathbb{F}_p[X]/(f_{p,k}))$ has at least n elements. So it is possible to select $a_{n,p} \in R$ such that within $\mathbb{F}_{p^{d_{p,k}}}$ it differs from $\alpha_1, \ldots, \alpha_{n-1}$. As a consequence all $a_{n,p} - \alpha_j$ are invertible within $\mathbb{F}_p[X]/(f_{p,k})$, and hence also within $\mathbb{F}_p[X]/(f_{p,k}^{\epsilon_{p,k}})$. Thus, $\Delta(\alpha_1, \ldots, \alpha_{n-1}, a_{n,p})$ is *non-zero* within $\mathbb{F}_p[X]/(f_{p,k}^{\epsilon_{p,k}})$, and therefore also non-zero modulo pR.

It remains to argue that $a_{n,p}$ may be chosen such that its first coordinate equals x_1. This is by adding a suitable rational integer multiple of a special

---

[7] It is given that $\tilde{g} = \sum_{i=1}^m \lambda_i F_i$ for some $\lambda_i \in \bar{\mathbb{Q}}[X_1, \ldots, X_m]$. There exists $\theta \in \bar{\mathbb{Q}}$ such that each of the coefficients of each of the λ_i's is in $\mathbb{Q}(\theta)$. Note that $\mathbb{Q}(\theta)$ is a $\mathbb{Q}$-vectorspace with basis $1, \theta, \ldots, \theta^{e-1}$ for some e. Consider the fraction field $L = \mathbb{Q}(X_1, \ldots, X_m)$. Similarly, $L(\theta)$ is an L-vectorspace with the same basis. Now consider an arbitrary non-zero coordinate of $\tilde{g}$ w.r.t. that basis. Then we have $g = \sum_{i=1}^m \mu_i F_i$ where $g \in \mathbb{Q}[X_1]$, respectively, $\mu_i \in \mathbb{Q}[X_1, \ldots, X_m]$, is this coordinate of $\tilde{g}$, respectively, of λ_i. Clearing denominators gives the desired result.

element $\delta_p \in R$ which has first coordinate 1 but that is 0 within $\mathbb{F}_{p^{d_{p,k}},k}$. We construct it below, and this finishes the proof.

For convenience, take $\bar{1}, \bar{X}, \dots, \bar{X}^{m-1} \in \mathbb{Z}[X]/(f)$ as the $\mathbb{Z}$-basis $e_1, \dots, e_m$ for R introduced earlier on. Let $c \in \mathbb{F}_p \setminus \{0\}$ be the constant coefficient of the irreducible polynomial $f_{p,k} \in \mathbb{F}_p[X]$ and $c^{-1} \in \mathbb{F}_p$ its inverse. Let $h(X) = h_0 + h_1 X + \cdots + h_{d_{p,k}} X^{d_{p,k}} \in \mathbb{Z}[X]$ have coefficients in $\{0, \dots, p-1\}$ such that modulo p it equals $c^{-1} f_{p,k}(X)$. Since f_p is reducible, h has degree smaller than m. Moreover it has constant coefficient $h_0 = 1$. Then define δ_p as $\overline{h(X)} \in R$. Indeed, its first coordinate is 1 and δ_p is clearly 0 within $\mathbb{F}_{p^{d_{p,k}},k}$.   □

5.3   A Generalization of Theorem 4

It is possible to give a generalization of Theorem 4 that applies to arbitrary orders (of non-zero discriminant), rather than only to integral extensions and which shows that the lower bound on n is tight if we require that not only Δ but all powers of Δ must have no non-trivial integral divisors. Consider for instance $f = X^2 + 1$ so that $R \simeq \mathbb{Z}[i]$ (the Gaussian integers). Then Theorem 4 promises a primitive set of size 2, while if fact there is a primitive set of size 3. Indeed, $\Delta(0, 1, i) = 1 + i$ has no non-trivial divisors; $\Delta(0, 1, i)^2 = (1 + i)^2 = 2i$ however has.

Definition 4 (*Radically* Primitive Elements and Sets). *A element δ in an order R is called* radically *primitive if the only rational integer divisors of any power of δ are 1 and -1, i.e., if $\delta^k \not\equiv 0$ modulo any prime $p \in \mathbb{Z}$, for all $k > 0$. And a set $\{\alpha_1, \dots, \alpha_n\}$ in R is called* radically primitive *if its Vandermonde determinant $\Delta(\alpha_1, \dots, \alpha_n)$ is radically primitive.*

Using similar but more general arguments as in the proof of Theorem 4, the following can be proved.

Theorem 6. *Let R be an order with discriminant $\Delta_{R/\mathbb{Z}} \neq 0$. For any prime $p \in \mathbb{Z}$ let $n(p) = \max_{\mathfrak{p}} |R/\mathfrak{p}|$, where $\mathfrak{p}$ ranges over all prime ideals $\mathfrak{p} \subseteq pR$ over p. Then the maximal cardinality for a radically primitive set in R is $\min_{p \ prime} n(p)$.*

5.4   Computational Complexity

Apart from having a small expansion factor, we would also like to exhibit that the number of black-box group operations is polynomial in the number of participants. This requires that the entries of the sharing matrix M are small (if we assume that d and e are sufficiently small as is the case for the constructions above). For an integral extension $R = \mathbb{Z}[X]/(f)$ this requires small coefficients of f and small coefficients of the evaluation points α_i when expressed as polynomials of minimal degree.

As mentioned in Section 4.2, for the scheme from [2] one method always works, namely picking an irreducible polynomial modulo p for every prime $p < n$ and using the Chinese Remainder Theorem to get a polynomial f over the integers. The coefficients of this polynomial are all smaller than $\prod_{p<n} p$, which corresponds to a bitlength linear in n. We cannot hope to find polynomials with coefficients that are much smaller than random CRT based polynomials and that are still irreducible modulo p for all $p < n$. The evaluation points that are used have minimal coefficients (either 0 or 1).

For our new construction, the set of suitable polynomials f is a proper superset of those employed by [2]. This means that we could take f as constructed above with coefficients whose sizes are linear in n. Unfortunately, the proof of existence of a primitive set $\alpha_1, \ldots, \alpha_n$ for a suitable f does not guarantee any reasonable bound on the size of the coefficients: the main problem in the proof occurs around the place where Hilbert's Nullstellensatz is invoked.

However, practical experiments indicate that f and the α_i's can in fact be chosen in such a way that their coefficients are within $\{-1, 0, 1\}$, which makes our scheme computationally more efficient by a factor n then the scheme from [2] (and any other scheme). Indeed, Fig. 1 shows polynomials f of degree m up to 12, and thus suitable for n up to 2^{12}, that allow the following primitive sets: choose $\alpha_1, \ldots, \alpha_n$ as (residue classes modulo f of) polynomials with coefficients in $\{0, 1\}$ and degree less than m such that α_i evaluates to $i-1$ at point 2 (similar as described in Section 4.2 for the β_i's in the scheme from [2]).

m	sample f	m	sample f
2	$X^2 - X - 1$	8	$X^8 + X^4 - X^3 + X - 1$
3	$X^3 - X - 1$	9	$X^9 + X^4 - 1$
4	$X^4 - X - 1$	10	$X^{10} - X^3 + X^2 + X - 1$
5	$X^5 - X^3 - X^2 + X + 1$	11	$X^{11} - X^5 + X^3 + X^2 - 1$
6	$X^6 - X - 1$	12	$X^{12} + X^6 - X^5 - X^4 - X^3 - X + 1$
7	$X^7 - X^3 + X^2 + X - 1$		

Fig. 1. Polynomials f that allow binary α_i's

We have been searching for suitable polynomials f with *minimal* residual degree $\deg(f - X^m)$, and that the polynomials found have rather small residual degree. This suggests that there is no shortage of suitable polynomials at all. As an aside, if we assume the existence of a suitable $\{-1, 0, 1\}$-polynomial for every n, then it can always be found in polynomial time.[8]

The best implementation of the scheme from [2] is given by Stam [14] using multi-exponentiation techniques. The achieved sharing complexity is $\tilde{O}(n^3)$ and the reconstruction complexity $\tilde{O}(n^{1+\lg 3})$ group operations. It appears to be hard to further improve the complexity of the scheme from [2], as the scheme

---

[8] In time $\tilde{O}(n^{3\lg 3})$: there are $O(n^{\lg 3})$ candidate polynomials f. Each candidate can be checked by computing the product of all non-zero $\{-1, 0, 1\}$-polynomials modulo f. There are $O(n^{\lg 3})$ factors in this product and the size of the coefficients in any step is also bounded by $\tilde{O}(n^{\lg 3})$. Note that for $n = 2^{12}$ this polynomial upper bound is already close to practically infeasible.

seems to be bound to an f with n-bit coefficients, and thus the module scalar-multiplication of a "small" number in $R = \mathbb{Z}[X]/(f)$ (meaning represented by a $\{-1, 0, 1\}$-polynomial of degree $< m$) with an element in G^m requires $\Theta(mn)$ group operations. That is where our complexity improvement stems from: since we can choose f with constant coefficients, a module scalar-multiplication with a small number requires only $O(m^2)$ group operations, and we achieve a sharing complexity of $\tilde{O}(n^2)$ and a reconstruction complexity of $\tilde{O}(n^{\lg 3})$ group operations.[9] (The $\lg(n)$-factors hidden by the $\tilde{O}$-notation have exponent at most 2).

The conclusion is that for reasonable values of n (namely for n up to 4096) our scheme is considerably more efficient than the scheme from [2] (and any other black-box secret sharing scheme). Furthermore, the evidence indicates that this is true for *any* n.

6   A Tight Lower Bound for the Randomness

In this final section we prove that our new black-box secret sharing scheme is not only optimal with regard to the expansion factor but also with regard to the randomness complexity. Specifically, we prove a lower bound of $t \cdot \lg(n) - O(t)$ for the randomness complexity of *binary linear* secret sharing schemes, which immediately implies the same bound for black-box secret sharing schemes. Recall that the randomness complexity of our scheme is $t \cdot \lceil \lg(n) \rceil$.

Recently, King proved the lower bound $\lg\big(n \cdot (n-1) \cdots (n-t+2)\big)$ [9, Theorem 12], which, using similar techniques as we do, can also be shown to be $t \cdot \lg(n) - O(t)$. However, the proposed proof assumes that the number of rows in any black-box secret sharing scheme $M \in \mathbb{Z}^{d \times e}$ is lower bounded by $d \geq n \lg(n)$, while in fact the best known lower bound is $d \geq n \lg(n+3) - n$ (see Theorem 2). Note that the bound $d \geq n \lg(n)$ used by King is widely conjectured to hold and sharpening the known lower bound to this conjectured lower bound is an interesting open problem.

Recall that a linear secret sharing scheme over a finite field F is defined along the lines of Definition 1 and 2, except that $\mathbb{Z}$ is replaced by F and G is restricted to $G = F$. In the following, e denotes the Euler number e ≈ 2.718.

Theorem 7. *For arbitrary $t, n \in \mathbb{Z}$ with $0 < t < n - 1$, the randomness complexity ρ of any binary linear secret sharing scheme $M \in \mathbb{F}_2^{d \times e}$, and thus in particular of any black-box secret sharing scheme $M \in \mathbb{Z}^{d \times e}$, for n and t satisfies $\rho > t \cdot \lg n - (1 + \lg e)t$.*

Proof. First of all, the bound for black-box secret sharing immediately follows from the bound on binary linear secret sharing, as any black-box secret sharing scheme $M \in \mathbb{Z}^{d \times e}$ reduced modulo 2 results in a binary linear secret sharing scheme.

---

[9] The exponent $\lg 3$ results from the fact that Δ (respectively $\Delta(\beta)$ in the scheme from [2]) can be replaced by its square-free part, which is the product of distinct polynomials of degree less than $m \approx \lg n$ with coefficients in $\{-1, 0, 1\}$, of which there exist $3^m \approx n^{\lg 3}$.

Consider a binary linear secret sharing scheme $M \in \mathbb{F}_2^{d \times e}$ for t and n as in the claim. Without loss of generality we may assume that the rows of M_i are linearly independent for any i. Also, by the lower bound on the expansion factor from Theorem 2, which also applies to binary linear schemes, we may assume that, say, M_n consists of $d_n \geq \lceil \lg(n+3) \rceil - 1$ rows (respectively $d_n \geq \lceil \lg(n) \rceil$ in case $t = 1$). Furthermore, as $t > 0$, $\boldsymbol{\varepsilon} = (1, 0, \ldots, 0)$ is not in the space spanned by the rows in M_n. Altogether this implies that, essentially by a basis change, M can be brought into a form where M_n consists of the $(d_n \times d_n)$-identity-matrix padded with zeroes to its left, while still being a binary linear secret sharing scheme for n and t. Consider now the labeled matrix $M' \in \mathbb{F}_2^{(d-d_n) \times (e-d_n)}$ by removing M_n as well as the last d_n columns of M (i.e. the columns that overlap with the identity matrix embedded in M_n). The labeling (of the remaining rows) is left unchanged. It is not hard to see that M' is a binary linear secret sharing scheme for $n' = n - 1$ and $t' = t - 1$. This procedure can be applied iteratively t times, resulting in a secret sharing scheme for $n - t$ participants and threshold 0 (which may have randomness complexity 0). The total number of rows removed during this process, and thus the randomness complexity of the original secret sharing scheme M is $\rho \geq \sum_{i=0}^{t-2} \left(\lceil \lg(n+3-i) \rceil - 1 \right) + \lceil (\lg(n-t+1) \rceil$. Using Stirling's bounds

$$\sqrt{2\pi}\, n^{n+1/2}\, e^{-n+1/(12n)} < n! < \sqrt{2\pi}\, n^{n+1/2}\, e^{-n+1/(12n+1)}$$

for factorials, we get

$$
\begin{aligned}
\rho &\geq \sum_{i=0}^{t-2} \left(\lceil \lg(n+3-i) \rceil - 1 \right) + \lceil (\lg(n-t+1) \rceil \\
&\geq \sum_{i=0}^{t-1} \lg(n-i) - t + 1 = \lg \prod_{i=0}^{t-1}(n-i) - t + 1 = \lg \frac{n!}{(n-t)!} - t + 1 \\
&> \lg \frac{n^{n+1/2}\, e^{-n+1/(12n)}}{(n-t)^{(n-t)+1/2}\, e^{-(n-t)+1/(12(n-t)+1)}} - t + 1 \\
&> \lg \frac{n^{n+1/2}\, e^{-n+1/(12n)}}{n^{(n-t)+1/2}\, e^{-(n-t)+1/(12(n-t)+1)}} - t + 1 \\
&= t \lg n - \left(t + \frac{1}{12(n-t)+1} - \frac{1}{12n} \right) \lg e - t + 1 \\
&> t \lg n - (1 + \lg e) t \qquad\qquad\qquad\qquad\qquad\qquad\qquad\qquad \square
\end{aligned}
$$

7 Concluding Remarks

From a practical point of view, the proposed black-box secret sharing scheme is essentially optimal with respect to its expansion factor (and its randomness complexity) and it is reasonably efficient for practical values of n: there seems to be little room for improvement (besides maybe squeezing the constant in the computational complexity). From a theoretical point of view, there are still a few open ends: First of all, we only have evidence but no proof that the proposed black-box secret sharing scheme is computationally efficient for large n. Furthermore, the question about the minimal achievable expansion factor is still not entirely solved, there is still a gap of (at most) 2 between the expansion factor achieved by the proposed scheme and the known lower bound; and we know that for certain parameters our construction is not optimal: it is for instance an easy exercise to construct a black-box secret sharing scheme for $t = 1$ and $n = 3$ with expansion factor $5/3$ (in contrast to 2, achieved by the proposed generic construction). Finally, all (reasonably good) black-box secret sharing schemes (for arbitrary t and n) are based on the framework discussed in Section 3. It would be interesting to discover completely new approaches.

Acknowledgements

The authors owe many thanks to H.W. Lenstra, jr. for contributing Theorem 6 and its proof to this work, and for his kind permission to include it in this paper; Theorem 4 (as well as its proof) is an adaptation to a special case of this more general theorem. Part of this work was done while Cramer was employed at Aarhus University and while Stam was a visitor under the Marie-Curie Program there. Part of this work was also done while Cramer was visiting the Centre de Recerca Matemàtica (CRM) in Bellaterra, Spain.

References

1. G. R. Blakley. Safeguarding cryptographic keys. In *Proc. National Computer Conference '79*, volume 48 of *AFIPS Proceedings*, pages 313–317, 1979.
2. R. Cramer and S. Fehr. Optimal black-box secret sharing over arbitrary Abelian groups. In M. Yung, editor, *Advances in Cryptography—Crypto'02*, volume 2442 of *Lecture Notes in Computer Science*, pages 272–287. Springer-Verlag, 2002.
3. R. Cramer, S. Fehr, Y. Ishai, and E. Kushilevitz. Efficient multi-party computation over rings. In E. Biham, editor, *Advances in Cryptography—Eurocrypt'03*, volume 2656 of *Lecture Notes in Computer Science*, pages 596–613. Springer-Verlag, 2003.
4. Y. Desmedt and Y. Frankel. Threshold cryptosystem. In G. Brassard, editor, *Advances in Cryptography—Crypto'89*, volume 435 of *Lecture Notes in Computer Science*, pages 307–315. Springer-Verlag, 1990.
5. Y. Desmedt, B. King, W. Kishimoto, and K. Kurosawa. A comment on the efficiency of secret sharing scheme over any finite abelian group. In C. Boyd and E. Dawson, editors, *ACISP'97*, volume 1438 of *Lecture Notes in Computer Science*, pages 391–402. Springer-Verlag, 1998.
6. S. Fehr. *Secure Multi-Player Protocols: Fundamentals, Generality, and Efficiency.* PhD thesis, University of Århus, 2003.
7. Y. Frankel, P. Gemmell, P. MacKenzie, and M. Yung. *Optimal resilience proactive public-key cryptosystems.* In: *Proceedings of FOCS '97*, IEEE Press, pp. 384–393, 1997.
8. M. Karchmer and A. Wigderson. On span programs. In *Proceedings of the Eigth Annual Structure in Complexity Theory Conference*, pages 102–111. IEEE Computer Society Press, 1993.

9. B. King. A Comment on Group Independent Threshold Sharing. In K. Chae and M. Yung, editors, *WISA'03*, volume 2908 of *Lecture Notes in Computer Science*, pages 425–441. Springer-Verlag, 2004.

10. B. S. King. *Some Results in Linear Secret Sharing*. PhD thesis, University of Wisconsin-Milwaukee, 2000.

11. S. Lang. *Algebra, 3rd ed.* Addison-Wesley Publishing Company, 1997.

12. A. Shamir. How to share a secret. *Communications of the ACM*, 22(11):612–613, 1979.

13. V. Shoup. Practical threshold signatures. In B. Preneel, editor, *Advances in Cryptography—Eurocrypt'00*, volume 1807 of *Lecture Notes in Computer Science*, pages 207–220. Springer-Verlag, 2000.

14. M. Stam. *Speeding up Subgroup Cryptosystems*. PhD thesis, Technische Universiteit Eindhoven, 2003.

15. D. Stinson and R. Wei. *Bibliography on Secret Sharing Schemes*. http://www.cacr.math.uwaterloo.ca/~dstinson/ssbib.html, 2003.

Secure Computation Without Authentication

Boaz Barak[1], Ran Canetti[2], Yehuda Lindell[3], Rafael Pass[4,*], and Tal Rabin[2]

[1] IAS
boaz@ias.edu
[2] IBM Research
{canetti, talr}@watson.ibm.com
[3] Bar-Ilan University, Israel
lindell@cs.biu.ac.il
[4] MIT
pass@csail.mit.edu

Abstract. In the setting of secure multiparty computation, a set of parties wish to jointly compute some function of their inputs. Such a computation must preserve certain security properties, like privacy and correctness, even if some of the participating parties or an external adversary collude to attack the honest parties. Until this paper, *all* protocols for general secure computation assumed that the parties can communicate reliably via authenticated channels. In this paper, we consider the feasibility of secure computation without *any* setup assumption.

We consider a completely unauthenticated setting, where all messages sent by the parties may be tampered with and modified by the adversary (without the honest parties being able to detect this fact). In this model, it is not possible to achieve the same level of security as in the authenticated-channel setting. Nevertheless, we show that meaningful security guarantees *can* be provided. In particular, we define a relaxed notion of what it means to "securely compute" a function in the unauthenticated setting. Then, we construct protocols for securely realizing any functionality in the stand-alone model, *with no setup assumptions whatsoever*. In addition, we construct universally composable protocols for securely realizing any functionality in the common reference string model (while still in an unauthenticated network). We also show that our protocols can be used to provide conceptually simple and unified solutions to a number of problems that were studied separately in the past, including *password-based authenticated key exchange* and *non-malleable commitments*.

1 Introduction

In the setting of secure multiparty computation, a set of parties with private inputs wish to jointly compute some function of their inputs in a secure way. Loosely speaking, the security requirements are that nothing is learned from the protocol other than the output (*privacy*), and that the output is distributed according to the prescribed functionality (*correctness*). These security properties must be guaranteed even when some subset of the parties and/or an external

---

* Research supported by an Akamai Presidential Fellowship.

V. Shoup (Ed.): Crypto 2005, LNCS 3621, pp. 361–377, 2005.

adversary maliciously and actively attack the protocol with the aim of compromising the honest parties' security.

Since the introduction of this problem in the 1980's [25,17,2,10], the research in this area has not subsided. The area has produced hundreds of papers that deal with many aspects of the problem. Works have included the definitional issues of secure computation, protocols with low round and communication complexity, protocols that rely on a wide variety of computational assumptions, lower bounds, security under composition and much much more.

Interestingly, one assumption that has appeared in *all* of the works in the field of secure computation until now is that *authenticated channels exist between the parties*. That is, it has always been assumed that the participating parties can communicate reliably with each other, without adversarial interference. In particular, the adversary is unable to send messages in the name of honest parties, or modify messages that the honest parties send to each other. There seem to be two main reasons that this assumption was always considered. First, the common belief was that these channels can easily be achieved, either through a physical designated channel connecting every pair of parties, or more realistically, via the deployment of a public-key infrastructure that can be used for implementing secure digital signatures. Second, it was assumed that no meaningful security guarantees can be provided in a distributed setting, unless honest parties can reliably communicate with each other.

Despite the above common belief, in real life the assumption that authenticated channels can be easily achieved is actually very problematic. It is clear that physical channels are generally unrealistic. In addition, a *fully deployed* public-key infrastructure is also far from reach. That is, although we can typically expect that most servers have an appropriate certificate for digital signatures, it is unreasonable today to require every participant (client) to also have one. This observation leads us to the following natural question:

What security can be obtained in a network without any authentication mechanism?

As we have seen, this question has important ramifications regarding the usefulness of secure multiparty protocols in real-world settings. However, it is also of great *theoretical* interest. In general, the theory of cryptography aims at understanding what tasks can be securely solved and under what (complexity and other) assumptions. Considered in this light, it is most natural to examine what security can be achieved in a setting with no setup assumptions whatsoever. In addition to highlighting the borders of feasibility and infeasibility for secure computation, answering this question enhances our understanding of the role of authentication in secure computation (detailed discussion follows).

Security Without Authenticated Channels. For simplicity, we begin by considering the important case of *two-party* protocols in an unauthenticated network. An immediate but important observation is that an adversary in such a network can simply "disconnect" the honest parties completely, and engage in separate executions with each of the two parties. Such an attack is unavoidable since there is no authentication between the parties. Therefore, the parties have no way of distinguishing the case that they interact with each other from the case that they each interact separately with a third party (in this case, the adversary). Given that this is an inherent limitation, our aim is to guarantee that this is the *only* attack that the adversary can carry out. More specifically, our notion of security guarantees that the adversary is limited to pursuing one of the two following strategies:

1. *Message relaying:* In this strategy, the adversary honestly relays the communication between the two parties, allowing them to perform their computation as if they were communicating over an authenticated channel.
2. *Independent executions:* In this strategy, the adversary intercepts the communication between the parties and engages in "independent" executions with each of them. That is, for parties A and B, the adversary can run an execution with A while playing B's role, and an execution with B while playing A's role. The security guarantee here is that the adversary is unable to make one execution depend on the other. Rather, the adversary must essentially choose an input for each execution and then run each execution as if it was running by itself. We remark that such "full" independence is actually impossible to achieve because the adversary can always run a complete execution with one of the parties, and then subsequently use the output it already received in order to choose its input for the execution with the other party. Therefore, our security definition guarantees that the *only* dependence the adversary can achieve is due to running the executions sequentially and choosing its input in the second execution after receiving its output from the first.

When considering the two-party case, the above security notion is a direct extension of the notion of *non-malleability*, introduced by Dolev, Dwork and Naor [11]. The work of [11] considered the specific tasks of encryption, commitments and zero-knowledge proofs. Here, we generalize these ideas to the more general concept of two-party (and multiparty) computation.

The same line of reasoning can be applied to analyze what is possible also in the case of multi-party protocols. Specifically, an adversary can always partition the honest parties into *disjoint* subsets. Then, given this partition, the adversary can run separate (and independent) executions with each subset in the partition, where in an execution with a given subset of honest parties H, the adversary plays the roles of all the parties outside of H. We guarantee that this is the only attack the adversary can carry out. In particular, we consider an adversary who interacts with a set of parties who are each willing to run a single execution (with each other). Our definition then states that although the adversary can actually run many executions with subsets of parties, it is guaranteed that:

1. The subsets of honest parties are disjoint,
2. Once a subset of honest parties is chosen, it is fixed for the duration of the protocol, and
3. The only dependence between the executions is due to the capability of the adversary to run the executions sequentially and choose its inputs as a function of the outputs from executions that have already terminated.

We remark that within each subset of parties, the execution that takes place is actually the same as when there are authenticated channels.

1.1 The Main Result

Our main result is a general proof of feasibility for stand-alone computation in the unauthenticated network setting. That is, we show that it is possible to securely compute any functionality according to the above security guarantees, even in a network with *no setup assumptions whatsoever*. This is in contrast to the widely held belief that authenticated channels, or some other setup, is a necessity for obtaining meaningful security. As an unusual step, before discussing the definition in more detail, we will first present the high-level idea behind our protocol. We feel that presenting the results in this order actually makes them easier to understand, especially because our protocol is in fact very simple.

It is clear that in order to run any of the known protocols for secure computation, authenticated channels are required. Our protocol for the unauthenticated setting is therefore comprised of two stages. In the first stage, some authenticated channels are set up. Then, in the second stage, a secure protocol is run on top of these authenticated channels. The basic idea of the protocol is:

Stage 1 – link initialization: In this stage, each party P_i generates a pair of signing and verification keys (s_i, v_i) and sends the signature verification key v_i to all other parties. In addition, after receiving verification keys v_j from all other parties, P_i signs on the series of all keys received (with its secret signing key s_i) and sends the signature σ_i to all other parties. Finally, each P_i checks that the signature it generated and all the signatures that it received refer to the same set of verification keys.

The idea behind this step is as follows. Let P_i and P_j be honest parties, let v_i be the verification key sent by P_i to P_j, and let v_j be the key sent by P_j to P_i. Since these keys are sent over unauthenticated channels, there is no guarantee that P_i will actually receive v_i and not some $v_i' \neq v_i$ generated by the adversary (and likewise for P_j). However, *if P_i and P_j do receive each other's real keys*, then they can set up a secure channel between them. In particular P_i has a verification key v_j associated with a secret signing-key known only to P_j, and vice versa. Thus, digital signatures can be used in a standard way in order to achieve authenticated communication between P_i and P_j. We note that if P_i received v_j (i.e., the key sent by P_j), then it will only continue if P_j received the exact same set of keys as P_i. This is guaranteed by the fact that the parties also sign on all the keys that they received. Thus, if P_j received different keys than P_i, then its signature σ_j will not include the same keys P_i received. When P_i receives the signature from P_j, it will therefore detect that adversarial interference has taken place, and so will abort. (Note that by our assumption here P_i already received v_j as generated by P_j, and so the adversary cannot hand P_i any other signature without breaking the signature scheme.)

From the above, we have that at the end of this stage, if P_i and P_j received each other's keys, then they have a secure bidirectional channel between them *and* they received the same set of verification keys. In contrast, if this is not the case, then we are guaranteed that their views of the verification keys are different. As we will show, this actually defines a *partition* of the honest parties so that **(1)** within each partition all of the honest parties hold each others' verification keys (and so all have mutually authenticated channels), and **(2)** the honest parties in different partitions have different views of the verification keys.

Stage 2 – secure computation using the generated links: In this stage, the parties run a protocol on top of the authenticated channels generated in the link initialization phase. The basic idea is to force the executions of the different subsets of honest parties, as defined by the above partition, to be *independent*. In order to do this, we view the series of verification keys as a *session identifier*. Then, we run a protocol for the authenticated channels model that is secure under concurrent composition, and guarantees independence between executions with different session identifiers. (We need security under concurrent composition, because different executions with different subsets may be run concurrently.)

As can be seen from the above protocol, and as we have discussed above, the only power provided to the adversary here is the ability to partition the honest parties into disjoint subsets and run separate executions with each set. (This adversarial "attack" can be carried out on *any* protocol in the unauthenticated model, and is not due to a weakness of our protocol.) We therefore model security by allowing the adversary to carry out such an "attack" in the ideal model as well. However, rather than modifying the standard ideal model, our basic definition of security is actually the same as in the standard model with authenticated channels and no honest majority. Then, the additional power awarded to the adversary here is modeled by modifying the definition of the functionality that is to be computed. That is, for any functionality $\mathcal{F}$ to be realized by two or more parties, we define a relaxed version of $\mathcal{F}$ called split-$\mathcal{F}$, or s$\mathcal{F}$, which is an *interactive functionality* and works as follows. Functionality s$\mathcal{F}$ lets the adversary define disjoint sets of parties, called **authentication sets**. Then, a separate and independent instance of the original functionality $\mathcal{F}$ is invoked for each authentication set. In an ideal execution of $\mathcal{F}$ for a given authentication set, functionality s$\mathcal{F}$ also allows the adversary to play the roles of all the honest parties not in the set (i.e., providing their inputs and receiving their outputs). In the two-party case, the adversary can either choose a *single* authentication set containing both parties (and then it cannot do anything more than in the authenticated channels model), or it can choose two authentication sets, each containing a single party (and so it must run an independent and separate execution with each party).

Theorem 1 (unauthenticated stand-alone computation): *Assume the existence of collision-resistant hash functions and enhanced trapdoor permutations,*[1] *and consider the stand-alone model with no setup whatsoever. Then, for any probabilistic polynomial-time multiparty functionality $\mathcal{F}$ there exists a protocol that securely computes the split functionality s$\mathcal{F}$, in the presence of static, malicious adversaries.*

Theorem 1 holds irrespective of the number of corrupted parties. In particular, this means that no honest majority is assumed (and therefore fairness and output delivery are not guaranteed, as is standard for this setting). We stress that unlike the setting of authenticated channels, here it would not help even if we did assume that

---

[1] See [19, Appendix C.1] for the definition of *enhanced* trapdoor permutations.

a large fraction of the parties are honest. This is due to the fact that the adversary can always choose all the subsets to be small, thereby ensuring an honest minority in each execution (and making it impossible to prevent an adversarial early abort).

We stress that although Theorem 1 constitutes a "general feasibility result", the security guarantee obtained is *far weaker* than that of the authenticated channels model. For example, agreement-type problems cannot be solved in this model, and indeed the split-functionality formalization explicitly removes all flavor of agreement (note that honest parties in different subsets run independent executions and so clearly cannot agree on anything).

Concurrency in a Stand-Alone World. From the above informal description of our protocol we see that the *stand-alone setting* with unauthenticated channels implicitly enables the adversary to run *concurrent executions* with different sets of honest parties. Thus, the protocol that is used in the second stage must be secure under concurrent composition. However, an important observation here is that when there are n honest parties, the adversary can force at most n concurrent executions (because the sets are disjoint and each honest party runs only once). It therefore follows that we only need security under *bounded* concurrency, which is fortunately much easier to achieve. (See [21] for impossibility results for the setting of unbounded concurrency, in contrast to the feasibility results of [20,23] and specifically for our use [24] when the concurrency is bounded.)

Entity Authentication Versus Session Authentication. One interesting corollary of our results is a more explicit distinction between entity authentication and session authentication. Entity authentication relates to a situation where a party A can verify that messages that it received in the name of party B were indeed sent by B. In contrast, session authentication relates to the fact that a party A establishes an authenticated channel with some other fixed party within a protocol execution or session. Party A does not know the identity of the party with whom it holds the channel; however, it knows that if the party is honest, then the adversary cannot interfere with any messages that are sent on the channel. This distinction is not new, and appears already in [11]. However, our results make it more explicit. Indeed, in the first stage of our protocol, we carry out a "session authentication protocol". Then, in the second stage, secure computation is carried out on top of this. By including entity authentication into the secure protocol of the second stage (or equivalently into the functionality being computed), we obtain an explicit separation of session authentication from entity authentication. This separation enables the entity authentication to be carried out *on top* of the session authentication, and in many different ways. Specifically, within the same execution, different parties may use different authentication mechanisms like passwords, digital signatures, interactive authentication protocols and so on.

1.2   Additional Results and Applications

The above result is of interest due to the fact that it requires no setup assumption whatsoever. However, it only holds for the rather limited stand-alone model. In this section, we briefly discuss some extensions and applications of this result. Formal statements and proofs of these results will appear in the full version of this paper.

UC Protocols Without Authenticated Channels. Universal composability (UC) is a definition of security with the property that any protocol that is UC-secure is guaranteed to remain secure under concurrent general composition [4] (i.e., when it is run many times concurrently with arbitrary other protocols). As in the stand-alone model, all UC-secure protocols until today assumed the existence of authenticated channels. We therefore extend our results to this setting.

We first note that in this setting, there is no hope of succeeding without setup assumptions. This is due to the fact that broad impossibility results for obtaining UC security have been demonstrated, even when there are authenticated channels [5,4,7]. We therefore consider the feasibility of obtaining UC-security without authenticated channels, but in the common reference string (CRS) model, where it is assumed that all parties have access to a single string that was chosen according to some predetermined distribution. In the CRS model and assuming authenticated channels, it has been shown that UC-secure protocols exist for essentially every functionality [8]. We combine our "link-initialization" protocol, as described above, with the protocol of [8] in order to achieve UC-secure protocols that compute essentially any split functionality $s\mathcal{F}$ in the CRS model with unauthenticated channels.

This combination of setup assumptions may seem strange. However, first note that at the very least, our result reduces the setup assumptions required for obtaining UC-security. More importantly, we argue that the assumption regarding a CRS is incomparable to that of authenticated channels. On the one hand, the generation of a CRS requires global trust of a stronger nature than that required for authenticated channels. On the other hand, it requires that only one string is generated and posted on some "secure bulletin board". In contrast, setting up authenticated channels essentially requires that all parties obtain a certificate for digital signatures. We also note that a common reference string by itself does not provide the means for parties to authenticate themselves to each other. Indeed, it is impossible to construct authenticated channels from unauthenticated ones even in the CRS model.

Partially Authenticated Networks. So far, we have discussed the completely unauthenticated setting, and have contrasted it to the standard completely authenticated setting. However, the most realistic setting is actually that of a *partially authenticated network*, where some of the parties have authenticated links and others do not. In addition, the authentication on these links may be unidirectional or bidirectional. For example, consider the case that only some of the parties have certificates for public (signature) keys as part of an implemented public-key infrastructure. Current protocols guarantee nothing in this setting. However, this is the real setting of the Internet today. We should be able to use a secure auction protocol, even if the only party who has a certificate is the auctioneer (in this case, all parties can obtain authenticated communication from the auctioneer, but that is all). In the full version of this paper, we show how to use our results in order to obtain secure computation in a partially authenticated network, while utilizing the authenticated links that do exist.

Password-Based Authenticated Key-Exchange. One problem that has received much attention, and is cast in the setting without authenticated channels, is that of password-based key exchange. Our results can also be applied to this problem. First, note that our definitional framework provides a way of modeling the problem easily within the setting of secure computation. Specifically, we define a functionality $\mathcal{F}$ as follows. Each party provides an input; if the inputs are equal, then $\mathcal{F}$ provides each with a long random value; if the inputs are not equal, then $\mathcal{F}$ hands $\bot$ to each party. Of course, the inputs we are referring to here are the parties' secret passwords.

The functionality $\mathcal{F}$, as defined, does not enable the adversary to make on-line password guesses, which is possible in password based key-exchange schemes. However, the transformation of $\mathcal{F}$ to its split functionality $\mathsf{s}\mathcal{F}$ provides this exact capability. Thus, the problem of securely computing $\mathsf{s}\mathcal{F}$ is exactly the problem of obtaining secure password-based authenticated key exchange. In particular, if the adversary plays a message relay strategy, then the parties will succeed in obtaining a shared secret key. In contrast, if the adversary runs independent executions, then the adversary will obtain exactly two password guesses. Furthermore, if the adversary guesses incorrectly, then the parties will obtain $\bot$. We note that this definition is essentially the same as that proposed in [6].

Our result therefore yields a conceptually simple framework and definition for solving this problem. Furthermore, we improve on previous solutions as follows. First, applying our first theorem we obtain secure password-based authenticated key-exchange in a setting with no setup assumptions. The only previously-known protocols to achieve this (without using random oracles) are [14,22]. Comparing our result to [14,22] we have the following advantages. First, we obtain a stronger security guarantee for the parties. Specifically, we guarantee exactly two password guesses per execution, rather than a constant or even polynomial number of guesses. Furthermore, these guesses are *explicit* (see [6] for a discussion about why this is advantageous). Second, our solution directly generalizes to password authentication protocols for *multiple* parties (whereas previous solutions only work for two parties). We note that like [14,22], our password-based protocol for the model with no setup assumptions (beyond the passwords themselves) is only secure if the same password is not used in concurrent executions of the protocol.

In addition to the above, we can apply our UC-secure protocol and obtain UC-secure password-based authenticated key-exchange in the common reference string model. This problem was previously considered by [6], who present highly efficient protocols based on specific assumptions. In contrast, we obtain protocols less efficient protocols, based on general assumptions. In addition, we can also extend our result to the setting of adaptive adversaries.

Alternative Authentication Mechanisms. Passwords are just one mechanism for authenticating parties. Due to the generality of our result which demonstrates that any function can be securely computed, we can obtain secure protocols for other, non-standard ways for parties to authenticate each other. The only requirement for accommodating these methods is that they can be described by an efficient functionality (and thus can be incorporated into stage 2 of the protocol). For example, we can accommodate "fuzzy" authentication where parties are authenticated if they pass at least k out of n "authentication tests", such as remembering the names of at least three of your childhood friends. Our solutions can also work in the case where parties are considered authenticated if they can perform some non-trivial computational task, like the "proof of work" in the anti-spam work of [12]. Finally, our protocols can be used to obtain "anonymous authentication" where two or more parties wish to authenticate themselves to each other based on useful data which they hold, as in the case of peer-to-peer and overlay networks.

Non-malleable Commitments. We remark that non-malleable commitments [11] can be obtained using our results in a similarly simple manner. Namely, define $\mathcal{F}$ to be a non-interactive (and potentially malleable) commitment function. Then, a protocol that securely computes $s\mathcal{F}$ constitutes a non-malleable commitment. This protocol does not improve on other known results. Nevertheless, it demonstrates the power of our general framework.

2   Split Functionalities

Due to the lack of space in this abstract, we will not present the definitions of secure computation. We refer the reader to [19, Chapter 7] for motivation and definitions. We note that we consider *reactive functionalities* here; see [4, Full version] for a formal discussion of this notion. Informally, the setting of secure computation with reactive functionalities is very similar to that of the more familiar "secure function evaluation". In the setting of secure function evaluation, an ideal model is defined where all parties send there inputs to a trusted party who computes the output and sends it back. When considering reactive functionalities, the only difference is that inputs and outputs can be provided interactively and at different stages. Thus, the trusted party interacts with the honest parties *and the adversary* multiple times, as specified by the code of the functionality.

In this section, we define what it means to realize an ideal functionality in an unauthenticated network without any setup. Before doing so, we remark that an unauthenticated network is formally modeled by having all communication go via the adversary. Thus, when a party P_i wishes to send a message m to P_j, it essentially just hands the tuple (P_i, P_j, m) to the adversary. It is then up to the adversary to deliver whatever message it wishes to P_j. We also remark that in the ideal model that we consider here, the communication between the honest parties and the trusted party remains ideally private and authenticated. Thus, the only change is to the real model.

We now proceed to the definition. As we have mentioned above, defining security in the unauthenticated model essentially involves defining a class of "split functionalities" that specifies the code of the trusted party in an ideal execution.[2] As we have mentioned, this class of functionalities enables the ideal-model adversary to split the honest parties into disjoint sets, called authentication

---

[2] As we have mentioned, this set of instructions for the trusted party could be incorporated into the definition of the ideal model. Equivalently, we have chosen to leave the ideal-model unchanged, and instead modify the functionality to be realized.

sets, in an adaptive way. The parties in each authentication set H then run a separate ideal execution with the trusted party. However, each such execution has the property that the adversary plays the roles of all the parties not in H (i.e., the parties that complete H to the full set of parties). Our specific formulation below provides three important guarantees:

1. An authentication set must be fixed before any computation in the set begins (and thus an authentication set cannot be chosen on the basis of the inputs of the honest parties in that set);
2. The computation within each set is secure in the standard sense (as in the case that authenticated channels are assumed);
3. The computation in a set is independent of the computations in other sets, except for the inputs provided by the adversary, which can be correlated to the outputs that it has received from computations with other authentication sets that have already been completed.

We now proceed to formalize the above. Let $\mathcal{F}$ be an ideal functionality. We define the relaxation of $\mathcal{F}$, called split-$\mathcal{F}$ or s$\mathcal{F}$, in Figure 1. We note that the functionality is slightly more involved than what is needed for the stand-alone case. The additional complications are included so that the same functionality will also be useful for the UC setting.

The Split Functionality s$\mathcal{F}$ – Explanation. In the initialization stage of the functionality, the adversary adaptively chooses subsets of honest parties H (the adaptivity relates to the fact that an authentication set can be chosen and a full execution completed, before the next authentication set is chosen). The adversary can choose any subsets that it wishes under the following constraints: First, the subsets must be *disjoint*. Second, the adversary must choose a *unique* session identifier sid_H for each authentication set H.

In the computation stage of the functionality s$\mathcal{F}$, each set H is provided with a different and independent copy of $\mathcal{F}$. This means that each set H essentially runs a separate ideal execution of $\mathcal{F}$. In each such execution, the parties $P_i \in H$ provide their own inputs, and the adversary provides the inputs for all $P_j \notin H$. This reflects the fact that in each execution, the roles of the parties *outside* of the authentication set are played by the adversary. Similarly, the parties $P_i \in H$ all receive their specified outputs as computed by their copy of $\mathcal{F}$. However, the adversary receives all of its own outputs, as well as the outputs of the parties $P_j \notin H$ (as is to be expected, since it plays the role of all of these parties in the execution). We stress that there is no interaction whatsoever between the different copies of $\mathcal{F}$ run by s$\mathcal{F}$.

The Functionality s$\mathcal{F}$ – Remarks:

1. The requirement that the authentication sets are disjoint guarantees that all the parties in an authentication set have consistent views of the interaction. In particular, each party participates in only one execution, and this is consistent with the other parties in its set.
2. s$\mathcal{F}$ requires the adversary to provide a unique identifier, sid_H for each authentication set. This identifier is used to differentiate between the various

---

Functionality s$\mathcal{F}$

For parties $P_1, \ldots, P_n$ and a given $\mathcal{F}$, functionality s$\mathcal{F}$ proceeds as follows:

Initialization:

1. Upon receiving a message (Init, sid) from a party P_i, send (Init, P_i) to the adversary.
2. Upon receiving a message (Init, sid, P_i, H, sid_H) from the adversary, verify that party P_i previously sent (Init, sid), that the list H of party identities includes P_i, and that for all previously recorded sets H', it holds that either **(1)** H and H' are disjoint and $sid_H \neq sid_{H'}$, or **(2)** $H = H'$ and $sid_H = sid_{H'}$. If any condition fails then do nothing. Otherwise, record the pair (H, sid_H), send (Init, sid, sid_H) to P_i, and initialize a new instance of the original functionality $\mathcal{F}$ with session identifier sid_H. Let $\mathcal{F}_H$ denote this instance of $\mathcal{F}$.

Computation:

1. Upon receiving a message (Input, sid, v) from party P_i, find the set H such that $P_i \in H$, and forward the copy of the functionality $\mathcal{F}_H$ the message v from P_i. If no such H is found then ignore the message.
2. Upon receiving a message (Input, sid, H, P_j, v) from the adversary, if $\mathcal{F}_H$ is initialized and $P_j \notin H$, then forward v to $\mathcal{F}_H$ as if coming from party P_j. Otherwise, ignore the message.
3. When a copy $\mathcal{F}_H$ generates an output v for party $P_i \in H$, functionality s$\mathcal{F}$ sends v to P_i. When the output is for a party $P_j \notin H$ or for the adversary, s$\mathcal{F}$ sends the output to the adversary.

Fig. 1. The split version of ideal functionality $\mathcal{F}$

copies of $\mathcal{F}$. Furthermore, this identifier is outputted explicitly to all the parties in this set. This is an important security guarantee: while the parties do not know, of course, which are the authentication sets, they have "evidence" of the set they are in. In particular, a global entity that sees the outputs of all parties can determine the authentication sets from the outputs alone. In a sense, this forces the adversary in the ideal process to mimic the same partitioning to authentication sets as in the protocol execution.

3. The above formalization of s$\mathcal{F}$ assumes for simplicity that the number and identities of the parties is known in advance. However, this requirement is not essential and neither the number of parties nor their identities need to be known in advance. Furthermore, they can be determined adaptively by the adversary as the computation proceeds. In this case, the only difference is that each party needs to receive the set of parties with which it should interact as part of its first input.

3   Obtaining Split Authentication

In this section, we show how to securely implement a link initialization phase. We proceed in two steps. First, we present an ideal functionality $\mathcal{F}_{\mathrm{SA}}$ that captures the property of authentication *within* an authentication set. Next, we present a simple protocol that UC-securely computes the $\mathcal{F}_{\mathrm{SA}}$ functionality *in the bare model,* without any setup. In Section 4 we will use the $\mathcal{F}_{\mathrm{SA}}$ functionality in order to obtain secure protocols for any split functionality s$\mathcal{F}$.

3.1   The Split Authentication Functionality $\mathcal{F}_{\mathrm{SA}}$

The split authentication functionality $\mathcal{F}_{\mathrm{SA}}$ is essentially a functionality that enables parties in the same authentication set to communicate in a reliable way. In particular, if the adversary wishes to deliver a message m to a party P_j with an alleged sender P_i, then $\mathcal{F}_{\mathrm{SA}}$ proceeds as follows:

1. If the authentication set H of P_j is not yet determined (i.e., P_j does not appear in any set H), then the delivery request is ignored. Otherwise:
2. If P_i is *not* in the same authentication set as P_j, then m is delivered as requested, regardless of whether it was actually sent by P_i.
3. If P_i and P_j *are* in the same authentication set, then the message is delivered to P_j only if it was sent by P_i and not yet delivered.

Formally, $\mathcal{F}_{\mathrm{SA}}$ is the split functionality of the functionality $\mathcal{F}_{\mathrm{AUTH}}$ defined in Figure 2 (we note that $\mathcal{F}_{\mathrm{AUTH}}$ here is a "multiple-session extension" of the $\mathcal{F}_{\mathrm{AUTH}}$ functionality defined in [4]). In other words, we define $\mathcal{F}_{\mathrm{SA}} = s\mathcal{F}_{\mathrm{AUTH}}$.

Functionality $\mathcal{F}_{\mathrm{AUTH}}$

$\mathcal{F}_{\mathrm{AUTH}}$ interacts with an adversary and parties $P_1, \ldots, P_n$ as follows:

1. Upon receiving (**send**, sid, P_i, P_j, m) from P_i, send (P_i, P_j, m) to the adversary and add (P_i, P_j, m) to an (initially empty) list $\mathcal{W}$ of waiting messages. Note that the same entry can appear multiple times in the list.
2. Upon receiving (**deliver**, sid, P_i, P_j, m) from the adversary, if there is a message $(P_i, P_j, m) \in \mathcal{W}$ then remove it from $\mathcal{W}$ and send (**received**, sid, P_i, P_j, m) to P_j. Otherwise do nothing.

Fig. 2. The authentication functionality $\mathcal{F}_{\mathrm{AUTH}}$

3.2   Realizing $\mathcal{F}_{\mathrm{SA}}$

In this section, we present a simple protocol for securely computing $\mathcal{F}_{\mathrm{SA}}$ in the bare model without any setup. The protocol that we present is actually UC-secure. This is important for two reasons. First, it is useful for achieving the extension of our results to the UC setting. Second, it enables us to claim that it remains secure even when run concurrently with any other protocol. This will be important in our final protocol (presented in Section 4) where the protocol for computing $\mathcal{F}_{\mathrm{SA}}$ is run together with the protocol of [23].

Our protocol uses a signature scheme that is existentially unforgeable against chosen message attacks as in [15] and is reminiscent of the technique used in [11] to construct non-malleable encryption. On a high-level our protocol also resembles the Byzantine Agreement protocol of [13] (although the goal and the actual protocol is very different). The main idea of the protocol has already been described in the introduction. We therefore proceed directly to its description.

Protocol 1   I. Link Initialization: *Upon input* (Init, *sid*), *each party* P_i *proceeds as follows:*

1. P_i *chooses a key pair* (VK_i, SK_i) *for the signature scheme.*
2. P_i *sends* VK_i *to all parties* P_j. (*Recall that in an unauthenticated network, sending* m *to* P_j *only means that the message* (P_i, P_j, m) *is given to the adversary.*)
3. P_i *waits until it receives keys from every* P_j, *for* $j \in [n], j \neq i$. (*Recall that these keys are actually received from the adversary and do not necessarily correspond to keys sent by other parties.*) *Denote by* VK_{i_j} *the key that* P_i *received from* P_j *and denote* $VK_{i_i} = VK_i$. *Now, let* $VK_{i'_1}, \ldots, VK_{i'_n}$ *be the same set of keys* $VK_{i_1}, \ldots, VK_{i_n}$ *arranged in ascending lexicographic order. If there are two keys that are the same, then* P_i *halts. Otherwise,* P_i *defines* $sid_i = \langle VK_{i'_1}, \ldots, VK_{i'_n} \rangle$.
4. P_i *computes* $\sigma_i = \mathsf{Sign}_{SK_i}(sid_i)$ *and sends* $\alpha_i = (sid_i, \sigma_i)$ *to all parties* P_j.
5. P_i *waits until it receives an* α_j *message from every* P_j, *for* $j \in [n], j \neq i$. *Denote by* $\alpha_{i_j} = (sid_{i_j}, \sigma_{i_j})$ *the pair that* P_i *received from* P_j *and denote* $\alpha_{i_i} = \alpha_i$. *Then,* P_i *checks that for every* j, $\mathsf{Verify}_{VK_{i_j}}(sid_{i_j}, \sigma_{i_j}) = 1$ *and that* $sid_{i_1} = sid_{i_2} = \cdots = sid_{i_n}$. *If all of these checks pass, then* P_i *outputs* (Init, *sid*, sid_i).

II. Authenticating Messages:

1. P_i *initializes a counter* c *to zero.*
2. *When* P_i *has input* (send, *sid*, P_i, P_j, m), *meaning that it wishes to send a message* m *to* P_j, *then it signs on* m *together with* sid_i, *the recipient identity, and the counter value. That is,* P_i *computes* $\sigma = \mathsf{Sign}_{SK_i}(sid_i, m, P_j, c)$, *sends* (P_i, m, c, σ) *to* P_j, *and increments* c.
3. *Upon receiving a message* (P_j, m, c, σ) *allegedly from* P_j, *party* P_i *first verifies that* c *did not appear in a message received from* P_j *in the past. It then verifies that* σ *is a valid signature on* (sid_i, m, P_i, c), *using the verification key* VK_{i_j}. *If the verification succeeds, then it outputs* (received, *sid*, P_j, P_i, m).

We have the following theorem:

Theorem 2 *Assume that the signature scheme used in Protocol 1 is existentially secure against chosen message attacks. Then, Protocol 1 securely computes the* $\mathcal{F}_{SA}$ *functionality under the UC-definition in the presence of malicious, adaptive adversaries, and in the bare model with no setup whatsoever.*

Proof Sketch: We show that for any adversary $\mathcal{A}$ there exists an ideal-process adversary (i.e., a simulator) $\mathcal{S}$ such that no environment $\mathcal{Z}$ can tell with non-negligible probability whether it is interacting with parties running Protocol 1 and adversary $\mathcal{A}$, or with $\mathcal{F}_{SA}$ and simulator $\mathcal{S}$. The simulator $\mathcal{S}$ internally invokes $\mathcal{A}$ and perfectly simulates the honest parties interacting with $\mathcal{A}$. Then, when an honest party P_i in the internal simulation by $\mathcal{S}$ completes its link initialization phase and computes sid_i, simulator $\mathcal{S}$ determines the set H of P_i to be the set of parties for which sid_i contain their "authentic" verification keys. Next, when $\mathcal{A}$ delivers a signed message to some P_i, simulator $\mathcal{S}$ asks $\mathcal{F}_{SA}$ to deliver the message to P_i in the ideal process only if the internally simulated honest party would accept the signature, according to the protocol specification. More specifically, $\mathcal{S}$ locally runs an interaction between $\mathcal{A}$ and simulated copies of all the parties. In addition:

1. All messages from the external $\mathcal{Z}$ to $\mathcal{S}$ are forwarded to the internal $\mathcal{A}$, and all messages that $\mathcal{A}$ wishes to send to $\mathcal{Z}$ are externally forwarded by $\mathcal{S}$ to $\mathcal{Z}$.
2. Whenever $\mathcal{S}$ receives a message that an honest party P_i sent an (Init, sid) message to $\mathcal{F}_{SA}$, simulator $\mathcal{S}$ simulates the actions of an honest P_i in the link initialization phase of Protocol 1.
3. Whenever an internally simulated party P_i completes the link initialization phase with sid_i, simulator $\mathcal{S}$ determines the set H_i to be the set of honest parties P_j such that the authentic verification key sent by P_j is included in sid_i. (Recall that $\mathcal{S}$ internally runs all the honest parties, so it can do this.) $\mathcal{S}$ then checks that for all previously computed sets H, it holds that either:
   - H_i and H are disjoint and $sid_{H_i} \neq sid_H$, or
   - $H_i = H$ and $sid_{H_i} = sid_H$.

   If this holds, then $\mathcal{S}$ sends (Init, sid, P_i, H_i, sid_i) to $\mathcal{F}_{SA}$. Otherwise, $\mathcal{S}$ halts and outputs fail1.
4. Whenever $\mathcal{S}$ receives a message (send, sid, P_i, P_j, m) from $\mathcal{F}_{SA}$ where P_i is honest, simulator $\mathcal{S}$ simulates the actions of an honest P_i sending a message m in the authentication phase of Protocol 1.
5. Whenever an internally simulated party P_i outputs (received, sid, P_j, P_i, m) in the simulation, $\mathcal{S}$ works as follows. If P_j is corrupted, then $\mathcal{S}$ instructs P_j to send an appropriate send message to $\mathcal{F}_{SA}$. Likewise, if P_j is not in the same authentication set as P_i, then $\mathcal{S}$ sends the appropriate send message to $\mathcal{F}_{SA}$ itself. Then, $\mathcal{S}$ sends $\mathcal{F}_{SA}$ the message (deliver, P_j, P_i, m), instructing it to deliver m to P_i from P_j.[3] If the request is not fulfilled then $\mathcal{S}$ halts and outputs fail2.
6. Whenever $\mathcal{A}$ corrupts a party P_i, simulator $\mathcal{S}$ hands $\mathcal{A}$ the state of the internally simulated P_i.

It is straightforward to verify that as long as $\mathcal{S}$ does not output fail1 or fail2, the view of $\mathcal{Z}$ in the ideal-model is identical to its view in a real execution of Protocol 1. (This is due to the fact that unless a fail occurs, $\mathcal{S}$ just mimics the actions of the honest parties. In addition, the local outputs of the honest parties in the internal simulation correspond exactly to the outputs of the actual honest parties in the ideal model.) It therefore suffices to show that $\mathcal{S}$ outputs a fail message with at most negligible probability.

We first show that $\mathcal{S}$ outputs a fail1 message with at most negligible probability. Below, we refer only to honest parties in the authentication sets because $\mathcal{S}$ never includes corrupted parties in these sets. There are three events that could cause a fail1 message:

1. *There exist two honest parties P_i and P_j for whom $\mathcal{S}$ defines sets H_i and H_j such that $H_i = H_j$, and yet $sid_i \neq sid_j$:* In order to see that this event cannot occur with non-negligible probability, notice that $\mathcal{S}$ only defines sets H_i and H_j for parties that conclude the Link Initialization portion of the protocol and places P_i and P_j in the same set if they received each others "authentic"

---

[3] Note that if P_j is honest and is in the same authentication set as P_i then $\mathcal{S}$ does not begin with a send message, but rather immediately sends a deliver message.

verification keys. By the signatures sent at the end of link initialization phase, it follows that either at least one of the parties aborts, or the adversary forged a signature relative to either P_i or P_j's verification key, or P_i and P_j both conclude with the same sid. (We note that the reduction here to the security of the signature scheme is straightforward.)

2. *There exist two sets $H_i \neq H_j$ that are not disjoint:* Let $P_i \in H_i \cap H_j$ be an honest party. Then, using the same arguments as above, except with negligible probability, P_i must have the same sid as all the honest parties in H_i *and* all the honest parties in H_j. Thus, all of the parties in $H_i \cup H_j$ have the same sid. Since this sid is comprised of the parties verification keys, it must hold that all parties in $H_i \cup H_j$ received each other's authentic verification keys. By the construction of S, it therefore holds that $H_i = H_j$.

3. *There exist two sets $H_i \neq H_j$, and yet $sid_i = sid_j$:* We have already seen that by the construction of S, if $sid_i = sid_j$ then $H_i = H_j$.

It remains to show that S outputs fail2 with at most negligible probability. This occurs if S sends a $(\texttt{deliver}, P_j, P_i, m)$ message to $\mathcal{F}_{\text{SA}}$ where P_i is honest, and the message is not actually delivered to P_i. By the definition of $\mathcal{F}_{\text{SA}}$ (and in general split functionalities), this can only occur if P_j is honest, and P_i and P_j are in the same authentication set H. (We ignore trivialities here like the case that H is not defined.) In order to see this, notice that if P_j is corrupted, then S first instructs it to send a send message to $\mathcal{F}_{\text{SA}}$ and so S's deliver message would not be ignored. The same is true in the case that P_i and P_j are *not* in the same authentication set (because then S first sends the send message itself). Now, if P_i and P_j *are* in the same authentication set, then they both hold each others "authentic" verification keys (as shown above). Furthermore, the deliver message of S is only ignored if P_j did not previously send an appropriate send message to $\mathcal{F}_{\text{SA}}$. This implies that S did not generate a signature on (P_i, m, c) in the internal simulation (see step 4 of the simulation by S), and yet P_i received a valid signature on this message. Thus, it follows that $\mathcal{A}$ must have forged a signature relative to the honest P_j's key. As above, such an adversary can be used to break the signature, and the actual reduction is straightforward. We conclude that the views of $\mathcal{Z}$ in the two interactions are statistically close. ∎

4   General Functionalities in the Stand-Alone Model

In this section, we prove the following theorem:

Theorem 3 (Theorem 1 – restated): *Assume the existence of collision-resistant hash functions and enhanced trapdoor permutations, and consider the stand-alone model with no setup whatsoever. Then, for any probabilistic polynomial-time multiparty functionality $\mathcal{F}$ there exists a protocol that securely computes the split functionality $s\mathcal{F}$, in the presence of static, malicious adversaries.*

Theorem 3 is obtained by combining Protocol 1 for securely computing $\mathcal{F}_{\text{SA}}$ with the protocol of [23] for securely computing any functionality in the setting of bounded-concurrency. Recall that in this model, there is an a priori bound on the number of protocol executions that can take place. As we have remarked above, in the setting considered here, we know that at most n concurrent executions can take place in a stand-alone execution with n parties in the unauthenticated model. Therefore, bounded concurrency suffices.

Our protocol for securely computing any n-party split functionality $\mathsf{s}\mathcal{F}$ works by first running the link initialization stage of Protocol 1, and obtaining a session identifier sid from this phase. Then, the protocol of [23] for securely computing $\mathcal{F}$ (under n-bounded concurrent composition) is executed, using the identifier sid and authenticating all messages sent and received as described in Protocol 1.[4]

The intuition behind the security of this protocol is that $\mathcal{F}_{\mathrm{SA}}$ guarantees that all the honest parties in a given authentication set H are essentially connected via pairwise authenticated channels. Thus, the execution of the protocol of [23] in our unauthenticated setting is the same as an execution of the protocol of [23] in the *authenticated channels model*, where the participating parties are comprised of the honest parties in H and $n - |H|$ corrupted parties. Now, in the unauthenticated model (by the definition of $\mathsf{s}\mathcal{F}$), the adversary is allowed to play the role of the $n - |H|$ parties not in H. Therefore, the above protocol suffices for securely computing the split functionality $\mathsf{s}\mathcal{F}$.

We note that the protocol of [23] relies on the existence of collision-resistant hash functions and enhanced trapdoor permutations. Furthermore, given any parameter m that is polynomial in the security parameter (and, in particular, setting m to equal the number of parties n), it is possible to obtain a protocol that remains secure for up to m concurrent executions, where in each execution any subset of the parties may participate. We note that these subsets may overlap in an arbitrary way, and security is still guaranteed. This point is *crucial* for our above use of the protocol. Namely, in order to prove security we actually consider a *virtual network* of $2n$ parties $P_1, \ldots, P_{2n}$ where all parties $P_{n+1}, \ldots, P_{2n}$ are corrupted. Then, for any authentication set $H \subseteq \{P_1, \ldots, P_n\}$ we consider an execution of the protocol of [23] with the subset of parties comprised of every $P_i \in H$, and every P_{n+j} for $P_j \notin H$. Note that this defines a subset of exactly n parties, where every party *not* in H is controlled by the adversary, as required. The important point to note now, however, is that some P_{n+j} may participate in many different executions of the protocol of [23]. It is therefore crucial that [23] remains secure when arbitrary subsets of parties run the protocol.

References

1. B. Barak. How to Go Beyond the Black-box Simulation Barrier. In *42nd FOCS*, pages 106–115, 2001.
2. M. Ben-Or, S. Goldwasser and A. Wigderson. Completeness Theorems for Non-cryptographic Fault-Tolerant Distributed Computations. In *20th STOC*, pages 1–10, 1988.

---

[4] More formally, our protocol works in the $\mathcal{F}_{\mathrm{SA}}$-hybrid model, and uses the authentication mechanism provided by $\mathcal{F}_{\mathrm{SA}}$ in order to run the protocol of [23].

3. R. Canetti. Security and Composition of Multiparty Cryptographic Protocols. *Journal of Cryptology*, 13(1):143–202, 2000.
4. R. Canetti. Universally Composable Security: A New Paradigm for Cryptographic Protocols. In *42nd FOCS*, pages 136–145, 2001. Full version available at http://eprint.iacr.org/2000/067.
5. R. Canetti and M. Fischlin. Universally Composable Commitments. In *CRYPTO 2001*, Springer-Verlag (LNCS 2139), pages 19–40, 2001.
6. R. Canetti, S. Halevi, J. Katz, Y. Lindell and P. MacKenzie. Universally Composable Password-Based Key Exchange. In *EUROCRYPT 2005*, Springer-Verlag (LNCS 3494), pages 404–421, 2005.
7. R. Canetti, E. Kushilevitz and Y. Lindell. On the Limitations of Universally Composable Two-Party Computation without Set-up Assumptions . In *EUROCRYPT '03*, Springer-Verlag (LNCS 2656), pages 68–86, 2003.
8. R. Canetti, Y. Lindell, R. Ostrovsky and A. Sahai. Universally Composable Two-Party and Multi-Party Secure Computation. In *34th STOC*, pp. 494–503, 2002.
9. R. Canetti and T. Rabin. Universal Composition with Joint State. In *CRYPTO 2003*, Springer-Verlag (LNCS 2729), pages 265–281, 2003.
10. D. Chaum, C. Crepeau and I. Damgard. Multiparty Unconditionally Secure Protocols. In *20th STOC*, pages 11–19, 1988.
11. D. Dolev, C. Dwork and M. Naor. Non-malleable Cryptography. *SIAM Journal on Computing*, 30(2):391–437, 2000.
12. C. Dwork and M. Naor. Pricing via Processing or Combating Junk Mail. In *CRYPTO'92*, Springer-Verlag (LNCS 740), pages 139–147, 1992.
13. M. Fitzi, D. Gottesman, M. Hirt, T. Holenstein and A. Smith. Detectable Byzantine Agreement Secure Against Faulty Majorities. *21st PODC*, pp. 118–126, 2002.
14. O. Goldreich and Y. Lindell. Session-Key Generation Using Human Passwords Only. In *CRYPTO 2001*, Springer-Verlag (LNCS 2139), pages 408–432, 2001.
15. S. Goldwasser, S. Micali and R. Rivest. A Digital Signature Scheme Secure Against Adaptive Chosen-Message Attacks. *SIAM J. on Computing*, 17(2):281–308, 1988.
16. S. Goldwasser, S. Micali and C. Rackoff. The Knowledge Complexity of Interactive Proof-Systems. *SIAM Journal on Computing*, 18(1):186–208, 1989.
17. O. Goldreich, S. Micali and A. Wigderson. How to Play Any Mental Game. In *19th STOC*, pages 218–229, 1987.
18. O. Goldreich. *Foundations of Cryptography – Vol. 1*. Cambridge Univ. Press, 2001.
19. O. Goldreich. *Foundations of Cryptography – Vol. 2*. Cambridge Univ. Press, 2004.
20. Y. Lindell. Bounded-Concurrent Secure Two-Party Computation Without Setup Assumptions. In *35th STOC*, pages 683–692, 2003.
21. Y. Lindell. Lower Bounds for Concurrent Self Composition. In *1st TCC*, Springer-Verlag (LNCS 2951), pages 203–222, 2004.
22. M. Nguyen and S. Vadhan. Simpler Session-Key Generation from Short Random Passwords. In *1st TCC*, Springer-Verlag (LNCS 2951), pages 428–445, 2004.
23. R. Pass. Bounded-Concurrent Secure Multi-Party Computation with a Dishonest Majority. In *36th STOC*, pages 232–241, 2004.
24. R. Pass and A. Rosen Bounded-Concurrent Secure Two-Party Computation in a Constant Number of Rounds. In *44th FOCS*, pages 404–413, 2003.
25. A.C. Yao. How to Generate and Exchange Secrets. *27th FOCS*, pp. 162–167, 1986.

Constant-Round Multiparty Computation Using a Black-Box Pseudorandom Generator

Ivan Damgård[1,*] and Yuval Ishai[2,**]

[1] Aarhus University
ivan@daimi.au.dk
[2] Technion
yuvali@cs.technion.ac.il

Abstract. We present a constant-round protocol for general secure multiparty computation which makes a *black-box* use of a pseudorandom generator. In particular, the protocol does not require expensive zero-knowledge proofs and its communication complexity does not depend on the computational complexity of the underlying cryptographic primitive. Our protocol withstands an active, adaptive adversary corrupting a minority of the parties. Previous constant-round protocols of this type were only known in the semi-honest model or for restricted classes of functionalities.

1 Introduction

General secure computation is often perceived as being inherently impractical. One valid reason for this perception is the fact that all current protocols either require many rounds of interaction (e.g., [20, 5, 27, 35, 13, 22]), or alternatively require only a constant number of rounds but make use of expensive zero-knowledge proofs for each gate of the circuit being computed (e.g., [40, 4, 28, 6, 26, 34, 25]). Indeed, in all constant-round protocols from the literature, players need to provide zero-knowledge proofs for statements that involve the computation of a pseudorandom generator or other cryptographic primitives on which the "semi-honest" version of the protocol relies. Thus, these protocols make a *non-black-box* use of their underlying cryptographic primitives. We stress that this holds for all settings of secure computation with security against malicious parties, both in the two-party and in the multi-party case. The only exceptions to this general state of affairs are unconditionally secure protocols that apply to restricted classes of functionalities such as NC^1 [2] or protocols that require an exponential amount of computation [3].

In this work we consider the setting of multiparty computation with an honest majority, and present a general constant-round protocol that makes a *black-*

---

* Supported by BRICS, Basic research in Computer Science, Center of the Danish National Research Foundation and FICS, Foundations in Cryptography and Security, funded by the Danish Natural Sciences Research Council.

** Research supported by Israel Science Foundation grant 36/03.

V. Shoup (Ed.): Crypto 2005, LNCS 3621, pp. 378–394, 2005.

box use of a pseudorandom generator.[1] Similarly to all general constant-round protocols from the literature, our protocol relies on Yao's garbled circuit technique [40], which was later adapted to the multi-party setting by Beaver, Micali, and Rogaway [4]. The latter "BMR protocol" requires players to verifiably secret-share seeds to a PRG as well as the outputs of the PRG on these seeds. To ensure that this is done correctly, the protocol makes a non-black-box use of the PRG by requiring players to prove via (distributed) zero-knowledge that the shared seeds are consistent with the shared PRG outputs. We get around this problem by modifying the basic structure of the BMR protocol, using "distributed symmetric encryption" and error-correction to replace the zero-knowledge proofs. Before providing a more detailed account of our results we give some more background to put them in context.

Black-Box Reductions in Cryptography. Most reductions between cryptographic primitives are (fully) black box, in the sense that they implement a primitive A by using some other primitive B as an oracle, without depending on the implementation details of B. Moreover, the security proof of such reductions is also black-box in the sense that an adversary breaking A can be used as an oracle in order to break the underlying primitive B. (See [36] for a more detailed definition and discussion.) In contrast, a non-black-box reduction can use the "code" of B when implementing A. Most examples for non-black-box reductions in cryptography are ones in which the construction of A requires parties to prove in zero-knowledge statements that involve the computation of the underlying primitive B. For instance, the construction of an identification scheme from a one-way function [14] makes a non-black-box use of the one-way function. A rich line of work, originating from [24], uses oracle separations to rule out the existence of (various forms of) black-box reductions. Most notably, it is shown in [24] that there is no black-box reduction from key agreement to a one-way function. A common interpretation of such results is that they rule out the existence of "practically feasible" reductions. Indeed, all known examples for non-black-box reductions in cryptography involve a considerable overhead. In the context of cryptographic protocols, this overhead typically involves not only local computation but also communication: the *communication* complexity of the protocol B depends on the *computational* complexity of the underlying primitive A.[2] Our work provides further demonstration for the usefulness of distinguishing between the two types of reductions.

Constant-Round Secure Computation. The question of implementing secure computation in a constant number of rounds has attracted a considerable amount of attention. The first general constant-round protocol for secure two-party computation was given by Yao [40]. Yao's original protocol considered only the case of semi-honest parties; an extension to the case of malicious parties (equivalently,

---

[1] Here we assume the standard model of secure point-to-point channels.

[2] In some cases it is possible to reduce the communication overhead by using communication-efficient zero-knowledge *arguments* (cf. [31]). However, this approach would make the computational overhead even higher.

an *active* adversary) was given by Lindell [28]. While Yao's original protocol makes a black-box use of the underlying primitives (a pseudorandom generator and oblivious transfer), the protocol from [28] relies on the methodology of Goldreich, Micali, ad Wigderson [20] and thus makes a non-black-box use of these primitives. Recently, Katz and Ostrovsky obtained a two-party protocol with an optimal exact round complexity [25].

An extension of Yao's protocol to the case of multiparty computation with an honest majority was given by Beaver, Micali, and Rogaway [4] (see also [37, 39]). Similarly to the two-party case, the BMR protocol makes a black-box use of a PRG in the semi-honest case and a non-black-box use of a PRG in the malicious case. (Because of the honest majority assumption, the protocol does not need to rely on OT.) Constant-round multiparty protocols withstanding a *dishonest* majority were recently obtained by Katz et al. [26] and by Pass [33]. (The latter protocol also achieves bounded concurrent security, extending a previous two-party protocol of Pass and Rosen [34].) These protocols are only proved secure with respect to a non-adaptive adversary and allow the adversary to prevent honest parties from receiving any output (even when corrupting a minority of the parties). Like the two-party protocols, these protocols follow the GMW methodology and thus make a non-black-box use of the underlying primitives.

Finally, there has also been a considerable amount of work on *unconditionally* secure constant-round multiparty computation in the case of an honest majority (e.g. [2, 15, 23, 9]). Unfortunately, all known protocols in this setting can only be efficiently applied to restricted classes of functions such as NC^1 or non-deterministic logspace.

1.1 Our Results

We consider the model of computationally secure multiparty computation against an active, adaptive adversary corrupting up to $t < n/2$ players. Our default network model assumes secure point-to-point channels and the availability of broadcast (see more on that later). As stated above, our main result is a new "black-box feasibility" result. Specifically, we construct the first general constant-round protocol which makes a black-box use of a PRG (equivalently, using [21], a black-box use of a one-way function). Since much of our motivation comes from the goal of making secure computation more efficient, we also attempt to minimize the amount of interaction and communication required by our protocols. To this end, it is convenient to cast the protocols in the following "client-server" framework.

The Client-Server Model. We divide the players into "input clients" who provide inputs, "output clients" who receive outputs, and "servers" who perform the actual computation. The security of the protocol should hold as long as at most t servers are corrupted, regardless of the number of corrupted clients. These three sets of players need not be disjoint, hence this is a strict generalization of the standard MPC framework in which all parties play all three roles. It also represents a likely scenario for applying MPC in practice, using specialized (but untrusted) servers to perform the bulk of the work. We stress again that this is just a refinement of the standard model. The main advantage of this refinement, besides conceptual clarity, is that it allows to decouple the number of "consumers" from the required "level of security". (The latter depends on the number of servers and the security threshold.) For instance, we can have just two clients and many servers (which may be viewed a distributed implementation of two-party computation), or a very large number of clients and only few servers. The latter might be the most realistic setting for secure computations involving inputs from many players.

Linear Preprocessing. We present our main protocol in two stages. First, we present a protocol in what we call the "linear preprocessing model". In this model, it is assumed that there is a trusted setup phase where a dealer can provide clients and servers with linearly-correlated resources, e.g., Shamir-shares of random secrets. Then we use standard subprotocols for emulating the trusted setup in the plain model. The linear preprocessing model is motivated by the pseudorandom secret-sharing technique of [11]: when the number of servers is small, linear preprocessing can be emulated using a "once and for all" setup phase in which (roughly $\binom{n}{t}$) replicated and independent seeds are given to the players. Following this setup, the players can locally generate the required correlated shares without further interaction.[3]

Our Protocols. Our main protocol in the linear preprocessing model requires only two communication rounds when $t < n/5$. In the first round each input client broadcasts its masked inputs to the servers, and in the second round the servers send to each output client a total of $O(n^2|C|k)$ bits, where $|C|$ is the size of the circuit being computed and k is a security parameter. In the plain model, one can obtain similar protocols at the cost of a higher communication complexity and additional rounds of interaction. When $t < n/5$, it suffices to use 3 rounds of interaction by relying on a VSS protocol from [17]. Alternatively, it is possible to tolerate $t < n/3$ or even $t < n/2$ malicious servers at the expense of further increasing the communication and the (constant) number of rounds.

In the case of computing a randomized functionality which has no inputs, the 2-round protocol in the linear preprocessing model becomes totally non-interactive when combined with pseudorandom secret-sharing. That is, to securely compute such a functionality it suffices for each server to send a single message to each output client, without using any broadcasts. Such non-interactive protocols can be used to obtain efficient distributed implementations of a trusted dealer in a wide range of applications.

On the Use of Broadcast. As in most of the MPC literature, our network model assumes the availability of broadcast as an atomic primitive. However, using the (expected) constant-round broadcast protocol of Feldman and Micali [16, 29], our protocols can be turned into (expected) constant-round protocols also in the point to point model. Concerning the communication complexity, since it is

---

[3] The method of [11] was only proved to be secure in the case of a non-adaptive adversary. Thus, when relying on pseudorandom secret sharing our protocol loses its provable adaptive security.

possible to implement our protocol so that the number of broadcasts involved is independent of $|C|$ (using the techniques of [22]), one can get the same (amortized) communication complexity in the point-to-point model. Moreover, in the typical scenario where the number of servers is small, even a "brute-force" simulation of the broadcasts will not have a major impact on efficiency.

Organization. The remainder of the paper is organized as follows. In Section 2 we define our security model and preprocessing models, and present some standard subprotocols in these models. Our main protocol and its variations are presented in Section 3, where the underlying distributed encryption idea is highlighted in Section 3.3. Due to our elaborate use of techniques from previous works (mostly in the context of *information-theoretic* multiparty computation), we omit some of the low-level details and assume the reader's familiarity with standard MPC techniques from the literature. Some discussions and extensions (e.g., the case $t < n/2$) were omitted for lack of space and can be found in the full version.

2   Preliminaries

The Model. We consider a system consisting of several players, who interact in synchronous rounds via authenticated secure point-to-point channels and a broadcast medium. Players can be designated three different roles: *input clients* who hold inputs, *output clients* who receive outputs, and *servers* who may be involved in the actual computation. As discussed in Section 1.1, this is just a generalization of the standard model, where each party can play all three roles. We denote the number of servers by n. The functionalities we wish to compute only receive inputs from input clients and only provide outputs to output clients. For simplicity we will only explicitly consider deterministic functionalities providing all output clients with the same output, though an extension of our results to the general case is straightforward.[4] (In contrast, we will employ sub-protocols that compute randomized functionalities and provide servers and input clients with outputs as well.)

We assume by default an active, adaptive, rushing adversary corrupting at most t servers. (There is no restriction on the number of corrupted clients.) We refer the reader to, e.g., [7] for the standard definition of security in this model.

Our protocols will employ secret-sharing over a finite field $K = \mathrm{GF}(2^k)$, where k is a security parameter that will be used as the length of a seed to a PRG. Slightly abusing notation, each server P_i is assigned a unique nonzero value $i \in K$.

2.1   Linear Preprocessing

As discussed in Section 1.1, it will be convenient to describe and analyze our protocols in the *linear preprocessing model*, where we allow some restricted trusted

---

[4] Standard reductions from the general case to this special case involve interaction between output clients. This can be avoided by directly generalizing the protocol to the randomized multi-output case.

setup as described below. The protocols can then be converted to the *plain model*, where no setup assumptions or preprocessing are allowed, at the price of some efficiency loss.

In the linear preprocessing model, we assume a dealer who initially gives to each player a set of values in K or in its subfield $GF(2)$. The values distributed by the dealer are restricted to be "linearly correlated". Specifically, the dealer picks a random codeword in a linear code defined over K or over $GF(2)$, and then hands to each player a subset of the coordinates in the codeword. It is public which subsets are used, but the values themselves are private. This procedure can be repeated multiple times, possibly using different linear codes.

Of course, we do not expect that such a dealer would exist in practice. This is only a convenient abstraction, that can be formalized as an ideal functionality. We later separately look at how such a dealer may be implemented.

Note that in Shamir's secret sharing scheme, shares are computed as linear functions of the secret and random elements chosen by the dealer. We may therefore assume that the dealer can give to players Shamir shares of a random secret or of 0. More concretely, we assume that the following subroutines are available. When they are called in our protocol descriptions that follow, this should be taken to mean that the players retrieve from their preprocessed material values as specified below.

RandSS(t)   Each server P_i obtains $f(i)$, where f is a random polynomial over K of degree at most t.

RandSS$_0$(t)   Same as RandSS(t), except that f is subject to $f(0) = 0$.

RandSS$_{bin}$(t)   Same as RandSS(t), except that f is subject to the restriction that $f(0)$ is either 0 or 1. Note that this correlation pattern is linear over $GF(2)$.

RandSS^P(t)   Same as RandSS(t), except that player P additionally receives f.

RandSS$^P_{bin}$(t)   Same as RandSS$_{bin}$(t), except that P additionally receives f.

As discussed in Section 1.1, the linear preprocessing model is motivated by the pseudorandom secret-sharing technique from [11] (see also [19]). When the number of servers is small, a "once and for all" setup is sufficient for enabling players to execute any number of calls to the above subroutines without having to communicate.

One can emulate the linear preprocessing model in the plain model using constant-round interaction between players. This is trivial for a passive adversary, and can be done for an active adversary based on standard verifiable secret sharing schemes from the literature (e.g., [5, 12, 10]). In particular, [12] shows how to build VSS from any linear secret sharing scheme, and this can conveniently be used to implement RandSS$_{bin}$(t) using VSS over $GF(2)$.

Our protocols will invoke the following variants of VSS as subroutines.

VSS^P(t)   Player P has a value $s \in K$ as private input. The goal is for each server to receive a Shamir share of s. Using linear preprocessing, this only requires a single round of broadcast: in the setup phase we invoke RandSS^P(t). Then the value $r = f(0)$ can be computed by P, and P broadcasts $z = s - r$. Each server P_i takes $z + f(i)$ to be his private output.

$\mathsf{VSS}^P_{bin}(t)$ Player P has private input a value $b \in \mathrm{GF}(2)$. The goal is for each server to receive a Shamir share of b (computed over K). This can be implemented similarly to $\mathsf{VSS}^P(t)$, replacing $\mathsf{RandSS}^P(t)$ by $\mathsf{RandSS}^P_{bin}(t)$.

2.2   Secure Computation of Low-Degree Polynomials

In the linear preprocessing model, we now show how to securely compute the following functionality, which will be useful later.

The functionality is defined by $Q()$, a degree d polynomial over K in l variables $x_1, ..., x_l$. Each input client is to supply values for some of the variables, the others are to be chosen at random by the functionality. For each server P_i we define an index set D_i; these sets are mutually disjoint and designate subsets of the random inputs to $Q()$. If $j \in D_i$, the functionality will output x_j to P_i. Finally, the functionality will output $Q(x_1, ..., x_l)$ to the output clients, as well as Shamir shares of this value to the servers.

The functionality is specifically designed to fit into our protocol for computing Yao-garbled circuits to be presented later. In particular, some of the random values x_j will be used as encryptions keys. Each such key has to be known to exactly one server, and this is the reason why the functionality outputs some of the x_j's to the servers. This functionality, denoted by $F_{Q,D_1,...,D_n}$, is more precisely defined as follows:

1. In the first round, it receives from each honest input client the x_j's this client supplies. In addition, it receives from the honest input clients and servers a set of values of the form produced in the linear preprocessing model. More precisely, these additional inputs take the following form:

   - For each x_j that is supplied by input client I, a set of values for $\mathsf{RandSS}^I(t)$ (as determined by a polynomial f_j).
   - For each x_j that is random, a set of values for $\mathsf{RandSS}(t)$ (as determined by a polynomial f_j).
   - For each x_j that is random and where $j \in D_i$, a set of values for $\mathsf{RandSS}^{P_i}(t)$ (as determined by a polynomial f_j).
   - A set of values for $\mathsf{RandSS}_0(dt)$ (as determined by a polynomial f_0).

   The functionality computes the shares and polynomials that all the above results in for the corrupt players and outputs this to the adversary. For instance, for every x_j supplied by corrupt input client I, this will be the polynomial $f_j()$. Note that, due to our assumed constraint on the number of corrupt servers, the honest players' information is enough to determine this information for the corrupt players. Also, for each x_j supplied by an honest input client, it sends $x_j - f_j(0)$ to the adversary. (No information is sent to honest players in this round.)

2. In round 2, the functionality receives from each corrupt input client the x_j's that it is responsible for. For each $j \in D_i$, the functionality will output x_j to P_i. It then outputs to each server a Shamir share of the value $Q(x_1, \ldots, x_l)$

generated by the polynomial $Q(f_1(), f_2(), ..., f_l()) + f_0()$ (this will be a univariate polynomial of degree at most dt). Finally, the functionality outputs $Q(x_1, \ldots, x_l)$ to all output clients.

We securely implement the above functionality in the linear preprocessing model using the following standard protocol:[5]

1. We do the following for each x_j: if x_j is supplied by input client I, execute $\mathsf{VSS}^I(t)$ where I uses x_j as his private input. The communication implied by this is the only communication in the first round.

   If x_j is random, execute $\mathsf{RandSS}(t)$. If x_j is random and $j \in D_i$, execute $\mathsf{RandSS}^{P_i}(t)$. In all cases, a set of shares of x_j is obtained. Let $x_{j,i}$ be the share of x_j obtained by server P_i. We execute $\mathsf{RandSS}_0(dt)$, creating shares of a degree dt polynomial that evaluates to 0 in 0. Let z_i be the share obtained by server P_i.

2. In the second round, each server P_i sends $Q(x_{1,i}, \ldots, x_{l,i}) + z_i$ to each output client. Each output client considers the values he receives as points on a degree dt polynomial f, reconstructs the polynomial (applying error-correction in the active adversary case) and outputs $f(0)$. Each server P_i outputs $Q(x_{1,i}, \ldots, x_{l,i}) + z_i$, and the values x_j for which $j \in D_i$.

We now show the security of this protocol using Canetti's UC framework [8]. We only show this for environments that supply inputs of correct form as specified above. This is sufficient, since we will only use the functionality in conjunction with the linear preprocessing, which is assumed to produce values of the right form.

Theorem 1. *There exists a 2-round protocol computing $F_{Q,D_1,\ldots,D_n}$, the protocol is secure for all environments that supply inputs for honest players as specified in the description of $F_{Q,D_1,\ldots,D_n}$. Furthermore, the protocol is secure for an adaptive adversary corrupting at most t servers and an arbitrary number of clients. For a passive adversary, we assume $dt < n$, for an active adversary we need $(d + 2)t < n$. The communication complexity involves each output client receiving n field elements and each input client broadcasting its (masked) inputs.*

Proof. To prove security, we describe the required simulator (ideal model adversary), which as usual works by running an internal copy of the real-life adversary. In the first round, we receive a set of polynomials and shares from the ideal functionality, which we pass on to the adversary. In particular, this includes, for each corrupt input client, a random polynomial $f_j()$ of degree at most t, for each x_j this client supplies. When the adversary broadcasts a values r_j on behalf of the client, we compute $x_j = f_j(0) - r_j$ and give x_j as input to the ideal functionality. We also received from the ideal functionality $f_j(0) - x_j$ for each x_j supplied by an honest client. We use these values to simulate the broadcasts of honest clients. Note that for each corrupt server P_i, the information received from the

---

[5] For our purposes the degree d of Q will be no larger than 3, hence we will not consider optimizations that apply to a larger d.

functionality now defines a share $x_{j,i}$ of each x_j, and that it also defines a share z_i of the degree dt sharing of 0.

In the second round, the ideal functionality sends $Q(x_1, ..., x_l)$ to the simulator (we assume the adversary has corrupted at least one output client, otherwise the simulation becomes trivial). For each corrupt output client, we use the value $Q(x_1, ..., x_l)$ and the shares of it known by corrupt servers to interpolate a random polynomial f of degree at most dt with $f(0) = Q(x_1, ..., x_l)$ and $f(i) = Q(x_{1,i}, ..., x_{l,i}) + z_i$, for each corrupt P_i, where the $x_{j,i}, z_i$ are the previously defined values for P_i. Then for each honest server P_i, we claim $f(i)$ as the value sent by P_i.

To establish adaptive security, we now show how to reconstruct the history of the players if they are corrupted after the protocol. Earlier corruptions are handled by truncating the reconstruction procedure.

If an input client or a server is corrupted after the protocol, we learn all his input, and pass this on to the adversary. This already determines his view of the protocol, and is consistent with what the adversary already knows by definition of the ideal functionality.

If an output client is corrupted after the protocol, we need not produce new values, as all output clients receive the same set of messages from honest servers, and these were already produced earlier.

It is straightforward to verify that this simulation leads to perfect indistinguishability between the real and ideal process. Namely, the only item that is produced using different algorithms in the two is the polynomial that determines $Q(x_1, ..., x_l)$. However, it is in both cases a random polynomial of degree at most dt under the constraints that it is consistent with corrupt servers' shares and its value at 0 is $Q(x_1, ..., x_l)$.   □

The protocol can be easily extended to secure computation in parallel of several low-degree polynomials, on some set of inputs, where some inputs may go to several polynomials. While it is not clear that this is implied by the composition theorem (because of the overlapping inputs) it can be shown by a trivial extension of the above proof. One can also modify the protocol so that only shares of $Q(x_1, ..., x_l)$ are computed for the servers, and not the value itself, by simply not sending the shares to the output clients.

3   Constant-Round MPC Using a Black-Box PRG

In this section we present our main protocol. We start in Section 3.1 by describing a variant of Yao's garbled circuit technique on which we rely. Then, in Section 3.2 we sketch the BMR approach for computing a garbled circuit in a distributed way. Finally, in Sections 3.3 and 3.4 we describe our modified approach.

3.1   The Basic Garbled Circuit Technique

Loosely speaking, the garbled circuit technique allows to represent a circuit C on ℓ input bits by an "encrypted" circuit $E(C)$ along with ℓ pairs of random keys, such that given $E(C)$ and the ℓ keys corresponding to a specific input $b_1, \ldots, b_\ell$, one can efficiently compute the output $C(b_1, \ldots, b_\ell)$, but this is the only information about the inputs that can be learned. Yao designed a method for generating such encrypted circuits, and used it to obtain a general constant-round two-party protocol for semi-honest parties. (See [30] for a formal proof of security of Yao's protocol and [32, 1] for other variants of this technique.) This protocol was generalized to the multi-party case by Beaver, Micali and Rogaway [4]. The circuit encryption process can be done in parallel for every gate in C, yielding a constant-round protocol for secure function evaluation.

We now describe how the basic technique for garbling a circuit works, by specifying how a trusted functionality could prepare an encrypted circuit and input keys as above. We will later show various protocols for implementing this functionality.

Without loss of generality, we assume the function to be computed is described as a Boolean circuit C with 2-input NAND gates. Let the number of wires in the circuit be W. We number the wires from 0 to $W - 1$. For simplicity, we assume the circuit produces just a single output bit, to be learnt by all output clients, where this bit corresponds to the last wire, number $W - 1$. Each input wire w has a bit b_w assigned to it, where each such bit is supplied by an input client. We will be using an index $0 \leq j \leq 2W - 1$, where the values $2w, 2w + 1$ are assigned to wire w.

We assume we have available a secure secret-key encryption scheme $E_S()$, where S is a k-bit key. We need to assume that the cryptosystem is semantically secure as long as each key is used on at most $2z$ messages, of length $k + 1$ bits each, where z is the maximal fan-out in C.

To compute the garbled circuit, we get as input the bits b_w for each input wire, and then proceed as follows (see below for intuition):

- For every wire w, choose a random bit λ_w (masking the true value of the wire) and random keys S_{2w}, S_{2w+1}.
- For every gate g in C, do the following: suppose g has input wires α, β and output wire γ. Define the following values

$$
\begin{aligned}
a_g^{00} = S_{2\gamma + \delta_g^{00}} &\; ; \; \delta_g^{00} = (\lambda_\alpha \ nand \ \lambda_\beta) \oplus \lambda_\gamma \\
a_g^{01} = S_{2\gamma + \delta_g^{01}} &\; ; \; \delta_g^{01} = (\lambda_\alpha \ nand \ \bar{\lambda_\beta}) \oplus \lambda_\gamma \\
a_g^{10} = S_{2\gamma + \delta_g^{10}} &\; ; \; \delta_g^{10} = (\bar{\lambda_\alpha} \ nand \ \lambda_\beta) \oplus \lambda_\gamma \\
a_g^{11} = S_{2\gamma + \delta_g^{11}} &\; ; \; \delta_g^{11} = (\bar{\lambda_\alpha} \ nand \ \bar{\lambda_\beta}) \oplus \lambda_\gamma
\end{aligned}
$$

Define $A_g^{cd} = (a_g^{cd}, \delta_g^{cd})$, for $c, d \in \{0, 1\}$. We compute the encryptions

$$
\begin{aligned}
&E_{S_{2\alpha}}(E_{S_{2\beta}}(A_g^{00})) \\
&E_{S_{2\alpha}}(E_{S_{2\beta+1}}(A_g^{01})) \\
&E_{S_{2\alpha+1}}(E_{S_{2\beta}}(A_g^{10})) \\
&E_{S_{2\alpha+1}}(E_{S_{2\beta+1}}(A_g^{11}))
\end{aligned}
$$

– We output, for each g, the 4 encryptions as above along with the mask λ_{W-1} of the output wire - this is the encrypted circuit which we denote by $E(C)$. We also output, for each input wire w, the values $b_w \oplus \lambda_w$ and $S_{2w+(b_w \oplus \lambda_w)}$ - these are the encrypted inputs.

A word about the underlying intuition behind this: assume we knew all the inputs and did an ordinary computation of the circuit. This would result in assigning a bit b_w to every wire w. Instead, we get to know exactly one of the two encryption keys that are assigned to each wire, namely the key $S_{2w+(b_w \oplus \lambda_w)}$ and the bit $b_w \oplus \lambda_w$, and this is ensured for all input wires initially. We can think of this information as an encrypted representation of the bit b_w. This also means that by making λ_{W-1} public, we reveal the output bit, and only that bit.

The idea behind making the individual gates work in this scenario is to encrypt the keys and bits that might be assigned to the gate's output wire under keys assigned to input wires in such a way that players will be able to decrypt the "correct" key and bits for the output wire, and only this information. For instance, suppose that some gate g in the circuit has input wires α, β and output wire γ. If the information known for the input wires is $S_{2\alpha+c}, c$ and $S_{2\beta+d}, d$ for bits c, d, then the bit that should be revealed for output wire is $\delta_g^{cd} = ((c \oplus \lambda_\alpha) \text{ nand } (d \oplus \lambda_\beta)) \oplus \lambda_\gamma$, and so the key that should revealed is $S_{2\gamma+\delta_g^{cd}}$. The idea is therefore to encrypt these two values under $S_{2\alpha+c}$ and $S_{2\beta+d}$, for all 4 values of c, d.

Anyone who is given encrypted circuit and inputs can compute the output by the following *local circuit evaluation procedure*: for each input wire w, the key $S_{2w+(b_w \oplus \lambda_w)}$ and the bit $b_w \oplus \lambda_w$ are given. There will now be a number of gates, for which a key and a bit are known for both input wires. Let g be such a gate, say with input wires α, β and output wire γ. Since we know $b_\alpha \oplus \lambda_\alpha$ and $b_\beta \oplus \lambda_\beta$, we know which of the encryptions associated to g we can decrypt, namely those where both involved keys are known. We decrypt and obtain as result a key and a bit, which are easily seen to be $S_{2\gamma+(b_\gamma \oplus \lambda_\gamma)}$ and the bit $b_\gamma \oplus \lambda_\gamma$. Continuing this way, we will obtain a key and a bit $b_{W-1} \oplus \lambda_{W-1}$ for the output wire $W-1$. Since we also know λ_{W-1}, we can compute the output bit b_{W-1}.

3.2 The BMR Protocol

The protocol from [4] can be seen as a concrete proposal for an encryption scheme E as required for the garbled circuit technique and a protocol for computing $E(C)$. Their scheme assumes a pseudorandom generator G taking as input a k-bit seed (such a generator can be constructed from any one-way function [21]). For a seed s, the output of $G(s)$ is split into k-bit blocks where the j'th block is denoted by $G(s)_j$.

A key S in the encryption scheme consists of n subkeys $S = (s^1, s^2, \ldots, s^n)$, each of which is k bits long, and where initially s^i is known only to P_i. An element $m \in K$ is encrypted under S as $E_S(m) = m \oplus G(s^1)_j \oplus \ldots \oplus G(s^n)_j$, assuming m is the j'th k-bit string we encrypt under S.

Assuming m and each s^i have been secret shared among the servers, we can securely compute the encryption by having server i locally compute and secret share $G(s^i)_j$. We can then use linearity of the secret sharing to get shares of $E_S(m)$, and send these shares to the output clients. This will work, and makes only a black-box use of G, if the adversary is passive. But if he is active, each server needs to prove in zero-knowledge that he has computed and secret shared $G(s^i)_j$ correctly. In general, this requires generic zero-knowledge techniques, which means we no longer make a black-box use of G, and also leads to a major loss of efficiency.

3.3   Our Distributed Encryption Scheme

We now suggest a different encryption scheme for the garbled circuit technique, allowing to avoid the use of zero-knowledge proofs in the case of an active adversary. We assume as before a pseudorandom generator G which expands a k bit seed. A key S consists again of n subkeys $S = (s^1, s^2, \ldots, s^n)$ where initially s^i is known only to P_i.

Consider now a situation where a message $m \in K$ has been secret shared among the n servers using a polynomial of degree d, where $t \le d < n$. Let m_i denote the share of m given to P_i. To encrypt such a message under a key S, we will let each server encrypt the share he knows under his part of the key (expanded by G).

We define $E_S^j(m) = (G(s^1)_j \oplus m_1, \ldots, G(s^n)_j \oplus m_n)$. Having received the parts of the ciphertext $E_S^j(m)$ from the servers and given the key S, one can decrypt each share and reconstruct m from the shares, where error correction[6] is used to recover m if the adversary has actively corrupted some of the servers. The following lemma is straightforward.

Lemma 1. *The above distributed encryption scheme has the following properties:*

- *If an adversary is given up to t of the s^i's, and E_S^j is used on at most one message m, the encryption keeps m semantically secure.*
- *If the adversary is passive, and an honest output client is given S and receives $E_S^j(m)$ from the servers, he can decrypt correctly if $d < n$.*
- *If the adversary is active, and an honest output client is given S and receives $E_S^j(m)$ from the servers, he can decrypt correctly if $d + 2t < n$.*

This generalizes in a straightforward way to cases where two keys U, V are used. We write $E_U^i(E_V^j(m)) = (G(u^1)_i \oplus G(v^1)_j \oplus m_1, \ldots, G(u^n)_i \oplus G(v^n)_j \oplus m_n)$.

In the following, we will be encrypting several elements in K under the same key. This is done in the natural way, by using a fresh part of the output from G for each new element.

3.4   Distributed Computation of a Garbled Circuit

We now apply the distributed encryption idea for securely computing a garbled circuit using a black-box PRG. The protocol takes place in the linear preprocess-

---

[6] In the case $t < n/2$ the subkeys and the message will be distributed using authenticated shares, in which case the decryption will involve a correction of *erasures* rather than errors.

ing model, and will use the subroutines and protocols described in Sections 2.1 and 2.2.

1. In round 1, for each wire $w = 0..W-1$ the servers execute $\mathsf{RandSS}_{bin}(t)$ to create shares of the secret wire masks λ_w's. Also, for $i = 1..n, j = 0..2W-1$, they execute $\mathsf{RandSS}^{P_i}(t)$ to create shares of the subkeys s_j^i, such that s_j^i is known to P_i. Finally, for each input bit b_w held by input client I_j, the players execute $\mathsf{VSS}^{I_j}(t)$ (i.e., with I_j as dealer) and b_w as shared secret. Thus the only communication in round 1 consists of broadcasts done in the VSS subroutines.

2. In round 2, the servers first do some local computation.
   - For each input wire w and $i = 1..n$, each server locally computes a random share of the value $s_{2w+(b_w \oplus \lambda_w)}^i$. Note that since we work over a field of characteristic 2, this value can be written as a degree 2 polynomial, namely $(1 + b_w + \lambda_w)s_{2w}^i + (b_w + \lambda_w)s_{2w+1}^i$. We can therefore compute shares of the value $s_{2w+(b_w \oplus \lambda_w)}^i$ defined by a random degree $2t$ polynomial and send all these shares to the output clients, using the protocol from Section 2.2.
   - For each input wire w, the servers compute shares of the value $b_w \oplus \lambda_w$ and send them to the output clients.
   - For each gate g in the circuit, suppose the two inputs and output wire are wires α, β, γ, respectively. Then for each $i = 1..n$, the servers locally compute random shares of the values

$$
\begin{aligned}
a_g^{00,i} &= s_{2\gamma+\delta_g^{00}}^i \;;\; \delta_g^{00} = (\lambda_\alpha \; nand \; \lambda_\beta) \oplus \lambda_\gamma \\
a_g^{01,i} &= s_{2\gamma+\delta_g^{01}}^i \;;\; \delta_g^{01} = (\lambda_\alpha \; nand \; \bar{\lambda}_\beta) \oplus \lambda_\gamma \\
a_g^{10,i} &= s_{2\gamma+\delta_g^{10}}^i \;;\; \delta_g^{10} = (\bar{\lambda}_\alpha \; nand \; \lambda_\beta) \oplus \lambda_\gamma \\
a_g^{11,i} &= s_{2\gamma+\delta_g^{11}}^i \;;\; \delta_g^{11} = (\bar{\lambda}_\alpha \; nand \; \bar{\lambda}_\beta) \oplus \lambda_\gamma
\end{aligned}
$$

   Note that these values can be written as degree 3 polynomials in the already shared values, for instance, $a_g^{00,i} = (\lambda_\alpha \lambda_\beta + \lambda_\gamma)s_{2\gamma}^i + (1 + \lambda_\alpha \lambda_\beta + \lambda_\gamma)s_{2\gamma+1}^i$. We can therefore use the protocol from Section 2.2 to locally compute these random shares (without sending them to the output clients).

3. Let $a_g^{cd} = (a_g^{cd,1}, ..., a_g^{cd,n})$, for $c, d \in \{0, 1\}$. (This vector of n subkeys replaces the single key a_g^{cd} in the basic garbled circuit construction from Section 3.1). Define $A_g^{cd} = (A_g^{cd}, \delta_g^{cd})$. The servers can now reveal to the output clients encryptions of the form $E_{S_{2\alpha+c}}(E_{S_{2\beta+d}}(A_g^{cd}))$ using the distributed encryption scheme from Section 3.3. (Note that both the data we need to encrypt and the encryption subkeys are already shared in the required form.)

4. The output clients now apply the local circuit evaluation procedure described in Section 3.1, replacing ordinary decryption with distributed decryption (Section 3.3).

Theorem 2 (Black-box constant-round protocol in linear preprocessing model). *In the linear preprocessing model, there is a general 2-round MPC* *protocol making a black-box use of a pseudorandom generator. The protocol tolerates an active, adaptive adversary corrupting $t < n/5$ servers and an arbitrary number of clients. [7] The communication complexity for computing a circuit C involves $O(n^2|C|k)$ bits sent to each output client, and each input client must broadcast its (masked) inputs to the n servers.*

Proof sketch: Formally speaking, we want to prove that the above protocol realizes a functionality F_C that accepts inputs $b_1, ..., b_\ell$ from the input clients and then outputs $C(b_1, ..., b_\ell)$ to all output clients.

Let $F_{low-degree}$ be an extended version of $F_{Q,D_1,...,D_n}$, computing all polynomials of degree 2, 3 and shares we compute in the protocol π as described above (see the remarks following Theorem 1). Let $\pi^{F_{low-degree}}$ be the protocol we obtain by replacing in the natural way steps 1 and 2 in π by a call to $F_{low-degree}$. By Theorem 1 and the composition theorem, to show security of π, it is sufficient to show security of $\pi^{F_{low-degree}}$.

We now describe a black-box simulator for this protocol. The simulator proceeds by running internally copies of the linear preprocessing functionality, $F_{low-degree}$ and the (initially) honest players. The internal copies of honest players are called *virtual honest players*. They will be given 0's as input instead of the real values of their b_i's (which are unknown to the simulator). Otherwise all these internal entities proceed according to the protocol. There are only two differences between this and the real process: When the adversary specifies input bits to $F_{low-degree}$, this in particular fixes values of the b_i's for the corrupt players. The simulator then sends these bits to F_C. Second, when we get $C(b_1, ..., b_\ell)$ from F_C and construct the encrypted circuit, we assign a number of ciphertexts to the output gate g, exactly one of which will be decrypted in the local evaluation procedure. The simulator will put the bit $\lambda_{W-1} \oplus C(b_1, ..., b_\ell)$ as plaintext inside this encryption. This is done as follows: the degree $3t$ polynomial that defines this bit is of the form $g() + z()$ where $z()$ is a random degree $3t$ polynomial with $z(0) = 0$. We then change $z()$ to a random $z'()$ of the same degree, but such that $g(0) + z'(0)$ is that value we want and $z'()$ is consistent with the shares of corrupt players. This change introduces no inconsistencies in the view of the virtual players, since $z()$ is used for nothing else than randomizing g. The simulated execution now results in output $C(b_1, ..., b_\ell)$ which is consistent with what the ideal functionality gives to the honest clients.

It remains to be described how the simulator handles corruptions. We describe how the simulator will reconstruct the view of a newly corrupted player. We assume the player is corrupted after the protocol terminates. For earlier corruptions, the reconstruction procedure is truncated appropriately. The general idea is that the simulator already has the views of the virtual honest players, including the one the adversary now wants to corrupt. We then modify this information so it becomes consistent with what we learn as a result of the corruption, without changing what the adversary already knows.

---

[7] In the full paper, we give an optimized version of Theorem 1 for the case of a passive adversary, allowing us to prove the above theorem for $t < n/2$ in this case.

If an input client is corrupted, we learn his input bit(s), say b_w. We already broadcasted a value r_w related to this, and the virtual client has from the pre-processing a polynomial f_w with $f_w(0) = r_w$. We then change f_w to f'_w, so f'_w is random of degree at most t, subject to $f'_w(0) = b_w \oplus r_w$, and $f_w(c) = f'_w(c)$ for all corrupt P_c. We now want to claim that the virtual honest players used the polynomial $g'_w() = f'_w() + r_w$ in the further computation, instead of $g_w() = f_w() + r_w$ we used so far. Note that, to compute the garbled inputs, the protocol computes a set of degree-2 polynomials. Consider one of them, say $Q(b_w, \lambda_w, s_{2w}, s_{2w+1})$. Say the last 3 variables are shared using polynomials $g_2(), g_3(), g_4()$. Before corruption, we had $Q(0, \lambda_w, s_{2w}, s_{2w+1})$ shared using a univariate polynomial of the form $Q(g_w(), g_2(), g_3(), g_4()) + z_w()$, where $z_w()$ is random of degree $2t$ and $z_w(0) = 0$. Define $z'_w()$ by

$$Q(g_w(), g_2(), g_3(), g_4()) + z_w() = Q(g'_w(), g_2(), g_3(), g_4()) + z'_w()$$

and change $z_w()$ to $z'_w()$. This will change honest virtual server's shares, but not the corrupted server's shares, by construction of $f'_w(), g'_w()$. It will also preserve the data sent to output clients. We now give to the adversary the updated view of the virtual client.

If a server or an output client is corrupted, note that this does not result in any new data learnt from F_C. We can therefore give the current view of the virtual output client or server to the adversary.

This concludes the description of the simulator. The intuition of the analysis of the simulation is that all plaintext data are identically distributed in simulation as in real execution, the difference lies in the data that remain encrypted, and this cannot be detected efficiently by semantic security of the encryption. □

3.5   The Plain Model

To implement the above protocol in the plain model, we need to emulate the procedures for generating random shared secrets. The cost of the resulting protocol is dominated by the cost of emulating $O(n|C|)$ invocations of (different variants of) RandSS(t). In the semi-honest case, each invocation of RandSS(t) can be implemented in a straightforward way by letting each player distribute a random secret and output the sum of the shares it received. (In fact, it suffices that $t + 1$ players share secrets.) In the malicious case, one could use a similar procedure based on any standard constant-round VSS protocol from the literature (e.g., the one from [5]). In fact, using the 2-round VSS protocol from [17], one can obtain a 3-round protocol in the plain model (assuming $t < n/5$).

We can weaken the assumption on t in the active case to $t < n/3$ by replacing the non-interactive polynomial evaluation protocol from Theorem 1 by an interactive one (e.g., using [5]). The resulting protocol will have a larger (but still constant) number of rounds and a higher communication complexity. In the full version, we sketch how to use the VSS and multiplication protocol from [10] to further extend the feasibility result to the case $t < n/2$. This is based on two observations: first, a variant of our distributed encryption scheme can be used to encrypt values that have been shared under any VSS with a non-interactive reconstruction protocol. Second, by requiring that all values in the computation as well as shares of these values are VSS'ed, we can obtain a multiplication protocol that is guaranteed to terminate in a constant number of rounds, even for the case of $t < n/2$. Thus, our main feasibility result in the plain model is the following:

Theorem 3 (Black-box constant-round protocol in plain model). *In the plain model, there is a general constant-round MPC protocol making a black-box use of a pseudorandom generator. The protocol tolerates an active, adaptive adversary corrupting $t < n/2$ servers and an arbitrary number of clients.*

References

[1] B. Applebaum, Y. Ishai, and E. Kushilevitz. Computationally private randomizing polynomials and their applications. In *Proc. 20th Conference on Computational Complexity*, 2005.

[2] J. Bar-Ilan and D. Beaver. Non-cryptographic fault-tolerant computing in a constant number of rounds. In *Proc. 8th ACM PODC*, pages 201–209, 1989.

[3] D. Beaver, J. Feigenbaum, J. Kilian, and P. Rogaway. Security with low communication overhead (extended abstract). In *Proc. of CRYPTO '90*.

[4] D. Beaver, S. Micali, and P. Rogaway. The round complexity of secure protocols (extended abstract). In *Proc. of 22nd STOC*, pages 503–513, 1990.

[5] M. Ben-Or, S. Goldwasser, and A. Wigderson. Completeness theorems for non-cryptographic fault-tolerant distributed computation. In *Proc. of 20th STOC*, pages 1–10, 1988.

[6] C. Cachin, J. Camenisch, J. Kilian, and J. Muller. One-round secure computation and secure autonomous mobile agents. In *Proceedings of ICALP' 00*, 2000.

[7] R. Canetti. Security and composition of multiparty cryptographic protocols. In *J. of Cryptology*, 13(1), 2000.

[8] R. Canetti. Ran Canetti. Universally Composable Security: A New Paradigm for Cryptographic Protocols. FOCS 2001: 136-145.

[9] R. Cramer and I. Damgård. Secure distributed linear algebra in a constant number of rounds. In *Proc. Crypto 2001*.

[10] R. Cramer, I. Damgård, S. Dziembowski, M. Hirt, and T. Rabin. Efficient Multiparty Computations Secure Against an Adaptive Adversary. In *Proc. EURO-CRYPT* 1999, pages 311-326.

[11] R. Cramer, I. Damgård, and Y. Ishai. Share conversion, pseudorandom secret-sharing and applications to secure computation. In *Proc. of second TCC*, 2005.

[12] R. Cramer, I. Damgård, and U. Maurer. General secure multi-party computation from any linear secret-sharing scheme. In *Proc. of EUROCRYPT '00*, LNCS 1807, pp. 316-334, 2000.

[13] R. Cramer, I. Damgård, and J. Nielsen. Multiparty computation from threshold homomorphic encryption. In *Proc. of EUROCRYPT '01*, LNCS 2045, pp. 280-299, 2001.

[14] U. Feige, A. Fiat, and A. Shamir. Zero-Knowledge Proofs of Identity. J. Cryptology 1(2): 77-94 (1988).

[15] Uri Feige, Joe Kilian, and Moni Naor. A minimal model for secure computation (extended abstract). In *Proc. 26th STOC*, pages 554–563. ACM, 1994.

[16] P. Feldman and S. Micali. An Optimal Algorithm for Synchronous Byzantine Agreement. *SIAM. J. Computing*, 26(2):873–933, 1997.

[17] R. Gennaro, Y. Ishai, E. Kushilevitz and T. Rabin. The Round Complexity of Verifiable Secret Sharing and Secure Multicast. In *Proceedings of the 33rd ACM Symp. on Theory of Computing (STOC '01)*, pages 580-589, 2001.

[18] R. Gennaro, Y. Ishai, E. Kushilevitz and T. Rabin. On 2-round secure multiparty computation. In *Proc. Crypto '02*.

[19] N. Gilboa and Y. Ishai. Compressing cryptographic resources. In *Proc. of CRYPTO '99*.

[20] O. Goldreich, S. Micali, and A. Wigderson. How to play any mental game (extended abstract). In *Proc. of 19th STOC*, pages 218–229, 1987.

[21] J. Håstad, R. Impagliazzo, L. A. Levin, and M. Luby. A pseudorandom generator from any one-way function. *SIAM J. Comput.*, 28(4):1364–1396, 1999.

[22] M. Hirt and U. M. Maurer. Robustness for Free in Unconditional Multi-party Computation. CRYPTO 2001: 101-118.

[23] Y. Ishai and E. Kushilevitz. Randomizing polynomials: A new representation with applications to round-efficient secure computation. In *Proc. 41st FOCS*, pp. 294–304, 2000.

[24] R. Impagliazzo and S. Rudich. Limits on the provable consequences of one-way permutations. Proceedings of 21st Annual ACM Symposium on the Theory of Computing, 1989, pp. 44 – 61.

[25] J. Katz and R. Ostrovsky. Round-Optimal Secure Two-Party Computation. In *CRYPTO 2004*, pages 335-354.

[26] J. Katz, R. Ostrovsky, and A. Smith. Round Efficiency of Multi-party Computation with a Dishonest Majority. In *EUROCRYPT 2003*, pages 578-595.

[27] J. Kilian. Founding cryptography on oblivious transfer. In *Proc. 20th STOC*, pages 20–31, 1988.

[28] Y. Lindell. Parallel Coin-Tossing and Constant-Round Secure Two-Party Computation. *J. Cryptology* 16(3): 143-184 (2003). Preliminary version in Crypto 2001.

[29] Y. Lindell, A. Lysyanskaya, and T. Rabin. Sequential composition of protocols without simultaneous termination. In *Proc. PODC 2002*, pages 203-212.

[30] Y. Lindell and B. Pinkas. A Proof of Yao's Protocol for Secure Two-Party Computation. Cryptology ePrint Archive, Report 2004/175, 2004.

[31] M. Naor and K. Nissim. Communication preserving protocols for secure function evaluation. In *Proc. STOC 2001*, pages 590-599.

[32] M. Naor, B. Pinkas, and R. Sumner. Privacy preserving auctions and mechanism design. In *Proc. 1st ACM Conference on Electronic Commerce*, pages 129–139, 1999.

[33] R. Pass. Bounded-concurrent secure multi-party computation with a dishonest majority. In *Proc. STOC 2004*, pages 232-241.

[34] R. Pass and A. Rosen. Bounded-Concurrent Secure Two-Party Computation in a Constant Number of Rounds. FOCS 2003.

[35] T. Rabin and M. Ben-Or. Verifiable Secret Sharing and Multiparty Protocols with Honest Majority. In *Proc. 21st STOC*, pages 73–85. ACM, 1989.

[36] O. Reingold, L. Trevisan, and S. P. Vadhan. Notions of Reducibility between Cryptographic Primitives. *TCC 2004*: 1-20.

[37] P. Rogaway. *The Round Complexity of Secure Protocols*. PhD thesis, MIT, June 1991.

[38] A. Shamir. How to share a secret. *Commun. ACM*, 22(6):612–613, June 1979.

[39] S. R. Tate and K. Xu. On garbled circuits and constant round secure function evaluation. CoPS Lab Technical Report 2003-02, University of North Texas, 2003.

[40] A. C. Yao. How to generate and exchange secrets. In *Proc. 27th FOCS*, pp. 162–167, 1986.

Secure Computation of Constant-Depth Circuits with Applications to Database Search Problems*

Omer Barkol and Yuval Ishai

Computer Science Department, Technion
{omerb, yuvali}@cs.technion.ac.il

Abstract. Motivated by database search problems such as partial match or nearest neighbor, we present secure multiparty computation protocols for constant-depth circuits. Specifically, for a constant-depth circuit C of size s with an m-bit input x, we obtain the following types of protocols.

- In a setting where $k \geq \operatorname{poly}\log(s)$ servers hold C and a client holds x, we obtain a protocol in which the client privately learns $C(x)$ by communicating $\tilde{O}(m)$ bits with each server.
- In a setting where x is arbitrarily distributed between $k \geq \operatorname{poly}\log(s)$ parties who all know C, we obtain a secure protocol for evaluating $C(x)$ using $O(m \cdot \operatorname{poly}(k))$ communication.

Both types of protocols tolerate $t = k/\operatorname{poly}\log(s)$ dishonest parties and their computational complexity is nearly linear in s. In particular, the protocols are optimal "up to polylog factors" with respect to communication, local computation, and minimal number of participating parties.

We then apply the above results to obtain sublinear-communication secure protocols for natural database search problems. For instance, for the partial match problem on a database of n points in $\{0,1\}^m$ we get a protocol with $k \approx \frac{1}{2} \log n$ servers, $\tilde{O}(m)$ communication, and nearly linear server computation. Applying previous protocols to this problem would either require $\Omega(nm)$ communication, $\tilde{\Omega}(m)$ servers, or super-polynomial computation.

1 Introduction

As networking becomes a common tool and data can be accessible to all, many applications require distributed access of clients to data servers over the web and other network environments. Once the search is not locally performed, privacy might become a major concern. This motivates the problem of *privacy-preserving database search*, allowing clients to search a database without revealing their search queries to the servers storing the database. Since the databases being searched might be very large, it is desirable to obtain privacy-preserving search protocols whose communication complexity is sublinear in the database size.

---

* Research supported by Israel Science Foundation grant 36/03.

V. Shoup (Ed.): Crypto 2005, LNCS 3621, pp. 395–411, 2005.

The above problem was extensively studied within the context of *private information retrieval* (PIR) [11]. The goal of PIR is to allow a client to privately retrieve the ith item (say, a bit) from a database stored in one or more servers. PIR can be used as a building block for more complex database search operations. Using PIR to probe a data structure representing the database, one can obtain sublinear-communication private protocols for problems such as keyword search [10, 13] or approximate nearest neighbor search (e.g., using [21]).[1] Unfortunately, many natural database search problems that arise in practice are not known to have efficient data structures, namely ones which provide the guarantee that each query can be answered by making few probes into the data structure. For instance, all known algorithms for the *partial match* problem (aka "keyword search with wildcards") require either the number of probes to be nearly linear in the database size or the data structure to be exponential in the length of database entries (see, e.g., the best algorithms known for partial match by Charikar et al. [9] and the survey by Miltersen [22]). Hence, the generic PIR-based approach is not useful for this problem. The same holds for many other natural and useful search problems, including Boolean information retrieval (supporting "advanced Google search" functionality), exact nearest neighbor search, and others.

Our point of departure is the observation that most practical database search operations can be efficiently implemented using *constant depth circuits*. (By default, we allow circuits to use AND, OR, NOT, and XOR gates with unbounded fan-in and fan-out.) That is, it is possible for the server to represent its database as a (large) constant-depth circuit C and for the client to independently represent its query as a (small) input x, such that $C(x)$ returns the answer to the client's query on the server's database. Given this observation, it suffices to obtain protocols for securely evaluating a constant-depth circuit C held by the server on an input x held by a client, such that the communication complexity of the protocol is dominated by the size of x rather than by the size of C. Unfortunately, no protocols of this type are known. Using the current toolbox of techniques for secure two-party computation, one can either obtain protocols whose communication complexity is (at least) linear in the circuit size [30, 17] or ones whose computational complexity is exponential in the input size [24].

A good solution to the above problem would imply a major breakthrough in the theory of secure computation. (A small step in this direction, resolving the case in which C is a 2-DNF formula, was very recently made in [6].) In the current work we consider a relaxed setting where (few) different servers hold copies of C, and the client's privacy should be protected against every individual server or collusion of servers of some bounded size t. This "data replication" scenario is the one originally considered in the context of PIR [11]. It is arguably becoming more and more relevant to practice, with the widespread use of peer-to-peer networks, distributed file backup and web caching systems, and other forms of replicated data. Moreover, in this setting it is possible to avoid the use of

---

[1] The PIR-based approach was generalized in [24] to turn *arbitrary* sublinear-communication protocols into private ones. However, the resulting protocols generally require a super-polynomial amount of computation.

expensive "cryptographic" computations (e.g., modular exponentiations) which would make the protocols computationally infeasible in practice. Our goal in this setting is to simultaneously obtain nearly-optimal communication (of the order of $|x|$) and local computation (of the order of $|C|$), while minimizing the number of servers.

In addition to the client-servers scenario discussed above, we also consider the complexity of evaluating constant-depth circuits in a multiparty setting, without any data replication assumptions. Here the circuit C specifies the functionality to be computed (hence it is known to *all* parties) and the input x is arbitrarily partitioned between the parties. In this case, the best known techniques from the multiparty computation literature would either require linear communication in $|C|$ [5] or an exponential amount of computation and $\tilde{\Omega}(|x|)$ parties [3]. Our goal, as before, is to simultaneously obtain nearly optimal communication and computation while minimizing the required number of parties (alternatively, maximizing the security threshold).

1.1 Our Results

We obtain communication-efficient protocols for securely evaluating constant-depth circuits in both the client-servers setting and the multiparty setting discussed above. Our main protocols are optimal "up to polylog factors" with respect to all three parameters of interest: communication, computation, and number of participating parties. Furthermore, the protocols typically require a minimal amount of interaction, consisting of only two communication rounds (a single round of queries and answers in the client-servers case). Since the number of servers or participating parties is the most crucial resource, we also attempt to optimize the multiplicative constants involved, as was done in the context of PIR. In the case of depth 2 circuits (which in particular suffices for capturing general DNF or CNF evaluation and secure partial match), the number of servers can be as low as $\frac{1}{2}\log_2 |C|$ while maintaining (essentially) optimal communication and computation.

We now provide a more detailed account of our results. We let C denote a circuit of size s and depth c with an m-bit input x. In the client-servers setting, each of k servers holds C and a client who holds x should privately learn $C(x)$. In the multiparty setting the circuit C is known to all k parties, where x is distributed between them, and (by default) they all privately learn $C(x)$. Let t denote a security threshold ($t = 1$ by default). We obtain the following two main types of protocols:

- In the client-servers setting we obtain protocols where the communication complexity and the client's computation complexity are $\tilde{O}(m)$ per server, the computation complexity of each server is $\tilde{O}(s)$, and $k = O(t \cdot \log^{c-1}(s))$.
- In the multiparty setting we obtain a protocol for $k = O(\log^{c-1}(s))$ parties in which the communication complexity is $O(m \cdot \text{poly}(k))$, the computation complexity of each party is $\tilde{O}(s)$, and the protocol resists $t = \Omega(k/\log^{c-1}(s))$ dishonest parties.

All these protocols are secure in an information-theoretic sense.

We then apply the above general results and optimize them to obtain efficient secure protocols for database search problems. For instance, for the partial match problem on a database of n points in $\{0,1\}^m$, we get a protocol with $k \approx \frac{1}{2}\log n$ servers, $\tilde{O}(m)$ communication, and nearly linear server computation.

1.2   Overview of Techniques

The main technical tool we use in obtaining the above results is a compact representation of constant-depth circuits by probabilistic low-degree multivariate polynomials. We rely on techniques that have been part of a large body of work proving lower bounds on the size of constant-depth circuits (originating from [25, 27]), tailoring them to our different goals. Additional tools we employ are ϵ-biased generators [23] and randomizing polynomials [19].

Randomizing polynomial provide a secure reduction of a "complex" functionality $f(x)$ to a low-degree randomized functionality $p(x, r)$. (Here r represents "private" randomness chosen by the functionality; both x and r count towards the degree.) Such reductions are motivated by the fact that most standard protocols for secure function evaluation can handle low-degree functionalities very efficiently. Specifically, the main motivation for introducing randomizing polynomials in [19] was the fact that evaluating low-degree polynomials requires few rounds of interaction. The current motivation is different: We are mainly interested in minimizing the communication complexity. We exploit the fact that the amount of communication required for evaluating a vector of low-degree polynomials is dominated by the length of the vector (i.e., the number of outputs), rather than by the description size of the polynomials. Thus, our goal is to construct short vectors of low-degree randomizing polynomials representing constant-depth circuits.

In all previous constructions of randomizing polynomials from the literature, the output length of p is at least linear in the representation size of f, even when f outputs only a single bit. For instance, in [19] it is shown how to construct a vector of degree-3 randomizing polynomials whose length is quadratic in the size of a branching program computing f. In our case, both the output length and the amount of private randomness must be sublinear in the circuit size. To this end we define the more general notion of a randomizing polynomials collection (RPC), which introduces public randomness in addition to the private randomness r. Specifically, an RPC is defined by a collection of polynomial vectors $p_\rho(x, r)$, where the "key" ρ is viewed as public randomness and thus does not count towards the degree. We say that the RPC $p_\rho(x, r)$ represents the function f if: (1) it is possible to recover $f(x)$ from $p_\rho(x, r)$ with a negligible failure probability (over the choices of ρ and r), and (2) the output of $p_\rho(x, r)$ gives (essentially) no additional information about x, even given the knowledge of the public randomness ρ. The usual notion of randomizing polynomials corresponds to the special case in which ρ is empty.

An RPC representation for f naturally gives rise to secure protocols for f, similarly to the case of standard randomizing polynomials. Thus, our goals reduce to constructing "good" RPCs for constant-depth circuits. The degree of the

RPC corresponds to the minimal number of participating parties (alternatively, maximal security threshold), and thus serves as our main optimization goal. In addition to minimizing the degree, we wish to optimize both the output length and the amount of randomness, and in particular require them to be sublinear in the circuit size.

Our main RPC construction proceeds in three stages. The first and main stage applies a variant of the techniques of Razborov and Smolensky [25, 27] to create a *short* vector of low-degree randomized polynomials that uses a large amount of public randomness (but no private randomness) in order to reduce the degree of f. This representation guarantees that $f(x)$ can be reconstructed from the outputs of the polynomials with overwhelming probability, yet these outputs might reveal additional information about x. In the second stage we reduce the amount of public randomness by using ϵ-biased generators [23, 1]. Finally, we eliminate the extra information about x revealed by the polynomials using previous constructions of degree-3 randomizing polynomials [19]. This stage introduces a small amount of private randomness and only incurs a minor increase to the degree.

Organization. The remainder of the paper is organized as follows. In Section 2 we give some definitions, and in particular define the notion of RPCs. In Section 3 we describe our main RPC construction for constant-depth circuits and in Section 4 we apply it to obtain secure protocols in both the client-servers setting and the multiparty setting. Finally, Section 5 discusses applications to concrete database search problems.

2   Preliminaries

2.1   Circuits

We represent functions using Boolean circuits with unbounded fan-in and fan-out, as defined below. A circuit C is a labelled directed acyclic graph. The nodes with no incoming edges are labelled with variables (x_i), their negations $(\bar{x}_i)$, or constants (0 or 1). All other nodes are called *gates* and are labelled with some operator. Our default basis of operators includes AND, OR, NOT, and XOR. The nodes from which there are edges to a gate g are called the inputs of g. We refer to the number of such inputs as the *fan-in* of g. In the full version we also consider a generalization of XOR gates to MOD_{p^e} gates, where p is prime and $e \geq 1$ is an integer. Such a gate outputs 1 *iff* the sum of its inputs is 0 mod p^e (for $p^e = 2$ this is a NOT $-$ XOR gate).

The *size* of a circuit is the number of edges. Its *depth* is the length of the longest path from a variable node to an output node, where intermediate NOT and XOR gates do not count towards the depth. Nodes with no outgoing edges are called the *output* nodes. We denote by $C(x)$ the output of C on input x, and say that C computes a function f if $C(x) = f(x)$ for all inputs x.

We will focus on the case of *constant-depth* circuits. By this we refer to (polynomial-time uniform) families of circuits whose depth is bounded by some constant c, independently of the input length. Such circuits over the basis AND, OR, NOT (resp., AND, OR, NOT, MOD_{p^e}) correspond to the complexity class AC^0 (resp., $AC^0(p^e)$). Note that, using De-Morgan's law, one can eliminate AND gates without increasing the depth.

The case of depth-2 circuits will be of particular interest. An *n-term DNF* formula is a depth-2 circuit computing the disjunction (OR) of n conjunctions (AND) of literals. For instance, $(x_1 \wedge \bar{x}_2) \vee (x_2 \wedge \bar{x}_3 \wedge x_4)$ is a 2-term DNF formula.

2.2 Secure Computation

We consider two different scenarios for secure computation: a client-servers scenario, which may be viewed as a distributed form of two-party computation, and the standard multi-party scenario. We begin by recalling the latter.

Multi-party Setting. In the multi-party setting there are k parties, each holding an input x_i to a functionality f. By default, we consider *deterministic, single-output* functionalities; that is, the output $f(x_1, \ldots, x_k)$ should be learned by all parties. Generalization to randomized, multiple-output functionalities is straightforward.

Our protocols in this setting satisfy standard definitions for secure multiparty computation from the literature [7, 8, 18]. In fact, all our protocols are secure in an *information-theoretic* sense, assuming the availability of secure point-to-point channels. More specifically, our protocols will be *statistically secure*, where security is parameterized by a (statistical) security parameter σ which is given to all parties as an additional input.

We will distinguish between security in the *semi-honest* model (capturing "honest-but-curious" players or a passive adversary) and security in the *malicious* model (capturing an active adversary). In the latter case, we assume the availability of broadcast. In both cases, we allow the adversary to adaptively corrupt up to t parties.

Client-Servers Setting. Our client-servers model generalizes the model for information-theoretic PIR introduced in [11]. In this model there is a client (or *user*) $\mathcal{U}$ who holds an input x of length m, and k servers $\mathcal{S}_1, \ldots, \mathcal{S}_k$ who all hold the same input C of length s. The goal is for the client to learn the value $f(C, x)$, for some publicly known function f, while keeping its input x hidden from any collusion of t servers. We will be particularly interested in the case where the servers hold (a description of) a constant-depth circuit C and the client holds an input x to this circuit. In this case, f will be a *universal* function defined by $f(C, x) = C(x)$.

All of our protocols in this setting will only require a single round of interaction in which the client sends a query to each server and receives an answer in return. The protocol is ϵ-correct if the client's output is correct except with error probability bounded by ϵ. By default, ϵ should be exponentially small in the security parameter σ.

Similarly to the case of PIR, our default security requirement only considers the privacy of the client. We say that the protocol is *t-private* if any collusion of t servers can learn nothing about the client's input x.

We will also consider enhanced client-servers protocols that additionally protect the privacy of the servers' input. In such a protocol, the client should learn essentially nothing about C except the output $f(C, x)$. More specifically, the protocol is said to be *δ-server-private* (with respect to f) if the view of the client can be simulated, up to statistical distance of δ, based on its input and output alone. (See full version for a formal definition.) We require $\delta(\sigma) = 2^{-\Omega(\sigma)}$ by default. Following the terminology that was used in the context of PIR [16], we refer to protocols that satisfy this additional server privacy requirement as being *symmetrically private*. To enable server privacy without direct interaction between the servers, it is required to allow the servers to share a common random string (CRS) [16].

The above security requirements induce three levels of security for client-servers protocols: (1) basic security, providing client-privacy only; (2) symmetric privacy with respect to a semi-honest client; and (3) symmetric privacy with respect to a malicious client.

2.3 Randomizing Polynomials

We generalize the notion of randomizing polynomials from [19] and consider what we call *collections* of randomizing polynomials. Before describing our generalization, we review the original notion of randomizing polynomials.

Randomizing polynomials represent a function f using a vector of multivariate polynomials over a finite field. (In this work, the underlying field will be $GF(2)$ by default.) Each polynomial has two types of inputs: ordinary inputs x and random inputs r. A randomizing polynomials vector will usually be denoted by $p(x, r)$. Note that p is a *vector* of polynomials which all act on the same variables x, r. The vector $p(x, r)$ is said to represent a function f if its output distribution is "equivalent" to the output of f in the following sense. First, given $p(x, r)$ it is possible to recover $f(x)$ (without knowing r). In the other direction, given $f(x)$ alone it is possible to sample from the output distribution $p(x, r)$ induced by a uniform choice of r (without knowing x).

For the purpose of allowing more compact representations, we generalize the notion of randomizing polynomials by considering *collections* of randomizing polynomials. Let $p_\rho(x, r)$ denote a collection of polynomial vectors, indexed by a key ρ. When ρ is picked at random, we refer to it as *public randomness*, whereas r is referred to as *private randomness*. We will say that $p_\rho(x, r)$ represents a function $f(x)$ if the following two properties hold: (1) it is possible to recover $f(x)$ from the output of $p_\rho(x, r)$ (except for a negligible failure probability over the choices of ρ and r), and (2) the output of $p_\rho(x, r)$ gives (essentially) no additional information about x *even given the knowledge of the public randomness ρ*. These properties guarantee that the secure computation of f can be reduced to that of p_ρ, where ρ is a *public* random string chosen independently of the inputs.

Definition 1. (Randomizing Polynomials Collection (RPC)).   *Let*
$p_\rho(x,r) = (p_{1_\rho}(x,r),\ p_{2_\rho}(x,r),\ldots,p_{l_\rho}(x,r))$ *be a vector of l polynomials over
the input $x = (x_1,\ldots,x_m)$, the private random input r, and the public random
input ρ. All polynomials are over a finite field F, where $F = \mathrm{GF}(2)$ by default.
We say that $p_\rho(x,r)$ is an ϵ-correct, δ-private randomizing polynomials collection (RPC) for $f(x)$ if the following holds.*

- *(ϵ-correctness) There exists a reconstruction algorithm $\mathcal{R}$ such that for every
  input x, $Pr_{r,\rho}[\mathcal{R}(p_\rho(x,r)) \neq f(x)] \leq \epsilon$, where r and ρ are chosen uniformly
  and independently. Note that reconstruction should not depend on ρ. (Intu-
  itively, correctness should hold for all but a negligible fraction of the ρ's.)*
- *(δ-privacy) There exists a simulator $\mathcal{M}$ such that for every input x,*

$$SD[(\rho, \mathcal{M}(f(x))), (\rho, p_\rho(x,r))] \leq \delta,$$

*where r and ρ are chosen uniformly and independently at random and SD
denotes statistical distance. Note that the simulator is not given ρ yet the
simulation should also be successful when considered jointly with ρ. This
implies that the output distribution of $p_\rho(x,r)$ should be essentially the same
given almost any fixed ρ.*

*The length of $p_\rho(x,r)$ is l. Its degree is the maximal degree of a polynomial in
the vector, taking into account only the input variables x and the private random
variables r. We refer to $|\rho|$ as the* public randomness complexity *and to $|r|$ as
the* private randomness complexity.

Universal RPC. We will sometimes want to represent each function f in a class
$\mathcal{F}$ by an RPC, such that all RPCs in the class share the same simulator and
reconstruction algorithms. In such a case, we say that the class of RPCs is
universal for the function class $\mathcal{F}$. Our main RPC construcsion will be *fully
universal:* the same reconstruction algorithm and simulator can be applied for
all functions f. (Of course, the RPC itself varies from one function to another.)
This feature will be useful for obtaining protocols in the client-servers model,
where a circuit held by the servers is evaluated on an input held by the client.

Polynomial Collection (PC). We will also consider RPCs which do not need
to satisfy the privacy requirement. (In fact, such collections will serve as an
intermediate step in constructing RPCs.) In this case, there is no need for private
randomness. We will refer to this relaxed type of RPC as a *polynomial collection*
(PC) and denote it by $p_\rho(x)$. A PC of length 1 will also be referred to as a
randomized polynomial. Note that the identity function $p(x) = x$ defines a trivial
PC with no public randomness. However, we will be interested in constructing
universal PCs (whose reconstruction algorithm does not depend on f), and in
particular ones in which the output length is sublinear in the input length.

Randomizing polynomials for branching programs. We will rely on an efficient
representation of branching programs by randomizing polynomials.

Lemma 1. *[20] Suppose $f(x)$ can be computed by a branching program of size ℓ. Then, f can be represented by a vector $p(x, r)$ of degree-3 (perfectly correct and private) randomizing polynomials of length $O(\ell^2)$ and randomness complexity $O(\ell^2)$. Moreover, the degree of p in the x variables is 1.*

2.4 ϵ-Biased Generators

The communication complexity of some of our protocols will depend on the *randomness complexity* of the underlying RPCs. This calls for the use of pseudo-randomness. It turns out that the pseudo-random generators we need are only required to fool linear distinguishers. Thus, we will rely on the following standard notion of ϵ-biased generators [23].

Definition 2. (ϵ-biased generator) *A function $G : \{0,1\}^\ell \rightarrow \{0,1\}^{n(\ell)}$ is an ϵ-biased generator (for some bias function $\epsilon(\ell)$) if for all sufficiently large ℓ and all linear functions $L : \mathrm{GF}(2)^{n(\ell)} \rightarrow \mathrm{GF}(2)$, we have*

$$| \Pr[L(G(U_\ell)) = 1] - \Pr[L(U_{n(\ell)}) = 1]| \leq \epsilon(\ell)$$

By default, the function $\epsilon(\ell)$ is required to be negligible.

3 Low-Degree RPCs for Constant Depth Circuits

In this section we present our main constructions of low-degree PCs and RPCs for constant-depth circuits. The high level idea is to simulate the given circuit in a gate-by-gate fashion, going from the inputs to the output, where each such simulation step does not add much to the degree and does not create a big error.

For simplicity, we assume that the circuit has a single output and thus computes a boolean function; a generalization to the non-boolean case is straightforward. We also restrict the attention to $AC^0(2)$ circuits (a generalization to $AC^0(p^e)$ circuits appears in the full version). Finally, we may assume without loss of generality that the circuit contains only OR, XOR, NOT gates and that the output gate is OR.

A central "gadget" in the construction is the following representation of the OR function by a single randomized polynomial, namely a PC of length 1.

OR *construction with parameter γ.* Given an OR gate with t inputs, let R be a random $\gamma \times t$ matrix over $\mathrm{GF}(2)$, and define the randomized polynomial:

$$p_R(x_1, \ldots, x_t) = 1 - \prod_{i=1}^{\gamma} (1 - (\sum_{j=1}^{t} R_{i,j} x_j)). \tag{1}$$

If $\mathrm{OR}(x) = 0$ then so is $p_R(x)$, while if $\mathrm{OR}(x) = 1$ then the probability of every inner product (the sum) to result in 0 or 1 is equal. Thus, $p_R(x) = \mathrm{OR}(x)$ except with probability $2^{-\gamma}$ over the choice of R, and we have the following.

Lemma 2. *An* OR *gate with t inputs has a $2^{-\gamma}$-correct PC representation over* GF(2) *of length 1 and degree γ.*

Notice that in this case we only have a one-sided error; however, applying the OR gadget within the general construction will generally result in a two-sided error. We now proceed to the case of a general circuit C.

Basic construction with parameter σ. Given a circuit C of size s, we define a randomized polynomial $p_\rho^g(x)$ for every gate g of C, so that the PC representing the circuit is the polynomial defined for the output gate. The polynomial p^g is defined inductively as follows. An input is represented by a deterministic polynomial corresponding to its straightforward arithmetization (e.g., $\bar{x}_i$ is represented by $1 - x_i$). If g is an OR gate, then p^g is defined by applying the above OR construction with $\gamma = \log s + \sigma$ to the polynomials representing its inputs. This step introduces new public randomness. Finally, if g is a XOR or a NOT gate, then p^g is naturally defined in terms of the polynomials representing the input gates (e.g., their summation in case of XOR). The degree of the output polynomial is bounded by the maximal degree of a polynomial representing a gate (as a function of its inputs) to the power of the depth of the circuit. Using union bound on the error probability of the representation[2] we have the following:

Lemma 3. *Given a circuit C with m inputs, one output, size s and depth c, the basic construction with parameter σ produces a $2^{-\sigma}$-correct PC representation for C over* GF(2) *of length 1, degree at most $(\log s + \sigma)^c$, and public randomness complexity $O(s(\log s + \sigma))$.*

The parameters of the above PC representation leave much to be desired. First, the degree depends on σ which will generally be larger than $\log s$; moreover, even for depth-2 circuits (capturing the important case of DNF) the degree grows *quadratically* with $\log s + \sigma$. As we shall see, one can make the degree linear in $\log s$ (and independent of σ) in the depth-2 case. Finally, the public randomness complexity is very large.

We will start by reducing the amount of randomness via the use of ϵ-biased generators. We use here the *powering construction* by Alon et al.:

Lemma 4. *[1] There exists an efficient ϵ-biased generator $G : \{0,1\}^{2\ell} \to \{0,1\}^t$ with $(t-1)2^{-\ell}$-bias.*

We use the above generator to produce the matrix R from the OR construction (1). To ensure independence between rows, we use a separate seed for every row. This gives the following:

Lemma 5. *Let $0 < \epsilon < \frac{1}{2}$. An* OR *gate of t inputs has a PC representation over* GF(2) *of length 1, degree γ, $(\frac{1}{2} + \epsilon)^\gamma$-correctness, and public randomness complexity $2\gamma \lceil \log \frac{t}{\epsilon} \rceil$.*

---

[2] The random matrices of the different gates are not considered independent in this analysis and thus the same matrix can be "recycled".

We turn to the question of optimizing the degree, our most crucial parameter. The main observation is that one can reduce the error probability in the basic construction by repeating it σ times *in parallel*, using independent randomness in each copy. This will result in a universal PC of length σ from which the output can be recovered, except with $2^{-\Omega(\sigma)}$ error probability, by applying some fixed *threshold* function. We will then enhance this PC into a universal RPC at a minor additional cost.

Improved construction with parameter σ. The improved construction is similar to the basic construction with the following changes:

- OR gates are represented using the construction of Lemma 5.
- The output gate is assigned a special parameter γ_o when Lemma 5 is applied.
- The representation is repeated σ times in parallel, producing a PC of length σ with threshold as its reconstruction function.
- If privacy is required, a randomizing polynomials representation of a threshold function is applied to the σ outputs of the PC, producing an RPC.

Theorem 1. *Let σ be a security parameter. Given a circuit C with m inputs, one output, size s and depth c, the improved construction yields representations of C by:*

- *a PC over* GF(2) *of length σ, degree $d_{PC} = \lceil(\log s + 3)\rceil^{c-1}$, $2^{-\Omega(\sigma)}$- correctness, and public randomness complexity $O(\sigma \log^2 s)$;*
- *an RPC over* GF(2) *of length $O(\sigma^4)$, degree $d_{PC} + 2$, $2^{-\Omega(\sigma)}$-correctness, $2^{-\Omega(\sigma)}$-privacy, public randomness complexity $O(\sigma \log^2 s)$, and private randomness complexity $O(\sigma^4)$.*

The above representations are universal, *i.e., their reconstruction algorithm and simulator do not depend on the circuit C.*

Proof. We instantiate the improved construction outlined above with the following parameters. Consider first the case where no ϵ-biased generators are used to reduce public randomness. In this case, we choose $\gamma = \log s + 2$ and $\gamma_o = 1$. By Lemma 2 each randomized polynomial representing an internal OR gate of C errs with at most $2^{-\gamma}$ probability. Using a union bound, the probability that at least one of them errs is bounded by $s \cdot 2^{-\gamma} \leq \frac{1}{4}$. Since $\gamma_o = 1$, the output will be 1 with probability at most $\frac{1}{4} \cdot \frac{1}{2} = \frac{1}{8}$ if $f(x) = 0$ and at least $\frac{3}{4} \cdot \frac{1}{2} = \frac{3}{8}$ if $f(x) = 1$. (Here we assume that an independent random matrix R is used for the top gate; this has no impact on the asymptotic complexity.)

Consider the PC obtained by concatenating σ copies of the above construction, using independent randomness in each copy. The degree remains as before. By Chernoff's bound, we have a (universal) $2^{-\Omega(\sigma)}$-correct reconstruction algorithm that outputs 1 if at least $\sigma/4$ out of the σ outputs evaluate to 1.

To optimize the amount of public randomness, we choose the bias parameter to be $\epsilon = \frac{1}{s}$. The improved construction is then applied with parameters $\gamma = \left\lceil (\log s + 2) / \log \frac{1}{\frac{1}{2}+\epsilon} \right\rceil$ and $\gamma_o = 1$. The resulting PC has degree $\gamma_o \cdot (\gamma)^{(c-1)}$, which, for sufficiently large s, is bounded by $\lceil (\log s + 3) \rceil^{c-1}$. The amount of randomness for each randomized polynomial in the PC is the amount of randomness needed for a single OR gate which, by Lemma 5, is $O(\gamma \log \frac{s}{\epsilon}) = O(\log^2 s)$.

The output of the above PC representation $p_\rho(x)$ might reveal additional information about x, other than what follow from $C(x)$. To this end, we apply a randomizing polynomials representation of a $(\sigma/4)$-threshold function to the σ outputs of p_ρ. Since any threshold function on σ bits can be computed by a branching program of size $O(\sigma^2)$, Lemma 1 guarantees a representation by degree-3, perfectly correct and private randomizing polynomials $P(\tilde{x}, r)$ where both the length and the private randomness complexity are $O(\sigma^4)$. Moreover, the degree in the $\tilde{x}$ variables is 1. Applying this construction with $p_\rho(x)$ as an input produces an RPC $\tilde{P}_\rho(x, r) = P(p_\rho(x), r)$ with the required parameters.

Note that the perfect simulator (resp., reconstruction algorithm) of P can serve as a universal $2^{-\Omega(\sigma)}$-private simulator (resp., $2^{-\Omega(\sigma)}$-correct reconstruction algorithm) for the above RPC. This follows from the fact that the simulation and reconstruction of P are perfect when conditioned on the event that the choice of ρ does not lead $p_\rho(x)$ to err.   □

We note that above PC and RPC constructions implicitly define efficient evaluation algorithms whose complexity is nearly linear in the circuit size s (defined as the number of *wires*). Moreover, for the purpose of bounding the degree, one can take s to be the number of OR gates in the circuit. Thus, for the important special case where C is an n-term DNF formula, we get a PC (resp., RPC) of degree $\log n + O(1)$ and length σ (resp., $O(\sigma^4)$).

4   Secure Computation of Constant Depth Circuits

In this section we describe the application of low-degree representations to communication-efficient secure computation. Combined with the results of the previous section, we will get efficient protocols for constant-depth circuits.

We will separately consider the client-servers model and the multiparty model, both defined in Section 2.2. In the basic client-servers setting, where only the privacy of the client is guaranteed, it will suffice to use an underlying PC representation of the function we wish to compute. In multi-party setting, as well as the client-servers setting with server privacy, we will need to rely on the stronger RPC representation.

The protocols we describe rely on standard techniques for securely evaluating low-degree polynomials, previously used in the contexts of information-theoretic secure multi-party computation [5, 2, 12, 19] and private information retrieval [11, 16, 4]. We only sketch the high-level structure of the protocols and the parameters they achieve. Further details can be found in the full version.

Throughout this section, assume F to be an extension field of GF(2) having more than k elements, where k is the number of servers or parties. Most of the protocols will employ Shamir's secret-sharing [26] over F in order to securely compute polynomials over its subfield GF(2). We let t denote a security threshold, where $t = 1$ by default.

4.1   The Client-Servers Setting

In the general definition of client-servers computation given in Section 2.2, the client holds an input x, the servers hold an input C, and the client wishes to learn $f(C, x)$ for some publicly known function f. It will be convenient for our purpose to focus on the case where C represents a circuit, x is an m-bit input to this circuit, and f is the *universal* function defined by $f(C, x) = C(x)$. We assume that C is taken from some known class $\mathcal{C}$, typically the class of depth-c, size-s circuits. Thus, the problem we consider is that of allowing the client to privately learn the value of a circuit $C \in \mathcal{C}$ held by the servers on its secret input x.

We start with the case of client-privacy only. In this case, a universal degree-d PC representation for $\mathcal{C}$ gives rise to the following simple protocol. Suppose $k > dt$. Let $p_\rho(x)$ denote be a PC representing the servers' circuit C. The client secret-shares each of its input bits between the servers, using the t-private Shamir's scheme over F. In parallel, it picks ρ at random and sends it to all k servers. Each server replies to the client with the output of p_ρ on the m-tuple of shares it received.[3] The client recovers $C(x)$ by first recovering the output y of p_ρ (since $k > dt$, this can be done by polynomial interpolation) and then applying the (universal) reconstruction algorithm of $\mathcal{C}$. The error probability of the protocol is the same as that of p_ρ. Combining the above protocol with the representation obtained by Theorem 1 we have the following.

Theorem 2. *Let f be the universal circuit evaluation function $f(C, x) = C(x)$. Suppose k servers hold a circuit C of size s, depth c, m inputs and a single output, and the client holds an assignment $x \in \{0, 1\}^m$. Then for every integer $t \geq 1$ there exists a t-private client-servers protocol with the following parameters:*

- *The number of servers is $k = t \cdot (\log s + O(1))^{c-1}$.*
- *The communication consists of $O(m \log k + \sigma \log^2 s)$ bits sent from the client to each server and $O(\sigma)$ bits sent in return.*
- *The computation of each party is nearly linear in its input length (up to factors of σ and $\log k$).*

In the special case of n-term DNF we can substitute in the above $c = 2$ and $s = n$ (see remark following Theorem 1). Thus, we get a protocol with $\log n + O(1)$ servers, $\tilde{O}(m)$ communication, and $\tilde{O}(mn)$ server computation.

Using a technique of Woodruff and Yekhanin [29], it is possible to reduce the number of servers by a factor of 2 without substantially increasing the *total* communication. Specifically, in the resulting protocol the total communication is $O(m\sigma \log m \log^{c-1} s)$ bits per server, and the computation of each server is $\tilde{O}(sm)$. The following is implicit in [29]:

Lemma 6. *Let $t \geq 1$ be an integer, and let $p(x)$ be an m-variate polynomial of degree d, over a field of a prime order $q > dt + 1$. Then, there exists a t-private*

---

[3] In fact, it suffices for each server to send back just a single bit by projecting the answers from F back to GF(2) (cf. [4], Lemma 3).

client-servers protocol to compute $p(x)$ (where p is held by the servers and x by the client), with $k = \lceil \frac{dt+1}{2} \rceil$ servers, queries of m field elements per server, and answers of $m + 1$ field elements per server.

The protocol from [29] is similar to the basic client-servers protocol described above, and in particular the client's queries have the same structure. The improvement comes from the fact that each server replies not only with the value of p on the point queried by the client, but also with the values of its m partial derivatives. In order to apply Lemma 6 in our context, we should emulate computation of $p_\rho(x)$ over GF(2) using computation over a field of a large prime order q. Since the value of $p_\rho(x)$ when computed *over the integers* is bounded by $2^{O(\log^{c-1} s \log m)}$, it suffices to use a prime q of the latter size. This yields the communication complexity specified above. In particular, in the case of n-term DNF we get a protocol with $\frac{1}{2} \log n + O(1)$ servers and nearly optimal communication and computation.

Symmetrically private client-servers protocols. We now briefly discuss an extension of the above results to the case where server privacy is also a requirement. (Further details can be found in the full version.) In this case, the stronger RPC representation should replace the previous RP representation. If the circuit C held by the servers is represented by a (universal) RPC $p_\rho(x, r)$, the protocol will let the client securely evaluate the polynomial vector $p_\rho(x, r)$, where both r and ρ are taken from the CRS shared by the servers. (Note that here it is crucial that r remain hidden from the client.) Unlike the previous case of client privacy only, here the polynomial evaluation protocol should prevent the client from learning anything other than the output of p_ρ. The simple polynomial evaluation protocol described before fails to achieve this property. In the case of a semi-honest client, it is easy to eliminate the extra information from the servers' answers by masking them with (private) randomness from the CRS. The case of a malicious client is more difficult, since the queries it sends to the servers might not be well formed. For instance, the client might share values from $F \setminus \{0, 1\}$ or send shares that are not of the right degree. In the full version we describe a client-servers protocol for fully secure polynomial evaluation that resists a malicious client without requiring additional rounds of interaction and with a very minor complexity overhead. The protocol relies on the conditional disclosure methodology of Gertner et al. [16] and can also be used to obtain better t-private SPIR protocols than those implied by [16]. Applying this machinery, we get an enhanced version of Theorem 2 which provides server privacy against a malicious client, with the main additional cost of increasing the answer size from σ to $\sigma^{O(1)}$ (reflecting the larger length of the RPC).

4.2   The Multi-party Setting

Here we consider the standard MPC setting in which an m-bit string x is partitioned between k parties, who all want to learn the output of a publicly known function $f(x)$. Given an RPC $p_\rho(x, r)$ representing f, we view p_ρ as a randomized low-degree ideal functionality in which r is a "private" random input and ρ

is a "public" random input. Both r and ρ are independently chosen by the functionality; the difference between them is that ρ is additionally revealed to the adversary. (Recall that the separation between ρ and r is motivated by the goal of minimizing the *degree* of the RPC, which in the current context will correspond to the minimal required number of parties.) The reconstruction algorithm of the RPC now defines a non-interactive reduction from f to p_ρ: to securely compute $f(x)$ the parties first securely compute $p_\rho(x, r)$ and then locally apply its reconstruction algorithm.

We turn to the question of efficiently computing the randomized functionality induced by an RPC p_ρ of degree d. We require the communication complexity to be dominated by the length of the inputs and outputs and not by the description size of p. This can be done by using standard MPC techniques (cf. [5, 3, 15]). In the semi-honest case, a simple two-round protocol for $k > dt$ parties is described in [19]. In the malicious case, we can use the following constant-round protocol for $k > (d + 2)t$ parties. First, the parties apply a VSS protocol (e.g., from [5]) to create degree-t shares of x along with shares of *secret* random values r, ρ and shares of 0 (the latter are created by locally adding shares contributed by different players). Using the protocol from [14] the above requires only two rounds. One also needs to ensure that the shared values are elements of the subfield GF(2) rather than general elements of F; this can be done (without additional interaction) using a method from [12]. Next, the players reveal ρ which now defines a *public* polynomial vector p_ρ of degree d. Finally, the players recover $p_\rho(x, r)$: this is done by having each player locally evaluate p_ρ on its shares of x, r, mask the resulting shares with the shares of 0 created in the first phase, and communicate the result to all other players. The players can now recover the output of p_ρ via local error-correction. Combining this protocol with Theorem 1, we have:

Theorem 3. *Let $t \geq 1$ be a security threshold, and C be a circuit of size s and depth c whose m input bits are partitioned between $k \geq \Omega(t \cdot \log^{c-1} s)$ players. Then, there exists a constant-round, statistically t-secure protocol computing $C(x)$ with communication complexity $O(m \cdot poly(k))$.*

5   Constant-Depth Circuits for Search Applications

In this section we demonstrate the usefulness of our general results in the context of database search problems. The goal is to translate the servers' database into a constant-depth circuit and the client's query into a corresponding assignment in a way that would optimize the complexity of our general constructions. For the following problems assume the database consists of n points in $\{0, 1\}^m$:

- *The partial match decision problem* supports queries in $\{0, 1, *\}^m$, where the symbol $*$ is interpreted as "don't care". An answer should be 1 *iff* there is a point in the database that agrees with the query in all indices that are not $*$. We encode the database as an n-term DNF over the $2m$ variables $x_{1,0}, x_{1,1}, \ldots x_{m,0}, x_{m,1}$. The DNF is constructed by assigning a term for every point in the database (e.g., 1011 translates into $x_{1,1} \wedge x_{2,0} \wedge x_{3,1} \wedge x_{4,1}$). The query is then translated by applying a two-bit encoding for each symbol (0 by 10, 1 by 01 and $*$ by 11). As noted in Section 3, for n-term DNF we get a PC (resp., RPC) of degree $\log n + O(1)$, length σ (resp., $\sigma^{O(1)}$) and $2m$ variables. Using Theorem 2, we get a client-servers protocol with $\log n + O(1)$ servers, $\tilde{O}(m)$ communication and $\tilde{O}(nm)$ server computation. One can reduce the number of servers to $\frac{1}{2} \log n$ as discussed in Section 4.1.

- *The partial match search problem* is similar to the decision version except that a matching point of the database should be retrieved. A naive construction gives a depth-4 circuit, paying a lot on tie breaking between matching points. Using the Valiant-Vazirani lemma [28], the database can be translated to a (probabilistic) circuit with $O(m^2)$ inputs, $\tilde{O}(m)$ outputs, and size $\tilde{O}(nm^2)$.
- *The nearest neighbor search problem* is a problem where the point with the smallest Hamming distance to the query should be retrieved. The database can be translated into an $AC^0(p)$ circuit for $p > m$ with m inputs, $O(m^2)$ outputs, size $O(nm^2)$ and depth 3. The use of MOD_p enables efficient constant-depth computation of Hamming distance.

In the full version we describe optimized constructions of constant-depth circuits for the problems mentioned above as well as for other problems.

Acknowledgement. We would like to thank Ronny Roth for helpful comments.

References

[1] N. Alon, O. Goldreich, J. Hastad, and R. Peralta. Simple construction of almost k-wise independent random variables. *Random Structures and Algorithms*, 3(1):289–304, 1992. Preliminary version in FOCS '90.

[2] D. Beaver and J. Feigenbaum. Hiding instances in multioracle queries. In *Proc. 7th STACS*, pages 37–48, 1990.

[3] D. Beaver, J. Feigenbaum, J. Kilian, and P. Rogaway. Security with low communication overhead. In *Proc. 10th CRYPTO*, pages 62–76, 1990.

[4] A. Beimel and Y. Ishai. Information-theoretic private information retrieval: A unified construction. In *Proc. 28th ICALP*, pages 912–926, 2001.

[5] M. Ben-Or, S. Goldwasser, and A. Wigderson. Completeness theorems for non-cryptographic fault-tolerant distributed computation. In *Proc. 20th STOC*, 1988.

[6] D. Boneh, E.J. Goh, and K. Nissim. Evaluating 2-DNF formulas on ciphertexts. In *Proc. 2nd TCC*, pages 325–341, 2005.

[7] R. Canetti. Security and composition of multiparty cryptographic protocols. *J. Cryptology*, 13(1):143–202, 2000.

[8] Ran Canetti. Universally composable security: A new paradigm for cryptographic protocols. In *Proc. 42st FOCS*, pages 136–145, 2001.

[9] M. Charikar, P. Indyk, and R. Panigrahy. New algorithms for subset query, partial match, orthogonal range searching and related problems. In *Proc. 29th ICALP*, pages 451–462, 2002.

[10] B. Chor, N. Gilboa, and M. Naor. Private information retrieval by keywords. Technical report, Department of Computer Science, Technion, 1997.

[11] B. Chor, O. Goldreich, E. Kushilevitz, and M. Sudan. Private information retrieval. In *Proc. 36th FOCS*, pages 41–50, 1995.

[12] R. Cramer, I. Damgård, and U. Maurer. General secure multy-party computation from any linear secret-sharing scheme. In *Proc. of EUROCRYPT*, 2000.

[13] M.J. Freedman, Y. Ishai, B. Pinkas, and O. Reingold. Keyword search and oblivious pseudorandom functions. In *Proc. 2nd TCC*, pages 303–324, 2005.

[14] R. Gennaro, Y. Ishai, E. Kushilevitz and T. Rabin. The Round Complexity of Verifiable Secret Sharing and Secure Multicast. In *Proc. 33rd STOC*, 2001.

[15] R. Gennaro, M.O. Rabin, and T. Rabin. Simplified VSS and fact-track multiparty computations with applications to threshold. In *Proc. 17th PODC*, 1998.

[16] Y. Gertner, Y. Ishai, E. Kushilevitz, and T. Malkin. Protecting data privacy in private information retrieval schemes. *J. of Computer and Systems Sciences*, 60, 2000. Preliminary version in STOC '98.

[17] O. Goldreich, S. Micali, and A. Wigderson. How to play any mental game. In *Proc. 19th STOC*, pages 218–229, 1987.

[18] Oded Goldreich. *Foundations of Cryptography: Basic Applications*. Cambridge University Press, 2004.

[19] Y. Ishai and E. Kushilevitz. Randomizing polynomials: A new representation with applications to round-efficient secure computation. In *Proc. 41st FOCS*, pages 294–304, 2000.

[20] Y. Ishai and E. Kushilevitz. Perfect constant-round secure computation via perfect randomizing polynomials. In *Proc. 29th ICALP*, pages 244–256, 2002.

[21] E. Kushilevitz, R. Ostrovsky, and Y. Rabani. Efficient search for approximate nearest neighbor in high dimensional spaces. In *Proc. 30th STOC*, 1998.

[22] P.B. Miltersen. Cell probe complexity–a survey. In *Pre-Conference Workshop on Advances in Data Structures at the 19th Conference on Foundations of Software Technology and Theoretical Computer Science*, 1999.

[23] J. Naor and M. Naor. Small-bias probability spaces: Efficient constructions and applications. *SIAM J. Comput.*, 22(4):838–856, 1993.

[24] M. Naor and K. Nissim. Communication preserving protocols for secure function evaluation. In *Proc. 33rd STOC*, pages 590–599, 2001.

[25] A. Razborov. Lower bounds for the size of circuits of bounded depth with basis (AND, XOR). *Math. Notes of the Academy of Science of the USSR*, 41(4):333–338, 1987.

[26] A. Shamir. How to share a secret. *Communication of the ACM*, 22(11):612–613, 1979.

[27] R. Smolensky. Algebric methods in the theory of lower bound for boolean circuit complexity. In *Proc. 19th STOC*, pages 77–82, 1987.

[28] L.G. Valiant and V.V. Vazirani. NP is as easy as detecting unique solutions. *Theoretical Computer Science*, 47:85–93, 1986. Preliminary version in STOC '85.

[29] D. Woodruff and S. Yekhanin. A geometric approach to information-theoretic private information retrieval. In *Electronic Colloquium on Computational Complexity (ECCC)*, 2005. Report TR05-009. To appear in CCC 2005.

[30] A. C. Yao. How to generate and exchange secrets. In *Proc. 27th FOCS*, 1986.

Analysis of Random Oracle Instantiation Scenarios for OAEP and Other Practical Schemes

Alexandra Boldyreva[1] and Marc Fischlin[2,*]

[1] College of Computing, Georgia Institute of Technology,
801 Atlantic Drive, Atlanta, GA 30332, USA
aboldyre@cc.gatech.edu
www.cc.gatech.edu/~aboldyre
[2] Institute for Theoretical Computer Science, ETH Zürich, Switzerland
marc.fischlin@inf.ethz.ch
www.fischlin.de

Abstract. We investigate several previously suggested scenarios of instantiating random oracles (ROs) with "realizable" primitives in cryptographic schemes. As candidates for such "instantiating" primitives we pick perfectly one-way hash functions (POWHFs) and verifiable pseudorandom functions (VPRFs). Our analysis focuses on the most practical encryption schemes such as OAEP and its variant PSS-E and the Fujisaki-Okamoto hybrid encryption scheme. We also consider the RSA Full Domain Hash (FDH) signature scheme. We first show that some previous beliefs about instantiations for some of these schemes are not true. Namely we show that, contrary to Canetti's conjecture, in general one cannot instantiate either one of the two ROs in the OAEP encryption scheme by POWHFs without losing security. We also confirm through the FDH signature scheme that the straightforward instantiation of ROs with VPRFs may result in insecure schemes, in contrast to regular pseudorandom functions which can provably replace ROs (in a well-defined way). But unlike a growing number of papers on negative results about ROs, we bring some good news. We show that one can realize *one* of the two ROs in a variant of the PSS-E encryption scheme and *either one* of the two ROs in the Fujisaki-Okamoto hybrid encryption scheme through POWHFs, while preserving the IND-CCA security in both cases (still in the RO model). Although this partial instantiation in form of substituting only one RO does not help to break out of the random oracle model, it yet gives a better understanding of the necessary properties of the primitives and also constitutes a better security heuristic.

1 Introduction

The random oracle (RO) model, introduced by Fiat and Shamir [15] and refined by Bellare and Rogaway [4], has been suggested as a trade-off between provable

---

* Part of the work done while both authors were at the University of California, San Diego. The second author was supported by the Emmy Noether Programme Fi 940/1-1 of the German Research Foundation (DFG).

security and practical requirements for efficiency. Schemes and proofs in this nowadays well-established model make the idealized assumption that all parties have oracle access to a truly random function. Availability of such a random oracle often allows to find more efficient solutions than in the standard model. In practice, it is then assumed that the idealized random function is instantiated through a "good" cryptographic hash function, like SHA-1 or a variation thereof.

The random oracle methodology has gained considerable attention as a design method. Numerous cryptographic schemes proven secure in the RO model have been proposed and some of them are implemented and standardized. The best known example is presumably the OAEP encryption scheme [5,18]. However, even though a RO-based scheme instantiated with a "good" hash function is usually believed to remain secure in the standard model, proofs in the RO model do not technically guarantee this, but merely provide some evidence of security.

Moreover, several recent works [10,21,23,2,19] raised concerns by proving that the random oracle model is not sound. Here lack of soundness refers to the situation when a scheme allows a security proof in the random oracle model but any instantiation of the scheme with any real function family is insecure in the standard model. Such schemes are called "uninstantiable" in [2]. While these results are certainly good reminders about the gap between the RO model and the standard model, the defenders of the RO model and practitioners are assured by the fact that most uninstantiable schemes involve somewhat esoteric examples, in terms of either a construction or sometimes with respect to a security goal.

TOWARDS INSTANTIATING RANDOM ORACLES FOR PRACTICAL SCHEMES. In this work we continue to study security of instantiated schemes designed in the RO model. But unlike the aforementioned works we turn our attention to the most practical cryptographic schemes such as OAEP encryption, the full domain hash (FDH) signature scheme, hybrid encryption schemes obtained via Fujisaki-Okamoto transform [17] and the PSS-E encryption scheme, an OAEP variant due to Coron et al. [12]. Our goal is different, too. We do not show that these schemes are uninstantiable (this would be really bad news). It also seems unrealistic to instantiate these schemes such that they are still efficient and provably secure in the standard model (though this would be great news). Rather, we investigate several possible instantiation scenarios for to these practical schemes somewhere in between.

As candidates for substituting random oracles we consider two primitives with known constructions whose security definitions capture various strong properties of the ideal random oracles, and which have actually been suggested as possible instantiations of random oracles [9,13]. These are the perfectly one-way hash functions (POWHFs) [9,24] and verifiable pseudorandom functions (VPRFs) [22].

The notion of perfectly one-way hash functions has been suggested by Canetti [9] (and was originally named "oracle hashing") to identify and realize useful properties of random oracles. POWHFs are special randomized collision-resistant one-way functions which hide all information about preimages. Canetti [9], and subsequently [24,16], gave several constructions of such POWHFs, based on specific number-theoretic and on more general assumptions. Usually, these POWHFs satisfy another property that requires the output look random, even to an adversary who knows "a little" about the inputs. We will refer to such POWHFs as pseudorandom. In [9] it is proved that a hybrid encryption scheme of Bellare and Rogaway [4] secure against chosen-plaintext attacks (IND-CPA secure) can be securely instantiated with a pseudorandom POWHF, and Canetti conjectured that one could also replace one of the two random oracles in OAEP by a POWHF without sacrificing security against chosen-ciphertext attacks (IND-CCA security) in the RO model.

Verifiable pseudorandom functions have been proposed by Micali et al. in [22]. They resemble pseudorandom functions in that their outputs look random. But their outputs also include proofs that allow verifying the correctness of the outputs with respect to a previously announced public key. In contrast to POWHFs, which are publicly computable given the inputs, VPRFs involve a secret key and therefore their global usage requires the participation of a third party or a device with a tamper-proof key. It is folklore that a secure RO scheme instantiated with a PRF implemented by a third party, will remain secure in the standard model. As suggested in [13] an application scenario for VPRFs, that lowers the amount of trust put on the third party, is a trusted third party implementing a VPRF, say, through a web interface. Now the correctness of the given image can be verified with the consistency proof, and this can be done locally, without further interactions with the third party. We note that this scenario is suitable mostly for digital signatures and not encryption schemes, as the third party has to know the inputs.

NEGATIVE RESULTS. In this work we show that the above intuition about securely replacing random oracles by the aforementioned primitives may be incorrect. We first disprove Canetti's [9] conjecture for the OAEP encryption scheme [5] saying that one can instantiate one of the two RO in the OAEP scheme without losing security (still in the RO model). Recall that, in the OAEP scheme with a (partial one-way) trapdoor permutation f, a ciphertext is of the form $C = f(s||t)$ for $s = G(r) \oplus M||0^k$ and $t = r \oplus H(s)$ for random r. For the security proof of OAEP it is assumed that both G and H are modeled as random oracles.

We prove that, with respect to general (partial one-way) trapdoor permutations f, one cannot replace either of the two random oracles G, H in OAEP by arbitrary pseudorandom POWHFs without sacrificing chosen-ciphertext security. Our negative result follows Shoup's idea to identify weaknesses in the original OAEP security proof [26], and holds relative to a malleable trapdoor function oracle from which a specific function f is derived. Yet, unlike [26], we consider *partial* one-way functions f which suffice to prove OAEP to be IND-CCA in the random oracle model [18]. Our construction also requires to come up with a malleable yet pseudorandom POWHF. We note that our impossibility result is not known to hold for the special case of the RSA function $f : x \mapsto x^e \bmod N$, yet indicates that further assumptions about the RSA function may be necessary to replace one of the random oracles by a POWHF.

The idea for OAEP can be also applied to the Full Domain Hash (FDH) signature scheme, where signatures are of the form $S = f^{-1}(H(M))$. Transferring our OAEP result shows that for a specific class of trapdoor permutations f the instantiation of the RO H through a POWHF can result in an insecure implementation. But here we also show that FDH becomes insecure when H is instantiated the obvious way with a VPRF, even for any trapdoor permutation f such as RSA. By obvious we mean that the pseudorandom value $H(M)$ and its correctness proof π is concatenated with the signature S, such that one can verify the signature's validity by verifying π and checking that $f(S) = H(M)$. Note that VPRFs already provide secure signatures directly, so substituting the random oracle by a VPRF in a signature scheme seems to be moot. However, our goal is to see if VPRFs are a good instantiation in general. Second, one might want additional properties of the signature scheme which FDH gives but not the VPRF, e.g., if used as a sub-protocol in Chaum's blind signature scheme [11]. We note that, independently of our work, [14] obtained a related result about FDH signatures, showing that *any* instantiation of H fails relative to a specific trapdoor function oracle f (whereas our result holds for arbitrary trapdoor functions such as RSA but for a specific instantiation candidate).

POSITIVE RESULTS. Our results show that the RO model is very demanding and even functions with extremely strong properties often cannot securely replace random oracles. However this does not mean that no real function family can be securely used in place of any random oracle. As mentioned, Canetti [9] for example shows how to instantiate an IND-*CPA* secure encryption scheme through POWHFs. Accordingly, we look beyond our negative results and present some positive results, but this time for IND-*CCA* secure encryption schemes.

We first show the following positive results for a variation of the PSS-E encryption scheme introduced by Coron et al. [12]. In the original PSS-E encryption scheme ciphertexts are given by $C = f(\omega||s)$ for $\omega = H(M||r)$ and $s = G(\omega) \oplus M||r$. The PSS transform has been originally proposed by Bellare and Rogaway in the RSA-based signature scheme with message recovery [6]. Coron et al. showed that PSS is a universal transform in that it can also be used for RSA-based encryption for random oracles G, H, achieving chosen-ciphertext security as an alternative to OAEP.

Here we consider a variation PSS-I, where ciphertexts have the form $(f(\omega), s)$ for $\omega = H(M||r)$ and $s = G(\omega) \oplus M||r$, i.e., where the s-part is moved outside of the trapdoor permutation. We prove that for any trapdoor function f the random oracle G can be *instantiated (hence the name PSS-I) with a pseudorandom POWHF such that the scheme remains IND-CCA secure (in the RO model). Interestingly, this also comes with a weaker assumption about the function f. While the original PSS-E scheme has been proven secure for *partial* one-way trapdoor permutations, our scheme PSS-I (with the G-instantiation through a POWHF) works for *any* trapdoor permutation f. A similar observation was made in [20] for OAEP. Concerning the substitution of the H-oracle (even if G is assumed to be a random oracle) we were neither able to prove or disprove that this oracle can be instantiated by some primitive with known construction. We remark that this result about PSS-I is in sharp contrast to OAEP where neither oracle can be replaced by such a POWHF.

As an example where we can replace two random oracles (individually) we discuss the Fujisaki-Okamoto transformation [17] for combining asymmetric and symmetric encryption schemes, where a ciphertext is given by $C = (\mathcal{E}_{\mathrm{asym}}(pk, \sigma; H(\sigma, M)), \mathcal{E}_{\mathrm{sym}}(G(\sigma), M))$ for random σ. It provides an IND-CCA secure hybrid encryption under weak security properties of the two encryption schemes (for random oracles G, H). We show that the scheme remains IND-CCA secure in the RO model if the oracle G is instantiated with a pseudorandom POWHF. We also show that one can instantiate oracle H through a POWHF (for random oracle G) but this requires a strong assumption about the joint security of the POWHF and the asymmetric encryption scheme. Hence, for the Fujisaki-Okamoto transformation both random oracles can be instantiated separately (albeit under a very strong assumption in case of the H oracle).

Our technical results do not mean that one scheme is "more" or "less" secure than the other one, just because one can substitute one random oracle by a primitive like POWHFs. In our positive examples there are usually two random oracles and, replacing one, the resulting scheme is still cast in the random oracle model. Yet, we believe that attenuating the assumption is beneficial, as substituting even one oracle by more "down-to-earth" cryptographic primitives gives a better understanding of the required properties, and it also provides a better heuristic than merely assuming that the hash function behaves as a random oracle.

ORGANIZATION. We give the basic definitions of the two primitives, POWHFs and VPRFs, in Section 2. In Section 3 we show our negative result about instantiating one of the random oracles in OAEP through a POWHF. We then show that in Section 4 that PSS-I admits such an instantiation for one oracle. Section 5 presents the Fujisaki-Okamoto transformation as an example of a scheme where we can replace both random oracles by POWHFs. The FDH scheme and its instantiation through VPRFs are discussed in Section 6.

2   Preliminaries

If x is a binary string, then $|x|$ denotes its length, and if $n \geq 1$ is an integer, then $|n|$ denotes the length of its binary encoding, meaning the unique integer ℓ such that $2^{\ell-1} \leq n < 2^{\ell}$. The string-concatenation operator is denoted "$\|$". If S is a set then $x \xleftarrow{\$} S$ means that the value x is chosen uniformly at random from S. More generally, if D is a probability distribution on S then $x \xleftarrow{D} S$ means that the value x is chosen from set S according to D. If $\mathcal{A}$ is a randomized algorithm with a single output then $x \xleftarrow{\$} \mathcal{A}(y, z, \dots)$ means that the value x is assigned the output of $\mathcal{A}$ for input $(y, z, \dots)$. We let $[\mathcal{A}(y, z, \dots)]$ denote the set of all points having positive probability of being output by A on inputs y, z, etc. A (possibly probabilistic) algorithm is called efficient if it runs in polynomial time in the input length (which, in our case, usually refers to polynomial time in the security parameter).

In the full version of the paper [8] we recall the definitions of asymmetric encryption schemes, their security against chosen-plaintext attacks (IND-CPA security) and chosen-ciphertext attacks (IND-CCA security), of deterministic symmetric encryption schemes, also known as data encapsulation mechanisms or one-time symmetric encryption schemes, and their IND-CPA security (that is a weaker notion than the standard IND-CPA security), and of digital signature schemes and their security against existential unforgeability under chosen-message attacks. For simplicity we give all definitions in the standard model. To extend these definitions to the random oracle model, all algorithms including the adversary get oracle access to one or more random functions $G, H, \ldots$, drawn from the set of all mappings from domain A_k to some range B_k (possibly distinct for different oracles). Here, the parameter k and therefore the domain and the range are usually determined by the cryptographic scheme in question.

2.1 Perfectly One-Way Hash Functions

Perfectly one-way hash functions describe (probabilistic) collision-resistant hash functions with perfect one-wayness. The latter refers to the strong secrecy of a preimage x, even if some additional information about x besides the hash value are known. For this purpose [9] introduces the notion of a function hint which captures these side information. One assumes, though, that it is infeasible to recover the entire value x from hint(x), else the notion becomes trivial. More formally, a (possibly randomized) function hint: $\{0,1\}^{m(k)} \to \{0,1\}^{n(k)}$, where m, n are polynomials, is *uninvertible* with respect to a probability distribution $\mathcal{X} = (\mathcal{X}_k)_{k \in \mathbb{N}}$ if for any probabilistic polynomial-time adversary $\mathcal{I}$ and x taken from $\mathcal{X}_k$, the probability $\Pr\left[\,\mathcal{I}(1^k, \mathrm{hint}(x)) = x\,\right]$ is negligible in k.

In the sequel we usually restrict ourselves to efficient and sufficiently smooth distributions. That is, a probability distribution $\mathcal{X} = (\mathcal{X}_k)_{k \in \mathbb{N}}$ is efficient if it can be computed in polynomial time in k; it is *well-spread* if the min-entropy of $\mathcal{X}$ is superlogarithmic in k.

Definition 1. *[Perfectly One-Way Hash Function] Let $\mathcal{K}$ be an efficient key generation algorithm that takes input 1^k for $k \in \mathbb{N}$ and outputs a function key K of length $l(k)$; let $\mathcal{H}$ be an efficient evaluation algorithm that takes a function key K, input $x \in \{0,1\}^{m(k)}$ and randomness $r \in Coins(K)$ for some fixed polynomial $m(k)$ and returns a hash value $y \in \{0,1\}^{n(k)}$; let $\mathcal{V}$ be an efficient verification algorithm that takes a function key K, an input $x \in \{0,1\}^{m(k)}$ and a hash value $y \in \{0,1\}^{n(k)}$ and outputs a decision bit. The tuple $\mathsf{POWHF} = (\mathcal{K}, \mathcal{H}, \mathcal{V})$ is called a perfectly one-way hash function (with respect to the well-spread, efficient distribution $\mathcal{X} = (\mathcal{X}_k)_{k \in \mathbb{N}}$ and the uninvertible function hint) if the following holds:*

1. *Completeness: For any $k \in \mathbb{N}$, any key $K \in [\mathcal{K}(1^k)]$, any $r \in Coins(K)$, any $x \in \{0,1\}^{m(k)}$ we have $\mathcal{V}(K, x, \mathcal{H}(K, x, r)) = 1$.*
2. *Collision-resistance: For every efficient adversary $\mathcal{C}$ the following holds. For $k \in \mathbb{N}$ pick $K \xleftarrow{\$} \mathcal{K}(1^k)$ and let $(x, x', y) \xleftarrow{\$} \mathcal{C}(K)$. Then $\Pr\left[\,\mathcal{V}(K, x, y) = 1 \wedge \mathcal{V}(K, x', y) = 1 \wedge x \neq x'\,\right]$ is negligible in k.*

3. *Perfect one-wayness (with respect to $\mathcal{X}$, hint): For any efficient adversary $\mathcal{A}$ with binary output the following random variables are computationally indistinguishable:*
   - *Let $K \xleftarrow{\$} \mathcal{K}(1^k)$, $r \xleftarrow{\$} Coins(K)$, $x \xleftarrow{\mathcal{X}_k} \{0,1\}^{m(k)}$.*
     Output $(K, x, \mathcal{A}(K, \text{hint}(x), \mathcal{H}(K, x, r)))$.
   - *Let $K \xleftarrow{\$} \mathcal{K}(1^k)$, $r \xleftarrow{\$} Coins(K)$, $x, x' \xleftarrow{\mathcal{X}_k} \{0,1\}^{m(k)}$.*
     Output $(K, x, \mathcal{A}(K, \text{hint}(x), \mathcal{H}(K, x', r)))$.

The perfectly one-way hash function may have the following additional properties:

4. *Public randomness: $\mathcal{H}$ can be written as $\mathcal{H}(K, x, r) = (r, \mathcal{H}^{pr}(K, x, r))$ for another function $\mathcal{H}^{pr}: \{0,1\}^{l(k)} \times \{0,1\}^{m(k)} \times Coins(K) \rightarrow \{0,1\}^{n(k)-|r|}$ for any $k \in \mathbb{N}$, any $K \in [\mathcal{K}(1^k)]$, any $x \in \{0,1\}^{m(k)}$ and any $r \in Coins(K)$.*
5. *Pseudorandomess (with respect to $\mathcal{X}$, hint): The function acts a pseudorandom generator such that the following random variables are computationally indistinguishable:*
   - *Let $K \xleftarrow{\$} \mathcal{K}(1^k)$, $r \xleftarrow{\$} Coins(K)$, $x \xleftarrow{\mathcal{X}_k} \{0,1\}^{m(k)}$.*
     Output $(K, \text{hint}(x), \mathcal{H}(K, x, r))$.
   - *Let $K \xleftarrow{\$} \mathcal{K}(1^k)$, $x \xleftarrow{\mathcal{X}_k} \{0,1\}^{m(k)}$, and $U \xleftarrow{\$} \{0,1\}^{n(k)}$.*
     Output $(K, \text{hint}(x), U)$.

As pointed out in [9] the notion of an uninvertible function is weaker than the one of a one-way function. For example, $\text{hint}(\cdot) = 0$, which reveals no information about x, is uninvertible but not one-way. We call this function the *trivial uninvertible function*. In fact, several constructions of POWHF based on the Decisional Diffie-Hellman assumption [9] and on more general assumptions like one-way permutations and regular hash functions [9,24,16] have been suggested in the literature. They are provably *pseudorandom* POWHFs with respect to trivial uninvertible function hint. For other uninvertible functions hint they are conjectured to remain secure, yet a formal proof is missing.

In this paper we will mostly consider perfectly one way function families with public randomness as this is a way to ensure correct function re-computation on the same input by different parties, needed for some encryption schemes functionality. All previous constructions [9,24,16] have been designed to meet this notion. For simplicity we will often use the notation $y \leftarrow \mathcal{H}_K(x, r)$ for $y \leftarrow \mathcal{H}(K, x, r)$ and $y \xleftarrow{\$} \mathcal{H}_K(x)$ for $r \xleftarrow{\$} Coins(K), y \leftarrow \mathcal{H}(K, x, r)$, and we often define a hash function with public randomness by just specifying $\mathcal{H}^{pr}$.

2.2   Verifiable Pseudorandom Functions

A verifiable pseudorandom function, defined in [22], is a pseudorandom function with an additional public key allowing to verify consistency of values. Any value for which one has not seen the proof should still look random:

Definition 2. *[Verifiable Pseudorandom Function] Let $\mathcal{K}$ be an efficient key generation algorithm that takes input 1^k for $k \in \mathbb{N}$ and outputs a function* *key and a verification key (fk, vk); let $\mathcal{H}$ be an efficient evaluation algorithm that takes the key fk, input $x \in \{0,1\}^*$ and returns the output $y \in \{0,1\}^{n(k)}$ and a proof $\pi \in \{0,1\}^{l(k)}$ for some fixed polynomials l, n; let $\mathcal{V}$ be an efficient verification algorithm that takes vk, x, y and π and returns a bit. The triple VPRF $= (\mathcal{K}, \mathcal{H}, \mathcal{V})$ is called a verifiable pseudorandom function if the following holds:*

1. *Completeness: For any $(vk, fk) \in [\mathcal{K}(1^k)]$, $x \in \{0,1\}^*$ and $(y, \pi) \in [\mathcal{H}(fk, x)]$, $\mathcal{V}(vk, x, y, \pi) = 1$.*
2. *Uniqueness: There exists a negligible function $\nu(\cdot)$ such that for any $(vk, fk) \in [\mathcal{K}(1^k)]$, any $x \in \{0,1\}^*$, $y_0 \neq y_1 \in \{0,1\}^{n(k)}$, $\pi_0, \pi_1 \in \{0,1\}^{l(k)}$ we have $\Pr[\mathcal{V}(vk, x, y_b, \pi_b) = 1] \leq \nu(k)$ for either $b = 0$ or $b = 1$.*
3. *Pseudorandomness: For any efficient algorithm $\mathcal{A}$ that has access to an oracle and the following experiment*

    Experiment $\mathbf{Exp}_{\text{VPRF},\mathcal{A}}^{\text{vprf-ind}}(1^k)$

    $b \xleftarrow{\$} \{0,1\}$

    $(fk, vk) \xleftarrow{\$} \mathcal{K}(1^k)$

    $(x, state) \xleftarrow{\$} \mathcal{A}^{\mathcal{H}(fk, \cdot)}$ *where x has never been submitted to oracle $\mathcal{H}(fk, \cdot)$*

    If $b = 0$ then $(y, \pi) \xleftarrow{\$} \mathcal{H}(fk, x)$ else $y \xleftarrow{\$} \{0,1\}^{n(k)}$ EndIf

    $d \xleftarrow{\$} \mathcal{A}^{\mathcal{H}(fk, \cdot)}(y, state)$ *where x has never been submitted to oracle $\mathcal{H}(fk, \cdot)$*

    the difference $\Pr\left[\mathbf{Exp}_{\text{VPRF},\mathcal{A}}^{\text{vprf-ind}}(1^k) = b\right] - 1/2$ is negligible in k.

3   (In)Security of OAEP Instantiations

Here we show that, for general trapdoor permutations, instantiating any of the two random oracles in OAEP with a pseudorandom POWHF does not yield a secure scheme.

3.1   OAEP Encryption Scheme

We first recall the OAEP encryption scheme [5]. It is parameterized by integers k, k_0 and k_1 (where k_0, k_1 are linear in k) and makes use of a trapdoor permutation family F with domain and range $\{0,1\}^k$ and two random oracles

$$G \colon \{0,1\}^{k_0} \to \{0,1\}^{k-k_0} \quad \text{and} \quad H \colon \{0,1\}^{k-k_0} \to \{0,1\}^{k_0}$$

The message space is $\{0,1\}^{k-k_0-k_1}$. The scheme $\text{OAEP}^{G,H}[F] = (\mathcal{EK}, \mathcal{E}, \mathcal{D})$ are defined as follows:

- $\mathcal{EK}(1^k)$: Pick a permutation f from F at random. Let pk specify f and let sk specify f^{-1}.
- $\mathcal{E}(pk, M)$: Compute $r \xleftarrow{\$} \{0,1\}^{k_0}$, $s \leftarrow (m\|0^{k_1}) \oplus G(r)$ and $t \leftarrow r \oplus H(s)$. Output $C \leftarrow f(s\|t)$.
- $\mathcal{D}(sk, C)$: Compute $s\|t \leftarrow f^{-1}(C)$, $r \leftarrow t \oplus H(s)$ and $M \leftarrow s \oplus G(r)$. If the last k_1 bits of M are zeros, then return the first $k - k_0 - k_1$ bits of M. Otherwise, return $\perp$.

The encryption scheme $OAEP^{G,H}[F]$ is proven to be IND-CCA secure in the RO model if the underlying permutation family F is partial one-way [18]. Partial one-wayness is a stronger notion than one-wayness; for the definitions see [18].

3.2   Insecurity of Instantiating the G-Oracle in OAEP with POWHFs

We first consider the OAEP scheme where the G-oracle is instantiated with a pseudorandom POWHF. Informally, a key specifying an instance of POWHF becomes a part of the public key and each invocation of the G-oracle is replaced with the function evaluation, such that in the encryption algorithm a new randomness for the function evaluation is picked and becomes part of the ciphertext, and in the decryption algorithm the function is re-computed using the given randomness. More formally:

Let $POWHF = (\mathcal{K}, \mathcal{G}, \mathcal{V})$, where $\mathcal{K}: \{1^k | k \in \mathbb{N}\} \rightarrow \{0,1\}^k$, $\mathcal{G}: \{0,1\}^k \times \{0,1\}^{k_0} \times Coins(K) \rightarrow \{0,1\}^{k-k_0}$ and $\mathcal{V}: \{0,1\}^k \times \{0,1\}^{k_0} \times \{0,1\}^{k-k_0} \rightarrow \{0,1\}$, be a perfectly one-way pseudorandom hash function with public randomness. An *instantiation of the G-oracle* in the $OAEP^{G,H}[F]$ encryption scheme with $POWHF = (\mathcal{K}, \mathcal{G}, \mathcal{V})$ results in the following encryption scheme $OAEP^{POWHF,H}[F] = (\mathcal{EK}, \mathcal{E}, \mathcal{D})$

- $\mathcal{EK}(1^k)$: Pick a random permutation f on $\{0,1\}^k$ and sample a POWHF key $K \stackrel{\$}{\leftarrow} \mathcal{K}(1^k)$. Let pk specify f and also contain K, and let sk specify f^{-1} and also contain K.
- $\mathcal{E}(pk, M)$: Pick randomness $r \stackrel{\$}{\leftarrow} \{0,1\}^{k_0}$ for encryption and $r_\mathcal{G} \stackrel{\$}{\leftarrow} Coins(K)$ for the POWHF. Compute $y \leftarrow \mathcal{G}_K^{pr}(r, r_\mathcal{G})$, $s \leftarrow (M \| 0^{k_1}) \oplus y$ and $t \leftarrow r \oplus H(s)$. Let $C \leftarrow f(s \| t)$ and output $(r_\mathcal{G}, C)$.
- $\mathcal{D}(sk, (r_\mathcal{G}, C))$: Compute $s \| t \leftarrow f^{-1}(C), r \leftarrow t \oplus H(s), M \leftarrow s \oplus \mathcal{G}^{pr}(r, r_\mathcal{G})$. If the last k_1 bits of M are zeros, then return the first $k - k_0 - k_1$ bits of M. Otherwise, return $\perp$.

We note that for simplicity we assume that $r_\mathcal{G}$, the randomness output by $\mathcal{G}_K$, is a public part of the ciphertext. If it was possible to tamper this value $r_\mathcal{G}$ into $r'_\mathcal{G}$ for a given ciphertext, such that this yields the same hash value, $\mathcal{G}_K^{pr}(r, r_\mathcal{G}) = \mathcal{G}_K^{pr}(r, r'_\mathcal{G})$, then it would be obviously easy to mount a successful chosen-ciphertext attack. To prevent such attacks one can in principle demand that such collisions for the hash function are infeasible to find —most known constructions [9,24,16] have this additional property— or one can protect $r_\mathcal{G}$ by some other means. We do not complicate the instantiation here, as our attack already succeeds without changing $r_\mathcal{G}$, e.g., the attack would even work if $r_\mathcal{G}$ was encrypted (separately or inside f) or authenticated.

INTUITION. Before we present our results in detail we provide some intuition. First we construct malleable POWHFs, i.e., for which $\mathcal{G}_K(x, r) \oplus \Delta = \mathcal{G}_K(x \oplus \delta, r)$ for some δ, Δ. We show how to construct such primitives in [8]. Our construction assumes that one-way permutations exist and employs the pseudorandom function tribe ensembles of [16] (which are one possibility to build

POWHFs). Assume that either RO in the $\text{OAEP}^{G,H}[F]$ encryption scheme is instantiated with such a POWHF. Here F is a partial one-way trapdoor permutation family. Now given the challenge ciphertext $C^* = f(s^*\|t^*)$ of some message M_b where f is an instance of F, an adversary $\mathcal{A}$ can find δ, Δ such that $C = f((s^*\|t^*) \oplus \delta)$ is a valid encryption of $M_b \oplus \Delta$, and given the decryption of this ciphertext one can easily compute M_b.

The only problem is that, although flipping bits by penetrating the POWHF is easy by construction, how can $\mathcal{A}$ compute $f((s^*\|t^*) \oplus \delta)$ without knowing $s^*\|t^*$? Here we use the idea of Shoup [26] about the existence of XOR-malleable trapdoor permutations which allow such modifications. We note that the attack is not known to work for OAEP with the RSA trapdoor family, but it nevertheless shows that security may fail in general if a RO is instantiated with a POWHF.

Our approach is somewhat similar to the attacks Shoup used to show that for a XOR-malleable one-way trapdoor permutation family F the encryption scheme $\text{OAEP}^{G,H}[F]$ is *not* IND-CCA secure in the RO model. However, Shoup's attack does not work if F is partial one way, and, moreover, for such F the scheme $\text{OAEP}^{G,H}[F]$ has been proven IND-CCA secure in the RO model [18]. Our attacks work even if F is partial one way.

Theorem 1. *Let* $\text{POWHF}' = (\mathcal{K}', \mathcal{G}', \mathcal{V}')$ *be a pseudorandom POWHF with public randomness (with respect to the uniform distribution and some uninvertible function* hint*). Then there exists a pseudorandom* $\text{POWHF} = (\mathcal{K}, \mathcal{G}, \mathcal{V})$ *with public randomness (with respect to the uniform distribution and* hint*) and an oracle relative to which there is a partial one-way permutation family* F*, such that* $\text{OAEP}^{\text{POWHF},H}[F]$*, an instantiation of the G-oracle in the* $\text{OAEP}^{G,H}[F]$ *encryption scheme with* POWHF*, is* not *IND-CCA in the RO model.*

Recall that we can assume that POWHF is malleable in the sense that $\mathcal{G}_K^{\text{pr}}(x, r) \oplus 1\|0^{n-1} = \mathcal{G}_K^{\text{pr}}(x \oplus 1\|0^{m-1}, r)$ for all k, x, r (we show how to construct such POWHFs form the given POWHF' in [8]). We now define a compliant XOR-malleable permutation family. We slightly strengthen the original definition of Shoup [26].

Definition 3. *A permutation family* F *is XOR-malleable if there exists an efficient algorithm* U*, such that on inputs a random instance permutation* f *from* F *with domain* $\{0,1\}^k$ *and* $f(t)$ *for random* $t \in \{0,1\}^k$ *and any* $\delta \in \{0,1\}^k$*, algorithm* $U(f, f(t), \delta)$ *outputs* $f(t \oplus \delta)$ *with non-negligible probability (in* k*).*

Even though Shoup uses a weaker definition of XOR-malleability, where U's success probability is also over the random choice of $\delta \in \{0,1\}^k$, his proof in [26] is also valid for the stronger Definition 3 with fixed δ:

Fact 1 ([26]). *There exists an oracle relative to which XOR-malleable one-way trapdoor permutations exist.*

Now we are ready to prove the theorem of the insecure instantiation of the G-oracle in OAEP. We present the formal proof in [8]. The idea is to construct the trapdoor permutation family F as $f(s\|t) = f'_{\text{left}}(s)\|f'_{\text{right}}(t)$ for random instances $f'_{\text{left}}, f'_{\text{right}}$ of the malleable family F'. Then an adversary $\mathcal{A}$ gets a challenge ciphertext $(r^*_{\mathcal{G}}, C^*_{\text{left}} \| C^*_{\text{right}})$ of one of two messages M_0, M_1, and invokes U to modify the right part to $C_{\text{right}} \leftarrow U(f'_{\text{right}}, C^*_{\text{right}}, 1\|0^{k_0-1})$. Submitting the ciphertext $(r^*_{\mathcal{G}}, C^*_{\text{left}} \| C_{\text{right}})$ to the decryption oracle is a valid ciphertext for the message $M_b \oplus 1\|0^{k-k_0-k_1-1}$ because for

$$(C^*_{\text{left}} \| C^*_{\text{right}}) = (f'_{\text{left}}(s^*)\|f'_{\text{right}}(t^*)), \ s^* = M_b\|0^{k_0} \oplus \mathcal{G}^{\text{pr}}_K(r^*, r^*_{\mathcal{G}}), \ t^* = r^* \oplus H(s^*)$$

we have:

$$
\begin{aligned}
C_{\text{right}} &= f'_{\text{right}}\big(t^* \oplus 1\|0^{k_0-1}\big) = f'_{\text{right}}\big((r^* \oplus 1\|0^{k_0-1}) \oplus H(s^*)\big) \\
C^*_{\text{left}} &= f'_{\text{left}}(s^*) = f'_{\text{left}}\big(M_b\|0^{k_0} \oplus \mathcal{G}^{\text{pr}}_K(r^*, r^*_{\mathcal{G}})\big) \\
&= f'_{\text{left}}\big((M_b\|0^{k_0} \oplus 1\|0^{k-k_0-1}) \oplus (\mathcal{G}^{\text{pr}}_K(r^*, r^*_{\mathcal{G}}) \oplus 1\|0^{k-k_0-1})\big) \\
&= f'_{\text{left}}\big((M_b\|0^{k_0} \oplus 1\|0^{k-k_0-1}) \oplus \mathcal{G}^{\text{pr}}_K(r^* \oplus 1\|0^{k_0-1}, r^*_{\mathcal{G}})\big)
\end{aligned}
$$

The answer of the decryption oracle now allows to determine the bit b easily.

3.3   Insecurity of Instantiating the H-Oracle in OAEP with POWHFs

For substituting the H-oracle we obtain a similar insecurity result as for the case of G. However, the proof (presented in [8]) is slightly different as we have to transform both ciphertext parts.

Theorem 2. *Let* $\mathsf{POWHF}' = (\mathcal{K}', \mathcal{H}', \mathcal{V}')$ *be a pseudorandom POWHFs with public randomness (with respect to the uniform distribution and some uninvertible function* hint*). Then there exists a pseudorandom* $\mathsf{POWHF} = (\mathcal{K}, \mathcal{H}, \mathcal{V})$ *with public randomness (with respect to the uniform distribution and* hint*), and there exists an oracle relative to which there is a partial one-way permutation family* F, *such that* $OAEP^{G,\mathsf{POWHF}}[F] = (\mathcal{EK}, \mathcal{E}, \mathcal{D})$, *an instantiation of the H-oracle in the* $OAEP^{G,H}[F]$ *encryption scheme with* POWHF, *is not IND-CCA in the RO model.*

4   Security of PSS-I Encryption Instantiations

In this section we show a positive result, allowing to replace one of the random oracles in our PSS-E variation, called PSS-I, by a pseudorandom POWHF. We were unable to prove or disprove that one can replace the other oracle in PSS-I.

4.1   The PSS-I Encryption Scheme

Coron et al. [12] suggested that the transformation used by the PSS signature scheme [6] can also be used for encrypting with RSA. Here we consider the following variation PSS-I. This scheme is parameterized by integers k, k_0 and k_1 (where k_0, k_1 are linear in k) and makes use of an instance of a trapdoor permutation family with domain and range $\{0,1\}^k$ (and it can be easily adapted for other domains like $\mathbf{Z}_N^*$ for the RSA permutation). The scheme also uses two random oracles

$$G\colon \{0,1\}^{k_1} \to \{0,1\}^{k-k_1} \quad \text{and} \quad H\colon \{0,1\}^{k-k_1} \to \{0,1\}^{k_1}.$$

The message space is $\{0,1\}^{k-k_0-k_1}$. The scheme PSS-I$^{G,H}[F]$ is given by the following algorithms:

- $\mathcal{EK}(1^k)$: Pick a random permutation f on $\{0,1\}^{k_1}$. Let pk specify f and let sk specify f^{-1}.
- $\mathcal{E}(pk, M)$: Compute $r \xleftarrow{\$} \{0,1\}^{k_0}$, $\omega \leftarrow H(M\|r)$ and $s \leftarrow G(\omega) \oplus (M\|r)$. Compute $C \leftarrow f(\omega)$ and output (C, s).
- $\mathcal{D}(sk, (C, s))$: Compute $\omega \leftarrow f^{-1}(C)$, $M\|r \leftarrow s \oplus G(\omega)$. If $\omega = H(M\|r)$ then return M. Otherwise, return $\perp$.

In the original PSS-E scheme [12] one computes f over both $\omega\|s$. We remark that our version here seems to be less secure than the original scheme at first, as the value s is now given in the clear. However, it nonetheless allows us to securely replace oracle G by a POWHF which we were unable to do in the original scheme. Moreover, we can prove security of our instantiation with respect to arbitrary trapdoor permutations, whereas the original scheme required partial one-way trapdoor permutations.

4.2   Instantiating the G-Oracle in PSS-I with POWHFs

An *instantiation of the G-oracle* in the PSS-I$^{G,H}[F]$ encryption scheme with a pseudorandom perfectly one-way hash function POWHF $=$ $(\mathcal{K}, \mathcal{G}, \mathcal{V})$ with public randomness results in the following encryption scheme PSS-I$^{\text{POWHF},H}[F] = (\mathcal{EK}, \mathcal{E}, \mathcal{D})$

- $\mathcal{EK}(1^k)$: Pick a random permutation f on $\{0,1\}^{k_1}$ and sample a POWHF key $K \xleftarrow{\$} \mathcal{K}(1^k)$ and randomness $r_\mathcal{G} \xleftarrow{\$} \text{Coins}(K)$. Let pk specify f and also contain $K, r_\mathcal{G}$, and let sk specify f^{-1} and also contain $K, r_\mathcal{G}$.
- $\mathcal{E}(pk, M)$: Pick randomness $r \xleftarrow{\$} \{0,1\}^{k_0}$ for the encryption algorithm and compute $\omega \leftarrow H(M\|r)$. Compute $s \leftarrow \mathcal{G}_K^{\text{pr}}(\omega, r_\mathcal{G}) \oplus (M\|r)$ and $C \leftarrow f(\omega)$. Output (C, s).
- $\mathcal{D}(sk, (C, s))$: Compute $\omega \leftarrow f^{-1}(C)$, $M\|r \leftarrow s \oplus \mathcal{G}_K^{\text{pr}}(\omega, r_\mathcal{G})$. If $\omega = H(M\|r)$ then return M. Otherwise, return $\perp$.

It is noteworthy that the randomness of the POWHF becomes part of the public key and is therefore fixed for each ciphertext. While this seems strange at first, it becomes clear in in light of the role of the randomness in POWHFs. Originally, POWHFs were designed to meet a stronger security requirement [9,24], demanding pairs $(\mathcal{G}(x, r_1), \mathcal{G}(x, r_2))$ for a single random x to be indistinguishable from pairs $(\mathcal{G}(x, r_1), \mathcal{G}(x', r_2))$ for independent samples x, x'. This of course requires that the randomness r_1, r_2 is chosen independently for each function evaluation, else distinguishing would be easy. However, security of PSS-I relies on pseudo-randomness of the corresponding function family and does not require the above security property. Accordingly, putting the randomness for the function family in the public key does not compromise security of the encryption scheme.

Theorem 3. *Let F be a trapdoor permutation family and let* POWHF $=$ $(\mathcal{K}, \mathcal{G}, \mathcal{V})$ *be a pseudorandom POWHF with public randomness, where pseudoran-domness holds with respect to the uniform distribution on and the uninvertible function* $\mathsf{hint}(x) = (f, f(x))$ *for random f drawn from F. Then PSS-I*$^{\mathsf{POWHF}, H}[F]$ *is IND-CCA secure in the RO model.*

The proof is delegated to [8]. We note that our proof does not make use of the collision-resistance of the POWHF. This is because the preimage ω of the POWHF is uniquely determined by the additional trapdoor function value $f(\omega)$ anyway. Hence, a pseudorandom generator for which distinguishing the output from random is infeasible, even if given $\mathsf{hint}(\omega)$, would actually suffice in this setting. In particular, such a generator G can be built in combination with the trapdoor permutation f via the Yao-Blum-Micali construction [27,7]. Namely, let f be of the form $f(x) = g^n(x)$ for a trapdoor permutation g and define $G(x) = (\mathsf{hb}(x), \mathsf{hb}(g(x)), \ldots, \mathsf{hb}(g^{n-1}(x)))$ through the hardcore bits hb. Then the output of G is still pseudorandom, even given $f(x)$.

5   Security of Instantiating the Fujisaki-Okamoto Transformation

Fujisaki and Okamoto [17] suggested a general construction of hybrid encryption schemes in the random oracle model. It is based on two random oracles, G and H. Here we show that one can replace G by a pseudorandom POWHF and still obtain a secure scheme (for a random oracle H). We then prove, under a somewhat non-standard assumption, that one can also replace H by a POWHF to obtain a secure scheme for a random oracle G.

5.1   Fujisaki-Okamoto Scheme

The Fujisaki-Okamoto construction is based on an asymmetric encryption scheme AS $= (\mathcal{EK}_{\mathrm{asym}}, \mathcal{E}_{\mathrm{asym}}, \mathcal{D}_{\mathrm{asym}})$ and a deterministic symmetric encryp-tion scheme SS $= (\mathcal{EK}_{\mathrm{sym}}, \mathcal{E}_{\mathrm{sym}}, \mathcal{D}_{\mathrm{sym}})$, as well as two random oracles G, H. For parameter $k \in \mathbb{N}$ let $\mathrm{Coins}_{\mathrm{asym}}(k)$ and $\mathrm{MsgSp}_{\mathrm{asym}}(k)$ denote the set of ran-dom strings and the message space of the asymmetric encryption scheme, and $\mathrm{Keys}_{\mathrm{sym}}(k)$ and $\mathrm{MsgSp}_{\mathrm{sym}}(k)$ denote the key and message space of the symmetric encryption scheme. Let

$$G\colon \mathrm{MsgSp}_{\mathrm{asym}}(k) \to \mathrm{Keys}_{\mathrm{sym}}(k) \quad \text{and} \quad H\colon \{0,1\}^k \times \{0,1\}^* \to \mathrm{Coins}_{\mathrm{asym}}(k)$$

The message space is $\mathrm{MsgSp}_{\mathrm{sym}}(k)$. The encryption scheme $\mathrm{FO}^{G,H}$ is given by the following algorithms:

- $\mathcal{EK}(1^k)$: Run $\mathcal{EK}_{\mathrm{asym}}(1^k)$ to generate a key pair (sk, pk).
- $\mathcal{E}(pk, M)$: Pick $\sigma \overset{\$}{\leftarrow} \mathrm{MsgSp}_{\mathrm{asym}}(k)$, compute
  $C_{\mathrm{asym}} \leftarrow \mathcal{E}_{\mathrm{asym}}(pk, \sigma; H(\sigma, M))$ and $C_{\mathrm{sym}} \leftarrow \mathcal{E}_{\mathrm{sym}}(G(\sigma), M)$. Output $C = (C_{\mathrm{asym}}, C_{\mathrm{sym}})$.
- $\mathcal{D}(sk, C)$: For $C = (C_{\mathrm{asym}}, C_{\mathrm{sym}})$ compute $\sigma \leftarrow \mathcal{D}(sk, C_{\mathrm{asym}})$,
  $M \leftarrow \mathcal{D}_{\mathrm{sym}}(G(\sigma), C_{\mathrm{sym}})$. Recompute $c \leftarrow \mathcal{E}_{\mathrm{asym}}(pk, \sigma; H(\sigma, M))$ and output M if $c = C_{\mathrm{asym}}$, else return $\perp$.

Security of this conversion has been shown under the assumption that the symmetric encryption scheme is IND-CPA (and that the symmetric encryption algorithm is deterministic), and that the public-key encryption scheme is one-way and γ-uniform, which roughly means that ciphertexts are almost uniform. Here we make different, yet "natural" assumptions about the encryption schemes, as specified below.

5.2   Instantiating the G-Oracle

An *instantiation of the G-oracle* in the Fujisaki-Okamoto scheme through a perfectly one-way hash function $\mathrm{POWHF} = (\mathcal{K}, \mathcal{G}, \mathcal{V})$ with public randomness, denoted by $\mathrm{FO}^{\mathrm{POWHF}, H}$, works as follows:

- $\mathcal{EK}(1^k)$: Run $\mathcal{EK}_{\mathrm{asym}}(1^k)$ to generate a key pair (sk, pk). Pick $K \overset{\$}{\leftarrow} \mathcal{K}(1^k)$
  and $r \overset{\$}{\leftarrow} \mathrm{Coins}_{\mathcal{G}}(k)$. Output $((sk, K, r), (pk, K, r))$.
- $\mathcal{E}((pk, K, r), M)$: Pick $\sigma \overset{\$}{\leftarrow} \mathrm{MsgSp}_{\mathrm{asym}}(k)$, compute $C_{\mathrm{asym}} \leftarrow \mathcal{E}_{\mathrm{asym}}(pk, \sigma,$
  $H(\sigma, M))$ and $C_{\mathrm{sym}} \leftarrow \mathcal{E}_{\mathrm{sym}}(\mathcal{G}^{\mathrm{pr}}(K, \sigma, r), M)$. Output $C = (C_{\mathrm{asym}}, C_{\mathrm{sym}})$.
- $\mathcal{D}((sk, K, r), C)$: For $C = (C_{\mathrm{asym}}, C_{\mathrm{sym}})$ compute $\sigma \leftarrow \mathcal{D}_{\mathrm{asym}}(sk, C_{\mathrm{asym}})$,
  $M \leftarrow \mathcal{D}_{\mathrm{sym}}(\mathcal{G}^{\mathrm{pr}}(K, \sigma, r), C_{\mathrm{sym}})$. Recompute $c \leftarrow \mathcal{E}_{\mathrm{asym}}(pk, \sigma; H(\sigma, M))$ and
  output M if $c = C_{\mathrm{asym}}$, else return $\perp$.

We note that we use the same trick as in the PSS-I case before and put the randomness r of the POWHF into the public key. See the remarks there for further discussion.

Theorem 4. *Let* AS *and* SS *be IND-CPA asymmetric and symmetric encryption schemes, where $\mathcal{E}_{sym}$ is deterministic. Let* POWHF $= (\mathcal{K}, \mathcal{G}, \mathcal{V})$ *be a pseudorandom POWHF with public randomness (with respect to the uniform distribution on* $(\mathrm{MsgSp}_{asym}(k))_{k \in \mathbb{N}}$ *and the trivial uninvertible function* hint*). Then the instantiation of the G-oracle in the Fujisaki-Okamoto scheme,* $\mathrm{FO}^{\mathrm{POWHF}, H}$, *is IND-CCA in the random oracle model.*

The proof is in [8]. Recall that such POWHF as in the claim can be built from any one-way permutation. We can thus instantiate the G-oracle under this condition. In fact, the proof actually shows that regular one-wayness (instead of perfect one-wayness) is sufficient for the pseudorandom POWHF, where for any efficient algorithm $\mathcal{A}$ the probability that $\mathcal{A}$ returns x on input $(K, \mathrm{hint}(x), \mathcal{H}(K, x, r))$ for $K \overset{\$}{\leftarrow} \mathcal{K}(1^k)$, $r \overset{\$}{\leftarrow} \mathrm{Coins}(K)$, $x \overset{x_k}{\leftarrow} \{0, 1\}^{m(k)}$, is negligible. Clearly, perfect one-wayness implies regular one-wayness.

5.3 Instantiating the H-Oracle

Instantiating the H-oracle is technically more involved and requires a strong assumption about the combination of the POWHF and the public-key encryption scheme. Our construction also requires a stronger (yet mild) assumption about the symmetric encryption scheme.

Before presenting our assumptions we first define the H-instantiation of the Fujisaki -Okamoto transformation. We call the encryption scheme below an *instantiation of the H-oracle* in the Fujisaki-Okamoto scheme, $\mathrm{FO}^{G,\mathsf{POWHF}}$, through a pseudorandom and strongly collision-resistant $\mathsf{POWHF} = (\mathcal{K}, \mathcal{H}, \mathcal{V})$:

- $\mathcal{EK}(1^k)$: Run $\mathcal{EK}_{\mathrm{asym}}(1^k)$ to generate a key pair (sk, pk). Generate $K \xleftarrow{\$} \mathcal{K}(1^k)$ and $r \xleftarrow{\$} \mathrm{Coins}_{\mathcal{H}}(k)$ for POWHF. Output (sk, K, r) and (pk, K, r).
- $\mathcal{E}((pk, K, r), M)$: Pick $\sigma \xleftarrow{\$} \mathcal{EK}_{\mathrm{sym}}(1^k)$, compute $\omega \leftarrow \mathcal{H}^{\mathrm{pr}}(K, \sigma \| M, r)$ and $C_{\mathrm{asym}} \leftarrow \mathcal{E}_{\mathrm{asym}}(pk, \sigma, \omega)$ and $C_{\mathrm{sym}} \leftarrow \mathcal{E}_{\mathrm{sym}}(G(\sigma), M)$. Output $C = (C_{\mathrm{asym}}, C_{\mathrm{sym}})$.
- $\mathcal{D}((sk, K, r), C)$: For $C = (C_{\mathrm{asym}}, C_{\mathrm{sym}})$ compute $\sigma \leftarrow \mathcal{D}(sk, C_{\mathrm{asym}})$, $M \leftarrow \mathcal{D}_{\mathrm{sym}}(G(\sigma), C_{\mathrm{sym}})$. Recompute $c \leftarrow \mathcal{E}_{\mathrm{asym}}(pk, \sigma; \mathcal{H}^{\mathrm{pr}}(K, \sigma \| M, r))$ and output M if $c = C_{\mathrm{asym}}$, else return $\perp$.

To show that this instantiation is secure we need the following additional assumption about the symmetric encryption scheme. We assume that the symmetric encryption scheme provides *integrity of ciphertexts* (INT-CTXT) [3], i.e., for any efficient adversary $\mathcal{B}$ let $\kappa \xleftarrow{\$} \mathcal{EK}_{\mathrm{sym}}(1^k)$, $C \xleftarrow{\$} \mathcal{B}^{\mathcal{E}_{\mathrm{sym}}(\kappa, \cdot)}(1^k)$ and let $M \xleftarrow{\$} \mathcal{D}_{\mathrm{sym}}(\kappa, C)$. Then the probability that $M \neq \perp$ and that C has never been submitted by $\mathcal{B}$ to its oracle $\mathcal{E}_{\mathrm{sym}}(\kappa, \cdot)$ is negligible. This INT-CTXT property can be accomplished for example by the encrypt-then-MAC paradigm [3]. We remark that this additional property, together with the IND-CPA security of the asymmetric encryption scheme, does not necessarily imply IND-CCA security of hybrid schemes; it is easy to construct counterexamples.

For our instantiation we also need a very strong assumption about the combination of POWHF and the public-key encryption scheme $(\mathcal{EK}_{\mathrm{asym}}, \mathcal{E}_{\mathrm{asym}}, \mathcal{D}_{\mathrm{asym}})$. That is, we assume that the following random variables are indistinguishable for any efficient message distribution $\mathcal{M}$ (which also outputs some information *state* about the sampling process):

- Let $(sk, pk) \xleftarrow{\$} \mathcal{EK}_{\mathrm{asym}}(1^k)$, $K \xleftarrow{\$} \mathcal{K}(1^k)$, $r \xleftarrow{\$} \mathrm{Coins}_G(k)$ and $(M, state) \xleftarrow{\$} \mathcal{M}(pk, K, r)$. Pick $\sigma \xleftarrow{\$} \mathrm{MsgSp}_{\mathrm{asym}}(k)$ and compute $\omega \leftarrow \mathcal{H}^{\mathrm{pr}}(K, \sigma \| M, r)$ and $C_{\mathrm{asym}} \leftarrow \mathcal{E}_{\mathrm{asym}}(pk, \sigma, \omega)$. Output $(pk, K, r, state, C_{\mathrm{asym}})$.
- Let $(sk, pk) \xleftarrow{\$} \mathcal{EK}_{\mathrm{asym}}(1^k)$, $K \xleftarrow{\$} \mathcal{K}(1^k)$, $r \xleftarrow{\$} \mathrm{Coins}_G(k)$ and $(M, state) \xleftarrow{\$} \mathcal{M}(pk, K, r)$. Pick $\sigma \xleftarrow{\$} \mathrm{MsgSp}_{\mathrm{asym}}(k)$ and $\omega \xleftarrow{\$} \mathrm{Coins}_{\mathrm{asym}}$ and compute $C_{\mathrm{asym}} \leftarrow \mathcal{E}_{\mathrm{asym}}(pk, \sigma, \omega)$. Output $(pk, K, r, state, C_{\mathrm{asym}})$.

We call this the *POWHF-encryption assumption for* POWHF *and* AS.

Informally, if one views the POWHFs as a pseudorandom generator, the assumption basically says that encrypting the seed σ of a pseudorandom generator with the pseudorandom output ω is indistinguishable from an encryption of the seed with independent randomness. Note that this assumption would be false in general if one is also given ω in clear (which is either pseudorandom or truly random). For example, for ElGamal encryption $(g^\omega, pk^\omega \cdot \sigma)$ one could easily recover σ if given ω (by dividing out pk^ω in the right part), and try to recompute ω through the pseudorandom generator applied to σ. However, if one is not given ω then such generic attacks (in the sense of [25]) fail.

Note also that our POWHF-encryption assumption is certainly not stronger than assuming that the pseudorandom generator is perfect and given by a random oracle. On the contrary, our result shows that seeing the adversary's queries to function H is not necessary to simulate attacks and to prove security. This holds, of course, as long as G is still a random oracle and the simulator learns the queries to this oracle. The proof of the following theorem is in [8]. Similar to the G-case the proof shows that regular one-wayness is enough for the pseudorandom POWHF.

Theorem 5. *Let* AS *and* SS *be IND-CPA public-key and private-key encryption schemes where $\mathcal{E}_{sym}$ is deterministic. Let* POWHF $= (\mathcal{K}, \mathcal{H}, \mathcal{V})$ *be a pseudorandom POWHF with public randomness (with respect to the uniform distribution and the trivial uninvertible function). Assume further that the symmetric encryption scheme provides integrity of ciphertexts and that the POWHF-encryption assumption holds for* POWHF *and* AS. *Then the instantiation of the H-oracle in the Fujisaki-Okamoto transformation, $FO^{G,POWHF}$, yields an IND-CCA encryption scheme in the random oracle model.*

6   (In)Security of FDH Signature Scheme Instantiations

In this section we consider the Full Domain Hash (FDH) signature scheme which is provably secure in the random oracle model if the associated permutation is one-way. We show that replacing the random oracle by a verifiable pseudorandom function does not necessarily yield a secure instantiation. For sake of concreteness we explain our negative result for the RSA case. The result can be transferred, mutatis mutandis, to other trapdoor permutations.

We note that one can easily transfer our negative result about OAEP (Theorems 1 and 2) to show that the FDH instantiated with a POWHF is insecure with respect to a specific trapdoor permutation oracle. But our result here for the VPRFs works for any trapdoor permutation, including RSA for example.

FULL DOMAIN HASH SIGNATURE SCHEME AND INSTANTIATION WITH VPRFS. Due to lack of space we omit the formal description of the well-known Full-domain hash (FDH) signature scheme [4] Basically, a signature S for a message M is given as $S = f^{-1}(H(M))$ and verification requires checking $f(S) = H(M)$. An instantiation of the FDH scheme with VPRF $= (\mathcal{K}, \mathcal{H}, \mathcal{V})$ is the following signature scheme $FDH^{VPRF}[F] = (\mathcal{SK}, \mathcal{S}, \mathcal{V})$:

- $\mathcal{SK}(1^k)$: pick a random permutation f on D_k from F, pick $(fk, vk) \xleftarrow{\$} \mathcal{K}(1^k)$. Let pk specify f and contain vk and let sk specify f^{-1} and contain vk.

- $\mathcal{S}^{\mathcal{H}(fk,\cdot)}(sk, M)$: $(y, \pi) \xleftarrow{\$} \mathcal{H}(fk, M)$, $S \leftarrow f^{-1}(y)$. Output (S, π, y).
- $\mathcal{V}^{\mathcal{V}(fk,\cdot)}(pk, M, (S, \pi))$: If $f(S) = y$ and $\mathcal{V}(vk, M, y, \pi) = 1$ then return 1, else return 0

It is important to note that in the attack the adversary is only given access to the signature oracle but not to the VPRF oracle. Although the application as a third-party web interface providing such values indicate that the adversary can get additional VPRF values, our result even holds in the setting where the adversary is denied such values.

ON THE INSECURITY OF RSA-FDH WITH VPRFs. A special case is the RSA-FDH signature scheme (and its instantiation through a VPRF) where f, f^{-1} are given by the RSA function $x \mapsto x^e \bmod N$ and its inverse $y \mapsto y^d \bmod N$. Here we consider the case *with large prime exponents* where the RSA exponent e has to be a prime of $(k+1)$ bits and therefore larger than the k-bit modulus N. We denote this function by $\text{RSA}_{\text{large-exponent}}$. According to the recent result about deterministic primality testing [1], this prerequisite allows to verify deterministically that a pair (N, e) really constitutes a permutation. We also remark that this RSA version is not known to be weaker than RSA with other exponents.

For the RSA-FDH scheme we construct a "bad" VPRF such that, when instantiated with this VPRF, RSA-FDH becomes insecure:

Theorem 6. *Suppose VPRFs exist. Then there exists a verifiable pseudorandom function* VPRF $= (\mathcal{K}, \mathcal{H}, \mathcal{V})$ *such that* $FDH^{\text{VPRF}}[RSA_{\text{large-exponent}}]$ *is subject to existential forgeries in chosen-message attacks.*

The basic idea is that the "bad" VPRF (which exists if any VPRF exists) itself will reveal signatures for free as part of the correctness proof. Thus, giving the signature oracle the right message will force the signer to query the VPRF at the right input which, in turn, allows to forge signatures. We prove this formally in [8] .

Acknowledgements

We thank Victor Shoup for clarifications on [26] and the anonymous reviewers of Crypto 2005 for useful comments.

References

1. M. Agrawal, N. Kayal, and N. Saxena. Primes is in P. http://www.cse.iitk.ac.in/news/primality.html, 2002.
2. M. Bellare, A. Boldyreva, and A. Palacio. An uninstantiable random-oracle-model scheme for a hybrid-encryption problem. In *Eurocrypt 2004*, volume 3027 of *LNCS*. Springer, 2004.
3. M. Bellare and C. Namprempre. Authenticated encryption: Relations among notions and analysis of the generic composition paradigm. In *ASIACRYPT 2000*, volume 1976 of *LNCS*. Springer, 2000.

4. M. Bellare and P. Rogaway. Random oracles are practical: a paradigm for designing efficient protocols. In *CCS '93*. ACM, 1993.
5. M. Bellare and P. Rogaway. Optimal asymmetric encryption – how to encrypt with RSA. In *Eurocrypt '94*, volume 950, 1995.
6. M. Bellare and P. Rogaway. The exact security of digital signatures: How to sign with RSA and Rabin. In *Eurocrypt '96*, volume 1070 of *LNCS*. Springer, 1996.
7. M. Blum and S. Micali. How to generate cryptographically strong sequences of pseudorandom bits. *SIAM Journal of Computing*, 13:850–864, 1984.
8. A. Boldyreva and M. Fischlin. Analysis of random-oracle instantiation scenarios for OAEP and other practical schemes. Full version of this paper. Available at http://www.cc.gatech.edu/~aboldyre/publications.html.
9. R. Canetti. Towards realizing random oracles: Hash functions that hide all partial information. In *CRYPTO '97*, volume 1294 of *LNCS*. Springer, 1997.
10. R. Canetti, O. Goldreich, and S. Halevi. The random oracle methodology, revisited. In *STOC '98*. ACM, 1998.
11. D. Chaum. Blind signatures for untraceable payments. In *CRYPTO '82*, 1983.
12. J.-S. Coron, M. Joye, D. Naccache, and .P. Paillier. Universal padding schemes for RSA. In *CRYPTO 2002*, volume 2442. Springer, 2002.
13. Y. Dodis. Efficient construction of (distributed) verifiable random functions. In *PKC 2003*, volume 2567 of *LNCS*. Springer, 2003.
14. Y. Dodis, R. Oliveira, and K. Pietrzak. On the generic insecurity of full-domain hash. In *CRYPTO 2005*, LNCS, 2005.
15. A. Fiat and A. Shamir. How to prove yourself: Practical solutions to identification and signature schemes. In *Crypto '86*, volume 263 of *LNCS*. Springer, 1986.
16. M. Fischlin. Pseudorandom function tribe ensembles based on one-way permutations: Improvements and applications. In *Eurocrypt '99*, volume 1592 of *LNCS*. Springer, 1999.
17. E. Fujisaki and T. Okamoto. Secure integration of asymmetric and symmetric encryption schemes. In *CRYPTO '99*, volume 1666 of *LNCS*, 1999.
18. E. Fujisaki, T. Okamoto, D. Pointcheval, and J. Stern. RSA-OAEP is secure under the RSA assumption. In *CRYPTO 2001*, volume 2139 of *LNCS*. Springer, 2001.
19. S. Goldwasser and Y. T. Kalai. On the (in)security of the Fiat-Shamir paradigm. In *FOCS 2003*. IEEE, 2003.
20. K. Kobara and H. Imai. OAEP++: A very simple way to apply OAEP to deterministic ow-cpa primitives. *Cryptology ePrint Archive, Report 2002/130.*, 2002.
21. U. Maurer, R. Renner, and C. Holenstein. Indifferentiability, impossibility results on reductions, and applications to the random oracle methodology. In *TCC 2004*, volume 2951 of *LNCS*. Springer, 2004.
22. S. Micali, M. Rabin, and S. Vadhan. Verifiable random functions. In *FOCS 1999*. IEEE, 1999.
23. J. Nielsen. Separating random oracle proofs from complexity theoretic proofs: The non-committing encryption case. In *CRYPTO 2002*, volume 2442 of *LNCS*. Springer, 2002.
24. D. Micciancio R. Canetti and O. Reingold. Perfectly one-way probabilistic hash functions. In *STOC '98*. ACM, 1998.
25. V. Shoup. Lower bounds for discrete logarithms and related problems. In *Eurocrypt '97*, volume 1233 of *LNCS*. Springer, 1997.
26. V. Shoup. OAEP reconsidered. In *CRYPTO 2001*, volume 2139 of *LNCS*. Springer, 2001.
27. A. Yao. Theory and applications of trapdoor functions. In *FOCS 1982*, pages 80–91. IEEE, 1982.

Merkle-Damgård Revisited: How to Construct a Hash Function

Jean-Sébastien Coron[1], Yevgeniy Dodis[2,*], and Cécile Malinaud[3], and Prashant Puniya[2,**]

[1] University of Luxembourg
coron@clipper.ens.fr
[2] New-York University
{dodis, puniya}@cs.nyu.edu
[3] Gemplus Card International
cecile.malinaud@normalesup.org

Abstract. The most common way of constructing a hash function (*e.g.*, SHA-1) is to iterate a compression function on the input message. The compression function is usually designed from scratch or made out of a block-cipher. In this paper, we introduce a new security notion for hash-functions, stronger than collision-resistance. Under this notion, the arbitrary length hash function H must behave as a random oracle when the fixed-length building block is viewed as a random oracle or an ideal block-cipher. The key property is that if a particular construction meets this definition, then any cryptosystem proven secure assuming H is a random oracle remains secure if one plugs in this construction (still assuming that the underlying fixed-length primitive is ideal). In this paper, we show that the current design principle behind hash functions such as SHA-1 and MD5 — the (strengthened) Merkle-Damgård transformation — does not satisfy this security notion. We provide several constructions that provably satisfy this notion; those new constructions introduce minimal changes to the plain Merkle-Damgård construction and are easily implementable in practice.

1 Introduction

RANDOM ORACLE METHODOLOGY. The random oracle model has been introduced by Bellare and Rogaway as a "paradigm for designing efficient protocols" [4]. It assumes that all parties, including the adversary, have access to a public, truly random hash function H. This model has been proven extremely useful for designing simple, efficient and highly practical solutions for many problems. From a theoretical perspective, it is clear that a security proof in the random oracle model is only a heuristic indication of the security of the system when instantiated with a particular hash function, such as SHA-1 [16] or MD5 [18].

* Supported by NSF CAREER Award CCR-0133806 and TC Grant No. CCR-0311095.
** Supported by NSF Cybertrust/DARPA Grant No. CNS-0430425.

V. Shoup (Ed.): Crypto 2005, LNCS 3621, pp. 430–448, 2005.

In fact, many recent "separation" results [11,26,19,2,12,15] illustrated various cryptographic systems secure in the random oracle model but completely insecure for *any* concrete instantiation of the random oracle (even by a *family* of hash functions). Nevertheless, these important separation results do not seem to directly attack any of the concrete, widely used cryptosystems (such as OAEP [6] and PSS [5] as used in the PKCS #1 v2.1 standard [27]) which rely on "secure hash functions". Moreover, we hope that such particular systems are in fact *secure when instantiated with a "good" hash function*. In the random oracle model, instead of making a highly non-standard (and possibly unsubstantiated) assumption that "my system is secure with this H" (e.g., H being SHA-1), one proves that the system is at least secure with an "ideal" hash function H (under standard assumptions). Such formal proof in the random oracle model is believed to indicate that there are no structural flaws in the design of the system, and thus one can heuristically hope that no such flaws will suddenly appear with a particular, "well designed" function H. *But can we say anything about the lack of structural flaws in the design of H itself?*

BUILDING RANDOM ORACLES.   On the first glance, it appears that nothing theoretically meaningful can be said about this question. Namely, we know that mathematically a concrete function H is not a random oracle, so to prove that H is "good" we need to directly argue the security of our system with this given H. And the latter task is usually unmanageable given our current tools (e.g., "realizable" properties of H such as collision-resistance, pseudorandomness or one-wayness are usually not enough to prove the security of the system). However, we argue that there is a significant gap in this reasoning. Indeed, most systems abstractly model H as a function from $\{0,1\}^*$ to $\{0,1\}^n$ (where n is proportional to the security parameter), so that H can be used on some arbitrary input domain. On the other hand, in practice such *arbitrary-length* hash functions are built by first heuristically constructing a *fixed-length* building block, such as a fixed-length compression function or a block cipher, and then iterating this building block in some manner to extend the input domain arbitrarily. For example, SHA-1, MD5, as well as all the other hash function we know of, are constructed by applying some variant of the Merkle-Damgård construction to an underlying compression function $f : \{0,1\}^{n+\kappa} \to \{0,1\}^n$ (see Figure 5):

Function $H(m_1, \ldots, m_\ell)$:
   let $y_0 = 0^n$ (more generally, some fixed IV value can be used)
   for $i = 1$ to ℓ do $y_i \leftarrow f(y_{i-1}, m_i)$
   return y_ℓ

When the number of κ-bit message blocks ℓ is not fixed, one essentially appends an extra block $m_{\ell+1}$ containing the binary representation $\langle |m| \rangle$ of the length of the message (prepended by 1 and a string of 0's in order to make everything a multiple of κ; the exact details will not matter for our discussion). The fixed-length compression function f can either be constructed from scratch or made out of a block-cipher E via the Davies-Meyer construction (see [31] and Figure 9): $f(x,y) = E_y(x) \oplus x$. For example, the SHA-1 compression function was designed specifically for hashing, but a block-cipher can nevertheless be derived from it, as illustrated in [20].

OUR MAIN QUESTION. Given such particular and "structured" design of our hash function H,— which is actually the design used in practice,— we argue that there exists a missing link in the claim that no structural flaws exist in the design of our system. Indeed, we only know that no such flaws exist when H was modeled as a "monolithic" random oracle, and not as an iterated hash function built from some smaller building block. As since the real implementation of H as an iterated hash function has much more structure than a random monolithic hash function would have, maybe this structure could somehow invalidate the security proof in the random oracle model? To put this into a different perspective, all the ad-hoc (and hopefully "secure") design effort for widely used hash functions, such as SHA-1 and MD5, has been placed into the design of the fixed-length building block f (or E). On the other hand, even if f (or E) were assumed to be ideal, the current proofs in the random oracle model do not guarantee the security of the resulting system when such iterated hash function H is used!

Let us illustrate our point on a well known example. A common suggestion to construct a MAC algorithm is to simply include a secret key k as part of the input of the hash function, and take for example $\text{MAC}(k, m) = H(k \| m)$. It is easy to see that this construction is secure when H is modeled as a random oracle [4], as no adversary can output a MAC forgery except with negligible probability. However, this MAC scheme is completely insecure for any Merkle-Damgård construction considered so far (including Merkle-Damgård strengthening used in current hash functions such as SHA-1, and any of the 64 block-cipher based variants of iterative hash-functions considered in [29,9]), no matter which (ideal) compression function f (or a block cipher E) is used. Namely, given $\text{MAC}(k, m) = H(k \| m)$, one can extend the message m with any single arbitrary block y and deduce $\text{MAC}(k, m \| y) = H(k \| m \| y)$ without knowing the secret key k (even with Merkle-Damgård strengthening, one could still forge the MAC by more or less setting $y = \langle |m| \rangle$, where the actual block depends on the exact details of the strengthening). This (well known) example illustrates that the construction of a MAC from an iterated hash function requires a specific analysis, and cannot be derived from the security of this MAC with a monolithic hash function H. On the other hand, while the Merkle-Damgård transformation and its variants have been intensively studied for many "realizable" properties such as collision-resistance [13,25,29,9], pseudorandomness [8], unforgeability [1,24] and randomness extraction [14], it is clear that these analyses are insufficient to argue its applicability for the purposes of building a hash function which can be modeled as a random oracle, since the latter is a considerably stronger security notion (in fact unrealizable in the standard model). For a simple concrete example, the Merkle-Damgård strengthening is easily seen to preserve collision-resistance when instantiated with a collision-resistant compression function, while we just saw that it does not work to yield a random oracle or even just a variable-length MAC, and this holds even if the underlying compression function is modeled as a random oracle.

OUR GOALS. Summarizing the above discussion, our goal is two-fold. First, we would like to give a formal *definition* of what it means to implement an arbitrary-length random oracle H from a fixed-length building block f or E. The key property of this definition should be the fact that if a particular construction of H from f (or E) meets this definition, then *any application* proven secure assuming H is a random oracle would remain secure if we plug in our construction (although still assuming that the underlying fixed-length primitive f or E was ideal). In other words, we can safely use our implementation of H as if we were using a monolithic random oracle H. We remark that this means that our definition should not just preserve the pseudorandomness properties of H, but also all the other "tricks" present in the random oracle model, such as "programmability" and "extractability". For example, we could try to set $H(x) = f(h(x))$, where f is a fixed-length random oracle and h is a collision-resistant hash function (not viewed as a random oracle). While pseudorandom, this simple implementation is clearly not "extractable": for example, given output $z = f(h(x))$ for some unknown x, we can only "extract" the value $h(x)$ (by observing the random oracle queries made to f), but then have no way of extracting x itself from $h(x)$ (indeed, we will show a direct attack on this implementation in Section 3.1). This shows that the security definition we need is an interesting and non-trivial task of its own, especially if we also want it to be simple, natural and easy to use.

Second, while the definition we seek should not be too specific to some variant of the Merkle-Damgård transformation, we would like to give secure constructions which resemble what is done in practice as much as possible. Unfortunately, we already argued that the current design principle behind hash functions such as SHA-1 and MD5 — the (strengthened) Merkle-Damgård transformation — will *not* be secure for our ambitious goal. Therefore, instead of giving new and practically unmotivated constructions, our secondary goal is to come up with *minimal* and *easily implementable in practice* changes to the plain Merkle-Damgård construction, which would satisfy our security definition.

OUR RESULTS. First, we give a satisfactory definition of what it means to implement an arbitrary-length random oracle H from a fixed length primitive g (where g is either an ideal compression function f, or a an ideal block cipher E). Our definition is based on the indifferentiability framework of Maurer et al. [23]. This framework enjoys the desired closure property we seek, and is very intuitive and easy to state.

Having a good security definition, we provide several provable constructions. We start by giving three modifications to the (insecure) plain Merkle-Damgård construction which yield a secure random oracle H taking *arbitrary-length* input, from a compression function viewed as a random oracle taking *fixed-length* input. This result can be viewed as a secure *domain extender* for the random oracle, which is an interesting result of independent interest. We remark that domain extenders are well studied for such primitives as collision-resistant hash functions [13,25], pseudorandom functions [8], MACs [1,24] and universal one-way hash functions [7,30]. Although the above works also showed that some variants of Merkle-Damgård yield secure domain extenders for the corresponding primitive in question, these results are not sufficient to claim a domain extender for the random oracle.

Our secure modifications to the plain Merkle-Damgård construction are the following. (1) *Prefix-Free Encoding* : we show that if the inputs to the plain MD construction are guaranteed to be *prefix-free*, then the plain MD construction is secure. (2) *Dropping Some Output Bits* : we show that by dropping a non-trivial number of output bits from the plain MD chaining, we get a secure random oracle H even if the input is not encoded in the prefix-free manner. (3) *Using NMAC construction* (see Figure 8a): we show that by applying an independent hash function g to the output of the plain MD chaining (as in the NMAC construction [8]), then once again we get a secure construction of an arbitrary-length random oracle H, in the random oracle model for f and g. (4) *Using HMAC Construction* (see Figure 8b): we show a slightly modified variant of the NMAC construction allowing us to conveniently build the function g from the compression function f itself (as in [8] when going from NMAC to HMAC)! In this latter variant, one implements a secure hash function H by making two *black-box calls to the plain Merkle-Damgård construction* (with the same fixed IV and a given compression function f): first on $(\ell+1)$-block input $0^\kappa m_1 \ldots m_\ell$, getting an n-bit output y, and then on one-block κ-bit input y' (obtained by either truncating or padding y depending on whether or not $\kappa > n$), getting the final output.

However, in practice most hash-function constructions are block-cipher based, either explicitly as in [29] or implicitly as for SHA-1. Therefore, we consider the question of designing an arbitrary-length random oracle H from an ideal block cipher E, specifically concentrating on using the Merkle-Damgård construction with the Davies-Meyer compression function $f(x, y) = E_y(x) \oplus x$, since this is the most practically relevant construction. We show that all of the four fixes to the plain MD chaining which worked when f was a fixed-length random oracle, are still secure (in the ideal cipher model) when we plug in $f(x, y) = E_y(x) \oplus x$ instead. Specifically, we can either use a prefix-free encoding, or drop a non-trivial number of output bits (when possible), or apply an independent random oracle g to the output of plain MD chaining, or use the optimized HMAC construction which allows us to build this function g from the ideal cipher itself.

2   Definitions

In this section, we introduce the main notations and definitions used throughout the paper. Our security notion for secure hash-function is based on the notion of indifferentiability of systems, introduced by Maurer *et al.* in [23]. This is an extension of the classical notion of indistinguishability, when one or more oracles are publicly available, such as random oracles or ideal ciphers. This notion is based on ideas from the Universal Composition framework introduced by Canetti in [10] and on the model of Pfitzmann and Waidner [28]. The indifferentiability notion in [23] is given in the framework of random systems providing interfaces to other systems, but equivalently we use this notion in the framework of Interactive Turing Machines (as in [10]).

We define an *ideal primitive* as an algorithmic entity which receives inputs from one of the parties and deliver its output immediately to the querying party. The ideal primitives that we consider in this paper are random oracles and ideal ciphers. A *random oracle* [4] is an ideal primitive which provides a random output for each new query. Identical input queries are given the same answer. An *ideal cipher* is an ideal primitive that models a random block-cipher E : $\{0,1\}^\kappa \times \{0,1\}^n \to \{0,1\}^n$. Each key $k \in \{0,1\}^\kappa$ defines a random permutation $E_k = E(k,\cdot)$ on $\{0,1\}^n$. The ideal primitive provides oracle access to E and E^{-1}; that is, on query $(0,k,m)$, the primitive answers $c = E_k(m)$, and on query $(1,k,c)$, the primitive answers m such that $c = E_k(m)$.

We now proceed to the definition of indifferentiability [23] :

Definition 1. *A Turing machine C with oracle access to an ideal primitive $\mathcal{G}$ is said to be $(t_D, t_S, q, \varepsilon)$ indifferentiable from an ideal primitive $\mathcal{F}$ if there exists a simulator S, such that for any distinguisher D it holds that :*

$$\left| \Pr\left[D^{C,\mathcal{G}} = 1 \right] - \Pr\left[D^{\mathcal{F},S} = 1 \right] \right| < \varepsilon$$

The simulator has oracle access to $\mathcal{F}$ and runs in time at most t_S. The distinguisher runs in time at most t_D and makes at most q queries. Similarly, $C^{\mathcal{G}}$ is said to be (computationally) indifferentiable from $\mathcal{F}$ if ε is a negligible function of the security parameter k (for polynomially bounded t_D and t_S).

As illustrated in Figure 1, the role of the simulator is to simulate the ideal primitive $\mathcal{G}$ so that no distinguisher can tell whether it is interacting with C and $\mathcal{G}$, or with $\mathcal{F}$ and S; in other words, the output of S should look "consistent" with what the distinguisher can obtain from $\mathcal{F}$. Note that the simulator does not see the distinguisher's queries to $\mathcal{F}$; however, it can call $\mathcal{F}$ directly when needed for the simulation.

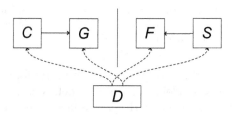

Fig. 1. The indifferentiability notion: the distinguisher D either interacts with algorithm C and ideal primitive $\mathcal{G}$, or with ideal primitive $\mathcal{F}$ and simulator S. Algorithm C has oracle access to $\mathcal{G}$, while simulator S has oracle access to $\mathcal{F}$

In the rest of the paper, the algorithm C will represent the construction of an iterative hash-function (such as the Merkle-Damgård construction recalled in the introduction). The ideal primitive $\mathcal{G}$ will represent the underlying primitive used to build the hash-function. $\mathcal{G}$ will be either a random oracle (when the compression function is modelled as a random oracle), or an ideal block-cipher (when the compression function is based on a block-cipher). The ideal primitive $\mathcal{F}$ will represent the random oracle that the construction C should emulate. Therefore, one obtains the following setting : the distinguisher has oracle access to both the block-cipher and the hash-function, and these oracles are implemented in one of the following two ways: either the block-cipher E is chosen at random and the hash-function C is constructed from it, or the hash-function H is chosen at random and the block-cipher is implemented by a simulator S with oracle access to H. Those two cases should be indistinguishable, that is the distinguisher should not be able to tell whether the block-cipher was chosen at random and the iterated hash-function constructed from it, or the hash-function was chosen at random and the block-cipher then "tailored" to match that hash-function.

It is shown in [23] that if $C^{\mathcal{G}}$ is indifferentiable from $\mathcal{F}$, then $C^{\mathcal{G}}$ can replace $\mathcal{F}$ in any cryptosystem, and the resulting cryptosystem is at least as secure in the $\mathcal{G}$ model as in the $\mathcal{F}$ model. For example, if a block-cipher based iterative hash function is indifferentiable from a random oracle in the ideal cipher model, then the iterative hash-function can replace the random oracle in any cryptosystem, and the resulting cryptosystem remains secure in the ideal cipher model if the original scheme was secure in the random oracle model.

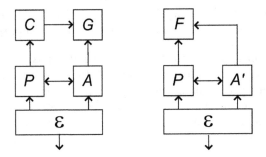

Fig. 2. The environment $\mathcal{E}$ interacts with cryptosystem $\mathcal{P}$ and attacker $\mathcal{A}$. In the $\mathcal{G}$ model (left), $\mathcal{P}$ has oracle access to C whereas $\mathcal{A}$ has oracle access to $\mathcal{G}$. In the $\mathcal{F}$ model, both $\mathcal{P}$ and $\mathcal{A}'$ have oracle access to $\mathcal{F}$

We use the definition of [23] to specify what it means for a cryptosystem to be at least as secure in the $\mathcal{G}$ model as in the $\mathcal{F}$ model. A cryptosystem is modelled as an Interactive Turing Machine with an interface to an adversary $\mathcal{A}$ and to a public oracle. The cryptosystem is run by an *environment* $\mathcal{E}$ which provides a binary output and also runs the adversary. In the $\mathcal{G}$ model, cryptosystem $\mathcal{P}$ has oracle access to C whereas attacker $\mathcal{A}$ has oracle access to $\mathcal{G}$. In the $\mathcal{F}$ model, both $\mathcal{P}$ and $\mathcal{A}$ have oracle access to $\mathcal{F}$. The definition is illustrated in Figure 2.

Definition 2. *A cryptosystem is said to be at least as secure in the $\mathcal{G}$ model with algorithm C as in the $\mathcal{F}$ model, if for any environment $\mathcal{E}$ and any attacker $\mathcal{A}$ in the $\mathcal{G}$ model, there exists an attacker $\mathcal{A}'$ in the $\mathcal{F}$ model, such that*

$$\left| \Pr\left[\mathcal{E}(\mathcal{P}^{C}, \mathcal{A}^{\mathcal{G}}) = 1 \right] - \Pr\left[\mathcal{E}(\mathcal{P}^{\mathcal{F}}, \mathcal{A}'^{\mathcal{F}}) = 1 \right] \right|$$

is a negligible function of the security parameter k. Similarly, a cryptosystem is said to be computationally at least as secure, etc., if $\mathcal{E}$, A and A' are polynomial-time in k.

The following theorem from [23] shows that security is preserved when replacing an ideal primitive by an indifferentiable one :

Theorem 1. *Let $\mathcal{P}$ be a cryptosystem with oracle access to an ideal primitive $\mathcal{F}$. Let C be an algorithm such that $C^{\mathcal{G}}$ is indifferentiable from $\mathcal{F}$. Then cryptosystem $\mathcal{P}$ is at least as secure in the $\mathcal{G}$ model with algorithm C as in the $\mathcal{F}$ model.*

Proof. We only provide a proof sketch; see [23] for a full proof. Let $\mathcal{P}$ be any cryptosystem, modelled as an Interactive Turing Machine. Let $\mathcal{E}$ be any environment, and A be any attacker in the $\mathcal{G}$ model. In the $\mathcal{G}$ model, $\mathcal{P}$ has oracle access to C whereas A has oracle access to ideal primitive $\mathcal{G}$; moreover environment $\mathcal{E}$ interacts with both $\mathcal{P}$ and A. This is illustrated in Figure 3 (left part).

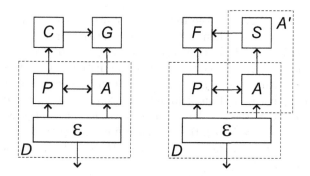

Fig. 3. Construction of attacker A' from attacker A and simulator S

Since $C^{\mathcal{G}}$ is indifferentiable from $\mathcal{F}$ (see Figure 1), one can replace $(C, \mathcal{G})$ by $(\mathcal{F}, S)$ with only a negligible modification of the environment's output distribution. As illustrated in Figure 3, by merging attacker A and simulator S, one obtains an attacker A' in the $\mathcal{F}$ model, and the difference in $\mathcal{E}$'s output distribution is negligible. □

3   Domain Extension for Random Oracles

In this section, we show how to construct an iterative hash-function indifferentiable from a random oracle, from a compression function viewed as a random oracle. We start with two simple and intuitive constructions that do not work.

3.1 $H(x) = f(h(x))$ for Random Oracle f and Collision-Resistant One-Way Hash-Function h

One could hope to emulate a random oracle (with arbitrary-length input) by taking :

$$C^f(x) = f(h(x))$$

where $f : \{0,1\}^n \to \{0,1\}^n$ is modelled as a random oracle and $h : \{0,1\}^* \to \{0,1\}^n$ is any collision-resistant one-way hash-function (not modelled as a random oracle). However, we show that such C^f is not indifferentiable from a random oracle; namely, we construct a distinguisher that can fool any simulator.

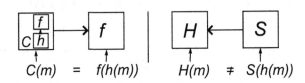

Fig. 4. The simulator cannot output $H(m)$ since it only receives $h(m)$ and cannot recover m from $h(m)$

As illustrated in Figure 4, the distinguisher first generates an arbitrary m and computes $u = h(m)$. Then it queries $v = f(u)$ to random oracle f and queries $z = C^f(m)$ to C^f. It then checks that $z = v$ and outputs 1 in this case, and 0 otherwise. It is easy to see that the distinguisher always output 1 when interacting with C^f and f, but outputs 0 with overwhelming probability when interacting with H and any simulator S. Namely, when the distinguisher interacts with H and S, the simulator only receives $u = h(m)$; therefore, in order to output v such that $v = H(m)$, the simulator must either recover m from $h(m)$ (and then query $H(m)$) or guess the value of $H(m)$, which can be done with only negligible probability.

3.2   Plain Merkle-Damgård Construction

We show that the plain Merkle-Damgård construction (see Figure 5) fails to emulate a random oracle (taking arbitrary-length input) when the compression function f is viewed as a random oracle (taking fixed-length input). For simplicity, we only consider the usual Merkle-Damgård variant, although the discussion easily extends to the strengthened variant which appends the message length $\langle |m| \rangle$ at the last block :

> **Function** $MD^f(m_1, \ldots, m_\ell)$:
>   let $y_0 = 0^n$ (more generally, some fixed IV value can be used)
>   for $i = 1$ to ℓ do $y_i \leftarrow f(y_{i-1}, m_i)$
>   return $y_\ell \in \{0,1\}^n$.

where for all i, $|m_i| = \kappa$ and $f : \{0,1\}^{n+\kappa} \to \{0,1\}^n$.

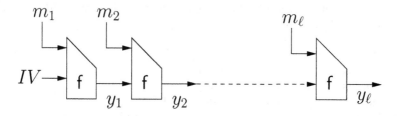

Fig. 5. The plain Merkle-Damgård Construction

We have already mentioned in introduction a counter-example based on MAC. Namely, we showed that $MAC(k, m) = H(k\|m)$ provides a secure MAC in the random oracle model for H, but is completely insecure when H is replaced by the previous Merkle-Damgård construction MD^f, because of the message extension attack. In the following, we give a more direct refutation based on the definition of indifferentiability, using again the message extension attack.

We consider only one-block messages or two-block messages. For such messages, we have that $MD^f(m_1) = f(0, m_1)$ and $MD^f(m_1, m_2) = f(f(0, m_1), m_2)$. We build a distinguisher that can fool any simulator as follows. The distinguisher first makes a MD^f-query for m_1 and receives $u = MD^f(m_1)$. Then it makes a query for $v = f(u, m_2)$ to random oracle f. The distinguisher then makes a MD^f-query for (m_1, m_2) and eventually checks that $v = MD^f(m_1, m_2)$; in this case it outputs 1, and 0 otherwise. It is easy to see that the distinguisher always outputs 1 when interacting with MD^f and f. However, when the distinguisher interacts with H and S (who must simulate f), we observe that S has no information about m_1 (because S does not see the distinguisher's H-queries). Therefore, the simulator cannot answer v such that $v = H(m_1, m_2)$, except with negligible probability.

3.3   Prefix-Free Merkle-Damgård

In this section, we show that if the inputs to the plain MD construction are guaranteed to be prefix-free, then the plain MD construction is secure. Namely, prefix-free encoding enables to eliminate the message expansion attack described previously. This "fix" is similar to the fix for the CBC-MAC [3], which is also insecure in its plain form. Thus, the plain MD construction can be safely used for any application of the random oracle H where the length of the inputs is fixed or where one uses domain separation (e.g., prepending $0, 1, \ldots$ to differentiate between inputs from different domains). For other applications, one must specifically ensure that prefix-freeness is satisfied.

A prefix-free code over the alphabet $\{0, 1\}^\kappa$ is an efficiently computable injective function $g : \{0, 1\}^* \to (\{0, 1\}^\kappa)^*$ such that for all $x \neq y$, $g(x)$ is not a prefix of $g(y)$. Moreover, it must be easy to recover x given only $g(x)$. We provide two examples of prefix-free encodings. The first one consists in prepending the message size in bits as the first block. The last block is then padded with the bit one followed by zeroes.

Function $g_1(m)$:
    let N be the message length of m in bits.
    write m as $(m_1, \ldots, m_\ell)$ where for all i, $|m_i| = \kappa$
        and with the last block m_ℓ padded with 10^r.
    let $g_1(m) = (\langle N \rangle, m_1, \ldots, m_\ell)$ where $\langle N \rangle$ is a κ-bit binary encoding of N.

An important drawback of this encoding is that the message length must be known in advance; this can be a problem for streaming applications in which a large message must be processed on the fly. Our second encoding g_2 does not suffer from this drawback, but requires to waste one bit per block of the message :

Function $g_2(m)$:
    write m as $(m_1, \ldots, m_\ell)$ where for all i, $|m_i| = \kappa - 1$
        and with the last block m_ℓ padded with 10^r.
    let $g_2(m) = (0|m_1, \ldots, 0|m_{\ell-1}, 1|m_\ell)$.

Given any prefix-free encoding g, we consider the following construction of the iterative hash-function pf-MD$_g^f$: $\{0,1\}^* \rightarrow \{0,1\}^n$, using the Merkle-Damgård hash-function MD^f : $(\{0,1\}^\kappa)^* \rightarrow \{0,1\}^n$ defined previously.

Function pf-MD$_g^f(m)$:
    let $g(m) = (m_1, \ldots, m_\ell)$
    $y \leftarrow \text{MD}^f(m_1, \ldots, m_\ell)$
    return y

Theorem 2. *The previous construction is (t_D, t_S, q, ϵ)-indifferentiable from a random oracle, in the random oracle model for the compression function, for any t_D, with $t_S = \ell \cdot \mathcal{O}(q^2)$ and $\epsilon = 2^{-n} \cdot \ell^2 \cdot \mathcal{O}(q^2)$, where ℓ is the maximum length of a query made by the distinguisher D.*

Proof. Due to lack of space, we only provide a proof sketch for a particular prefix-free encoding which has a simpler proof; the proof for any prefix-free encoding will be provided in the full version of this paper.

The particular prefix-free encoding that we consider consists in adding the message-length as part of the input of f; moreover, the index of the current block is also included as part of the input of f, so that f can be viewed as an independent random oracle for each block m_i. Specifically, we construct an iterative hash-function C^f : $(\{0,1\}^\kappa)^* \rightarrow \{0,1\}^n$ from a compression function $f : \{0,1\}^{n+\kappa+2 \cdot t} \rightarrow \{0,1\}^n$ as follows :

Function $C^f(m_1, \ldots, m_\ell)$:
    let $y_0 = 0^n$
    for $i = 1$ to ℓ do $y_i \leftarrow f(y_{i-1}, m_i, \langle \ell \rangle, \langle i \rangle)$
    return y_ℓ

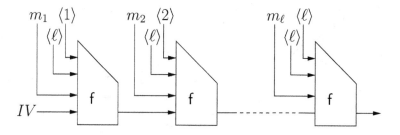

Fig. 6. Merkle-Damgård with a particular prefix-free encoding

where for all i, $|m_i| = \kappa$. The string $\langle \ell \rangle$ is a t-bit binary encoding of the message length ℓ, and $\langle i \rangle$ is a t-bit encoding of the block index. The construction is shown in Figure 6.

In the following, we show that C^f is indifferentiable from a random oracle, in the random oracle model for f. Since the block-length ℓ is part of the input of the compression function f, we have that C^f behaves independently for messages of different length. Therefore, we can restrict ourselves to messages of fixed length ℓ, i.e. it suffices to show that for all ℓ, the construction C^f with message length ℓ is indifferentiable from random oracle $H_\ell : (\{0,1\}^\kappa)^\ell \to \{0,1\}^n$.

We consider for all $1 \leq j \leq \ell$ the function $C_j^f : (\{0,1\}^\kappa)^j \to \{0,1\}^n$ outputting the intermediate value y_j in C^f. From the definition of C^f, we have for all $2 \leq j \leq \ell$:

$$C_j^f(m_1, \ldots, m_j) = f(C_{j-1}^f(m_1, \ldots, m_{j-1}), m_j, \langle \ell \rangle, \langle j \rangle) \qquad (1)$$

We provide a recursive proof that for all j, the construction C_j^f is indifferentiable from a random oracle. The result for C^f will follow for $j = \ell$. The property clearly holds for $j = 1$. Assuming now that it holds for $j - 1$, we show that it holds for j. We use the following lemma :

Lemma 1. *Let $h_1 : \{0,1\}^a \to \{0,1\}^n$ and $h_2 : \{0,1\}^{n+\kappa} \to \{0,1\}^n$. The construction $R^{h_1,h_2} = h_2(h_1(x), y)$ is indifferentiable from a random oracle, in the random oracle model for h_1 and h_2.*

Replacing C_{j-1}^f by h_1 and $f(\cdot, \langle \ell \rangle, \langle j \rangle)$ by h_2 in equation (1), one then obtains that C_j^f is indifferentiable from a random oracle (see Figure 7 for an illustration).

We now proceed to the proof of lemma 1; due to lack of space, we only provide a proof sketch. One must construct a simulator S such that interacting with $(R, (h_1, h_2))$ is indistinguishable from interacting with (H, S), where H is a random oracle. Our simulator is defined as follows:

Simulator S :
On h_1-query x, return a random $v \in \{0,1\}^n$.
On h_2-query (v', y), check if $v' = h_2(x')$ for some previously queried x'.
   In this case, query (x', y) to H and output $H(x', y)$.
   Otherwise return a random output.

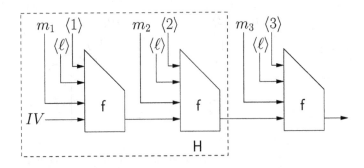

Fig. 7. The output of first two blocks is replaced by a random oracle using Lemma 1

The distinguisher either interacts with $(R, (h_1, h_2))$ or with (H, S). We denote by F the event that a collision occurs for h_1, that is $h_1(x) = h_1(x')$ for some distinct queries x, x'. We denote by F' the event that the distinguisher makes a h_2-query (v', y) such that $v' = h_1(x)$ and (x, y) was previously queried to R, but x was never queried directly to h_1 by the distinguisher. We claim that conditioned on the complement of $F \vee F'$, the simulation of S is perfect (see the full paper for a complete justification). The distinguishing probability is then at most $\Pr[F \vee F']$; for a distinguisher making at most q queries, this gives:

$$\Pr[F \vee F'] \le \frac{2q^2}{2^n}$$

which shows a negligible distinguishing probability.   □

3.4   The Chop Solution

In this section, we show that by removing a fraction of the output of the plain Merkle-Damgård construction MD^f, one obtains a construction indifferentiable from a random oracle. This "fix" is similar to the method used by Dodis et al. [14] to overcome the problem of using plain MD chaining for randomness extraction from high-entropy distributions, and to the suggestion of Lucks [22] to increase the resilience of plain MD chaining to multi-collision attacks. It is also already used in practice in the design of hash functions SHA-348 and SHA-224 [17] (both obtained by dropping some output bits from SHA-512 and SHA-256). Here we show that by dropping a non-trivial number of output bits from the plain MD chaining, one gets a secure random oracle H even if the input is not encoded in the prefix-free manner. For example, such dropping prevents the "extension" attacks we saw in the MAC application, since the attacker cannot guess the value of the dropped bits, and cannot extend the output of the MAC to a valid MAC of a longer message.

Formally, given a compression function $f : \{0, 1\}^{n+\kappa} \to \{0, 1\}^n$, the new construction chop-MD_s^f is defined as follows:

Function chop-MD$_s^f(m)$:
    let $m = (m_1, \ldots, m_\ell)$
    $y \leftarrow \text{MD}^f(m_1, \ldots, m_\ell)$
    return the first $n - s$ bits of y.

Theorem 3. *The chop-MD$_s^f$ construction is (t_D, t_S, q, ϵ) indifferentiable from a random oracle, for any t_D, with $t_S = \ell \cdot \mathcal{O}(q^2)$ and $\epsilon = 2^{-s} \cdot \ell^2 \cdot \mathcal{O}(q^2)$. Here ℓ is the maximum length of a query made by the distinguisher D.*

While really simple, the drawback of this method is that its exact security is proportional to $q^2 2^{-s}$, where s is the number of chopped bits and q is the number of oracle queries. Thus, to achieve adequate security level the value of s has to be relatively high, which means that short-output hash functions such as SHA-1 and MD5 cannot be fixed using this method. However, functions such as SHA-512 can naturally be fixed (say, by setting $s = 256$).

3.5   The NMAC and HMAC Constructions

The NMAC construction [8], which is the basis of the popular HMAC construction, applies an *independent* hash function g to the output of the plain MD chaining. It has been shown very valuable in the design of MACs [8], and recently also randomness extractors [14]. Here we show that if g is modelled as another fixed-length random oracle *independent* from the random oracle f (used for the compression function), then once again one gets a secure construction of an arbitrary-length random oracle H, even if plain MD chaining is applied without prefix-free encoding. Intuitively, applying g gives another way to hide the output of the plain MD chaining, and thus prevent the "extension" attack described earlier.

Formally, given $f : \{0,1\}^{n+\kappa} \rightarrow \{0,1\}^n$ and $g : \{0,1\}^n \rightarrow \{0,1\}^{n'}$, the function NMAC^{f,g} is defined as (see Figure 8a):

Function NMAC$^{f,g}(m)$:
    let $m = (m_1, \ldots, m_\ell)$
    $y \leftarrow MD^f(m_1, \ldots, m_\ell)$
    $Y \leftarrow g(y)$
    return Y

Theorem 4. *The construction NMAC^{f,g} is (t_D, t_S, q, ϵ) indifferentiable from a random oracle for any t_D, $t_S = \ell \cdot \mathcal{O}(q^2)$ and $\epsilon = 2^{-\min(n, n')} \ell^2 \mathcal{O}(q^2)$, in the random oracle model for f and g, where ℓ is the maximum message length queried by the distinguisher.*

To practically instantiate this suggestion, we would like to implement f and g from a single compression function. This problem is analogous to the problem in going from NMAC to HMAC in [8], although our solution is slightly different. One simple way for achieving this is to use domain separation: e.g., by prepending 0 for calls to f and 1 — for calls to g. However, with this modeling we are effectively using the prefix-free encoding mapping $m_1 m_2 \ldots m_\ell$ to $0 m_1 0 m_2 \ldots 0 m_\ell 1 0^\kappa$, which appears slightly wasteful. Additionally, this also forces us to go into the lower-level implementation details for the compression function, which we would like to avoid. Instead, our solution consists in applying two *black-box calls to the plain Merkle-Damgård construction* MD^f (with the same f and IV) : first to the input $0^\kappa m_1 \ldots m_\ell$, getting an n-bit output y, and again to κ-bit y', where y' is defined from y as follows (see Figure 8b):

> **Function** $\text{HMAC}^f(m)$:
>   let $m = (m_1, \ldots, m_\ell)$
>   let $m_0 = 0^\kappa$
> $y \leftarrow \text{MD}^f(m_0, m_1, \ldots, m_\ell)$
>   if $n < \kappa$ then $y' \leftarrow y \parallel 0^{\kappa - n}$
>   else $y' \leftarrow y|_\kappa$
> $Y \leftarrow \text{MD}^f(y')$
>   return Y

Intuitively, we are almost using the NMAC construction with $g(y) = f(IV, y')$ (where y' is obtained from y as above), except we prepend a fixed block $m_0 = 0^\kappa$ to our message. This latter tweak is done to ensure that there are no interdependencies between using the same IV on y' and the first message block (which would have been under adversarial control had we not prepended m_0). Indeed, it is very unlikely that "high-entropy" y' will ever be equal to $m_0 = 0^\kappa$, so the analysis for NMAC can be easily extended for this optimization.

Theorem 5. *The construction $HMAC^f$ is (t_D, t_S, q, ϵ) indifferentiable from a random oracle for any t_D, $t_S = \ell \cdot \mathcal{O}(q^2)$ and $\epsilon = 2^{-\min(n,\kappa)} \cdot \ell^2 \cdot \mathcal{O}(q^2)$, in the random oracle model for f, where ℓ is the maximum message length queried by the distinguisher.*

4   Constructions Using Ideal Cipher

In practice, most hash-function constructions are block-cipher based, either explicitly as in [29] or implicitly as for SHA-1. Therefore, we consider the question of designing an arbitrary-length random oracle H from an ideal block cipher $E : \{0,1\}^\kappa \times \{0,1\}^n \to \{0,1\}^n$, specifically concentrating on using the Merkle-Damgård construction with the Davies-Meyer compression function $f(x, y) = E_y(x) \oplus x$ (see Figure 9), since this is the most practically relevant construction. We notice that the question of designing a *collision-resistant* hash function H from an ideal block cipher was explicitly considered by Preneel, Govaerts and Vandewalle in [29], and latter formalized and extended by Black, Rogaway and Shrimpton [9]. Specifically, the authors of [9] actually considered 64 block-cipher variants of the Merkle-Damgård transform (which included the Davies-Meyer variant among them), and formally showed that exactly 20 of these variations

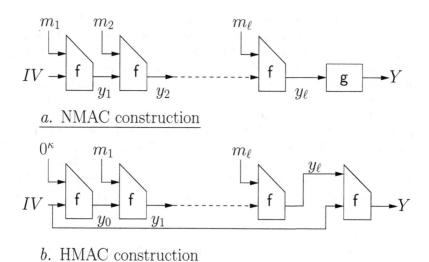

a. NMAC construction

b. HMAC construction

Fig. 8. The NMAC and HMAC constructions

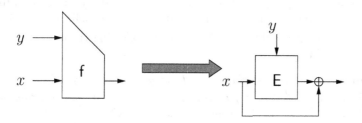

Fig. 9. The Davies-Meyer Compression function

(including the Davies-Meyer variant) are collision-resistant when the block cipher E is modeled as an ideal cipher. However, while our work will also model E
as an ideal cipher, our security goal is considerably stronger than mere collision-
resistance. Indeed, we already pointed out that none of the 64 variants above
can withstand the "extension" attack on the MAC application, even with the
Merkle-Damgård strengthening. And even when restricting to a fixed number
of blocks ℓ (which invalidates the "extension" attack), collision-resistance is
completely insufficient for our purposes. For example, the authors of [9] show
the collision-resistance when using the plain MD chaining with fixed IV and
compression function $f(x, y) = E_y(x)$. On the other hand, it is easy to see
that this method does not provide a secure random oracle H according to our
definition.

From a different direction, if we could show that the Davies-Meyer compression function $f(x, y) = E_y(x) \oplus x$ is a secure random oracle when E is an ideal
block-cipher, then we could directly apply any of the three fixes discussed above.
Unfortunately, this is again not the case: intuitively, the above construction al-

lows anybody to compute x from $f(x, y) \oplus x$ and y (since $x = E_y^{-1}(f(x,y) \oplus x)$), which should not be the case if f was a true random oracle. Thus, we need a direct proof to argue the security of the Davies-Meyer construction. Luckily, using such direct proofs we indeed argue that all of the fixes to the plain MD chaining which worked when f was a fixed-length random oracle, are still secure when $f(x, y) = E_y(x) \oplus x$ is used instead. Namely, we can either use a prefix-free encoding, or drop a non-trivial number of output bits, or apply an independent random oracle g to the output of plain MD chaining. With respect to this latter fix, we also show that we can implement this independent g using the ideal cipher itself, similarly to the case with an ideal compression function f.

Formally, given a block-cipher $E : \{0,1\}^\kappa \times \{0,1\}^n \to \{0,1\}^n$, the plain Merkle-Damgård hash-function with Davies-Meyer's compression function is defined as :

> **Function** $\mathrm{MD}^E(m_1, \ldots, m_\ell)$:
>   let $y_0 = 0^n$ (more generally, some fixed *IV* value can be used)
>   for $i = 1$ to ℓ do $y_i \leftarrow E_{m_i}(y_{i-1}) \oplus y_{i-1}$
>   return $y_\ell \in \{0,1\}^n$.

where for all i, $|m_i| = \kappa$. The block-cipher based iterative hash-functions pf-MD_g^E, chop-MD_s^E, NMAC_g^E and HMAC^E are then defined as in section 3, using MD^E instead of MD^f. The proof of the following theorem is given in the full version of this paper.

Theorem 6. *The block-cipher based constructions* pf-MD_g^E, *chop-MD_s^E,* NMAC_g^E *and* HMAC^E *are* (t_D, t_S, q, ϵ)-indifferentiable from a random oracle, in the ideal cipher model for E, for any t_D and $t_S = \ell \cdot \mathcal{O}(q^2)$, with $\epsilon = 2^{-n} \cdot \ell^2 \cdot \mathcal{O}(q^2)$ for pf-MD_g^E, $\epsilon = 2^{-s} \cdot \ell^2 \cdot \mathcal{O}(q^2)$ for chop-MD_s^E, $\epsilon = 2^{-\min(n,n')} \cdot \ell^2 \cdot \mathcal{O}(q^2)$ for NMAC_g^E and $\epsilon = 2^{-\min(\kappa,n)} \cdot \ell^2 \cdot \mathcal{O}(q^2)$ for HMAC^E. Here ℓ is the maximum message length queried by the distinguisher.*

5   Conclusion

In this paper, we pointed the attention of the cryptographic community to the gap between assuming an arbitrary-length random oracle H and assuming a fixed-length ideal building block for H such as a fixed-length compression function or a block cipher. We then provided a formal definition which suffices to eliminate this gap, noticed that the current iterative hash functions like SHA-1 and MD5 do not satisfy our security notion, and showed several practically motivated, easily implementable and provably secure fixes to the plain Merkle-Damgård transformation. Specifically, one can either ensure that all the inputs appear in the prefix-free form, or drop a nontrivial number of the output bits (if the output of the hash function is long enough to allow it), or, — when the above methods are not applicable — apply an independent fixed-length hash function to the output, which, as we illustrated, can be conveniently implemented using the corresponding building block itself.

An interesting open problem is to provide a construction in the opposite direction, that is, a construction that securely realizes an ideal block-cipher (or a random permutation) from a random oracle. One could use the Luby-Rackoff construction of a pseudo-random permutation from a pseudo-random function [21], but the major difference is that here the adversary has oracle access to the inner functions. One can show that at least six rounds are required to securely realize a random permutation from a random oracle (which should be contrasted with the secret-key case where four rounds are necessary and sufficient [21]), but we were not able to find a proof that six or more rounds would be sufficient.

Acknowledgments. We would like to deeply thank Victor Shoup for his invaluable contribution to all aspects of this work. We also thank the anonymous referees for many useful comments.

References

1. J. H. An, M. Bellare, *Constructing VIL-MACs from FIL-MACs: Message Authentication under Weakened Assumptions*, CRYPTO 1999, pages 252-269.
2. Mihir Bellare, Alexandra Boldyreva and Adriana Palacio. An Uninstantiable Random-Oracle-Model Scheme for a Hybrid-Encryption Problem. Proccedings of Eurocrypt 2004.
3. M. Bellare, J. Kilian, and P. Rogaway. The Security of Cipher Block Chaining. In *Crypto '94*, pages 341–358, 1994. LNCS No. 839.
4. M. Bellare and P. Rogaway, *Random oracles are practical : a paradigm for designing efficient protocols.* Proceedings of the First Annual Conference on Computer and Commmunications Security, ACM, 1993.
5. M. Bellare and P. Rogaway, *The exact security of digital signatures - How to sign with RSA and Rabin.* Proceedings of Eurocrypt'96, LNCS vol. 1070, Springer-Verlag, 1996, pp. 399-416.
6. M. Bellare and P. Rogaway, *Optimal Asymmetric Encryption*, Proceedings of Eurocrypt'94, LNCS vol. 950, Springer-Verlag, 1994, pp. 92–111.
7. M. Bellare and P. Rogaway, *Collision-Resistant Hashing: Towards Making UOWHFs Practical*, In *Crypto '97*, LNCS Vol. 1294.
8. M. Bellare, R. Canetti, and H. Krawczyk, *Pseudorandom Functions Re-visited: The Cascade Construction and Its Concrete Security*, In Proc. 37th FOCS, pages 514-523. IEEE, 1996.
9. J. Black, P. Rogaway, T. Shrimpton, *Black-Box Analysis of the Block-Cipher-Based Hash-Function Constructions from PGV*, in Advances in Cryptology - CRYPTO 2002, California, USA.
10. R. Canetti, *Universally Composable Security: A New Paradigm for Cryptographic Protocols*, proceedings of the 42nd Symposium on Foundations of Computer Science (FOCS), 2001. Cryptology ePrint Archive, Report 2000/067, http://eprint.iacr.org/.
11. R. Canetti, O. Goldreich and S. Halevi, *The random oracle methodology, revisited*, STOC' 98, ACM, 1998.
12. Ran Canetti, Oded Goldreich and Shai Halevi. On the random oracle methodology as applied to Length-Restricted Signature Schemes. In *Proceedings of Theory of Cryptology Conference*, pp. 40–57, 2004.

13. I. Damgård, *A Design Principle for Hash Functions*, In Crypto '89, pages 416-427, 1989. LNCS No. 435.

14. Y. Dodis, R. Gennaro, J. Håstad, H. Krawczyk, and T. Rabin, *Randomness Extraction and Key Derivation Using the CBC, Cascade and HMAC Modes*, Advances in Cryptology - CRYPTO, August 2004.

15. Y. Dodis, R. Oliveira, K. Pietrzak, *On the Generic Insecurity of the Full Domain Hash*, Advances in Cryptology - CRYPTO, August 2005.

16. FIPS 180-1, *Secure hash standard*, Federal Information Processing Standards Publication 180-1, U.S. Department of Commerce/N.I.S.T., National Technical Information Service, Springfield, Virginia, April 17 1995 (supersedes FIPS PUB 180).

17. National Institute of Standards and Technology (NIST). Secure hash standard. FIPS 180-2. August 2002.

18. RFC 1321, *The MD5 message-digest algorithm*, Internet Request for Comments 1321, R.L. Rivest, April 1992.

19. Shafi Goldwasser and Yael Tauman. On the (In)security of the Fiat-Shamir Paradigm. In *Proceedings of the 44th Annual IEEE Symposium on Foundations of Computer Science* (2003), 102-114.

20. H. Handschuh and D. Naccache, *SHACAL*, In B. Preneel, Ed., First Open NESSIE Workshop, Leuven, Belgium, November 13-14, 2000

21. M. Luby and C. Rackoff, *How to construct pseudo-random permutations from pseudo-random functions*, SIAM J. Comput., Vol. 17, No. 2, April 1988.

22. Stefan Lucks. *Design Principles for Iterated Hash Functions*, available at E-Print Archive, http://eprint.iacr.org/2004/253.

23. U. Maurer, R. Renner, and C. Holenstein, *Indifferentiability, Impossibility Results on Reductions, and Applications to the Random Oracle Methodology*, Theory of Cryptography - TCC 2004, Lecture Notes in Computer Science, Springer-Verlag, vol. 2951, pp. 21-39, Feb 2004.

24. Ueli Maurer and Johan Sjodin. *Single-key AIL-MACs from any FIL-MAC*, In *ICALP 2005*, July 2005.

25. R. Merkle, *One way hash functions and DES*, Advances in Cryptology, Proc. Crypto'89, LNCS 435, G. Brassard, Ed., Springer-Verlag, 1990, pp. 428-446.

26. Jesper Buus Nielsen. Separating Random Oracle Proofs from Complexity Theoretic Proofs: The Non-Committing Encryption Case. In *Advances in Cryptology - Crypto 2002 Proceedings* (2002), 111 -126

27. PKCS #1 v2.1, *RSA Cryptography Standard (draft)*, document available at www.rsa security.com/rsalabs/pkcs.

28. B. Pfitzmann and M. Waidner, *A model for asynchronous reactive systems and its application to secure message transmission*. In IEEE Symposium on Security and Privacy, pages 184-200. IEEE Computer Society Press, 2001.

29. B. Preneel, R. Govaerts and J. Vandewalle, *Hash Functions Based on Block Ciphers: A Synthetic Approach*, in Advances in Cryptology - CRYPTO '93,, Santa Barbara, California, USA.

30. V. Shoup, *A composition theorem for universal one-way hash functions*, In *Eurocrypt '00*, pp. 445–452, LNCS Vol. 1807.

31. R. Winternitz, *A secure one-way hash function built from DES*, in Proceedings of the IEEE Symposium on Information Security and Privacy, pages 88-90. IEEE Press, 1984.

On the Generic Insecurity of the Full Domain Hash

Yevgeniy Dodis[1,*], Roberto Oliveira[2], and Krzysztof Pietrzak[3,**]

[1] New York University
dodis@cs.nyu.edu
[2] IBM T.J. Watson Research Center
riolivei@us.ibm.com
[3] ETH Zürich
pietrzak@inf.ethz.ch

Abstract. The *Full-Domain Hash* (FDH) signature scheme [3] forms one the most basic usages of random oracles. It works with a family $\mathcal{F}$ of trapdoor permutations (TDP), where the signature of m is computed as $f^{-1}(h(m))$ (here $f \in_{\mathcal{R}} \mathcal{F}$ and h is modelled as a random oracle). It is known to be existentially unforgeable for any TDP family $\mathcal{F}$ [3], although a much tighter security reduction is known for a restrictive class of TDP's [10,14] — namely, those induced by a family of claw-free permutations (CFP) pairs. The latter result was shown [11] to match the best *possible* "black-box" security reduction in the random oracle model, irrespective of the TDP family $\mathcal{F}$ (e.g., RSA) one might use.

In this work we investigate the question if it is possible to instantiate the random oracle h with a "real" family of hash functions $\mathcal{H}$ such that the corresponding schemes can be proven secure *in the standard model*, under some natural assumption on the family $\mathcal{F}$. Our main result *rules out* the existence of such instantiations for *any* assumption on $\mathcal{F}$ which (1) is satisfied by a family of random permutations; and (2) does not allow the attacker to invert $f \in_{\mathcal{R}} \mathcal{F}$ on an a-priori unbounded number of points. Moreover, this holds even if the choice of $\mathcal{H}$ can arbitrarily depend on f. As an immediate corollary, we rule out instantiating FDH based on general claw-free permutations, which shows that in order to prove the security of FDH in the standard model one must utilize significantly more structure on $\mathcal{F}$ than what is sufficient for the best proof of security in the random oracle model.

1 Introduction

FULL DOMAIN HASH. Dating back to Diffie-Hellman [13], the simplest classical suggestion for the design of digital signature schemes is to set the signature of the message m to be $\sigma = f^{-1}(m)$, where f comes from a family of trapdoor permutations (TDP) $\mathcal{F}$ such RSA. Unfortunately, this simple scheme is existentially

* Supported by NSF CAREER Award CCR-0133806 and TC Grant No. CCR-0311095.
** Supported by the Swiss National Science Foundation, project No. 200020-103847/1.

V. Shoup (Ed.): Crypto 2005, LNCS 3621, pp. 449–466, 2005.

forgeable (even under no message attack), since any σ happens to be the signature of $m = f(\sigma)$. A folklore suggestion to fix this problem, which is the basis of several existing standards such as PKCS #1 [1], is to hash the message before inverting f: namely, to set $\sigma = f^{-1}(h(m))$ for a carefully chosen hash function h. This invalidates the trivial existential forgery above and seems to work well in practice for a "crazy" enough h, such as SHA-1. This signature scheme is commonly called *Full Domain Hash* (FDH), and yields one of the simplest and most practical signature schemes known.

From a theoretical point of view, however, one can wonder if *it is possible to formally prove the security of FDH for some TDP f and hash function h?*

RANDOM ORACLE MODEL. Partially motivated by this question, in their seminal paper Bellare and Rogaway [3] introduced the *random oracle (RO) model* as a "paradigm for designing efficient protocols". It mathematically models h as a truly random function, which is freely available to all the parties including the adversary. In particular, under this idealized assumption Bellare and Rogaway formally confirmed the intuition of practitioners that the FDH signature scheme is existentially unforgeable in the RO model, for *any* TDP family $\mathcal{F}$. In fact, this was one of the first applications of the so called "random oracle methodology". Namely, one first formally analyzes and proves the security of a scheme like FDH in the RO model, and then practically *instantiates* this abstract scheme by replacing the ideal hash function h by some "real" implementation (such as SHA-1, or, more abstractly, some family of "real" functions $\mathcal{H}$), heuristically hoping that no security flaws will suddenly appear in the standard model. Therefore, it is clearly of fundamental importance to understand under which conditions one can *provably* instantiate the random oracles in the standard model. In particular, in this work we will concentrate on the FDH signature scheme, which, as we said, is one of the most basic and important applications of random oracles. Before addressing it in more detail, however, let us summarize what is known about this scheme in the RO model.

FDH IN RO MODEL. As we mentioned, Bellare and Rogaway showed that FDH is existentially unforgeable in the RO model, for any TDP family $\mathcal{F}$. On the other hand, a much tighter security reduction in the random oracle model was subsequently found by [10,14] for a special class of TDP's: namely, those induced by a family of *claw-free permutation (CFP) pairs*[1] which luckily includes all popular families such as RSA. Moreover, Coron [11] subsequently showed that the above tighter reduction from CFP-induced TDP's is *optimal*, as long as the reduction treats the adversary as a "black-box" and irrespective of which particular TDP family $\mathcal{F}$ is used (e.g., even with RSA one cannot find a better black-box reduction in the RO model).

OUR GOAL. As we see, in the RO model very weak assumptions on the function family $\mathcal{F}$ are sufficient to prove the security of FDH: in fact, a single (although

---

[1] Such families consist of pairs of functions (f, g) for which is it infeasible to find a "claw" (x, y) satisfying $f(x) = g(y)$. One get an induced TDP family by taking f and "ignoring" g.

"ideal") hash function h simultaneously works for all such $\mathcal{F}$. Unfortunately, it is not hard to see that previously studied "realizable" properties of random oracles, such as collision-resistance, pseudorandomness (even verifiable; see [5]) or perfect one-wayness [9] are not sufficient *in general* to implement the random oracle h, even for specific function families $\mathcal{F}$ (i.e., one can come up with an artificial counter-example family $\mathcal{H}$ which nevertheless satisfies the given property but for which the FDH scheme is insecure with $\mathcal{F}$). On the other hand, many practitioners strongly believe that for most "real" TDP families $\mathcal{F}$ there should probably exist "good enough" hash functions like SHA-1 which would make FDH with $\mathcal{F}$ secure. Therefore, our main question is to examine *for which TDP families $\mathcal{F}$ can we provably instantiate FDH in the standard model?*

INITIAL ATTEMPT. Let us make this question more precise. Given a function family $\mathcal{F}$, we are trying to design a hash family $\mathcal{H}$, such that a random h sampled from $\mathcal{H}$ will make FDH secure. Clearly, $\mathcal{H}$ should be allowed to depend on $\mathcal{F}$ (since assuming otherwise seems to place unfair restrictions on the signature designer). In fact, we also want to allow $\mathcal{H}$ to depend on a specific function f sampled from $\mathcal{F}$ (and whatever public information is associated with such f). For example, if $\mathcal{F}$ is induced by a family of claw-free permutation pairs (which, as we know, is very beneficial in the RO model), a random member f from $\mathcal{F}$ is sampled by choosing a random pair (f, g) from the CFP family, and then "ignoring" g. In this case it seems natural that the signature designer might want to use both f and g in designing the hash function h. For example, setting $h = g$ results in a signature scheme $f^{-1}(g(m))$ which is provably unforgeable under *no message attack*. Although the latter task can be easily achieved by other means (e.g., making h to be a random constant), this shows a potential utility one might get by using g in a less obvious manner.

Thus, the ambitious question would be to characterize the TDP families $\mathcal{F}$ for which one can choose an efficient $\mathcal{H}$ (depending of f) which would make FDH secure. Unfortunately, this seems to be an extremely difficult question given our current state-of-the-art knowledge. In particular, even for specific families such as RSA we do not seem to be able to say anything more meaningful than making a tautological assumption of the form "SHA-1 makes a good RSA-based FDH signature scheme".

OUR APPROACH. Instead, we will ask a slightly more general question: which *security assumptions on $\mathcal{F}$* are sufficient to instantiate FDH in the standard model. For example, can we match the RO result stating that any TDP family can be instantiated? And, if not, maybe more restricted CFP-induced families can? Or maybe some other elegant assumption on $\mathcal{F}$ will be sufficient?

While a positive answer to these kind of questions would be even harder and more remarkable than the ambitious question asked about specific families $\mathcal{F}$ like RSA, the extra generality will allow us to get a *meaningful negative result*, which we believe is still very important. In particular, it will allow us to further realize the differences between the standard and the random oracle model. For example, we will see that being induced by a family of CFP's *by itself* is insufficient to instantiate FDH, in contrast to the RO model, where *nothing beyond this prop-*

erty is likely to be of any extra advantage! Additionally, looking at the current general proofs of security of FDH in the RO model, it seems reasonable to hope that even in the standard model some natural and relatively general assumption on $\mathcal{F}$ might be sufficient for the proof to "go through" (with an appropriately chosen $\mathcal{H}$). In this regard, our approach allows us to further understand which security assumptions on $\mathcal{F}$ will certainly be *insufficient* (by themselves) to try instantiating FDH. In particular, if a given set of properties of $\mathcal{F}$ will be consistent with some assumption that we formally rule out, then more properties are needed. For example, it easily follows from our result below that one cannot instantiate FDH even if we assume $\mathcal{F}$ to be one-way against *any* distribution of "super-logarithmic" entropy. This is an *extremely* strong assumption that might appear quite useful for FDH upon the first look: for example, a similar assumption was recently utilized by Wee [28] to *successfully* obfuscate "equality" queries in the standard model, which was previously known only in the random oracle model [24]. Yet, we show that this assumption is insufficient for FDH.

OUR MODELING. A bit more formally, we will study the question if there exists a *black-box reduction* (see [21]) from a given security assumption on $\mathcal{F}$ (such as being one-way or induced by CFP family, etc.) to the security of FDH. This means that all the relevant parties — adversary and "challenger" for the assumption (see below), potential forger for FDH, *as well as the designer of the hash family $\mathcal{H}^2$* — should work given oracle access to f (and possibly even f^{-1}; see below). While seemingly restrictive, we believe this captures the essence of what it means to instantiate FDH given *any* $\mathcal{F}$ satisfying a given security assumption. Indeed, allowing non-black-box access to $\mathcal{F}$ essentially maps us back to the original "beyond-the-reach" question, where the designer of $\mathcal{H}$ can use some "extra" properties of $\mathcal{F}$ which do not follow from the security assumption alone. For example, we do not know how to show the insecurity of RSA-based FDH when the designer of the scheme chooses h to be SHA-1. In fact, most practitioners actually hope that the resulting scheme is secure!

In our modeling, a given security assumption is formalized by a "game" G between the adversary A and the *challenger* C. At the start of the game, a random f is chosen from $\mathcal{F}$ (possibly with some other public information), after which A and C engage in some protocol using *oracle access to f*, and the end of which C output 1 if the adversary has won and 0 otherwise. For example, in the one-wayness game defining plain TDP's the challenger simply asks A to invert $f(x)$, for a random x of its choice. Similarly, in the "claw-free" game defining $\mathcal{F}$ induced by some CFP family, C simply waits for A to provide a claw (x, y), where A can have oracle access to both f and its "twin" permutation g. Many other assumptions can be put in this framework as well.

Given such an abstract game G, we can look at the corresponding class of black-box permutation families $\mathcal{F}$ for which no polynomial time adversary can win with non-negligible probability. To argue a separation result for a given game G, we must essentially (see below) show that there exist a black-box family $\mathcal{F}$ such that (1) $\mathcal{F}$ is "black-box" secure with respect to G, but (2) $\mathcal{F}$ cannot be

---

[2] As we stated, it seems very restrictive not to allow such a dependency.

instantiated for the use in FDH, for any polynomial size circuit family $\mathcal{H}$ (which is allowed to depend on $\mathcal{F}$, but in a "black-box" manner).

OUR MAIN RESULT. Our main result is pretty general: we show than no game G between A and C can lead to an "instantiable" security assumption on $\mathcal{F}$, provided that a family of *truly random permutations* satisfies the security of G. Intuitively, it rules out all the assumptions involving "inverting" f on more or less arbitrary inputs (since random permutations are very hard to invert), or finding some inputs to f whose images satisfy some non-trivial relation (e.g., x and y such that $f(x) = f(y) \oplus 1$), etc. In fact, our main results extends even to games where the challenger is allowed to invert f to the attacker, as long as this is done for an a-priori bounded number of times.[3] To state this result differently, *any assumption on $\mathcal{F}$ which (1) is satisfied by a family of random permutations; and where (2) the challenger does not invert f on an a-priori unbounded number of points, is insufficient to instantiate FDH in the standard model.*

Thus, to generically instantiate FDH one must assume a property on $\mathcal{F}$ which is *not* satisfied by random permutations, such as being "homomorphic" or "self-reducible".

OTHER RESULTS. As special cases, we rule out such instantiations based on plain TDP's, as well the sub-class of TDP's induced by CFP's, since both of those are easily seen to be satisfied by random permutations. In particular, this shows that more assumptions on $\mathcal{F}$ are needed in the standard model than what is sufficient for the best reduction in the RO model, giving yet another separation between the standard the the RO model (see related work below). As another interesting corollary, we notice that many cryptographic primitives such as collision-resistant hash functions, trapdoor commitments and even general signature schemes follow — in a black-box manner — from the existence of CFP families. Our separation result therefore shows that even assuming the existence of all these powerful primitives is not sufficient to build an "FDH-like" signature scheme (in a black-box manner), despite the fact that general, "non-FDH-like" signature schemes can be built! For example, there seem to be a "price to pay" for insisting on *inverting a trapdoor permutation* on the hash of the message, as opposed to applying to it any secure signature scheme on short messages: the latter is *provably secure* as long as the hash is collision-resistant (this is the famous "hash-then-sign" paradigm), while we show that much stronger assumptions seem to be required for the former.

We remark that our main impossibility result uses the full power of the chosen message attack, since our FDH breaker is allowed to ask more signing queries than the description of the hash function h. If we restrict our attention to the class of general TDP's (as opposed to *all* hard games satisfied by random permutations), we also strengthen our separation and show that there is no black-box reduction from the security of TDP's to the security of FDH even as

---

[3] Essentially, for a number of times slightly smaller than the number of signing queries the FDH forger is allowed to make. Without this restriction, one can define games modeling tautological assumptions of the form "SHA-1 makes FDH secure for a given $\mathcal{F}$" (which are trivially instantiable by setting h equal to SHA-1).

a *one-time* signature scheme, as long as the message space is super-polynomial in the security parameter.[4]

OUR TECHNIQUES. In both of our results we use an elegant "two-oracle" observation of Hsiao and Reyzin [21] for showing general black-box separation results. Applied to our setting, they show that it is sufficient to design an oracle F for $\mathcal{F}$ and another "breaking" oracle G, such that G does not help the attacker to win the game G with F, but always helps the forger to break the security of FDH (even if $\mathcal{H}$ can depend of F but not on G). In both of our results we use a family of random permutations to model the oracle F for our TDP family $\mathcal{F}$. The oracles G, however, are very different.

For our general separation result we use a novel oracle G which takes a description of the hash function h, and will forge the FDH-like signature of a new message *only if* the attacker can "prove" that he has oracle access to the FDH signing oracle. Remarkably, the oracle G is designed in such a careful way that its addition is literally of "no use" to the attacker in *any* game G! So if G was secure with random permutations, addition of G will not change this fact. Yet, it clearly breaks any FDH instantiation $\mathcal{H}$, since the forger has a "real" access to the signing oracle, and thus can successfully utilize G.

On the other hand, the oracle G for our TDP-specific separation is very different and is based on to the corresponding oracle by Simon [27] used to separate collision-resistant hash functions from one-way permutations.[5] In essence, this oracle returns collisions for any length-decreasing function h (which could depend of f), but in a careful way which does not allow the attacker to invert f. On the other hand, any collision clearly makes FDH insecure against one-message attack, as both of the colliding messages have the same signature. The main technical difficulty we have to resolve here is the fact that Simon's oracle only covers length-decreasing function families $\mathcal{H}$ (in fact, it is completely useless for most length-increasing hash families). Therefore, we have to non-trivially extend it to allow one break FDH for arbitrary function families $\mathcal{H}$, and yet without suddenly helping the adversary to invert f at a random point.

RELATED WORK. Our work is related to several important results [7,26,20,8,2] showing that various schemes provably secure in the random oracle model cannot be securely instantiated in the standard model. Canetti, Goldreich and Halevi [7,8] gave concrete (although somewhat artificial) examples of general signature and encryption scheme with this property. Nielsen [26] considered the question of designing so called "non-committing encryption schemes" [6] capable of encrypting arbitrary number of messages, and showed that one cannot build such scheme *at all* in the standard model, although simple solutions in the random oracle model exist. Goldwasser and Tauman [20] concentrated on the soundness of the Fiat-Shamir heuristics [15], and showed a secure (although artificial) 3-round identification scheme which does not result in a secure signature scheme

---

[4] Otherwise, one can of course instantiate FDH by "hardwiring" an independent random challenge y_m to be the hash of m.

[5] For example, such oracle cannot be extended to cover CFP-induced TDP's, since it is known how to build collision-resistance hash functions from CFP's.

in the standard model, no matter how one implements the hash family. Finally, Bellare, Boldyreva and Palacio [2] showed a natural ElGamal-based key encapsulation mechanism for hybrid encryption which is secure in the random oracle model (for any symmetric-key component), but where for every real hash family one can come up with (artificial) symmetric-key encryption scheme making the overall hybrid scheme insecure.

We notice that an attractive feature of all these results as compared to our result, is that their separations are not black-box. However, our setting appears to be significantly more constrained as well. Intuitively, in all of the above results the syntax of the question allowed one enough freedom to adapt the scheme *after* the hash function h was chosen. While such adaptation was pretty non-trivial in each of the above works, our setting appears to be more restrictive. Namely, we must "commit" to a "real" TDP family $\mathcal{F}$ (possibly satisfying even more constraints), and then, given an arbitrary non-black-box function h *depending of* f, find some point m where we can invert $h(m)$! Without "reverse-engineering" such an h, the latter task seems quite hopeless to do (even using the signing oracle since it is hard to predict on which points it will invert f). Indeed, our black-box assumption essentially allows us to get a weak, but luckily sufficient "handle" to determine how h actually depends on f.

From a different perspective, our work naturally relates to a rich body of work on various black-box separations [22,27,18,23,17,16,11,14,21]. For example, we already pointed out how our breaking oracle for the case of general TDP's relates to the oracle of Simon [27], and how we use the simplified framework of Hsiao and Reyzin [21] to get our black-box separations. To the best of our knowledge, however, our work is the first to show a black-box separation result with respect to *instantiating random oracles in the standard model*, as opposed to separating different cryptographic assumptions from each other [22,27,18,21] or showing lower bounds on the efficiency or exact security of various "black-box" reductions [23,17,16,11,14].

Finally, we already mentioned a complimentary recent work of Boldyreva and Fischlin [5], who considered the question of instantiating random oracles in various scenarios, including FDH, by popular families of "realizable" hash functions, such as verifiable pseudorandom functions [25] (VRFs). In particular, they showed that such VRFs cannot generically instantiate FDH, no matter which TDP family $\mathcal{F}$ is used.

2   Preliminaries

BASIC DEFINITIONS AND NOTATION. For a set $\mathcal{X}$ we denote by $x \in_R \mathcal{X}$ a value chosen uniformly at random from $\mathcal{X}$. A function $\mu : \mathbb{N} \to [0, 1]$ is negligible if for any $c > 0$ there is an n_0 such that $\mu(n) \leq 1/n^c$ for all $n \geq n_0$. We write $\mathsf{negl}(\cdot)$ as a shorthand for a *negligible* function.

TM is a shorthand for Turing-machine. We use the standard definition of probabilistic polynomial-time TMs (pptTM for short) and pptTMs with oracle access (oppTM for short). We say that something can be efficiently computed (relative to an oracle O) if it can be computed by a pptTM (by a opptTM with oracle access to O).

TRAPDOOR PERMUTATIONS. A *trapdoor permutation family* (TDP) is a pair of efficient algorithms $(KeyGen, F)$. $KeyGen$ is probabilistic and on input 1^n generates a key/trapdoor pair $KeyGen(1^n) \rightarrow (pk, td)$ where $F(pk, \cdot)$ implements a permutation $f_{pk}(.)$ over $\{0,1\}^n$ and $F(td, \cdot)$ implements its inverse $f_{pk}^{-1}(.)$.

SECURITY OF TDPs. The standard security property for TDPs is *one-wayness* which says that inverting is hard *without* the trapdoor, i.e. for any pptTM A

$$\mathbf{Pr}_{KeyGen(1^n)\rightarrow(pk,td),x\in_R\{0,1\}^n}[A(f_{pk}(x),pk) = x] = \mathsf{negl}(n).$$

A stronger security property is *claw-freeness* which says that given two independently sampled permutations it is hard to find a collision, i.e. for any pptTM A

$$\mathbf{Pr}_{i\in\{1,2\}:KeyGen(1^n)\rightarrow(pk_i,td_i)}[A(pk_1,pk_2) = (x_1,x_2) \text{ where } f_{pk_1}(x_1) = f_{pk_2}(x_2)]$$
$$= \mathsf{negl}(n).$$

A TDP with this property is not a standard assumption, but it implies the following popular primitive.

CLAW-FREE PAIRS OF TRAPDOOR PERMUTATIONS. A *family of claw-free pairs of trapdoor permutations* (CFP) is a triple of efficient algorithms $(KeyGen, F, G)$ where $KeyGen$ is probabilistic and on input 1^n generates a key/trapdoor pair $KeyGen(1^n) \rightarrow (pk, td)$ for which $F(pk, \cdot)$ and $G(pk, \cdot)$ implement permutations $f_{pk}(.)$ and $g_{pk}(.)$ over $\{0,1\}^n$ respectively. $F(td, \cdot)$ and $G(td, \cdot)$ implement the inverses $f_{pk}^{-1}(.)$ and $g_{pk}^{-1}(.)$. The security property for CFPs requires that for any pptTM A

$$\mathbf{Pr}_{KeyGen(1^n)\rightarrow(pk,td)}[A(pk) = (x_1,x_2) \text{ where } f_{pk}(x_1) = g_{pk}(x_2)] = \mathsf{negl}(n).$$

HASH-FUNCTION. A family of hash-functions is a pair of efficient algorithms $(Index, H)$. $Index$ is probabilistic and on input 1^n generates an index $i \in \mathcal{I}_n$. For each $i \in \mathcal{I}_n$, $H(i, .)$ implements a function $h_i : \{0,1\}^* \rightarrow \{0,1\}^n$. A family of hash-functions is collision resistant if

$$\mathbf{Pr}_{Index(1^n)\rightarrow i}[A(1^n, i) = (x_1,x_2) \text{ where } x_1 \neq x_2 \text{ and } h_i(x_1) = h_i(x_2)]$$
$$= \mathsf{negl}(n).$$

FULL-DOMAIN HASH (FDH). The FDH signature-scheme based on a trapdoor permutation family $(KeyGen_F, F)$ and a family of hash functions $(Index, H)$ is defined as a triple of functions $(KeyGen_{FDH}, sign, verify)$ where for security parameter n

– $KeyGen_{FDH}(1^n)$ first runs $KeyGen_F(1^n) \rightarrow (pk, td)$ and $Index(1^n) \rightarrow i$.[6] It outputs the triple (pk, td, i). The public-key of the signature scheme is (pk, i) and the secret key is (td, i).

---

[6] One would probably choose the randomness for $Index$ and for $KeyGen_F$ independent here, but we make no such assumption. In particular, (pk, td) and i can be arbitrarily correlated.

- $sign(m, td, i)$, the signature of a message $m \in \{0,1\}^*$ is $f_{pk}^{-1} h_i(m)$ (i.e. computed as $F(td, H(i, m))$).
- $verify(\sigma, m, pk, i)$, the verification function evaluates to 1 (with the meaning that the signature is valid) iff $f_{pk}(\sigma) = h_i(m)$ and to 0 otherwise.

SECURITY OF FDH. A FDH signature scheme as above is secure against an existential forgery in a chosen message attack if for any opptTM A

$$\mathbf{Pr}_{KeyGen_{FDH}(1^n) \to (pk, td, i)} [A^{sign(., td, i)}(pk, i) \to (m, \sigma) \text{ where}$$
$$verify(\sigma, m, pk, i) = 1 \text{ and } A \text{ did not make the oracle query } m] = \mathsf{negl}(n). \quad (1)$$

This means that A cannot come up with a valid signature/message pair for a message that he had not already signed by the signing oracle.

GAME. A game is defined by two opptTMs, a prover A and a challenger C, which have a common communication tape over which they can exchange messages. The challenger finally output a decision bit. We say that A wins the game if this bit is 1 and denote this event by $\langle A, C \rangle \to 1$.

HARD GAME. An opptTM C as above defines a *hard* game if no opptTM A can win the game when the oracle is instantiated with t (where $t = t(n)$ is implicitly defined by C and can be polynomial in n) uniform random permutations $\pi_1, \ldots, \pi_t$ over $\{0,1\}^n$. I.e. C defines a hard game if for all opptTM A

$$\mathbf{Pr}[\langle A^{\pi_1(.), \ldots, \pi_t(.)}(1^n), C^{\pi_1(.), \ldots, \pi_t(.)}(1^n) \rangle \to 1] = \mathsf{negl}(n). \quad (2)$$

A TDP $(KeyGen, F)$ is secure for the hard game C if (2) is satisfied even if the random permutations are replaced with this TDP, i.e. for all pptTM A

$$\mathbf{Pr}_{\forall i=1...t:KeyGen(n) \to (pk_i, td_i)} [\langle A(pk_1, \ldots, pk_t), C^{F(pk_1, \cdot), \ldots, F(pk_t, \cdot)}(1^n) \rangle \to 1]$$
$$= \mathsf{negl}(n).$$

Hard games capture many natural security properties, in particular

- one-wayness: $C^{f(.)}(1^n)$ samples $x \in \{0,1\}^n$ uniformly at random and sends $f(x)$ to A. It outputs 1 iff it receives as the next message x.
- claw-freeness: $C^{f_1(.), f_2(.)}(1^n)$ just expects $x_1, x_2 \in \{0,1\}^n$ and outputs 1 iff $f_1(x_1) = f_2(x_2)$.

In the next section we will show that a TDP which is secure for *all* hard games cannot be black-box reduces the security of FDH. There does not exist a TDP which is secure for all hard games in the standard model,[7] but we show an impossibility result, and showing impossibility from such a hypothetic TDP implies

---

[7] Consider for example a game where C expects as input a circuit and then checks if the circuit computes the same value as its oracle on a few (n is enough) random inputs and outputs 1 only if this is the case. In the standard model A can always win this game by sending a circuit which computes C's oracle $F(pk, .)$. But this is a hard game as if the oracle is a random permutation, it will with high probability disagree with every polynomial size circuit on most inputs, and C will reject almost certainly.

impossibility for any assumption it implies. Then in section 4 we will extend the notion of hard games and give the challenger also access to inversion oracles $\pi_i^{-1}(.)$ which he may query at most polynomially many times (for some arbitrary but a priori fixed polynomial). With such games we can cover additional natural assumptions for TDPs, which will therefore be also insufficient to get a reduction to an FDH signature scheme.

3   No Reduction from Any Hard Game

Theorem 1. *There is no black-box reduction from a trapdoor permutations family which is secure for all hard games to a FDH signature scheme secure against chosen-message attacks.*

More precisely, given a TDP $(KeyGen, F)$ which is secure for all hard games and any hash function family $(Index, H)$, the security of the signature scheme $sign(m) = f_{pk}^{-1}h(m)$ (where $KeyGen(1^n) \rightarrow (pk, sk)$ and $Index(1^n) \rightarrow i$) cannot be black-box reduced from the security of the TDP. Here the hash function can use the TDP as a black-box and the randomness used for $KeyGen$ and $Index$ can be arbitrarily correlated. Moreover, if we let $s(n) = \max\{|h_i| : i \in Range(Index(1^n))\}$ denote an upper bound on the size of a description of the hash function used, then the theorem even holds if we restrict the number of chosen message queries to $s(n)$ and the size of the message-space of the signature scheme to $s(n) + 1$.

As corollaries we get that any assumption on TDPs which can be formulated as a hard game will not be enough to get a reduction to FDH, e.g.

Corollary 1. *There is no black-box reduction from claw-free pairs of trapdoor permutations to a FDH signature scheme secure against chosen-message attacks.*

Proof (of Theorem 1). Following [21] (Lemma 1), as to rule out black-box reductions, it is enough to prove that there are two oracles F and G such that the following holds:

1. There is an opptTM D such that D^F implements[8] TDP.
2. There is an opptTM A such that $A^{F,G}$ finds a forgery for any signature scheme of the form $sign(m) = f^{-1}(h^F(m))$ in a chosen message attack, where f is the TDP implemented by D^F and h is any oracle circuit.
3. There is no opptTM B where $B^{F,G}$ breaks the security of TDP implemented by D^F. This means that $B^{F,G}$ cannot win any hard game C instantiated with this TDP with non-negligible probability.

Points 2 and 3 will follow from Lemmas 1 and 2 below. The first point is satisfied by the definition of the oracle F we will give, which implements TDP. This F *alone* is trivially a *secure* implementation of TDP. We then define a *breaking oracle* G

---

[8] Here implement has a purely functional meaning and does not imply any security assumptions.

for which we will show that it can be used to break any FDH scheme based on the TDP implemented by F but not the security of the TDP itself. The oracle G will simply provide a forgery (for the message $m = 0$) to any signature scheme of the form $sign(m) = f^{-1}(h^F(m))$ (where $f \in F$ and h is any oracle circuit), *but only if* it can be sure that the requesting party can compute those signatures herself (e.g. because she has access to the signing oracle $sign(m) = f^{-1}(h^F(m))$ or knows the trapdoor for f). For this our G expects as input the values $f^{-1}(h^F(m))$ for $m = 1 \ldots \ell$, where $\ell = |h|$. This choice of ℓ should make it impossible for an adversary to hardwire the outputs of h on all the inputs requested to values where she can invert f. However, there would still be at least two ways in which an adversary could abuse the oracle G to break the security of TDP implemented by F.

1. She could define an h such that the output of h^F collides on (some of) the requested inputs. Say $h^F(i) = y$ for all $1 \leq i \leq \ell$ (where she knows $f^{-1}(y)$) and $h^F(0) = z$ (where z could be a challenge in the one-wayness game). As she can provide the requested signatures $f^{-1}(h^F(i))$ to G, G will output a forgery $w = f^{-1}(h^F(0)) = f^{-1}(z)$ and she wins the game! To avoid this our G will check if there is such a collision before providing the forgery. This will not affect the usability of G to provide forgeries, as having a collision for h^F one can compute a forgery without the help of G anyway.

2. She could use f in the definition of h^F in a clever way, for example by choosing an h where $h^F(m) = f(m)$ for $m \neq 0$ (then $f^{-1}(h^F(m)) = m$) and $h(0) = z$ (where z could be a challenge in the one-wayness game). Our G will prevent this by checking whether in the computation of h^F on any of the requested inputs, the oracle for f is queried on an input x where $f(x) = h^F(i)$ for $i, 1 \leq i \leq \ell$. Again, if this check fails we have a forgery as $x = f^{-1}(h^F(i))$.

We will show that the two above checks are not only necessary, but already sufficient to guarantee that G cannot be used to break the security of TDP implemented by F.

Definition of F (TDP secure for every hard game). The definition of the oracle F is straight forward. For any $n \in \mathbb{N}$ choose $2^n + 1$ permutations $f_{0,n}, \ldots, f_{2^n-1,n}$ and t_n at random. Now F is defined as[9]

- $F(td2pk, n, td) \rightarrow t_n(td)$
- $F(eval, n, pk, x) \rightarrow f_{pk,n}(x)$
- $F(invert, n, td, y) \rightarrow f_{pk,n}^{-1}(y)$

---

[9] With this F a TDP $(KeyGen, F)$ can be implemented as follows. $KeyGen(1^n)$ first samples a random trapdoor $td \in_R \{0,1\}^n$, then computes the corresponding public-key $F(td2pk, n, td) \rightarrow pk$ and outputs (pk, td). $F(pk, .)$ and $F(td, .)$ are computed by $F(eval, n, pk, .)$ and $F(invert, n, td, .)$ respectively. Informally the reason that this TDP is secure for every hard game follows from the fact that a permutation, chosen at random from a set of 2^n randomly chosen permutations, is computationally indistinguishable from a truly random permutation. But if there was a hard game that this TDP could win, we could turn it into a distinguisher.

Definition of G *(Breaking Oracle).* The oracle G takes as input $(n \in \mathbb{N}, k \in \{0,1\}^n, h \in \{0,1\}^*, V)$ where h is (the description of) an oracle circuit.[10] This can be seen as a request for an existential forgery for the signature scheme $sign(m) = f_{pk,n}^{-1}(h^{\mathsf{F}}(m))$. The vector $V = [v_1, \ldots, v_{|h|}]$ is a "proof" that the requesting party can compute those signatures herself. We say that G accepts the input if the input has the correct form (as above) and

1. $f_{pk,n}^{-1}(h^{\mathsf{F}}(i)) = v_i$ for all $i = 1, \ldots, |h|$.
2. $v_i \neq v_j$ for all $1 \leq i < j \leq |h|$.
3. $\{h^{\mathsf{F}}(1), \ldots, h^{\mathsf{F}}(|h|)\} \cap Y_{\mathsf{F}}^h = \emptyset$ where Y_{F}^h is defined as

$$Y_{\mathsf{F}}^h = \{f_{pk,n}(x) | \exists i, 1 \leq i \leq |h|, h^{\mathsf{F}}(i) \text{ makes the query } \mathsf{F}(eval, n, pk, x)\} \ (3)$$

If G accepts the input it outputs a forgery $f_{pk,n}^{-1}(h^{\mathsf{F}}(0))$ and $\perp$ otherwise.

G *Breaks any FDH Signature Scheme.* Now we will show that G breaks any FDH signature scheme based on F.

Lemma 1. *There is an* opptTM *A which outputs a forgery for any signature scheme* $sign(m) = f_{pk,n}^{-1}(h^{\mathsf{F}}(m))$ *with probability 1, i.e.*

$$\mathbf{Pr}[A^{\mathsf{F},\mathsf{G},sign(.)}(n, pk, h) \rightarrow (m, s) \quad where \quad s = f_{pk,n}^{-1}(h^{\mathsf{F}}(m))$$
$$and \quad sign(.) \text{ was not queried on input } m] = 1$$

Proof (of Lemma). A must only check if h satisfies conditions 2 and 3. If one of them is not satisfied, this directly gives a forgery, otherwise A can use G to get a forgery. More formally A does the following:

– Compute $h^{\mathsf{F}}(1), \ldots, h^{\mathsf{F}}(|h|)$, doing this also compute Y_{F}^h as in (3).
  - If any of the $h^{\mathsf{F}}(1), \ldots, h^{\mathsf{F}}(|h|)$ collide we have a forgery: If say $h^{\mathsf{F}}(i) = h^{\mathsf{F}}(j)$, then query $sign(i)$ and output the forgery $(j, sign(i))$.
  - If $\{h^{\mathsf{F}}(1), \ldots, h^{\mathsf{F}}(|h|)\} \cap Y_{\mathsf{F}}^h \neq \emptyset$, then we have found an x and an i satisfying $f_{pk,n}(x) = h^{\mathsf{F}}(i)$ and thus have a forgery as $sign(i) = f_{pk,n}^{-1}(h^{\mathsf{F}}(i)) = x$.
– If none of the above is the case, then call the oracle $sign$ on inputs $1, \ldots, |h|$ and let $V = [sign(1), \ldots, sign(|h|)]$. Now query G on input (n, pk, h, V) to get a forgery for the message $m = 0$. $\Diamond$

G *does not break the security of* F. In this section we will prove that F is a *secure* implementation of a family of claw-free trapdoor permutations, even when given access to G, i.e.

---

[10] Usually the hash function h is given as a TM and not as a circuit, but a TM can be simulated by a circuit whose size is only polynomial in the running time of the TM. In particular for every efficient h there is an $m \in \mathbb{N}$ and a circuit h_c such that $\forall i \in \{0,1\}^m : h_c(i) = h(i)$ and $|h_c| < 2^m$, moreover such h_c can be efficiently computed and is sufficient here.

Lemma 2. *With probability 1 (over the choice of* F*) for any* opptTM *B and any hard game* C *(with* $t = t(n)$ *implicitly defined by* C*)*

$$\mathbf{Pr}_{\forall i=1...t:KeyGen(n)\to(pk_i,td_i)}[(B^{\mathsf{F},\mathsf{G}}(pk_1,\ldots,pk_t), C^{f_{pk_1,n},\ldots,f_{pk_t,n}}(1^n)) \to 1]$$
$$= \mathsf{negl}(n). \tag{4}$$

Proof (Proof Sketch of Lemma). If the oracle G was not there, then (4) would follow from the fact that for a random pk, $f_{pk,n}$ is computationally indistinguishable from a random permutation and that the randomly chosen permutation t_n is one-way (thus one cannot get the trapdoor $t_n^{-1}(pk)$).

Now we must argue that the presence of the oracle G will not help to win any hard game. This is not so obvious, after all G provides forgeries $f_{pk,n}^{-1}(h^{\mathsf{F}}(0))$ for an h of our choice. But to learn such a forgery we must find an accepting input (see the definition of G) for G. From Lemma 3 below it now follows that B cannot find such an accepting input for a random pk and thus will not learn anything about the $f_{pk_i,n}$'s that he could not compute on its own.[11]   ◇

Lemma 3. *Let* f *be a random permutation on* $\{0,1\}^n$ *and* $c \geq 1$ *be a constant. For any oracle TM* A *which makes at most* n^c *oracle calls, we have (the probability is over the random permutation* f*)*

$$\mathbf{Pr}[A^f \to (h, x_1, \ldots, x_{|h|})] = \mathsf{negl}(n)$$

where $h, |h| \leq n^c$ *is an oracle circuit and the output satisfies the conditions*

1. $f^{-1}(h^f(i)) = x_i$ *for all* $i = 1, \ldots, |h|$.
2. $x_i \neq x_j$ *for all* $1 \leq i < j \leq |h|$.
3. $\{h^f(1), \ldots, h^f(|h|)\} \cap Y_f^h = \emptyset$ *where* Y_f^h *is defined as*

$$Y_f^h = \{f(x)| \exists i, 1 \leq i \leq |h|, h^f(i) \text{ makes the oracle query } x\}$$

Proof (of Lemma). Consider any oracle TM A where A^f makes n^c oracle queries. After having used up all his oracle queries A^f must come up with an output $(h, x_1, \ldots, x_{|h|})$ where h satisfies conditions 2 and 3. Below we prove that with overwhelming probability there does not even exist an h which satisfies conditions 2 and 3 *and* where A^f has made all the queries $x_1, \ldots, x_{|h|}$ satisfying condition 1. But in this case, even when choosing an h which satisfies conditions 2 and 3, A^f would still have to guess at least one x_i (i.e. $f^{-1}(h^f(i))$). The

---

[11] To make the proof and the statement of Lemma 3 simple (i.e. purely information theoretic), we will consider a computationally unbounded TM with oracle access to a truly random permutation which it can access a polynomial number of times, whereas Lemma 2 is about a opptTM and permutations chosen randomly from some family of exponential size. But as already mentioned, considering a random permutation is fine as a opptTM cannot distinguish a random permutation from $f_{pk,n}$ where $pk \in_R \{0,1\}^n$ anyway. And considering any computationally unbounded oracle TM (instead of only opptTMs) makes the lemma only stronger.

probability that it will guess correctly (i.e. this x_i will satisfy condition 1) is negligible.[12] We must now prove the above statement, i.e. that an h satisfying conditions 2 and 3 and where A^f made all the queries $x_1, \ldots, x_{|h|}$ satisfying condition 1 exists only with negligible probability. Let $X_f^A, |X_f^A| = n^c$ denote all oracle queries made by A^f, i.e.

$$X_f^A := \{x \mid A^f \text{ makes the oracle query } x\}.$$

Now consider any fixed oracle circuit $h, |h| \leq n^c$ which satisfies the conditions 2 and 3. Let $X_f^h = \{f^{-1}(y) \mid y \in Y_f^h\}$, i.e.

$$X_f^h := \{x \mid \exists i, 1 \leq i \leq |h|, h^f(i) \text{ makes the oracle query } x\}$$

and let

$$H := \{f^{-1}(h^f(1)), \ldots, f^{-1}(h^f(|h|))\}.$$

Condition 3 states that $f(H) \cap f(X_f^h) = \emptyset$, and as f is a permutation this is equivalent to

$$H \cap X_f^h = \emptyset.$$

Given X_f^h and conditioned on h^f satisfying condition 3, the set H is a random subset of $\{0, 1\}^n \setminus X_f^h$. If condition 2 is satisfied then $|H| = |h|$ moreover $|X_f^h| \leq |h|^2 \leq n^{2c}$. Now the probability that $H \subseteq X_f^A$ can be upper bounded as (here the probability is over the random permutation f and for a fixed h conditioned on h^f satisfying conditions 2 and 3)

$$\mathbf{Pr}[H \subseteq X_f^A] = \prod_{i=0}^{|H|-1} \frac{|X_f^A| - |X_f^A \cap X_f^h| - i}{2^n - i - |X_f^h|}$$

$$\leq \left(\frac{|X_f^A|}{2^n - n^c - |X_f^h|} \right)^{|H|} = \left(\frac{n^c}{2^n - 2n^{2c}} \right)^{|h|}.$$

By taking the union bound over all oracle circuits $h, |h| \leq n^c$ we can now upper bound the probability that there exists an h satisfying conditions 2 and 3 and where A^f knows all x_i satisfying condition 1 as

$$\sum_{|h|=1}^{n^c} 2^{|h|} \left(\frac{n^c}{2^n - 2n^{2c}} \right)^{|h|} \leq \left(\frac{2n^c}{2^n - 2n^{2c}} \right) = \mathsf{negl}(n)$$

where in the first step we assumed that the sum takes it maximum for $|h| = 1$ which holds for all sufficiently large n.   ◇ □

---

[12] It can easily be upper bounded by $1/(2^n - n^c - n^{2c})$: given the n^c oracle queries (not containing the query x_i) made by A^f and additionally the $\leq n^{2c}$ oracle queries made by h^f on inputs $1, \ldots, |h|$ (which will not contain the query x_i because of condition 3), x_i is a random variable with the uniform distribution over a set of size $\geq 2^n - n^c - n^{2c}$.

4   Hard Games with Inversions

In the last section we have seen that a TDP which is secure for all hard games (and thus has the one-wayness and claw-freeness security property) cannot be black-box reduced to a FDH signature scheme. In this section we will see that even a stronger notion of hard games does not allow for such a reduction. We extend the definition of a hard-game and allow (a limited number of) inversion queries.[13] To motivate this let us define one more security property for TDPs which can be modelled as such a game.

- A TDP has the *one-way with $q(.)$-inversions* security property if it is one-way, even with an oracle for f_{pk}^{-1} that can be used at most $q(n)$ times on any input except the challenge $f_{pk}(x)$, i.e.[14]

$$\mathbf{Pr}_{KeyGen(1^n) \to (pk,td), x \in_R \{0,1\}^n}[A^{f_{pk}^{-1}(\cdot)}(f_{pk}(x), pk) = x] = \mathsf{negl}(n).$$

HARD GAME WITH $q(.)$ INVERSIONS. An opptTM C defines a hard game with $q(.)$ inversions if for a random permutation π and all opptTM A

$$\mathbf{Pr}[\langle A^{\pi(\cdot)}(1^n), C^{\pi(\cdot), \pi^{-1}(\cdot)}(1^n)\rangle \to 1] = \mathsf{negl}(n) \qquad (5)$$

where C may query the $\pi^{-1}(.)$ oracle at most $q(n)$ times. A TDP $(KeyGen, F)$ is secure for a hard game C with $q(.)$ inversions if

$$\mathbf{Pr}_{KeyGen(n) \to (pk,td)}[\langle A(pk), C^{F(pk,\cdot), F(td,\cdot)}(1^n)\rangle \to 1] = \mathsf{negl}(n).$$

The one-way with $q(.)$ inversions property is captured by such a game as follows:

- $C^{f(\cdot), f^{-1}(\cdot)}(1^n)$ samples $x \in_R \{0,1\}^n$ and sends $f(x)$ to A. Now C answers at most $q(n)$ queries $z \in \{0,1\}^n$ where $z \neq f(x)$ with $f^{-1}(z)$. C accepts and outputs 1 if it receives x as the $(q(n)+1)$'th message.

Lemma 2 is easily seen *not* to extend to hard games with $q(.)$ inversions already for $q(n) = O(n)$.[15] But if in the definition of the breaking oracle G we increase the number of requested signatures from $|h|$ to $|h| + q(n)$, then it is again impossible to find an accepting input for G and Lemma 2 can be shown to hold even for hard games with $q(.)$ inversions (using a similar strengthening of Lemma 3).

---

[13] For clarity of exposition we will consider the case where C expects only one permutation oracle, i.e. $t = 1$.

[14] This property directly implies some others like security for the *known-target inversion problem* introduced in [4]. Here one gets $q(n) + 1$ random challenges to invert and may use an inversion oracle on arbitrary inputs $q(n)$ times, i.e. once less than the number of challenges.

[15] For example A could win the one-way with cn inversions game (for some constant c) as follows. On challenge $y = f_{pk}(x)$ let $h(x) = x \oplus y$ (the c must satisfy $|h| \leq cn$). Now use the cn inversion queries to C to find an accepting input for the breaking oracle G, which will then provide the forgery $s = f_{pk}^{-1}(h(0) = y)$. Send $s = x$ to C and win the game.

Theorem 2. *For any polynomially bounded function $q(.)$, there is no black-box reduction from a TDP family which is secure for all hard games with $q(.)$ inversions to a FDH signature scheme secure against chosen-message attacks.*[16]

As corollaries we get that any assumption on TDPs which can be formulated as such a game will not be enough to get a reduction to FDH, e.g.

Corollary 2. *For any polynomially bounded function $q(.)$, there is no black-box reduction from a TDP satisfying the one-way with $q(.)$ inversions security property to a FDH signature scheme secure against chosen-message attacks.*

Finally, let us remark that in Theorem 2 it is necessary to have $q(.)$ bounded by some fixed polynomial. As if one allows a superpolynomial $q(n) \in n^{\omega(1)}$ then a TDP which is secure for all hard games with $q(.)$ inversions *can* be black-box reduced to a secure FDH signature scheme (note that this has a priori no practical consequences as such TDPs do not exist in the standard model). The main observation here is that the "existential forgery in a chosen message attack" (1) can be seen as a game where the challenger plays the role of the signing oracle $sign(m, td, i) \to f_{pk}^{-1}(h_i(m))$ and finally accepts if it receives a forgery from the prover A. We have not yet defined which FDH signature scheme to use in the above game. This scheme can not be arbitrary as we must make sure that this game is actually a *hard* game (i.e. no efficient A can win it when the oracles are instantiated with random permutations), but it is not difficult to construct a secure FDH scheme from random permutations $\pi_1, \pi_2, \ldots$ with only the signer having access to inversion oracles. For example, for a message space restricted to $\{0,1\}^n$, $sign(m) = \pi_1^{-1}(\pi_2(m))$ will already do it.

5   No Reduction from Trapdoor Permutations

We conclude the paper by observing that "the plain TDP assumption" implies an extreme black-box security limitation for FDH: not even security against a *one-chosen-message attack* can be achieved.[17]

Theorem 3. *There is no black-box reduction from trapdoor permutation families to a full-domain hash scheme secure against one-chosen-message attacks.*

For space reasons, we leave the proof of this theorem to a full version of the paper, here only discussing the key choice in the proof: that of the oracle G that breaks FDH but not TDP (cf. the proof of Theorem 1; F is the same as before).
   G is partly based on the collision-finding oracle of Simon [27]. However, his "collision-finding" oracle only works for length-decreasing hash functions. To deal

---

[16] Moreover, if we let $s(n) = \max\{|h_i| : i \in Range(Index(1^n))\}$ denote an upper bound on the size of a description of the hash function used, then the theorem even holds if we restrict the number of chosen message queries to $s(n) + q(n)$ and the size of the message-space of the signature scheme to $s(n) + q(n) + 1$.

[17] A one-chosen-message attack is precisely an attack where at most one query to the signing oracle is allowed.

with arbitrary (potentially length-increasing) hash functions, we extend Simon's oracle to forge the FDH signature of a special input when no "good collision" to h was found: But we have to make sure that the inversion of $f_{pk,n}$ resulting from this forgery will not allow the attacker to invert $f_{pk,n}$ on its own challenge.

More specifically, G takes inputs of the form $(1^L, 1^t, \langle h \rangle, pk)$, where $\langle h \rangle$ is the description of a *deterministic* oracle TM. Such a query can be seen as a request for a forgery to signature scheme $sign(m) = f_{pk,n}^{-1}(h(m))$, here $n = |pk|$. G first checks if the running time of $h^{F_0}(x)$ is $> \lfloor t/2 \rfloor$ for some $x \in \{0,1\}^L$ and all potential choices $F = F_0$ for the oracle F; if so, it outputs $\perp$ and stops. Otherwise, $u \in_R \{0,1\}^L$ is chosen and $w \equiv h^F(u)$ is computed; then $v \in_R \{0,1\}^L$ is sampled *conditioned* on $h^F(v) = w$. If $u = v$, $|w| = n$ *and* $L \geq \mu(n)$, where $\mu(n) = \log^2(n)$, F outputs $(u, v, y, f_{pk,n}^{-1}(w), s, \text{inversion})$, where s describes the computations $h^F(u)$ and $h^F(v)$ (including all F-queries). Else, the output of G is $(u, v, w, s, \text{collision})$, with s as above.

It is easy to see that with this oracle G one can forge $sign(m) = f_{pk,n}^{-1}(h(m))$ for any efficient h: Just query $G(1^L, 1^t, \langle h \rangle, pk)$ (for appropriate L, t) to obtain u and v with $h(u) = h(v)$. If $u \neq v$, we can forge a signature for v by asking the signing oracle to sign u, which will also give a signature of v. If $u = v$, then G also outputs $f_{pk,n}^{-1}(h(u))$, which is a direct forgery (with no queries to its signing oracle).

More subtleties arise when showing that G does not help the adversary to invert F. In particular, they motivate the need for t and the "μ-test" in G. The former avoids that an adversary $A = A^{F,G}$ receives the result of more oracle queries than she would have time to compute. As for the μ-test, it avoids that the TDP is *inverted on specific inputs*, for it makes negligible the probability that A could use G to invert some specific y of interest (e.g., the challenge in the one-wayness game). Indeed, for this to happen (1) a random u should map to y; *and* (2) a random preimage v of y ($v \in h^{-1}(y)$) should be u again. Now, it is easy to see that the probability of this happening is negligible indeed:

$$\mathbf{Pr}[u \in h^{-1}(y)]\mathbf{Pr}[v = u \mid v \in h^{-1}(y)] = \frac{|h^{-1}(y)|}{2^L} \frac{1}{|h^{-1}(y)|} = 2^{-L} \leq 2^{-\mu(n)}.$$

$$(6)$$

This simple fact turns out to be ultimately responsible for G not breaking the TDP property. More details will be given in the full version.

References

1. PKCS #1 v2.1, *RSA Cryptography Standard (draft)*, document available at www.rsa security.com/rsalabs/pkcs.
2. Mihir Bellare, Alexandra Boldyreva and Adriana Palacio. An Uninstantiable Random-Oracle-Model Scheme for a Hybrid-Encryption Problem. *EUROCRYPT 04*, pp. 171–188.
3. Mihir Bellare and Phillip Rogaway. Random Oracles are Practical: A Paradigm for Designing Efficient Protocols. *ACM CCS 93*, pp. 62–73.

4. Mihir Bellare, Chanathip Namprempre, David Pointcheval and Michael Semanko. The One-More-RSA-Inversion Problems and the Security of Chaum's Blind Signature Scheme. *J. of Cryptology*, **16** (3), pp. 185–215, 2003.
5. Alexandra Boldyreva and Marc Fischlin. Analysis of Random Oracle Instantiation Scenarios for OAEP and Other Practical Schemes. *CRYPTO 05*.
6. Ran Canetti, Uri Feige, Oded Goldreich and Moni Naor. Adaptively Secure Multi-Party Computation. *STOC 96*, pp. 22–24.
7. Ran Canetti, Oded Goldreich and Shai Halevi. The Random Oracle Methodology, Revisited. *STOC 98*, pp. 209–218.
8. Ran Canetti, Oded Goldreich and Shai Halevi. On the Random Oracle Methodology as Applied to Length-Restricted Signature Schemes. *TCC 04*, pp. 40–57.
9. Ran Canetti, Daniele Micciancio and Omer Reingold. Perfectly One-Way Probabilistic Hash Functions. *STOC 98*, pp. 131–140.
10. Jean-Sébastian Coron. On the Exact Security of Full Domain Hash. *CRYPTO 00*, pp. 229–235.
11. Jean-Sébastian Coron. Optimal Security Proofs for PSS and other Signature Schemes. *EUROCRYPT 02*, pp. 272–287.
12. Ivan Damgård. Collision-Free Hash Functions and Public-Key Signature Schemes. *EUROCRYPT 87*, pp. 203-216.
13. Whitfield Diffie and Martin Hellman. New directions in cryptography. IEEE Transactions on Information Theory 22 (1976), pp. 644–654.
14. Yevgeniy Dodis and Leonid Reyzin. On the Power of Claw-Free Permutations. *SCN 02*, pp. 55–73.
15. Amos Fiat and Adi Shamir. How to prove yourself: Practical solutions to identification and signature problems. *CRYPTO 86*, pp. 186–194.
16. Rosario Gennaro, Yael Gertner and Jonathan Katz. Lower Bounds on the Efficiency of Encryption and Digital Signature Schemes. *STOC 03*, pp. 417–425.
17. Rosario Gennaro and Luca Trevisan. Lower Bounds on the Efficiency of Generic Cryptographic Constructions. *FOCS 00*, pp. 305–313.
18. Yael Gertner, Tal Malkin, and Omer Reingold. On the Impossibility of Basing Trapdoor Functions on Trapdoor Predicates. *FOCS 01*, pp. 126–135.
19. Yael Gertner, Sampath Kannan, Tal Malkin, Omer Reingold and Mahesh Viswanathan. The Relationship Between Public-Key Encryption and Oblivious Transfer. *FOCS 00*, pp. 325–335.
20. Shafi Goldwasser and Yael Tauman. On the (In)security of the Fiat-Shamir Paradigm. *FOCS 03*, pp. 102–114.
21. Chun-Yuan Hsiao and Leonid Reyzin. Finding Collisions on a Public Road, or do Secure Hash Functions Need Secret Coins? *CRYPTO 04*, pp. 92–105.
22. Russell Impagliazzo and Steven Rudich. Limits on the Provable Consequences of One-Way Permutations. *STOC 89*, pp. 44–61.
23. Jeong Han Kim, Daniel R. Simon and Prasad Tetali. Limits on the Efficiency of One-Way Permutation-Based Hash Functions. *FOCS 99*, pp. 535–542.
24. Ben Lynn, Manoj Prabhakaran and Amit Sahai. Positive Results and Techniques for Obfuscation. *EUROCRYPT 04*, pp. 20–39.
25. Silvio Micali, Michael Rabin and Salil Vadhan. Verifiable Random Functions. *FOCS 99*, pp. 120–130.
26. Jesper Buus Nielsen. Separating Random Oracle Proofs from Complexity Theoretic Proofs: The Non-Committing Encryption Case. *CRYPTO 02*, pp. 111–126.
27. Daniel Simon. Finding Collisions on a One-Way Street: Can Secure Hash Functions be Based on General Assumptions? *EUROCRYPT 98*, pp. 334–345.
28. Hoeteck Wee. On Obfuscating Point Functions. *STOC 05*, pp. 523–532.

New Monotones and Lower Bounds in Unconditional Two-Party Computation

Stefan Wolf and Jürg Wullschleger

Département d'Informatique et R.O.
Université de Montréal, Québec, Canada
{wolf, wullschj}@iro.umontreal.ca

Abstract. Since bit and string *oblivious transfer* and *commitment*, two primitives of paramount importance in secure two- and multi-party computation, cannot be realized in an unconditionally secure way for both parties from scratch, *reductions* to weak information-theoretic primitives as well as between different variants of the functionalities are of great interest. In this context, we introduce three independent *monotones*—quantities that cannot be increased by any protocol—and use them to derive lower bounds on the *possibility* and *efficiency* of such reductions. An example is the transition between different versions of oblivious transfer, for which we also propose a new protocol allowing to increase the number of messages the receiver can choose from at the price of a reduction of their length. Our scheme matches the new lower bound and is, therefore, optimal.

1 Introduction, Motivation, and Main Results

The advantage of *unconditional* or *information-theoretic* security—as compared to computational security—is that it does not depend on any assumption on an adversary's computing power or memory space, nor on the hardness of any computational problem. Its disadvantage, on the other hand, is that it cannot be realized simply from scratch. This is why *reductions* are of great interest and importance in this context: Which functionality can be realized from which other? If a reduction is possible in principle, what is the best efficiency, i.e., the minimum number of instances of the initial primitive required per realization of the target functionality?

Two tasks of particular importance in secure two-party computation are *oblivious transfer* and *bit commitment*. Both primitives are known to be impossible to realize from scratch in an unconditionally secure way for both parties by any (classical or even quantum) protocol. On the other hand, they *can* be realized from noisy channels [6], [7], weak versions of oblivious transfer [3], correlated pieces of information [18], or the assumption that one of the parties' memory space is limited.

For the same reason, reductions between different variants of oblivious transfer are of interest as well: chosen 1-out-of-2 oblivious transfer from Rabin oblivious transfer [5], string oblivious transfer from bit oblivious transfer [3], 1-out-of-n

V. Shoup (Ed.): Crypto 2005, LNCS 3621, pp. 467–477, 2005.

oblivious transfer from 1-out-of-2 oblivious transfer, oblivious transfer from A to B from oblivious transfer from B to A [8], [19], and so forth. A number of lower bounds in the context of such reductions have been given, based on information-theoretic arguments [9], [13].

With respect to information-theoretic reductions between cryptographic and information-theoretic functionalities, quantities which never increase during the execution of a protocol—so-called *monotones* [4]—are of great importance. In *key agreement*, for instance, two parties A and B can start with correlated pieces of information X and Y, respectively, and try to generate a secret key S by public communication such that an adversary E, who initially knows a third random variable Z, is virtually ignorant about S. It has been shown in [16] that the *intrinsic information* [14] of A's and B's entire knowledge, given E's, is a monotone, i.e., cannot increase. This immediately leads to the following bound on the size of the generated key: $H(S) \leq I(X; Y \downarrow Z)$.

The main results of our paper are the following.

Three monotones of unconditional two-party computation.
> In Section 3, we define three information-theoretic quantities (the underlying notions are introduced in Section 2) and prove them to be monotones: No protocol allows for increasing them.

Lower bounds for oblivious-transfer reductions.
> In Section 4.1, we derive a new lower bound on the efficiency of reductions from one variant of oblivious transfer to another, and of realizing oblivious transfer from shared correlated pieces of information.

Optimally trading message length for choice in oblivious transfer.
> In Section 4.2, we present a new protocol allowing for increasing the number of messages from which the receiver can choose at the price of a reduction of their length. Our lower bound shows that the protocol is optimal.

New error bounds for bit commitment.
> In Section 5, we show new lower bounds on the probability of failure of any protocol for bit commitment based on correlated pieces of information.

2   Preliminaries: Common and Dependent Parts

As a preparation, we introduce two notions, namely the *common part* $X \wedge Y$ and the *dependent parts* $X \searrow Y$ and $Y \searrow X$ of two random variables X and Y. In the context of cryptography, the notions have first been used in [10], [12], [18]. Both notions have appeared previously in other information-theoretic contexts [11], the latter under the name of *sufficient statistics*.

2.1   Common Part

Let X and Y be two random variables with joint distribution P_{XY}. Intuitively, the common part $X \wedge Y$ is the maximal element of the set of all random variables that can be generated both from X and from Y.

Definition 1. [18] *Let X and Y be random variables with (disjoint) ranges $\mathcal{X}$ and $\mathcal{Y}$ and distributed according to P_{XY}. Then $X \wedge Y$, the* common part *of X and Y, is constructed in the following way:*

- *Consider the bipartite graph G with vertex set $\mathcal{X} \cup \mathcal{Y}$, and where two vertices $x \in \mathcal{X}$ and $y \in \mathcal{Y}$ are connected by an edge if $P_{XY}(x, y) > 0$ holds.*
- *Let $f_X : \mathcal{X} \to 2^{\mathcal{X} \cup \mathcal{Y}}$ be the function that maps a vertex $v \in \mathcal{X}$ of G to the set of vertices in the connected component of G containing v. Let $f_Y : \mathcal{Y} \to 2^{\mathcal{X} \cup \mathcal{Y}}$ be the function that does the same for a vertex $w \in \mathcal{Y}$ of G.*
- *$X \wedge Y := f_X(X) = f_Y(Y)$.*

Note that $X \wedge Y$ is symmetric—i.e., $X \wedge Y \equiv Y \wedge X$ [1]. There exist functions f_X and f_Y with $X \wedge Y = f_X(X) = f_Y(Y)$. Hence, $X \wedge Y$ can be calculated both from X and from Y.

Lemma 1. [18] *For all X, Y, and $\overline{C}$ for which there exist functions $\overline{f}_X$ and $\overline{f}_Y$ such that $\overline{C} = \overline{f}_X(X) = \overline{f}_Y(Y)$ holds, there exists a function g with $\overline{C} = g(X \wedge Y)$.*

2.2 Dependent Part

Intuitively, the *dependent part* of X from Y, denoted $X \searrow Y$, is the minimal element of the set of all random variables K that can be generated from X and are such that $X \longleftrightarrow K \longleftrightarrow Y$ is a Markov chain.

Definition 2. [10] *Let X and Y be two random variables, and let $f(x) = P_{Y|X=x}$. The* dependent part *of X from Y is defined as $X \searrow Y := f(X)$.*

Lemma 2 shows that all of X that is dependent on Y is included in $X \searrow Y$, i.e., more formally, $I(X; Y | X \searrow Y) = 0$ holds or, equivalently, $X, X \searrow Y$, and Y form a Markov chain.

Lemma 2. [10] *For all X and Y, $X \longleftrightarrow (X \searrow Y) \longleftrightarrow Y$ is a Markov chain.*

On the other hand, there does not exist a random variable with the same properties that is "smaller" than $X \searrow Y$.

Lemma 3. [18] *Let X, Y, and $\overline{K}$ be random variables such that there exists a function $\overline{f}$ such that $\overline{K} = \overline{f}(X)$ and $X \longleftrightarrow \overline{K} \longleftrightarrow Y$ hold. Then there exists a function g with $X \searrow Y = g(\overline{K})$.*

---

[1] We say that two random variables A and B are equivalent, denoted by $A \equiv B$, if there exists a bijective function $g : \mathcal{A} \to \mathcal{B}$ such that $B = g(A)$ holds with probability 1.

3   Three Two-Party-Protocol Monotones

In this section we show that the following three quantities are monotones, i.e., cannot increase during the execution of any protocol based on (noiseless) communication and (lossless) processing (where X' and Y' are the random variables summarizing the entire information accessible to A and B, respectively):

$$H(Y' \searrow X'|X') \,,$$
$$H(X' \searrow Y'|Y') \,,$$
$$I(X'; Y'|X' \wedge Y') \,.$$

We first show that local randomness generation and data processing, and second, that noiseless bi-directional communication do not allow for increasing any of these quantities.

3.1   Invariance Under Randomness Generation and Data Processing

Lemma 4. *Let X, Y, and Z be random variables such that $X \longleftrightarrow Y \longleftrightarrow Z$ is a Markov chain. Then we have*

$$X \searrow [Y, Z] \equiv X \searrow Y \,.$$

Proof. We have $P_{YZ|X=x} = P_{Y|X=x}P_{Z|Y}$. Therefore, for all $x, x' \in \mathcal{X}$, the function $P_{YZ|X=x}$ is different from $P_{YZ|X=x'}$ if and only if $P_{Y|X=x}$ is different from $P_{Y|X=x'}$.   □

Lemma 5. *Let W, X, and Y be random variables such that $W \longleftrightarrow X \longleftrightarrow Y$ is a Markov chain. Then we have*

$$[W, X] \searrow Y \equiv X \searrow Y \,.$$

Proof. We have $P_{Y|W=w,X=x} = P_{Y|X=x}$. Therefore, for all $w, w' \in \mathcal{W}$ and $x, x' \in \mathcal{X}$, the function $P_{Y|W=w,X=x}$ is different from $P_{Y|W=w',X=x'}$ if and only if $P_{Y|X=x}$ is different from $P_{Y|X=x'}$.   □

Lemma 6. *Let X, Y, and Z be random variables such that $X \longleftrightarrow Y \longleftrightarrow Z$ is a Markov chain. Then we have*

$$X \wedge [Y, Z] \equiv X \wedge Y \,.$$

Proof. We have $P_{XYZ} = P_{XY}P_{Z|Y}$. Let us look at the connection graph between all the values x and (y, z) for which $P_X(x) > 0$ and $P_{YZ}(y, z) > 0$ hold. Then x and (y, z) are connected if and only if $P_{XYZ}(x, y, z) > 0$ holds. Since $P_{Z|Y}(z, y) > 0$, this holds if and only if $P_{XY}(x, y) > 0$ holds. Hence, $X \wedge [Y, Z] \equiv X \wedge Y$.   □

Theorem 1 shows that local data processing does not increase any of the quantities in question. It is a direct consequence of Lemmas 4, 5, and 6.

Theorem 1. *Let X, Y, and Z be random variables with $X \longleftrightarrow Y \longleftrightarrow Z$. Then we have*

$$H([Y, Z] \searpoint X | X) = H(Y \searpoint X | X) \ ,$$
$$H(X \searpoint [Y, Z] | [Y, Z]) = H(X \searpoint Y | Y) \ ,$$
$$I(X; [Y, Z] | X \wedge [Y, Z]) = I(X; Y | X \wedge Y) \ .$$

3.2   No Increase by Communication

We now show that the same holds with respect to noise-free communication between A and B. We first prove three lemmas.

Lemma 7. *Let X and Y be random variables and f a function. Then*

$$X \searpoint [Y, f(X)] \equiv [X \searpoint Y, f(X)] \ .$$

Proof. Let $h_1(X) := X \searpoint [Y, f(X)]$ and $h_2(X) := [X \searpoint Y, f(X)]$, and let $F = f(X)$. We have $P_{YF|X} = P_{Y|X}P_{F|X}$. For all x, x' with $h_1(x) = h_1(x')$, we have $P_{YF|X=x} = P_{YF|X=x'}$, which holds exactly if $P_{Y|X=x} = P_{Y|X=x'}$ and $f(x) = f(x')$ hold, which is equivalent to $h_2(x) = h_2(x')$. Hence, $X \searpoint [Y, f(X)] \equiv [X \searpoint Y, f(X)]$. □

Lemma 8. *Let X and Y be random variables and f a function. Then there exists a function g such that*

$$[Y, f(X)] \searpoint X = g([Y \searpoint X, f(X)]) \ .$$

Proof. Let $h_1(X, Y) := [Y, f(X)] \searpoint X$ and $h_2(X, Y) := [Y \searpoint X, f(X)]$. For all x, x', y, and y' with $h_2(x, y) = h_2(x', y')$, we have $P_{X|Y=y} = P_{X|Y=y'}$ and $f(x) = f(x')$. It follows $P_{X|Y=y, f(X)=f(x)} = P_{X|Y=y', f(X)=f(x)}$, and, hence, $h_1(x, y) = h_1(x', y')$. Therefore, there must exist a function g with $h_1 = g \circ h_2$. □

Lemma 9. *Let X, Y, and Z be random variables. There exists a function f such that*

$$X \wedge Y = f([X, Z] \wedge Y) \ .$$

Proof. $X \wedge Y$ can be calculated from X, and, hence, also from $[X, Z]$. The statement now follows from Lemma 1. □

Theorem 2 states that noiseless communication between the two parties cannot increase any of the quantities in question.

Theorem 2. *Let X and Y be two random variables and f a function. Then we have*

$$H([Y, f(X)] \searpoint X | X) \leq H(Y \searpoint X | X) \ ,$$
$$H(X \searpoint [Y, f(X)] | Y, f(X)) \leq H(X \searpoint Y | Y) \ ,$$
$$I(X; [Y, f(X)] | X \wedge [Y, f(X)]) \leq I(X; Y | X \wedge Y) \ .$$

Proof. Using Lemmas 7, 8, and 9, we obtain

$$H([Y, f(X)] \searbackslash X|X) \leq H([Y \searbackslash X, f(X)]|X)$$
$$= H(Y \searbackslash X|X)$$

$$H(X \searbackslash [f(X), Y]|f(X), Y) = H([X \searbackslash Y, f(X)]|f(X), Y)$$
$$= H(X \searbackslash Y|f(X), Y)$$
$$\leq H(X \searbackslash Y|Y)$$

$$I(X; [f(X), Y]|X \wedge [f(X), Y]) \leq I(X; [f(X), Y]|f(X), X \wedge Y)$$
$$= I(X; Y|f(X), X \wedge Y)$$
$$\leq I(X; Y|X \wedge Y)$$

$\square$

Corollary 1 is a direct consequence of Theorems 1 and 2.

Corollary 1. *Let X and Y be two parties' entire knowledge before, and X' and Y' after the execution of a protocol including local data processing and noiseless communication. Then we have*

$$H(X' \searbackslash Y'|Y') \leq H(X \searbackslash Y|Y),$$
$$H(Y' \searbackslash X'|X') \leq H(Y \searbackslash X|X),$$
$$I(X'; Y'|X' \wedge Y') \leq I(X; Y|X \wedge Y).$$

4   Oblivious Transfer: Lower Bounds and an Optimal Reduction

4.1   New Bounds on Oblivious-Transfer Reductions

In m-out-of-n k-string oblivious transfer, denoted $\binom{n}{m}$-OT^k, the sender inputs n k-bit messages out of which the receiver can choose to read m, but does not obtain any further information about the messages; the sender, on the other hand, does not obtain any information on the receiver's choice.

In [1], it has been shown that $\binom{2}{1}$-OT^1 is *equivalent* to pieces of information with a certain distribution (in other words, oblivious transfer can be pre-computed and stored). This result generalizes to $\binom{n}{m}$-OT^k in a straight-forward way. By determining the corresponding values of the three monotones derived in Section 3 we can, thus, obtain lower bounds on the reducibility between different variants of oblivious transfer. The bound of Theorem 4 is an improvement on an earlier bound by Dodis and Micali [9].

Theorem 3. *Assume that there exists a protocol for realizing unconditionally secure* $\binom{N}{M}$-OT^K *from distributed random variables* X *and* Y. *Then we have*

$$(N - M)K \leq H(X \searrow Y|Y) \, ,$$

$$\log \binom{N}{M} \leq H(Y \searrow X|X) \, ,$$

$$MK \leq I(X; Y|X \wedge Y) \, .$$

Proof. As mentioned, $\binom{N}{M}$-OT^K can be stored. More specifically, the corresponding random variables X' and Y' arise when $\binom{N}{M}$-OT^K is executed with random and independent inputs. We have $H(X' \searrow Y'|Y') = (N - M)K$, $I(X'; Y'|X' \wedge Y') = MK$, and $H(Y' \searrow X'|X') = \log \binom{N}{M}$. The assertion now follows from Corollary 1. □

Theorem 4. *Assume that there exists a protocol for realizing unconditionally secure* $\binom{N}{M}$-OT^K *from* t *instances of* $\binom{n}{m}$-OT^k. *Then we have*

$$t \geq \max \left(\frac{(N - M)K}{(n - m)k} \, , \, \frac{\log \binom{N}{M}}{\log \binom{n}{m}} \, , \, \frac{MK}{mk} \right) \, .$$

Proof. Since $\binom{n}{m}$-OT^k is *equivalent* to the pieces of information obtained when the primitive is used with random inputs, we can assume that A and B start the protocol with such random variables X_i and Y_i, respectively, for $i = 1, \ldots, t$. (The first step in this protocol can be to restore $\binom{n}{m}$-OT^k from the shared information.) We have $H(X_i \searrow Y_i|Y_i) = (n - m)k$, $I(X_i; Y_i|X_i \wedge Y_i) = mk$, and $H(Y_i \searrow X_i|X_i) = \log \binom{n}{m}$. For $X = [X_1, \ldots, X_t]$ and $Y = [Y_1, \ldots, Y_t]$, we have $H(X \searrow Y|Y) = t(n - m)k$, $I(X; Y|X \wedge Y) = tmk$, and $H(Y \searrow X|X) = t \log \binom{n}{m}$. Now we can apply Theorem 3, and the statement follows. □

For the special case where $M = m = 1$, the obtained bounds are shown in Figure 1.

$t \geq \ldots$	$K \geq k$	$K < k$
$N \geq n$	$\frac{(N-1)K}{(n-1)k}$	$\max \left(\frac{(N-1)K}{(n-1)k}, \frac{\log N}{\log n} \right)$
$N < n$	$\frac{K}{k}$	1

Fig. 1. The bounds for $M = m = 1$

4.2 Optimally Trading Message Length for Choice

We present a protocol allowing for increasing the number of messages sent in oblivious transfer if, at the same time, their length is reduced. The number of calls to the original oblivious transfer equals the lower bound of Theorem 4.

Let $n, k, t \in \mathbf{N}, t > 1, N = n^t$, and $K \leq k/n^{t-1}$. Protocol 1 reduces $\binom{N}{1}$-OT^K to t instances of $\binom{n}{1}$-OT^k.

Protocol 1. Let A's inputs be $x_0, \ldots, x_{N-1} \in \{0,1\}^K$, whereas B's choice is $c \in \{0, \ldots, N-1\}$. Let $c = \sum_{i=0}^{t-1} c_i n^i$, $c_i \in \{0, \ldots, n-1\}$.

1. A chooses $R_0^0, R_1^0, \ldots, R_{n-1}^0, R_0^1, \ldots, R_{n-1}^{t-1} \in_R \{0,1\}^k$.
2. A and B run $\binom{n}{1}$-OT^k t times. In round $i \in \{0, \ldots, t-1\}$, A inputs $R_0^i, \ldots, R_{n-1}^i$, and B inputs c_i. B receives Y_i.
3. A and B subdivide each string R_j^i and Y_i into n^{t-1} pieces of length $K = k/n^{t-1}$: $R_j^i = R_j^i(0) \| \cdots \| R_j^i(n^{t-1} - 1)$, $Y_i = Y_i(0) \| \cdots \| Y_i(n^{t-1} - 1)$.
4. For every $j \in \{0, \ldots, N-1\}$, let $j = \sum_{i=0}^{t-1} j_i n^i$ and $d_j = \sum_{i=0}^{t-2} (j_i + j_{t-1} \bmod n) n^i$. A sends $m_j = x_j \oplus R_{j_0}^0(d_j) \oplus \cdots \oplus R_{j_{t-1}}^{t-1}(d_j)$ to B.
5. B calculates $d_c = \sum_{i=0}^{t-1} (c_i + c_{t-1} \bmod n) n^i$ and outputs $y = m_c \oplus Y_0(d_c) \oplus \cdots \oplus Y_{t-1}(d_c)$.

Theorem 5. *Protocol 1 is a perfect reduction of $\binom{N}{1}$-OT^K to $\binom{n}{1}$-OT^k for $N = n^t, t > 1$, and $K \leq k/n^{t-1}$.*

Proof. If both players are honest, we have $Y_i = R_{c_i}^i$ for all $i \in \{0, \ldots, t-1\}$. Therefore,

$$
\begin{aligned}
y &= m_c \oplus Y_0(d_c) \oplus \cdots \oplus Y_{t-1}(d_c) \\
&= x_c \oplus m_c \oplus R_{c_0}^0(d_c) \oplus \cdots \oplus R_{c_{t-1}}^{t-1}(d_c) \oplus Y_0(d_c) \oplus \cdots \oplus Y_{t-1}(d_c) \\
&= x_c \ .
\end{aligned}
$$

A does not receive any messages, so she does not get any information about c.

It remains to be proven that B only gets information about one value sent by A, even if he is given all the other values. First of all, note that if two different j and j' take the same value d, then $j_i + j_{t-1} \equiv j_i' + j_{t-1}' \pmod{n}$ holds for all $i \in \{0, \ldots, t-1\}$. It follows $j_{t-1} \neq j_{t-1}'$, and, hence, $j_i \neq j_i'$ for all $i \in \{0, \ldots, t-1\}$. Therefore, every $R_j^i(d)$ is used at most once in Step 4. B has to choose a value c_i in every round, so he will always be able to reconstruct x_c for $c = \sum_{i=0}^t c_i n^i$. But for every other value $x_{c'}$, $c' \neq c$, he is missing at least one of the $R_{c_i'}^i(d_{c'})$ for $i \in \{0, \ldots, t-1\}$. This value is a one-time pad on $x_{c'}$ since it is not used anywhere else. Therefore, B does not get any information about any $x_{c'}$ for $c' \neq c$, even if he is given all the other values $x_{c''}$ for $c'' \neq c'$. □

5   Bit and String Commitment: Tight Lower Bounds

Unlike oblivious transfer, bit commitment that is *perfectly secure* for both parties is impossible to achieve even when they share correlated pieces of information

X and Y initially. Intuitively speaking, the reason is that if the commitment is perfectly hiding, there must exist, after the "commit message," an "open message" to be accepted by the receiver for any possible value one can commit to. Theorems 6 and 7 make this precise and explicit by giving lower bounds on the success probability of such cheating by the committer, depending on the distribution P_{XY}. Our bounds are improvements on similar bounds presented in [2] and [15].

Theorem 6. *Assume that a commitment protocol exists where the committer initially knows a random variable X and the receiver knows Y. If the protocol is perfectly hiding and the committer has committed to a value $v \in \mathcal{V}$, then the probability p_s that she succeeds in opening the commitment to a different value $v' \neq v$ is at least*

$$p_s \geq 2^{-H(Y \setminus X | X)} .$$

Proof. Note first that under the given assumptions, there must also exist a commitment protocol with the same security properties if the parties are given X and $Y \setminus X$, respectively, since the part of Y that is independent of X can be simulated by B because $X \longleftrightarrow Y \setminus X \longleftrightarrow Y$ is a Markov chain. As the protocol is perfectly hiding, there must exist, for every value v', an opening of the commitment to v' that B accepts. Let y' be the value maximizing $P_{Y|X=x}$. A then opens the commitment for v' in such a way that B accepts if his value y is equal to y', and this is successful if $y = y'$ indeed holds. The expected probability of this event is

$$
\begin{aligned}
E_X \left[2^{-H_\infty(Y \setminus X | X = x)} \right] &\geq 2^{-E_X[H_\infty(Y \setminus X | X = x)]} \\
&\geq 2^{-E_X[H(Y \setminus X | X = x)]} \\
&= 2^{-H(Y \setminus X | X)} .
\end{aligned}
$$

In the first step, we have used Jensen's inequality. □

Theorem 7. *Assume that a commitment protocol exists where the committer initially knows a random variable X and the receiver knows Y. If the protocol is perfectly hiding and the committer has committed to a value $v \in \mathcal{V}$, then the probability p_s that she succeeds in opening the commitment to a different value $v' \neq v$ is at least*

$$p_s \geq 2^{-H(X \setminus Y) + \log(|\mathcal{V}| - 1)} .$$

Proof. We can assume without loss of generality that the pieces of information known to the parties are $X \setminus Y$ and Y. Let the committer hold x and commit to $v \in \mathcal{V}$, and let $v' \neq v$. Since the protocol is perfectly hiding, there must exist $x' \in \mathcal{X}$ such that the commit message sent corresponds to the correct commitment for v'. The probability of correctly guessing this value x', maximized over all v', is at least

$$2^{-H_\infty(X \setminus Y)}(|\mathcal{V}| - 1) \geq 2^{-H(X \setminus Y) + \log(|\mathcal{V}| - 1)} .$$ □

Corollary 2. *Assume that a commitment protocol exists where the committer initially knows a random variable X and the receiver knows Y. If the protocol is perfectly hiding and the committer has committed to a value $v \in \mathcal{V}$, then the probability p_s that she succeeds in opening the commitment to a different value $v' \neq v$ is at least*

$$p_s \geq \max \left(2^{-H(Y \searrow X|X)} , \; 2^{-H(X \searrow Y)+\log(|\mathcal{V}|-1)} \right) .$$

The commitment protocol of [17] achieves this bound: Given a prime number q, we have $H(X \searrow Y) = 2\log q$, $H(Y \searrow X|X) = \log q$, and $|\mathcal{V}| = q$. It is perfectly hiding, and the "binding error probability" p_s is

$$p_s = 1/q = 2^{-H(Y \searrow X|X)} .$$

6   Concluding Remarks

We have presented three information-theoretic quantities with the property that no two-party protocol can increase them—so-called *monotones*. Based on these, we have derived new lower bounds on the possibility and efficiency of realizing oblivious transfer and bit commitment from pieces of correlated information, as well as on reductions between different versions of oblivious transfer. Finally, we have proposed a new protocol for such a reduction of the latter kind which is optimal.

We suggest as an open problem to find a general reduction of $\binom{N}{M}$-OT^K to $\binom{n}{m}$-OT^k which attains the given lower bound for *any* choice of the parameters. Furthermore, it would be interesting and useful to find similar monotones for *multi*-party protocols.

Acknowledgments

The authors thank Don Beaver, Claude Crépeau, Anderson Nascimento, and Renato Renner for interesting discussions on the subject of this paper, and three anonymous reviewers for their helpful comments on an earlier version. This work was supported by Canada's NSERC and Québec's FQRNT.

References

1. D. Beaver. Precomputing oblivious transfer. *Advances in Cryptology—Proceedings of CRYPTO '95*, LNCS, Vol. 963, pp. 97–109, Springer-Verlag, 1992.
2. C. Blundo, B. Masucci, D. R. Stinson, and R. Wei. Constructions and bounds for unconditionally secure non-interactive commitment schemes. *Designs, Codes, and Cryptography*, 26(1-3): 97–110, 2002.
3. G. Brassard, C. Crépeau, and S. Wolf. Oblivious transfers and privacy amplification. *Journal of Cryptology*, Vol. 16, No. 4, pp. 219–237, 2003.

4. N. J. Cerf, S. Massar, and S. Schneider. Multipartite classical and quantum secrecy monotones. *Phys. Rev. A*, Vol. 66, No. 042309, 2002.
5. C. Crépeau. *Correct and private reductions among oblivious transfers*. Ph. D. thesis, MIT, 1990.
6. C. Crépeau. Efficient cryptographic protocols based on noisy channels. *Advances in Cryptology—Proceedings of CRYPTO '97*, LNCS, Vol. 1233, pp. 306–317, Springer-Verlag, 1997.
7. C. Crépeau, K. Morozov, and S. Wolf. Efficient unconditional oblivious transfer from almost any noisy channel. *Proceedings of Fourth Conference on Security in Communication Networks (SCN) '04*, LNCS, Vol. 3352, pp. 47–59, Springer-Verlag, 2004.
8. C. Crépeau and M. Sántha. On the reversibility of oblivious transfer. *Advances in Cryptology—Proceedings of Eurocrypt '91*, LNCS, Vol. 547, pp. 106–113, Springer-Verlag, 1991.
9. Y. Dodis and S. Micali, Lower bounds for oblivious transfer reductions, *Advances in Cryptology—Proceedings of EUROCRYPT '99*, LNCS, Vol. 1592, pp. 42–55, Springer-Verlag, 1999.
10. M. Fitzi, S. Wolf, and J. Wullschleger. Pseudo-signatures, broadcast, and multi-party computation from correlated randomness. *Advances in Cryptology—Proceedings of CRYPTO '04*, LNCS, Vol. 3152, pp. 562–579, Springer-Verlag, 2004.
11. P. Gacs and J. Körner, Common information is far less than mutual information, *Probl. Contr. Inform. Theory*, Vol. 2, pp. 149–162, 1973.
12. H. Imai, J. Müller-Quade, A. Nascimento, and A. Winter. Rates for bit commitment and coin tossing from noisy correlation. *Proceedings of the IEEE International Symposium on Information Theory (ISIT '04)*, IEEE, 2004.
13. U. Maurer. Information-theoretic cryptography. *Advances in Cryptography—Proceedings of CRYPTO' 99*, LNCS, Vol. 1666, pp. 47–64, Springer-Verlag, 1999.
14. U. Maurer and S. Wolf. Unconditionally secure key agreement and the intrinsic conditional information. *IEEE Transactions on Information Theory*, Vol. 45, No. 2, pp. 499–514, 1999.
15. A.C.A. Nascimento, J. Müller-Quade, A. Otsuka, and H. Imai. Unconditionally secure homomorphic pre-distributed commitments. *Proceedings of AAECC 2003*, pp. 87–97, 2003.
16. R. Renner and S. Wolf. New bounds in secret-key agreement: the gap between formation and secrecy extraction. *Advances in Cryptography—Proceedings of EUROCRYPT 2003*, LNCS, Vol. 2656, pp. 562–577, Springer-Verlag, 2003.
17. R. L. Rivest. Unconditionally secure commitment and oblivious transfer schemes using private channels and a trusted initializer. Unpublished manuscript, 1999.
18. S. Wolf and J. Wullschleger. Zero-error information and applications in cryptography. *Proceedings of 2004 IEEE Information Theory Workshop (ITW 2004)*, 2004.
19. S. Wolf and J. Wullschleger. Oblivious transfer is symmetric. *Cryptology ePrint Archive*, Report 2004/336. http://eprint.iacr.org/2004/336, 2004.
20. S. Wolf and J. Wullschleger. Oblivious transfer and quantum non-locality. *Quantum Physics e-print Archive*, quant-ph/0502030, 2005.

One-Way Secret-Key Agreement and Applications to Circuit Polarization and Immunization of Public-Key Encryption[*]

Thomas Holenstein and Renato Renner

Department of Computer Science,
Swiss Federal Institute of Technology (ETH),
Zürich, Switzerland
{thomahol, renner}@inf.ethz.ch

Abstract. Secret-key agreement between two parties Alice and Bob, connected by an insecure channel, can be realized in an information-theoretic sense if the parties share many independent pairs of correlated and partially secure bits. We study the special case where only one-way communication from Alice to Bob is allowed and where, for each of the bit pairs, with a certain probability, the adversary has no information on Alice's bit. We give an expression which, for this situation, exactly characterizes the rate at which Alice and Bob can generate secret key bits.

This result can be used to analyze a slightly restricted variant of the problem of polarizing circuits, introduced by Sahai and Vadhan in the context of statistical zero-knowledge, which we show to be equivalent to secret-key agreement as described above. This provides us both with new constructions to polarize circuits, but also proves that the known constructions work for parameters which are tight.

As a further application of our results on secret-key agreement, we show how to immunize single-bit public-key encryption schemes from decryption errors and insecurities of the encryption, a question posed and partially answered by Dwork, Naor, and Reingold. Our construction works for stronger parameters than the known constructions.

1   Introduction

Consider two parties, Alice and Bob, connected by an authentic but otherwise fully insecure communication channel. It is well known that it is impossible for Alice and Bob to establish information-theoretically secure private communication (see [16,11]). In particular, they are unable to generate an unconditionally secure key. This changes dramatically if we additionally assume that Alice and Bob have access to some correlated randomness on which an adversary has only partial information.

The initial correlation shared by Alice and Bob can originate from various sources. For example, Wyner [20] and, subsequently, Csiszár and Körner [3] have

---

[*] Supported by the Swiss National Science Foundation, project no. 200020-103847/1.

V. Shoup (Ed.): Crypto 2005, LNCS 3621, pp. 478–493, 2005.

studied a scenario where Alice and Bob are connected by a noisy channel on which an adversary has only limited access. Maurer [11] (cf. also [1]) proposed to consider a setting where a satellite broadcasts uniform random bits with low signal intensity, such that Alice, Bob, and also Eve cannot receive them perfectly. It has been shown that, in both settings, Alice and Bob can indeed generate an information-theoretically secure key and thus communicate secretly.

In this paper, we study one-way secret-key agreement, i.e., we assume that only one-way communication from Alice to Bob is allowed. We fully analyze the case where Alice and Bob hold many independent pairs of correlated bits, and where the only secrecy guarantee is that, for each of these pairs, with a certain probability, the adversary has no information about Alice's value. It turns out that this particular kind of information-theoretic secret-key agreement has interesting applications, even in the context of computational cryptography.

1.1   Secret-Key Agreement

Previous Work: Information-theoretically secure secret-key agreement from correlated information has first been proposed by Maurer in [11]. He considered a setting where Alice, Bob, and Eve hold many independent realizations of correlated random variables X, Y, and Z, respectively, with joint probability distribution P_{XYZ}. The (two-way) *secret-key rate* $S(X; Y|Z)$, i.e., the rate at which Alice and Bob can generate secret-key bits per realization of (X, Y, Z), has further been studied in [1] and later in [12], where the *intrinsic information* $I(X; Y \downarrow Z)$ is defined and shown to be an upper bound on $S(X; Y|Z)$, which, however, is not tight [13].

For *one-way* communication, it is already implied by a result in [3] and has later been shown in [1] that the secret-key rate $S_\rightarrow(X; Y|Z)$ is given by the supremum of $H(U|ZV) - H(U|YV)$, taken over all possible random variables U and V obtained from X.[1] However, as this is a purely information-theoretic result, it does not directly imply that there exists an *efficient* key-agreement protocol.

Our Contributions: In Section 2, we show that $H(U|ZV) - H(U|YV)$ is the exact rate at which Alice and Bob can *efficiently* generate a secret key. The methods used to show this are not new, but as far as we know this result has not appeared anywhere else in the literature.

Furthermore, we study the class of distributions P_{XYZ} where X and Y are random variables over $\{0, 1\}$ with some bounded error $\Pr[X \neq Y]$, and where all that is known about Z is that, with a certain probability, it does not give any information on X.[2] This class will be important for our applications. Using novel techniques, we give an explicitly computable lower bound on the one-way

---

[1] This result is proven with respect to a slightly different definition of the secret-key rate than we use. For completeness, we thus provide a new proof for this.

[2] As the *exact* distribution of the initial randomness—especially the part held by Eve—is usually not known, it is natural to consider such classes.

secret-key rate as well as a tight characterization of the parameters for which one-way secret-key agreement is possible.

1.2   Circuit Polarization

Previous Work: In [17], Sahai and Vadhan introduced the promise problem *statistical difference*. This problem is defined for parameters α and β, $\alpha > \beta$ as follows: given two circuits which, on uniform random input, produce output distributed according to C_0 and C_1 with the promise that the statistical distance of the distributions is either bigger than α or smaller than β, decide which of the two is the case. If $\alpha^2 > \beta$, Sahai and Vadhan show (cf. also [18]) how to *polarize* such a pair of circuits, i.e., they give an efficient construction which takes a pair of circuits and outputs a pair of circuits such that, if the statistical distance of the initial pair was at least α to begin with, the statistical distance of the resulting distributions is very high (i.e., at least $1 - 2^{-k}$ for an arbitrary k), and if the statistical distance of the pair was at most β, then the resulting statistical distance is very small (i.e., at most 2^{-k}).

In order to achieve this only two operations are used, where one of them increases the statistical distance of the distributions at hand and the other reduces the distance. These operations share a certain similarity to operations used in secret-key agreement protocols (cf. [11] and [19]), and indeed, in [5], Dwork et al. note that their construction to immunize public-key encryption is inspired by [17].

Our Contributions: In this work, we make the connection anticipated in [5] explicit by showing that one-way secret-key agreement for the class of distributions given in Section 2.3 is equivalent to the task of circuit polarization, as long as one is restricted to black-box constructions (i.e., the description of the circuits given may not be used), only gives independent and uniform random inputs to the circuits, and directly outputs the samples of the circuits. These restrictions may seem quite strong at first, but the method given in [18] is of this form. Using our bounds for secret-key agreement, we show that such a polarization method *does only exist* if $\alpha^2 > \beta$, i.e., the bounds given in [18] are optimal for this class of constructions.

1.3   Immunization of Public-Key Encryption

Previous Work: Assume that a public-key encryption scheme for single bits is given, which has the property that the receiver may succeed in decrypting correctly only with probability $(1+\alpha)/2$, and also that a potential eavesdropper Eve may have probability up to $(1+\beta)/2$ to find the message, for some constants (or functions of a security parameter) α and β. In [5], the question was posed whether such a scheme can be used to get a public-key encryption scheme in the usual sense. Furthermore, the question was answered in the positive sense in two cases: if $\alpha^2 > c\beta$, for some absolute constant $c \gg 1$, a scheme is given. Also, for every constant $\beta < 1$ a construction which works for some constant $\alpha < 1$ is given. However, this construction is not very strong: for example, for $\beta = 1/2$, the constant α is about $1 - 2^{-15}$. Note that Dwork et al. make no attempt to optimize these constants.

In [7] a similar question was asked for key agreement where Alice and Bob may communicate an arbitrary number of rounds.

Our Contributions: Using a lemma from [7], we improve the result of [5] and show that, for constants α and β, immunizing such an encryption scheme is possible if $\alpha^2 > \beta$. Furthermore we show that, in a setting which is sufficiently black-box, this is optimal.

1.4   Notation

Throughout the paper, we use calligraphic letters (e.g. $\mathcal{X}$, $\mathcal{Y}$, $\mathcal{U}$) to denote sets. Uppercase letters (X, Y, U) are used to denote random variables, and lowercase letters denote values of these random variables.

For distributions P_X and P_Y over the same domain $\mathcal{X}$, we denote by $\|P_X - P_{X'}\| = \frac{1}{2} \sum_{x \in \mathcal{X}} |P_X(x) - P_{X'}(x)|$ the statistical distance between P_X and $P_{X'}$. If X and X' are the corresponding random variables we sometimes slightly abuse notation and write $\|X - X'\|$ instead.

The min-entropy (or Rényi entropy of order ∞) of a random variable X over $\mathcal{X}$ is defined as $H_\infty(X) := - \log(\max_{x \in \mathcal{X}} P_X(x))$, and the Rényi entropy of order zero is $H_0(X) := \log(|\{x \in \mathcal{X} | P_X(x) > 0\}|)$. More generally, the conditional Rényi entropies are

$$H_\infty(X|Y) := - \log\big(\max_{x \in \mathcal{X}, y \in \mathcal{Y}} P_{X|Y}(x|y)\big),$$

$$H_0(X|Y) := \log\big(\max_{y \in \mathcal{Y}} |\{x \in \mathcal{X} | P_{X|Y}(x|y) > 0\}|\big).$$

Additionally, we use the following smoothed versions of these entropy measures [14], which are defined for any $\varepsilon \geq 0$:

$$H_\infty^\varepsilon(X) := \max_{P_{X'} : \|P_X - P_{X'}\| \leq \varepsilon} H_\infty(X'),$$

$$H_\infty^\varepsilon(X|Y) := \max_{P_{X'Y'} : \|P_{XY} - P_{X'Y'}\| \leq \varepsilon} H_\infty(X'|Y').$$

For a random variable X, we write $U \leftarrow X$ if, for any other random variable Z, $U \leftrightarrow X \leftrightarrow Z$ is a Markov chain. It other words, one can think of U as being obtained from X by sending it through a channel without considering anything else.

2   One-Way Secret-Key Agreement

2.1   Notation and Definitions

A one-way secret-key agreement protocol has three important parameters, which are denoted by the same letters throughout the paper: the length m of the secret key produced, a security parameter k, and the number n of instances of the initial random variables used. It will be convenient in applications to assume that, for given m and k, n can be computed by a function $n(k,m)$.

Definition 1 (Protocol). *A one-way secret-key agreement (OW-SKA) proto-col on $\mathcal{X} \times \mathcal{Y}$ consists of the function $n(k,m) : \mathbb{N} \times \mathbb{N} \to \mathbb{N}$; a function fam-ily, called Alice, with parameters k and m, mapping n instances of X to a bit string $S_A \in \{0,1\}^m$ (the secret key) and a bit string $\Gamma \in \{0,1\}^*$ (the communica-tion); and a function family, called Bob, with parameters k and m, mapping Γ and n instances of Y to a bit string $S_B \in \{0,1\}^m$. The protocol is efficient if $n(k,m)$, Alice, and Bob can be computed by probabilistic Turing machines in time $\mathrm{poly}(k,m)$. The rate of the protocol is $\lim_{k\to\infty} \lim_{m\to\infty} \frac{n(k,m)}{m}$.*

The goal of secret-key agreement is to get a secure key (S_A, S_B), i.e., two strings which are likely to be equal and look like a uniform random string to Eve. We can define this as follows:

Definition 2 (Secure Key). *A pair (X,Y) over $\{0,1\}^m \times \{0,1\}^m$ of random variables is ε-secure with respect to Z if*

$$\|P_{XYZ} - P_{UU} \times P_Z\| \le \varepsilon,$$

where P_{UU} is the probability distribution over $\{0,1\}^m \times \{0,1\}^m$ given by

$$P_{UU}(x,y) = \begin{cases} 2^{-m} & \text{if } x = y \\ 0 & \text{otherwise.} \end{cases}$$

We say that a protocol is secure if it generates a 2^{-k}-secure key with respect to the information Eve has after the protocol execution, that is, the initial ran-domness $Z_1, \ldots, Z_n$ and the communication Γ. In some cases it is desirable to have a protocol which works for a class of distributions rather than for a sin-gle distribution (since one may not know the exact distribution of the random variables).

Definition 3 (Secure protocol). *A OW-SKA protocol on $\mathcal{X} \times \mathcal{Y}$ is secure on a probability distribution P_{XYZ} over $\mathcal{X} \times \mathcal{Y} \times \mathcal{Z}$ if, for any $k, m \in \mathbb{N}$, (S_A, S_B) is 2^{-k}-secure with respect to $(Z_1, \ldots, Z_{n(k,m)}, \Gamma)$.*

A protocol is secure on a set $\mathcal{P} = \{P_{XYZ}\}$ of tripartite probability distribu-tions if it is secure for every distribution $P_{XYZ} \in \mathcal{P}$.

This way, we can study the secret-key rate of classes of distributions, and also of single distributions.

Definition 4 (One-way secret-key rate). *The one-way secret key rate $S_\to(\mathcal{P})$ of a set $\mathcal{P} = \{P_{XYZ}\}$ of probability distributions is the supremum of the rate of any OW-SKA protocol which is secure on $\mathcal{P}$.*

We also write $S_\to(X; Y|Z)$ to denote the one-way secret-key rate of a single distribution, i.e., $S_\to(X; Y|Z) := S_\to(\{P_{XYZ}\})$.

2.2 A General Expression for the One-Way Secret-Key Rate

In this section, we derive a simple expression for the one-way secret-key rate of a general tripartite probability distribution. As mentioned in the introduction, Theorem 1 has already been known to hold for general (not necessarily efficient) protocols [3,1].

Theorem 1. *Let P_{XYZ} be a probability distribution. Then*

$$S_{\rightarrow}(X;Y|Z) = \sup_{V \leftarrow U \leftarrow X} H(U|ZV) - H(U|YV).$$

Moreover, the same identity holds if only efficient secret-key agreement protocols are considered.

For the (two-way) secret-key rate no comparable expression is known. We prove Theorem 1 in two steps: We first give an efficient protocol for any rate which is below $\sup_{V \leftarrow U \leftarrow X} H(U|ZV) - H(U|YV)$ (Theorem 2) and then show that no protocol can achieve a higher rate (Theorem 3).

The protocol is based on the following proposition. A proof can be found in [6]; the idea is to concatenate a random linear code with a Reed-Solomon code such that the decoding can be done in polynomial time.

Proposition 1. *For any memoryless channel and any rate s below the capacity it is possible to design codes $C : \mathcal{X}^{\ell} \to \mathcal{X}^n$ of growing length $\ell \to \infty$ with overall complexity (construction, encoding, and decoding) of order n^2 and decoding error probability $2^{-c_s \cdot n}$ where the constant c_s only depends on the channel and the rate s.*

Furthermore, we use the following from [15]:

Proposition 2. *Let P_{XYZ} be a probability distribution. For any $\varepsilon, \varepsilon' \geq 0$,*

$$H_{\infty}^{\varepsilon+\varepsilon'}(X|Y) \geq H_{\infty}^{\varepsilon}(XY) - H_0(Y) - \log\left(\frac{1}{\varepsilon'}\right)$$

$$H_{\infty}^{\varepsilon+\varepsilon'}(XY) \geq H_{\infty}^{\varepsilon}(X) + H_{\infty}^{\varepsilon'}(Y|X).$$

More generally, the statement still holds if all entropies are conditioned on some additional random variable Z.

Also, we use the following from [8].[3]

Proposition 3. *Let $(X_1, Y_1), \ldots, (X_n, Y_n)$ i.i.d. according to P_{XY}. Then,*

$$H_{\infty}^{\varepsilon}(X_1, \ldots, X_n|Y_1, \ldots, Y_n) \geq nH(X|Y) - 4\sqrt{n \log(1/\varepsilon)} \log(|\mathcal{X}|).$$

---

[3] Note that a non-quantitative version of this statements follows directly from the asymptotic equipartition property (see, e.g., [4]). A slightly different quantitative version can be found in [9].

Also, we need the left-over hash-lemma, first given in [9] (see also [2]). The function Ext used is a two-universal hash-function.

Proposition 4 (Left-Over Hash-Lemma). *Let X be a random variable over $\{0,1\}^n$. Let U^n and U^m be independent and uniform over $\{0,1\}^n$ and $\{0,1\}^m$, respectively. There exists an efficiently computable function $\mathrm{Ext} : \{0,1\}^n \times \{0,1\}^n \to \{0,1\}^m$ such that, if $H_\infty(X|Z) \geq m + 2\log(1/\varepsilon)$, then $\|(\mathrm{Ext}(X,U^n),U^n,Z) - (U^m,U^n,Z)\| \leq \varepsilon$.*

Lemma 1. *Let P_{XYZ} be an arbitrary tripartite probability distribution and let $r < H(X|Z) - H(X|Y)$. There exists a constant d_r (depending on P_{XYZ} and r) and an efficient OW-SKA protocol secure on P_{XYZ} such that $n \leq \max(m/r, k \cdot d_r)$.*

Proof. Let γ be such that $r + 3\gamma = H(X|Z) - H(X|Y)$. Let $\oplus$ be an arbitrary group operation over $\mathcal{X}$. For the channel which maps x to a pair $(X \oplus x, Y)$ (by choosing X and Y according to P_{XY}, this channel has capacity $H_0(X) - H(X|Y)$), we use Proposition 1 to get a code $\mathcal{C}$ with rate $s := H_0(X) - H(X|Y) - \gamma$.

Choose n such that $n \geq \frac{m}{r}$, $n \geq \frac{32k \log^2 |\mathcal{X}|}{\gamma^2}$, $n \geq \frac{2k}{c_s}$ and such that there exists a code of this length in the family guaranteed by Proposition 1. From the code of this length, Alice now choses a random word $C = (C_1, \ldots, C_n)$ and sends, for all i, $C_i \oplus X_i$ to Bob, who gets $(Y_i, C_i \oplus X_i)$. Using the property of the code, Bob can find the original codeword C with probability $1 - 2^{-c_s \cdot n} \geq 1 - 2^{-2k}$. Alice then sends a randomly chosen seed of a two-universal hash-function which maps the codeword to a string of length m. Both parties apply the hash-function and output S_A and S_B, respectively.

We show that Eve gets no information with probability 2^{-k}. For this, we set $\varepsilon := 2^{-2k}$. From Proposition 2 and using $H_\infty(C|Z^n) = ns$ (which follows from the fact that the codeword is chosen uniformly at random), we get

$$H_\infty^{2\varepsilon}(C|(X^n \oplus C)Z^n)$$

$$\geq H_\infty(C|Z^n) + H_\infty^\varepsilon(X^n \oplus C|CZ^n) - H_0(X^n \oplus C|Z^n) - \log\left(\frac{1}{\varepsilon}\right)$$

$$= ns + H_\infty^\varepsilon(X^n|Z^n) - nH_0(X) - 2k.$$

From Proposition 3 we get $H_\infty^\varepsilon(X^n|Z^n) \geq nH(X|Z) - 4\log(|\mathcal{X}|)\sqrt{2nk}$. Together, we obtain

$$H_\infty^{2\varepsilon}(C|(X^n \oplus C)Z^n)$$

$$\geq \underbrace{n(H(X|Z) - H(X|Y) - \gamma)}_{=n(r+2\gamma)} - \underbrace{4\log(|\mathcal{X}|)\sqrt{2nk}}_{=\sqrt{32nk \log^2 |\mathcal{X}|} \leq n\gamma} - 2k$$

$$\geq nr + n\gamma - 2k.$$

From Proposition 4 we see that it is possible to extract a secret key of length $nr + n\gamma - 6k > nr \geq m$ such that Eve gets no information except with probability $2\varepsilon + \varepsilon = 3 \cdot 2^{-2k} \leq 2^{-k}$. $\qquad\square$

Theorem 2. *Let P_{XYZ} be an arbitrary probability distribution and let r be a constant satisfying $r < \sup_{V \leftarrow U \leftarrow X} H(U|ZV) - H(U|YV)$. There exists a constant d_r and an efficient OW-SKA protocol which is secure on P_{XYZ} and uses at most $\max(m/r, k \cdot d_r)$ instances of the initial random variables.*

Proof. For any random variables U and V such that $V \leftarrow U \leftarrow Z$, Alice can compute an instance of U and V locally from an instance of X, and then send V over the channel to Bob (and Eve). The result then follows from Lemma 1.

Theorem 3. *Let P_{XYZ} be a probability distribution. Then*

$$S_\rightarrow(X;Y|Z) \le \sup_{V \leftarrow U \leftarrow X} H(U|ZV) - H(U|YV).$$

Proof (sketch). We show that $\sup_{V \leftarrow U \leftarrow X} H(U|ZV) - H(U|YV)$ does not increase by any step of a one-way key-agreement protocol. More precisely, it does not increase by local processing of either Alice or Bob, or sending a message from Alice to Bob. Furthermore, taking n copies of X, Y, and Z at most multiplies this quantity by n. Finally, if Alice and Bob share a secret key of length m, then this quantity is arbitrarily close to m (depending on k). Hence, the initial quantity is at least m.   □

Proof (Theorem 1). From Theorems 2 and 3.   □

2.3   The Secret Key Rate of a Class of Binary Distributions

In this section we study the one-way secret-key rate of a general class of distributions. Namely, for parameters α and β, we assume that Alice and Bob are given binary random variables X and Y which have the property that they are equal with probability at least $(1 + \alpha)/2$ (i.e., X and Y have correlation at least α). Furthermore, we assume that with probability $1 - \beta$, the random variable Z does not give any information about X. This class will also be of interest for Sections 3 and 4.

Definition 5. *Let $\mathcal{D}(\alpha, \beta)$ be the set of probability distributions P_{XYZ} over $\{0, 1\} \times \{0, 1\} \times \mathcal{Z}$ satisfying*

- $\Pr[X = 0] = \Pr[X = 1] = \frac{1}{2}$,
- $\Pr[X = Y] \ge \frac{1+\alpha}{2}$,
- *there exists an event $\mathcal{E}$ such that $H(X|Z\mathcal{E}) = 1$ and $\Pr[\mathcal{E}] \ge 1 - \beta$.*

It is not hard to see that we could similarly look at the distributions which satisfy $\|P_{Y|X=0} - P_{Y|X=1}\| \ge \alpha$ and $\|P_{Z|X=0} - P_{Z|X=1}\| \le \beta$, where Y does not have to be binary. This condition implies that Bob can apply a function to Y such that a distribution from $\mathcal{D}(\alpha, \beta)$ results. Furthermore, all distributions in $\mathcal{D}(\alpha, \beta)$ satisfy this characterization.

Some distributions in $\mathcal{D}(\alpha, \beta)$ have a higher secret-key rate than others, of course. We will see that the following distribution has the lowest secret-key rate of all distributions in $\mathcal{D}(\alpha, \beta)$. Intuitively, this distribution gives as much information to Eve as possible, and makes X and Y as independent as possible under the constraints of Definition 5. For a random variable X, let[4] $\mathbb{X}_\lambda(X)$ be the random variable describing the output of a binary symmetric channel taking input X, i.e., $P_{\mathbb{X}_\lambda(X)|X=0}(0) = P_{\mathbb{X}_\lambda(X)|X=1}(1) = \frac{1+\lambda}{2}$.

Definition 6. *For fixed α, β, the characteristic distribution P_{XYZ} of $\mathcal{D}(\alpha, \beta)$ is given by the following random process: we chose $X \in \{0,1\}$ uniformly at random. Then, Y is given as $\mathbb{X}_\alpha(X)$, and Z over $\{0, 1, \bot\}$ is given as the output of an erasure channel with symmetric error probability $1 - \beta$ on input X, i.e., $\Pr[Z = X] = \beta$, independently of X, and $\Pr[Z = \bot] = 1 - \beta$.*

We are now ready to formulate our main statement of this section, namely an easily computable expression for $S_\rightarrow(\mathcal{D}(\alpha, \beta))$:

Theorem 4. *For any α, β, let P_{XYZ} be the characteristic distribution of $\mathcal{D}(\alpha, \beta)$. Then,*

$$S_\rightarrow(\mathcal{D}(\alpha, \beta)) = \max_\lambda H(\mathbb{X}_\lambda(X)|Z) - H(\mathbb{X}_\lambda(X)|Y). \qquad (1)$$

In particular, if $\alpha^2 > \beta$ then $S_\rightarrow(\mathcal{D}(\alpha, \beta)) \geq \frac{1}{7}(\alpha^2 - \beta)^2$ and if $\alpha^2 \leq \beta$ then $S_\rightarrow(\mathcal{D}(\alpha, \beta)) = 0$.

Since the term in the maximum of (1) only involves random variables whose distribution is explicitly known (cf. Definition 6) we can get the following form of it (where $h(x)$ is the binary entropy function):

$$g_{\alpha,\beta}(\lambda) := H(\mathbb{X}_\lambda(X)|Z) - H(\mathbb{X}_\lambda(X)|Y)$$
$$= (1 - \beta) + \beta h\left(\frac{1+\lambda}{2}\right) - h\left(\frac{1+\alpha\lambda}{2}\right) \qquad (2)$$

In order to prove Theorem 4, we need a few properties of $g_{\alpha,\beta}$ (see also Fig. 1). As they can be obtained with standard tools from calculus, the (not very interesting) proof is omitted.

Lemma 2. *Let the function $g_{\alpha,\beta} : [-1, 1] \rightarrow \mathbb{R}$ be as in (2). If $\alpha^2 \leq \beta$, then $g_{\alpha,\beta}(\lambda) \leq 0$ for all $\lambda \in [-1, 1]$ and $g_{\alpha,\beta}$ is concave. If $\alpha^2 > \beta$, then $g_{\alpha,\beta}$ has one local minimum at $\lambda = 0$ with $g_{\alpha,\beta}(0) = 0$ and two local maxima at $-\lambda^+$ and λ^+, $\lambda^+ \in (0, 1]$ with $g_{\alpha,\beta}(-\lambda^+) = g_{\alpha,\beta}(\lambda^+) \geq \frac{1}{7}(\alpha^2 - \beta)^2$. Furthermore, $g_{\alpha,\beta}$ is concave in $[-1, -\lambda^+]$ and $[\lambda^+, 1]$.*

We first give an upper bound on $S_\rightarrow(X; Y|Z)$ for the distribution from Definition 6.

Lemma 3. *Let P_{XYZ} be the characteristic distribution of $\mathcal{D}(\alpha, \beta)$. Then, $S_\rightarrow (X; Y|Z) \leq \max_\lambda g_{\alpha,\beta}(\lambda)$, where $g_{\alpha,\beta}$ is defined by (2).*

---

[4] The symbol $\mathbb{X}$ is supposed to look like a binary symmetric channel, and can be pronounced as *noise*.

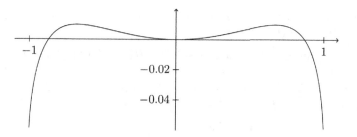

Fig. 1. Plot of $g_{\alpha,\beta}(\lambda)$ with $\alpha = 0.8$ and $\beta = 0.59$

Proof. We know that $S_{\rightarrow}(X;Y|Z) = \sup_{V \leftarrow U \leftarrow X} H(U|ZV) - H(U|YV)$ (Theorem 1). Let $P_{U|X}$ and $P_{V|U}$ be fixed channels. It is sufficient to show that $H(U|ZV) - H(U|YV) \leq \max_\lambda g_{\alpha,\beta}(\lambda)$.

We can rewrite $H(U|ZV) - H(U|YV)$ as

$$H(U|ZV) - H(U|YV) = H(UZV) - H(UYV) - (H(ZV) - H(YV))$$
$$= H(Z|UV) - H(Y|UV) - (H(Z|V) - H(Y|V)). \quad (3)$$

Consider now a fixed pair (u, v). Setting $\frac{1+\lambda_{uv}}{2} := \Pr[X{=}0|U{=}u, V{=}v]$ and $\frac{1+\lambda_v}{2} := \Pr[X{=}0|V{=}v]$, a straightforward computation yields:

$$H(Z|U{=}u, V{=}v) - H(Y|U{=}u, V{=}v) = h(\beta) + \beta h\left(\tfrac{1+\lambda_{uv}}{2}\right) - h\left(\tfrac{1+\alpha\lambda_{uv}}{2}\right)$$
$$H(Z|V{=}v) - H(Y|V{=}v) = h(\beta) + \beta h\left(\tfrac{1+\lambda_v}{2}\right) - h\left(\tfrac{1+\alpha\lambda_v}{2}\right).$$

Because $g_{\alpha,\beta}$ differs from these expressions only by a constant, together with (3) this gives

$$H(U|ZV) - H(U|YV) = \mathbf{E}_{uv}[g_{\alpha,\beta}(\lambda_{uv})] - \mathbf{E}_{v}[g_{\alpha,\beta}(\lambda_v)].$$

Using $\mathbf{E}_u[\lambda_{uv}] = \lambda_v$, where u is chosen according to the probability distribution $P_{U|V=v}$, we thus obtain

$$H(U|ZV) - H(U|YV) = \mathbf{E}_v\Big[\mathbf{E}_u[g_{\alpha,\beta}(\lambda_{uv})] - g_{\alpha,\beta}(\mathbf{E}_u[\lambda_{uv}])\Big].$$

For every fixed v, we can use Lemma 2 to obtain the following upper bound on the term in the expectation:

$$\mathbf{E}_u[g_{\alpha,\beta}(\lambda_{uv})] - g_{\alpha,\beta}(\mathbf{E}_u[\lambda_{uv}]) \leq \max_\lambda g_{\alpha,\beta}(\lambda) - g_{\alpha,\beta}(0) = \max_\lambda g_{\alpha,\beta}(\lambda),$$

which can now be inserted in the above expression. □

Next, we show that for every distribution in $\mathcal{D}(\alpha, \beta)$ we can achieve at least this rate by sending X over a fixed channel. As we want a protocol which works for *every* distribution in $\mathcal{D}(\alpha, \beta)$, it is important that this processing only depends on the parameters α and β.

Lemma 4. *Let α, β be fixed, $P_{XYZ} \in \mathcal{D}(\alpha, \beta)$, $g_{\alpha, \beta}$ as in (2), $\lambda \in [0,1]$. Then $H(\mathbb{X}_\lambda(X)|Z) - H(\mathbb{X}_\lambda(X)|Y) \geq g_{\alpha, \beta}(\lambda)$.*

Proof. Using a simple calculation we see that $H(\mathbb{X}_\lambda(X)|Z) \geq (1-\beta) + \beta h(\frac{1+\lambda}{2})$. To see that $H(\mathbb{X}_\lambda(X)|Y) \leq h(\frac{1+\alpha\lambda}{2})$, let B be a uniform random bit, which is independent of X and Y. Then we obtain $H(\mathbb{X}_\lambda(X)|Y) = H(\mathbb{X}_\lambda(X \oplus B)|Y \oplus B, B)$ $\leq H(\mathbb{X}_\lambda(X \oplus B)|Y \oplus B) = h(\frac{1+\alpha\lambda}{2})$. □

We are now ready to prove Theorem 4.

Proof (Theorem 4). From Theorem 1, Lemmata 2, 3, and 4. □

Furthermore, together with the results of the previous section, we conclude that for any α, β with $\alpha^2 > \beta$ there exists an efficient one-way secret-key agreement protocol secure on $\mathcal{D}(\alpha, \beta)$.

Corollary 1. *Let α, β be constant with $\alpha^2 > \beta$. There exists an efficient one-way secret-key agreement protocol with rate $(\alpha^2 - \beta)^2/8$ which is secure on $\mathcal{D}(\alpha, \beta)$.*

Proof. From Theorems 1 and 4.[5] □

3   Circuit Polarization

3.1   Polarization and Oblivious Polarization

Circuit polarization was introduced by Sahai and Vadhan in [17] in the context of statistical zero knowledge. It can be described as follows: assume that two circuits are given, which on uniform random input yield output distributions C_0 and C_1 over $\{0,1\}^\ell$, respectively. We look for an efficient method to *polarize* the circuits: if $\|C_0 - C_1\| \geq \alpha$, for some parameter α, the method should output circuits which are near disjoint, if $\|C_0 - C_1\| \leq \beta$, for some parameter β, then the method should output circuits which produce very close distributions.

In general, such a method uses a description of the circuits given. Here, we focus on methods which use the given circuits in a black-box manner, obliviously and with random input only.

Definition 7. *An* oblivious polarization method *for parameters α and β is a randomized algorithm which, on input k and b, outputs "query bits" $Q_b^1, \ldots, Q_b^n$ and a string R_b. For two distributions C_0 and C_1 it satisfies:*

$$\|C_0 - C_1\| \geq \alpha \implies \|(C_{Q_0^1}, \ldots, C_{Q_0^n}, R_0) - (C_{Q_1^1}, \ldots, C_{Q_1^n}, R_1)\| \geq 1 - 2^{-k}$$

$$\|C_0 - C_1\| \leq \beta \implies \|(C_{Q_0^1}, \ldots, C_{Q_0^n}, R_0) - (C_{Q_1^1}, \ldots, C_{Q_1^n}, R_1)\| \leq 2^{-k}.$$

The method is efficient if the algorithm runs in time polynomial in k.

---

[5] Technically speaking, Theorem 1 only guarantees that such a protocol exists for one single distribution, and in general the protocol *will* depend on the distribution at hand. Of course the protocol cannot depend on the distribution of $P_{Z|XY}$, but the distribution $P_{Y|X}$ can vary in $\mathcal{D}(\alpha, \beta)$, so we have to be careful. However, since the protocol just uses an error correcting code which is too strong for some distributions, it is easy to see that this is not a problem.

Note that the method given in [18] to polarize circuits is oblivious in this sense.[6] The method given to *invert* the statistical distance is not oblivious (and cannot possibly be).

3.2   Equivalence of Polarization and Secret-Key Agreement

The goal of this section is to prove that an oblivious polarization method for parameters α and β is equivalent to a secret-key agreement protocol (for a one bit key) secure on $\mathcal{D}(\alpha, \beta)$, as defined in Section 2.3.

Theorem 5. *There exists an oblivious polarization method for parameters α and β if and only if there exists a one-way secret-key agreement protocol secure on $\mathcal{D}(\alpha, \beta)$. Moreover, there exists an* efficient *oblivious polarization method if and only if there exists a protocol with efficient encoding (i.e., Alice is efficient).*

We prove Theorem 5 in both directions separately, and start by showing that a polarization method implies the existence of a one-way secret-key agreement protocol:

Lemma 5. *Let an oblivious polarization method for parameters α, β be given. Then there exists a one-way secret-key agreement protocol which is secure on $\mathcal{D}(\alpha, \beta)$. Furthermore, if the polarization method is efficient, then Alice is efficient.*

Proof. It is sufficient to show how to get a one-way secret-key agreement protocol for $m := 1$ bit.

The number of random variables $n := n(k, 1)$ the protocol uses is set to the number of queries produced by the polarization method. Alice first simulates the polarization method with input k and a uniform random bit b which yields R_b and $Q_b^1, \ldots, Q_b^n$. Subsequently, Alice sends R_b as well as $(X_1 \oplus Q_b^1, \ldots, X_n \oplus Q_b^n)$ as communication to Bob, and outputs b as secret bit.

We show that Bob can find b with high probability from the communication and Y^n (this may not necessarily be efficient). Since $P_{XYZ} \in \mathcal{D}(\alpha, \beta)$ the random variables $C_0 := (X, Y)$ and $C_1 := (1 \oplus X, Y)$ satisfy $\|C_0 - C_1\| \geq \alpha$. Furthermore, $Y_1, \ldots, Y_n$ and the communication gives Bob a sample of the distribution $(C_{Q_b^1}, \ldots, C_{Q_b^n}, R_b)$. The definition of the polarization method now implies that a statistical test can find b except with probability exponentially small in k.

Also the protocol is secure against Eve: consider the random variable $D_0 := (Z, X)$ and the random variable $D_1 := (Z, X \oplus 1)$. Here, $P_{XYZ} \in \mathcal{D}(\alpha, \beta)$ implies $\|D_0 - D_1\| \leq \beta$, and Eve sees exactly a sample of $(D_{Q_b^1}, \ldots, D_{Q_b^n}, R_b)$, which is independent of b except with probability exponentially small in k. □

On the other hand, a one-way secret-key agreement protocol yields a polarization method:

---

[6] In fact, R_0 and R_1 are empty in the method given.

Lemma 6. *Let a one-way secret-key agreement protocol secure on $\mathcal{D}(\alpha, \beta)$ be given. Then, there exists an oblivious polarization method for parameters α and β using $n(k, 1)$ copies of the given distribution. Furthermore, if Alice is efficient, then the polarization method is efficient.*

Proof. Throughout the proof we only need key agreement for one key bit and set $m := 1$. On input b and k, the polarization method first chooses random (uniform and independent) queries $Q_b^1, \ldots, Q_b^n$. Then Alice is simulated with random variables $X_1 := Q_b^1, \ldots, X_n := Q_b^n$, which yields communication C, and a secret bit S. The string R_b is then defined as $R_b := (C, S \oplus b)$.

We first show that $\|C_0 - C_1\| \geq \alpha$ implies that $\|(C_{Q_0^1}, \ldots, C_{Q_0^n}, R_0) - (C_{Q_1^1}, \ldots, C_{Q_1^n}, R_1)\|$ is exponentially close to 1. For this, it is enough to show how to find b from $(C_{Q_b^1}, \ldots, C_{Q_b^n}, R_b)$ with probability almost 1. $\|C_0 - C_1\| \geq \alpha$ implies that there exists a function y (a statistical test) such that setting $Y_i := y(C_{Q_b^i})$ gives $\Pr[Y_i = Q_b^i] \geq \frac{1+\alpha}{2}$. Thus we can use the decoding algorithm Bob of the secret key agreement protocol to reconstruct S with very high probability. Since $S \oplus b$ is also given, we can find b.

Now assume that $\|C_0 - C_1\| \leq \beta$. Consider the tripartite probability distribution P_{XYZ} where $X = Y$ is a uniform random bit, and $Z = C_X$. It is not hard to see that $P_{XYZ} \in \mathcal{D}(\alpha, \beta)$. Thus, in the one-way secret-key agreement protocol (using this distribution) Eve will see exactly a sample of $(C_{Q_b^1}, \ldots, C_{Q_b^n}, R_b)$ and the value $S \oplus b$. The properties of the protocol imply that this distribution is statistically independent (with high probability) of S. Furthermore, in the construction above only $S \oplus b$ depends on b, which implies the lemma. $\square$

Proof (Theorem 5). Follows from Lemmata 5 and 6.

Furthermore, since we know for which parameters α and β a protocol exists, we get:

Corollary 2. *There exists an (efficient) oblivious black-box polarization method for constant parameters α and β if and only if $\alpha^2 > \beta$.*

Proof. Using Theorem 4 and Corollary 1. Additionally, we observe that if $S_\rightarrow(X; Y|Z) = 0$ then no one-way secret-key agreement protocol can exist, since one could use it to get a positive rate. $\square$

As mentioned before such a polarization method was already given in [18]. However, it was unknown that this is tight for oblivious methods.

3.3   Further Improvements

Note that instead of using the code as guaranteed in Proposition 1, we could have used a random linear code in this application (where the code is chosen by Alice and a description is sent as communication). In this case, the resulting polarization method is very efficient, as only $k \cdot \text{poly}((\alpha^2 - \beta)^{-1})$ copies of the circuits are needed. If this method is used in a statistical zero-knowledge proof system however, the prover needs additional power since he needs to decode a random linear code.

Finally, a statistical zero knowledge proof for the promise problem statistical difference (with parameters α and β) can be realized as follows: the two given circuits are sampled obliviously and uniformly at random by the verifier, sending the samples to the prover. The information *which* circuit was sampled is used as random variables $X_1, \ldots, X_n$ in a one-way secret-key agreement protocol, whose communication is also sent to the prover. Now, if the given instance produces distributions with statistical distance at least α, then the prover gets the same information as Bob does, and he can prove this to the verifier by sending back the secret key. If the circuits produce distributions with statistical distance at most β, the prover gets the same information as Eve does, and cannot find the secret key. Thus, it can be useful to use protocols which yield more than one secret bit, as this immediately reduces the error of the zero-knowledge proof.

4   Immunizing Bit Encryption Schemes

In this section we study the implications of our work on the task of immunizing bit encryption schemes. Thus, we assume that a public-key encryption scheme for bits is given, which has a certain probability of being correct, and a certain security.

Definition 8. *A $(\alpha(k), \beta(k))$-secure public-key bit encryption scheme is a triple (G, E, D) of probabilistic polynomial time algorithms such that*

- *Algorithm G, on input 1^k produces a pair $(\mathsf{pk}, \mathsf{sk})$.*
- *For a random bit $b \in \{0, 1\}$, $\Pr[D_{\mathsf{sk}}(E_{\mathsf{pk}}(b)) = b] > \frac{1+\alpha(k)}{2}$, where the probability is over the randomness of G (giving the pair $(\mathsf{pk}, \mathsf{sk})$), E, D, and the choice of b.*
- *For any polynomial time algorithm A, and a uniform random bit b: $\Pr[A(\mathsf{pk}, E_{\mathsf{pk}}(b)) = b] < \frac{1+\beta(k)}{2}$, where the probability is over the randomness of A, G, E, and the choice of b.*

If such a scheme is (α, β)-secure for every function $1 - \alpha = \beta \in \frac{1}{\mathrm{poly}(k)}$, we say that it is a secure public-key encryption scheme.

We can combine information-theoretic and computational protocols to obtain the following:

Lemma 7. *Let $\alpha, \beta : \mathbb{N} \to [0, 1]$ be noticeable and computable in time $\mathrm{poly}(k)$. Let (G, E, D) be a (α, β)-secure public-key bit encryption scheme. If there exists an efficient one-way secret-key agreement protocol secure on $\mathcal{D}(\alpha, \beta)$, then there exists a secure public-key encryption scheme (G', E', D').*

The proof of this is very similar to the corresponding Lemma in [7]. Due to space constraints it is only sketched here.

Proof (sketch). We only need one key bit, and therefore we will set $m = 1$ throughout the proof. On input 1^k, algorithm G' then does $n(k, 1)$ invocations of algorithm G with input 1^k. This gives a key pair $(\mathsf{pk}', \mathsf{sk}')$, such that both the public- and the secret-key are n-tuples $\mathsf{pk}' = (\mathsf{pk}_1, \ldots, \mathsf{pk}_n)$ and $\mathsf{sk}' = (\mathsf{sk}_1, \ldots, \mathsf{sk}_n)$.

To encrypt a bit b with public key pk', Alice first encrypts n random bits $X_1, \ldots, X_n$ with the underlying scheme, i.e., X_i is encrypted with E_{pk_i}. It then uses the information-theoretic one-way secret-key agreement protocol, where the X_i are used as random variables. Let S_A be the resulting secret bit. The output of algorithm E' is then the encryption of $X_1, \ldots, X_n$, the communication of the information-theoretic one-way secret key agreement protocol, and $S_A \oplus b$.

It is easy to see that the communication together with the secret key suffices to decode the encrypted bit. Furthermore, the security of the protocol can be shown using a standard hybrid argument together with the uniform hard-core lemma given in [7] (see also [10]).   □

Lemma 7 together with Corollary 1 implies that a (α, β)-secure public-key cryptosystem can be used to get a secure public-key cryptosystem if $\alpha^2 > \beta$. For a limited class of reductions this is tight: a *strong black-box reduction* is a black-box reduction which allows Alice and Bob to use such a cryptosystem only in a way such that it can be modeled by an oracle where Alice and Bob obtain random bits X and Y, respectively, and an attacking algorithm obtains information Z.[7]

Theorem 6. *Let α and β be constants. There exists a strong black-box reduction from a (α, β)-secure public-key cryptosystem to a secure public-key cryptosystem if and only if $\alpha^2 > \beta$.*

Proof. If $\alpha^2 > \beta$, this is implied by Lemma 7 and Corollary 1.

Assume now that $\alpha^2 \leq \beta$, and assume that a reduction is given. It is easy to see that for suitably chosen random variables X, Y and Z an attacker can break every protocol in polynomial space from the information given. Consequently, by giving the attacker access to a PSPACE-complete oracle we can obtain a contradiction.   □

References

1. Rudolph Ahlswede and Imre Csiszàr. Common randomness in information theory and cryptography—part I: Secret sharing. *IEEE Transactions on Information Theory*, 39(4):1121–1132, 1993.
2. Charles H. Bennett, Gilles Brassard, Claude Crépeau, and Ueli Maurer. Generalized privacy amplification. *IEEE Transaction on Information Theory*, 41(6):1915–1923, 1995.

---

[7] As an example, this excludes the possibility of using the (α, β)-secure cryptosystem to obtain a one-way function.

3. Imre Csiszár and János Körner. Broadcast channels with confidential messages. *IEEE Transactions on Information Theory*, 22(6):644–654, 1978.

4. Thomas M. Cover and Joy A. Thomas. *Elements of Information Theory*. John Wiley & Sons, Inc., first edition, 1991. ISBN 0-471-06259-6.

5. Cynthia Dwork, Moni Naor, and Omer Reingold. Immunizing encryption schemes from decryption errors. In Christian Cachin and Jan Camenisch, editors, *EURO-CRYPT 2004*, volume 3027 of *LNCS*, pages 342–360, 2004.

6. Ilya I. Dumer. Concatenated codes and their multilevel generalizations. In V. S. Pless and W. C. Huffman, editors, *The Handbook of Coding Theory*, volume 2, chapter 23, pages 1191–1988. North-Holland, Elsevier, 1998.

7. Thomas Holenstein. Key agreement from weak bit agreement. In *Proceedings of the 37th STOC*, pages 664–673, 2005.

8. Thomas Holenstein and Renato Renner. On the smooth Rényi entropy of independently repeated random experiments. manuscript, 2005.

9. Russell Impagliazzo, Leonid A. Levin, and Michael Luby. Pseudo-random generation from one-way functions (extended abstract). In *Proceedings of the 21st STOC*, pages 12–24, 1989.

10. Russell Impagliazzo. Hard-core distributions for somewhat hard problems. In *36th FOCS*, pages 538–545, 1995.

11. Ueli Maurer. Secret key agreement by public discussion. *IEEE Transaction on Information Theory*, 39(3):733–742, 1993.

12. Ueli Maurer and Stefan Wolf. Unconditionally secure key agreement and the intrinsic conditional information. *IEEE Transaction on Information Theory*, 45(2):499–514, 1999.

13. Renato Renner and Stefan Wolf. New bounds in secret-key agreement: The gap between formation and secrecy extraction. In Eli Biham, editor, *EUROCRYPT 2003*, volume 2656 of *LNCS*, pages 562–577, 2003.

14. Renato Renner and Stefan Wolf. Smooth Rényi entropy and applications. In *Proceedings of 2004 IEEE International Symposium on Information Theory*, page 233. IEEE, 2004.

15. Renato Renner and Stefan Wolf. Simple and tight bounds for information reconciliation and privacy amplification. Manuscript, 2005.

16. Claude E. Shannon. Communication theory of secrecy systems. *Bell Systems Technical Journal*, 28:656–715, 1949.

17. Amit Sahai and Salil Vadhan. A complete promise problem for statistical zero-knowledge. In *The 38th FOCS*, pages 448–457, 1997.

18. Amit Sahai and Salil Vadhan. Manipulating statistical difference. In Panos Pardalos, Sanguthevar Rajasekaran, and José Rolim, editors, *DIMACS Series*, volume 43, pages 251–270, 1999.

19. Stefan Wolf. *Information-Theoretically and Computationally Secure Key Agreement in Cryptography*. PhD thesis, ETH Zürich, 1999.

20. Aaron D. Wyner. The wire-tap channel. *Bell Systems Technical Journal*, 54:1355–1387, 1975.

A Quantum Cipher with Near Optimal Key-Recycling

Ivan Damgård, Thomas Brochmann Pedersen*, and Louis Salvail*

BRICS**, FICS***, Dept. of Computer Science, University of Århus
{ivan, pede, salvail}@brics.dk

Abstract. Assuming an insecure quantum channel and an authenticated classical channel, we propose an unconditionally secure scheme for encrypting classical messages under a shared key, where attempts to eavesdrop the ciphertext can be detected. If no eavesdropping is detected, we can securely re-use the entire key for encrypting new messages. If eavesdropping is detected, we must discard a number of key bits corresponding to the length of the message, but can re-use almost all of the rest. We show this is essentially optimal. Thus, provided the adversary does not interfere (too much) with the quantum channel, we can securely send an arbitrary number of message bits, independently of the length of the initial key. Moreover, the key-recycling mechanism only requires one-bit feedback. While ordinary quantum key distribution with a classical one time pad could be used instead to obtain a similar functionality, this would need more rounds of interaction and more communication.

Keywords: Quantum cryptography, key-recycling, unconditional security, private-key encryption.

1 Introduction

It is well known that only assuming a quantum channel and an authenticated classical channel, Quantum Key Distribution (QKD) can be used to generate an unconditionally secure shared key between two parties. If we want to use this key for encrypting classical messages, the simplest way is to use it as a one-time pad. This way, an m-bit key can be used to encrypt no more than m message bits, since re-using the key would not be secure (without extra assumptions like in the bounded storage model[19,10,13]).

However, if we allow the same communication model for message transmission as for key exchange — which seems quite natural — an obvious question is whether we might gain something by using the quantum channel to transmit

---

 * Part of this research was funded by European projects PROSECCO and SECOQC.
 ** Basic Research in Computer Science (www.brics.dk),
    funded by the Danish National Research Foundation.
 *** Foundations in Cryptography and Security,
    funded by the Danish Natural Sciences Research Council.

V. Shoup (Ed.): Crypto 2005, LNCS 3621, pp. 494–510, 2005.

ciphertexts. The reason why this might be a good idea is that the ciphertext is now a quantum state, and so by the laws of quantum mechanics, the adversary cannot avoid affecting the ciphertext when trying to eavesdrop. We may therefore hope being able to detect — at least with some probability — whether the adversary has interacted with the ciphertext. Clearly, if we know he has not, we can re-use the entire key. Even if he has, we may still be able to bound the amount of information he can obtain on the key, and hence we can still re-use part of the key. Note that the authenticated classical channel is needed in such a scheme, in order for the receiver to tell the sender whether the ciphertext arrived safely, and possibly also to exchange information needed to extract the part of the key that can be re-used. Such a system is called a Quantum Key-Recycling Scheme (QKRS).

A possible objection against QKRS is that since it requires interaction, we might as well use QKD to generate new key bits whenever needed. However, in the model where the authenticated classical channel is given as a black-box (i.e., not implemented via a shared key) QKD requires at least three messages: the quantum channel must be used, and the authenticated channel must be used in both directions, since otherwise the adversary could impersonate one of the honest parties. Further, each move requires a substantial amount of communication (if N qubits were transmitted then the two classical moves require more than N classical bits each). Finally, N is typically larger than the length of the secret-key produced. Hence, if we can build a QKRS scheme that is efficient, particularly in terms of how much key material can be re-used, this may be an advantage over straightforward use of QKD.

From a more theoretical point of view, our work can be seen as a study of the recycling capabilities of quantum ciphers in general. In particular, how many key bits can be recycled, and how much feedback information must go from receiver to sender in order to guarantee the security of the recycled key? How do these capabilities differ from those of classical ciphers? In this paper we give precise answers to these questions.

The idea behind a QKRS originates from Bennett and Brassard during the early days of quantum cryptography[4]. Although they did not provide any fully satisfying solution or security proof, their approach to the problem is similar to our. More recently Leung studied recycling of quantum keys in a model where Alice and Bob are allowed three moves of interaction[12]. In this model, however, quantum key distribution can be applied. Leung also suggested that classical keys can be recycled when no eavesdropping is detected. In [16], a QKRS was proposed based on quantum authentication codes[2]. The key-recycling capabilities of their scheme can be described in terms of 2 parameters: the message length m and the security parameter ℓ. The scheme uses $2m + 2\ell$ bits of key, and is based on quantum authentication schemes that, as shown in [2], must always encrypt the message. The receiver first checks the authenticity of the received quantum state and then sends the result to the sender on the authenticated channel. Even when the receiver accepts, the adversary may still have obtained a small amount of information on the key. The receiver therefore also sends a universal hash function, and privacy amplification is used to extract from the original key a secure key of length $2m + \ell$. If the receiver rejects then a secure key of length $m + \ell$ can be extracted.

In this paper, we propose a QKRS for encrypting classical messages. Our QKRS is based on a new technique where we append a k-bit classical authentication tag to the message, and then encrypt the $n = m + \ell$-bit plaintext using the W_n-quantum cipher introduced in [8]. The authentication is based on universal hashing using an m-bit key. The cipher uses $2n = 2(m + \ell)$ bits of key, where $m + \ell$ bits are used as a one-time pad, and $m + \ell$ bits are used to select in which basis to send the result, out of a set of $2^{m+\ell}$ so called mutually unbiased bases. Thus, the entire key of the QKRS consists of $3m + 2\ell$ bits. The receiver decrypts and checks the authentication tag. If the tag is correct, we can show that the adversary has exponentially small information about the key, and the entire key can therefore be recycled. If the tag is incorrect, we can still identify $2m + \ell$ bits of the key, about which the adversary has no information, and they can therefore be re-used. Since this subset of bits is always the same, the receiver only needs to tell the sender whether he accepts or not.

Being able to recycle the entire key in case the receiver accepts is of course optimal. On the other hand, we can show that any QKRS must discard at least $m - 1$ bits of key in case the receiver rejects. Since m can be chosen to be much larger than ℓ, discarding $m + \ell$ bits, as we do, is almost optimal.

In comparison with earlier works, our technique completely eliminates the use of privacy amplification, and hence reduces the communication on the authenticated channel to a single bit. Moreover, we can recycle the entire key when the receiver accepts the authentication tag. Hence, in scenarios where interference from the adversary is not too frequent, our keys can last much longer than with previous schemes, even though we initially start with a longer key.

Our results differ from those of [16], since quantum authentication based QKRS do not guarantee the privacy of the authentication tag. Therefore, part of the key must be discarded even if the receiver accepts. Instead of quantum authentication, we use classical Wegman and Carter authentication codes[6] and a quantum encryption of classical messages[8] applied to both the message and the tag. This construction allows to recycle the entire authentication key securely.

The scheme we introduce can also be used as an authentication code for quantum messages. However, it requires a longer secret-key than the scheme in [2], but allows for recycling the authentication key entirely upon acceptance.

Our QKRS is composable since the security is expressed in terms of distance from uniform. The secret-keys and plaintexts are private when, from the adversary's point of view, they look like uniformly distributed random variables. This has been shown to provide *universal composability* in the quantum world[17].

We end this introduction with some remarks on the authenticated classical channel. Having such a channel given for free as a black-box may not be a realistic assumption, but it is well known that it can be implemented assuming

the players initially have a (short) shared key.[1] In this model, the distinction
between QKD and QKRS is not as clear as before, since we now assume an initial
shared key for both primitives. Indeed, our QKRS can be seen as an alternative
way to do QKD: we can form a message as the concatenation of new random
key bits to be output and a short key for implementing the next usage of the
authenticated channel. Having sent enough messages of this form successfully,
we can generate a much larger number of secure key bits than we started from.
Note that this is harder to achieve when using the earlier QKRS scheme since
bits of the original key are lost even in successful transmissions.

2 Preliminaries

2.1 Density Operators and Distance Measures

We denote by $S(\mathcal{H})$ the set of density operators on Hilbert space $\mathcal{H}$ (i.e. positive
operators σ such that $\text{tr}(\sigma) = 1$). In the following, $\mathcal{H}_n$ denotes the 2^n-dimensional
Hilbert space over $\mathbb{C}$, $\mathbb{1}_n$ denotes the $2^n \times 2^n$ identity operator, and $\mathbb{I}_n = 2^{-n}\mathbb{1}_n$
denotes the completely mixed state. The trace-norm distance between two quan-
tum states $\rho, \sigma \in S(\mathcal{H})$ is defined as:

$$D(\rho, \sigma) = \frac{1}{2} \text{tr}\left(|\rho - \sigma|\right),$$

where the right-hand side denotes half the sum over the absolute value of all
eigenvalues of $\rho - \sigma$. The trace-norm distance is a metric over the set of density operators in $S(\mathcal{H})$. In the following, we use the same notation as [17]. Let
(Ω, P) be a discrete probability space. A *random state* ρ is a function from Ω
to $S(\mathcal{H})$. This means that to $\omega \in \Omega$ corresponds the mixed state $\rho(\omega)$. To an
observer ignorant of the randomness $\omega \in \Omega$, the density operator described by
ρ is given by

$$[\rho] = \sum_{\omega \in \Omega} P(\omega)\rho(\omega).$$

For any event $\mathcal{E}$, the density operator described by ρ conditioned on $\mathcal{E}$ is given by

$$[\rho|\mathcal{E}] = \frac{1}{\Pr(\mathcal{E})} \sum_{\omega \in \mathcal{E}} P(\omega)\rho(\omega).$$

Classical random variables can also be represented as random states. Let X
be a random variable with range $\mathcal{X}$ and let $\mathcal{H}$ be a $\#\mathcal{X}$-dimensional Hilbert
space with orthonormal basis $\{|x\rangle\}_{x \in \mathcal{X}}$. The random state corresponding to X
is denoted by $\{X\} = |X\rangle\langle X|$ and $[\{X\}] = \sum_{x \in \mathcal{X}} P(x)|x\rangle\langle x|$ denotes its associated
density operator. Let $\rho \otimes \{X\}$ be a random state with a classical part $\{X\}$. The
corresponding density operator is given by

$$[\rho \otimes \{X\}] = \sum_{x \in \mathcal{X}} P(x)[\rho|X = x] \otimes |x\rangle\langle x|.$$

---

[1] Even in this case, QKD does something that is impossible classically, namely it
generates a shared key that is longer than the initial one.

If X is independent of ρ then $[\rho \otimes \{X\}] = [\rho] \otimes [\{X\}]$. Let X be a classical random variable with range $\mathcal{X}$ and let ρ be a random state. The *distance to uniform of X given ρ* is defined by

$$d(X|\rho) = D([\{X\} \otimes \rho], [\{U\}] \otimes [\rho]), \tag{1}$$

where U is a random variable uniformly distributed over $\mathcal{X}$.

2.2 Quantum Ciphers

A quantum encryption scheme for classical messages is the central part of any QKRS. Such schemes where introduced in [1], and further studied in [8], where their performances were analyzed against known-plaintext attacks. We adopt a similar definition here except that we allow for the encryption to provide only statistical instead of perfect privacy. As in [1,8], we model encryption under key k by an appropriate unitary operator E_k acting upon the message and a possible ancilla of any size initially in state $|0\rangle$. Decryption is simply done by applying the inverse unitary.

For convenience we will use the notation

$$\rho_x = \sum_{k \in \{0,1\}^n} 2^{-n} E_k |x\rangle\langle x| \otimes |0\rangle\langle 0| E_k^\dagger,$$

for the equal mixture of a plaintext $x \in \{0,1\}^m$ encrypted under all possible keys with uniform probability. A quantum cipher is private if, given a cipherstate, almost no information can be extracted about the plaintext.

Definition 1. *For a non-negative function $\epsilon(n)$, a $\epsilon(n)$-private (n,m)-quantum cipher is described by a set of 2^n unitary encryption operators $\{E_k\}_{k \in \{0,1\}^n}$, acting on a set of m-bit plaintexts and an arbitrary ancilla initially in state $|0\rangle$ such that,*

$$(\forall x, x' \in \{0,1\}^m)[D(\rho_x, \rho_{x'}) < \epsilon(n)].$$

If $\epsilon(n)$ is a negligible function of n we say that the scheme is statistically private.

The total mixture of ciphertexts associated with an ϵ-private (n,m)-quantum cipher with encryption operators $\{E_k\}_{k \in \{0,1\}^n}$ is

$$\xi = \sum_{k \in \{0,1\}^n} 2^{-n} \sum_{x \in \{0,1\}^m} 2^{-m} E_k |x\rangle\langle x| \otimes |0\rangle\langle 0| E_k^\dagger. \tag{2}$$

The next technical Lemma states that the total mixture of any ϵ-private quantum cipher is ϵ-close to any plaintext encryption under a random and private key.

Lemma 1. *Any ϵ-private (n,m)-quantum cipher satisfy that for all $x \in \{0,1\}^m$, $D(\xi, \rho_x) < \epsilon$.*

Proof. Simply observe that,

$$D(\xi, \rho_x) = D\left(2^{-m} \sum_{y \in \{0,1\}^m} \rho_y, \rho_x\right) \le \sum_{y \in \{0,1\}^m} 2^{-m} D(\rho_y, \rho_x) < \epsilon,$$

from the convexity of $D(\cdot, \cdot)$ and the ϵ-privacy of the quantum cipher.   □

2.3   Mutually Unbiased Bases

A set $\mathcal{B}_n = \{B_1, \ldots, B_{2^t}\}$ of 2^t orthonormal bases in a Hilbert space of dimension 2^n is said to be *mutually unbiased* (we abbreviate mutually unbiased bases set as MUBS) if for all $|u\rangle \in B_i$ and $|v\rangle \in B_j$ for $i \neq j$, we have $|\langle u|v\rangle| = 2^{-n/2}$. Wootters and Fields[20] have shown that there are MUBSs of up to $2^n + 1$ bases in a Hilbert space of dimension 2^n, and such sets are *maximum*. They also give a construction for a maximal MUBS in Hilbert spaces of prime-power dimensions. For $\mathcal{B}_n = \{B_b\}_{b \in \{0,1\}^t}$ a MUBS, $w \in \{0,1\}^n$, and $b \in \{0,1\}^t$, we denote by $|v_w^{(b)}\rangle$ the w-th state in basis $B_b \in \mathcal{B}_n$.

Lawrence, Brukner, and Zeilinger[11] introduced an alternative construction for maximal MUBSs based on algebra in the Pauli group. Their construction plays an important role in the security analysis of our QKRS. The method for constructing a maximal MUBS in $\mathcal{H}_n$ relies on a special partitioning of all Pauli operators in $\mathcal{H}_n$. These operators live in a vector space of dimension 4^n. Let $\Sigma = \{\sigma_x, \sigma_y, \sigma_z, \sigma_1\}$ (where $\sigma_1 = \mathbb{1}_1$) be the set of Pauli operators in $\mathcal{H}_1$. This set forms a basis for all one-qubit operators. A basis for operators on n qubits is constructed as follows for $i \in \{0, \ldots, 4^n - 1\}$:

$$O_i = \sigma_{\mu(1,i)}^1 \sigma_{\mu(2,i)}^2 \cdots \sigma_{\mu(n,i)}^n = \prod_{k=1}^{n} \sigma_{\mu(k,i)}^k, \qquad (3)$$

such that $\sigma_{\mu(k,i)}^k$ is an operator in Σ acting only on the k-th qubit. We use the convention $O_0 = \mathbb{1}_n$. The action of O_i on the k-th qubit is $\sigma_{\mu(k,i)}$ where $\mu(k,i) \in \{x, y, z, \mathbb{1}\}$. The basis described in (3) is orthogonal, $\operatorname{tr}(O_i O_j) = 2^n \delta_{i,j}$ where $i = j$ means that $\mu(k,i) = \mu(k,j)$ for any qubit k. Every Pauli operator O_i is such that $O_i^2 = \mathbb{1}_n$. Apart from the identity $\mathbb{1}_n$, all O_i's are traceless and have eigenvalues ± 1.

In [11], it is first shown how to partition the set of $4^n - 1$ non-trivial Pauli operators $\{O_i\}_{i=1}^{4^n-1}$ into $2^n + 1$ subsets, each containing $2^n - 1$ commuting members. Second, each such partitioning is shown to define a maximal MUBS. Let us denote by $P_\beta^b = |v_\beta^{(b)}\rangle\langle v_\beta^{(b)}|$ the projector on the β-th vector in basis B_b. Saying that $\mathcal{B}_n = \{B_i\}_i$ is a MUBS means that $\operatorname{tr}(P_\alpha^a P_\beta^b) = 2^{-n}$ when $a \neq b$ and $\operatorname{tr}(P_\beta^b P_{\beta'}^b) = \delta_{\beta,\beta'}$. Let $(\varepsilon_{b,\beta})_{b,\beta}$ be a $2^n \times 2^n$ matrix consisting of orthogonal rows, one of which is all $+1$, and the remaining ones all contain as many $+1$ as -1. The b-th partition contains Pauli operators $\{O_\beta^b\}_{\beta=1}^{2^n-1}$ such that

$$O_\beta^b = \sum_{\alpha=1}^{2^n} \varepsilon_{\beta,\alpha} P_\alpha^b. \qquad (4)$$

In the following, $(\varepsilon_{\beta,\alpha})_{\beta,\alpha}$ will always denote the operator $2^{n/2} H^{\otimes n}$ where $H^{\otimes n}$ is the n-qubit Hadamard transform (i.e. $\varepsilon_{\beta,\alpha} = (-1)^{\beta \cdot \alpha}$).

The number of partitions $\{O_\beta^b\}_\beta$ defined by (4) is $2^n + 1$ when constructed from a maximal MUBS. Each partition contains $2^n - 1$ operators after discarding the identity (they all contain the identity). Each of these operators is traceless and has ± 1 eigenvalues as for the Pauli operators. It is easy to verify that for $a \neq b$,

$$\mathrm{tr}\left(O_\alpha^a O_\beta^b\right) = \sum_{\mu,\nu} \varepsilon_{\alpha,\mu} \varepsilon_{\beta,\nu}\, \mathrm{tr}\left(P_\mu^a P_\nu^b\right) = 0. \tag{5}$$

Moreover,

$$\mathrm{tr}\left(O_\beta^b O_{\beta'}^b\right) = \sum_{\mu,\nu} \varepsilon_{\beta,\mu} \varepsilon_{\beta',\nu}\, \mathrm{tr}\left(P_\mu^b P_\nu^b\right) = \sum_\mu \varepsilon_{\beta,\mu} \varepsilon_{\beta',\mu} = 2^n \delta_{\beta,\beta'}. \tag{6}$$

It follows from (5) and (6) that all operators in (4) are unitarily equivalent to Pauli operators. This essentially shows that partitioning the Pauli operators the way we want is always possible.

It remains to argue that any such partitioning defines a maximal MUBS. Notice that partition $\{O_1^b, \ldots, O_{2^n-1}^b\}$ (i.e. without the identity O_0^b) defines a unique basis $\{P_\beta^b\}_\beta$ where

$$P_\beta^b = 2^{-n} \sum_\mu \varepsilon_{\mu,\beta} O_\mu^b. \tag{7}$$

It is not difficult to verify that $\mathrm{tr}(P_\beta^b P_{\beta'}^b) = \delta_{\beta,\beta'}$ and for $a \neq b$, $\mathrm{tr}(P_\beta^b P_\alpha^a) = 2^{-n}$ thus leading to a maximal MUBS.

In other words, there is a one-to-one correspondence between maximal MUBSs and the partitionings $\{\{O_\beta^b\}_\beta\}_b$ of the $4^n - 1$ Pauli operators (except the identity), acting on n qubits, into $2^n + 1$ partitions $\{O_\beta^b\}_\beta$ of $2^n - 1$ commuting members. Each partition is a subgroup of the n-qubit Pauli group and is generated by n of these operators. Any Pauli operator commutes with all other operators in the partition in which it is, and anti-commutes with exactly half of the operators, including the identity, in all other partitions. See [11] for more details.

2.4   The W_n-Cipher

In [8], quantum ciphers based on MUBSs were introduced and studied with respect to their secret-key uncertainty against known-plaintext attacks. Our QKRS, presented in Sect. 5.1, uses one of these ciphers, the W_n-cipher, as its main building block. The W_n-cipher is a $(2n, n)$-quantum cipher, that is, it encrypts n-bit classical messages with the help of a $2n$-bit secret-key. The W_n-cipher enjoys perfect privacy when the secret-key is perfectly private. It is easy to verify that the cipher is ϵ-private if the secret-key is only ϵ-close to uniform[17].

Let $\mathcal{B}_n = \{B_b\}_{b \in \{0,1\}^n}$ be a MUBS of cardinality 2^n for $\mathcal{H}_n$. Remember that $|v_w^{(b)}\rangle$ denotes the w-th basis state in basis $B_b \in \mathcal{B}$. The secret-key k for the W_n-cipher is conveniently written as $k = (z, b)$ where $z, b \in_R \{0,1\}^n$. Encryption according secret-key $k = (z, b)$ of message $x \in \{0,1\}^n$ consists in preparing the following state:

$$E_k|x\rangle = E_{(z,b)}|x\rangle = \left|v_{x \oplus z}^{(b)}\right\rangle \in B_b.$$

In other words, the encryption process first one-time pad message x with key z before mapping the resulting state to basis B_b. Encryption and decryption can be performed efficiently on a quantum computer[20,8].

3   Key-Recycling Schemes

A QKRS is an encryption scheme with authentication. In addition, there are two key-recycling mechanisms, $R_{ok}^{n,s}$ and $R_{no}^{n,t}$, allowing one to recycle part of the secret-key shared between Alice and Bob in case where the authentication succeeds and fails respectively. We model the recycling mechanism by privacy amplification. That is, $R_{ok}^{n,s}$ and $R_{no}^{n,t}$ are classes of hashing functions mapping the current key $k \in \{0,1\}^n$ into a recycled key $\hat{k}$ of length s and t respectively. In order to apply privacy amplification, an *authentic classical feedback chan-nel* is necessary for announcing Bob's random recycling function $R \in_R R_{ok}^{n,s}$ or $R \in_R R_{no}^{n,t}$ depending on the outcome of authentication. Alice and Bob then compute $\hat{k} = R(k)$ as their recycled secret-key. We do not allow further inter-action between Alice and Bob since otherwise quantum key distribution could take place between them allowing not only to recycle their secret-key but even to increase its length. Key-recycling should be inherently non-interactive from Bob to Alice since the authentication outcome should anyway be made available to Alice. For simplicity, we assume that the classical feedback channel between Bob and Alice is authenticated. In general, a small secret key could be used for providing classical message-authentication on the feedback channel.

Definition 2. *A* (n,m,s,t)-QKRS *is defined by a pair* $(\mathfrak{C}^{m,n}, (R_{ok}^{n,s}, R_{no}^{n,t}))$ *where*

- $\mathfrak{C}^{m,n}$ *is a* (m,n)-*quantum cipher, and*
- $(R_{ok}^{n,s}, R_{no}^{n,t})$ *is a key-recycling mechanism.*

In this paper, the privacy of the recycled key is characterized by its distance from uniform. In [17], it is shown that when the distance is negligible, the key behaves as a perfectly private key except with negligible probability. It follows that the application is composable provided the adversary is static[14,17,3].

For a QKRS to be secure, we require that even knowing the plaintext, the function R, and the authentication outcome, the adversary's view about the recycled key is at negligible distance from uniform. This should hold except for a negligible number of functions in $R_{ok}^{n,s}$ and $R_{no}^{n,t}$. Security against known plaintext attacks is an important property of good key-recycling mechanisms. Otherwise, extra conditions on the *a posteriori* probability distribution over plaintexts have to be enforced. In particular a recycled key could be compromised if a previous plaintext gets revealed to the adversary.

The adversary's view typically changes depending on whether the authentica-tion succeeds or fails. Let $\mathcal{A}_{ok}$ (resp. $\mathcal{A}_{no}$) be the event consisting in a successful (resp. unsuccessful) authentication. Conditioned on $\mathcal{A}_{ok}$, the adversary should have access only to very limited amount of information about the secret-key. The better the authentication scheme is, the more key material the recycling mechanism can handle. When $\mathcal{A}_{no}$ occurs, however, the adversary may hold the entire cipherstate. Let K be the random variable for the secret-key. Let $\rho(x)$ be the random state corresponding to the adversary's view on an encryption of classical message x using a random key. We denote by $[\rho_{ok}(x)] = [\rho(x)|\mathcal{A}_{ok}]$ and $[\rho_{no}(x)] = [\rho(x)|\mathcal{A}_{no}]$ the random state $\rho(x)$ conditioned on the event $\mathcal{A}_{ok}$ and $\mathcal{A}_{no}$ respectively.

Definition 3. *A key-recycling mechanism* $(\mathsf{R}_{ok}^{n,s}, \mathsf{R}_{no}^{n,t})$ *is* $(\delta_{ok}, \delta_{no})$-*indistinguishable if for all* $x \in \{0,1\}^m$:

1. $d(R(K)|\rho_{ok}(x) \otimes \{R\}) \leq \delta_{ok}$ *(where* $R \in_R \mathsf{R}_{ok}^{n,s}$*), and*
2. $d(R(K)|\rho_{no}(x) \otimes \{R\}) \leq \delta_{no}$ *(where* $R \in_R \mathsf{R}_{no}^{n,t}$*).*

For δ_{ok}, δ_{no} *negligible functions of* n*, we say that the key-recycling mechanism is statistically indistinguishable. The class of key-recycling functions* $\mathsf{R}_{ok}^{n,s}$ *or* $\mathsf{R}_{no}^{n,t}$ *is said to be* δ-*indistinguishable if condition 1 or 2 respectively holds relative to* δ.

Finally, a QKRS is secure if it is a private encryption scheme together with a statistically indistinguishable key-recycling mechanism. In general,

Definition 4. *A* (n, m, s, t)-*QKRS defined by* $(\mathfrak{C}^{m,n}, (\mathsf{R}_{ok}^{n,s}, \mathsf{R}_{no}^{n,t}))$ *is* $(\epsilon, \delta_{ok}, \delta_{no})$-*secure if*

1. $\mathfrak{C}^{m,n}$ *is* ϵ-*private,*
2. *when no eavesdropping occurs the key-recycling mechanism* $\mathsf{R}_{ok}^{n,s}$ *is used, and*
3. $(\mathsf{R}_{ok}^{n,s}, \mathsf{R}_{no}^{n,t})$ *is a* $(\delta_{ok}, \delta_{no})$-*indistinguishable key-recycling mechanism.*

If the scheme is such that ϵ, δ_{ok}, *and* δ_{no} *are all negligible functions of* n *then we say that the scheme is* statistically secure.

The efficiency of a QKRS is characterized by n, s and t. When authentication succeeds $n - s$ bits of secret-key must be thrown away while, when authentication fails, $n - t$ have to be discarded. Clearly, any purely classical key-recycling scheme must have $s, t \leq n - m$. This does not have to hold for quantum schemes. However, we show next that quantum schemes suffer the same restrictions as classical ciphers when authentication fails.

4   Upper Bound on Key-Recycling

In this section, we show that any statistically secure QKRS must discard as many key-bits as the length of the plaintext (minus one bit) when the authentication fails. In other words, when authentication fails no QKRS does better than the classical one-time-pad.

When authentication fails, the adversary may have kept the entire ciphertext and may know the plaintext $x \in \{0,1\}^m$. On the other hand, condition 2 in Definition 3 requires that the key-recycling mechanism satisfies $d(R(K)|\rho_{no}(x) \otimes \{R\}) \leq \delta(n)$ where $\delta(n)$ is negligible and $R \in_R \mathsf{R}_{no}^{n,t}$. Using (1), it follows that

$$D([\{R(K)\} \otimes \rho_{no}(x) \otimes \{R\}], [\{U\}] \otimes [\rho_{no}(x) \otimes \{R\}]) \leq \delta(n). \qquad (8)$$

The density operator $\rho_{\mathrm{no}}(\hat{k}, x, R) = [\boldsymbol{\rho}_{\mathrm{no}}(x)|R(K) = \hat{k}]$ corresponds to the adversary's view when the plaintext is x, the recycled key is $\hat{k} \in \{0,1\}^t$, and the privacy amplification function is $R \in \mathsf{R}_{\mathrm{no}}^{n,t}$. We have that,

$$\rho_{\mathrm{no}}(\hat{k}, x, R) = \sum_{k:R(k)=\hat{k}} \frac{1}{\#R^{-1}(\hat{k})} E_k|x\rangle\langle x| \otimes |0\rangle\langle 0|E_k^{\dagger}. \tag{9}$$

For convenience, we define $\rho_{\mathrm{no}}(\hat{k}, x) = \frac{1}{\#\mathsf{R}_{\mathrm{no}}^{n,t}} \sum_{R \in \mathsf{R}_{\mathrm{no}}^{n,t}} \rho_{\mathrm{no}}(\hat{k}, x, R) \otimes |R\rangle\langle R|$. If a key-recycling scheme is statistically indistinguishable then for a negligible function $\delta(n)$,

$$\delta(n) \geq d(R(K)|\boldsymbol{\rho}_{\mathrm{no}}(x) \otimes \{R\}) \tag{10}$$

$$= D\left(\sum_{\hat{k}} p_{\hat{K}}(\hat{k})\Big|\hat{k}\Big\rangle\Big\langle\hat{k}\Big| \otimes \rho_{\mathrm{no}}(\hat{k}, x), \mathbb{I}_t \otimes \sum_{\hat{k}} p_{\hat{K}}(\hat{k})\rho_{\mathrm{no}}(\hat{k}, x)\right) \tag{11}$$

$$\geq \frac{1}{\#\mathsf{R}_{\mathrm{no}}^{n,t}} \sum_R \sum_{\hat{k}} 2^{-n}\#R^{-1}(\hat{k})D(\rho_{\mathrm{no}}(\hat{k}, x, R), \rho_x), \tag{12}$$

where (10) follows by definition of statistical indistinguishability, and (11) is obtained using (8) and (9). The last step follows from the fact that $D(\rho, \sigma) = \max_{\{E_m\}_m} D(p(m), q(m))$ where the maximum is computed over all POVMs $\{E_m\}_m$ and $p(m) = \mathrm{tr}(\rho E_m)$, $q(m) = \mathrm{tr}(\sigma E_m)$ are probability distributions for the outcomes of $\{E_m\}_m$ when applied to ρ and σ respectively (see for example Theorem 9.1 in [15]). In order to get (12) from (11) one only has to consider a POVM that first measures R and $\hat{k}$ before performing the POVM $\{E_m'\}_m$ (depending on R and $\hat{k}$) on the residual state that satisfies $D(\rho_{\mathrm{no}}(\hat{k}, x, R), \rho_x) = d(p'(m), q'(m))$.

It can be shown that for $t \geq n - m + 2$, (12) implies the existence of $R \in \mathsf{R}_{\mathrm{no}}^{n,t}$ and $\hat{k}_0 \in \{0,1\}^t$ such that $\#R^{-1}(\hat{k}_0) \leq 2^{m-1}$ and $D(\rho_{\mathrm{no}}(\hat{k}_0, x, R), \rho_x) \leq c$ for any constant $0 < c \leq 1$. Moreover, since the cipher is statistically private, there exists a negligible function $\epsilon(n)$ such that,

$$D(\rho_{\mathrm{no}}(\hat{k}_0, x, R), \xi) \leq D(\xi, \rho_x) + D(\rho_x, \rho_{\mathrm{no}}(\hat{k}_0, x, R)) \leq \epsilon(n) + c. \tag{13}$$

On the other hand, an argument along the lines of the proof of Lemma IV.3.2 in [5] allows us to conclude that when $\#R^{-1}(\hat{k}_0) \leq 2^{m-1}$, $D(\rho_{\mathrm{no}}(\hat{k}_0, x, R), \xi) \geq 1/2$ which contradicts (13) when $c < 1/2$ and $\epsilon(n)$ is negligible (see Lemma 3 in [9]). Next Theorem, proven in [9], follows:

Theorem 1 (Key-Recycling Bound). *Any statistically secure (n, m, s, t)-QKRS is such that $t \leq n - m + 1$.*

We believe that a more careful analysis would show that statistically secure (n, m, s, t)-QKRS must satisfy $t \leq n - m$. Theorem 1 implies that in order to recycle more secret-key bits than any classical scheme, quantum ciphers must provide authentication. It is only when the authentication succeeds that a QKRS may perform better than classical ones.

5   A Near Optimal Quantum Key-Recycling Scheme

We introduce a QKRS, called W_nC_m, that recycles an almost optimal amount of key material. Moreover, the key-recycling mechanism does not use privacy amplification. Deterministic functions are sufficient to guarantee the statistical indistinguishability of the recycled key. The scheme is introduced in Sect. 5.1. In Sect. 5.2 we present an EPR-version of the scheme and we prove it secure. In Sect. 5.3 we reduce the security of W_nC_m to that of the EPR-version.

5.1   The Scheme

The W_nC_m-cipher encrypts a message together with its Wegman-Carter one-time authentication tag[6] using the W_n-cipher[8]. We need an authentication code constructed from XOR-universal classes of hash-functions:

Definition 5 ([6]). *An XOR-universal family of hash-functions is a set of functions $H_{m,\mu} = \{h_u : \{0,1\}^m \rightarrow \{0,1\}^\mu\}_u$ such that for all $a \neq b \in \{0,1\}^m$ and all $x \in \{0,1\}^\mu$, $\#\{h \in H_{m,\mu} | h(a) \oplus h(b) = x\} = \frac{\#H_{m,\mu}}{2^\mu}$.*

There exists an XOR-universal class of hash-functions $H_{m,\mu}^\oplus$ (for any $m \geq \mu$) that requires only m bits to specify and such that picking a function at random can be done efficiently.

For the transmission of m-bit messages, W_nC_m requires Alice and Bob to share a secret-key of size $N = 2n + m$ bits where $n = m + \ell(m)$, and $\ell(m) \in \Omega(m)$ is the size of the Wegman-Carter authentication tag. We denote secret-key k by the triplet: $k = (z, b, u)$ where $z, b \in \{0,1\}^n$ is the key for the W_n-cipher and $u \in \{0,1\}^m$ is the description of a random function $h_u \in H_{m,\ell(m)}^\oplus$. Encrypting message $x \in \{0,1\}^m$ is performed by first computing the Wegman-Carter one-time authentication tag $h_u(x)$. The message $(x, h_u(x)) \in \{0,1\}^n$ is then encrypted using the W_n-cipher with secret-key (z, b). Bob decrypts the W_n-cipher and verifies that a message of the form $(x, h_u(x))$ is obtained. Bob announces to Alice the outcome of the authentication using the authenticated feedback channel. When it is successful, Alice and Bob recycle the whole secret-key. If the authentication fails then Alice and Bob throw away the one-time-pad z. The remaining part (b, u) is entirely recycled. In other words, $R_{ok}^{N,s}$ is the identity with $s = N$ and $R_{no}^{N,t}$ is deterministic with $t = N - n = N - m - \ell(m)$.

It is almost straightforward to show that our key-recycling function is perfectly indistinguishable when authentication fails.

Lemma 2. *Let $N = 2n + m$ where $n = m + \ell(m), \ell(m) > 0$ be the key-length used in W_nC_m and let $R(z, b, u) = (b, u)$ for $z, b \in \{0,1\}^n$ and $u \in \{0,1\}^m$. The key-recycling mechanism $R_{no}^{N,N-n} = \{R\}$ is 0-indistinguishable.*

Proof. Since $\rho_{no}((b, u), x, R) = \mathbb{I}_n = \rho_{no}((b', u'), x, R)$ for all $(b, u), (b', u')$, and x, it easily follows that $d(R(K)|\rho_{no}(x) \otimes \{R\}) = 0$.   □

Since W_nC_m encrypts m-bit messages and recycles $N - n$ bits of key, the scheme is sub-optimal according Theorem 1. In the next sections, we see that W_nC_m remains statistically secure for any $\ell(m) \in \Omega(m)$. It follows that although sub-optimal, W_nC_m is *nearly* optimal.

Private-Key: $(z, b, u) \in_R \{0,1\}^{2n+m}$ where $n = m + \ell(m)$.

1. Alice creates the message $c = (x, h_u(x))$ where $h_u \in H^{\oplus}_{m,\ell(m)}$. She then encrypts this message with key (z, b) according to the W_n-cipher.
2. Bob decodes the received W_n-cipher with key (z, b) and gets $c' = (x', t')$. He then verifies the authentication tag $t' = h_u(x')$. Bob sends the result of the test to Alice through a classical authentic channel.
3. **[Key-Recycling]** If Bob accepts then Alice and Bob recycle the entire key (b, z, u). If Bob rejects then Alice and Bob recycle (b, u) and throw away $z \in \{0,1\}^n$.

Fig. 1. The $W_n C_m$

5.2   An EPR Variant of $W_n C_m$

We establish the security of the key-recycling mechanism in $W_n C_m$ when the authentication is successful. We prove this case using a Shor-Preskill argument[18] similar to the ones invoked in [16] and [2] for key-recycling and quantum authentication respectively.

We first define a variant of $W_n C_m$, called EPR-$W_n C_m$, using EPR-pairs and having access to an additional authenticated and private classical channel. The key-recycling mechanism of EPR-$W_n C_m$ can be proven secure more easily since it has access to more powerful resources. Second, we show that the security of $W_n C_m$ follows from the security of EPR-$W_n C_m$.

In EPR-$W_n C_m$, Alice and Bob initially share an n-bit key b, and an m-bit key u. They agree on 2^n mutually unbiased bases in $\mathcal{H}_n$, and a family of XOR-universal hash-functions $H^{\oplus}_{m,\mu} = \{h_u\}_{u \in \{0,1\}^m}$. As for $W_n C_m$, the key b is used to select in which of the bases of the MUBS the encryption will take place. The key u indicates the selection of the hash-function for authentication. The key z in EPR-$W_n C_m$ is not shared beforehand but will be implicitly generated by measuring the shared EPR-pairs. This corresponds to refreshing z before each round of EPR-$W_n C_m$.

In order for Alice to send classical message $x \in \{0,1\}^m$ to Bob, Alice and Bob proceeds as described in Fig. 2. The key-recycling mechanism of EPR-$W_n C_m$ only takes place when authentication succeeds. The quantum transmission in $W_n C_m$ is replaced by transmitting half of a maximally entangled state consisting of n EPR-pairs.

$$|\Psi\rangle = \sum_{x \in \{0,1\}^n} 2^{-n/2} |x\rangle^A |x\rangle^B = \sum_{x \in \{0,1\}^n} 2^{-n/2} \left|\xi_x^{(b)}\right\rangle^A \left|v_x^{(b)}\right\rangle^B, \qquad (14)$$

for some orthonormal basis $\{|\xi_x^{(b)}\rangle\}_x$.

Private-Key: $(b, u) \in_R \{0, 1\}^{n+m}$.

1. Alice prepares the n EPR-pairs in state $|\Psi\rangle^{AB}$.
2. Alice sends the B-register to Bob.
3. Bob acknowledges receiving the state using the classical authentic feedback channel.
4. Alice measures her A-register in basis $\{|\xi_c^{(b)}\rangle\}_{c \in \{0,1\}^n}$ (See (14)). On classical outcome c, she computes $z := c \oplus (x, h_u(x))$.
5. Alice sends z to Bob through the additional private and authenticated classical channel.
6. Bob measures his B-register in the b-th basis of the MUBS, gets outcome c', and computes $(x', t') = c' \oplus z$. Bob verifies that $t' = h_u(x')$ and announces the result to Alice through the classical authenticated feedback channel.
7. If Bob accepts, Alice and Bob recycle the whole key (b, u).

Fig. 2. The EPR-W_nC_m-cipher using an extra private and authentic classical channel

Any trace-preserving operator the adversary can apply to Bob's half EPR-pairs can be described in terms of the 4^n Pauli operators,

$$\hat{\rho} = \mathcal{E}(|\Psi\rangle\langle\Psi|) = \sum_{i=0}^{4^n-1} \sum_{j=0}^{4^n-1} c_i \overline{c_j} (\mathbb{1}_n \otimes O_i)|\Psi\rangle\langle\Psi|(\mathbb{1}_n \otimes O_j)^\dagger, \qquad (15)$$

where $O_0 = \mathbb{1}_n$. We can split (15) into the case where the error leaves the state untouched, and the case where the state is changed

$$\hat{\rho} = |c_0|^2 |\Psi\rangle\langle\Psi| + (1 - |c_0|^2)\rho_E^{b,u}, \qquad (16)$$

where $\rho_E^{b,u} = \sum_{(i,j)\neq(0,0)} \frac{c_i \overline{c_j}}{(1-|c_0|^2)} (\mathbb{1}_n \otimes O_i)|\Psi\rangle\langle\Psi|(\mathbb{1}_n \otimes O_j)^\dagger$, and $|c_0|^2$ is the probability that the state is left unchanged by $\mathcal{E}$.

The idea behind the security of the key-recycling mechanism is that an eavesdropper, performing any non-trivial action upon Bob's system, will fail authentication with high probability. Any eavesdropping strategy that remains undetected with a *not too small* probability is such that $|c_0|^2$ is at negligible distance from 1. This means that the ciphertext will be left untouched with probability essentially 1. In other words the probability of being detected is closely related to $1 - |c_0|^2$.

The probability that Bob will accept the authentication tag, when Alice and Bob share key (b, u) can be expressed by the observable projecting onto the space of states where Alice has her untouched EPR-halves, and Bob has anything that passes the authentication test:

$$\Pi_{\text{Acc}}^{b,u} = \sum_{z \in \{0,1\}^n} \sum_{\hat{x} \in \{0,1\}^m} \left|\xi_{e_{z,u}(x)}^{(b)}\right\rangle\left\langle\xi_{e_{z,u}(x)}^{(b)}\right| \otimes \left|v_{e_{z,u}(\hat{x})}^{(b)}\right\rangle\left\langle v_{e_{z,u}(\hat{x})}^{(b)}\right|, \qquad (17)$$

where $e_{z,u}(x) = z \oplus (x, h_u(x))$. The probability that Bob will accept the authentication, when using key (b, u), is $p_{Acc}^{b,u} = \text{tr}(\Pi_{Acc}^{b,u} \hat{\rho})$.

As mentioned in Sect. 2.3, all $4^n - 1$ Pauli operators (excluding the identity) are partitioned into $2^n + 1$ sets, each containing $2^n - 1$ commuting members. Each operator, O_i, appearing in (15), will be in one of the $2^n + 1$ partitions (i.e. which each forms a subgroup). In the partition or basis where an error operator O_i belongs, its action will leave all cipherstates unchanged. For each other 2^n basis b, O_i will anti-commute with exactly half the operators (including the identity). This means that in basis b, the action of O_i permutes the basis vectors. Since this permutation is independent of the authentication code, we can show that the probability for O_i to remain undetected is negligible when the class of Wegman-Carter authentication functions is XOR-universal. Let $\hat{\rho}_{Acc}^{b,u}$ be the normalized state conditioned on $\mathcal{A}_{ok}$ defined as,

$$\hat{\rho}_{Acc}^{b,u} = \frac{\Pi_{Acc}^{b,u} \hat{\rho} \Pi_{Acc}^{b,u}}{\text{tr}\left(\Pi_{Acc}^{b,u} \hat{\rho}\right)}. \tag{18}$$

We are going to estimate the average fidelity[2] of $\hat{\rho}_{Acc}^{b,u}$ to the ideal state $|\Psi\rangle\langle\Psi|$. To do so we split $\hat{\rho}$ according to (16) and use the concavity of the fidelity, $F(\hat{\rho}_{Acc}^{b,u}, |\Psi\rangle\langle\Psi|) \geq \frac{|c_0|^2}{p_{Acc}^{b,u}}$. Applying (16) to $p_{Acc}^{b,u}$, gives us

$$F(\hat{\rho}_{Acc}^{b,u}, |\Psi\rangle\langle\Psi|) \geq \frac{|c_0|^2}{|c_0|^2 + (1 - |c_0|^2) \text{tr}\left(\Pi_{Acc}^{b,u} \rho_E^{b,u}\right)}.$$

To lower bound the average fidelity, $\sum_{b,u} 2^{-n-m} F(\hat{\rho}_{Acc}^{b,u}, |\Psi\rangle\langle\Psi|)$. We split the sum into keys (bases and authentication keys) for which $\text{tr}(\Pi_{Acc}^{b,u} \rho_E^{b,u})$ is small, and keys for which this probability is large. We know from the previous argument, that the probability of accepting a non-trivial error will be small in most bases, and indeed the terms with $\text{tr}(\Pi_{Acc}^{b,u} \rho_E^{b,u})$ negligible compared to $|c_0|^2$ give the main contribution to the fidelity.

In summary, an undetected attack is almost always trivial since it corresponds to the case where no eavesdropping occurred. Next Theorem, proven in [9], gives the desired result.

Theorem 2. *For all adversary strategies for which* $p_{Acc} \geq 2^{-(n-m-2)/2+1}$,

$$\sum_{b\in\{0,1\}^n} \sum_{u\in\{0,1\}^m} 2^{-n-m} F\left(\hat{\rho}_{Acc}^{b,u}, |\Psi\rangle\langle\Psi|\right) \geq 1 - 2^{-\frac{n-m-2}{4}+1},$$

provided n *is sufficiently large.*

Let $\rho_{ok}^{epr}(x)$ be the random state corresponding to the adversary's view in EPR-$W_n C_m$ given $\mathcal{A}_{ok}$. Let $K = (B, U, Z)$ be the random variable describing the

---

[2] Where the fidelity $F(\hat{\rho}_{Acc}^{b,u}, |\Psi\rangle\langle\Psi|) = \langle\Psi|\hat{\rho}_{Acc}^{b,u}|\Psi\rangle$.

key $(b, u) \in \{0,1\}^n \times \{0,1\}^m$, and $z \in \{0,1\}^n$ computed from the measurement outcome. Using the same line of arguments as [3] (for completeness, the proof can be found in [9]), Theorem 2 implies that:

Theorem 3. *For all adversary strategies for which $p_{Acc} \geq 2^{-(n-m-2)/2+1}$,*

$$d(K|\rho_{\mathrm{ok}}^{\mathrm{epr}}(x) \otimes \{R\}) \leq 2^{-\frac{(n-m-2)}{8}+1},$$

provided n is sufficiently large.

5.3   Back to W_nC_m

We now show that Theorem 3 also applies to W_nC_m. Similarly to other Shor-Preskill arguments[18,2,16], we transform EPR-W_nC_m into W_nC_m by simple modifications leaving the adversary's view unchanged.

In Step 4 of EPR-W_nC_m, Alice measures her part of the entangled pair in order to extract $c \in \{0,1\}^n$. Instead, she could have measured already in Step 1 since the measurement commutes with everything the adversary and Bob do up to Step 4. Measuring half the EPR-pairs immediately after creating them is equivalent to Alice preparing $c \in_R \{0,1\}^n$ before sending $|v_c^{(b)}\rangle$ in Step 2.

Instead of picking $c \in_R \{0,1\}^n$ in Step 1, Alice could choose $z \in_R \{0,1\}^n$ at random before sending $|v_{z \oplus (x, h_u(x))}^{(b)}\rangle$ to Bob. All these modifications change nothing to the adversary's view.

Now, sending z through the private and authenticated classical channel in Step 5 becomes unnecessary if Alice and Bob share z before the start of the protocol (thus making z part of the key). We have now removed the need for the private and authenticated classical channel.

The resulting protocol is such that Bob first acknowledges receiving the cipher, then measures it, and finally replies with either accept or reject. The acknowledgment of Step 3 is unnecessary and can safely be postponed to Bob's announcement in Step 6. The EPR-W_nC_m-cipher has now been fully converted into the W_nC_m-cipher without interfering with the eavesdropper's view. It follows directly that Theorem 3 also applies to W_nC_m.

Theorem 3 shows that one use of the W_nC_m-cipher leaves the secret-key at negligible distance from uniform when it was initially 0-indistinguishable. In general, if a random variable K is at distance no more than ϵ from uniform then K behaves exactly like a uniform random variable except with probability at most ϵ[17]. Our main result follows:

Theorem 4 (Main Result). *Let $n = m + \ell(m)$. For all adversary strategies the W_nC_m-cipher used with an initial ϵ-indistinguishable private-key satisfies,*

1. *either $d(K|\rho_{\mathrm{ok}}(x) \otimes \{R\}) \leq \epsilon + 2^{-\frac{\ell(m)-2}{8}+1}$ or $p_{Acc} \leq 2^{-(\ell(m)-2)/2+1}$,*
2. *$d(K|\rho_{\mathrm{no}}(x) \otimes \{R\}) \leq \epsilon$,*

provided n is sufficiently large.

In other words, the key-recycling mechanism is statistically indistinguishable when $\ell(m) \in \Omega(m)$. It follows that, when starting from a statistically indistinguishable secret-key, key-recycling can take place exponentially many times without compromising the statistical indistinguishability of the resulting key. As mentioned in Sect. 3, Theorem 4 and the discussion in [17] imply that the W_nC_m-cipher is universally composable against static adversaries.

6   Conclusion and Open Questions

We have shown that the W_nC_m-cipher is an almost optimal key-recycling cipher with one-bit feedback. There are many possible improvements of our scheme. In this paper, we assume noiseless quantum communication. This is of course an unrealistic assumption. Our scheme can easily be made resistant to noise by encoding the quantum cipher using a quantum error-correcting code. Since a quantum error-correcting code is also a secret-sharing[7], it can be shown that when authentication succeeds almost no information about the cipherstate is available to the eavesdropper. On the other hand, if the eavesdropper gains information about the cipherstate then authentication will fail similarly to the case where no error-correction is used.

It would be interesting to show that the key recycling bound(i.e. Theorem 1) can be improved to $t \leq n - m$ (instead of $n - m + 1$) as for classical schemes. It is an open question whether there exists a QKRS achieving this upper bound.

It is also possible to allow for more key-recycling mechanisms associated to different output values for the authentication process. Such a generalized scheme would allow to recycle key-material as a function of the adversary's available information but would require more than one-bit feedback.

It is easy to see that the W_nC_m-cipher can be used as a re-usable quantum authentication scheme when authentication succeeds. Our construction (using MUBSs) is different than the ones based on purity testing codes[2] and may be of independent interest.

Acknowledgments. The authors would like to thank M. Horodecki, D. Leung, and J. Oppenheim for enlightening discussions. We are also grateful to the program committee for valuable comments and suggestions.

References

1. A. Ambainis, M. Mosca, A. Tapp, and R. de Wolf. Private quantum channels. In *Proceedings of the 41st IEEE Symposium on Foundations of Computer Science — FOCS 2000*, pages 547–553, 2000.
2. H. Barnum, C. Crépeau, D. Gottesman, A. Smith, and A. Tapp. Authentication of quantum messages. In *Proc. 43rd Annual IEEE Symposium on the Foundations of Computer Science (FOCS '02)*, pages 449–458, 2002.

3. M. Ben-Or, M. Horodecki, D. W. Leung, D. Mayers, and J. Oppenheim. The universal composable security of quantum key distribution. In J. Kilian, editor, *Theory of Cryptography — TCC 2005*, volume 3378 of *Lecture Notes in Computer Science*, pages 386–406. Springer-Verlag Heidelberg, 2005.

4. C. H. Bennett, G. Brassard, and S. Breidbart. Quantum cryptography II: How to re-use a one-time pad safely even if P = NP. Unpublished manuscript, 1982.

5. R. Bhatia. *Matrix Analysis*. Graduate Texts in Mathematics. Springer-Verlag, 1997.

6. J. L. Carter and M. N. Wegman. Universal classes of hash functions (extended abstract). In *Proceedings of the 9th ACM Symposium on Theory of Computing — STOC 1977*, pages 106–112. ACM Press, 1977.

7. R. Cleve, D. Gottesman, and H.-K. Lo. How to share a quantum secret. *Physical Review Letter*, 83(3):648–651, 1999.

8. I. B. Damgård, T. B. Pedersen, and L. Salvail. On the key-uncertainty of quantum ciphers and the computational security of one-way quantum transmission. In *Advances in Cryptology — EUROCRYPT 2004*, volume 3027 of *Lecture Notes in Computer Science*, pages 91–108. Springer-Verlag Heidelberg, 2004.

9. I. B. Damgård, T. B. Pedersen, and L. Salvail. A quantum cipher with near optimal key-recycling. rs RS-05-17, BRICS, Department of Computer Science, University of Aarhus, Ny Munkegade, DK-8000 Aarhus C, Denmark, 2005.

10. S. Dziembowski and U. M. Maurer. On generating the initial key in the bounded-storage model. In *Advances in Cryptology — EUROCRYPT 2004*, pages 126–137, 2004.

11. J. Lawrence, Č. Brukner, and A. Zeilinger. Mutually unbiased binary observable sets on N qubits. *Physical Review A*, 65(3), 2002.

12. D. W. Leung. Quantum vernam cipher. *Quantum Information and Computation*, 2(1):14–34, 2002.

13. C.-J. Lu. Encryption against storage-bounded adversaries from on-line strong extractors. *Journal of Cryptology*, 17(1):27–42, 2004.

14. J. B. Nielsen. Private communication, 2005.

15. M. A. Nielsen and I. L. Chuang. *Quantum Computation and Quantum Information*. Cambridge university press, 2000.

16. J. Oppenheim and M. Horodecki. How to reuse a one-time pad and other notes on authentication, encryption and protection of quantum information. Available at http://arxiv.org/abs/quant-ph/0306161.

17. R. Renner and R. König. Universally composable privacy amplification against quantum adversaries. In J. Kilian, editor, *Theory of Cryptography — TCC 2005*, volume 3378 of *Lecture Notes in Computer Science*, pages 407–425. Springer-Verlag Heidelberg, 2005.

18. P. W. Shor and J. Preskill. Simple proof of security of the BB84 quantum key distribution protocol. *Physical Review Letters*, 85:441–444, 2000.

19. S. P. Vadhan. On constructing locally computable extractors and cryptosystems in the bounded storage model. *Journal of Cryptology*, 17(1):43–77, 2004.

20. W. K. Wootters and B. D. Fields. Optimal state-determination by mutually unbiased measurements. *Annals of Physics*, 191(2):363–381, 1989.

An Efficient CDH-Based Signature Scheme with a Tight Security Reduction

Benoît Chevallier-Mames[1,2]

[1] Gemplus, Card Security Group,
La Vigie, Avenue du Jujubier, ZI Athélia IV,
F-13705 La Ciotat Cedex, France
[2] École Normale Supérieure, Département d'Informatique,
45 rue d'Ulm, F-75230 Paris 05, France
benoit.chevallier-mames@gemplus.com

Abstract. At EUROCRYPT '03, Goh and Jarecki showed that, contrary to other signature schemes in the discrete-log setting, the *EDL* signature scheme has a *tight* security reduction, namely to the Computational Diffie-Hellman (CDH) problem, in the Random Oracle (RO) model. They also remarked that *EDL* can be turned into an off-line/on-line signature scheme using the technique of Shamir and Tauman, based on chameleon hash functions.

In this paper, we propose a new signature scheme that also has a tight security reduction to CDH but whose resulting signatures are smaller than *EDL* signatures. Further, similarly to the Schnorr signature scheme (but contrary to *EDL*), our signature is naturally efficient on-line: no additional trick is needed for the off-line phase and the verification process is unchanged.

For example, in elliptic curve groups, our scheme results in a 25% improvement on the state-of-the-art discrete-log based schemes, with the same security level. This represents to date the most efficient scheme of any signature scheme with a tight security reduction in the discrete-log setting.

Keywords: Public-key cryptography, signature schemes, discrete logarithm problem, Diffie-Hellman problem, *EDL*.

1   Introduction

In a signature scheme, a party, called *signer*, generates a signature using his own private key so that any other party, called *verifier*, can check the validity of the signature using the corresponding signer's public-key. Following the IEEE P1363 standard [P1363], there are two main settings commonly used to build signature schemes: the integer factorization setting and the discrete logarithm setting.

A signature scheme should protect against *impersonation* of parties and *alteration* of messages. Informally, the security is assessed by showing that if an adversary can violate one of the two previous properties then the same adversary can

V. Shoup (Ed.): Crypto 2005, LNCS 3621, pp. 511–526, 2005.

also break the underlying cryptographic problem — for example, the integer factorization problem, the RSA problem [RSA78], the discrete logarithm problem or the Diffie-Hellman problem [DH76]. As the cryptographic problem is supposed to be intractable, no such adversary exists. This methodology for assessing the security is called *security reduction*. The "quality" of the reduction is given by the success probability of the adversary against a signature scheme to break the underlying intractable problem. A security reduction is said *tight* when this success probability is close to 1; otherwise it is said *close* or *loose* [MR02]. This notion of tightness is very important, and allows to distinguish between *asymptotic* security and *exact* security, the first one meaning that a scheme is secure *for sufficiently large parameters*, while the second one means that the underlying cryptographic problem is almost as hard to solve as the scheme to break.

The first efficient signature scheme *tightly* related to the RSA problem is due to Bellare and Rogaway [BR96]. The security stands in the Random Oracle (RO) model [BR93] where hash functions are idealized as random oracles. Their scheme, called RSA-PSS, appears in most recent cryptographic standards. Other RSA-based signature schemes shown to be secure in the standard model include [GHR99] and [CS00].

Amongst the signature schemes based on the discrete logarithm problem (or on the Diffie-Hellman problem), we quote the ElGamal scheme [ElG85], the Schnorr scheme [Sch91], and the Girault-Poupard-Stern scheme [Gir91, PS98]. The security of these schemes is assessed (in the RO model) thanks to the forking lemma by Pointcheval and Stern [PS96]. Basically, the idea consists in running the adversary twice with different hash oracles so that it eventually gets two distinct valid forgeries on the same message. The disadvantage of the forking lemma technique is that the so-obtained security reductions are loose.

Even if the security reductions are loose, those signature schemes present the nice feature that there are very efficient *on-line* [FS87] compared to RSA-based signature schemes. In the off-line phase, the signer precomputes a quantity (independent of the message) called a *coupon* that will be used in the on-line phase to produce very quickly a signature on an arbitrary message.

To date, the only signature scheme whose security is tightly related to the discrete logarithm problem or to the Diffie-Hellman problem (in the RO model) is *EDL*, a scheme independently considered in [CP92] and [JS99]. Indeed, at EUROCRYPT '03, Goh and Jarecki [GJ03] showed that the security of *EDL* can be reduced in a tight way to the Computational Diffie-Hellman (CDH) problem. Its on-line version as suggested in [GJ03] requires the recent technique by Shamir and Tauman [ST01] based on chameleon hash functions [KR00] and so is not as efficient as the aforementioned signature schemes: the resulting signatures are longer and the verification is slower.

It is to note that *EDL* was recently modified by Katz and Wang [KW03] into a scheme with shorter signatures and a tight security reduction but on a stronger assumption, namely the Decisional Diffie-Hellman (DDH) assumption. In the same paper, Katz and Wang also proposed an improvement to *EDL*, that uses a single bit instead of a long random, and which has a tight reduction to the CDH problem. The cost of this nice improvement is simply a decrease of the security parameter of one bit.

To finalize the related work part, we stress that the shortest signature scheme that is known today is a scheme of Boneh, Lynn and Shacham [BLS04]. This scheme is loosely related to the CDH problem, but gives very short signatures, as it consists in only one single group element. However, this scheme is limited to certain elliptic and hyper-elliptic curve groups, and so less general than *EDL*. Furthermore, the on-line version of the Boneh-Lynn-Shacham signature scheme requires the technique by Shamir and Tauman, which doubles the size of the signature, and hence is less interesting.

Our Contribution. In this paper, we firstly review the definition of *EDL*, its proof by Goh and Jarecki, and the scheme of Katz and Wang. Secondly, we propose a new signature scheme which, similarly to *EDL*, features a *tight* security reduction relatively to the CDH problem but whose resulting signatures are smaller than *EDL* signatures. Furthermore, contrary to *EDL*, no additional trick is needed to turn our signature scheme in an off-line/on-line version.

Notably, in elliptic curve settings, our scheme supersedes other discrete logarithm based schemes with same security level, as it uses signatures that are 25% smaller.

Organization of the Paper. The rest of this paper is organized as follows. In the next section, we give some background on signature schemes and provide a brief introduction to "provable" security. Then, in Section 3, we review the *EDL* signature scheme and its proof by Goh and Jarecki. Section 4 is the core of our paper. We describe our signature scheme, prove that its security is tightly related to CDH in the RO model and show how it outperforms *EDL*. Finally, we conclude in Section 5.

2 Definitions

In this section, we remind some background on signature schemes and on their security. We also define the Diffie-Hellman and the discrete logarithm problems. We then provide a brief introduction to provable security. Finally, we review the concept of *on-the-fly* signatures.

2.1 Signature Schemes

A signature scheme SIG = (GenKey, Sign, Verify) is defined by the three following algorithms:

- The *key generation algorithm* GenKey. On input 1^k, algorithm GenKey produces a pair (pk, sk) of matching public (verification) and private (signing) keys.
- The *signing algorithm* Sign. Given a message m in a set of messages $\mathcal{M}$ and a pair of matching public and private keys (pk, sk), Sign produces a signature σ. The signing algorithm can be probabilistic.

- The *verification algorithm* VERIFY. Given a signature σ, a message $m \in \mathcal{M}$ and a public key pk, VERIFY tests whether σ is a valid signature of m with respect to pk.

Several security notions have been defined about signature schemes, mainly based on the seminal work of Goldwasser, Micali and Rivest [GMR84, GMR88]. It is now customary to ask for the impossibility of existential forgeries, even against adaptive chosen-message adversaries:

- An *existential forgery* is a new message-signature pair, valid and generated by the adversary. The corresponding security notion is called *existential un-forgeability* (EUF).
- The verification key is public, including to the adversary. But more information may also be available. The strongest kind of information is definitely formalized by the *adaptive chosen-message attacks* (CMA), where the attacker can ask the signer to sign any message of its choice, in an adaptive way.

As a consequence, we say that a signature scheme is *secure* if it prevents existential forgeries, even under adaptive chosen-message attacks (EUF-CMA). This is measured by the following success probability, which should be negligibly small, for any adversary $\mathcal{A}$ which outputs a valid signature σ on a message m that was never submitted to the signature oracle, within a "reasonable" bounded running-time and with at most q_s signature queries to the signature oracle:

$$\mathsf{Succ}_{\mathrm{SIG}}^{\mathsf{euf-cma}}(\mathcal{A}, q_s) = \Pr \left[\begin{array}{c} (\mathsf{pk}, \mathsf{sk}) \leftarrow \mathrm{GENKEY}(1^k), (m, \sigma) \leftarrow \mathcal{A}^{\mathrm{SIGN}(\mathsf{sk};\cdot)}(\mathsf{pk}) : \\ \mathrm{VERIFY}(\mathsf{pk}; m, \sigma) = \mathrm{TRUE} \end{array} \right] .$$

In the *random oracle model* [BR93], adversary $\mathcal{A}$ has also access to a hash oracle: $\mathcal{A}$ is allowed to make at most q_h queries to the hash oracle.

2.2   The Diffie-Hellman and the Discrete Logarithm Problems

The security of signature schemes relies on problems that are supposed intractable, such as the *Diffie-Hellman problem* [DH76] or the *discrete logarithm problem*.

Let $\mathbb{G}$ be a (multiplicatively written) abelian group. Given an element $g \in \mathbb{G}$ of prime order q, we let $G_{g,q} \subseteq \mathbb{G}$ denote the cyclic group generated by g, i.e., $G_{g,q} = \{g^i, i \in \mathbb{Z}_q\}$.

Let x be a random number in $\mathbb{Z}_q$. Define $y = g^x$. Being given (g, y), the discrete logarithm problem in $G_{g,q}$ is defined as finding the value of x. In this paper, the discrete logarithm of y w.r.t. g will be denoted as $DL_g(y) = x$. On the other hand, being given (g, y, g^a), for an unknown random number a in $\mathbb{Z}_q$, the (computational) Diffie-Hellman problem is defined as returning $g^{ax} = y^a$.

For cryptographic applications, group $G_{g,q}$ is chosen so that the problems are (supposed) hard. A classical example is to choose $G_{g,q} \subseteq \mathbb{F}_p^*$, where q divides $(p-1)$. Another widely used group family is the one of elliptic curves over finite fields [Mil85, Kob87, BSS99].

There are plenty of such signature schemes, including the schemes by ElGamal [ElG85], Girault-Poupard-Stern [Gir91, PS98], Schnorr [Sch91], and particulary the one we are interested in this paper, the *EDL* scheme [CP92, JS99, GJ03].

2.3   Security Reduction and Provable Security

Today, schemes are "proved" secure, using what is called a *reduction*. For this reason, some authors prefer to use the term of reductionist security (*e.g.,* [KM04]) instead of provable security.

Basically, the idea is to prove that a scheme is secure by exhibiting a machine (the so-called reduction) that uses a chosen-message attacker on a given signature scheme, in order to solve a hard cryptographic problem. In the standard model, the attacker is used by simulating signature queries on q_s chosen-messages. In addition, in the random oracle mode, the simulator also simulates hash queries on q_h chosen data.

Two classes of provably secure signature schemes can be distinguished. The first class of provable signature schemes proposes reductions that are said *loose*, as they can turn an attacker into a machine to solve the cryptographic problem asymptotically. The second class of provable signature schemes features so-called *tight* reductions, using the attacker to solve the problem with almost the same probability.

Of course, tightly secure schemes are the preferred ones, but there are just few of them. Notably, RSA-PSS and its derivatives are tightly related to the RSA problem [RSA78, BR96, Cor02], and Rabin-PSS is equivalent to the factorisation problem [Rab79]. For a long time, no *tightly* secure schemes were known, based on the Diffie-Hellman or discrete logarithm problems, but only loosely secure schemes, as their security was shown thanks to the *forking lemma* technique by Pointcheval and Stern [PS96]. Proved recently at EUROCRYPT '03, the *EDL* scheme is the first tight secure scheme, based on the computational Diffie-Hellman problem.

2.4   Signature with Coupons

Some signature schemes have the nice feature that one can precompute (off-line) some quantities, independent from the messages, called *coupons*, and use them in a very fast way to generate signatures once the message is received [FS87]. Such signature schemes are also known as *on-the-fly* signature schemes.

This coupon technique is very useful, especially in constrained environments such as smart cards and finds numerous applications. Most signature schemes based on discrete logarithm or Diffie-Hellman problems allow the use of coupons. However, as previously explained, they do not offer a tight security reduction. To our knowledge, the only exception is the *EDL* signature scheme using a technique proposed by Shamir and Tauman, based on chameleon hashes by Krawczyk and Rabin [ST01, KR00]. However, this use of chameleon hashes is at the price of a slower verification, as the verifier must compute chameleon hashes (which are multi-exponentiations) before verifying the signature.

3 The *EDL* Signature

3.1 The Scheme

The *EDL* signature scheme, independently proposed in [CP92, JS99], is defined as follows.

Global set-up: Let ℓ_p, ℓ_q, and ℓ_r denote security parameters.[1] Let also a cyclic group $G_{g,q}$ of order q, generated by g, where q is a ℓ_q-bit prime and the representation of the elements of $G_{g,q}$ is included in $\{0,1\}^{\ell_p}$. Finally, let two hash functions, $\mathcal{H} : \mathcal{M} \times \{0,1\}^{\ell_r} \to G_{g,q}$ and $\mathcal{G} : (G_{g,q})^6 \to \mathbb{Z}_q$.

Key generation: The private key is a random number $x \in \mathbb{Z}_q$. The corresponding public key is $y = g^x$.

Signature: To sign a message $m \in \mathcal{M}$, one first randomly chooses $r \in \{0,1\}^{\ell_r}$, and computes $h = \mathcal{H}(m,r)$ and $z = h^x$. Follows a proof of logarithm equality that $DL_h(z) = DL_g(y)$: for a random number $k \in \mathbb{Z}_q$, one computes $u = g^k$, $v = h^k$, $c = \mathcal{G}(g,h,y,z,u,v)$ and $s = k + cx \bmod q$. The signature on m is $\sigma = (z,r,s,c)$.

Verification: To verify a signature $\sigma = (z,r,s,c) \in G_{g,q} \times \{0,1\}^{\ell_r} \times (\mathbb{Z}_q)^2$ on a message $m \in \mathcal{M}$, one computes $h' = \mathcal{H}(m,r)$, $u' = g^s \, y^{-c}$ and $v' = h'^s \, z^{-c}$. The signature σ is accepted iff $c = \mathcal{G}(g,h',y,z,u',v')$.

In *EDL*, the only quantity that can be precomputed in off-line signature phase is u. The on-line part is so two hash function evaluations plus two modular exponentiations.

3.2 Security of *EDL*

In this section, we just remind that the security of *EDL* reduces to the security of the computational Diffie-Hellman problem. The proof basically shows that the *EDL* scheme is a proof that $DL_h(z) = DL_g(y) = x$. We refer to [GJ03] or to the full version of this paper [Che05] for more details.

Theorem 1 ([GJ03]). *Let $\mathcal{A}$ be an adversary which can produce, with success probability ε, an existential forgery under a chosen-message attack within time τ, after q_h queries to the hash oracles and q_s queries to the signing oracle, in the random oracle model. Then the computational Diffie-Hellman problem can be solved with success probability ε' within time τ', with*

$$\varepsilon' \geq \varepsilon - q_s \left(\frac{q_s + q_h}{q^2} + \frac{q_s + q_h}{2^{\ell_r}} \right) - \frac{q_h}{q}$$

and

$$\tau' \lesssim \tau + (6q_s + q_h)\tau_0$$

where τ_0 is the time for an exponentiation in $G_{g,q}$.

---

[1] For normal use-cases, $\ell_r \leq \ell_q$.

3.3   Features of the *EDL* Signature

The *EDL* signature scheme is proven secure relatively to the computational Diffie-Hellman problem, with a tight reduction. Hence, its security is a strong point.

The scheme yields signatures of $(\ell_p + 2\ell_q + \ell_r)$ bits. This may appear somewhat long but actually it is not, given such a strong security.[2]

In its classical use, the scheme cannot be used with coupons, but, as noted by Goh and Jarecki, one can use the technique of [ST01] based on chameleon hash functions [KR00] to transform this signature into a signature with coupons, what we will call *EDL-CH* in the sequel. Producing a *EDL-CH* signature forgery is equivalent to produce a signature forgery in the regular *EDL* signature scheme, or to find a collision in the chameleon hash function. Hence, the natural way to get a signature with coupons and with a tight security reduction to the computational Diffie-Hellman problem is to use a chameleon hash function whose collision-resistance is also based on discrete logarithm or Diffie-Hellman problem (*e.g.,* $\mathcal{H}(m, r) = \mathcal{H}_0(g^m\, y^r)$, where $\mathcal{H}_0 : G_{g,q} \rightarrow G_{g,q}$ is a hash function). But the cost of this way to create coupons is a slower verification. Further, using the chameleon hash $\mathcal{H}(m, r) = \mathcal{H}_0(g^m\, y^r)$ implies that one needs to define random number $r \in \mathbb{Z}_q$ (and not in $\{0, 1\}^{\ell_r}$). This makes the *EDL-CH* signatures slightly longer: $(\ell_p + 3\ell_q)$ bits.

3.4   Katz-Wang Signature Scheme

In [KW03], Katz and Wang proposed two modifications of *EDL*, one which consists in a scheme with short signatures tightly based on the DDH assumption, and one another that uses signature shorter than *EDL* but keeps tightly related to the CDH problem. In this section, we briefly remind the second scheme.

The idea of Katz and Wang is to remove the *randomness* of r, and to replace it by *unpredictability*. Namely, r is replaced by a bit b that can only be computed by the signer (*e.g.,* b is the result of a PRF, under a secret key included in the signing key):[3] the signatures are then (z, s, c, b), and so are shorter than *EDL* signatures by 110 bits. The proof of *EDL* is then slightly modified for Katz-Wang scheme.

This modification gives a signature scheme with a signature length of $(\ell_p + 2\ell_q + 1)$ bits, and which is just one bit less secure than *EDL* when taking identical parameters. Unfortunately, in this scheme, only u can be computed off-line, and so the on-line part of the signature is two modular exponentiations in $G_{g,q}$.

---

[2] In [GJ03], the authors estimate that if the discrete logarithm problem is supposed to be infeasible for 1000-bit primes, the forking lemma's technique tells that Schnorr signatures are secure in a field modulo a 8000-bit prime.

[3] In other words, in *EDL*, signing few times the same message would result in different random numbers r, while doing the same with Katz-Wang scheme would give always the same bit b.

4   Our Signature Scheme

Looking at the description of *EDL*, we can see that basically two random values are used: k is used to generate a proof of knowledge of the discrete logarithm while r is used to ensure that the attacker cannot predict the value of h, that will be used during simulations.

More precisely, in *EDL*, h is taken equal to $\mathcal{H}(m, r)$, with a sufficiently large random number r. As RSA-PSS does in a certain sense, the goal is to avoid, with overwhelming probability, that the attacker requests the value of $\mathcal{H}(m, r)$ with a random number r that will afterwards appear during signature queries on m. Indeed, we want to build the $\mathcal{H}(m, r)$'s involved in signature simulations in a certain form and the $\mathcal{H}(m, r)$'s returned to direct queries (and susceptible to be used in the final forgery) in another form (see [GJ03] or [Che05] for more detail).

Our first idea is the following: *Why not trying to put the randomness of k inside $\mathcal{H}(m, \cdot)$ instead of using another random number r that increases the size of the signature?* Clearly, one cannot use $\mathcal{H}(m, k)$ directly, but $\mathcal{H}(m, u)$ looks promising (and appears to be secure, as proven in the full version of this paper). As a result, the size of the so-constructed signature is reduced.

Our second idea is the following: *Would it be possible to put m inside $\mathcal{G}(\cdot)$ rather than in $\mathcal{H}(\cdot)$, as done in [Sch91] or in [KW03]?* The goal here is to allow as many precomputations as possible. This trick does not apply to *EDL*, but when combined with the previously suggested technique, the answer appears to be positive.

Intuitively, using $z = \mathcal{H}(r)^x$ and putting m in $\mathcal{G}(\cdot)$ in *EDL* is insecure because an attacker could easily reuse a z returned by the signer, and so a simulator would not solve a CDH problem. On the contrary, in our construction, we will show that using $z = \mathcal{H}(u)^x$ remains secure, as an attacker could not reuse an $\mathcal{H}(u)^x$ returned by the signer, unless the discrete logarithm is revealed: indeed, u satisfies a certain relation ($u = g^s y^{-c}$) that cannot be given for two different c's for the same u without revealing the discrete logarithm.

In this section, we describe more formally our scheme and prove strictly the intuition that we have just given.

4.1   Description

Our scheme goes as follows:

Global set-up: Let ℓ_p and ℓ_q denote security parameters. Let also a cyclic group $G_{g,q}$ of order q, generated by g, where q is a ℓ_q-bit prime and the representation of the elements of $G_{g,q}$ is included in $\{0, 1\}^{\ell_p}$. Finally, let two hash functions, $\mathcal{H} : G_{g,q} \to G_{g,q}$ and $\mathcal{G} : \mathcal{M} \times (G_{g,q})^6 \to \mathbb{Z}_q$.

Key generation: The private key is a random number $x \in \mathbb{Z}_q$. The corresponding public key is $y = g^x$.

Signature: To sign a message $m \in \mathcal{M}$, one first randomly chooses $k \in \mathbb{Z}_q$, and computes $u = g^k$, $h = \mathcal{H}(u)$, $z = h^x$ and $v = h^k$. Next, one computes $c = \mathcal{G}(m, g, h, y, z, u, v)$ and $s = k + cx \bmod q$. The signature on m is $\sigma = (z, s, c)$.

Verification: To verify a signature $\sigma = (z, s, c) \in G_{g,q} \times (\mathbb{Z}_q)^2$ on a message $m \in \mathcal{M}$, one computes $u' = g^s y^{-c}$, $h' = \mathcal{H}(u')$, and $v' = h'^s z^{-c}$. The signature σ is accepted iff $c = \mathcal{G}(m, g, h', y, z, u', v')$.

As an advantage, our signatures are smaller than the *EDL*'s ones: they are only $(\ell_p + 2\ell_q)$-bit long. We still have to prove that the scheme is tightly related to the computational Diffie-Hellman problem, which is done in the next section — but assuming this for the moment, we can see that, using the numerical values of [GJ03], our scheme leads to a gain of $\ell_r = 111$ bits per signature.

4.2 Security of the Proposed Scheme

In this section, we reduce the security of the proposed scheme to the security of the computational Diffie-Hellman problem. The proof consists in showing that the proposed scheme is a proof that $DL_h(z) = DL_g(y) = x$.

> **Theorem 2.** *Let $\mathcal{A}$ be an adversary which can produce, with success probability ε, an existential forgery under a chosen-message attack within time τ, after q_h queries to the hash oracles and q_s queries to the signing oracle, in the random oracle model. Then the computational Diffie-Hellman problem can be solved with success probability ε' within time τ', with*
>
> $$\varepsilon' \geq \varepsilon - 2q_s \left(\frac{q_s + q_h}{q} \right)$$
>
> *and*
>
> $$\tau' \lesssim \tau + (6q_s + q_h)\tau_0$$
>
> *where τ_0 is the time for an exponentiation in $G_{g,q}$.*

Proof. We are given a group $G_{g,q}$ and a CDH challenge (g, g^x, g^a). We will use an attacker $\mathcal{A}$ against our signature scheme to solve this challenge, *i.e.*, to find g^{ax}. Our attacker $\mathcal{A}$, after $q_{\mathcal{H}}$ (resp. $q_{\mathcal{G}}$) hash queries to $\mathcal{H}$ (resp. $\mathcal{G}$) oracle and q_s signature queries, is able to produce a signature forgery with probability ε within time τ. We let $q_h = q_{\mathcal{H}} + q_{\mathcal{G}}$.

Attacker $\mathcal{A}$ is run with the following simulation:

Initialization: $\mathcal{A}$ is initialized with public key $y = g^x$ and public parameters $(g, q, G_{g,q})$.

Answering new $\mathcal{G}(m, g, h, y, z, u, v)$ query: The simulator returns a random number in $\mathbb{Z}_q$.

Answering new $\mathcal{H}(u)$ query: The simulator generates a random number $d \in \mathbb{Z}_q$, and returns $(g^a) g^d$. All queries u are stored in a list called U-List.

Answering signatures query on $m \in \mathcal{M}$: The simulator randomly generates $(\kappa, s, c) \in (\mathbb{Z}_q)^3$. Then, it computes $u = g^s y^{-c}$. If $\mathcal{H}(u)$ is already set, the simulator stops (Event 1). Else, the simulator sets $h = \mathcal{H}(u) = g^\kappa$ and computes $z = (g^x)^\kappa$ — remark that $DL_h(z) = DL_g(y)(= x)$. Finally, the simulator computes $v = h^s z^{-c}$. If $\mathcal{G}(m, g, h, y, z, u, v)$ is already set, the simulator stops and fails (Event 2). Else, the simulator sets $\mathcal{G}(m, g, h, y, z, u, v) = c$, and returns the valid signature (z, s, c). All u's computed during signature queries are stored in a list called Υ-List

As we can see, this simulator is valid and indistinguishable from an actual signer, except for some events:

- <u>Event 1</u>: As u is a random number in $G_{g,q}$, the probability that the $\mathcal{H}(u)$ is already set is less than $\frac{q_s + q_\mathcal{H}}{q}$, for one signature query. For q_s signature queries, the failure probability is thus upper bounded by $\frac{q_s \cdot (q_s + q_\mathcal{H})}{q}$.

- <u>Event 2</u>: From the simulation, the input tuples to the $\mathcal{G}$ oracle are of the form $(m, g, h, y, z, u, v) = (m, g, g^\kappa, y, y^\kappa, g^k, g^{\kappa k})$ for $k \in \mathbb{Z}_q$ and κ which is determined by the relation $h = \mathcal{H}(g^k) = g^\kappa$; but as Event 1 did not happened, h is absolutely unknown for the attacker, and so κ is a random integer of $\mathbb{Z}_q$. Then, the probability that $\mathcal{G}(m, g, h, y, z, u, v)$ is already set is less than $\frac{q_s + q_\mathcal{G}}{q^2}$. For q_s signature queries, the failure probability is thus upper bounded by $\frac{q_s \cdot (q_s + q_\mathcal{G})}{q^2} \leq \frac{q_s \cdot (q_s + q_\mathcal{G})}{q}$.

As a conclusion, except with a probability smaller than $\delta_{sim} = q_s\left(\frac{q_\mathcal{H} + 2q_s}{q}\right)$, the simulation is successful.

In other words, with a probability $\varepsilon_{sim} \geq \varepsilon - \delta_{sim}$, the attacker $\mathcal{A}$ is able to return a valid signature forgery $(\hat{z}, \hat{s}, \hat{c})$ on a message $\hat{m} \in \mathcal{M}$ that was never submitted to the signature oracle. The simulator deduces from this forgery the corresponding tuple $(\hat{u}, \hat{v}, \hat{h})$, by the following computations: $\hat{u} = g^{\hat{s}} y^{-\hat{c}}$, $\hat{h} = \mathcal{H}(\hat{u})$, and $\hat{v} = \hat{h}^{\hat{s}} \hat{z}^{-\hat{c}}$. Notably, if $\mathcal{H}(\hat{u})$ has not been queried to the $\mathcal{H}$ oracle by the attacker or set by the signature oracle, the simulator queries it to the $\mathcal{H}$ oracle itself. Hence, $\hat{u}$ is a member of U-List or a member of Υ-List.

Solving the CDH Challenge (g, g^x, g^a). At this step, once the forgery is returned by the attacker, there are two cases, contrary to the proof of *EDL*.

In the first case, $\hat{u}$ is a member of U-List. This is the case that corresponds to the only case of the proof of *EDL*. As in *EDL*, we write $\hat{u} = g^k$, $\hat{v} = \hat{h}^{k'}$ and $\hat{z} = \hat{h}^{x'}$, and we get, as the signature is valid, $k = \hat{s} - \hat{c}x \bmod q$ and $k' = \hat{s} - \hat{c}x' \bmod q$. Then, if $x \neq x'$, we have $\hat{c} = \mathcal{G}(\hat{m}, g, \hat{h}, y, \hat{h}^{x'}, g^k, \hat{h}^{k'}) = \frac{k-k'}{x'-x} \bmod q$. As the message $\hat{m}$ is new, $\mathcal{G}(\hat{m}, g, \hat{h}, y, \hat{h}^{x'}, g^k, \hat{h}^{k'})$ was not set during a signature query, and so we know that $DL_{\hat{h}}(\hat{z}) = DL_g(y)(= x)$, except with a probability $\frac{q_\mathcal{G}}{q}$. Apart this error, the simulator receives from the attacker a signature with $\hat{z} = \hat{h}^x$, and it knows d such that $\hat{h} = \mathcal{H}(\hat{u}) = (g^a) g^d$. Then the simulator can return the solution to the CDH challenge, which is $\hat{z} (g^x)^{-d}$. In this first case, the forgery is successfully used to solve the CDH challenge, except with a probability smaller than $\delta_1 = \frac{q_\mathcal{H}}{q}$.

In the second case, $\hat{u}$ is not a member of U-List, and so is a member of Υ-List. This case can happen, contrary to the *EDL* signature scheme, as there is no message in the input of $\mathcal{H}$, and so we can imagine that the attacker reuse a u that corresponds to a u of a signature given by the signature oracle. Then, the simulator can recover from its log files all quantities that correspond to this $u = \hat{u}$, i.e., h, v, z, s, c and m.

At this moment, we can see that we have $u = g^s y^{-c} = \hat{u} = g^{\hat{s}} y^{-\hat{c}}$. It is exactly the kind of hypothesis that is used by the forking lemma to prove a (loose) security. But here, this equality is not obtained by restarting the attacker (as it is done in the forking lemma), but just by construction. More precisely, we can recover easily the private key x, as far as $\hat{c} \neq c \bmod q$.

As the message $\hat{m}$ is new, $c \neq \hat{c}$ or a collision on $\mathcal{G}$ function happened, between a $\mathcal{G}$ returned the signature simulation and a $\mathcal{G}$ returned by a direct $\mathcal{G}$ query, which occurs with a probability smaller than $\frac{q_s \cdot q_{\mathcal{G}}}{q}$. Hence, except an error with a probability smaller than $\delta_2 = \frac{q_s \cdot q_{\mathcal{G}}}{q}$, we have $\hat{c} \neq c$, and so we can recover the private key x: equation $s - xc = \hat{s} - x\hat{c} \bmod q$ gives $x = \frac{s - \hat{s}}{c - \hat{c}} \bmod q$. We can see that this second case gives not only the solution to the CDH challenge, but also the solution to the discrete logarithm.

As a conclusion, we can see that in both cases, our simulator can transform the forgery given by the attacker into the solution to the CDH challenge.

Putting all together, the success probability ε' of our reduction satisfies $\varepsilon' \geq \varepsilon - \delta_{sim} - \max(\delta_1, \delta_2)$, which gives, using $q_{\mathcal{H}} + q_{\mathcal{G}} = q_h$,

$$\varepsilon' \geq \varepsilon - 2q_s \left(\frac{q_s + q_h}{q} \right)$$

and the running time τ' satisfies

$$\tau' \lesssim \tau + (6q_s + q_h)\tau_0 \ .$$

As we can see, our scheme is tight, as far as $\frac{q_s \cdot q_h}{q} \leq \frac{\varepsilon}{4}$. $\square$

4.3   Our Proposed Scheme with Coupons

Interestingly, our scheme allows what we call a *cost-free* use of coupons. By this, we mean that the signer is free to choose to use coupons or not: this choice of the signer does not affect the verifier as the verification step remains unchanged.

This is done in a very natural way: the signature step (cf. Section 4.1) is simply split into two steps.

Off-line signature: To create a new coupon, one randomly chooses $k \in \mathbb{Z}_q$ and computes $u = g^k$, $h = \mathcal{H}(u)$, $z = h^x$ and $v = h^k$. The coupon is the tuple (u, v, h, z, k).

On-line signature: To sign a message $m \in \mathcal{M}$, one uses a fresh coupon (u, v, h, z, k) and just computes $c = \mathcal{G}(m, g, h, y, z, u, v)$ and $s = k + cx \bmod q$. The signature on m is $\sigma = (z, s, c)$.

The verification step remains the same. This property is very useful as it allows the signer to precompute coupons and to sign on-line very quickly, namely, by just performing one hash function evaluation followed by one modular multiplication.[4]

As previously described, our scheme features a coupon size of $(4\ell_p + \ell_q)$ bits. This size can be reduced to $(3\ell_p + \ell_q)$ bits by not storing the value of h, i.e., a coupon is defined as (u, v, z, k). Then, $h = \mathcal{H}(u)$ is evaluated in the on-line step. This option turns out useful for memory constrained devices like smart cards.

An even more sophisticated solution that minimizes the size of the coupon is described in the full version of this paper [Che05].

4.4 Size of Parameters

In this section, we show how to set the values of ℓ_q and ℓ_p to attain a security level of 2^κ. Our analysis basically follows Goh and Jarecki's for *EDL*. Assuming we take the best $(q_h, q_s, \tau, \varepsilon)$-attacker against our scheme, he can find a forgery in an average time of $\frac{\tau}{\varepsilon}$. Letting $\tau = 2^n$ and $\varepsilon = 2^{-e}$, we get $\log_2(\frac{\tau}{\varepsilon}) = n + e = \kappa$, by definition of the security level of our scheme.

Furthermore, we can use this attacker, as shown in the proof of Section 4.2, to solve the CDH problem in a time of $\frac{\tau'}{\varepsilon'}$. We let $2^{\kappa'}$ denote the security level of the CDH in the subgroup $G_{g,q}$. By definition, we have $\kappa' \leq \log_2(\frac{\tau'}{\varepsilon'})$. Because of the $O(\sqrt{q})$ security for the discrete logarithm in $G_{g,q}$, we have $\ell_q \geq 2\kappa'$.

We use the cost of the evaluation of a hash function as the unit of time. Hence, $q_h \leq 2^n$. We suppose that τ_0 (the time for an exponentiation in $G_{g,q}$) is 100 times the time of a hash function evaluation. So, using $q_s \leq q_h$, we obtain that $\tau' \simeq 2^{n+7}$ and $\varepsilon' \gtrsim \varepsilon - \frac{4 q_s \cdot q_h}{q}$. As long as $q_s \leq 2^{\ell_q - e - 3 - n} = 2^{\ell_q - \kappa - 3}$ (e.g., $\kappa = 80$, $q_s \leq 2^{80}$, $q_h \leq 2^{80}$ and $\ell_q \geq 176$), we have $\varepsilon' \gtrsim 2^{-e-1}$. Then, $\log_2(\frac{\tau'}{\varepsilon'}) \lesssim n + 7 + e + 1 = \kappa + 8$. We finally obtain $\kappa \geq \kappa' - 8$.

For example, if the targeted security level is $\kappa = 80$, it is sufficient to use $\kappa' = 88$ (and hence $\ell_q \geq 176$). It proves that our scheme is very efficient in terms of signature size, as we can use the same subgroup $G_{g,q}$ as the one used by Goh and Jarecki for *EDL* and have the same security. One can remark that our scheme remains secure even if we limit q_s to 2^{80}, while in *EDL*, q_s was limited to 2^{30}, or the random number r was made appropriately longer.

4.5 Detailed Comparison with *EDL*, the Katz-Wang Scheme and Other Schemes

In this paragraph, we sum up the advantages of our scheme. Compared to *EDL*, our scheme features

---

[4] This is comparable to the fastest off-line/on-line signature schemes of Schnorr, Girault-Poupard-Stern or Poupard-Stern [Sch91, Gir91, PS98, PS99]. One would remark that Girault-Poupard-Stern scheme does not require a reduction modulo the group order, but yields longer signatures: this elegant technique can also be used in our scheme, to get an even faster on-line signature scheme at the price of longer signatures.

1. faster signatures with a cost-free use of coupons: the on-line part only requires one hash function evaluation followed by one modular multiplication in $\mathbb{Z}_q$, while in EDL, this phase consists of two hash function evaluations and two modular exponentiations in $G_{g,q}$;
2. same verification step efficiency;
3. shorter signatures of $\ell_r \geq 111$ bits: in a subgroup of $\mathbb{F}_p^*$, taking $\ell_p = 1024$ and $\ell_q = 176$, this represents an improvement of 7%. In the elliptic curve setting, the gain is even more sensible, as z can be represented with a length around $\ell_q = 176$, resulting in an improvement of 17%.

Compared to the Katz-Wang scheme, our scheme features

1. faster signatures with a cost-free use of coupons: the on-line part only requires one hash function evaluation followed by one modular multiplication in $\mathbb{Z}_q$, while in Katz-Wang signature scheme, this phase consists of two hash function evaluations and two modular exponentiations in $G_{g,q}$;
2. same verification step efficiency;
3. less significantly, shorter signatures of 1 bit and a security parameter greater of 1 bit;
4. smaller key size, as computing b by a PRF or in another way require an additional key, that should better not be related to the private key x.

Furthermore, as noticed in [KW03], the computation of an hash $\mathcal{H} : G_{g,q} \rightarrow G_{g,q}$ can be very long, namely it costs an exponentiation of $(\ell_p - \ell_q)$ bits, which is much longer than the two exponentiations in $G_{g,q}$. In our scheme, this hash computation is done off-line, contrary to EDL and Katz-Wang schemes.

Compared to the off-line/on-line version of EDL, EDL-CH, the off-line/on-line version of our scheme presents

1. faster and unchanged verification step (remember that EDL-CH relies on chameleon hashes, which requires additional exponentiations);
2. shorter signatures, *i.e.*, $\ell_q \geq 176$ bits less than EDL-CH; again, in a subgroup of $\mathbb{F}_p^*$, taking $\ell_p = 1024$ and $\ell_q = 176$, this represents an improvement of 11% and of 25% in the elliptic curve setting.

Finally, owing to its security tightness, our scheme fulfills or even improves most of the advantages of EDL that were presented by Goh and Jarecki, by comparison with other discrete-logarithm schemes, such as Schnorr signature, with same security level.

On the one hand, using our scheme in $G_{g,q} \subseteq \mathbb{F}_p^*$, we can use a field 8 times smaller and a subgroup of order twice smaller than in other discrete-logarithm schemes (as in EDL). Notably, it means that public keys are smaller by a factor of 8, private keys are smaller by a factor of 2. In this case, our signatures are about twice as long as other discrete-logarithm schemes.

On the other hand, in the elliptic curve setting, our public and private keys are smaller by a factor of 2 and our signatures are 25% smaller than in previously known schemes.

This clearly shows the advantages of the proposed scheme.

5 Conclusion

At EUROCRYPT '03, Goh and Jarecki gave a proof that the security of *EDL* is tightly related to the CDH problem, in the random oracle model. They also proposed to use the technique of Shamir and Tauman, based on chameleon hash functions, to get a version of *EDL* scheme with coupons: *EDL-CH*.

In this paper, we have proposed a new signature scheme which, similarly to *EDL*, features a *tight* security reduction relatively to the CDH problem but whose resulting signatures are smaller: if coupons are not used, we gain ℓ_r bits compared to *EDL* signatures; in the off-line/on-line version, we gain ℓ_q bits compared to *EDL-CH* signatures. Furthermore, contrary to *EDL*, no additional trick is needed to turn our signature scheme in an off-line/on-line version.

Our scheme represents to date the most efficient scheme of any signature scheme with a tight security reduction in the discrete-log setting.

Acknowledgements

The author would like to thank his careful PhD advisor, David Pointcheval, as well as Marc Joye for his attention and support in our research. Anonymous referees are also thanked for their precious remarks, and notably for corrections on our previous proofs.

The author thanks Jean-François Dhem, David Naccache and Philippe Proust, as well as Dan Boneh and Jonathan Katz for their comments.

References

[BR93]   M. Bellare and P. Rogaway. Random oracles are practical: A paradigm for designing efficient protocols. In *ACM Conference on Computer and Communications Security*, pages 62–73. ACM Press, 1993.

[BR96]   M. Bellare and P. Rogaway. The exact security of digital signatures: How to sign with RSA and Rabin. In U. Maurer, editor, *Advances in Cryptology – EUROCRYPT '96*, volume 1070 of *Lecture Notes in Computer Science*, pages 399–416. Springer-Verlag, 1996.

[BLS04] D. Boneh, B. Lynn, and H. Shacham. Short signatures from the weil pairing. *Journal of Cryptology*, 17(4):297–319, 2004.

[BSS99] I. Blake, G. Seroussi, and N.P. Smart. Elliptic Curves in Cryptography. Cambridge University Press, 1999.

[Che05] B. Chevallier-Mames. An Efficient CDH-based Signature Scheme With a Tight Security Reduction. Full version available from http://eprint.iacr.org/2005/035.

[Cor02] J.-S. Coron. Optimal security proofs for PSS and other signature schemes. In L.R. Knudsen, editor, *Advances in Cryptology – EUROCRYPT 2002*, volume 2332 of *Lecture Notes in Computer Science*, pages 272–287. Springer-Verlag, 2002.

[CP92]   D. Chaum and T.P. Pedersen. Wallet databases with observers. In E. Brickell, editor, *Advances in Cryptology – CRYPTO '92*, volume 740 of *Lecture Notes in Computer Science*, pages 89–105. Springer-Verlag, 1992.

[CS00]   R. Cramer and V. Shoup. Signature scheme based on the strong RSA assumption. *ACM Transactions on Information and System Security*, 3(3):161–185, 2000.

[DH76]   W. Diffie and M.E. Hellman. New directions in cryptography. *IEEE Transactions on Information Theory*, IT-22(6):644–654, 1976.

[ElG85] T. ElGamal. A public key cryptosystem and a signature scheme based on discrete logarithms. *IEEE Transactions on Information Theory*, IT-31(4):469–472, 1985.

[FS87]   A. Fiat and A. Shamir. How to prove yourself: Practical solutions to identification and signature problems. In A.M. Odlyzko, editor, *Advances in Cryptology – CRYPTO '86*, volume 263 of *Lecture Notes in Computer Science*, pages 186–194. Springer-Verlag, 1987.

[GHR99] R. Gennaro, S. Halevi, and T. Rabin. Secure hash-and-sign signatures without the random oracle. In M. Bellare, editor, *Advances in Cryptology – EUROCRYPT '99*, volume 1592 of *Lecture Notes in Computer Science*, pages 123–139. Springer-Verlag, 1999.

[Gir91] M. Girault. An identity-based identification scheme based on discrete logarithms modulo a composite number. In I.B. Damgård, editor, *Advances in Cryptology – EUROCRYPT '90*, volume 473 of *Lecture Notes in Computer Science*, pages 481–486. Springer-Verlag, 1991.

[GJ03]   E.-J. Goh and S. Jarecki. A signature scheme as secure as the Diffie-Hellman problem. In E. Biham, editor, *Advances in Cryptology – EUROCRYPT 2003*, Lecture Notes in Computer Science, pages 401–415. Springer-Verlag, 2003.

[GMR84] S. Goldwasser, S. Micali, and R. Rivest. A "paradoxical" solution to the signature problem. In *Proceedings of the 25th FOCS*, pages 441–448. IEEE, 1984.

[GMR88] S. Goldwasser, S. Micali, and R. Rivest. A digital signature scheme secure against adaptive chosen message attacks. *SIAM Journal of Computing*, 17(2):281–308, 1988.

[JS99]   M. Jakobsson and C.P. Schnorr. Efficient oblivious proofs of correct exponentiation. In B. Preneel, editor, *Communications and Multimedia Security – CMS '99*, volume 152 of *IFIP Conference Proceedings*, pages 71–86. Kluver, 1999.

[Kob87] N. Koblitz. Elliptic curve cryptosystems. *Mathematics of Computation*, vol. 48, pp. 203-209, 1987.

[KM04]   N. Koblitz and A. Menezes. Another look at "provable security". Cryptology ePrint Archive, Report 2004/152, 2004. http://eprint.iacr.org/.

[KR00]   H. Krawczyk and T. Rabin. Chameleon signatures. In *Symposium on Network and Distributed System Security – NDSS 2000*, pages 143–154. Internet Society, 2000.

[KW03]   J. Katz and N. Wang. Efficiency improvements for signature schemes with tight security reductions. In *ACM Conference on Computer and Communications Security*, pages 155–164. ACM Press, 2003.

[MR02]   S. Micali and L. Reyzin. Improving the exact security of digital signatre schemes. *Journal of Cryptology*, 15(1):1–18, 2002.

[Mil85] V. Miller. Use of elliptic curves in cryptography. In H. C. Williams, editor, *Advances in Cryptology – CRYPTO '85*, Lecture Notes in Computer Science, pages 417–426. Springer-Verlag, 1986.

[P1363] IEEE P1363. IEEE Standard Specifications for Public-Key Cryptography. IEEE Computer Society, August 2000.

[PS96]   D. Pointcheval and J. Stern. Security proofs for signature schemes. In U. Maurer, editor, *Advances in Cryptology – EUROCRYPT '96*, volume 1070 of *Lecture Notes in Computer Science*, pages 387–398. Springer-Verlag, 1996.

[PS98]   G. Poupard and J. Stern. Security analysis of a practical "on the fly" authentication and signature generation. In K. Nyberg, editor, *Advances in Cryptology – EUROCRYPT '98*, volume 1403 of *Lecture Notes in Computer Science*, pages 422–436. Springer-Verlag, 1998.

[PS99]   G. Poupard and J. Stern. On the fly signatures based on factoring. In *ACM Conference on Computer and Communications Security*, pages 37–45. ACM Press, 1999.

[Rab79]   M.O. Rabin. Digital signatures and public-key functions as intractable as factorization. Technical Report MIT/LCS/TR-212, MIT Laboratory for Computer Science, January 1979.

[RSA78]   R.L. Rivest, A. Shamir, and L.M. Adleman. A method for obtaining digital signatures and public-key cryptosystems. *Communications of the ACM*, 21(2):120–126, 1978.

[Sch91]   C.-P. Schnorr. Efficient signature generation by smart cards. *Journal of Cryptology*, 4(3):161–174, 1991.

[ST01]   A. Shamir and Y. Tauman. Improved online/offline signature schemes. In J. Kilian, editor, *Advances in Cryptology – CRYPTO 2001*, Lecture Notes in Computer Science, pages 355–367. Springer-Verlag, 2001.

Improved Security Analyses for CBC MACs

Mihir Bellare[1], Krzysztof Pietrzak[2], and Phillip Rogaway[3]

[1] Dept. of Computer Science & Engineering, University of California San Diego,
9500 Gilman Drive, La Jolla, CA 92093, USA
mihir@cs.ucsd.edu
www-cse.ucsd.edu/users/mihir
[2] Dept. of Computer Science, ETH Zürich, CH-8092 Zürich Switzerland
pietrzak@inf.ethz.ch
[3] Dept. of Computer Science, University of California, Davis, California, 95616, USA;
and Dept. of Computer Science, Faculty of Science, Chiang Mai University,
Chiang Mai 50200, Thailand
rogaway@cs.ucdavis.edu
www.cs.ucdavis.edu/~rogaway/

Abstract. We present an improved bound on the advantage of any q-query adversary at distinguishing between the CBC MAC over a random n-bit permutation and a random function outputting n bits. The result assumes that no message queried is a prefix of any other, as is the case when all messages to be MACed have the same length. We go on to give an improved analysis of the encrypted CBC MAC, where there is no restriction on queried messages. Letting m be the block length of the longest query, our bounds are about $mq^2/2^n$ for the basic CBC MAC and $m^{o(1)}q^2/2^n$ for the encrypted CBC MAC, improving prior bounds of $m^2q^2/2^n$. The new bounds translate into improved guarantees on the probability of forging these MACs.

1 Introduction

SOME DEFINITIONS. The CBC function CBC_π associated to a key $\pi\colon \{0,1\}^n \to \{0,1\}^n$ takes as input a message $M = M^1 \cdots M^m$ that is a sequence of n-bit blocks and returns the n-bit string C^m computed by setting $C^i = \pi(C^{i-1} \oplus M^i)$ for each $i \in [1..m]$, where $C^0 = 0^n$. Consider three types of attacks for an adversary given an oracle: atk = eq means all queries are exactly m blocks long; atk = pf means they have at most m blocks and no query is a prefix of any another; atk = any means the queries are arbitrary distinct strings of at most m blocks. Let $\mathbf{Adv}^{\text{atk}}_{\text{CBC}}(q, n, m)$ denote the maximum advantage attainable by any q-query adversary, mounting an atk attack, in distinguishing whether its oracle is CBC^π_n for a random permutation π on n bits, or a random function that outputs n bits. We aim to upper bound this quantity as a function of n, m, q.

PAST WORK AND OUR RESULTS ON CBC. Bellare, Kilian and Rogaway [2] showed that $\mathbf{Adv}^{\text{eq}}_{\text{CBC}}(q, n, m) \leq 2m^2q^2/2^n$. Maurer reduced the constant 2 to 1 and provided a substantially different proof [13]. Petrank and Rackoff [15] showed

V. Shoup (Ed.): Crypto 2005, LNCS 3621, pp. 527–545, 2005.

Construct	atk	Previous bound	Our bound
CBC	pf	$m^2 q^2 / 2^n$ [2,13,15]	$mq^2/2^n \cdot (12 + 8m^3/2^n)$
ECBC	any	$2.5\,m^2 q^2 / 2^n$ [7]	$q^2/2^n \cdot (d'(m) + 4m^4/2^n)$

Fig. 1. Bounds on $\mathbf{Adv}^{\mathsf{pf}}_{\mathrm{CBC}}(q, n, m)$ and $\mathbf{Adv}^{\mathsf{any}}_{\mathrm{ECBC}}(q, n, m)$, assuming $m \leq 2^{n/2-1}$

that the same bounds hold (up to a constant) for $\mathbf{Adv}^{\mathsf{pf}}_{\mathrm{CBC}}(q, n, m)$. In this paper we show that $\mathbf{Adv}^{\mathsf{pf}}_{\mathrm{CBC}}(q, n, m) \leq 20mq^2/2^n$ for $m \leq 2^{n/3}$. (The result is actually a little stronger. See Fig. 1.) This implies the same bound holds for $\mathbf{Adv}^{\mathsf{eq}}_{\mathrm{CBC}}(q, n, m)$.

CONTEXT AND DISCUSSION. When $\pi = E(K, \cdot)$, where $K \in \mathcal{K}$ is a random key for blockcipher $E \colon \mathcal{K} \times \{0,1\}^n \to \{0,1\}^n$, the function CBC_π is a popular message authentication code (MAC). Assuming E is a good pseudorandom permutation (PRP), the dominant term in a bound on the probability of forgery in an atk-type chosen-message attack is $\mathbf{Adv}^{\mathsf{atk}}_{\mathrm{CBC}}(q, n, m)$, where q is the sum of the number of MAC-generation and MAC-verification queries made by the adversary (cf. [1]). Thus the quality of guarantee we get on the security of the MAC is a function of how good an upper bound we can prove on $\mathbf{Adv}^{\mathsf{atk}}_{\mathrm{CBC}}(q, n, m)$.

It is well known that the CBC MAC is insecure when the messages MACed have varying lengths (specifically, it is forgeable under an any-attack that uses just one MAC-generation and one MAC-verification query, each of at most two blocks) so the case atk = any is not of interest for CBC. The case where all messages MACed have the same length (atk = eq) is the most basic one, and where positive results were first obtained [2]. The case atk = pf is interesting because one way to get a secure MAC for varying-length inputs is to apply a prefix-free encoding to the data before MACing it. The most common such encoding is to include in the first block of each message an encoding of its length.

We emphasize that our results are about CBC_π for a random permutation $\pi \colon \{0,1\}^n \to \{0,1\}^n$, and not about CBC_ρ for a random function $\rho \colon \{0,1\}^n \to \{0,1\}^n$. Since our bounds are better than the cost to convert between a random n-bit function and a random n-bit permutation using the switching lemma [2], the distinction is significant. Indeed for the prefix-free case, applying CBC over a random function on n bits is known to admit an attack more effective than that which is ruled out by our bound [6].

ENCRYPTED CBC. The ECBC function $\mathrm{ECBC}_{\pi_1, \pi_2}$ associated to permutations π_1, π_2 on n bits takes a message M that is a multiple of n bits and returns $\pi_2(\mathrm{CBC}_{\pi_1}(M))$. Define $\mathbf{Adv}^{\mathsf{atk}}_{\mathrm{ECBC}}(q, n, m)$ analogously to the CBC case above (atk $\in \{\mathsf{any}, \mathsf{eq}, \mathsf{pf}\}$). Petrank and Rackoff [15] showed that $\mathbf{Adv}^{\mathsf{any}}_{\mathrm{ECBC}}(q, n, m) \leq 2.5\,m^2 q^2/2^n$. A better bound, $\mathbf{Adv}^{\mathsf{eq}}_{\mathrm{ECBC}}(q, n, m) \leq q^2/2^n \cdot (1 + cm^2/2^n + cm^6/2^{2n})$ for some constant c, is possible for the atk = eq case based on a lemma of Dodis et al. [9], but the point of the ECBC construction is to achieve any-security. We improve on the result of Petrank and Rackoff to show that $\mathbf{Adv}^{\mathsf{any}}_{\mathrm{ECBC}}(q, n, m) \leq q^2/2^n \cdot (d'(m) + 4m^4/2^n)$ where $d'(m)$ is the maximum, over all $m' \leq m$, of the number of divisors of m'. (Once again see Fig. 1.) Note that the function $d'(m) \approx m^{1/\ln\ln(m)}$ grows slowly.

The MAC corresponding to ECBC (namely $\text{ECBC}_{\pi_1,\pi_2}$ when $\pi_1 = E(K_1, \cdot)$ and $\pi_2 = E(K_2, \cdot)$ for random keys $K_1, K_2 \in \mathcal{K}$ of a blockcipher $E: \mathcal{K} \times \{0,1\}^n \to \{0,1\}^n$) was developed by the RACE project [5]. This MAC is interesting as a natural and practical variant of the CBC MAC that correctly handles messages of varying lengths. A variant of ECBC called CMAC was recently adopted as a NIST-recommended mode of operation [14]. As with the CBC MAC, our results imply improved guarantees on the forgery probability of the ECBC MAC under a chosen-message attack, but this time of type any rather than merely pf, and with the improvement being numerically more substantial.

MORE DEFINITIONS. The collision-probability $\mathbf{CP}^{\text{atk}}_{n,m}$ of the CBC MAC is the maximum, over all pairs of messages (M_1, M_2) in an appropriate atk-dependent range, of the probability, over random π, that $\text{CBC}_\pi(M_1) = \text{CBC}_\pi(M_2)$. For atk = any the range is any pair of distinct strings of length a positive multiple of n but at most mn; for atk = pf it is any such pair where neither string is a prefix of the other; and for atk = eq it is any pair of distinct strings of exactly mn bits. The *full collision probability* $\mathbf{FCP}^{\text{atk}}_{n,m}$ is similar except that the probability is of the event $C_2^{m_2} \in \{C_1^1, \ldots, C_1^{m_1}, C_2^1, \ldots, C_2^{m_2-1}\}$ where, for each $b \in \{1,2\}$, we have $C_b^i = \pi(C_b^{i-1} \oplus M_b^i)$ for $m_b = |M_b|/n$ and $i \in [1..m_b]$ and $C_b^0 = 0^n$. Note that these definitions do not involve an adversary and in this sense are simpler than the advantage functions considered above.

REDUCTIONS TO FCP AND CP. By viewing ECBC as an instance of the Carter-Wegman paradigm [18], one can reduce bounding $\mathbf{Adv}^{\text{atk}}_{\text{ECBC}}(q,n,m)$ (for atk $\in \{$any, eq, pf$\}$) to bounding $\mathbf{CP}^{\text{atk}}_{n,m}$ (see [7], stated here as Lemma 3). This simplifies the analysis because one is now faced with a combinatorial problem rather than consideration of a dynamic, adaptive adversary.

The first step in our analysis of the CBC MAC is to provide an analogous reduction (Lemma 1) that reduces bounding $\mathbf{Adv}^{\text{pf}}_{\text{CBC}}(q,n,m)$ to bounding $\mathbf{FCP}^{\text{pf}}_{n,m}$. Unlike the case of ECBC, the reduction is not immediate and does not rely on the Carter-Wegman paradigm. Rather it is proved directly using the game-playing approach [4,16].

BOUNDS ON FCP AND CP. Black and Rogaway [7] show that $\mathbf{CP}^{\text{any}}_{n,m} \leq 2(m^2 + m)/2^n$. Dodis, Gennaro, Håstad, Krawczyk, and Rabin [9] show that $\mathbf{CP}^{\text{eq}}_{n,m} \leq 2^{-n} + cm^2/2^{2n} + cm^3/2^{3n}$ for some absolute constant c. (The above-mentioned bound on $\mathbf{Adv}^{\text{eq}}_{\text{ECBC}}(q,n,m)$ is obtained via this.) We build on their techniques to show (cf. Lemma 4) that $\mathbf{CP}^{\text{any}}_{n,m} \leq 2d'(m)/2^n + 8m^4/2^{2n}$. Our bound on $\mathbf{Adv}^{\text{any}}_{\text{ECBC}}(q,n,m)$ then follows. We also show that $\mathbf{FCP}^{\text{pf}}_{n,m} \leq 8m/2^n + 8m^4/2^{2n}$. Our bound on $\mathbf{Adv}^{\text{pf}}_{\text{CBC}}(q,n,m)$ then follows.

We remark that the security proof of RMAC [11] had stated and used a claim that implies $\mathbf{CP}^{\text{any}}_{n,m} \leq 12m/2^n$, but the published proof was wrong. Our Lemma 4 both fixes and improves that result.

FURTHER RELATED WORK. Other approaches to the analysis of the CBC MAC and the encrypted CBC MAC include those of Maurer [13] and Vaudenay [17], but they only obtain bounds of $m^2 q^2 / 2^n$.

2   Definitions

NOTATION. The empty string is denoted ε. If x is a string then $|x|$ denotes its length. We let $B_n = \{0,1\}^n$. If $x \in B_n^*$ then $|x|_n = |x|/n$ denotes the number of n-bit blocks in it. If $X \subseteq \{0,1\}^*$ then $X^{\leq m}$ denotes the set of all non-empty strings formed by concatenating m or fewer strings from X and X^+ denotes the set of all strings formed by concatenating one or more strings from X. If $M \in B_n^*$ then M^i denotes its i-th n-bit block and $M^{i \to j}$ denotes the string $M^i \| \cdots \| M^j$, for $1 \leq i \leq j \leq |M|_n$. If S is a set equipped with some probability distribution then $s \xleftarrow{\$} S$ denotes the operation of picking s from S according to this distribution. If no distribution is explicitly specified, it is understood to be uniform.

We denote by $\mathrm{Perm}(n)$ the set of all permutations over $\{0,1\}^n$, and by $\mathrm{Func}(n)$ the set of all functions mapping $\{0,1\}^*$ to $\{0,1\}^n$. (Both these sets are viewed as equipped with the uniform distribution.) A blockcipher E (with block-length n and key-space $\mathcal{K}$) is identified with the set of permutations $\{E_K : K \in \mathcal{K}\}$ where $E_K : \{0,1\}^n \to \{0,1\}^n$ denotes the map specified by key $K \in \mathcal{K}$. The distribution is that induced by a random choice of K from $\mathcal{K}$, so $f \xleftarrow{\$} E$ is the same as $K \xleftarrow{\$} \mathcal{K}$, $f \leftarrow E_K$.

SECURITY. An adversary is a randomized algorithm that always halts. Let $\mathcal{A}_{q,n,m}^{\mathsf{atk}}$ denote the class of adversaries that make at most q oracle queries, where if $\mathsf{atk} = \mathsf{eq}$, then each query is in B_n^m; if $\mathsf{atk} = \mathsf{pf}$, then each query is in $B_n^{\leq m}$ and no query is a prefix of another; and if $\mathsf{atk} = \mathsf{any}$ then each query is in $B_n^{\leq m}$. We remark that the adversaries considered here are computationally unbounded. In this paper we always consider deterministic, stateless oracles and thus we will assume that an adversary never repeats an oracle query. We also assume that an adversary never asks a query outside of the implicitly understood domain of interest.

Let $F \colon D \to \{0,1\}^n$ be a set of functions and let $A \in \mathcal{A}_{q,n,m}^{\mathsf{atk}}$ be an adversary, where $\mathsf{atk} \in \{\mathsf{eq}, \mathsf{pf}, \mathsf{any}\}$. By "$A^f \Rightarrow 1$" we denote the event that A outputs 1 with oracle f. The advantage of A (in distinguishing an instance of F from a random function outputting n bits) and the advantage of F are defined, respectively, as

$$\mathbf{Adv}_F(A) = \Pr[f \xleftarrow{\$} F \colon A^f \Rightarrow 1] - \Pr[f \xleftarrow{\$} \mathrm{Func}(n) \colon A^f \Rightarrow 1] \quad \text{and}$$
$$\mathbf{Adv}_F^{\mathsf{atk}}(q, n, m) = \max_{A \in \mathcal{A}_{q,n,m}^{\mathsf{atk}}} \{ \mathbf{Adv}_F(A) \} \, .$$

Note that since $\mathcal{A}_{q,n,m}^{\mathsf{eq}} \subseteq \mathcal{A}_{q,n,m}^{\mathsf{pf}} \subseteq \mathcal{A}_{q,n,m}^{\mathsf{any}}$, we have

$$\mathbf{Adv}_F^{\mathsf{eq}}(q, n, m) \ \leq \ \mathbf{Adv}_F^{\mathsf{pf}}(q, n, m) \ \leq \ \mathbf{Adv}_F^{\mathsf{any}}(q, n, m) \, . \tag{1}$$

CBC AND ECBC. Fix $n \geq 1$. For $M \in B_n^m$ and $\pi\colon B_n \to B_n$ then define $\mathrm{CBC}_\pi^M[i]$ inductively for $i \in [0..m]$ via $\mathrm{CBC}_\pi^M[0] = 0^n$ and $\mathrm{CBC}_\pi^M[i] = \pi(\mathrm{CBC}_\pi^M \oplus M^i)$ for $i \in [1..m]$. We associate to π the CBC MAC function $\mathrm{CBC}_\pi\colon B_n^+ \to B_n$ defined by $\mathrm{CBC}_\pi(M) = \mathrm{CBC}_\pi^M[m]$ where $m = |M|_n$. We let $\mathrm{CBC} = \{\mathrm{CBC}_\pi\colon \pi \in \mathrm{Perm}(n)\}$. This set of functions has the distribution induced by picking π uniformly from $\mathrm{Perm}(n)$.

To functions $\pi_1, \pi_2\colon B_n \to B_n$ we associate the encrypted CBC MAC function $\mathrm{ECBC}_{\pi_1,\pi_2}\colon B_n^+ \to B_n$ defined by $\mathrm{ECBC}_{\pi_1,\pi_2}(M) = \pi_2(\mathrm{CBC}_{\pi_1}(M))$ for all $M \in B_n^+$. We let $\mathrm{ECBC} = \{\mathrm{ECBC}_{\pi_1,\pi_2}\colon \pi_1, \pi_2 \in \mathrm{Perm}(n)\}$. This set of functions has the distribution induced by picking π_1, π_2 independently and uniformly at random from $\mathrm{Perm}(n)$.

COLLISIONS. For $M_1, M_2 \in B_n^*$ we define the *prefix predicate* $\mathsf{pf}(M_1, M_2)$ to be true if either M_1 is a prefix of M_2 or M_2 is a prefix of M_1, and false otherwise. Note that $\mathsf{pf}(M, M) = \mathsf{true}$ for any $M \in B_n^*$. Let

$$\mathcal{M}_{n,m}^{\mathsf{eq}} = \{(M_1, M_2) \in B_n^m \times B_n^m : M_1 \neq M_2\},$$
$$\mathcal{M}_{n,m}^{\mathsf{pf}} = \{(M_1, M_2) \in B_n^{\leq m} \times B_n^{\leq m} : \mathsf{pf}(M_1, M_2) = \mathsf{false}\}, \quad \text{and}$$
$$\mathcal{M}_{n,m}^{\mathsf{any}} = \{(M_1, M_2) \in B_n^{\leq m} \times B_n^{\leq m} : M_1 \neq M_2\}.$$

For $M_1, M_2 \in B_n^+$ and $\mathsf{atk} \in \{\mathsf{eq}, \mathsf{pf}, \mathsf{any}\}$ we then let

$$\mathbf{CP}_n(M_1, M_2) = \Pr[\pi \xleftarrow{\$} \mathrm{Perm}(n): \mathrm{CBC}_\pi(M_1) = \mathrm{CBC}_\pi(M_2)]$$
$$\mathbf{CP}_{n,m}^{\mathsf{atk}} = \max_{(M_1,M_2)\in\mathcal{M}_{n,m}^{\mathsf{atk}}} \left\{ \mathbf{CP}_n(M_1, M_2) \right\}.$$

For $M_1, M_2 \in B_n^+$ we let $\mathbf{FCP}_n(M_1, M_2)$ (the full collision probability) be the probability, over $\pi \xleftarrow{\$} \mathrm{Perm}(n)$, that $\mathrm{CBC}_\pi(M_2)$ is in the set

$$\{\mathrm{CBC}_\pi^{M_1}[1], \ldots, \mathrm{CBC}_\pi^{M_1}[m_1], \mathrm{CBC}_\pi^{M_2}[1], \ldots, \mathrm{CBC}_\pi^{M_2}[m_2 - 1]\}$$

where $m_b = |M_b|_n$ for $b = 1, 2$. For $\mathsf{atk} \in \{\mathsf{eq}, \mathsf{pf}, \mathsf{any}\}$ we then let

$$\mathbf{FCP}_{n,m}^{\mathsf{atk}} = \max_{(M_1,M_2)\in\mathcal{M}_{n,m}^{\mathsf{atk}}} \left\{ \mathbf{FCP}_n(M_1, M_2) \right\}.$$

3   Results on the CBC MAC

We state results only for the $\mathsf{atk} = \mathsf{pf}$ case; results for $\mathsf{atk} = \mathsf{eq}$ follow due to (1). To bound $\mathbf{Adv}_{\mathrm{CBC}}^{\mathsf{pf}}(q, n, m)$ we must consider a dynamic adversary that adaptively queries its oracle. Our first lemma reduces this problem to that of bounding a more "static" quantity whose definition does not involve an adversary, namely the full collision probability of the CBC MAC. The proof is in Section 5.

Lemma 1. *For any* $n, m, q,$

$$\mathbf{Adv}_{\mathrm{CBC}}^{\mathsf{pf}}(q, n, m) \leq q^2 \cdot \mathbf{FCP}_{n,m}^{\mathsf{pf}} + \frac{4mq^2}{2^n}. \quad \blacksquare$$

The next lemma bounds the full collision probability of the CBC MAC. The proof is given in Section 8.

Lemma 2. *For any n, m with $m^2 \leq 2^{n-2}$,*

$$\mathbf{FCP}_{n,m}^{\mathsf{pf}} \leq \frac{8m}{2^n} + \frac{8m^4}{2^{2n}} \, . \quad \blacksquare$$

Combining the above two lemmas we bound $\mathbf{Adv}_{\mathrm{CBC}}^{\mathsf{pf}}(q, n, m)$:

Theorem 1. *For any n, m, q with $m^2 \leq 2^{n-2}$,*

$$\mathbf{Adv}_{\mathrm{CBC}}^{\mathsf{pf}}(q, n, m) \leq \frac{mq^2}{2^n} \cdot \left(12 + \frac{8m^3}{2^n}\right) \, . \quad \blacksquare$$

4   Results on the Encrypted CBC MAC

Following [7], we view ECBC as an instance of the Carter-Wegman paradigm [18]. This enables us to reduce the problem of bounding $\mathbf{Adv}_{\mathrm{ECBC}}^{\mathsf{atk}}(q, n, m)$ to bounding the collision probability of the CBC MAC, as stated in the next lemma. A proof of the following is provided in [3].

Lemma 3. *For any $n, m, q \geq 1$ and any $\mathsf{atk} \in \{\mathsf{eq}, \mathsf{pf}, \mathsf{any}\}$,*

$$\mathbf{Adv}_{\mathrm{ECBC}}^{\mathsf{atk}}(q, n, m) \leq \frac{q(q-1)}{2} \cdot \left(\mathbf{CP}_{n,m}^{\mathsf{atk}} + \frac{1}{2^n}\right) \, . \quad \blacksquare$$

Petrank and Rackoff [15] show that

$$\mathbf{Adv}_{\mathrm{ECBC}}^{\mathsf{any}}(q, n, m) \leq 2.5 \, m^2 q^2 / 2^n \, . \tag{2}$$

Dodis *et al.* [9] show that $\mathbf{CP}_{n,m}^{\mathsf{eq}} \leq 2^{-n} + cm^2 \cdot 2^{-2n} + cm^6 \cdot 2^{-3n}$ for some absolute constant c. Combining this with Lemma 3 leads to

$$\mathbf{Adv}_{\mathrm{ECBC}}^{\mathsf{eq}}(q, n, m) \leq \frac{q^2}{2^n} \cdot \left(1 + \frac{cm^2}{2^n} + \frac{cm^6}{2^{2n}}\right) \, .$$

However, the case of $\mathsf{atk} = \mathsf{eq}$ is not interesting here, since the point of ECBC is to gain security even for $\mathsf{atk} = \mathsf{any}$. To obtain an improvement for this, we show the following, whose proof is in Section 7:

Lemma 4. *For any n, m with $m^2 \leq 2^{n-2}$,*

$$\mathbf{CP}_{n,m}^{\mathsf{any}} \leq \frac{2d'(m)}{2^n} + \frac{8m^4}{2^{2n}}$$

where $d'(m)$ is the maximum, over all $m' \leq m$, of the number of positive numbers that divide m'. $\quad \blacksquare$

The function $d'(m)$ grows slowly; in particular, $d'(m) < m^{0.7/\ln\ln(m)}$ for all sufficiently large m [10, Theorem 317]. We have verified that $d'(m) \leq m^{1.07/\ln\ln m}$ for all $m \leq 2^{64}$ (and we assume for all m), and also that $d'(m) \leq \lg^2 m$ for all $m \leq 2^{25}$.

Combining the above with Lemma 3 leads to the following:

Theorem 2. *For any* n, m, q *with* $m^2 \leq 2^{n-2}$,

$$\mathbf{Adv}_{\text{ECBC}}^{\text{any}}(q, n, m) \leq \frac{q^2}{2^n} \cdot \left(d'(m) + \frac{4m^4}{2^n} \right) . \quad \blacksquare$$

5 Bounding FCP Bounds CBC (Proof of Lemma 1)

The proof is by the game-playing technique [2,4]. Let A be an adversary that asks exactly q queries, $M_1, \ldots, M_q \in B_{\bar{n}}^{\leq m}$, where no queries M_r and M_s, for $r \neq s$, share a prefix in B_n^+. We must show that $\mathbf{Adv}_{\text{CBC}}(A) \leq q^2 \cdot \mathbf{FCP}_{n,m}^{\text{pf}} + 4mq^2/2^n$.

Refer to games D0–D7 as defined in Fig. 2. Sets $\text{Dom}(\pi)$ and $\text{Ran}(\pi)$ start off as empty and automatically grow as points are added to the domain and range of the partial function π. Sets $\overline{\text{Dom}}(\pi)$ and $\overline{\text{Ran}}(\pi)$ are the complements of these sets relative to $\{0,1\}^n$. They automatically shrink as points join the domain and range of π. We write boolean values as 0 (false) and 1 (true), and we sometimes write **then** as a colon. The flag *bad* is initialized to 0 and the map π is initialized as everywhere undefined. We now briefly explain the sequence.

D1: Game D1 faithfully simulates the CBC MAC construction. Instead of choosing a random permutation π up front, we fill in its values as-needed, so as to not to create a conflict. Observe that if *bad* = 0 following lines 107–108 then $\widehat{C}_s^{m_s} = C_s^{m_s}$ and so game D1 always returns $C_s^{m_s}$, regardless of *bad*. This makes clear that $\Pr[A^{\text{D1}} \Rightarrow 1] = \Pr[\pi \xleftarrow{\$} \text{Perm}(n) : A^{\text{CBC}_\pi} \Rightarrow 1]$. **D0:** Game D0 is obtained from game D1 by omitting line 110 and the statements that immediately follow the setting of *bad* at lines 107 and 108. Thus this game returns the random n-bit string $C_s^{m_s} = \widehat{C}_s^{m_s}$ in response to each query M_s, so $\Pr[A^{\text{D0}} \Rightarrow 1] = \Pr[\rho \xleftarrow{\$} \text{Func}(n) : A^{\rho} \Rightarrow 1]$. Now games D1 and D0 have been defined so as to be syntactically identical except on statements that immediately follow the setting of *bad* to true or the checking if *bad* is true, so the fundamental lemma of game-playing [4] says us that $\Pr[A^{\text{D1}} \Rightarrow 1] - \Pr[A^{\text{D0}} \Rightarrow 1] \leq \Pr[A^{\text{D0}} \text{ sets } bad]$. As $\mathbf{Adv}_{\text{CBC}}(A) = \Pr[A^{\text{CBC}_\pi} \Rightarrow 1] - \Pr[A^{\rho} \Rightarrow 1] = \Pr[A^{\text{D1}} \Rightarrow 1] - \Pr[A^{\text{D0}} \Rightarrow 1]$, the rest of the proof bounds $\mathbf{Adv}_{\text{CBC}}(A)$ by bounding $\Pr[A^{\text{D0}} \text{ sets } bad]$.

D0→D2: We rewrite game D0 as game D2 by dropping the variable $\widehat{C}_s^{m_s}$ and using variable $C_s^{m_s}$ in its place, as these are always equal. We have that $\Pr[A^{\text{D0}} \text{ sets } bad] = \Pr[A^{\text{D2}} \text{ sets } bad]$. **D2→D3:** Next we eliminate line 209 and then, to compensate, we set *bad* any time the value $X_s^{m_s}$ or $C_s^{m_s}$ would have been accessed. This accounts for the new line 303 and the new disjunct on lines 310. To compensate for the removal of line 209 we must also set *bad* whenever C_s^i, chosen at line 204, happens to be a prior value $C_r^{m_r}$. This is done at line 306. We have that $\Pr[A^{\text{D2}} \text{ sets } bad] \leq \Pr[A^{\text{D3}} \text{ sets } bad]$. **D3→D4:** Next we remove the

On the sth query $F(M_s)$ Game D1	On the sth query $F(M_s)$ Game D2
100 $m_s \leftarrow \lvert M_s \rvert_n, \quad C_s^0 \leftarrow 0^n$	200 $m_s \leftarrow \lvert M_s \rvert_n, \quad C_s^0 \leftarrow 0^n$
101 **for** $i \leftarrow 1$ **to** $m_s - 1$ **do**	201 **for** $i \leftarrow 1$ **to** $m_s - 1$ **do**
102 $\quad X_s^i \leftarrow C_s^{i-1} \oplus M_s^i$	202 $\quad X_s^i \leftarrow C_s^{i-1} \oplus M_s^i$
103 $\quad$ **if** $X_s^i \in \mathrm{Dom}(\pi)$ **then** $C_s^i \leftarrow \pi(X_s^i)$	203 $\quad$ **if** $X_s^i \in \mathrm{Dom}(\pi)$ **then** $C_s^i \leftarrow \pi(X_s^i)$
104 $\qquad$ **else** $\pi(X_s^i) \leftarrow C_s^i \overset{\$}{\leftarrow} \overline{\mathrm{Ran}}(\pi)$	204 $\qquad$ **else** $\pi(X_s^i) \leftarrow C_s^i \overset{\$}{\leftarrow} \overline{\mathrm{Ran}}(\pi)$
105 $X_s^{m_s} \leftarrow C_s^{m_s-1} \oplus M_s^{m_s}$	205 $X_s^{m_s} \leftarrow C_s^{m_s-1} \oplus M_s^{m_s}$
106 $\widehat{C}_s^{m_s} \leftarrow C_s^{m_s} \overset{\$}{\leftarrow} \{0,1\}^n$	206 $C_s^{m_s} \overset{\$}{\leftarrow} \{0,1\}^n$
107 **if** $C_s^{m_s} \in \mathrm{Ran}(\pi)$: $bad \leftarrow 1, C_s^{m_s} \overset{\$}{\leftarrow} \overline{\mathrm{Ran}}(\pi)$	207 **if** $X_s^{m_s} \in \mathrm{Dom}(\pi) \vee C_s^{m_s} \in \mathrm{Ran}(\pi)$
108 **if** $X_s^{m_s} \in \mathrm{Dom}(\pi)$: $bad \leftarrow 1, C_s^{m_s} \leftarrow \pi(X_s^{m_s})$	208 $\quad$ **then** $bad \leftarrow 1$
109 $\pi(X_s^{m_s}) \leftarrow C_s^{m_s}$	209 $\pi(X_s^{m_s}) \leftarrow C_s^{m_s}$
110 **if** bad **then return** $C_s^{m_s}$	210 **return** $C_s^{m_s}$
111 **return** $\widehat{C}_s^{m_s}$	

On the sth query $F(M_s)$ Game D3	On the sth query $F(M_s)$ Game D4
300 $m_s \leftarrow \lvert M_s \rvert_n, \quad C_s^0 \leftarrow 0^n$	400 $m_s \leftarrow \lvert M_s \rvert_n, \quad C_s^0 \leftarrow 0^n$
301 **for** $i \leftarrow 1$ **to** $m_s - 1$ **do**	401 **for** $i \leftarrow 1$ **to** $m_s - 1$ **do**
302 $\quad X_s^i \leftarrow C_s^{i-1} \oplus M_s^i$	402 $\quad X_s^i \leftarrow C_s^{i-1} \oplus M_s^i$
303 $\quad$ **if** $(\exists r < s)(X_s^i = X_r^{m_r})$: $bad \leftarrow 1$	403 $\quad$ **if** $(\exists r < s)(X_s^i = X_r^{m_r})$: $bad \leftarrow 1$
304 $\quad$ **if** $X_s^i \in \mathrm{Dom}(\pi)$ **then** $C_s^i \leftarrow \pi(X_s^i)$	404 $\quad$ **if** $X_s^i \in \mathrm{Dom}(\pi)$ **then** $C_s^i \leftarrow \pi(X_s^i)$
305 $\qquad$ **else** $\pi(X_s^i) \leftarrow C_s^i \overset{\$}{\leftarrow} \overline{\mathrm{Ran}}(\pi)$,	405 $\qquad$ **else** $\pi(X_s^i) \leftarrow C_s^i \overset{\$}{\leftarrow} \overline{\mathrm{Ran}}(\pi)$
306 $\qquad\quad$ **if** $(\exists r < s)(C_s^i = C_r^{m_r})$: $bad \leftarrow 1$	406 $X_s^{m_s} \leftarrow C_s^{m_s-1} \oplus M_s^{m_s}$
307 $X_s^{m_s} \leftarrow C_s^{m_s-1} \oplus M_s^{m_s}$	407 **if** $X_s^{m_s} \in \mathrm{Dom}(\pi) \vee$
308 $C_s^{m_s} \overset{\$}{\leftarrow} \{0,1\}^n$	408 $\quad (\exists r < s)(X_s^{m_s} = X_r^{m_r})$ **then** $bad \leftarrow 1$
309 **if** $X_s^{m_s} \in \mathrm{Dom}(\pi) \vee C_s^{m_s} \in \mathrm{Ran}(\pi) \vee$	409 $C_s^{m_s} \overset{\$}{\leftarrow} \{0,1\}^n$
310 $\quad (\exists r < s)(X_s^{m_s} = X_r^{m_r} \vee C_s^{m_s} = C_r^{m_r})$	410 **return** $C_s^{m_s}$
311 $\quad$ **then** $bad \leftarrow 1$	
312 **return** $C_s^{m_s}$	

500 **for** $s \leftarrow 1$ **to** q **do** Game D5	600 $\pi \overset{\$}{\leftarrow} \mathrm{Perm}(n)$ Game D6
501 $\quad C_s^0 \leftarrow 0^n$	601 **for** $s \in [1..q]$ **do**
502 $\quad$ **for** $i \leftarrow 1$ **to** $\mathrm{m}_s - 1$ **do**	602 $\quad C_s^0 \leftarrow 0^n$
503 $\qquad X_s^i \leftarrow C_s^{i-1} \oplus \mathrm{M}_s^i$	603 $\quad$ **for** $i \leftarrow 1$ **to** $\mathrm{m}_s - 1$ **do**
504 $\qquad$ **if** $(\exists r < s)(X_s^i = X_r^{\mathrm{m}_r})$: $bad \leftarrow 1$	604 $\qquad X_s^i \leftarrow C_s^{i-1} \oplus \mathrm{M}_s^i$
505 $\qquad$ **if** $X_s^i \in \mathrm{Dom}(\pi)$ **then** $C_s^i \leftarrow \pi(X_s^i)$	605 $\qquad C_s^i \leftarrow \pi(X_s^i)$
506 $\qquad\quad$ **else** $\pi(X_s^i) \leftarrow C_s^i \overset{\$}{\leftarrow} \overline{\mathrm{Ran}}(\pi)$	606 $\quad X_s^{\mathrm{m}_s} \leftarrow C_s^{\mathrm{m}_s-1} \oplus \mathrm{M}_s^{\mathrm{m}_s}$
507 $\quad X_s^{\mathrm{m}_s} \leftarrow C_s^{\mathrm{m}_s-1} \oplus \mathrm{M}_s^{\mathrm{m}_s}$	607 $bad \leftarrow (\exists (r,i) \neq (s,\mathrm{m}_s))\,[X_r^i = X_s^{\mathrm{m}_s}]$
508 $\quad$ **if** $(\exists r < s)\,(X_s^{\mathrm{m}_s} \in \mathrm{Dom}(\pi) \vee$	
509 $\qquad X_s^{\mathrm{m}_s} = X_r^{\mathrm{m}_r})$ **then** $bad \leftarrow 1$	

700 $\pi \overset{\$}{\leftarrow} \mathrm{Perm}(n)$ Game D7
701 $C_1^0 \leftarrow C_2^0 \leftarrow 0^n$
702 **for** $i \leftarrow 1$ **to** m_1 **do**
703 $\quad X_1^i \leftarrow C_1^{i-1} \oplus \mathrm{M}_1^i, \quad C_1^i \leftarrow \pi(X_1^i)$
704 **for** $i \leftarrow 1$ **to** m_2 **do**
705 $\quad X_2^i \leftarrow C_2^{i-1} \oplus \mathrm{M}_2^i, \quad C_2^i \leftarrow \pi(X_2^i)$
706 $bad \leftarrow X_2^{\mathrm{m}_2} \in \{X_1^1, \ldots, X_1^{\mathrm{m}_1},$
707 $\qquad\qquad X_2^1, \ldots, X_2^{\mathrm{m}_2-1}\}$

Fig. 2. Games D0–D7 used in the proof of Lemma 1

test $(\exists r < s)(C_s^i = C_r^{m_r})$ at line 306, the test if $C_s^{m_s} \in \text{Ran}(\pi)$ at line 309, and the test for $C_s^{m_s} = C_r^{m_r}$ at line 310, bounding the probability that *bad* gets set due to any of these three tests. To bound the probability of *bad* getting set at line 306: A total of at most mq times we select at line 305 a random sample C_s^i from a set of size at least $2^n - mq \ge 2^{n-1}$. (We may assume that $mq \le 2^{n-1}$ since the probability bound given by our lemma exceeds 1 if $mq > 2^{n-1}$.) The chance that one of these points is equal to any of the at most q points $C_r^{m_r}$ is thus at most $2mq^2/2^n$. To bound the probability of *bad* getting set by the $C_s^{m_s} \in \text{Ran}(\pi)$ test at line 309: easily seen to be at most $mq^2/2^n$. To bound the probability of *bad* getting set by the $C_s^{m_s} = C_r^{m_r}$ test at line 310: easily seen to be at most $q^2/2^n$. Overall then, $\Pr[A^{D3} \text{ sets } bad\,] \le \Pr[A^{D4} \text{ sets } bad\,] + 4mq^2/2^n$.

D4→D5: The value $C_s^{m_s}$ returned to the adversary in response to a query in game D4 is never referred to again in the code and has no influence on the game and the setting of *bad*. Accordingly, we may think of these values as being chosen up-front by the adversary who, correspondingly, makes an optimal choice of message queries $M_1, \ldots, M_q$ so as to maximize the probability that *bad* gets set in game D4. Queries $M_1, \ldots, M_q \in B_n^{\le m}$ are prefix-free (meaning that no two strings from this list share a prefix $P \in B_n^+$) and the strings have block lengths of $m_1, \ldots, m_q$, respectively, where each $m_i \le m$. We fix such an optimal vector of messages and message lengths in passing to game D5, so that $\Pr[A^{D4} \text{ sets } bad\,] \le \Pr[\text{D5 sets } bad\,]$. The adversary has effectively been eliminated at this point.

D5→D6: Next we postpone the evaluation of *bad* and undo the "lazy defining" of π to arrive at game D6. We have $\Pr[\text{D5 sets } bad\,] \le \Pr[\text{D6 sets } bad\,]$.
D6→D7: Next we observe that in game D6, some pair r, s must contribute at least an average amount to the probability that *bad* gets set. Namely, for any $r, s \in [1 .. q]$ where $r \ne s$ define $bad_{r,s}$ as

$$(X_s^{m_s} = X_r^i \text{ for some } i \in [1 .. m_r]) \vee (X_s^{m_s} = X_s^i \text{ for some } i \in [1 .. m_s - 1])$$

and note that *bad* is set at line 607 iff $bad_{r,s} = 1$ for some $r \ne s$, and so there must be an $r \ne s$ such that $\Pr[\text{D6 sets } bad_{r,s}] \ge (1/q(q-1)) \Pr[\text{D6 sets } bad\,]$. Fixing such an r, s and renaming $M_1 = M_r$, $M_2 = M_s$, $m_1 = m_r$, and $m_2 = m_s$, we arrive at game D7 knowing that

$$\Pr[\text{D6 sets } bad\,] \le q^2 \cdot \Pr[\text{D7 sets } bad\,]. \tag{3}$$

Now $\Pr[\text{D7 sets } bad\,] = \mathbf{FCP}_n(M_1, M_2) \le \mathbf{FCP}_{n,m}^{\text{pf}}$ by the definition of FCP and the fact that π is a permutation. Putting all the above together we are done.

6   A Graph-Based Representation of CBC

In this section we describe a graph-based view of CBC computations and provide some lemmas that will then allow us to reduce the problem of upper bounding the collision probabilities $\mathbf{CP}_{n,m}^{\text{any}}$ and $\mathbf{FCP}_{n,m}^{\text{pf}}$ to combinatorial counting problems.

We fix for the rest of this section a blocklength $n \ge 1$ and a pair of distinct messages $M_1 = M_1^1 \cdots M_1^{m_1} \in B_n^{m_1}$ and $M_2 = M_2^1 \cdots M_2^{m_2} \in B_n^{m_2}$ where $m_1, m_2 \ge 1$. We let $\ell = \max(m_1, m_2)$.

---

algorithm Perm2Graph(M_1, M_2, π) $// M_1 \in B_n^{m_1}, M_2 \in B_n^{m_2}, \pi \in \mathrm{Perm}(n)$
   $\sigma(0) \leftarrow 0^n, \quad \nu \leftarrow 0, \quad E \leftarrow \emptyset$
   for $b \leftarrow 1$ **to** 2 **do**
     $v \leftarrow 0$
     for $i \leftarrow 1$ **to** m_b **do**
      if $\exists w$ s.t. $(v, w) \in E$ **and** $L((v, w)) = M_b^i$ **then** $v \leftarrow w$
      else if $\exists w$ s.t. $\pi(\sigma(v) \oplus M_b^i) = \sigma(w)$ **then**
         $E \leftarrow E \cup \{(v, w)\}, \; L((v, w)) \leftarrow M_b^i, \; v \leftarrow w$
        else $\nu \leftarrow \nu + 1, \; \sigma(\nu) \leftarrow \pi(\sigma(v) \oplus M_b^i),$
          $E \leftarrow E \cup \{(v, \nu)\}, \; L((v, \nu)) \leftarrow M_b^i, \; v \leftarrow \nu$
   return $G \leftarrow ([0..\nu], \; E, \; L)$

---

algorithm Graph2Profs(G) $// G \in \mathcal{G}(M_1, M_2), M_1 \in B_n^{m_1}, M_2 \in B_n^{m_2}$
   $\mathrm{Prof}_1 \leftarrow \mathrm{Prof}_2 \leftarrow \mathrm{Prof}_3 \leftarrow (), \; V' \leftarrow \{0\}, \; E' \leftarrow \emptyset$
   for $b \leftarrow 1$ **to** 2 **do**
     for $i \leftarrow 1$ **to** m_b **do**
      if $\exists w \in V'$ s.t. $V_b^i(G) = w$ **then**
        if $b = 1$ **then** $p \leftarrow (w, i)$ **else** $p \leftarrow (w, m_1 + i)$
        $\mathrm{Prof}_1 \leftarrow \mathrm{Prof}_1 \parallel p$
        if $(V_b^{i-1}(G), w) \notin E'$ **then** $\mathrm{Prof}_2 \leftarrow \mathrm{Prof}_2 \parallel p$
        if $\mathrm{CYCLE}_G(V', E', V_b^{i-1}(G), w) = 0$ **then** $\mathrm{Prof}_3 \leftarrow \mathrm{Prof}_3 \parallel p$
      $V' \leftarrow V' \cup \{V_b^i(G)\}, \; E' \leftarrow E' \cup \{(V_b^{i-1}(G), V_b^i(G))\}$
   return $(\mathrm{Prof}_1, \mathrm{Prof}_2, \mathrm{Prof}_3)$

---

algorithm Prof2Graph(A) $// A = ((i_1, t_1), \ldots, (i_a, t_a)) \in \mathrm{Prof}_2(M_1, M_2)$
   $V \leftarrow \{0\}, \; E \leftarrow \emptyset, \; c \leftarrow 1, \; v_0^1 \leftarrow v_0^2 \leftarrow \nu \leftarrow 0$
   for $b \leftarrow 1$ **to** 2 **do**
     for $i \leftarrow 1$ **to** m_b **do**
      if $i = t_c$ **then** $v_i^b \leftarrow i_c, \; c \leftarrow c + 1$ **else** $\nu \leftarrow \nu + 1, \; v_i^b \leftarrow \nu$
      $E \leftarrow E \cup \{(v_{i-1}^b, v_i^b)\}, \; L((v_{i-1}^b, v_i^b)) \leftarrow M_b^i$
   return $G \leftarrow ([0..\nu], E, L)$

Fig. 3. The first algorithm above builds the structure graph $G_\pi^{M_1, M_2}$ associated to M_1, M_2 and a permutation $\pi \in \mathrm{Perm}(n)$. The next associates to $G \in \mathcal{G}(M_1, M_2)$ its type-1, type-2 and type-3 collision profiles. The last algorithm constructs a graph from its type-2 collision profile $A \in \mathrm{Prof}_2(M_1, M_2)$.

STRUCTURE GRAPHS. To M_1, M_2 and any $\pi \in \mathrm{Perm}(n)$ we associate the *structure graph* $G_\pi^{M_1, M_2}$ output by the procedure Perm2Graph (permutation to graph) of Fig. 3. The structure graph is a directed graph (V, E) together with an edge-labeling function $L \colon E \to \{M_1^1, \ldots, M_1^{m_1}, M_2^1, \ldots, M_2^{m_2}\}$, where $V = [0..\nu]$ for some $\nu \leq m_1 + m_2 + 1$. To get some sense of what is going on here, let

$$C_\pi^{M_1, M_2} = \{\mathrm{CBC}_\pi^{M_1}[i] : 0 \leq i \leq m_1\} \cup \{\mathrm{CBC}_\pi^{M_2}[i] : 0 \leq i \leq m_2\}.$$

Note that due to collisions the size of the set $C_\pi^{M_1, M_2}$ could be strictly less than the maximum possible size of $m_1 + m_2 + 1$. The structure graph $G_\pi^{M_1, M_2}$ has vertex set $V = [0..\eta]$ where $\eta = |C_\pi^{M_1, M_2}|$. Associated to a vertex $v \in V$ is a label $\sigma(v) \in C_\pi^{M_1, M_2}$, with $\sigma(0) = 0^n$. (This label is constructed by the code but not part of the final graph.) An edge from a to b with label x exists in the structure graph iff $\pi(\sigma(a) \oplus x) = \sigma(b)$.

Let $\mathcal{G}(M_1, M_2) = \{G_\pi^{M_1, M_2} : \pi \in \mathrm{Perm}(n)\}$ denote the set of all structure graphs associated to messages M_1, M_2. This set has the probability distribution induced by picking π at random from $\mathrm{Perm}(n)$.

We associate to $G = (V, E, L) \in \mathcal{G}(M_1, M_2)$ sequences $V_b^0, \ldots, V_b^{m_b} \in V$ that for $b = 1, 2$ are defined inductively as follows: set $V_b^0 = 0$ and for $i \in [1..m_b]$ let V_b^i be the unique vertex $w \in V$ such that there is an edge $(V_b^{i-1}, w) \in E$ with $L(e) = M_b^i$. Note that this defines the following walks in G:

$$0 = V_1^0 \xrightarrow{M_1^1} V_1^1 \xrightarrow{M_1^2} V_1^2 \longrightarrow \cdots \longrightarrow V_1^{m_1} \xrightarrow{M_1^{m_1}} V_1^{m_1} \text{ and}$$

$$0 = V_2^0 \xrightarrow{M_2^1} V_2^1 \xrightarrow{M_2^2} V_2^2 \longrightarrow \cdots \longrightarrow V_2^{m_2-1} \xrightarrow{M_2^{m_2}} V_2^{m_2} .$$

If $G = G_\pi^{M_1, M_2}$ then observe that $\sigma(V_b^i) = \mathrm{CBC}_\pi^{M_1, M_2}[i]$ for $i \in [0..m_b]$ and $b = 1, 2$, where $\sigma(\cdot)$ is the vertex-labeling function defined by $\mathsf{Perm2Graph}(\pi)$. We emphasize that V_b^i depends on G (and thus implicitly on M_1 and M_2), and if we want to make the dependence explicit we will write $V_b^i(G)$.

COLLISIONS. We use the following notation for sequences. If $s = (s_1, \ldots, s_k)$ is a sequence then $|s| = k$; $y \in s$ iff $y = s_i$ for some $i \in [1..k]$; $s \parallel x = (s_1, \ldots, s_k, x)$; and $(\ )$ denotes the empty sequence. For $G = (V, E) \in \mathcal{G}$, $E' \subseteq E$, $V' \subseteq V$ and $a, b \in V$ we define $\mathrm{CYCLE}_G(V', E', a, b) = 1$ if adding edge (a, b) to graph $G' = (V', E')$ closes a cycle of length at least four with directions of edges on the cycle alternating. Formally, $\mathrm{CYCLE}_G(V', E', a, b) = 1$ iff there exists $k \geq 2$ and vertices $a = v_1, v_2, \ldots, v_{2k-1}, v_{2k} = b \in V'$ such that $(v_{2i-1}, v_{2i}) \in E'$ for all $i \in [1..k]$, $(v_{2i+1}, v_{2i}) \in E'$ for all $i \in [1..k-1]$, and $(b, a) \in E$. To a graph $G \in \mathcal{G}$ we associate sequences $\mathsf{Prof}_1(G), \mathsf{Prof}_2(G), \mathsf{Prof}_3(G)$ called, respectively, the type-1, type-2 and type-3 collision profiles of G. They are returned by the algorithm $\mathsf{Graph2Profs}$ (graph to collision profiles) of Fig. 3 that refers to the predicate CYCLE_G we have just defined. We say that G has a type-a (i, t)-collision $(a \in \{1, 2, 3\})$ if $(i, t) \in \mathsf{Prof}_a(G)$. Type-3 collisions are also called accidents, and type-1 collisions that are not accidents are called induced collisions. We let $\mathsf{col}_i(G) = |\mathsf{Prof}_i(G)|$ for $i = 1, 2, 3$.

Lemma 5. Let $n \geq 1$, $M_1 \in B_n^{m_1}$, $M_2 \in B_n^{m_2}$, $\ell = \max(m_1, m_2)$. Let $H \in \mathcal{G}(M_1, M_2)$ be a structure graph. Then

$$\Pr[G \xleftarrow{\$} \mathcal{G}(M_1, M_2) : G = H] \leq \frac{1}{(2^n - m - m')^{\mathsf{col}_3(H)}} \leq \frac{1}{(2^n - 2\ell)^{\mathsf{col}_3(H)}} . \blacksquare$$

The lemma builds on an unpublished technique from [8,9]. A proof is given in [3]. For $i = 1, 2, 3$ let $\mathsf{Prof}_i(M_1, M_2) = \{\mathsf{Prof}_i(G) : G \in \mathcal{G}(M_1, M_2)\}$. Note that if $A = ((w_1, t_1), \ldots, (w_a, t_a)) \in \mathsf{Prof}_2(M_1, M_2)$ then $1 \leq t_1 < \cdots < t_a \leq m_1 + m_2$ and $w_i < t_i$ for all $i \in [1..a]$. Algorithm $\mathsf{Prof2Graph}$ (collision profile to graph) of Fig. 3 associates to $A \in \mathsf{Prof}_2(M_1, M_2)$ a graph in a natural way. We leave the reader to verify the following:

Lemma 6. Prof2Graph(Prof$_2(G)$) = G *for any* $G \in \mathcal{G}(M_1, M_2)$.   ∎

This means that the type-2 collision profile of a graph determines it uniquely. Now for $i = 1, 2, 3$ and an integer $a \geq 0$ we let $\mathcal{G}_i^a(M_1, M_2) = \{G \in \mathcal{G}(M_1, M_2) : \text{col}_i(G) = a\}$ and $\text{Prof}_i^a(M_1, M_2) = \{A \in \text{Prof}_i(M_1, M_2) : |A| = a\}$

Lemma 7. *Let* $n \geq 1$, $M_1 \in B_n^{m_1}$, $M_2 \in B_n^{m_2}$, $\ell = \max(m_1, m_2)$, *and assume* $\ell^2 \leq 2^{n-2}$. *Then*

$$\Pr[G \xleftarrow{\$} \mathcal{G}(M_1, M_2) : \text{col}_3(G) \geq 2] \leq \frac{8\ell^4}{2^{2n}} . \quad \blacksquare$$

Proof. By Lemma 5

$$\Pr[G \xleftarrow{\$} \mathcal{G}(M_1, M_2) : \text{col}_3(G) \geq 2]$$

$$= \sum_{a=2}^{\ell} \sum_{H \in \mathcal{G}_3^a(M_1, M_2)} \Pr[G \xleftarrow{\$} \mathcal{G}(M_1, M_2) : G = H]$$

$$\leq \sum_{a=2}^{\ell} \frac{|\mathcal{G}_3^a(M_1, M_2)|}{(2^n - 2\ell)^a} .$$

Since every type-3 collision is a type-2 collision, $|\mathcal{G}_3^a(M_1, M_2)| \leq |\mathcal{G}_2^a(M_1, M_2)|$. By Proposition 6, $|\mathcal{G}_2^a(M_1, M_2)| = |\text{Prof}_2^a(M_1, M_2)|$. Now $|\text{Prof}_2^a(M_1, M_2)| \leq (\ell(\ell+1)/2)^a \leq \ell^{2a}$, so we have

$$\sum_{a=2}^{\ell} \frac{|\mathcal{G}_3^a(M_1, M_2)|}{(2^n - 2\ell)^a} \leq \sum_{a=2}^{\ell} \frac{\ell^{2a}}{(2^n - 2\ell)^a} .$$

Let $x = \ell^2/(2^n - 2\ell)$, and observe that the assumption $\ell^2 \leq 2^{n-2}$ made in the lemma statement implies that $x \leq 1/2$. Thus the above is

$$\sum_{a=2}^{\ell} x^a = x^2 \cdot \sum_{a=0}^{\ell-2} x^a \leq x^2 \cdot \sum_{a=0}^{\infty} x^a \leq 2x^2 = \frac{2\ell^4}{(2^n - 2\ell)^2} \leq \frac{8\ell^4}{2^{2n}} ,$$

where the last inequality used the fact that $\ell \leq 2^{n-2}$.   ∎

Let P denote a predicate on graphs. Then $\phi_{M_1, M_2}[P]$ will denote the set of all $G \in \mathcal{G}_3^1(M_1, M_2)$ such that G satisfies P. (That is, it is the set of structure graphs G having exactly one type-3 collision and satisfying the predicate.) For example, predicate P might be $V_1^{m_1}(\cdot) = V_2^{m_2}(\cdot)$ and in that case $\phi_{M_1, M_2}[V_1^{m_1} = V_2^{m_2}]$ is $\{G \in \mathcal{G}_3^1(M_1, M_2) : V_1^{m_1}(G) = V_2^{m_2}(G)\}$.

Note that if G has exactly one accident then $\text{Prof}_2(G) = \text{Prof}_3(G)$, meaning the accident was both a type-2 and a type-3 collision. We will use this below. In this case when we talk of an (i, t)-accident, we mean a type-2 (i, t)-collision.

Finally, let $\text{in}_G(v)$ denote the in-degree of a vertex v in a structure graph G.

7 Bounding $\mathbf{CP}^{\mathsf{any}}_{n,m}$ (Proof of Lemma 4)

In this section we prove Lemma 4, showing that $\mathbf{CP}^{\mathsf{any}}_{n,\ell} \leq 2d'(\ell)/2^n + 8\ell^4/2^{2n}$ for any n, ℓ with $\ell^2 \leq 2^{n-2}$, thereby proving Lemma 4.

Lemma 8. *Let $n \geq 1$ and $1 \leq m_1, m_2 \leq \ell$. Let $M_1 \in B_n^{m_1}$ and $M_2 \in B_n^{m_2}$ be distinct messages and assume $\ell^2 \leq 2^{n-2}$. Then*

$$\mathbf{CP}^{\mathsf{any}}_{n,\ell}(M_1, M_2) \leq \frac{2 \cdot |\phi_{M_1,M_2}[V_1^{m_1} = V_2^{m_2}]|}{2^n} + \frac{8\ell^4}{2^{2n}} . \; \blacksquare$$

Proof. With the probability over $G \xleftarrow{\$} \mathcal{G}(M_1, M_2)$, we have:

$$
\begin{aligned}
\mathbf{CP}_n(M_1, M_2) \\
&= \Pr\left[V_1^{m_1} = V_2^{m_2} \right] \\
&= \Pr\left[V_1^{m_1} = V_2^{m_2} \wedge \mathsf{col}_3(G) = 1 \right] + \Pr\left[V_1^{m_1} = V_2^{m_2} \wedge \mathsf{col}_3(G) \geq 2 \right] \quad (4) \\
&\leq \frac{|\phi_{M_1,M_2}[V_1^{m_1} = V_2^{m_2}]|}{2^n - 2\ell} + \frac{8\ell^4}{2^{2n}} \quad (5) \\
&\leq \frac{2 \cdot |\phi_{M_1,M_2}[V_1^{m_1} = V_2^{m_2}]|}{2^n} + \frac{8\ell^4}{2^{2n}} . \quad (6)
\end{aligned}
$$

In (4) above we used that $\Pr\left[V_1^{m_1} = V_2^{m_2} \wedge \mathsf{col}_3(G) = 0 \right] = 0$ as $V_1^{m_1} = V_2^{m_2}$ with $M_1 \neq M_2$ implies that there is at least one accident. In (5) we first used Lemma 5, and then used Lemma 7. In (6) we used the fact that $\ell \leq 2^{n-2}$, which follows from the assumption $\ell^2 \leq 2^{n-2}$. $\quad\blacksquare$

Next we bound the size of the set that arises above:

Lemma 9. *Let $n, \ell \geq 1$ and $1 \leq m_2 \leq m_1 \leq \ell$. Let $M_1 \in B_n^{m_1}$ and $M_2 \in B_n^{m_2}$ be distinct messages. Then*

$$|\phi_{M_1,M_2}[V_1^{m_1} = V_2^{m_2}]| \leq d'(\ell) . \quad \blacksquare$$

Putting together Lemmas 8 and 9 completes the proof of Lemma 4.

Proof (Lemma 9). Let $k \geq 0$ be the largest integer such that M_1, M_2 have a common suffix of k blocks. Note that $V_1^{m_1} = V_2^{m_2}$ iff $V_1^{m_1-k} = V_2^{m_2-k}$. Thus, we may consider M_1 to be replaced by $M_1^{1 \to m_1-k}$ and M_2 to be replaced by $M_2^{1 \to m_2-k}$, with m_1, m_2 correspondingly replaced by $m_1 - k, m_2 - k$ respectively. We now have distinct messages M_1, M_2 of at most ℓ blocks each such that either $m_2 = 0$ or $M_1^{m_1} \neq M_2^{m_2}$. (Note that now m_2 could be 0, which was not true before our transformation.) Now consider three cases. The first is that $m_2 \geq 1$ and M_2 is a prefix of M_1. This case is covered by Lemma 10. (Note in this case it must be that $m_1 > m_2$ since M_1, M_2 are distinct and their last blocks are different.) The second case is that $m_2 = 0$ and is covered by Lemma 11. (In this case, $m_1 \geq 1$ since M_1, M_2 are distinct.) The third case is that $m_2 \geq 1$ and M_2 is not a prefix of M_1. This case is covered by Lemma 12. $\quad\blacksquare$

Lemma 10. *Let $n \geq 1$ and $1 \leq m_2 < m_1 \leq \ell$. Let $M_1 \in B_n^{m_1}, M_2 \in B_n^{m_2}$. Assume M_2 is a prefix of M_1 and $M_1^{m_1} \neq M_2^{m_2}$. Then $|\phi_{M_1,M_2}[V_1^{m_1} = V_2^{m_2}]| \leq d'(\ell)$.* ∎

Proof. Because M_2 is a prefix of M_1 we have that $V_2^{m_2} = V_1^{m_2}$, and thus $|\phi_{M_1,M_2}[V_1^{m_1} = V_2^{m_2}]| = |\phi_{M_1,M_2}[V_1^{m_2} = V_1^{m_1}]|$. We now bound the latter.

Let $G \in \mathcal{G}_3^1(M_1, M_2)$. Then $V_1^{m_1}(G) = V_1^{m_2}(G)$ iff $\exists t \geq m_2$ such that G has a type-2 $(t, V_1^{m_2}(G))$-collision. (This is also a type-3 $(V_1^{m_2}(G), t)$-collision since G has exactly one accident.) To see this note that since there was at most one accident, we have $\mathrm{in}_G(V_1^i(G)) \leq 1$ for all $i \in [1..m_1]$ except one, namely the i such that $V_1^i(G)$ was hit by the accident. And it must be that $i = m_2$ since $V_1^{m_2}(G)$ has in-going edges labeled $M_1^{m_2}$ and $M_1^{m_1}$, and these edges cannot be the same as $M_1^{m_1} \neq M_1^{m_2}$.

Let $c \geq 1$ be the smallest integer such that $V_1^{m_2+c}(G) = V_1^{m_2}(G)$. That is, we have a cycle $V_1^{m_2}(G), V_1^{m_2+1}(G), \ldots, V_1^{m_2+c}(G) = V_1^{m_2}(G)$. Now, given that there is only one accident and $V_1^{m_2}(G) = V_1^{m_1}(G)$, it must be that $m_1 = m_2 + kc$ for some integer $k \geq 1$. (That is, starting from $V_1^{m_2}(G)$, one traverses the cycle k times before reaching $V_1^{m_1}(G) = V_1^{m_2}(G)$.) This means that c must divide $m_1 - m_2$. But $|\phi_{M_1,M_2}[V_1^{m_2} = V_1^{m_1}]|$ is at most the number of possible values of c, since this value uniquely determines the graph. So $|\phi_{M_1,M_2}[V_1^{m_2} = V_1^{m_1}]| \leq d(m_1 - m_2)$, where $d(s)$ is the number of positive integers $i \leq s$ such that i divides s. But $d(m_1 - m_2) \leq d'(\ell)$ by definition of the latter. ∎

Lemma 11. *Let $n \geq 1$ and $1 \leq m_1 \leq \ell$. Let $M_1 \in B_n^{m_1}$, let $M_2 = \varepsilon$ and let $m_2 = 0$. Then $|\phi_{M_1,M_2}[V_1^{m_1} = V_2^{m_2}]| \leq d'(\ell)$.* ∎

Proof. Use an argument similar to that of Lemma 10, noting that $V_{m_1}^0(G) = V_1^0(G)$ implies that $\mathrm{in}_G(V_1^0(G)) \geq 1$. ∎

Lemma 12. *Let $n \geq 1$ and $1 \leq m_2 \leq m_1 \leq \ell$. Let $M_1 \in B_n^{m_1}, M_2 \in B_n^{m_2}$. Assume M_2 is not a prefix of M_1 and $M_1^{m_1} \neq M_2^{m_2}$. Then $|\phi_{M_1,M_2}[V_1^{m_1} = V_2^{m_2}]| \leq 1$.* ∎

Proof. Let $p \in [0..m_2 - 1]$ be the largest integer such that $M_1^{1 \to i} = M_2^{1 \to i}$ for all $i \in [1..p]$. Then $V_1^i = V_2^i$ for $i \in [1..p]$ and $V_1^{p+1} \neq V_2^{p+1}$. Now to have $V_1^{m_1} = V_2^{m_2}$ we need an accident. Since $M_1^{m_1} \neq M_2^{m_2}$ and there is only one accident, the only possibility is that this is a $(V_1^{m_1}, m_1 + m_2)$-collision. Thus, there is only one way to draw the graph. ∎

8   Bounding $\mathbf{FCP}_{n,\ell}^{\mathsf{pf}}$ (Proof of Lemma 2)

In this section we show that $\mathbf{FCP}_{n,\ell}^{\mathsf{pf}} \leq 8\ell/2^n + 8\ell^4/2^{2n}$ for any n, ℓ with $\ell^2 \leq 2^{n-2}$, thereby proving Lemma 2. Recall that $\mathsf{pf}(M_1, M_2) = \mathsf{false}$ iff M_1 is not a prefix of M_2 and M_2 is not a prefix of M_1. The proof of the following is similar to the proof of Lemma 8 and is omitted.

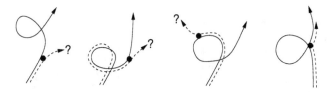

Fig. 4. Some shapes where the M_1-path (solid line) makes a loop. In the first three cases the M_1-path passes only once through V_1^p (the dot), and we see that we cannot draw the M_2-path such that $V_2^{m_2} \in \{V_1^{p+1}, \ldots, V_1^{m_1}\}$ without a second accident in any of those cases. In the last graph $V_2^{m_2} \in \{V_1^{p+1}, \ldots, V_1^{m_1}\}$, but there also $V_1^p \in \{V_1^0, \ldots, V_1^{p-1}, V_1^{p+1}, \ldots, V_1^{m_1}\}$.

Lemma 13. *Let $n \geq 1$ and $1 \leq m_1, m_2 \leq \ell$. Let $M_1 \in B_n^{m_1}, M_2 \in B_n^{m_2}$ with $\mathsf{pf}(M_1, M_2) = \mathsf{false}$. Assume $\ell^2 \leq 2^{n-2}$. Then*

$$\mathbf{FCP}_{n,\ell}^{\mathsf{pf}}(M_1, M_2) \leq \frac{2 \cdot \left|\phi_{M_1, M_2}[V_2^{m_2} \in \{V_1^1, \ldots, V_1^{m_1}, V_2^1, \ldots, V_2^{m_2-1}\}]\right|}{2^n} + \frac{8\ell^4}{2^{2n}}. \quad \blacksquare$$

Next we bound the size of the set that arises above:

Lemma 14. *Let $n, \ell \geq 1$ and $1 \leq m_1, m_2 \leq \ell$. Let $M_1 \in B_n^{m_1}, M_2 \in B_n^{m_2}$ with $\mathsf{pf}(M_1, M_2) = \mathsf{false}$. Then*

$$\left|\phi_{M_1, M_2}[V_2^{m_2} \in \{V_1^1, \ldots, V_1^{m_1}, V_2^1, \ldots, V_2^{m_2-1}\}]\right| \leq 4\ell. \quad \blacksquare$$

Putting together Lemmas 13 and 14 completes the proof of Lemma 2.

We denote by $\mathsf{cpl}(M_1, M_2)$ the number of blocks in the longest common block-prefix of M_1, M_2. That is, $\mathsf{cpl}(M_1, M_2)$ is the largest integer p such that $M_1^i = M_2^i$ for all $i \in [1..p]$. Define the predicate $\mathsf{NoLoop}(G)$ to be true for structure graph $G \in \mathcal{G}_2^1(M_1, M_2)$ iff $V_1^0(G), \ldots, V_1^{m_1}(G)$ are all distinct and also $V_2^0(G), \ldots, V_2^{m_2}(G)$ are all distinct. Let Loop be the negation of NoLoop.

Proof (Lemma 14). Let $p = \mathsf{cpl}(M_1, M_2)$. Since $\mathsf{pf}(M_1, M_2) = \mathsf{false}$, it must be that $p < m_1, m_2$ and $M_1^{p+1} \neq M_2^{p+1}$. Note then that $V_1^i = V_2^i$ for all $i \in [0..p]$ but $V_1^{p+1} \neq V_2^{p+1}$. Now we break up the set in which we are interested as

$$\phi_{M_1, M_2}[V_2^{m_2} \in \{V_1^1, \ldots, V_1^{m_1}, V_2^1, \ldots, V_2^{m_2-1}\}]$$
$$= \phi_{M_1, M_2}[V_2^{m_2} \in \{V_2^1, \ldots, V_2^{m_2-1}\}] \cup \phi_{M_1, M_2}[V_2^{m_2} \in \{V_1^{p+1}, \ldots, V_1^{m_1}\}].$$

Lemma 15 implies that $\left|\phi_{M_1, M_2}[V_2^{m_2} \in \{V_2^1, \ldots, V_2^{m_2-1}\}]\right| \leq m_2$ and Lemma 17 says that $\left|\phi_{M_1, M_2}[V_2^{m_2} \in \{V_1^{p+1}, \ldots, V_1^{m_1}\} \wedge \mathsf{NoLoop}]\right| \leq m_1$. It remains to bound $\left|\phi_{M_1, M_2}[V_2^{m_2} \in \{V_1^{p+1}, \ldots, V_1^{m_1}\} \wedge \mathsf{Loop}]\right|$. We use a case analysis, which is illustrated in Fig. 4. The condition Loop means that either the M_1- or the M_2-path (or both) must make a loop. If the M_1-path makes a loop then we can only draw the M_2-path such that $V_2^{m_2} \in \{V_1^{p+1}, \ldots, V_1^{m_1}\}$ if the loop goes twice through V_1^p. The same argument works if only the M_2-path makes a loop. Thus

$$\phi_{M_1, M_2}[V_2^{m_2} \in \{V_1^{p+1}, \ldots, V_1^{m_1}\} \wedge \mathsf{Loop}] \subseteq \mathcal{S}_1 \cup \mathcal{S}_2$$

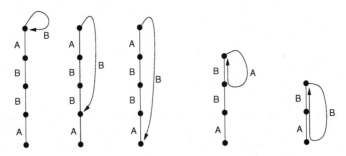

Fig. 5. An example for the proof of Lemma 15 with $m_1 = 5$ and $M_1 = A\|B\|B\|A\|B$ for distinct $A, B \in \{0,1\}^n$. Here we have $N_5 = 5 - \mu_1(M_1^5) + 1 = 5 - \mu_1(B) + 1 = 5 - 3 + 1 = 3$ and $N_4 = \mu_1(M_1^5) - \mu_1(M_1^{4\rightarrow5}) = \mu_1(B) - \mu_1(A\|B) = 3 - 2 = 1$ and $N_3 = \mu_1(M_1^{4\rightarrow5}) - \mu_1(M_1^{3\rightarrow5}) = \mu_1(A\|B) - \mu_1(B\|A\|B) = 2 - 1 = 1$ and $N_2 = N_1 = 0$. The first three graphs show the N_5 cases, the fourth and the fifth graph show the single cases for N_4 and N_3.

where

$$S_1 = \phi_{M_1,M_2}[V_1^p \in \{V_1^0, \ldots, V_1^{p-1}, V_1^{p+1}, \ldots, V_1^{m_1}\}]$$
$$S_2 = \phi_{M_1,M_2}[V_2^p \in \{V_2^0, \ldots, V_2^{p-1}, V_2^{p+1}, \ldots, V_2^{m_2}\}] .$$

Lemma 16 says that $|S_1| \leq m_1$ and $|S_2| \leq m_2$. Putting everything together, the lemma follows as $2(m_1 + m_2) \leq 4\ell$. ∎

Lemma 15. *Let* $n, m_1, m_2 \geq 1$. *Let* $M_1 \in B_n^{m_1}$, $M_2 \in B_n^{m_2}$ *with* $\mathsf{pf}(M_1, M_2) = $ *false. Then for* $b \in \{1, 2\}$,

$$\left|\phi_{M_1,M_2}[V_b^{m_b} \in V_b^0, V_b^1, \ldots, V_b^{m_b-1}\}]\right| = m_b \quad ∎$$

Proof. We prove the claim for $b = 1$ and then briefly discuss how to extend the proof to $b = 2$. If $V_1^{m_1} \in \{V_1^0, \ldots, V_1^{m_1-1}\}$ then there must be a (V_1^i, j)-accident for some $i \in [0..m_1 - 1]$ and $j \in [i + 1..m_1]$ and then induced collisions in steps $j + 1$ to m_1. Thus $V_1^{j+k} = V_1^{i+k}$ for all $k \in [0..m_1 - j]$. For $j \in [1..m_1]$ let N_j be the number of structure graphs $G \in \mathcal{G}_2^1(M_1, M_2)$ such that $V_1^{m_1}(G) \in \{V_1^0(G), \ldots, V_1^{m_1-1}(G)\}$ and there is a $(V_1^i(G), j)$-accident for some $i \in [0..j-1]$. Then

$$\left|\phi_{M_1,M_2}[V_1^{m_1} \in \{V_1^0, \ldots, V_1^{m_1-1}\}]\right| = \sum_{j=1}^{m_1} N_j .$$

Let $\mu_1(S)$ denote the number of block-aligned occurrences of the substring S in M_1. (For example, $\mu_1(A \| B) = 2$ if $M_1 = A \| B \| B \| \| A \| B$ for some distinct $A, B \in \{0, 1\}^n$.) It is possible to have a (V_1^i, m_1)-accident for any $i \in [0..m_1 - 1]$ for which $M_1^i \neq M_1^{m_1}$ (cf. Fig. 5) and thus $N_{m_1} = m_1 - \mu_1(M_1^{m_1}) + 1$. It is possible to have a $(V_1^i, m_1 - 1)$-accident and also have $V_1^{m_1} \in \{V_1^0, \ldots, V_1^{m_1-1}\}$ for any $i \in [0..m_1 - 2]$ for which $M_1^i \neq M_1^{m_1-1}$ and $M_1^{i+1} = M_1^{m_1}$ and thus

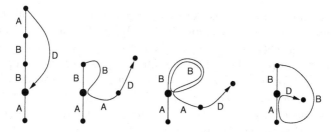

Fig. 6. An example for the proof of Lemma 16 with $m_1 = 5, M_1 = A\|B\|B\|A\|D$ and $r = 1$, where $A, B, D \in \{0,1\}^n$ are distinct. (The large dot is $V_1^r = V_1^1$.) Here we have $N_r = m - r = \mu_2(M_1^1) = N_1 = m_1 - 1 - \mu_2(M_1^1) = 5 - 1 - \mu_2(A) = 5 - 1 - 1 = 3$. Those cases correspond to the first three graphs in the figure. The fourth graph corresponds to $N_{r-1} = N_0 = \mu_2(\star \| M_1^{1 \to r}) = \mu_2(\star \| A) = 1$.

$N_{m_1-1} = \mu_1(M_1^{m_1}) - \mu_1(M_1^{m_1-1 \to m_1})$. In general for $j \in [1..m_1 - 1]$ we have $N_j = \mu_1(M_1^{j+1 \to m_1}) - \mu_1(M_1^{j \to m_1})$. Using cancellation of terms in the sum we have

$$\sum_{j=1}^{m_1} N_j = m_1 + 1 - \mu_1(M_1^{1 \to m_1}) = m_1$$

which proves the lemma for the case $b = 1$. For $b = 2$ we note that we can effectively ignore the part of the graph related to M since it must be a straight line, and thus the above counting applies again with the (V_1^i, j)-accident now being a $(V_2^i, m_1 + j)$-accident and M_1, m_1 replaced by M_2, m_2 respectively. ∎

Next we have a generalization of Lemma 15.

Lemma 16. *Let $n, m_1, m_2 \geq 1$. Let $M_1 \in B_n^{m_1}, M_2 \in B_n^{m_2}$ with $\mathsf{pf}(M_1, M_2) =$ false. Then for $b \in \{1, 2\}$ and any $r \in [0..m_b]$,*

$$\left|\phi_{M_1,M_2}[V_b^r \in \{V_b^0, \ldots, V_b^{r-1}, V_b^{r+1}, \ldots, V_b^{m_b}\}]\right| \leq m_b . \quad ∎$$

Proof. We prove it for the case $b = 1$. (The case $b = 2$ is analogous.) By Lemma 15 we have $|\phi_{M_1,M_2}[V_1^r \in \{V_1^0, \ldots, V_1^{r-1}\}]| = r$. It remains to show that

$$\left|\phi_{M_1,M_2}[V_1^r \in \{V_1^{r+1}, \ldots, V_1^{m_1}\} \wedge V_1^r \notin \{V_1^0, \ldots, V_1^r\}]\right| \leq m_1 - r .$$

We may assume that $V_1^i \neq V_1^j$ for all $0 \leq i < j \leq r - 1$, as otherwise we have already used up our accident and there's no way to get $V_1^r \in \{V_1^{r+1}, \ldots, V_1^{m_1}\}$ any more. If $V_r^\in \{V_1^{r+1}, \ldots, V_1^{m_1}\}$ then there is a (V_1^j, i)-accident for some $0 \leq j \leq r < i$. For $j \in [0..r]$ let N_j be the number of structure graphs $G \in \mathcal{G}_2^1(M_1, M_2)$ such that $V_1^r(G) \in \{V_1^{r+1}(G), \ldots, V_1^{m_1}(G)\}$, $V_1^r(G) \notin \{V_1^0(G), \ldots, V_1^r(G)\}$ and there is a (V_1^j, i)-accident for some $i \in [r+1..m_1]$. Then

$$\left|\phi_{M_1,M_2}[V_1^r \in \{V_1^{r+1}, \ldots, V_1^{m_1}\} \wedge V_1^r \notin \{V_1^0, \ldots, V_1^r\}]\right| = \sum_{j=0}^{r} N_j .$$

Let $\mu_2(S)$ be the number of block-aligned occurrences of the substring S in $M_1^{r+1\to m_1}$, and adopt the convention that $\mu_2(M_1^0) = 0$. Since we can only have an (V_1^r, j)-accident when $M_1^j \neq M_1^r$ we have $N_r = m - r - \mu_2(M_1^r)$. For $i > r$, a (V_1^r, i)-accident is possible and will result in $V_1^r \in \{V_1^{r+1}, \ldots, V_1^{m_1}\}$ only if $M_1^{i \to i+1} = X \| M_r$ for some $X \neq M_1^{r-1}$. Now with $\star$ being a wildcard standing for an arbitrary block we have $N_{r-1} = \mu_2(\star \| M_1^r) - \mu_2(M_1^{r-1 \to r})$. In general, for $j \in [1..r-1]$ we have $N_j = \mu_2(\star \| M_1^{j+1\to r}) - \mu_2(M_1^{j\to r})$ and $N_0 = \mu_2(\star \| M_1^{1\to r})$. Now, as $\mu_2(\star \| S) \leq \mu_2(S)$ for any S, we get

$$\sum_{j=0}^{r} N_j \leq m_1 - r . \ \blacksquare$$

The proof of the following is in [3].

Lemma 17. *Let* $n, m_1, m_2 \geq 1$. *Let* $M_1 \in B_n^{m_1}$, $M_2 \in B_n^{m_2}$ *with* $\mathsf{pf}(M_1, M_2) = $ *false. Let* $p = \mathsf{cpl}(M_1, M_2)$. *Then*

$$\left| \phi_{M_1, M_2}[V_2^{m_2} \in \{V_1^{p+1}, \ldots, V_1^{m_1}\} \wedge \mathsf{NoLoop}] \right| \leq m_1 . \qquad \blacksquare$$

Acknowledgments

Bart Preneel was the first we heard to ask, back in 1994, if the m^2 term can be improved in the CBC MAC bound of $m^2 q^2/2^n$.

Bellare was supported by NSF grants ANR-0129617 and CCR-0208842, and by an IBM Faculty Partnership Development Award. Pietrzak was supported by the Swiss National Science Foundation, project No. 200020-103847/1. Rogaway carried out most of this work while hosted by the Department of Computer Science, Faculty of Science, Chiang Mai University, Thailand. He is currently hosted by the School of Information Technology, Mae Fah Luang University, Thailand. He is supported by NSF grant CCR-0208842 and a gift from Intel Corp.

References

1. M. Bellare, O. Goldreich, and A. Mityagin. The power of verification queries in message authentication and authenticated encryption. Cryptology ePrint Archive: Report 2004/309.
2. M. Bellare, J. Kilian, and P. Rogaway. The security of the cipher block chaining message authentication code. *Journal of Computer and System Sciences* (JCSS), vol. 61, no. 3, pp. 362–399, 2000. Earlier version in *Crypto '94*.
3. M. Bellare, K. Pietrzak, and P. Rogaway. Improved security analyses for CBC MACs. Full version of this paper. Available via authors' web pages.
4. M. Bellare and P. Rogaway. The game-playing technique. Cryptology ePrint Archive: Report 2004/331.

5. A. Berendschot, B. den Boer, J. Boly, A. Bosselaers, J. Brandt, D. Chaum, I. Damgård, M. Dichtl, W. Fumy, M. van der Ham, C. Jansen, P. Landrock, B. Preneel, G. Roelofsen, P. de Rooij, and J. Vandewalle. *Final Report of Race Integrity Primitives.* Lecture Notes in Computer Science, vol. 1007, Springer-Verlag, 1995

6. R. Berke. On the security of iterated MACs. Diploma Thesis, ETH Zürich, August 2003.

7. J. Black and P. Rogaway. CBC MACs for arbitrary-length messages: the three-key constructions. *Advances in Cryptology – CRYPTO '00*, Lecture Notes in Computer Science Vol. 1880, M. Bellare ed., Springer-Verlag, 2000.

8. Y. Dodis. Personal communication to K. Pietrzak. 2004.

9. Y. Dodis, R. Gennaro, J. Håstad, H. Krawczyk, and T. Rabin. Randomness extraction and key derivation using the CBC, Cascade, and HMAC modes. *Advances in Cryptology – CRYPTO '04*, Lecture Notes in Computer Science Vol. 3152 , M. Franklin ed., Springer-Verlag, 2004.

10. G. Hardy and E. Wright. *An Introduction to the Theory of Numbers.* Oxford University Press, 1980.

11. E. Jaulmes, A. Joux, and F. Valette. On the security of randomized CBC-MAC beyond the birthday paradox limit: a new construction. *Fast Software Encryption '02*, Lecture Notes in Computer Science Vol. 2365 , J. Daemen, V. Rijmen ed., Springer-Verlag, 2002.

12. J. Kilian and P. Rogaway. How to protect DES against exhaustive key search (an analysis of DESX). *Journal of Cryptology*, vol. 14, no. 1, pp. 17–35, 2001. Earlier version in *Crypto '96*.

13. U. Maurer. Indistinguishability of random systems. *Advances in Cryptology – EUROCRYPT '02*, Lecture Notes in Computer Science Vol. 2332, L. Knudsen ed., Springer-Verlag, 2002.

14. National Institute of Standards and Technology, U.S. Department of Commerce, M Dworkin, author. Recommendation for block cipher modes of operation: the CMAC mode for authentication. NIST Special Publication 800-38B, May 2005.

15. E. Petrank and C. Rackoff. CBC MAC for real-time data sources. *Journal of Cryptology*, vol. 13, no. 3, pp. 315–338, 2000.

16. V. Shoup. Sequences of games: a tool for taming complexity in security proofs. Cryptology ePrint report 2004/332, 2004.

17. S. Vaudenay. Decorrelation over infinite domains: the encrypted CBC-MAC case. *Communications in Information and Systems* (CIS), vol. 1, pp. 75–85, 2001.

18. M. Wegman and L. Carter. New classes and applications of hash functions. *Symposium on Foundations of Computer Science* (FOCS), pp. 175–182, 1979.

HMQV: A High-Performance Secure Diffie-Hellman Protocol (Extended Abstract)⋆

Hugo Krawczyk

IBM T.J.Watson Research Center,
PO Box 704, Yorktown Heights, NY 10598, USA
hugo@ee.technion.ac.il

Abstract. The MQV protocol of Law, Menezes, Qu, Solinas and Vanstone is possibly the most efficient of all known authenticated Diffie-Hellman protocols that use public-key authentication. In addition to great performance, the protocol has been designed to achieve a remarkable list of security properties. As a result MQV has been widely standardized, and has recently been chosen by the NSA as the key exchange mechanism underlying *"the next generation cryptography to protect US government information"*.

One question that has not been settled so far is whether the protocol can be proven secure in a rigorous model of key-exchange security. In order to provide an answer to this question we analyze the MQV protocol in the Canetti-Krawczyk model of key exchange. Unfortunately, we show that MQV fails to a variety of attacks in this model that invalidate its basic security as well as many of its stated security goals. On the basis of these findings, we present HMQV, a carefully designed variant of MQV, that provides the same superb performance and functionality of the original protocol but for which all the MQV's security goals can be formally proved to hold in the random oracle model under the computational Diffie-Hellman assumption.

We base the design and proof of HMQV on a new form of "challenge-response signatures", derived from the Schnorr identification scheme, that have the property that both the challenger and signer can compute the *same* signature; the former by having chosen the challenge and the latter by knowing the private signature key.

1 Introduction

The classic Diffie-Hellman (DH) key-exchange protocol that marked the birth of modern cryptography has since been one of the main pillars of both theory and practice of cryptography. While the basic protocol as originally proposed (i.e., two parties $\hat{A}$ and $\hat{B}$ exchange values g^x and g^y, and compute a secret shared key as g^{xy}) is believed to be secure against an eavesdropping-only attacker, the

---

⋆ This is a very partial and informal version of the full paper available at
http://eprint.iacr.org/2005/

V. Shoup (Ed.): Crypto 2005, LNCS 3621, pp. 546–566, 2005.

quest for an "authenticated Diffie-Hellman" protocol that resists active, man-in-the-middle, attacks has resulted in innumerable ad-hoc proposals, many of which have been broken or shown to suffer from serious weaknesses. Fortunately, with the development of rigorous security models for key exchange in the last years, we are now in a much better position to judge the security of these protocols as well as to develop designs that provably withstand realistic active attacks.

In addition to the need for sound security, the many practical applications of key exchange have driven designers to improve on the performance cost associated with authentication mechanisms, especially those based on public key. One ambitious line of investigation, initiated by Matsumoto, Takashima and Imai in 1986 [31], is to design DH protocols whose communication is identical to the basic DH protocol (i.e., no explicit authentication added except for the possible transmission of PK certificates), yet they are implicitly authenticated by the sole ability of the parties to compute the resultant session key (i.e., rather than agreeing on the key g^{xy}, the parties would agree on a key that combines g^x, g^y with their public/private keys). Not only can this approach generate protocols that are very efficient communication-wise, but the combination of authentication with the key derivation procedure can potentially result in significant computational savings. For these reasons, several of these "implicitly authenticated" protocols have been standardized by major national and international security standards.

Of these protocols, the most famous, most efficient and most standardized is the MQV protocol of Law, Menezes, Qu, Solinas and Vanstone [33,30]. This protocol has been standardized by many organizations, e.g. [2,3,20,21,35], and has recently been announced by the US National Security Agency (NSA) as the key exchange mechanism underlying "the next generation cryptography to protect US government information" (which includes the protection of "classified or mission critical national security information") [36]. Indeed, MQV appears to be a remarkable protocol that not only is the most efficient and versatile authenticated DH protocol in existence, but it has also been designed to satisfy an impressive array of security goals.

Yet, in spite of its attractiveness and success, MQV has so far eluded any formal analysis in a well-defined model of key exchange. The present work was initially motivated by the desire to provide such an analysis. Our findings, however, have been disappointing: we found that when formally studied virtually none of the stated MQV goals can be shown to hold (specifically, we carried this study in the computational key exchange model of Canetti and Krawczyk [11]). This raises clear concerns about the security of the protocol and triggers a natural question: Do we have a replacement for MQV with the same superb performance and versatility but for which the MQV security goals can be guaranteed in a well analyzed, provable way?

The main contribution of this paper is in identifying the various analytical shortcomings of the MQV design and proposing a "hashed variant" of the protocol, which we call HMQV, that provides the same (almost optimal) performance

---

Key Computation in the MQV and HMQV Protocols

Both protocols: $\hat{A}$ and $\hat{B}$ exchange $X = g^x, Y = g^y$ (via a basic DH run)

$\hat{A}$ computes $\sigma_{\hat{A}} = (YB^e)^{x+da}$, $\hat{B}$ computes $\sigma_{\hat{B}} = (XA^d)^{y+eb}$

Both parties set $K = H(\sigma_{\hat{A}}) = H(\sigma_{\hat{B}})$

MQV: $d = \bar{X}$, $e = \bar{Y}$ ($\bar{X}$ and $\bar{Y}$ are defined in the text)

HMQV: $d = \bar{H}(X, \hat{B})$, $e = \bar{H}(Y, \hat{A})$

Fig. 1. Computation of the session key K in each of the two protocols ($A = g^a$ and $B = g^b$ are $\hat{A}$'s and $\hat{B}$'s public keys, respectively.)

---

of MQV but also delivers, in a provable way, the original security goals of MQV (and even more).

Organization. Due to space limitations most of this proceedings version is devoted to a high-level informal description of our results (Sections 2 and 3). The full version of the paper [28] contains a detailed presentation of our results and their proofs. The only technical section in this extended abstract is Section 4 which presents XCR signatures, the main technical tool developed here as a basis for the proof of the HMQV protocol (see the end of that section for a 1-paragraph rationale of HMQV's design). We end with some discussion of related work in Section 5 and concluding remarks in Section 6.

Note on Groups and Notation. All the protocols and operations discussed in this paper assume a cyclic group G of prime order q generated by a generator g. We denote by $|q|$ the bit length of q (i.e., $|q| = \lceil \log_2 q \rceil$), and use this quantity as an implicit security parameter. The parameters G, g and q are assumed to be fixed and known in advance to the parties (this is usually the case in practice, e.g., [19]; alternatively, one could include these values in certificates, etc.).
We use the multiplicative representation of group operations but our treatment is equally applicable to additive (prime order) groups such as elliptic curves.
In our protocols, public keys (denoted by upper case letters) are elements in the group G, and the private keys (denoted by the corresponding lower case letters) are elements in Z_q. For example, to a public key $A = g^a$ corresponds a private key a. The party having A as its public key will be denoted by $\hat{A}$ (in general, the "hat notation" is used to denote the identities of parties in the protocol, possibly including the party's PK certificate).

2   The MQV Protocol and Its Security Shortcomings

The communication in the MQV protocol is identical to the basic unauthenticated DH protocol except that the identities $\hat{A}, \hat{B}$ may include a public-key certificate. The computation of the session key is shown in Figure 1 where: party $\hat{A}$

possesses a long-term private key $a \in Z_q$ and corresponding public key $A = g^a$, $\hat{B}$'s private/public key pair is $(b, B = g^b)$, and the ephemeral DH values exchanged in the protocol are $X = g^x, Y = g^y$ (x, y chosen by $\hat{A}, \hat{B}$, respectively). The computation of the session key also uses the values $d = \bar{X}$ and $e = \bar{Y}$, where $\bar{X} = 2^\ell + (X \bmod 2^\ell)$ and $\bar{Y} = 2^\ell + (Y \bmod 2^\ell)$ for $\ell = |q|/2$. The computation of the session key by $\hat{A}$ (and similarly by $\hat{B}$) involves the exponentiations $X = g^x$, B^e, and $(YB^e)^{x+da}$. Note, however, that e is of length $|q|/2$ and hence B^e counts as "half exponentiation" (i.e. half the number of modular multiplication relative to a regular exponentiation of g). Also, note that $X = g^x$ can be pre-computed. This sums up to an impressive performance: same communication as the basic DH protocol and just half exponentiation more than the basic protocol, i.e. *a mere 25% increase in computation to achieve an authenticated exchange!* This is significantly better than any of the proven DH protocols that rely on digital signatures or public key encryption for authentication (which involve more expensive operations and increased bandwidth), and is also the most efficient of the implicitly-authenticated DH protocols (the closest are the "Unified Model" protocols [8,23] that require three full exponentiations and offer substantially less security features – see Section 5).

2.1   Stated Security Goals of the MQV Protocol

The designers of MQV clearly, albeit informally, stated the security goals behind the MQV design (see [33,30] and related publications). This includes the resistance to a variety of explicit attack strategies such as guessing attacks, impersonation attacks, known-key attacks, key-compromise impersonation (KCI) attacks, and the provision of perfect forward secrecy (PFS).

While resistance to guessing attacks and impersonation attacks are basic and obvious security requirements, it is worth expanding on the meaning of the other attacks. They all represent realizations of the same fundamental security principle: a good security system is not one that denies the possibility of failures but rather one designed to confine the adverse effects of such failures to the possible minimum.

In the case of known key attacks, one is concerned with the realistic possibility that some session-specific information, such as a session key (or the ephemeral secrets that led to the computation of that key), will leak to an attacker. This can happen in a variety of ways ranging from the simple mishandling of information to a temporary break-in into a computer system or the malicious action of an insider. In this case, one does not expect the exposed session to remain secure, but a well-designed key-exchange protocol needs to guarantee that such a failure will *only* affect the specific compromised session. Other sessions, by the same or other parties, should not be endangered by this leakage. The resistance to known-key attacks enforces other basic security principles as well; most importantly, that keys from different sessions should be fully "computationally independent" (i.e., from learning one session key nothing can be implied about the value of other session keys).

The properties of PFS and KCI resistance are also concerned with limiting the effects of eventual failures, in this case the disclosure of long-term keys. Clearly, the discovery by an attacker $\mathcal{M}$ of the long-term authentication key of party $\hat{A}$ allows $\mathcal{M}$ to impersonate $\hat{A}$ and establish its own sessions in the name of $\hat{A}$. A protocol is said to have PFS if session keys established (and deleted from memory) before the compromise of the long-term key cannot be recovered (even with the use of this key). In the case of KCI, the question is whether the knowledge of $\hat{A}$'s private key allows $\mathcal{M}$ not only to impersonate $\hat{A}$ to others but also *to impersonate other, uncorrupted, parties to $\hat{A}$*. A protocol that prevents this form of "reverse impersonation" is said to resist KCI attacks. In other words, in such a protocol the only way $\mathcal{M}$ can take advantage of the knowledge of $\hat{A}$'s private key is by active impersonation of $\hat{A}$. Any session established by $\hat{A}$, without being actively controlled by $\mathcal{M}$, remains secure. Resistance to KCI attacks is a very significant security property that has added much to the attractiveness of MQV as it is not offered by other implicitly-authenticated protocols, such as the unified-model protocols of [8,23] (see Section 5), that use the static DH key g^{ab} for authentication (this key functions as a long-term shared key and hence cannot resist a KCI attack).

2.2   Weaknesses of the MQV Protocol

In spite of the ambitious security goals described above, it turns out that when casting these goals in a well-defined formal setting as the one in [11], the MQV protocol falls short of delivering most of its intended security. Due to page limitations we only present a summary of these findings here; please refer to [28] for a detailed account (which also includes a succinct description of the formal model from [11]).

Group Representation Attacks. We first observe that MQV's security is strongly susceptible to the specific way the group elements are represented in the protocol. We show how some representations render the protocol totally insecure. While ordinary group representations may not have such an extreme effect on the security of the protocol, this result shows that any attempt at proving MQV would need to involve restricted group representations. Moreover, the inherent weaknesses of the protocol discussed below show that the protocol cannot be proven secure even for specific groups.

UKS Attacks. We study the vulnerability of the protocol to "unknown key share" (UKS) attacks which were also listed as a security consideration in MQV. A successful UKS attack [16] is one in which two parties compute the same session key but have different views of who the peer to the exchange was (this attack represents both an authentication failure as well as a vulnerability to known-key attacks). Originally, it was thought that MQV (at least when the registrants of public keys are required to prove "possession" of the corresponding private keys) was immune to these attacks; later it was shown by Kaliski [24] that even with such proofs of possession MQV fails to a UKS attack. Since then it has been believed that augmenting the protocol with a "key confirmation" step (which adds a third message to the protocol) would solve the problem. Here we show that this is *not* the case. Indeed, the 3-message variant of MQV is still vulnerable to this form of attack if the attacker can access ephemeral secret session-state information for sessions other than the session being attacked.

Lack of PFS. MQV does not provide Perfect Forward Secrecy (PFS). This, however, is not just a failure of MQV but it's an inherent limitation of implicitly-authenticated 2-message protocols based on public-key authentication. Indeed no such protocol can provide PFS. We present a generic attack against any such protocol where an active attacker $\mathcal{M}$ causes the establishment of a session key K at party $\hat{A}$ with peer $\hat{B}$ such that a later corruption of $\hat{B}$ (even after K was erased) allows $\mathcal{M}$ to find K.

KCI Attacks. Since MQV is susceptible to basic authentication attacks even when the private key of the victim is not known to the attacker, then KCI resistance cannot be satisfied. Yet, it is interesting to see explicit KCI attacks that take advantage of the knowledge of such private key. We show such an attack against MQV in the case that the attacker has access to the ephemeral values σ from which the session key is computed. This serves to motivate two design principles in HMQV: (i) the essential role of the hashing of σ for session key derivation (in MQV this hashing is recommended but separated from the core specification of the protocol [30])[1]; and (ii) the care required in handling ephemeral information (that may be learned by the attacker in some situations).

Prime-Order Checks. Our description in Figure 1 omitted an element from the session-key computation in MQV: In case where the group G generated by g is a subgroup of a larger group G' (which is the case in typical mod p and elliptic curve groups), MQV specifies that the key be computed as $K = H(\sigma)$ for $\sigma = (\sigma_{\hat{A}})^h = (\sigma_{\hat{B}})^h$ ($\sigma_{\hat{A}}$ and $\sigma_{\hat{B}}$ as defined in Figure 1) where h is the co-factor $|G'|/|G|$. This measure is used to ensure that the value σ belong to the group G, and has been added to MQV as a safeguard against potential attacks resulting from the lack of explicit authentication of the DH values, e.g., small-group attacks. We note, however, that this addition is of no help against the vulnerabilities mentioned above (and as we will see is not needed to provide security in HMQV). Moreover, lacking a proof that the above counter-measure really works, several standards defining MQV, as well as various descriptions of the protocol in the literature, often specify (or at least recommend) that the parties to MQV explicitly check that the DH value presented by the peer is of prime order q. This adds a costly extra exponentiation to each peer and takes significantly from the almost-optimal performance of MQV. As we will see, HMQV "provably dispenses" of the need for this costly check.

---

[1] The MQV paper is somewhat ambiguous about the need to hash σ (see the end of Sections 1 and 5 in [30]). In particular, this hashing is not viewed as essential to the security of the protocol but as a possible safeguard against potential weak bits in σ (which is not the source of weakness here, but rather the malleability of σ is.)

3   The HMQV Protocol and Its Proven Security

The HMQV protocol ("H" is for Hash) is a simple but powerful variant of MQV. As in MQV, its communication is identical to the basic DH exchange with the possible addition of certificates. The computation of the session key, shown in Figure 1, differs from MQV's in the computation of the values d and e which involves the hashing of the party's own DH value and the peer's identity. The output of this hash is $\ell = |q|/2$ bits. In addition, HMQV *mandates* the hashing of the values $\sigma_{\hat{A}} = \sigma_{\hat{B}}$ into k-bit keys where k is the length of the desired session key. We denote the hash function with ℓ bits of output by $\bar{H}$ and the one with k bits by H. In practice the same hash function can be used with different output lengths. (As a mnemonic, the bar in $\bar{H}$ indicates that the output of the function is used as an exponent).

From this description one can see that HMQV preserves the outstanding performance of MQV (both in terms of communication and computation). At the same time, *HMQV overcomes all the mentioned security shortcomings of MQV to the largest possible extent in a 2-message protocol.* We prove that in the random oracle model [5], and under the Computational Diffie-Hellman (CDH) assumption [15], the protocol is secure in the Canetti-Krawczyk security model [11]. In particular, this establishes the security of the protocol against impersonation attacks, known-key attacks, and UKS attacks. We also prove the resistance of HMQV to KCI attacks (which we formally define) under the same assumptions.

Furthermore, HMQV enjoys an additional performance advantage in that it provably dispenses of the need for costly prime-order tests on the DH values transmitted in the protocol. Indeed, our proof shows that the only way an attacker can benefit from the choice of rogue DH values is by choosing these to be zero, and thus a simple non-zero check is all is required (hence, there is no need for prime-order tests or for the co-factor h used in MQV).

Regarding forward secrecy, we said earlier that PFS cannot be achieved by any implicitly authenticated 2-message protocol, including HMQV. Yet, the following limited forward secrecy property holds for HMQV: any session key established without the active intervention of the attacker (except for eavesdropping the communication) is guaranteed to be irrecoverable by the attacker once the session key is erased from memory. This is the case even if the attacker knew the private keys of both peers when the session was established.

For applications that require full PFS we present a 3-message variant of HMQV which adds a third message and a MAC computation by each party and guarantees full PFS. This 3-message protocol, called HMQV-C, also provides "key confirmation" to both parties, i.e., the assurance that the assumed peer indeed participated in the protocol and that it computed the same session key. Another advantage of HMQV-C is that it can be proven secure in the stronger *universally composable* (UC) KE security model of [13] which ensures the security of key-exchange protocols when run concurrently with other applications. We note that while HMQV-C requires an extra message, its computational cost is essentially the same as HMQV as the MAC computation is negligible relative to the exponentiation cost. On the other hand, [23] note that 2-message symmetric protocols such as the basic HMQV allow for simultaneous initiation of a session by both $\hat{A}$ and $\hat{B}$, a desirable property in some network settings.

Another variant of HMQV is a one-pass authenticated key-exchange protocol in which $\hat{A}$ sends a single message to $\hat{B}$ after which both parties share a secret key. We show also this protocol to be secure (under CDH and in the random oracle model) in a security model adapted from [11] to one-pass protocols (the only difference is that we cannot prevent the adversarial replay of a message from $\hat{A}$ to $\hat{B}$ and, of course, cannot provide PFS). In particular, this one-pass protocol provides the functionality of public-key based deniable authentication as well as an authenticated CCA encryption scheme (in the random oracle model) in a more efficient way than existing alternatives.

An important security consideration not discussed by the authors of MQV is the resilience of the protocol to the disclosure of the secret exponent x corresponding to an ephemeral (session-specific) DH value $X = g^x$. This is a prime concern for any Diffie-Hellman protocol since many applications will boost protocol performance by pre-computing ephemeral pairs $(x, X = g^x)$ for later use in the protocol (this may apply to low-power devices as well as to high-volume servers). In this case, however, these stored pairs are more vulnerable to leakage than long-term static secrets (the latter may be stored in a hardware-protected area while the ephemeral pairs will be typically stored on disk and hence more available to a temporary break or to a malicious user of the system). We prove that HMQV's security is preserved even in the presence of the leakage of ephemeral secret DH exponents (see Section 5 for comparison to other work). For this property (and only for it) we need to resort to two strong assumptions: Gap Diffie-Hellman [37] and Knowledge of Exponent (KEA1) [14,18,4]; in return we get a guarantee that not even the session key computed using the exposed exponent is compromised by this leakage.

We end by noting an important property of our analysis: all results in this paper hold under a strong adversarial model in which the attacker is allowed to register arbitrary public key for all corrupted parties (and at any time during the protocol). This may include a public key that is identical, or related, to the public key of another (possibly uncorrupted) party; in particular, the attacker may not know the private key corresponding to the public key it chose. In practical terms this means that the security of our protocols does not depend on the certification authority requiring registrants of public keys to prove knowledge of the corresponding private keys. This is important since in many practical settings such "proofs of possession" are not required or performed by the CA (for contrast, see the comparison with [23] in Section 5).

4   Exponential Challenge-Response Signatures

Here we introduce the Exponential Challenge-Response (XCR) Signature Scheme which is the main building block used in the design and analysis of the HMQV protocol. As in a regular digital signature scheme, in a challenge-response signature scheme a signer has a pair of private and public keys used for generation and verification, respectively, of signatures. However, in contrast to regular signatures, challenge-response signatures are inherently interactive and require the recipient (i.e., the verifier) of a signature to issue a *challenge* to the signer before the latter can generate the signature on a given message. A *secure* challenge-response signature scheme needs to guarantee that no one other than the legitimate signer be able to generate a signature that will convince the challenger to accept it as valid (in particular, a signature is not only message-specific but also challenge-specific). On the other hand, we are only interested to ensure verifiability of the signature by the challenger, and thus we make no assumptions or requirements regarding the transferability, or verifiability by a third party, of the signature. Moreover, in the scheme described below the party that chooses the challenge can always generate a signature, on any message, which is valid with respect to that particular challenge. What is even more important for our application (and differentiates our scheme from other interactive signatures) is the fact that the verifier can compute, using the challenge, the *same signature string* as the signer.

While the above description may serve as a basis for a general definition of challenge-response signatures, we omit here such a general treatment in favor of a more focused description of the specific challenge-response signature used in this work. In particular, the definition of security is simplified by tailoring it to this specific scheme.

As before, we use g to denote a generator of a group G of prime order q, and $\bar{H}$ to denote a hash function that outputs $\ell = |q|/2$ bits. Our results require the following assumption (our treatment of polynomial-time, asymptotics, etc. is very informal, and uses $|q|$ as an implicit security parameter).

The CDH Assumption.   For two elements $U = g^u, V = g^v$ in G we denote by $CDH(U, V)$ the result of applying the Diffie-Hellman computation (wrt to generator g) to U and V, i.e., $CDH(U, V) = g^{uv}$. An algorithm is called a CDH solver for G if it takes as input pairs of elements (U, V) in G and a generator g of G and outputs the Diffie-Hellman result $CDH(U, V)$ wrt g. We say that the Computational Diffie-Hellman (CDH) assumption holds in the group $G = \langle g \rangle$ if for all probabilistic polynomial-time CDH solvers for G, the probability that on a pair (U, V), for $U, V \in_R G$, the solver computes the correct value $CDH_g(U, V)$ is negligible.

4.1   Definition of the XCR Signature Scheme

Definition 1. The exponential challenge-response (XCR) signature scheme. *The signer in a XCR scheme, denoted by $\hat{B}$, possesses a private key $b \in_R Z_q$ and a public key $B = g^b$. A verifier (or challenger), denoted $\hat{A}$, provides a message m for signature by $\hat{B}$ together with a challenge X which $\hat{A}$ computes as $X = g^x$ for $x \in_R Z_q$ (x is chosen, and kept secret, by $\hat{A}$). The signature of $\hat{B}$ on m using challenge X is defined as a pair $(Y, X^{y+\bar{H}(Y,m)b})$, where $Y = g^y$ and $y \in_R Z_q$ is* chosen by $\hat{B}$. The verifier $\hat{A}$ accepts a signature pair (Y, σ) as valid (for message m and with respect to challenge $X = g^x$) if and only if it holds that $Y \neq 0$ and $(YB^{\bar{H}(Y,m)})^x = \sigma$).

Notation: For message m, challenge X, and value Y we define $XSIG_{\hat{B}}(Y, m, X)$ $\stackrel{\text{def}}{=} X^{y+\bar{H}(Y,m)b}$ (i.e., $XSIG_{\hat{B}}$ denotes the second element in an XCR signature pair).

Relation Between XCR and Schnorr's Scheme. The main motivation for introducing the XCR scheme comes from its use in our design and analysis of the HMQV protocol [28]. By now, however, it may be illustrative to motivate this scheme via its relation to the Schnorr's identification scheme from which the XCR scheme is derived. We sketch this relation next. Schnorr's (interactive) identification scheme consists of a proof of knowledge of the discrete logarithm b for a given input $B = g^b$. Let $\hat{B}$ denote the prover in this scheme (that possesses b) and $\hat{A}$ the verifier (that is given the input B). The basic Schnorr's identification consists of three messages: (i) $\hat{B}$ chooses $y \in_R Z_q$ and sends $Y = g^y$ to $\hat{A}$; (ii) $\hat{A}$ responds with a random value $e \in_R Z_q$; and (iii) $\hat{B}$ sends $\hat{A}$ the value $s = y + eb$. $\hat{A}$ accepts if and only if $g^s = YB^e$ holds. This protocol is a Arthur-Merlin zero-knowledge proof of knowledge (of b) for an honest verifier $\hat{A}$ (i.e., one that chooses e uniformly at random). Therefore, it can be transformed via the Fiat-Shamir methodology into a signature scheme, namely $\text{SIG}_{\hat{B}}(m) = (Y, y + \bar{H}(Y, m)b)$, that is provably secure in the random oracle model [38].

Now consider the following 4-message variant of Schnorr's protocol in which a first message from $\hat{A}$ to $\hat{B}$ is added. In this first message $\hat{A}$ sends to $\hat{B}$ a value $X = g^x$. Then the 3 messages from Schnorr's scheme follow, except that in message (iii) (the fourth message in the modified protocol) rather than sending $s = y + eb$ to $\hat{A}$, $\hat{B}$ sends $S = X^s$. $\hat{A}$ accepts if and only if $S = (YB^e)^x$. It can be shown that this protocol is a proof of the "ability" of $\hat{B}$ to compute $CDH(B, X)$ for any value $X \in G$. Moreover, the protocol is zero-knowledge against a verifier $\hat{A}$ that chooses e at random (while X may be chosen arbitrarily). Now, note that applying the Fiat-Shamir transformation to this protocol one obtains the challenge-response signature XCR.[2] This also explains why we use the term "exponential" in naming the XCR scheme: it refers to the replacement of $s = y + eb$ in the Schnorr scheme with X^s in the last message of the protocol.

Next we establish our security requirement from the XCR scheme.

Definition 2. Security of the XCR signature scheme. *We say that the XCR challenge-response signature scheme is secure if no polynomial-time machine $\mathcal{F}$ can win the game in Figure 2 with non-negligible probability.*

---

[2] Note that if in Schnorr's protocol one chooses e in $\{0,1\}^\ell$ (rather than from Z_q) the protocol remains valid except that the soundness is limited by $2^{-\ell}$. This is the basis for our choice of $\ell = |q|/2$ as the output length of $\bar{H}(Y, m)$, namely, a trade-off between efficiency and security. See Remark 1 for a more accurate discussion.

Forger $\mathcal{F}$ in Definition 2

1. $\mathcal{F}$ is given values B, X_0 where $B, X_0 \in_R G$.
2. $\mathcal{F}$ is given access to a signing oracle $\hat{B}$ (representing a signer $\hat{B}$ with private key b and public key B) which on query (X, m) outputs a pair $(Y, XSIG_{\hat{B}}(Y, m, X))$ where $Y = g^y, y \in_R Z_q$, is chosen by $\hat{B}$ afresh with each query.
3. $\mathcal{F}$ is allowed a polynomial number of queries to $\hat{B}$ where the queries are chosen (possibly adaptively) by $\mathcal{F}$.
4. $\mathcal{F}$ halts with output "fail" or with a *guess* in the form of a triple (Y_0, m_0, σ).

$\mathcal{F}$'s guess is called a (successful) *forgery* if the following two conditions hold:

   (a) The pair (Y_0, σ) is a valid XCR signature of $\hat{B}$ on message m_0 with respect to challenge X_0 (i.e., $Y_0 \neq 0$ and $\sigma = XSIG_{\hat{B}}(Y_0, m_0, X_0)$; also note that the value of X_0 is the one received by $\mathcal{F}$ as input).

   (b) The pair (Y_0, m_0) did not appear in any of the responses of $\hat{B}$ to $\mathcal{F}$'s queries.

We say that $\mathcal{F}$ wins the game (or simply *forges*) if it outputs a successful forgery.

Fig. 2. Forgery Game for XCR Signatures

Remarks on Definition 2

1. Note that in order to be successful the forger has to use the input X_0 in its forgery. This captures the fact that XCR signatures are only unforgeable with respect to a challenge not chosen by the attacker.
2. According to our definition, if $\hat{B}$ outputs a signature (Y, σ) on a message m wrt challenge X, and the forger can find a signature (Y', σ') for the same m wrt the same challenge X but with $Y' \neq Y$, then we consider $\mathcal{F}$ successful (namely, finding a second signature for the same message is considered a valid forgery). In some sense, we look at these signatures as signing both m and Y. This property may not be essential in other applications of challenge-response signatures but it is crucial for the application to HMQV security in this paper.
3. We do not ask that $\mathcal{F}$ will always output good forgeries; it can output "fail" or even invalid triples. The only requirement is that with non-negligible probability (over the distribution of inputs to $\mathcal{F}$, the choices by the random oracle $\bar{H}$, the coins of $\mathcal{F}$, and the coins of $\hat{B}$) $\mathcal{F}$ output a successful forgery.
4. Note that we only restricted Y_0 to be non-zero. In particular, we require no check that Y_0 be of prime order q (only that it represents a non-zero element for which the group operation is defined). This is an important aspect of XCR signatures. In particular, the requirement to run a prime-order test would translate into an additional exponentiation for each party in the HMQV protocol, thus degrading significantly the "almost optimal" performance of the protocol.

4.2   Proof of Unforgeability for the XCR Signature Scheme

The following theorem states the security of the XCR scheme in the random oracle model under the CDH assumption; it constitutes the basis for the proof of security of protocol HMQV.

Theorem 1. *Under the CDH assumption, the* XCR *signature scheme is secure (according to Definition 2) in the random oracle model.*

Proof. Given an efficient and successful forger $\mathcal{F}$ against the XCR signature scheme (i.e., $\mathcal{F}$ wins the forgery game from Definition 2 with non-negligible probability), we build an efficient solver $\mathcal{C}$ for the CDH problem, namely, $\mathcal{C}$ gets as input a pair of random elements U, V in G, and outputs the value $CDH(U, V)$ with non-negligible probability. Unsuccessful runs of $\mathcal{C}$ may end with "fail" or just the wrong value of $CDH(U, V)$. Using results by Maurer and Wolf [32] and Shoup [40] such a "faulty CDH solver" can be transformed, using the self-reducibility properties of the CDH problem, into an efficient algorithm that solves CDH for every input U, V with only negligible probability of error.

Algorithm $\mathcal{C}$ is presented in Figure 3, and it follows a mostly standard argument for Fiat-Shamir type signatures. The idea is that if $\mathcal{F}$ can succeed in forging a signature with a pair (Y_0, m_0) and a given value $\bar{H}(Y_0, m_0)$ output by the function $\bar{H}$, then $\mathcal{F}$ is likely to succeed also when $\bar{H}(Y_0, m_0)$ is set to a *different* random value. Using this property, we construct $\mathcal{C}$ such that after running $\mathcal{F}$ twice, $\mathcal{C}$ obtains (with non-negligible probability) two forgeries with the same pair (Y_0, m_0) but different values of $\bar{H}(Y_0, m_0)$. Now, using these two forgeries $\mathcal{C}$ is able to compute $CDH(U, V)$.

Examining the specification of $\mathcal{C}$ in more detail, first note that in the run of $\mathcal{F}$ by $\mathcal{C}$ all queries to the signer $\hat{B}$ are answered by $\mathcal{C}$ without knowledge of the private key b, and without access to an actual signing oracle for $\hat{B}$. Instead all these answers are simulated by $\mathcal{C}$ in steps S1-S3. It is easy to see that this is a perfect simulation of the XCR signature generation algorithm under private key b except for the following deviation that happens with negligible probability: In step S3 of the simulation, $\mathcal{C}$ does not complete the run of $\mathcal{F}$ if the value (Y, m) was queried earlier from $\bar{H}$. However, since the value Y, as generated by $\mathcal{C}$, is distributed uniformly over G and chosen independently of previous values in the protocol, then the probability that the point (Y, m) was queried earlier from $\bar{H}$ is at most Q/q where Q is an upper bound on the number of queries to $\bar{H}$ that occur in a run of $\mathcal{C}$.

Therefore, the probability of $\mathcal{F}$ outputting a successful forgery in the run under $\mathcal{C}$ is the same, up to a negligible difference, as in a real run of $\mathcal{F}$, and therefore non-negligible. In particular, when such a successful forgery is output by $\mathcal{F}$ then conditions F1 and F2 checked by $\mathcal{C}$ necessarily hold. Condition F3 also holds except for probability $2^{-\ell}$, i.e., the probability that $\mathcal{F}$'s forgery is correct when it did not query $\bar{H}(Y_0, m_0)$. To see this, note that if one fixes the pair (Y_0, m_0) and the challenge X_0, then the signature produced with $e = \bar{H}(Y_0, m_0)$ is necessarily different than the signature produced with $e' = \bar{H}(Y_0, m_0)$ if $e \neq e' \pmod{q}$.

Building a CDH solver $\mathcal{C}$ from an XCR forger $\mathcal{F}$

Setup. Given a successful XCR-forger $\mathcal{F}$ we build an algorithm $\mathcal{C}$ to solve the CDH problem. The inputs to $\mathcal{C}$ are random values $U = g^u, V = g^v$ in G. $\mathcal{C}$'s goal is to compute $CDH(U,V) = g^{uv}$.

$\mathcal{C}$'s actions. $\mathcal{C}$ sets $B = V$ and $X_0 = U$, and runs the forger $\mathcal{F}$ on input (B, X_0) against a signer $\hat{B}$ with public key B. $\mathcal{C}$ provides $\mathcal{F}$ with a random tape and provides random answers to the $\bar{H}$ queries generated in the run (if the same $\bar{H}$ query is presented more than once $\mathcal{C}$ answers it with the same response as in the first time).

Each time $\mathcal{F}$ queries $\hat{B}$ for a signature on values (X, m) chosen by $\mathcal{F}$, $\mathcal{C}$ answers the query for $\hat{B}$ as follows (note that $\mathcal{C}$ does not know b):

S1. Chooses $s \in_R Z_q, e \in_R \{0,1\}^\ell$.
S2. Sets $Y = g^s/B^e$.
S3. Sets $\bar{H}(Y, m) = e$ (if $\bar{H}(Y, m)$ was defined by a previous query to $\bar{H}$, $\mathcal{C}$ aborts its run and outputs "fail").
$\mathcal{C}$ responds to $\mathcal{F}$'s query with the signature pair (Y, X^s)

When $\mathcal{F}$ halts $\mathcal{C}$ checks whether the three following conditions hold:

F1. $\mathcal{F}$ output a guess (Y_0, m_0, σ), $Y_0 \neq 0$.
F2. The pair (Y_0, m_0) was not used as the (Y, m) pair in any of the signatures generated by $\hat{B}$.
F3. The value $\bar{H}(Y_0, m_0)$ was queried from the random oracle $\bar{H}$.

If the three conditions hold, then $\mathcal{C}$ proceeds to the "repeat experiment" below; in all other cases $\mathcal{C}$ halts and outputs "fail".

The repeat experiment. $\mathcal{C}$ runs $\mathcal{F}$ again for a second time under the same input (B, X_0) and using the same coins for both $\mathcal{C}$ and $\mathcal{F}$. The difference between the two runs is in the way in which $\mathcal{C}$ answers the $\bar{H}$ queries during the second run. Specifically, all queries to $\bar{H}$ performed before the $\bar{H}(Y_0, m_0)$ query are answered identically as in the first run. The query $\bar{H}(Y_0, m_0)$, however, is answered with a new independent value $e' \in_R \{0,1\}^\ell$. Subsequent queries to $\bar{H}$ are also answered at random from $\{0,1\}^\ell$, independently of the responses provided in the first run.

Output. If at the end of the second run, $\mathcal{F}$ outputs a guess (Y_0, m_0, σ') (with same (Y_0, m_0) as in the first run) and $e \neq e'$, then $\mathcal{C}$ computes the value $W = (\sigma/\sigma')^{(e-e')^{-1}}$ and outputs W as its guess for $CDH(U, V)$; otherwise $\mathcal{C}$ outputs "fail".

Fig. 3. Proof of Theorem 1: Reduction from CDH to XCR forgeries

---

Hence, the probability that $\mathcal{F}$ will guess the right signature without querying $\bar{H}(Y_0, m_0)$ is at most as the probability of guessing the value of $\bar{H}(Y_0, m_0)$, i.e., $2^{-\ell}$. Since conditions F1-F3 determine the run of the "repeat experiment" then the simultaneous probability that $\mathcal{F}$ outputs a correct forgery in a run under $\mathcal{C}$ AND that $\mathcal{C}$ executes the "repeat experiment" in that run is non-negligible.

Now, using the Forking Lemma from [38] (our setting differs slightly from [38] in the use of a challenge, yet this does not affect the validity and applicability of the lemma), we obtain that the probability that in the "repeat experiment" $\mathcal{F}$ will output a correct forgery for the pair (Y_0, m_0) given that $\mathcal{F}$ did so in the first run is non-negligible. Moreover, in such a case we are guaranteed that the forgeries in the first and second runs use *different* values e, e' of $\bar{H}(Y_0, m_0)$.

We now proceed to show that in the case that both the first run and the repeat experiment end up with two valid forgeries for the same pair (Y_0, m_0), and $e \neq e'$ (which happens with probability $1 - 2^{-\ell}$), then the value W computed by $\mathcal{C}$ equals $CDH(X_0, B)$. Indeed, a simple computation shows that, if $Y_0 \neq 0$ as necessary for a valid forgery, then writing $X_0 = g^{x_0}$ we have:

$$W = \left(\frac{\sigma}{\sigma'}\right)^{\frac{1}{e-e'}} = \left(\frac{(YB^e)^{x_0}}{(YB^{e'})^{x_0}}\right)^{\frac{1}{e-e'}} = (B^{(e-e')x_0})^{\frac{1}{e-e'}} = B^{x_0} = CDH(X_0, B).$$

Now, since X_0 and B are, respectively, the inputs U and V provided to $\mathcal{C}$, then we get that in this case (which happens with non-negligible probability) $\mathcal{C}$ has successfully computed $CDH(U, V)$.   □

Remark 1. (Number of bits in $\bar{H}(Y, m)$). Let ℓ be the number of bits in the output of $\bar{H}(Y, m)$. Clearly, the smaller ℓ the more efficient the signature scheme is; on the other hand, a too small ℓ implies a bad security bound (since once the exponent $\bar{H}(Y, m)$ is predictable the signature scheme is insecure). But how large a ℓ do we need for security purposes? Here we see that setting $\ell = \frac{1}{2}|q|$, as we specified for XCR signatures (and for its application to the HMQV protocol), provides the right performance-security trade-off. In order to assess the level of security that ℓ provides to the XCR signatures (and consequently to HMQV), we note that there are two places in the above proof where this parameter ℓ enters the analysis. One is when bounding the probability that the attacker could forge a signature with parameters (Y_0, m_0) without querying $\bar{H}$ on this pair. As we claimed the probability in this case is $2^{-\ell}$ (or $1/\sqrt{q}$ using the fact that we defined $\ell = |q|/2$). The other use of ℓ is in the proof of the Forking Lemma by Pointcheval and Stern [38]. When written in terms of XCR signatures Lemma 9 and Theorem 10 from [38] show that given a forger against XCR signatures that works time T, performs Q queries to $\bar{H}$ and forges with probability ε, one can build a CDH solver that runs expected time $c\frac{Q}{\varepsilon}T$ provided that $\varepsilon \geq \frac{c'Q}{2^\ell}$ (c, c' are constants). Now, since we know how to build CDH solvers that run time $\sqrt{q}$ (e.g., Shanks algorithm) then the above analysis tells us something significant only when $c\frac{Q}{\varepsilon}T \ll \sqrt{q}$, in particular $Q/\varepsilon \ll \sqrt{q}$. From this and the condition $\varepsilon \geq c'Q/2^\ell$ we get that we need $Q/\varepsilon < \min\{\sqrt{q}, 2^\ell\}$. Since this is the only constraint on ℓ we see that choosing ℓ such that $2^\ell > \sqrt{q}$ does not add to the security of the scheme, and therefore setting $\ell = \frac{1}{2}|q|$ provides the best trade-off between security and performance. (Note that the same length consideration applies to the parameter e in the modified Schnorr's identification scheme described following Definition 1). We also comment, independently from the above considerations on ℓ, that the above constraint $Q < \varepsilon\sqrt{q}$ also guarantees that the simulation error in step S3 of Figure 3 (which we showed in the proof of Theorem 1 to be at most Q/q) is no more than $1/\sqrt{q}$.

Remark 2. (A non-interactive XCR *variant.)* XCR signatures can be made non-interactive, but verifier-specific, by putting $X = A$, where A is a public key of the verifier. In this case the signature will be a pair (Y, t) where t is a MAC tag computed on the signed message using a key derived by hashing $XSIG_{\hat{A}}(Y, \text{"}\hat{A}\text{"}, A)$. This provides for a very efficient non-interactive verifier-specific deniable authentication mechanism. It does *not* provide for a universally-verifiable non-repudiable signature.

Remark 3. (HCR and DSS signatures.) We do not know whether XCR signatures remain secure if the exponent y corresponding to a value Y used by $\hat{B}$ in a signature is revealed to the forger (note that in this case the simulation steps S1-S3 in Figure 3 do not work). On the other hand, if one modifies the definition of XCR such that the $XSIG_{\hat{B}}$ component is replaced with a hash of this value (note that the signature is still verifiable by the challenger) then one obtains a signature scheme in which revealing y does not help the forger. More precisely, in [28] we study these signatures, which we call HCR, in detail and show that under the Gap Diffie-Hellman and KEA1 assumptions they are unforgeable in the random oracle model even if y is revealed to the attacker. As a result, HCR signatures provide for a more secure alternative to DSS signatures as they resolve the main DSS vulnerability by which the disclosure of a single ephemeral exponent (i.e., k in the component $r = g^k$ of a DSS signature) suffices to reveal the signature key. On the other hand, HCR signatures are verifier-specific and require interaction (or the possession of a public key by the verifier as in Remark 2), and do not provide for third-party verifiability (a property that may be a bug or a feature of HCR depending on the application).

4.3   Dual XCR Signatures (DCR)

An important property of XCR signatures is that the challenger (having chosen the challenge) can compute the signature by itself. Here we show how to take advantage of this property in order to derive a related challenge-response signature scheme (which we call the "dual XCR scheme", or DCR for short) with the property that any two parties, $\hat{A}, \hat{B}$, can interact with each other with the dual roles of challenger and signer, and each produce a signature that no third party can forge. Moreover, *and this is what makes the scheme essential to the HMQV protocol,* the resultant signatures by $\hat{A}$ and by $\hat{B}$ have the *same value.* (More precisely, they have the same *XSIG* component.)

Definition 3. The dual (exponential) challenge-response (DCR) signature scheme. *Let $\hat{A}, \hat{B}$ be two parties with public keys $A = g^a, B = g^b$, respectively. Let m_1, m_2 be two messages. The* dual XCR *signature (DCR for short) of $\hat{A}$ and $\hat{B}$ on messages m_1, m_2, respectively, is defined as a triple of values: X, Y and $DSIG_{\hat{A}, \hat{B}}(m_1, m_2, X, Y) \stackrel{\text{def}}{=} g^{(x+da)(y+eb)}$, where $X = g^x, Y = g^y$ are challenges* *chosen by $\hat{A}$ and $\hat{B}$, respectively, and the symbols d and e denote $\bar{H}(X, m_1)$ and $\bar{H}(Y, m_2)$, respectively.*

As said, a fundamental property of a DCR signature is that after exchanging the values X and Y (with x and y chosen by $\hat{A}$ and $\hat{B}$, respectively), both $\hat{A}$ and $\hat{B}$ can compute (and verify) the **same** signature $DSIG_{\hat{A},\hat{B}}(m_1, m_2, X, Y)$. This can be seen from the identities:

$$DSIG_{\hat{A},\hat{B}}(m_1, m_2, X, Y) = g^{(x+da)(y+eb)} = (YB^e)^{x+da} = (XA^d)^{y+eb}$$

Moreover, as shown next the attacker cannot feasibly compute this signature.

The Security of DCR Signatures. Roughly speaking, a dual signature is an XCR signature by $\hat{A}$ on message m_1, under challenge YB^e, and at the same time an XCR signature by $\hat{B}$ on message m_2, under challenge XA^d. More precisely, since the values d and e are determined during the signature process (via the possibly adversarial choice of messages m_1, m_2), then we will say that a DCR signature of $\hat{B}$ is secure (with respect to A) if no efficient attacker can win, with non-negligible probability, the game of Figure 2 with the following modifications. In step 2, the queries to $\hat{B}$ are of the form (X, m, m_1) and the signature by $\hat{B}$ is the pair $(Y, XSIG_{\hat{B}}(Y, m, XA^d))$ where Y is chosen by $\hat{B}$ and $d = \bar{H}(X, m_1)$. A successful forgery is a quadruple (Y_0, m_0, m_1, σ) where $\sigma = XSIG_{\hat{B}}(Y_0, m_0, X_0 A^d)$, $Y_0 \neq 0$, the pair (Y_0, m_0) satisfies the validity requirement (b) from Figure 2, and m_1 is an arbitrary message chosen by $\mathcal{F}$. We say that the dual signature of $\hat{B}$ is secure if it is secure *with respect to any value $A = g^a$ not chosen by the attacker.*

Theorem 2. *Let $\hat{A}, \hat{B}$ be two parties with public keys $A = g^a, B = g^b$, resp. Under the CDH assumption, the DCR signature of $\hat{B}$ with respect to A is secure even if the forger is given the private key a of $\hat{A}$ (but not the private key of $\hat{B}$).*

Proof. The same proof of unforgeability of XCR (Theorem 1) works here with a modified computation of W as specified below. First note that since the DCR signature of $\hat{B}$ now involves the value $d = \bar{H}(X_0, m_1)$, where m_1 is a message that $\mathcal{F}$ may choose at will, then the value of m_1 chosen by $\mathcal{F}$ before the repeat experiment may differ from the value of m_1 chosen during the repeat experiment. In this case we get two different values, d, d', used in the σ and σ' signatures. Also, note that even with a single value of d the specification of W in the proof of Theorem 1 would result in the value $(XA^d)^b$ rather than X^b (as required in order to solve the CDH problem on inputs $U = X, V = B$). We deal with these two issues by redefining W as follows (here we use the fact that a, the private key of $\hat{A}$, is known to $\mathcal{C}$):

$$W = \left(\frac{\sigma/(YB^e)^{da}}{\sigma'/(YB^{e'})^{d'a}} \right)^{\frac{1}{e-e'}} \tag{1}$$

The rest of the proof remains unchanged. $\square$

HMQV in a Nutshell. The HMQV protocol consists of an exchange between parties $\hat{A}$ and $\hat{B}$ of DH values $X = g^x$ and $Y = g^y$ that serve as challenges from which both parties compute the dual XCR signature $DSIG_{\hat{A},\hat{B}}(\text{``}\hat{A}\text{''}, \text{``}\hat{B}\text{''}, X, Y) = g^{(x+da)(y+eb)}$. The session key is then derived by hashing this value. In this way the signature itself need not be transmitted: it is the uniqueness of the signature that ensures a common derived value for the session key, and it is the *ability to compute the key (equivalently, the signature)* that provides for a proof that the exchange was carried by the alleged parties $\hat{A}$ and $\hat{B}$. Moreover, since the messages m_1, m_2 on which the signature is computed are the identities of the peers, both parties get assurance that the key they computed is uniquely bound to the correct identities (this is essential to avoid some authentication failures such as the UKS attacks). We end by noting that while the casting of the HMQV design in terms of DCR signatures is the main conceptual contribution of our work, showing that this idea indeed works for proving the security of the protocol turns out to be technically challenging (see [28] for the gory details).

5 Related Work

Implicitly-authenticated DH protocols were first studied in the work of Matsumoto, Takashima and Imai [31] in 1986. Since then this line of research generated many protocols, many of which suffer from various weaknesses. See [9,34,7,8] for some surveys which also include the discussion of desirable security goals for these protocols as well as some of the shortcomings of specific proposals. Two works that study such protocols in a formal model are those of Blake-Wilson et al [8] and Jeong et al [23]. They both treat very similar protocols referred to in the literature as the "unified model". In these protocols, parties $\hat{A}$ and $\hat{B}$ use their public keys g^a, g^b to generate a shared key g^{ab} that they then use to authenticate a DH exchange (c.f., [26]).

The variant studied in [8] is shown to be open to interleaving and known-key attacks and hence insecure (unfortunately, this variant has been widely standardized [2,3,20]). One main flaw of this protocol is that it does not explicitly authenticate (or includes under the key derivation hashing) the ephemeral DH values exchanged by the parties. [23] studies the version in which the DH values are included under the key derivation and shows this protocol to be secure in the random oracle model. However, the protocol does not provide resistance to KCI and is open to a UKS attack if the CA does not enforce a proof-of-posession check at time of certificate issuance.[3] Lack of KCI is one aspect of a more substantial drawback of these protocols, namely, the use of the keys g^{ab} as long-term shared keys between the parties; these keys become particularly vulnerable when cached for efficiency.

---

[3] In [23], this protocol is also claimed to enjoy perfect forward secrecy (PFS), but what they actually show is a weaker and non-standard notion (see Section 3) where PFS holds only for sessions created without active intervention by the attacker. As we pointed out, lack of (full) PFS is an inherent limitation of any 2-message implicitly-authenticated DH exchange, including the 2-message protocols from [23].

In contrast, HMQV is a significantly stronger protocol which, in particular, does not use g^{ab} as a long-term key, does not require (even for efficiency) to cache this value, and even if the value of g^{ab} is ever learned by the attacker it is of no help for impersonating either $\hat{A}$ or $\hat{B}$, or for learning anything about their session keys. On top of all its security advantages (which hold without relying on proofs of possession performed by CAs or prime-order tests performed by the parties), HMQV is more efficient than the unified model protocols that take 3 exponentiations per-party.

Finally, we mention the works of Shoup [41] and Jeong et al [23] that present 2-message authenticated DH exchanges with *explicit* authentication (via signatures and MAC, respectively) that they show to satisfy the security definitions from [41,6] in the standard (non-random-oracle) model. In these protocols, however, it is sufficient for the attacker to learn a single ephemeral exponent x of a DH value g^x exchanged between parties $\hat{A}$ and $\hat{B}$ to be able to impersonate $\hat{A}$ to $\hat{B}$ indefinitely, and without ever having to learn $\hat{A}$'s or $\hat{B}$'s private keys. This is a serious security weakness which violates the basic principle that the disclosure of ephemeral session-specific information should not compromise other sessions. The reason that these protocols could be proven secure in [41,23] is that the models of key exchange security considered in these works do not allow the attacker to find any session-specific information beyond the session key itself. The above vulnerability, however, indicates that such models are insufficient to capture some realistic attack scenarios. In contrast, the model of [11], used as the basis for our analysis, captures such attacks via state-reveal queries (see [28]).

6   Concluding Remarks

The results in this paper show vulnerabilities of MQV to known-key and other attacks (in the case of the 3-message variant of MQV some of these vulnerabilities depend on the ability of the attacker to access ephemeral state information for incomplete sessions). The extent to which these weaknesses are exploitable in practice depends of course on the application, the computing and communication environment, threat model, etc. These results do *not* mean that applications already using MQV are necessarily insecure in their specific environments. At the same time, the identified weaknesses should not be dismissed as theoretical only, especially when considering that MQV has been (and is being) standardized as a key-exchange protocol for use in heterogeneous and unknown scenarios (including the highly sensitive applications such as those announced by the NSA [36]). This is particularly true when, as shown here, these weaknesses are not inherent to the problem being solved nor to the formal analytical setting.

Indeed, HMQV provides the same functionality with the same (or even better) performance of MQV while enjoying a full proof of security.[4] Two caveats regarding this proof are the use of the idealized random oracle methodology and

---

[4] We hope that standard bodies will take into account provability of protocols when selecting or revising protocols for standardization.

the significant (though polynomially bounded) reduction cost. Other proven protocols (such as SKEME [26], ISO [22,11] and SIGMA [27,12]), while less efficient, enjoy less expensive reductions and do not directly require random oracles. We note, however, that dispensing of random oracles in these protocols requires the use of the more expensive (and seldom used in practice) signature and encryption schemes that do not rely on random oracles, and also requires the stronger DDH assumption. Certainly, coming up with a protocol that offers the many attractive security and performance properties of HMQV and does not rely on the random oracle model in its analysis is an important open question.

We end by stressing that in spite of the weaknesses demonstrated here, the MQV protocol contains some remarkable ideas without which the design of HMQV would have not been possible. Nor would this design have been possible without the rigorous examination of these ideas in a formal framework such as the one in [11]. The design of HMQV is a demonstration of the strength of "proof-driven designs" which guide us in choosing the necessary design elements of the protocol while dispensing of unnecessary "safety margins". As a result, one obtains solutions that are not only cryptographically sound but are also more efficient.

References

1. M. Abdalla, M. Bellare and P. Rogaway, "The Oracle Diffie-Hellman Assumptions and an Analysis of DHIES", *CT-RSA 2001*. LNCS 2020, 2001.
2. American National Standard (ANSI) X9.42-2001, Public Key Cryptography for the Financial Services Industry: Agreement of Symmetric Keys Using Discrete Logarithm Cryptography.
3. American National Standard (ANSI) X9.63: Public Key Cryptography for the Financial Services Industry: Key Agreement and Key Transport using Elliptic Curve Cryptography.
4. M. Bellare and A. Palacio, "The Knowledge-of-Exponent Assumptions and 3-round Zero-Knowledge Protocols", *Crypto'04*, LNCS 3152, 2004, pp. 273–289.
5. M. Bellare and P. Rogaway, "Random Oracles are Practical: A Paradigm for Designing Efficient Protocols", *First ACM Conference on Computer and Communications Security*, pp. 62–73, 1993.
6. M. Bellare and P. Rogaway, "Entity authentication and key distribution", *Crypto'93*, LNCS 773, 1994, pp. 232-249.
7. S. Blake-Wilson and A. Menezes, "Authenticated Diffie-Hellman Key Agreement Protocols", *Proceedings of SAC '99*, LNCS Vol. 1556, 1999.
8. S. Blake-Wilson, D. Johnson and A. Menezes, "Key exchange protocols and their security analysis," *6th IMA International Conf. on Cryptography and Coding*, 1997.
9. C. Boyd and A. Mathuria, *Protocols for Authentication and Key Establishment*, Springer, 2003.
10. R. Canetti, "Universally Composable Security: A New paradigm for Cryptographic Protocols", *42nd FOCS*, 2001.
11. Canetti, R., and Krawczyk, H., "Analysis of Key-Exchange Protocols and Their Use for Building Secure Channels", Eurocrypt'2001, LNCS Vol. 2045. Full version in: http://eprint.iacr.org/2001/040.

12. Canetti, R., and Krawczyk, H., "Security Analysis of IKE's Signature-based Key-Exchange Protocol", *Crypto 2002.* LNCS Vol. 2442.
13. Canetti, R., and Krawczyk, H., "Universally Composable Notions of Key Exchange and Secure Channels", *Eurocrypt 02*, 2002. Full version available at http://eprint.iacr.org/2002/059.
14. I. Damgård, "Towards Practical Public Key Systems Secure Against Chosen Ciphertext Attacks", Crypto'91, LNCS Vol. 576.
15. W. Diffie and M. Hellman, "New Directions in Cryptography", IEEE Trans. Info. Theor. 22, 6 (Nov 1976), pp. 644–654.
16. W. Diffie, P. van Oorschot and M. Wiener, "Authentication and authenticated key exchanges", *Designs, Codes and Cryptography*, 2, 1992, pp. 107–125.
17. C. Dwork, M. Naor and A. Sahai, "Concurrent Zero-Knowledge", *STOC'98*, pp. 409–418.
18. Hada, S. and Tanaka, T., "On the Existence of 3-round Zero-Knowledge Protocols", Crypto'98, LNCS Vol. 1462.
19. D. Harkins and D. Carrel, "The Internet Key Exchange (IKE)", RFC 2409, Nov. 1998.
20. IEEE 1363-2000: Standard Specifications for Public Key Cryptography.
21. ISO/IEC IS 15946-3 "Information technology – Security techniques – Cryptographic techniques based on elliptic curves – Part 3: Key establishment", 2002.
22. ISO/IEC IS 9798-3, "Entity authentication mechanisms — Part 3: Entity authentication using asymmetric techniques", 1993.
23. Ik Rae Jeong, Jonathan Katz, Dong Hoon Lee, "One-Round Protocols for Two-Party Authenticated Key Exchange", ACNS 2004: 220-232
24. B. Kaliski, "An unknown key-share attack on the MQV key agreement protocol", *ACM Transactions on Information and System Security (TISSEC)*, Vol. 4 No. 3, 2001, pp. 275–288.
25. J. Katz, "Efficient and Non-Malleable Proofs of Plaintext Knowledge and Applications", EUROCRYPT'03, LNCS 2656.
26. H. Krawczyk, "SKEME: A Versatile Secure Key Exchange Mechanism for Internet,", *1996 Internet Society Symposium on Network and Distributed System Security*, pp. 114-127, Feb. 1996.
27. H. Krawczyk, "SIGMA: The 'SiGn-and-MAc' Approach to Authenticated Diffie-Hellman and Its Use in the IKE Protocols", *Crypto '03*, LNCS No. 2729. pp. 400–425, 2003.
28. H. Krawczyk, "HMQV: A High-Performance Secure Diffie-Hellman Protocol" (full version). http://eprint.iacr.org/2005/
29. H. Krawczyk, "On the Security of Implicitly-Authenticated Diffie-Hellman Protocols", work in progress.
30. L. Law, A. Menezes, M. Qu, J. Solinas, and S. Vanstone, "An efficient Protocol for Authenticated Key Agreement", *Designs, Codes and Cryptography, 28, 119-134, 2003.*
31. T. Matsumoto, Y. Takashima, and H. Imai, "On seeking smart public-key distribution systems", *Trans. IECE of Japan*, 1986, E69(2), pp. 99-106.
32. U. Maurer and S. Wolf, "Diffie-Hellman oracles", CRYPTO '96, LNCS, vol. 1109. 1996, pp. 268-282.
33. A. Menezes, M. Qu, and S. Vanstone, "Some new key agreement protocols providing mutual implicit authentication", *Second Workshop on Selected Areas in Cryptography (SAC 95)*, pp. 22–32, 1995.
34. A. Menezes, P. Van Oorschot and S. Vanstone, "Handbook of Applied Cryptography," CRC Press, 1996.

35. NIST Special Publication 800-56 (DRAFT): Recommendation on Key Establishment Schemes. Draft 2, Jan. 2003.

36. "NSAs Elliptic Curve Licensing Agreement", presentation by Mr. John Stasak (Cryptography Office, National Security Agency) to the IETF's Security Area Advisory Group, Nov 2004.
    http://www.machshav.com/~smb/saag-11-2004/NSA-EC-License.pdf

37. T. Okamoto and D. Pointcheval, "The Gap-Problems: A New Class of Problems for the Security of Cryptographic Schemes", *PKC 2001*, LNCS 1992, 2001.

38. D. Pointcheval and J. Stern, "Security Arguments for Digital Signatures and Blind Signatures", *J.Cryptology* (2000) 13:361-396.

39. M.O. Rabin, "Digitalized Signatures", Foundations of Secure Computing, DeMillo-Dobkins-Jones-Lipton, editors, 155-168, Academic Press, 1978.

40. V. Shoup, "Lower Bounds for Discrete Logarithms and Related Problems", Eurocrypt'97, LNCS 1233, pp. 256-266.

41. V. Shoup, "On Formal Models for Secure Key Exchange", Theory of Cryptography Library, 1999. http://philby.ucsd.edu/cryptolib/1999/99-12.html.

Author Index

Lecture Notes in Computer Science

For information about Vols. 1–3508

please contact your bookseller or Springer